CHEMICAL GENOMICS

Advances in chemistry, biology, and genomics coupled with laboratory automation and computational technologies have led to the rapid emergence of the multidisciplinary field of chemical genomics. This edited text with contributions from experts in the field discusses evolving concepts, essential techniques, and a wide range of applications to help further the study of chemical genomics. The beginning chapters provide an overview of the basic principles of chemical biology and chemical genomics. This is followed by a technical section that describes the sources of small molecule chemicals; the basics of high-throughput screening technologies; various bioassays for biochemical-, cellular-, and organism-based screens; and computational approaches. The final chapters connect the chemical genomics field with personalized medicine and the druggable genome for future discovery of new therapeutics. A resource section is included in selected chapters for further information.

This book will be valuable to researchers, professionals, and students in many fields besides chemical genomics, including biology, biomedicine, chemistry, and drug discovery.

Haian Fu is a professor of Pharmacology, Hematology, and Medical Oncology at Emory University School of Medicine. He is the Director of the Emory Chemical Biology Discovery Center and leader of the Discovery and Developmental Therapeutics Program at the Winship Cancer Institute of Emory University. Dr. Fu serves as a Steering Committee member for the NIH National Cancer Institute Chemical Biology Consortium and is co-founder of the International Chemical Biology Society. His research focuses on protein-protein interactions in signal transduction pathways that control cell survival and death in normal and cancer cells, as well as targeting these interactions for chemical biology studies and drug discovery. He serves on the editorial board of several scientific journals, including *Current Chemical Genomics*, and edited *Protein-Protein Interactions: Methods and Applications* (2004). Dr. Fu has received various honors, including the PhRMA Faculty Development Award, the Burroughs Wellcome Fund New Investigator Award, the GRA Distinguished Investigator Award, and the Georgia Cancer Coalition Distinguished Cancer Scholar. He is also a recipient of the Award for Excellence in Mentorship from the Emory Graduate Division of Biological and Biomedical Sciences (2000), the Outstanding Postdoctoral Mentor Award from Emory University (2011), and the Emory Pharmacology Teaching Excellence Award (2006).

CHEMICAL GENOMICS

Edited by

Haian Fu

Emory University School of Medicine

Shaftesbury Road, Cambridge CB2 8EA, United Kingdom

One Liberty Plaza, 20th Floor, New York, NY 10006, USA

477 Williamstown Road, Port Melbourne, VIC 3207, Australia

314–321, 3rd Floor, Plot 3, Splendor Forum, Jasola District Centre, New Delhi – 110025, India

103 Penang Road, #05–06/07, Visioncrest Commercial, Singapore 238467

Cambridge University Press is part of Cambridge University Press & Assessment, a department of the University of Cambridge.

We share the University's mission to contribute to society through the pursuit of education, learning and research at the highest international levels of excellence.

www.cambridge.org
Information on this title: www.cambridge.org/9780521889483

First published 2012

A catalogue record for this publication is available from the British Library

Library of Congress Cataloging-in-Publication data
Chemical genomics / [edited by] Haian Fu.
p. cm.
Includes bibliographical references and index.
ISBN 978-0-521-88948-3 (hardback)
1. Chemogenomics. I. Fu, Haian.
QH438.4.C45C45 2012
615′.7–dc23 2011015783

ISBN 978-0-521-88948-3 Hardback

Contents

Contributors

Nicholas Aberle
Department of Molecular, Cellular, and Developmental Biology
Yale University
New Haven, Connecticut

Douglas S. Auld
Novartis Institutes for Biomedical Research
Center for Proteomic Chemistry
Cambridge, Massachusetts

Sandra Bartoli
Menarini Ricerche
Pomezia
Rome
Italy

Andreas Bender
Leiden/Amsterdam Center for Drug Research
Gorlaeus Laboratories
Leiden University
Leiden
The Netherlands

Denzil Bernard
Cancer Center and Departments of Internal Medicine, Pharmacology, and Medicinal Chemistry
University of Michigan
Ann Arbor, Michigan

Brenda Bondesen
Department of Pharmacology and Emory Chemical Biology Discovery Center
Emory University
Atlanta, Georgia

Mark M. Bouzyk
AKESOgen, Inc
Norcross, Georgia

Craig M. Crews
Department of Molecular, Cellular, and Developmental Biology
Yale University
New Haven, Connecticut

Monica Diaz-Gavilan
Department of Chemistry
University of Cambridge
Cambridge
England

Ray Dingledine
Department of Pharmacology and Emory Chemical Biology Discovery Center
Emory University
Atlanta, Georgia

Yuhong Du
Department of Pharmacology and Emory Chemical Biology Discovery Center
Emory University
Atlanta, Georgia

Bruce S. Edwards
University of New Mexico Center for Molecular Discovery
Cancer Research and Treatment Center
Albuquerque, New Mexico

Peter Eimon
Zygogen, LLC
Atlanta, Georgia

Daniela Fattori
Menarini Ricerche
Pomezia
Rome
Italy

Robert A. Field
Department of Biological Chemistry
The John Innes Centre
Norwich Research Park
Norwich
England

James V. Follen
Harvard Medical School
Harvard University
Boston, Massachusetts

Haian Fu
Department of Pharmacology and
Emory Chemical Biology Discovery Center
Emory University
Atlanta, Georgia

Warren R. J. D. Galloway
Department of Chemistry
University of Cambridge
Cambridge
England

Jonathan J. Havel
Program in Molecular and Systems Pharmacology
Emory University
Atlanta, Georgia

Irena Ivnitski-Steele
University of New Mexico Center for Molecular
Discovery
Cancer Research and Treatment Center
Albuquerque, New Mexico

Hualiang Jiang
Shanghai Institute of Materia Medica
Chinese Academy of Sciences
Shanghai, China

Margaret A. Johns
Emory Chemical Biology Discovery Center
Emory University
Atlanta, Georgia

Paul A. Johnston
University of Pittsburgh Drug Discovery Institute and
Department of Pharmaceutical Sciences
University of Pittsburgh
Pittsburgh, Pennsylvania

Iestyn Lewis
Department of Pharmacology and
Emory Chemical Biology Discovery Center
Emory University
Atlanta, Georgia

Brian Leyland-Jones
Winship Cancer Institute
Emory University
Atlanta, Georgia

Honglin Li
School of Pharmacy
East China University of Science and Technology
Shanghai
China

Xiaofeng Liu
School of Pharmacy
East China University of Science and Technology
Shanghai
China

Ryan E. Looper
Department of Chemistry
University of Utah
Salt Lake City, Utah

Clinton Maddox
Southern Research Institute
Drug Discovery Division, High Throughput Screening
Center
Birmingham, Alabama

Akihisa Matsuyama
Chemical Genomics Research Group/Chemical Genetics
Laboratory
RIKEN Advanced Science Institute
Japan

Thomas Mayer
Department of Medicine
University of Massachusetts Medical School
Worcester, Massachusetts

Cheryl L. Meyerkord
Department of Pharmacology and
Emory Chemical Biology Discovery Center
Emory University
Atlanta, Georgia

Jaeki Min
Department of Chemistry and
Emory Chemical Biology Discovery Center
Emory University
Atlanta, Georgia

Teresa Moran
Catalan Institute of Oncology
Hospital Germans Trias i Pujol
Badalona (Barcelona)
Spain

Kieron M. G. O'Connell
Department of Chemistry
University of Cambridge
Cambridge
England

Ronald J. Quinn
Eskitis Institute
Griffith University
Brisbane, Queensland
Australia

Brian Revennaugh
Department of Pharmacology and
Emory Chemical Biology Discovery Center
Emory University
Atlanta, Georgia

Rafael Rosell
Catalan Institute of Oncology
Hospital Germans Trias i Pujol
Badalona (Barcelona)
Spain

Stewart P. Rudnicki
Harvard Medical School
Harvard University
Boston, Massachusetts

Andreas Russ
Department of Biochemistry
University of Oxford
Oxford
England

Virginia M. Salas
Department of Biology
Northern New Mexico College
Española, New Mexico

Stephan Schürer
Department of Pharmacology, Miller School of Medicine
Center for Computational Science
University of Miami
Miami, Florida

Eduard A. Sergienko
Conrad Prebys Center for Chemical
Genomics
Sanford-Burnham Medical Research Institute
La Jolla, California

Caroline E. Shamu
Harvard Medical School
Harvard University
Boston, Massachusetts

Larry A. Sklar
University of New Mexico Center for Molecular
Discovery
Cancer Research and Treatment Center
Albuquerque, New Mexico

Melinda I. Sosa
Southern Research Institute
Drug Discovery Division, High Throughput Screening
Center
Birmingham, Alabama

Richard J. Spandl
Department of Chemistry
University of Cambridge
Cambridge
England

David R. Spring
Department of Chemistry
University of Cambridge
Cambridge
England

Antonella Squarcia
Menarini Ricerche
Pomezia
Rome
Italy

J. Jacob Strouse
University of New Mexico Center for Molecular
Discovery
Cancer Research and Treatment Center
Albuquerque, New Mexico

Zurab Surviladze
University of New Mexico Center for Molecular
Discovery
Cancer Research and Treatment Center
Albuquerque, New Mexico

Weining Tang
Winship Cancer Institute
Emory University
Atlanta, Georgia

Miguel Taron
Catalan Institute of Oncology
Hospital Germans Trias i Pujol
Badalona (Barcelona)
Spain

Pahk Thepchatri
Department of Chemistry
Emory University
Atlanta, Georgia

Gemma L. Thomas
Department of Chemistry
University of Cambridge
Cambridge
England

Natasha Thorne
NIH Chemical Genomics Center
National Institutes of Health
Bethesda, Maryland

Gregory P. Tochtrop
Department of Chemistry
Case Western Reserve University
Cleveland, Ohio

Nicola J. Tolliday
Broad Institute
Cambridge, Massachusetts

Matthew L. Tomlinson
School of Biological Sciences
University of East Anglia
Norwich Research Park
Norwich
England

Shaomeng Wang
Cancer Center and Departments of Internal Medicine, Pharmacology, and Medicinal Chemistry
University of Michigan
Ann Arbor, Michigan

Grant N. Wheeler
School of Biological Sciences
University of East Anglia
Norwich Research Park
Norwich
England

Arron S. Xu
Biomarker Center of Excellence
Covance Laboratories, Inc.
Greenfield, Indiana

Jie Xu
Electro-Optical Systems Laboratory
Georgia Tech Research Institute
Atlanta, Georgia

Chao-Yie Yang
Cancer Center and Departments of Internal Medicine, Pharmacology, and Medicinal Chemistry
University of Michigan
Ann Arbor, Michigan

Yoko Yashiroda
Chemical Genomics Research Group/Chemical Genetics Laboratory
RIKEN Advanced Science Institute
Japan

Minoru Yoshida
Chemical Genomics Research Group/Chemical Genetics Laboratory
RIKEN Advanced Science Institute
Japan

Mingyue Zheng
Shanghai Institute of Materia Medica
Chinese Academy of Sciences
Shanghai
China

Preface

Chemical Genomics provides a starting point for scientists and students who are interested in the discovery and application of active chemical compounds for the interrogation of biological systems and improvement of human health in the genomics era. Despite the clear realization of the immense power of chemical strategies in biology and medicine, the discovery of effective small molecule modulators for a particular biological target or phenotype remains a daunting challenge. While "omics" studies have revealed an expansive landscape of various biological systems and disease models, only a limited number of defined, biologically active compounds are available for functional exploration and for therapeutic discovery. Meanwhile, an expanding research area that uses information from genomes to develop chemical modulators for biology and therapeutics is rapidly developing due to major advances in chemistry, biology, and genomics, coupled with laboratory automation and computational technologies. This new field, chemical genomics, holds promise for transforming biological research and medicine using the power of chemistry.

The book *Chemical Genomics* offers a collection of insightful contributions from international experts in the field and aims to further the study of chemical genomics. It covers a range of essential topics, including the basic principles of chemical biology and chemical genomics (Section One: Overviews), the fundamentals of chemical libraries (Section Two: Molecules for Chemical Genomics), the essentials of high-throughput screening and high-content screening (Section Three: Basics of High-Throughput Screening), and the design of innovative biochemical, cellular, and organismal bioassays and computational screening approaches (Section Four: Chemical Genomics Assays and Screens). The final section (Section Five: Chemical Genomics and Medicine) includes chapters that link chemical genomics with the movement toward personalized medicine and the druggable genome of the future. In selected chapters, a resource section is presented for further information. My hope is that this book provides an introduction to the principles of chemical genomics; serves as a resource for the associated technologies; and gives readers general concepts, useful vocabulary, practical techniques, and broad, forward-thinking perspectives.

My deepest appreciation goes to the contributors of each chapter for their dedication, commitment, and exceptional patience! Together, we have now an introductory book for the field of chemical biology and chemical genomics.

Haian Fu

SECTION ONE

OVERVIEWS

CHAPTER 1

Harnessing the Power of Chemistry for Biology and Medicine

Cheryl L. Meyerkord
Haian Fu

Small molecule natural products and synthetic compounds have long been developed and housed by pharmaceutical companies and biotechnology firms for drug discovery. In the past, academic scientists studied important biological processes and drug targets while industry developed drugs directed to these well-studied and validated targets. Recently, two major paradigm shifts in the field of small molecule drug discovery have occurred: 1) an expanded role for academia in drug discovery, and 2) an increased focus on developing small molecules as chemical tools for basic research. Academic institutions have now moved into the arena of small molecule drug discovery by establishing various centers or institutes with high-throughput screening (HTS) and medicinal chemistry capabilities. At the same time, bioactive small molecule compounds are being identified and characterized to serve not only as drug leads but also as molecular tools, or chemical probes, to elucidate principles of biological processes. These transitions are primarily driven by several major developments: 1) Complete sequencing and analysis of the human genome and subsequent comparative and disease genomics have revealed a large number of potential new drug targets, which also provides enormous opportunities for functional investigations; 2) Advances in combinatorial chemistry have made small molecule compounds, which once were exclusively proprietary property of private sectors, commercially available to academic institutions; 3) The cost of acquiring robots for laboratory automation and informatics has been significantly reduced, thus making these essential resources accessible to academic investigators; and 4) Recognition of the tremendous potential of these unique opportunities and subsequent investment by various academic institutions and the National Institutes of Health (NIH) has catalyzed this transition from an almost exclusively industrial endeavor to an academic pursuit. Today, the concept that academic institutions have the ability to discover active small molecule compounds and use them to decipher the function of important biological processes has become a reality. These advances have greatly accelerated the development of the chemical biology field and have allowed scientists to harness the power of chemistry to elucidate biological processes and transform medicine.

Chemical biology is a discipline that combines the principles and practices of chemistry, biology, high-throughput assay and screening technology, and informatics to discover small molecule modulators to understand the basis of life. With recent advances in genomics, this discipline is now capable of using chemical tools to address complex biological questions at the genome level and is expected to rapidly transform biomedical research and the future of patient care.

1. CHEMICAL PROBES FOR THE STUDY OF BIOLOGY AND GENOMES

1.1. *Advantages of using small molecules as chemical probes.* Biological approaches, such as traditional genetics and recent RNA interference, are indispensible for elucidating fundamental principles of life and identifying structural elements involved in the complex networks of biological systems. However, there are certain unique advantages to using cell-active small organic compounds. For instance, 1) cell-permeable small molecules are applicable to genetically intractable organisms; 2) these compounds allow experiments to be performed in a dose-dependent and reversible manner, thus providing unparalleled temporal control over target gene products under native physiologic conditions; 3) compounds with desired specificity can be used to modulate an individual function of a multifunctional protein; and 4) bioactive compounds that regulate disease-related targets could be further developed as leads for drug discovery. Indeed, such small molecules have been successfully used in biological studies and have shown remarkable potential for post–genome era investigations.

1.2. *From chemical biology to chemical genomics.* Several terminologies have appeared in the literature to describe three intrinsically interrelated fields at the center of the chemistry-biology continuum: chemical biology, chemical genetics, and chemical genomics. Chemical biology is a term used to describe a general discipline that uses chemical compounds as tools for functional and mechanistic investigation of biological systems. Chemical genetics uses individual small molecules in place of random mutagens or mutations, as in traditional genetics, to discover the gene/protein of interest (forward genetics) or the phenotype and function of a gene/protein of interest (reverse genetics) (see Chapters 2 and 3 for detailed discussions). On a genome-wide scale, chemical genomics aims to systematically explore the interactions between small molecules and biological systems for basic research and therapeutic development. Armed with a wealth of genomics information, chemical genomics seeks 1) to identify and characterize bioactive small molecules through HTS or high-content screening (HCS) supported by informative bioassays and reiterative chemical optimization, and 2) to use these bioactive compounds as chemical tools to study the vast biological target space and to discover novel therapeutics. In addition, chemical genomics seeks to understand the impact of a target-specific compound at the genome level through the use of systematic genomics tools, such as global expression profiling and systems biology. This truly multidisciplinary field employs principles and methodologies from chemistry, biology, genomics, and informatics and is powered by computational tools and laboratory automation-enabled HTS technologies.

1.3. *Discovery of bioactive chemical probes.* A number of experimental and computational approaches are employed to identify bioactive compounds for chemical biology and chemical genomics. Among these approaches, a primary HTS or HCS assay in a microplate format is often used to screen a library of small molecule chemicals for positive hits that give rise to a desired effect. A chemical library with a small number of molecules, such as the Library of Pharmacologically Active Compounds (LOPAC1280, Sigma-Aldrich), is frequently used in a pilot screen for assay validation, which is then followed by screening of a larger-scale library. The size and type of the selected chemical library depend on the nature of the assay, target class, and resources available. Positive hits from HTS or HCS are identified and ranked with the aid of informatics tools and confirmed by biologically relevant secondary assays. Next, the nature of the chemical/biological target interaction is determined. Structural clustering of confirmed positives identifies characteristics responsible for bioactivity. Iterative chemical synthesis is then coupled with bioassays and structure-activity information to produce an optimized probe compound with the desired activity and specificity. As an alternative, computational approaches are also used for virtual screening to identify active chemicals in conjunction with a defined bioassay. These active probe compounds are next used to systematically examine the function of biological targets or processes and their mechanisms of action. For drug discovery, additional pharmacologic parameters are used to filter compounds with desired bioavailability, pharmacokinetics, and toxicology profiles. Details on chemical libraries, essentials of HTS and HCS, and various assay technologies are described in Sections 2-4 of this book. Hopefully, scientists eventually will be able to associate each gene product with at least one active compound and to annotate genomes using the language of chemistry in such a way as to uncover new principles of biology and molecular bases of diseases.

2. CHALLENGES AND OPPORTUNITIES

2.1. *Searching for biologically relevant chemical space.* The identification of active small molecules in an HTS, HCS, or virtual screening campaign relies on the presence of certain "chemical space" in a library that is complementary to explored biological target space (Figure 1.1). Chemical space, or "the chemical universe," encompasses the set of all possible molecular structures and is characterized by multiple molecular dimensions. Such dimensions, or chemical descriptors, within this space include surface area, charge, hydrophobicity, and number of hydrogen bond donors or acceptors. The size and complexity of a particular chemical space largely depends on which set of dimensions or chemical descriptors one chooses to use. Thus, it is not surprising that the estimated number of molecular structures with drug-like characteristics that can possibly be made varies greatly, from 10^{18} to 10^{200}. The widely cited number of chemicals with drug-like properties that could potentially be synthesized is in excess of 10^{60}, according to Bohacek. Interestingly, the Beilstein database indicates that only about 10^{7} molecules already have been made. Thus, only a minute fraction of chemical space is currently accessible, of which only an even smaller fraction can interact with biological systems. It certainly is a daunting challenge to even try to attempt to fully populate the vast regions of chemical space. Although combinatorial chemistry has increased the number of small molecule compounds, we are far from achieving the chemical diversity that is necessary for effective interaction with biological systems. While it is vital to develop ground-breaking and innovative synthetic chemistry strategies and to effectively use natural product–based scaffolds as unique templates for new chemistry, biology-guided synthesis offers immediate opportunities to populate biologically important chemical space.

Regions of chemical space in which biologically active compounds reside are termed biologically

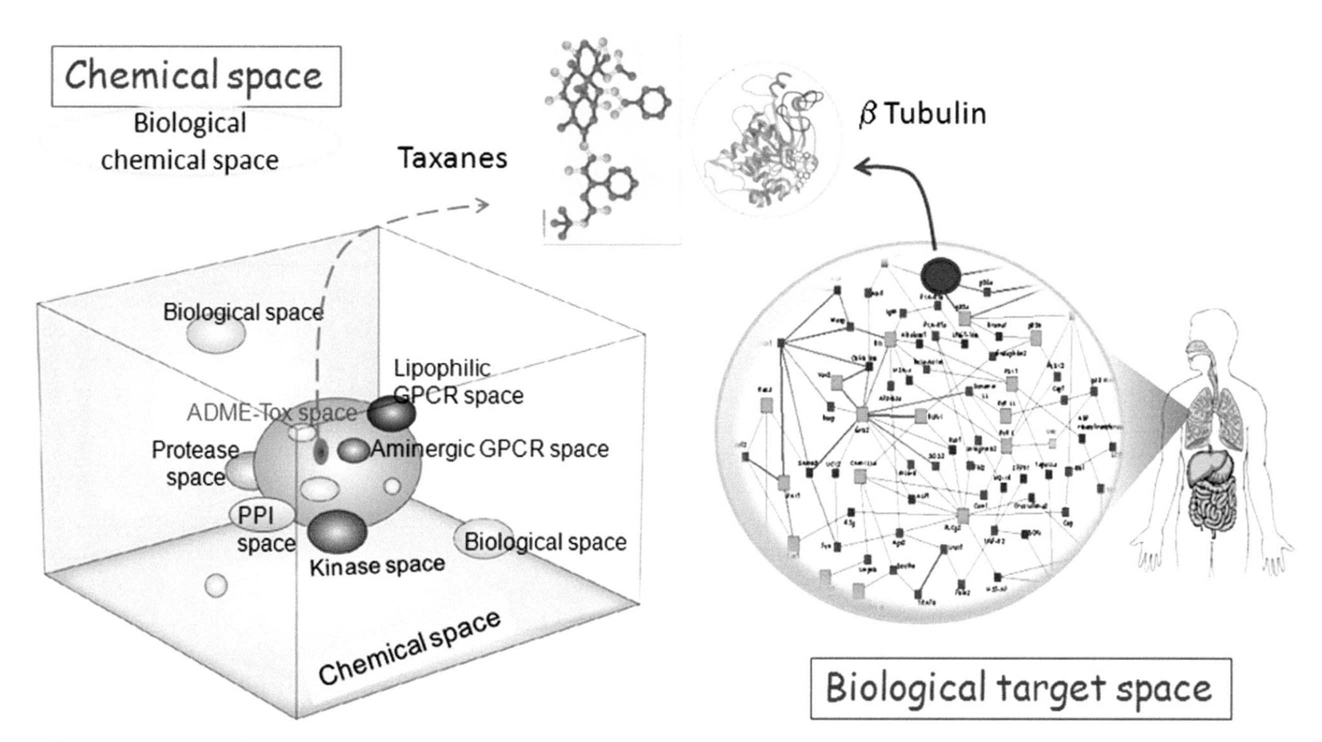

FIGURE 1.1:** Interaction of chemical space with biological systems. Chemical genomics seeks to populate biological target space with small molecule probes to elucidate the basis of life. Among all of the small molecule organic compounds that are possible to synthesize (chemical space; left box), some possess properties that allow them to interact with biological systems (right panel). These compounds reside in particular regions of chemical space termed biological chemical space (such as discreet chemical space that interacts with kinases, proteases, GPCRs, or PPI interfaces). Some of these compounds, such as taxanes, have additional favorable pharmacologic properties allowing them to serve as therapeutic agents (drugs). Taxanes specifically interact with and stabilize β tubulin in human target cells and act as effective anticancer agents. Thus, identification of bioactive compounds that target biological systems will lead to the population of biologically relevant chemical space. Some of these compounds may possess additional properties allowing them to serve as drugs and occupy therapeutic chemical space. The contents of this figure were adapted by permission from Macmillan Publishers Ltd (Smith, A. (2002) Exploring chemical space.* **Nature Reviews Drug Discovery 1, 95.) and Elsevier (Holstein, S. A. and Hohl, R. J. (2009) Lung Cancer. In* **Pharmacology and therapeutics: Principles to Practice** *(p. 924) Philadelphia: Saunders.)*

relevant chemical space, or biological chemical space. Small molecule compounds occupying biological chemical space are defined by a wide range of molecular descriptors, including effect on enzyme activity, ability to cause phenotypic changes, and pharmacologic parameters. Thus, these biological descriptors can be classified as bioactivity, phenotypic, or pharmacologic descriptors. Bioactivity descriptors used to define compounds in chemical space include conventional enzyme-modulating activities, such as the ability to inhibit or activate a protein kinase or protease. For cell-based assays, biologically active compounds can be defined by activity that results in a change of phenotype or a biological reporter readout, that is, phenotypic descriptors. For probing cell biology and drug discovery, cell permeability, accumulation, and bioavailability are additional parameters that are used to define compounds – that is, pharmacologic descriptors. To ensure effective application of compounds, it is essential that compounds are able to enter cells and reach the action site to exert their effect. Thus, HTS or HCS bioassays that give rise to meaningful biological or pharmacologic readouts provide informative molecular descriptors to assign compounds into different regions of chemical space. Therefore, chemical synthesis directed by results of bioassays offers opportunities for effective development of bioactivity-rich libraries and for the identification and optimization of biologically active chemicals. The decoding of the human genome has illuminated vast expanses of biological target space, thereby allowing innovative assay design in revealing and expanding biologically relevant chemical space.

2.2. *Innovative bioassay platforms for expanded biological descriptors.* As a result of recent biological and technological advances, novel bioassay platforms also have evolved. Traditionally, widely used HTS assays are limited to a few classes of biological molecules and processes, such as various enzyme reactions, ligand-binding assays, G protein-coupled receptors, and transcription factor reporters. This is largely due to their therapeutic importance and tractability for HTS operations. Although these bioassays remain powerful and popular in drug and probe discovery, new advances in high-throughput technologies have revolutionized the small molecule discovery process. It is now possible to monitor sophisticated biological events in an automated fashion. For instance, automated imaging systems enable the visualization of molecular movement at the subcellular level in a microplate format for HCS. Such HCS assays are widely used to monitor a broad spectrum of biological processes with multiple readouts, such as the subcellular localization of signaling molecules or transcription factors, neurite outgrowth, and stem cell differentiation. State-of-the-art flow cytometry technology has led to highly efficient multiplexed HTS. In addition, the application of various label-free biosensor technologies makes targets that were once intractable for HTS, such as complex pathways in primary cells, tractable. It is now feasible to screen for bioactive compounds that modulate a defined molecular event or a phenotypic change in cell- or whole organism–based systems. These phenotypic and label-free biosensor assays expand the repertoire of practical biological descriptors for classifying bioactive chemicals.

In parallel, advances in genomics and subsequent biological insights have dramatically expanded the scope of molecular targets and biological processes available for interrogation by small molecules. In essence, advances in our understanding of biology have widened the biological chemical space. Traditional assays, such as enzyme assays, have evolved and now have been applied to newly discovered enzymes or new functions of a known enzyme. For example, new members of the protease class, such as caspases in apoptosis and inflammatory signaling pathways, have been identified from basic research. The opportunity to find small molecules that target these proteases defines new biologically relevant chemical space. Furthermore, it is increasingly recognized that biological processes are mediated and dictated by a network of macromolecular interactions, mainly protein-protein interactions (PPIs). A change in a particular PPI may lead to functional alterations of biological systems. Thus, protein-protein interaction interfaces, with their own unique physicochemical characteristics, define novel biological target space that can be interrogated with various chemicals. Technologies that examine PPI by HTS and HCS are rapidly evolving to allow for efficient discovery of small molecule PPI modulators. Chemicals that interfere with these surface interactions, which reside in PPI chemical space, are expected to have a significant impact on biology.

2.3. *Unexplored biological target space*. Although enzymes, receptors, channel proteins, transcription factors, and PPI interfaces continue to dominate the bioassay and HTS/HCS field, new genomics-driven target areas have begun to emerge. Novel opportunities have arisen from the dissection of genome structure and function, which has revealed a vast, unexplored biological target space for small molecule probe and drug discovery. Interestingly, fewer than 2% of the 3 billion base pairs of the human genome are used to encode functional genes. A better understanding of epigenetic mechanisms has rendered bioassays and HTS/HCS approaches applicable to the discovery of small molecule epigenetic modulators. For example, the elucidation of the histone code makes it possible to screen for small molecules that control this mechanism and regulate the gene expression machinery. The genome also encodes a large number of functional small RNAs, such as microRNAs. Understanding how these small RNAs participate in and regulate cellular processes and how they themselves are controlled will lead to the design of new bioassays to identify compounds to selectively control small RNA-mediated functions. In addition, large numbers of repetitive DNA sequences exist that are critical for chromatin structure and important for cellular functions and organism development. Furthermore, sequences with totally unknown biological functions still remain. It is envisioned that insights into these new biological targets will lead to novel ways of designing HTS or HCS assays for small molecule modulator discovery. Conversely, phenotype-based, innovative HTS or HCS assays may help us to unveil functions of uncharacterized segments of the genome through the identification and use of bioactive chemical probes.

2.4. *Accelerated collision of chemical and biological target space*. Advances in genomics and biology drive bioassay design, chemical probe identification, and drug discovery. Identified bioactive molecules in turn provide unparalleled chemical tools that will enable the accelerated discovery of fundamental biological principles. Similar to the manner in which the human genome project accelerated the way we address biological questions and mechanisms of diseases, the chemical genomics pursuit to populate each gene product and its interfaces with bioactive compounds is expected to revolutionize the process of biological functional studies at the cellular and organism levels. These advances make drug discovery a natural and integrated part of biological investigations.

It certainly is a formidable challenge to define at least one bioactive compound for each gene product in the genome. At the current stage, however, we have to be selective in choosing target areas and biological systems with fundamental importance in biology and/or therapeutic potential. In this regard, comparative genomics and pathogen genomics have begun to provide some guidance for target prioritization. For example, the Cancer Genome Atlas program sponsored by the National Cancer Institute (NCI) aims to complete the sequencing of patient tumor genomes from a number of cancer types, including brain, lung, and ovarian cancers. This initiative has identified networks of genes critical for tumorigenesis and progression as well as those required to control normal cell growth. Developing bioassays to identify chemical probes for these identified gene products may lead to chemical libraries that will be used to populate important disease-related biological space. It is expected that promising bioactive compounds could be further developed into the next generation of targeted cancer therapeutic agents. The collision of chemical and biological space, accelerated by chemical genomics, is expected to power intimate collaborations between academia and industry, result in truly innovative therapeutic development, and transform the future of basic and clinical research.

3. RESOURCES

3.1. *Industrial experience.* Through their intensive drug discovery and development activities, pharmaceutical companies have accumulated valuable experience in the identification of bioactive compounds and their progression through the pipeline from hits to leads to drugs. Experienced pharmaceutical scientists and their evolving practices in assay development, HTS, and medicinal chemistry are invaluable resources for harnessing the power of chemistry for biology and medicine. Major professional societies garner industrial talents with increasing academic participation, including the Society for Biomolecular Sciences (SBS; http://www.sbsonline.org) and the Association for Laboratory Automation (ALA; http://www.labautomation.org). Recently, SBS and ALA have merged as sections of the newly formed Society for Laboratory Automation and Screening (SLAS; http://www.sbs-alamerger.org/) – an inclusive worldwide organization dedicated to advancing scientific research and discovery through laboratory automation and screening technology. Each section will continue to pursue its current mission while collectively addressing the SLAS mission: to provide forums for education and information exchange to encourage the study of, and improve the science and practice of, laboratory automation and screening. Their Web sites contain educational materials, such as SBS reference libraries with a collection of screening and drug discovery–related terminologies. Eli Lilly and Company and the NIH Center for Translational Therapeutics have collaborated to create another valuable resource, the *Assay Guidance Manual Version 5.0*, which is freely available online (http://assay.nih.gov/assay/index.php/Table_of_Contents).

3.2. *Academic resources.* Another major resource for studying chemical genomics comes from academic centers and facilities devoted to accelerating chemical probe and small molecule drug discovery. As of 2011, there were eighty-five such centers and facilities from all over the world listed on the SLAS Web site (http://www.slas.org/screeningFacilities/facilityList.cfm), including the Harvard Institute of Chemistry and Cell Biology (ICCB)-Longwood Screening Facility and the High-Throughput Drug Screening Facility of Memorial Sloan-Kettering Cancer Center. Distinct expertise and various capabilities can be found at different centers. In recognition of the unique historical opportunities and the potential of using small molecules for biology and medicine, government research agencies have launched major new initiatives. These initiatives include the US NIH Molecular Libraries and Imaging initiative (MLI) (http://nihroadmap.nih.gov/molecularlibraries/), the Genome Canada Chemical Biology Project (http://www.genomecanada.ca/), the European Society of Combinatorial Sciences (http://www.combichem.org/), European Infrastructure of Open Screening Platforms for Chemical Biology (http://www.eu-openscreen.eu/), the Chinese Chemical Biology initiative (http://www.nature.com/nchembio/journal/v4/n9/full/nchembio0908-515.html), and the Japanese Society for Chemical Biology (http://www.jscb.jp/index_e.html). As a part of the Roadmap Initiative (http://nihroadmap.nih.gov/), the NIH MLI established the Molecular Libraries Screening Centers Network (MLSCN) (http://mli.nih.gov/mli/secondary-menu/mlscn/) as a pilot program and the subsequent Molecular Libraries Probe Production Centers Network (MLPCN) (http://mli.nih.gov/mli/mlpcn/). This program encourages investigators to submit HTS or HCS assays to stimulate the discovery of novel small molecule tools for chemical biology. These screening centers house collections of compound libraries, various HTS and HCS platforms, and specialized probe discovery and/or medicinal chemistry capabilities. The MLSCN and MLPCN programs are intended to complement, rather than duplicate, drug discovery efforts carried out by the pharmaceutical industry. The results of the MLSCN and MLPCN projects are deposited in public domains, with PubChem as a major portal. PubChem (http://pubchem.ncbi.nlm.nih.gov) is an open-access data repository system housing detailed protocols of bioassays screened in the network and an array of associated chemical and biological data for a large NIH chemical library. In 2009, the NCI launched the Chemical Biology Consortium (CBC) (http://dctd.cancer.gov/CurrentResearch/ChemicalBioConsortium.htm). This consortium focuses on the discovery and development of novel, innovative small molecule cancer therapeutics for unmet medical needs. Institutes participating in these national initiatives are listed below along with their Web sites, when available.

- Broad Institute Comprehensive Screening Center, Cambridge, Mass. (MLPCN; http://www.broadinstitute.org/chembio/mlpcn)
- Emory Chemical Biology Discovery Center, Atlanta, Ga. (CBC, MLSCN; http://www.pharm.emory.edu/ECBDC/)
- Small Molecule Discovery Center at the University of California, San Francisco, San Francisco, Calif. (CBC; http://smdc.ucsf.edu/about/index.htm)
- Georgetown University Medical Center, Washington, D.C. (CBC; http://gumc.georgetown.edu/)
- Johns Hopkins Ion Channel Center, Baltimore, Md. (MLPCN; http://www.jhicc.org/)
- Kansas Specialized Chemistry Center, Lawrence, Kan. (MLPCN; http://www.scc.ku.edu/)
- MLSCN Center at Columbia University, New York, N.Y. (MLSCN; http://genomecenter.columbia.edu/)
- New Mexico Center for Molecular Discovery, Albuquerque, N.M. (MLSCN, MLPCN; http://nmmlsc.health.unm.edu/)
- The NIH Center for Translational Therapeutics (NCTT), Bethesda, Md. (CBC, MLSCN, MLPCN; http://www.nctt.nih.gov/)
- North Carolina Comprehensive Chemical Biology Center, Chapel Hill, N.C. (CBC; http://pharmacy.unc.edu/labs/center-for-integrative-chemical-biology-and-drug-discovery)
- The Penn Center for Molecular Discovery, Philadelphia, Pa. (MLSCN; http://www.seas.upenn.edu/~pcmd/)
- Sanford-Burnham Conrad Prebys Center for Chemical Genomics, La Jolla, Calif. (CBC, MLSCN, MLPCN; http://www.sanfordburnham.org/default.asp?contentID=902)
- Scripps Research Institute Molecular Screening Center, La Jolla, Calif., and Jupiter, Fla. (MLSCN, MLPCN; http://mlpcn.florida.scripps.edu/)
- Southern Research Institute, Birmingham, Ala. (CBC, MLSCN, MLPCN; http://www.southernresearch.org/)
- SRI International, Menlo Park, Calif. (CBC; http://www.sri.com/)
- The University of Minnesota Chemical Diversity Center, Minneapolis, Minn. (CBC; http://next.cancer.gov/discoveryResources/cbc_minnesota.htm)
- The University of Pittsburgh Chemical Diversity Center/University of Pittsburgh Specialized Application Center, Pittsburgh, Pa. (CBC, MLSCN; http://pmlsc.pitt.edu/)
- Vanderbilt Chemical Diversity Center, Nashville, Tenn. (CBC, MLSCN, MLPCN; http://www.vanderbilt.edu/mlscn/Templates/index.htm, http://www.mc.vanderbilt.edu/centers/mlpcn/index.html)

3.3. *Databases and publications.* In addition to information that can be obtained from various organizations and institutions, a plethora of information is readily available on the Internet. Internet resources include databases, software, protocol and assay guidance, and literature, including journal articles and books. Web sites containing online databases include Blueprint's Small-Molecule Interaction Database (SMID) (http://smid.blueprint.org), ChemBank (http://chembank.broadinstitute.org), Chembionet (http://www.chembionet.info/), ChemDB (http://cdb.ics.uci.edu), Chemical Entities of Biological Interest (ChEBI) (http://www.ebi.ac.uk/chebi/), ChemMine (http://bioweb.ucr.edu/ChemMineV2/), FDA drug databases (Drugs@FDA, http://www.accessdata.fda.gov/scripts/cder/drugsatfda/; the Orange Book, http://www.accessdata.fda.gov/scripts/cder/ob/default.cfm); and PubChem (http://pubchem.ncbi.nlm.nih.gov). The NIH Center for Translational Therapeutics's Web site contains links to free software (http://nctt.nih.gov/27543665), as well as assay and protocol guidance as mentioned earlier. Current Protocols in Chemical Biology (http://cda.currentprotocols.com/WileyCDA/CPTitle/isbn-0470559268.html) details some tested methods. A number of scientific publications offer platforms for effective communications among chemical biologists or anyone interested in the field. Journals with a key focus on chemical biology include the following:

- *ACS Chemical Biology* (http://pubs.acs.org/journal/acbcct)
- *BMC Chemical Biology* (http://www.biomedcentral.com/bmcchembiol/)
- *Chemical Biology & Drug Design* (http://onlinelibrary.wiley.com/journal/10.1111/(ISSN)1747-0285)
- *Chemistry & Biology* (http://www.cell.com/chemistry-biology/)
- *Current Chemical Biology* (http://www.bentham.org/ccb/index.htm)
- *Current Chemical Genomics* (http://www.bentham.org/open/tochgenj/)

- *Current Opinion in Chemical Biology* (http://www.current-opinion.com/JCHL/about.htm?jcode=jchl)
- *Highlights in Chemical Biology* (http://www.rsc.org/publishing/journals/cb/Index.asp)
- *Journal of Chemical Biology* (http://www.springerlink.com/content/120990/)
- *Nature Chemical Biology* (http://www.nature.com/nchembio/index.html)

The resources in section 3 of this chapter provide a starting point for readers. Additional resources related to a particular subject can be found in subsequent chapters.

SUGGESTED READINGS

Aguero, F., Al-Lazikani, B., Aslett, M., Berriman, M., Buckner, F. S., Campbell, R. K., Carmona, S., Carruthers, I. M., Chan, A. W., Chen, F., et al. (2008). Genomic-scale prioritization of drug targets: the TDR Targets database. *Nat Rev Drug Discov 7*, 900–907.

Crews, C. M. (2010). Targeting the undruggable proteome: the small molecules of my dreams. *Chem Biol 17*, 551–555.

Dobson, C. M. (2004). Chemical space and biology. *Nature 432*, 824–828.

Lipinski, C., and Hopkins, A. (2004). Navigating chemical space for biology and medicine. *Nature 432*, 855–861.

Paolini, G. V., Shapland, R. H., van Hoorn, W. P., Mason, J. S., and Hopkins, A. L. (2006). Global mapping of pharmacological space. *Nat Biotechnol 24*, 805–815.

Schreiber, S. L. (2005). Small molecules: the missing link in the central dogma. *Nat Chem Biol 1*, 64–66.

Smukste, I., and Stockwell, B. R. (2005). Advances in chemical genetics. *Annu Rev Genomics Hum Genet 6*, 261–286.

Wells, J. A., and McClendon, C. L. (2007) Reaching for high-hanging fruit in drug discovery at protein-protein interfaces. *Nature 450*, 1001–1009.

CHAPTER 2

Exploring Biology with Small Organic Molecules

Nicholas Aberle
Craig M. Crews

1. FROM CHEMICAL GENETICS TO CHEMICAL GENOMICS

Small molecules have long been recognized as an invaluable part of human medicine. Well before chemistry, biology, and chemical genomics were established as scientific disciplines, people routinely used plants to treat and prevent illnesses. Not until the advent of modern science, however, were we able to isolate and identify the active ingredients that conferred on the flora around us such a vast array of effects. Along with synthetic and semisynthetic molecules, these naturally occurring compounds have formed the cornerstone of today's therapeutics and have provided a way of improving our understanding of fundamental biological processes.

Exploration of these processes using chemical genetics requires certain basic elements common to all studies: 1) a small molecule with good target specificity, 2) a protein of interest, 3) a phenotype that is being investigated, and 4) an assay or screen that will bring these elements together to provide useful information (Figure 2.1).

Investigations often start with a small molecule that produces a specific phenotype, and the objective is to identify the protein that binds to the molecule and causes the phenotype by virtue of having its function modulated. This is referred to as *forward chemical genetics*, as it operates in a similar fashion to traditional forward genetics – by identifying the gene responsible for the phenotype. In *reverse chemical genetics*, the starting point is a protein of interest, and the endeavor focuses on identifying a small molecule that modulates its function and then observing the phenotype that is produced. This has parallels to reverse genetics, but here the abrogation of function is achieved through use of a small molecule rather than through genetic manipulation.

Given this description of chemical genetics, what *is* chemical genomics? Consistent with the difference between traditional genetics and genomics, chemical genomics can be thought of as the application of chemical genetic strategies to the genome as a whole. The sheer number of proteins encoded by the roughly twenty thousand genes in the human genome [1] highlights the overwhelming nature of this field: Discovering a small molecule ligand that can selectively modulate each function of each of these proteins represents a major challenge for scientists. Progress toward this goal will rely on advancing the techniques already in use in chemical genetics and adapting these techniques to the larger scale of chemical genomics.

The necessary adaptation revolves around a basic guiding principle: smaller, faster, cheaper [2], or, in other words, obtaining as much information as possible as quickly as possible in the most cost-effective way. It simply is not practical to probe ten thousand proteins in the same way you would probe one or two proteins. Similarly, synthesizing and testing one hundred thousand molecules (or more!) is a different proposition from performing a total synthesis of one natural product and testing its biological activity. Although the principles often remain the same, the details must be adjusted. Miniaturization and automation become indispensable allies.

The chapters that follow describe in detail many of the advances that have been made toward miniaturization and automation, such as high-throughput screening (HTS) and diversity-oriented synthesis (DOS), so these are discussed only briefly here. The remainder of this chapter focuses on the basics of chemical genetics to provide an understanding of the fundamentals that are then adapted to the greater goal of chemical genomics.

2. ORIGINS OF SMALL MOLECULES FOR CHEMICAL GENOMICS

As one might imagine, the small molecules used in a given assay are of utmost importance, and it is worth contemplating where these molecules come from to create a context within which we might consider how the field of chemical genomics has advanced.

2.1. Natural Products

Natural products typically are secondary metabolites: chemicals that are not directly involved in the growth or

FORWARD CHEMICAL GENETICS

Observe phenotype *Identify small molecule* *Determine protein target*

REVERSE CHEMICAL GENETICS

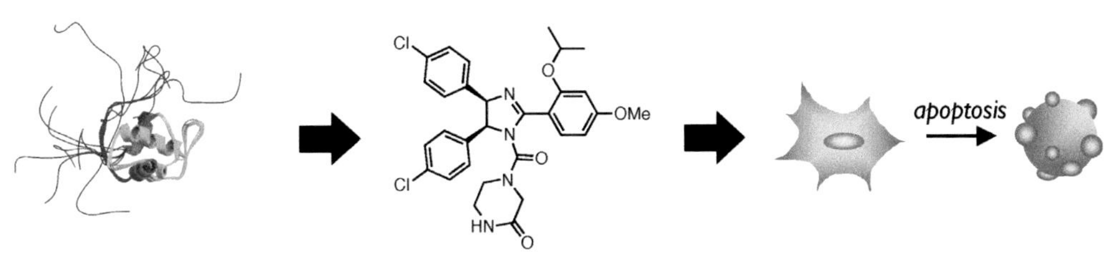

FIGURE 2.1: ***Schematics of forward and reverse chemical genetics. The forward chemical genetics example shows the discovery of fumagillin as a potent antibiotic, later found to inhibit angiogenesis through its interaction with the protein MetAP-2. In the reverse chemical genetics example, the important regulatory protein Mdm2 was found to have its function modulated by the small molecule called nutlin, which caused apoptotic cell death. [Fumagillin phenotype image reprinted by permission from Macmillan Publishers Ltd: Nature, 348, 555, copyright 1990; Mdm2 structure obtained from Protein Data Bank (1z1m)]***

development of the organism that produces them. Primary metabolites, essential for development, are broadly shared across a variety of species, but secondary metabolites are generally species specific and initially were considered to be waste products with no useful function. It was then discovered that these molecules controlled less fundamental processes, such as aspects of physical appearance and defense from pests and disease. With their origins in biological systems ranging from terrestrial bacteria to marine sponges, many of these small molecule secondary metabolites have been found to interact with human proteins, giving rise to the potential for therapeutic utility and use as biological probes.

A statistical analysis of approved therapeutics by Newman et al. highlights the importance of natural products as potential drugs [3]. Although only 6% of small molecule new chemical entities (NCEs) registered with the U.S. Food and Drug Administration (FDA) between 1981 and 2002 were actually natural products, a further 43% of NCEs either were derivatives of natural products or incorporated a naturally occurring pharmacophore. Despite providing such a significant proportion of NCEs over the past two decades, natural product research has been scaled down by large pharmaceutical companies [4].

Developments in the fields of structural biology, drug design, and combinatorial chemistry have fueled this shift in focus. Armed with the ability to either 1) make large numbers of compounds rapidly, followed by HTS, or 2) use structural information of target proteins to design a target-specific ligand, drug companies moved away from the somewhat serendipitous natural product field despite its many past successes. Several observers have drawn a connection between this shift and the recent decrease in annual NCE numbers [3–5]. It is worth noting, however, that advances in screening technology are equally applicable to natural product drug discovery and that use of combinatorial chemistry does not preclude the use of natural product scaffolds as the starting point for diversification (see notes 27–30 and associated text see page 14). Despite the advances just mentioned, early random synthetic libraries are failing to live up to expectations in terms of producing new drugs, prompting a resurgent interest in natural products [6].

Beyond their role as therapeutics, natural products have made significant contributions to the study of biological processes. As a leading paradigm of forward chemical genetics, mechanism of action (MOA) studies on naturally occurring compounds have enabled the elucidation

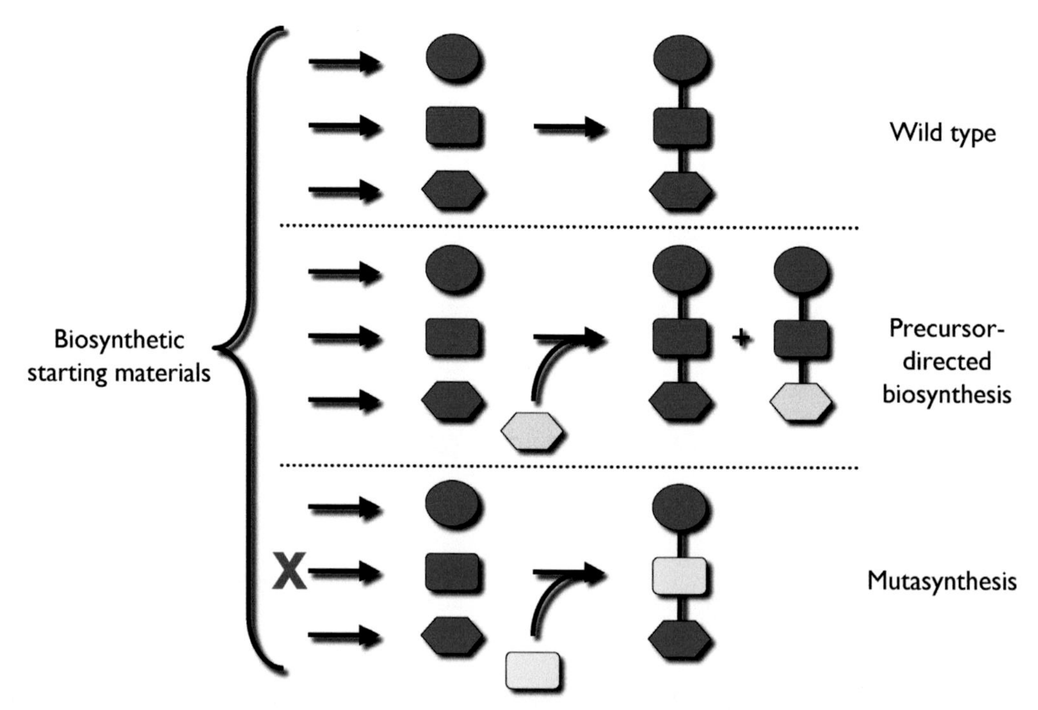

FIGURE 2.2: ***Biosynthesis of small molecules. In a wild type strain, the natural fragments are incorporated into the cartoon product shown. When an alternative fragment is fed to the organism in precursor-directed biosynthesis, both the original product and a new product can be formed. Using mutasynthesis, production of the original product is blocked, such that the only product formed is that incorporating the artificial fragment.***

of many important aspects of biology, examples of which are given later. Their application to larger-scale investigations is somewhat limited, however, by the rate at which the molecules themselves can be isolated, purified, and characterized. Many organisms produce only small numbers of compounds; others produce large numbers of similar compounds that can be difficult to separate through chromatography. Advances in separation techniques have expedited the purification process, as has bioassay-guided fractionation, which involves isolating individual compounds only from fractions that exhibit the desired activity.

One approach that has been adopted to combat any lack of structural diversity from a particular organism relies on intervening in the biosynthesis to encourage the organism to produce different compounds [7, 8] (Figure 2.2). The most straightforward examples of this are referred to as *precursor-directed biosynthesis*, in which synthetically prepared precursors are fed into the natural product–producing organism, such that these precursors are incorporated into analogs of the natural product through the usual operation of the biosynthetic pathway. As a result, a bacterium that previously produced only one compound might now be used to produce a library of derivatives based on the feeding of a range of starting materials. Genetic engineering of bacteria has created a subset of precursor-directed biosynthesis: The biosynthetic mechanism that produces the original natural product is blocked at one stage via deletion of the appropriate genes, meaning that feeding experiments are not competing with production of the original natural product, typically resulting in improved yields of the new compounds. This technique, known as *mutasynthesis* (for recent reviews, see references [7, 8]), is more powerful than standard precursor-directed biosynthesis (in which potentially only small quantities of new natural products may be obtained along with larger amounts of the original natural product). However, mutasynthesis can require significant effort to mutate the strain appropriately, although this is becoming easier as technology advances [9]. These biosynthetic strategies can help increase the number of "natural" products available for testing, but they do not bypass the need to isolate and purify the compounds.

Numerous recent applications of this methodology have been reported with the use of well-known natural products (Figure 2.3). For example, a thorough investigation of the biosynthetic gene cluster of rapamycin (**1**) [10], a potent immunosuppressant, enabled subsequent preparation of a range of derivatives using mutasynthesis [11, 12] and other analogs using precursor-directed biosynthesis [13]. Similarly, the proteasome inhibitor salinosporamide (**2**) has had both of its side chains modified with various artificial replacements, resulting in lower general toxicity without disrupting its ability to inhibit the proteasome [14, 15].

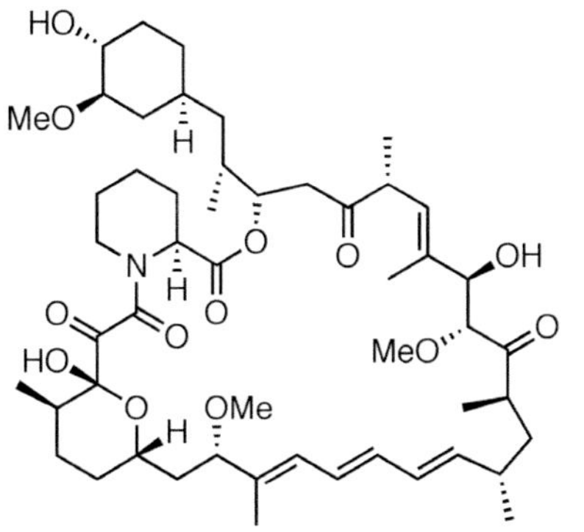

FIGURE 2.3: ***Structures of rapamycin (1) and salinosporamide (2).***

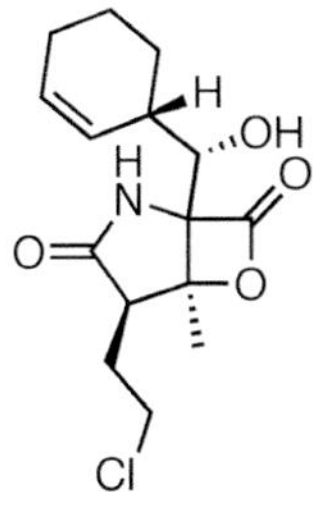

A related approach to increasing the number and diversity of available natural products is combinatorial biosynthesis [16]. This differs from mutasynthesis in that it takes genes encoding enzymes from various biosynthetic pathways and assembles them to form a new biosynthetic cluster that can produce novel compounds. Combinatorial biosynthesis shares many of the difficulties of mutasynthesis, such as the uncertainty regarding how to assemble the biosynthetic enzymes in a way that will produce compounds, but it is also a powerful way of generating novel structures.

Natural products have repeatedly demonstrated their utility as tools for studying biological systems, but the role of synthetic molecules should not be overlooked.

2.2. Synthetic Molecules

As alluded to earlier, the early combinatorial libraries did not provide the bountiful biologically active compounds that the technology promised to deliver [17]. Although technology had advanced to facilitate the rapid and impressive preparation of very large collections of molecules [18], these molecules typically differed only at their extremities, retaining the same core framework and providing a library more suited to probing a structure-activity relationship (SAR) than to sampling a broad chemical space; the same criticisms could be leveled at precursor-directed biosynthesis. To address this shortcoming, DOS, a new paradigm in library synthesis, began to take hold.

This is diametrically opposed to the well-established field of total synthesis or target-oriented synthesis, which aims to start from a broad base and converge into a single product (typically a natural product), although it often proceeds in a linear fashion. By contrast, DOS starts from a narrow base and aims to create high levels of chemical diversity as efficiently as possible (Figure 2.4). It routinely makes use of multicomponent reactions that can rapidly introduce diversity [19]; solid-phase synthesis techniques to facilitate separation of products; branching points from which

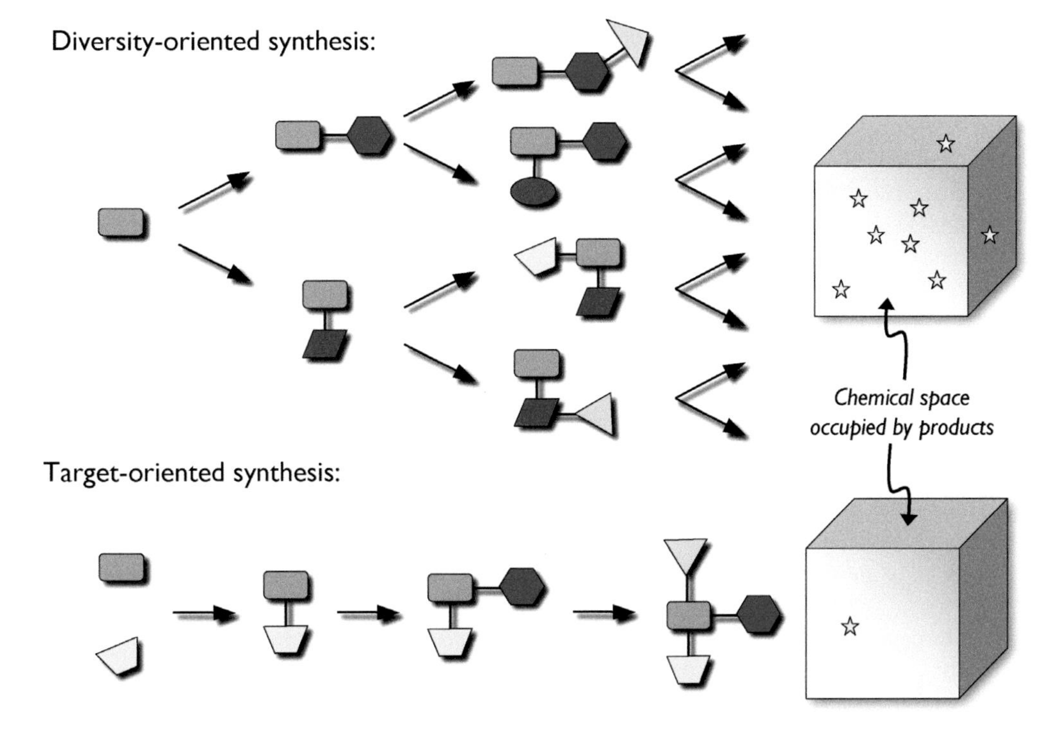

FIGURE 2.4: Diversity-oriented synthesis enables rapid preparation of skeletally diverse compounds, which occupy a wide range of chemical space. The increasing diversity introduced with each step is shown in contrast to the more linear and focused target-oriented synthesis.

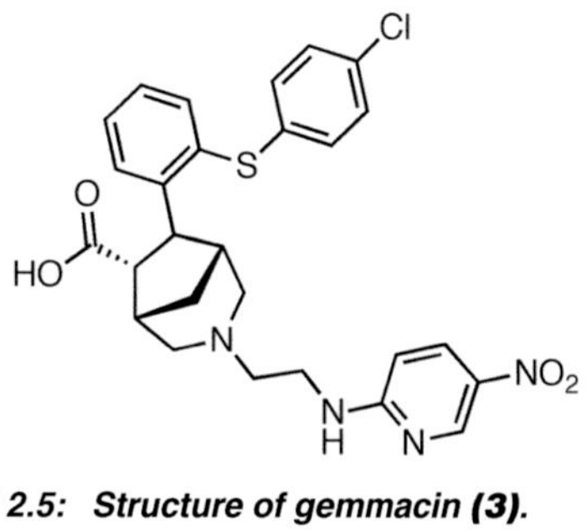

FIGURE 2.5: Structure of gemmacin (3).

a single substrate can undergo various reactions (rather than simply reacting with one type of functional group); and, perhaps most crucially, a forward synthetic analysis as opposed to the usual retrosynthetic analysis [20–22].

Some selected examples demonstrate the power of DOS and its potential to influence chemical genomics. Schreiber and co-workers have reported a method for the preparation of twelve distinct molecular skeletons from a single starting point: enamine-containing dihydropyridines and dihydroisoquinolines [23]. These building blocks are usually too unstable to be useful, but they were surprisingly stable when attached to a solid support. The authors found that these enamines could undergo a range of reactions (alkylations, reductions, and various cycloadditions) that allowed them to generate significant structural diversity.

In another excellent report, a relatively small library of 242 novel compounds was synthesized by the Spring group and tested for antibacterial activity [24]. Requiring between one and four steps to prepare, the compounds in the library represented eighteen natural scaffolds starting from twelve α,β-unsaturated acyl imidazolidinones, which in turn were obtained from a single, immobilized β-ketophosphonate. Molecules in the collection also contained a significant number of stereocenters, a feature of natural products that was absent from compounds in early combinatorial libraries. The most promising of these compounds, dubbed gemmacin (Figure 2.5, **3**), displayed potency comparable to erythromycin against *Staphylococcus aureus* and improved activity against methicillin-resistant strains of *S. aureus*. It is thought to exert its effect via disruption of the cell membrane. The molecule also showed broad growth inhibition against other Gram-positive bacteria, including vancomycin-resistant enterococci, but only limited inhibition of Gram-negative bacteria growth. The DOS of gemmacin allowed access to a structure that is unique among known antibacterials, indicating that it could form the basis of a new class of much-needed therapeutics.

In another DOS approach, Porco and co-workers have developed a strategy for converting densely functionalized Michael adducts into a range of carbon skeletons [25]. The adducts, obtained enantioselectively from a range of 1,3-diketones and β-nitrostyrenes, display three or four different functional groups that can be selectively "paired" with one another using controlled reaction conditions, thus providing access to a diverse set of fused and bridged ring systems.

Another development in the area of library preparation is termed *biology-oriented synthesis* (BIOS). Following a similar philosophy to Wender's function-oriented synthesis [26], Waldmann and colleagues have argued that natural products represent a prevalidated starting point for discovery of biologically relevant molecules [17] (see also [27]) and that, if the appropriate chemical space is investigated, the collection of molecules required to obtain useful hits can be relatively small compared to many DOS libraries. Using natural product–inspired libraries of synthetic molecules containing as few as fifty members, these researchers have identified compounds with low micromolar IC_{50} values (concentration at which inhibition is 50%) against a range of targets [28–30]. Although these libraries do not exhibit the same chemical diversity as typical DOS libraries, they are still capable of achieving the ultimate aim of discovering new modulators of protein function.

Another class of small molecule that finds use in chemical genetic studies and has the potential to be transferred to chemical genomics is that of specifically designed compounds. These molecules usually are not naturally occurring and are rarely synthesized through the generation of a large random library, as in DOS or BIOS. Instead, they are carefully designed so that a particular biological system may be studied closely. A specific enzyme inhibitor may be useful for studying the function of the protein it inactivates, but such compounds may be the result of random library synthesis. Computer-aided drug design has contributed to many medicinal chemistry campaigns and will make valuable contributions to chemical genomics [31], but the designed molecules of particular interest here incorporate multiple moieties that perform different roles in the cell. Although small molecules that inhibit protein-protein interactions, for example, are of interest as potential therapeutic agents [32–34], molecules that induce a controlled and specific protein-protein interaction are also useful in the manipulation and study of a network of protein function. These bifunctional molecules, often termed *chemical inducers of dimerization* (CID), that induce a particular biological effect have great potential in chemical genomic studies.

Early work in this area relied on the specific interaction between the natural product FK506 (Figure 2.6, **4**) and the protein FKBP12 (FK506 binding protein 12). Formation of the FK506-FKBP12 complex causes the recruitment of the protein phosphatase calcineurin, providing a ternary complex and highlighting the utility of FK506 as a heterodimerizer. Subsequent research developed an artificial dimer of FK506 termed FK1012 that was capable of recruiting two FKBP12 proteins. Proteins of interest could then be dimerized using genetic fusion to FKBP12 followed by treatment with FK1012. A structurally similar natural product,

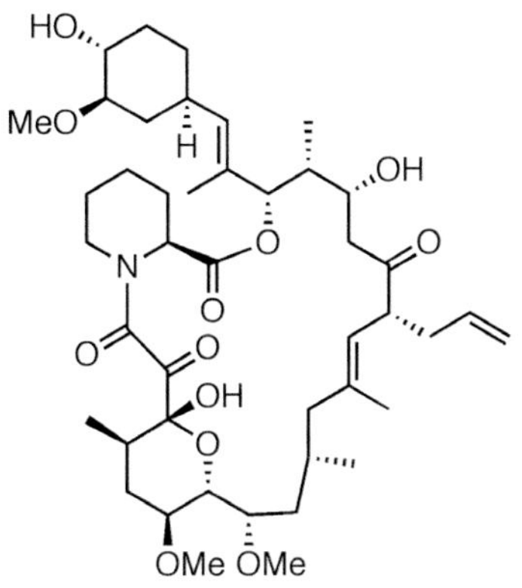

4 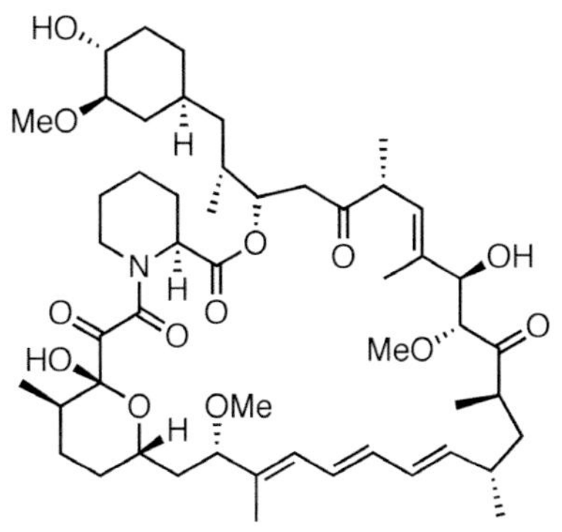 1

FIGURE 2.6: ***Structure of FK506 (4) and rapamycin (1).***

rapamycin (**1**), contains the same FKBP12-binding domain as FK506 and has another domain that binds to the mTOR protein. By creating fusions of FKBP12 and mTOR to two different proteins, rapamycin is thus capable of inducing controlled heterodimerization. This widely studied and well-characterized system has found a plethora of applications and has been extensively reviewed [35–37].

Bifunctional molecules outside the FK506-rapamycin paradigm have also been used to study many biological systems [38]. These studies often rely on taking two small molecules with known activities and linking them together via a simple chemical tether. Some are designed to have a particular therapeutic effect, such as improving the selectivity of a known toxin, and others are purely intended for use as probes of cellular processes. An example of this type of research that is of particular interest in chemical genetics is the use of small molecules to degrade a specific protein via the ubiquitin-proteasome pathway, providing an illustration of reverse chemical genetics.

Ubiquitin is a small, highly conserved protein that attaches to exposed lysine residues on proteins marked for degradation, such as damaged or unfolded proteins. Polyubiquitination of a protein results in its degradation by the proteasome, part of regular homeostasis. Ubiquitination occurs through the function of ubiquitin-activating enzymes (E1), ubiquitin-conjugating enzymes (E2), and ubiquitin-protein ligases (E3). The E3 ligase binds directly to the targeted protein and transfers the ubiquitin monomer, and conjugation of further ubiquitin units provides a polyubiquitin chain that is recognized by the proteasome.

Proteolysis-targeting chimeric molecules (PROTACs) are heterobifunctional molecules that have been designed to bind at one end to a specific protein chosen for degradation and at the other to an E3 ligase [39–41] (Figure 2.7).

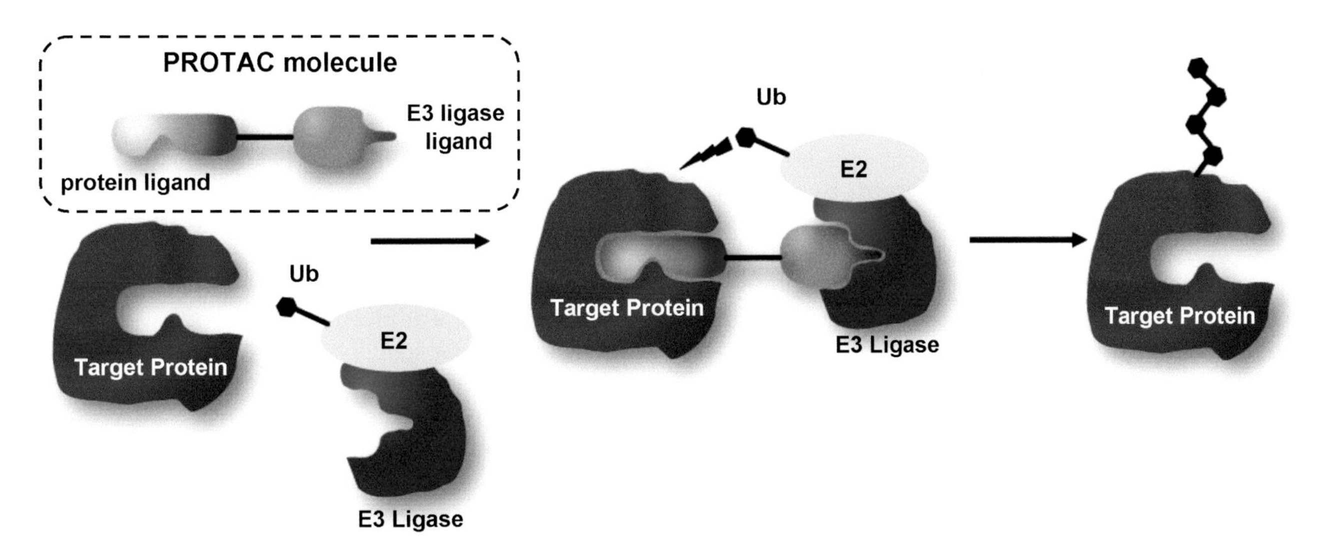

FIGURE 2.7: ***Proteolysis targeting chimeric molecules. The PROTAC, a heterobifunctional molecule that contains a protein-binding component and an E3 ligase binding component, mediates polyubiquitination of a target protein, marking it for degradation by the proteasome.***

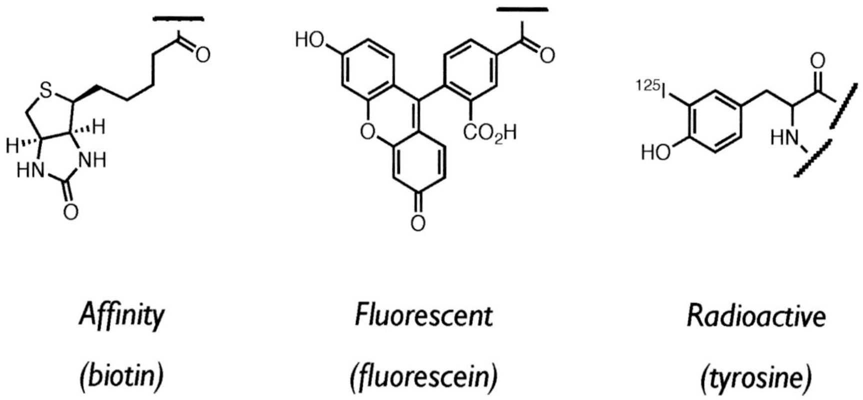

Affinity *Fluorescent* *Radioactive*

(biotin) *(fluorescein)* *(tyrosine)*

FIGURE 2.8: Common tags for biochemical approaches to determining mechanism of action.

This allows for controlled degradation of a protein, thus enabling detailed study of its function in a similar way to traditional reverse genetics. PROTACs are particularly useful where the protein of interest is essential for development, meaning that genetically engineered knockout mice do not survive long enough to be studied. Like all other chemical genetics strategies, PROTACs offer the advantage of being tunable and reversible.

Some bifunctional molecules require a level of design that makes them somewhat unsuitable for the larger scale of chemical genomics, despite their power in narrower chemical genetic studies. Other such molecules are more amenable to broader studies. Libraries of bifunctional molecules can indeed be prepared [42], and this type of compound collection may start to find increased use in the targeting of biological mechanisms.

3. EXPLORATION OF BIOLOGICAL PROCESSES

With this basic understanding of the origins of small molecules used for chemical genomics, we turn now to the other essential elements of a chemical genetic study: a protein of interest, a particular phenotype, and an assay that provides the desired information. As mentioned previously (see note 2), successful chemical genomics studies require the adaptation of techniques that have proven useful in chemical genetics. As such, this section provides an overview of these techniques, focusing initially on MOA studies and then providing a discussion of traditional screening methods and how these have developed toward genome-wide investigations.

3.1. Mechanism of Action Studies

Where a small molecule gives rise to a desired phenotype, such as inhibition of tumor cell growth, much can be learned by determining precisely how the molecule causes that phenotype. For example, basic disease biology can be illuminated, and the identification of a binding protein can potentially reveal a therapeutic target for future drugs. Such MOA studies have contributed a wealth of knowledge about biological processes and have enabled the identification of many new proteins through a powerful combination of synthetic chemistry and biochemical or genetic techniques.

Identification of a molecule's target binding protein using biochemical techniques typically requires chemical manipulation of the molecule. To track the protein-ligand complex, the molecule is usually tagged with something that provides a convenient way of monitoring its location, such as fluorescence, radioactivity, or an affinity reagent like biotin (Figure 2.8). Following this tag allows the isolation of the small molecule–protein complex, and the protein can then be characterized using mass spectrometry and sequence analysis. When using affinity reagents (unlike reporter tags like fluorescence and radioactivity), the target protein can be physically extracted from the complex cellular lysate, often referred to as a pull-down experiment (Figure 2.9).

The attachment of the tag can be complicated: If the affinity reagent (or other tag) is linked to the small molecule in a manner that disrupts the pharmacophore, the subsequent experiments to isolate the protein will fail because the small molecule will not bind its target protein. Also, derivatization of a small molecule ligand may not be synthetically trivial, often especially in the case of complex natural products. Total syntheses and SARs of the natural product being studied can greatly facilitate both the choice of site for attachment of the tag and the chemistry required for attachment itself, although a nonspecific and universal method for immobilization of small molecules onto beads has been reported [43]. Having synthesized the affinity reagent, it is essential to confirm that the derivative exhibits activity at least comparable with that of the parent molecule.

Some natural products bind covalently to their target protein, which means that the binding affinity between small molecule and protein will not be a concern in the isolation process. Whether a molecule binds covalently can

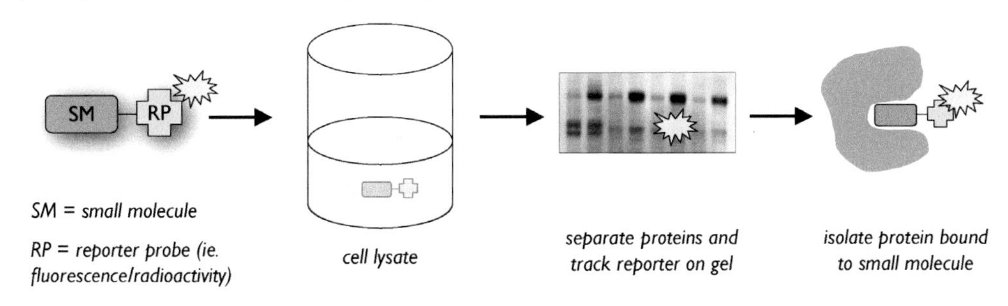

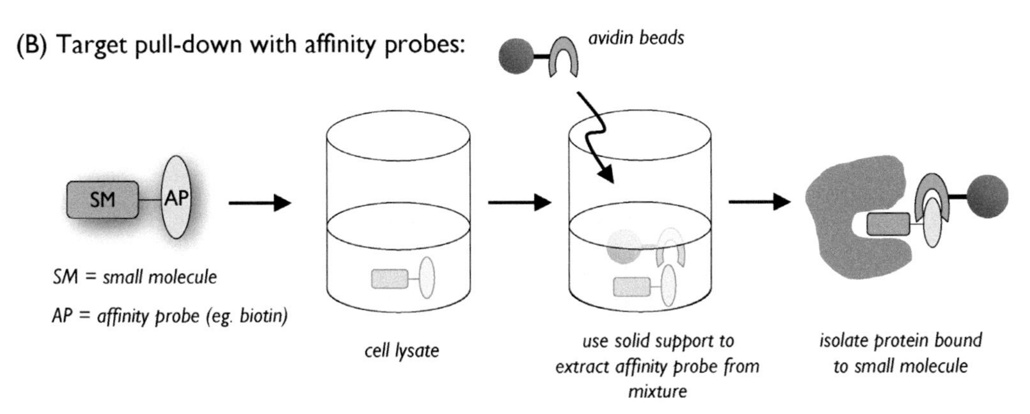

FIGURE 2.9: ***A.*** ***The reporter probe is used to track the small molecule–protein complex throughout the isolation process.*** ***B.*** ***The affinity probe enables the physical separation of the small molecule–protein complex from a cell lysate.***

often be predicted by simple examination of its structure (for example, if it contains reactive groups such as epoxides) but can only be proven experimentally. For molecules that bind only weakly to their target proteins, it is possible that the interaction will not survive the rigorous washing conditions needed to isolate the complex. One solution to this is to incorporate into the affinity reagent a moiety capable of photocrosslinking to the protein [44–46] (Figure 2.10). The small molecule brings the affinity reagent into close proximity to the protein, and upon exposure to UV light, the photocrosslinking moiety forms random covalent links to the protein – essentially a molecular stapler – so that the

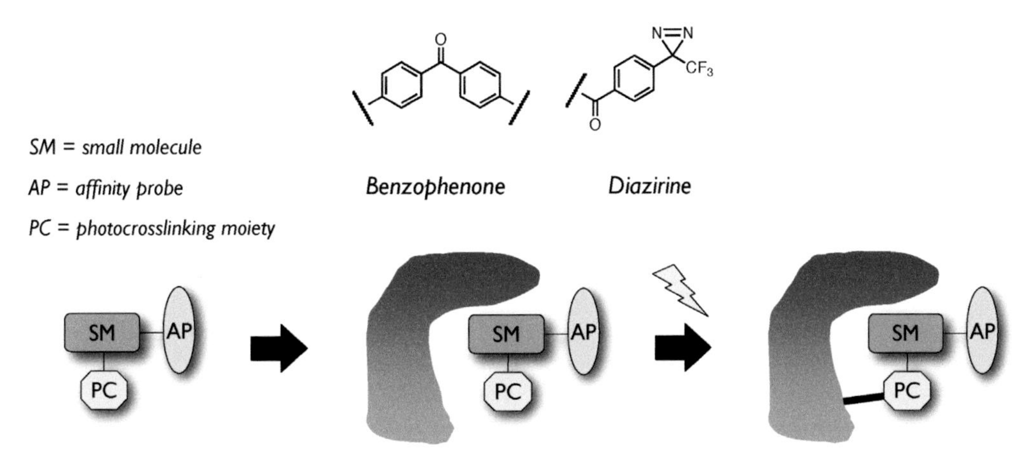

FIGURE 2.10: ***The utility of photocrosslinking. The small molecule, with appended affinity probe and photocrosslinking moiety, binds to its target protein. Upon irradiation with the appropriate light, an irreversible link forms between the protein and the small molecule, enabling isolation of the protein.***

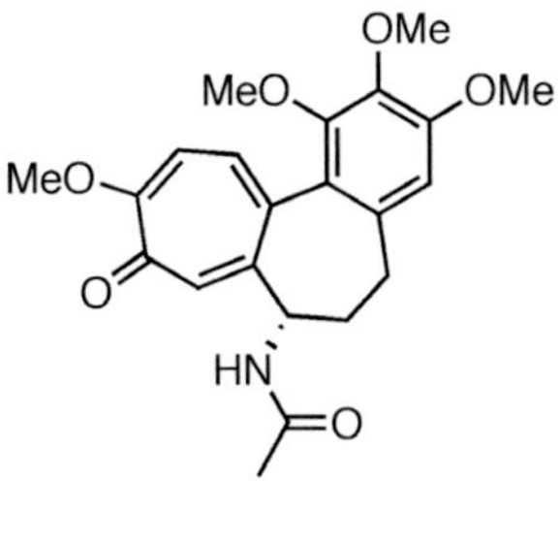

FIGURE 2.11: Mechanism of action studies. Structures of colchcine (5), taxol (6), epothilone B (7), monastrol (8), fumagillin (9), TNP-470 (10), reveromycin (11).

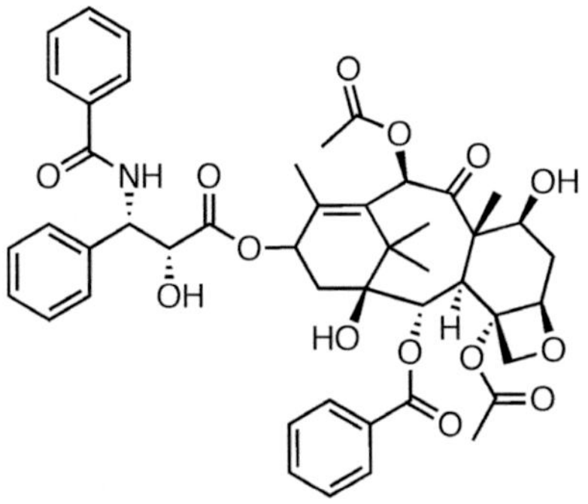

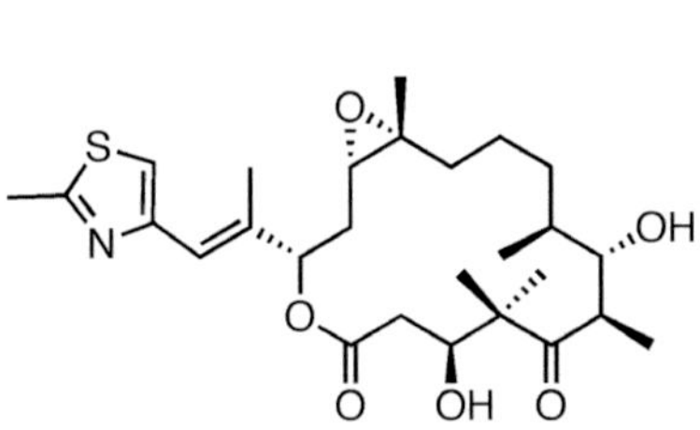

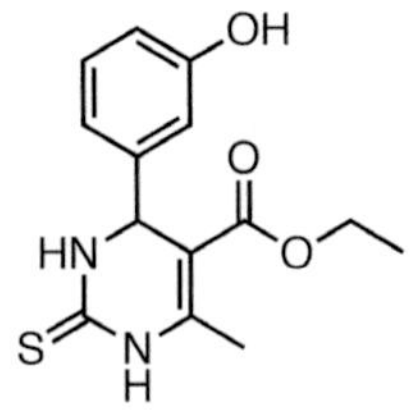

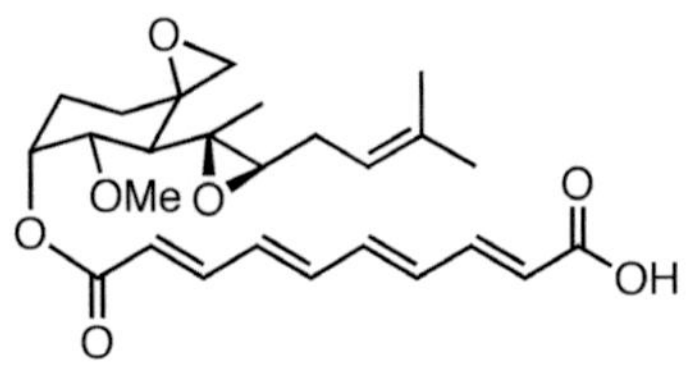

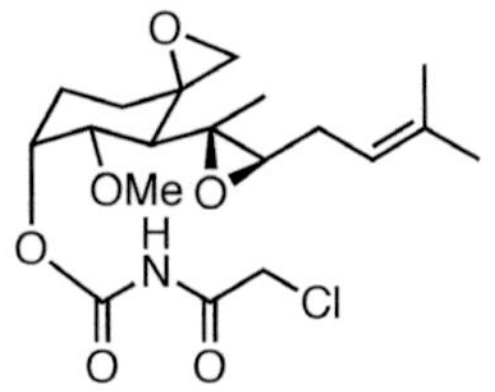

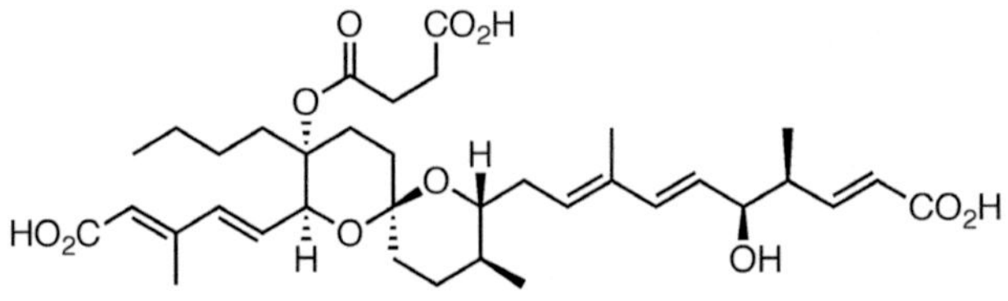

affinity handle may be used to extract the protein of interest from the cell lysate. Chemically induced crosslinking methods can also be used [47].

Several examples illustrate how some of these mode-of-action techniques have been applied to specific natural products and how useful biological information can be obtained (Figure 2.11).

A field in which small molecules have played a crucial role in developing our understanding of biology is the study of the cytoskeleton, a framework of filaments that provide cells with their shape and structure [48]. In early investigations, the alkaloid colchicine (**5**) was found to affect mitosis [49], but at the time, the composition of microtubules was not known. Identification of the natural product's binding protein was performed using radiolabeled colchicine [50–52] and revealed tubulin to be the subunit of these microtubule polymers [53]. Since then, modulators of this component of the cytoskeleton have found extensive use in studying cell division and in treating cancers. For example, the natural products taxol (**6**) [54], one of the best-known anti-cancer drugs, and epothilone (**7**) [55, 56] (derivatives of which are in clinical trials) both promote polymerization of microtubules.

However, many other structural proteins are important in the regulation of the cytoskeleton, and forward chemical genetic studies have been employed to find compounds that can affect these structures by interacting with proteins other than tubulin. In a cell-based assay, a collection of compounds was tested for their ability to induce mitotic arrest, with a counterscreen to exclude those that affected microtubules. A synthetic compound called monastrol (**8**) was observed to produce significant changes in mitotic spindle arrangement [57], exhibiting a phenotype previously recorded upon inhibition of the kinesin motor protein Eg5 (the *Xenopus laevis* homologue of human kinesin spindle protein). Biochemical confirmation that monastrol binds to Eg5 was obtained, and, intriguingly, crystallographic studies on KSP have shown that the small molecule binds in an allosteric pocket that is not visible in the crystal structure in the absence of monastrol [58]. This molecule has subsequently been used to study the role of Eg5 in mitosis [59], enhanced by the fact that the molecule is specific and exerts its effect reversibly.

Isolated from the fungus *Aspergillus fumigatus*, fumagillin (**9**) has been used for more than fifty years to treat a parasitic infection that affects honeybees. Of more interest to human health, it was subsequently discovered to inhibit angiogenesis – the formation of new blood vessels – which is required for important processes such as growth and wound healing but is also crucial for the progression of tumors. A synthetic derivative of fumagillin, TNP-470 (**10**) (also known as AGM-1470), was found to be fifty times more potent than the natural product [60], and, using the SAR knowledge from this study, a biotin-functionalized derivative of these was used to isolate the fumagillin binding protein, methionine aminopeptidase 2 (MetAP-2) [61, 62]. The binding mode was discovered by x-ray crystallography, which revealed a covalent link between the spirocyclic epoxide of fumagillin and histidine-231 of MetAP-2. Further studies on the effect of TNP-470 revealed that it induced expression of p53 and p21$^{CIP/WAF}$, a cyclin-dependent kinase inhibitor, exclusively in endothelial cells, ultimately causing G1 cell cycle arrest [63]. Although the precise mechanism by which fumagillin and TNP-470 exert their antiangiogenic activity is not fully clear, these molecules have implicated MetAP-2 as playing an important role in angiogenesis [63]. More recent studies have shown that MetAP-2 activity is required for noncanonical Wnt signaling (also inhibited by TNP-470) [64], thus suggesting a link between angiogenesis and this signaling pathway [65].

Genetic approaches to determining MOA have also shown great potential. Through mutagenesis, a cell line (or organism strain) can be found that is resistant to the effects of the molecule that causes the phenotype of interest. By sequencing the mutant and comparing it with the wild type, the responsible genetic mutation and the protein it encodes can be discovered. A conceptually related approach is to overexpress each gene and observe which mutants are unaffected by drug treatment, referred to as an overexpression screen. A major advantage of these methods is that chemical modification of the small molecule is not required. Furthermore, the targets identified are physiologically relevant. Two drawbacks of this approach are the need for biochemical confirmation of target proteins and the difficulty in generating the mutant library. However, genetic approaches do offer a powerful alternative to the biochemical methods discussed at the beginning of this section [66].

An early example that demonstrated the utility of genetic approaches to MOA studies involved the well-studied molecule rapamycin. Although the compound was already known to bind to FKBP, genetic studies in yeast were the first to identify two other genes, *TOR1* and *TOR2*, the protein products of which were required for rapamycin's toxicity [67].

The natural product reveromycin A (**11**) was discovered because of its ability to block the epidermal growth factor(EGF)-dependent responses of epidermal cells in mice and was known to induce G1 cell cycle arrest, but its precise biological target was not known until a mutant screen by Miyakawa and co-workers [68]. After confirming the ability of **11** to block protein synthesis in yeast, the authors generated a strain resistant to the drug. Subsequent cloning of this strain revealed the *ILS1* gene, encoding the protein IleRS (isoleucyl-tRNA synthetase) to be responsible for the resistance. Overexpression of *ILS1* also conferred resistance. IleRS is essential for growth and protein synthesis, consistent with the observed effect of reveromycin.

3.2. Approaches to Screening Small Molecules

Before MOA studies can be performed, however, compounds that possess the desired activity must be found. The speed with which small molecules can be obtained has prompted similar advances in the methods used to screen these molecules for biological activity. Traditional HTS has become increasingly miniaturized, progressing from 96-well plates to 384-well plates and even to 1536-well plates containing just microliters of solution [69]. The benefits of this miniaturization lie in both the decreased time taken per assay and the smaller quantities of expensive and potentially precious compounds and other reagents that are required. The sheer number of compounds being analyzed also demands a fully automated setup, and many robotic liquid handling systems are commercially available.

Like any other assay, an HTS assay requires significant development to produce meaningful results. Negative and (ideally) positive controls must be incorporated; the type of readout needs to be chosen from a number of options (e.g., fluorescence, absorbance, luminescence); any hits require validation, often using an alternative readout or a different assay altogether; the compounds must be checked to ensure that hits are not caused by decomposition products; and the results of the assay must be statistically significant (determined using the so-called Z' factor [70], which is related to the standard deviations of the positive and negative controls in the assay).

Many HTS campaigns fall into the category of reverse chemical genetics – a purified protein of interest is treated

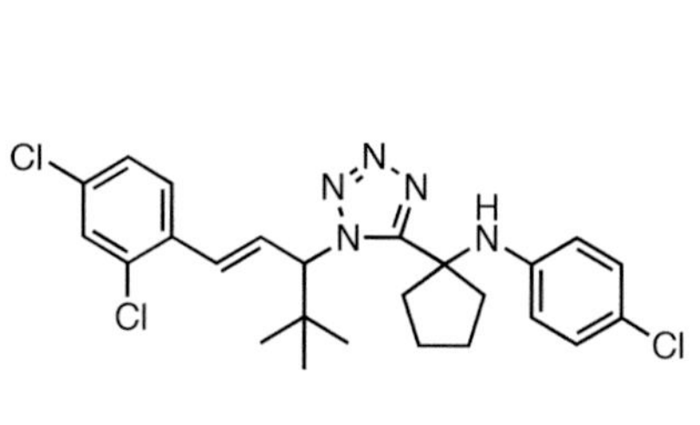

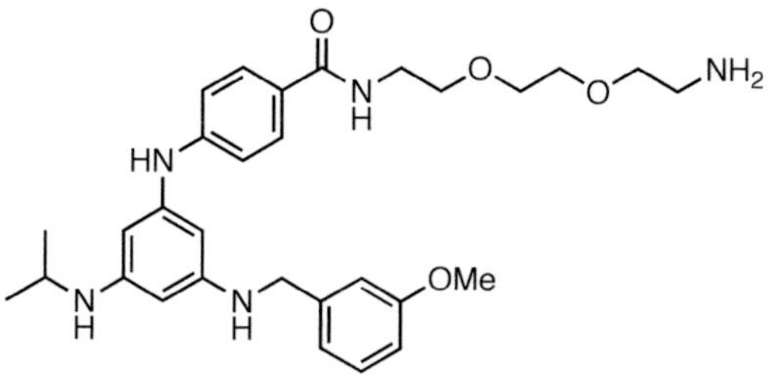

12 13

*FIGURE 2.12: Structures of 31N3 (**12**) and tagged triazine (**13**).*

in vitro with the compound collection in the hope of finding a molecule that binds to the protein. Cell-based assays can also be designed that give information specifically about the interaction of a small molecule with a particular protein: Reporter gene assays involve genetic manipulation of the cell line of interest, splicing into it the gene for a "reporter," often a fluorophore such as green fluorescent protein (GFP) or a luminescent substrate such as luciferase. The reporter is designed so that when the protein of interest is activated (or inactivated), downstream signaling causes the reporter gene to also be activated, thus giving rise to an observable readout. A powerful component of this technique is the intensity of the response, rendering it suitable to the required miniaturization. As with any assay, though, these reporter gene assays can result in both false positives and false negatives, such as enhancers or inhibitors of fluorescence. Therefore, an understanding of the shortcomings of an assay is crucial in the efficient pursuit of hits [71].

The ultimate quest of chemical genomics, identification of a small molecule modulator for every member of the proteome, will be hard to achieve by testing one protein at a time. For this reason, HTS is now shifting toward assays in which the compounds are tested in whole cells or even in model organisms such as zebrafish (see notes 72–76 and accompanying text) or *Caenorhabditis elegans* [72]. In this manner, each compound is tested against a wide range of proteins in their native environment, wherein changes in cell morphology or phenotype are used as the readout of the assay.

Phenotype-based assays are particularly powerful where disease mechanisms are either complex or not well understood, such as in neurodegenerative conditions like Alzheimer's disease and Parkinson's disease. Biologically useful compounds can still be identified based on the phenotype they produce, even though the specific proteins they target are not known to be involved in those diseases. Forward genetic studies can then be employed to determine the MOA, thus leading to greater understanding of disease pathways. Another advantage of phenotype-based assays is that they are closer to what might be considered a "real" biological setting in the sense that in vitro studies are a long way removed from a functioning organism.

The zebrafish (*Danio rerio*) has proven to be a popular organism for phenotype-based assays for a number of reasons [73]. First, it is a vertebrate, and its organs are quite similar to those of its human counterparts. Although results in zebrafish may not always correlate with identical activity in humans, it is a sufficiently close match for a model system. Second, zebrafish during the embryonic and larval stages are less than 2 mm long and thus fit very comfortably into 384-well plates. Third, they can be raised cheaply and efficiently, making them amenable to screening large compound collections. Finally, embryonic zebrafish are transparent. As a consequence, it is easy to monitor internal organs without needing to sacrifice the animal, and changes in phenotype can be readily observed. This physical characteristic also enhances the power of transgenic fluorescent markers. A further bonus in the use of zebrafish in chemical genomics studies is that they have been extensively investigated via traditional genetics: Random mutagenesis has been able to demonstrate the range of phenotypes that occur upon deletion of genes, providing a point of reference for studies looking at modulating the function of gene products.

An early example in this field was the screening of 1,100 synthetic compounds by Peterson et al. They looked at the effect of these molecules on four systems: the central nervous system, the cardiovascular system, pigmentation, and ear development [74]. Each well of a 96-well plate housed three zebrafish embryos, and visual inspection of the entire organism enabled a screening rate of approximately four hundred compounds per person per day. Aside from discovering a number of compounds that induced various phenotypes, the authors exploited the ease with which small molecules can provide temporal control over a system to study the timing of certain developmental processes. A molecule referred to as 31N3 (Figure 2.12, **12**) was found to affect, in a highly time-dependent manner, the formation of otoliths, bony structures of the inner ear that sense vibrations and assist in maintaining balance. Initial treatment with 31N3 does not affect otolith formation if the drug is washed out of the solution by fourteen hours post-fertilization (hpf). Also, if the drug is added *after* 26 hpf, it has no effect on development. When the drug is added

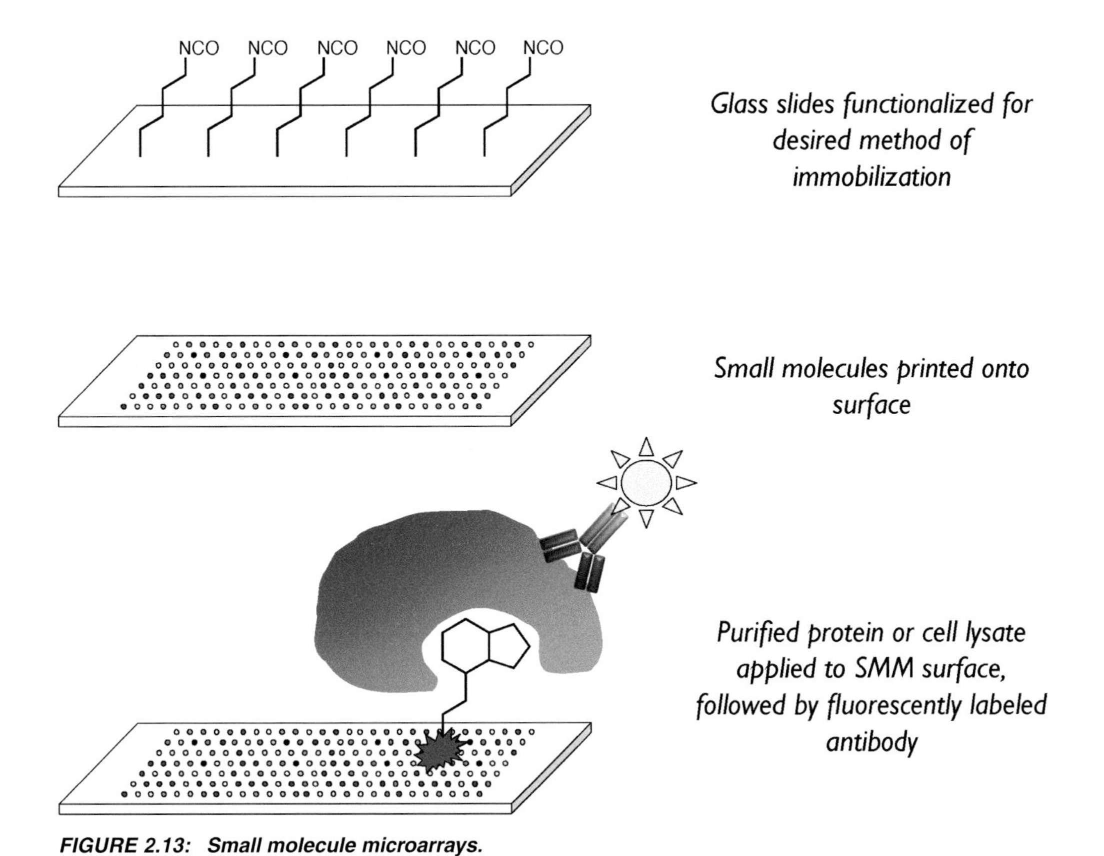

FIGURE 2.13: Small molecule microarrays.

at 8 hpf, however, no otoliths are observed; addition at 14 hpf often resulted in formation of otoliths in only one ear. This study thus paves the way for further investigation of the timing of these developmental processes.

An interesting combination of phenotype-based assay and identification of binding proteins has been performed in the Chang laboratory [75]. Their compound collection is a so-called "tagged" library of triazines, wherein each molecule is prepared to already contain a linker with a primary amine terminus, ready for immobilization onto a solid support for subsequent target identification studies. This tagged library approach also bypasses the hurdle of retrospectively finding a site on the molecule that does not affect biological activity, because the activity is discovered with the linker already in place [76]. More than 1,500 tagged triazines were screened in zebrafish embryos for effects on brain and eye development, culminating in the discovery of a modulator (**13**) of eye phenotypes with micromolar potency. Direct attachment of this compound onto beads facilitated the isolation of four binding proteins that have previously been shown by mutagenesis to be involved in causing changes in brain/eye phenotypes, thus providing validation of the assay. Far from reinventing the wheel, this report illustrates the great promise of phenotype-based assays in live organisms as well as that of tagged libraries.

A recent development that has further enhanced the power of phenotype-based assays is the use of automated monitoring equipment. Rather than rely on researchers to manually and painstakingly inspect each well in the screen of a several-thousand-compound library, it is now possible to fully automate this process. Sensitive cameras and software can recognize changes in a range of phenotypes even in whole organisms, allowing the testing of large numbers of compounds much more quickly than can be achieved by people staring down microscopes.

A final topic worth mentioning in this overview is small molecule microarrays (SMMs). In response to the small amount of each compound that is produced in a combinatorial library, miniaturization of assays was required. SMMs typically involve immobilization of very small amounts of compounds onto a functionalized glass slide, followed by probing with a labeled protein of interest (Figure 2.13). As an alternative to standard HTS assays, SMM offer an operationally simple, general binding assay. A pioneering example from the Schreiber laboratory [77] used robotic printing techniques to attach molecules in more than ten thousand positions on a single glass slide, using just nanoliters of solution, with full control over the spatial distribution. Fluorescently tagged known binding partners for these molecules confirmed the ability of immobilized small molecules to interact with proteins, thus validating the utility of the SMM approach.

A key aspect of SMMs is the chemistry used to immobilize the compounds [78]. Functionalizing the surface of the glass slide with a known functional group then allows chemoselective attachment of suitable molecules. For example, immobilized maleimides have been used to react with thiols [77], immobilized silyl chlorides react with alcohols [79], and immobilized alkynes can be used to trap molecules containing azides via Huisgen cycloaddition [80]. Steps toward universal immobilization strategies have been taken, some of which are similar to the photocrosslinking discussed above in MOA studies (Section 3.1). Glass surfaces can be functionalized with a diazirene, a photoreactive moiety that, upon UV irradiation, generates a highly reactive carbene that can be inserted into carbon-hydrogen and heteroatom-hydrogen bonds [81, 82]. Using this approach, the molecules to be immobilized need not contain any specific functional group, and the molecules will also (in theory) be linked to the surface through each possible combination of bonds, meaning that knowledge of the pharmacophore is not required. It is possible, however, for molecules to react in only a few positions, and to overcome this, a strategy has been reported that uses five different photoaffinity reagents, ensuring a better spread of immobilization sites [83]. These photoaffinity approaches have been criticized for producing too many false positives in SMM assays, and an alternative method has been described that provides attachment chemistry for essentially any nucleophile using immobilized isocyanates [84, 85].

Although the studies of SMM reported thus far [86] indicate their great potential to be useful in the identification of new ligands for modulating protein function, they are still very much in the development stage, with the inevitable discoveries of exciting new compounds yet to be made. In the move toward applications in chemical genomics, SMM have now been used to screen mammalian cells rather than purified proteins or cell lyates [87]. To achieve this, compounds were embedded in a biodegradable polymer ([D],[L]-lactide/glycolide copolymer) on the glass surface, which was then coated with cells. As the polymer degraded, compounds were slowly released, allowing them to interact with nearby cells. Changes in cell morphology indicated that the cell was growing above a molecule of interest, thus providing an easy, functional readout. SMM have also found application in proof-of-principle experiments in the screening of natural product extracts [88].

4. SUMMARY AND FUTURE DIRECTIONS

The quest of chemical genomics – to find a small molecule to modulate each function of every protein in the human genome – represents a holy grail of chemical biology. Gradually, it is coming closer to becoming a reality through significant advances in a number of fields. The ability of chemists to rapidly produce large numbers of diverse, biologically relevant, and stereochemically rich molecules is infinitely greater than it was just twenty years ago. Adaptation of traditional screening techniques by biologists has allowed us to quickly, efficiently, and accurately assess the biological function of these molecules. These two crucial elements have been united by greater automation, enhanced robotics, bioinfomatics, and even software.

The research environment has also shifted in a direction that will aid in this quest. Improved funding for research at the interface of chemistry and biology assists current investigations as well as the training of the next generation of chemical genomics researchers. Establishment of core screening capabilities at many major universities and at organizations such as the National Institutes of Health [89, 90] has given biologists access to large compound collections that they can test in their assays and has given chemists something to test their compounds against. In the past, providing chemists and biologists with a common goal has led to great scientific leaps, and the same will be true of chemical genomics.

REFERENCES

1. Clamp, M., Fry, B., Kamal, M., Xie, X., Cuff, J., Lin, M. F., Kellis, M., Lindblad-Toh, K., and Lander, E. S. (2007). Distinguishing protein-coding and noncoding genes in the human genome. *Proc Natl Acad Sci U S A 104*, 19428–19433.
2. MacBeath, G. (2001). Chemical genomics: what will it take and who gets to play? *Genome Biol 2*, COMMENT2005.
3. Newman, D. J., Cragg, G. M., and Snader, K. M. (2003). Natural products as sources of drugs over the period 1981–2002. *J Nat Prod 66*, 1022–1037.
4. Rouhi, A. M. (2003). Rediscovering natural products. *Chem Eng News 81*, 77–91.
5. Butler, M. S. (2005). Natural products to drugs: natural product derived compounds in clinical trials. *Nat Prod Rep 22*, 162–195.
6. Koehn, F. E., and Carter, G. T. (2005). The evolving role of natural products in drug discovery. *Nat Rev Drug Discov 4*, 206–220.
7. Kirschning, A., Taft, F., and Knobloch, T. (2007). Total synthesis approaches to natural product derivatives based on the combination of chemical synthesis and metabolic engineering. *Org Biomol Chem 5*, 3245–3259.
8. Kennedy, J. (2008). Mutasynthesis, chemobiosynthesis, and back to semi-synthesis: combining synthetic chemistry and biosynthetic engineering for diversifying natural products. *Nat Prod Rep 25*, 25–34.
9. Weissman, K. J. (2007). Mutasynthesis – uniting chemistry and genetics for drug discovery. *Trends Biotechnol 25*, 139–142.
10. Schwecke, T., Aparicio, J. F., Molnar, I., Konig, A., Khaw, L. E., Haydock, S. F., Oliynyk, M., Caffrey, P., Cortes, J., Lester, J. B., Bohm, G. A., Staunton, J. and Leadlay, P. F. (1995). The biosynthetic gene cluster for the polyketide immunosuppressant rapamycin. *Proc Natl Acad Sci U S A 92*, 7839–7843.
11. Gregory, M. A., Petkovic, H., Lill, R. E., Moss, S. J., Wilkinson, B., Gaisser, S., Leadlay, P. F., and Sheridan, R. M. (2005). Mutasynthesis of rapamycin analogues through the

manipulation of a gene governing starter unit biosynthesis. *Angew Chem Int Ed Engl 44*, 4757–4760.
12. Goss, R. J., Lanceron, S. E., Wise, N. J., and Moss, S. J. (2006). Generating rapamycin analogues by directed biosynthesis: starter acid substrate specificity of mono-substituted cyclohexane carboxylic acids. *Org Biomol Chem 4*, 4071–4073.
13. Graziani, E. I., Ritacco, F. V., Summers, M. Y., Zabriskie, T. M., Yu, K., Bernan, V. S., Greenstein, M., and Carter, G. T. (2003). Novel sulfur-containing rapamycin analogs prepared by precursor-directed biosynthesis. *Org Lett 5*, 2385–2388.
14. Eustaquio, A. S., and Moore, B. S. (2008). Mutasynthesis of fluorosalinosporamide, a potent and reversible inhibitor of the proteasome. *Angew Chem Int Ed Engl 47*, 3936–3938.
15. McGlinchey, R. P., Nett, M., Eustaquio, A. S., Asolkar, R. N., Fenical, W., and Moore, B. S. (2008). Engineered biosynthesis of antiprotealide and other unnatural salinosporamide proteasome inhibitors. *J Am Chem Soc 130*, 7822–7823.
16. Zhang, W., and Tang, Y. (2008). Combinatorial biosynthesis of natural products. *J Med Chem 51*, 2629–2633.
17. Breinbauer, R., Vetter, I. R., and Waldmann, H. (2002). From protein domains to drug candidates – natural products as guiding principles in the design and synthesis of compound libraries. *Angew Chem Int Ed Engl 41*, 2879–2890.
18. Tan, D. S., Foley, M. A., Shair, M. D., and Schreiber, S. L. (1998). Stereoselective synthesis of over two million compounds having structural features both reminiscent of natural products and compatible with miniaturized cell-based assays. *J Am Chem Soc 120*, 8565–8566.
19. Zhu, J., and Bienayme, H., eds. (2005). *Multicomponent Reactions* (Weinheim, Germany: Wiley-VCH).
20. Schreiber, S. L. (2000). Target-oriented and diversity-oriented organic synthesis in drug discovery. *Science 287*, 1964–1969.
21. Spring, D. R. (2003). Diversity-oriented synthesis; a challenge for synthetic chemists. *Org Biomol Chem 1*, 3867–3870.
22. Tan, D. S. (2005). Diversity-oriented synthesis: exploring the intersections between chemistry and biology. *Nat Chem Biol 1*, 74–84.
23. Taylor, S. J., Taylor, A. M., and Schreiber, S. L. (2004). Synthetic strategy toward skeletal diversity via solid-supported, otherwise unstable reactive intermediates. *Angew Chem Int Ed Engl 43*, 1681–1685.
24. Thomas, G. L., Spandl, R. J., Glansdorp, F. G., Welch, M., Bender, A., Cockfield, J., Lindsay, J. A., Bryant, C., Brown, D. F., Loiseleur, O., Rudyk, H., Ladlow, M., and Spring, D. R. (2008). Anti-MRSA agent discovery using diversity-oriented synthesis. *Angew Chem Int Ed Engl 47*, 2808–2812.
25. Comer, E., Rohan, E., Deng, L., and Porco, J. A., Jr. (2007). An approach to skeletal diversity using functional group pairing of multifunctional scaffolds. *Org Lett 9*, 2123–2126.
26. Wender, P. A., Verma, V. A., Paxton, T. J., and Pillow, T. H. (2008). Function-oriented synthesis, step economy, and drug design. *Acc Chem Res 41*, 40–49.
27. Cordier, C., Morton, D., Murrison, S., Nelson, A., and O'Leary-Steele, C. (2008). Natural products as an inspiration in the diversity-oriented synthesis of bioactive compound libraries. *Nat Prod Rep 25*, 719–737.
28. Noren-Muller, A., Reis-Correa, I., Jr., Prinz, H., Rosenbaum, C., Saxena, K., Schwalbe, H. J., Vestweber, D., Cagna, G., Schunk, S., Schwarz, O., Schiewe, H., and Waldmann, H. (2006). Discovery of protein phosphatase inhibitor classes by biology-oriented synthesis. *Proc Natl Acad Sci U S A 103*, 10606–10611.
29. Lessmann, T., Leuenberger, M. G., Menninger, S., Lopez-Canet, M., Muller, O., Hummer, S., Bormann, J., Korn, K., Fava, E., Zerial, M., Mayer, T. U., and Waldmann, H. (2007). Natural product-derived modulators of cell cycle progression and viral entry by enantioselective oxa Diels-Alder reactions on the solid phase. *Chem Biol 14*, 443–451.
30. Noren-Muller, A., Wilk, W., Saxena, K., Schwalbe, H., Kaiser, M., and Waldmann, H. (2008). Discovery of a new class of inhibitors of Mycobacterium tuberculosis protein tyrosine phosphatase B by biology-oriented synthesis. *Angew Chem Int Ed Engl 47*, 5973–5977.
31. Dean, P. M. (2007). Chemical genomics: a challenge for de novo drug design. *Mol Biotechnol 37*, 237–245.
32. Arkin, M. R., and Wells, J. A. (2004). Small-molecule inhibitors of protein-protein interactions: progressing towards the dream. *Nat Rev Drug Discov 3*, 301–317.
33. Chene, P. (2006). Drugs targeting protein-protein interactions. *ChemMedChem 1*, 400–411.
34. Wells, J. A., and McClendon, C. L. (2007). Reaching for high-hanging fruit in drug discovery at protein-protein interfaces. *Nature 450*, 1001–1009.
35. Gestwicki, J. E., and Marinec, P. S. (2007). Chemical control over protein-protein interactions: beyond inhibitors. *Comb Chem High Throughput Screen 10*, 667–675.
36. Clackson, T. (2006). Dissecting the functions of proteins and pathways using chemically induced dimerization. *Chem Biol Drug Des 67*, 440–442.
37. Clackson, T. (2007). Controlling protein-protein interactions using chemical inducers and disrupters of dimerization. In *Chemical Biology*, S. L. Schreiber, T. M. Kapoor, and G. Wess, eds. (Weinheim, Germany: Wiley-VCH).
38. Corson, T. W., Aberle, N., and Crews, C. M. (2008). Design and applications of bifunctional molecules: why two heads are better than one. *ACS Chem Biol 3*, 677–692.
39. Sakamoto, K. M., Kim, K. B., Kumagai, A., Mercurio, F., Crews, C. M., and Deshaies, R. J. (2001). Protacs: chimeric molecules that target proteins to the Skp1-Cullin-F box complex for ubiquitination and degradation. *Proc Natl Acad Sci U S A 98*, 8554–8559.
40. Schneekloth, J. S., Jr., Fonseca, F. N., Koldobskiy, M., Mandal, A., Deshaies, R., Sakamoto, K., and Crews, C. M. (2004). Chemical genetic control of protein levels: selective in vivo targeted degradation. *J Am Chem Soc 126*, 3748–3754.
41. Lee, H., Puppala, D., Choi, E.-Y., Swanson, H., and Kim, K.-B. (2007). Targeted degradation of the aryl hydrocarbon receptor by the PROTAC approach: a useful chemical genetic tool. *ChemBioChem 8*, 2058–2062.
42. Koide, K., Finkelstein, J. M., Ball, Z., and Verdine, G. L. (2001). A synthetic library of cell-permeable molecules. *J Am Chem Soc 123*, 398–408.
43. Kanoh, N., Honda, K., Simizu, S., Muroi, M., and Osada, H. (2005). Photo-cross-linked small-molecule affinity matrix for facilitating forward and reverse chemical genetics. *Angew Chem Int Ed Engl 44*, 3559–3562.
44. Kotzyba-Hibert, F., Kapfer, I., and Goeldner, M. (1995). Recent trends in photoaffinity labeling. *Angew Chem Int Ed Engl 34*, 1296–1312.

45. Fuwa, H., Takahashi, Y., Konno, Y., Watanabe, N., Miyashita, H., Sasaki, M., Natsugari, H., Kan, T., Fukuyama, T., Tomita, T., and Iwatsubo, T. (2007). Divergent synthesis of multifunctional molecular probes to elucidate the enzyme specificity of dipeptidic gamma-secretase inhibitors. *ACS Chem Biol 2*, 408–418.
46. Dorman, G., and Prestwich, G. D. (2000). Using photolabile ligands in drug discovery and development. *Trends Biotechnol 18*, 64–77.
47. Lim, H. S., Cai, D., Archer, C. T., and Kodadek, T. (2007). Periodate-triggered cross-linking reveals Sug2/Rpt4 as the molecular target of a peptoid inhibitor of the 19S proteasome regulatory particle. *J Am Chem Soc 129*, 12936–12937.
48. Peterson, J. R., and Mitchison, T. J. (2002). Small molecules, big impact: a history of chemical inhibitors and the cytoskeleton. *Chem Biol 9*, 1275–1285.
49. Levine, M. (1951). The action of colchicine on cell division in human cancer, animal, and plant tissues. *Ann N Y Acad Sci 51*, 1365–1408.
50. Borisy, G. G., and Taylor, E. W. (1967). The mechanism of action of colchicine. Binding of colchincine-3H to cellular protein. *J Cell Biol 34*, 525–533.
51. Taylor, E. W. (1965). The mechanism of colchicine inhibition of mitosis. I. Kinetics of inhibition and the binding of H3-colchicine. *J Cell Biol 25 (Suppl.)*, 145–160.
52. Weisenberg, R. C., Borisy, G. G., and Taylor, E. W. (1968). The colchicine-binding protein of mammalian brain and its relation to microtubules. *Biochemistry* 7, 4466–4479.
53. Mohri, H. (1968). Amino-acid composition of "tubulin" constituting microtubules of sperm flagella. *Nature 217*, 1053–1054.
54. Horwitz, S. B. (1994). Taxol (paclitaxel): mechanisms of action. *Ann Oncol 5 Suppl. 6*, S3–S6.
55. Altmann, K. H., Pfeiffer, B., Arseniyadis, S., Pratt, B. A., and Nicolaou, K. C. (2007). The chemistry and biology of epothilones – the wheel keeps turning. *ChemMedChem 2*, 396–423.
56. Donovan, D., and Vahdat, L. T. (2008). Epothilones: clinical update and future directions. *Oncology (Williston Park) 22*, 408–416; discussion 416, 421, 424 passim.
57. Mayer, T. U., Kapoor, T. M., Haggarty, S. J., King, R. W., Schreiber, S. L., and Mitchison, T. J. (1999). Small molecule inhibitor of mitotic spindle bipolarity identified in a phenotype-based screen. *Science 286*, 971–974.
58. Yan, Y., Sardana, V., Xu, B., Homnick, C., Halczenko, W., Buser, C.A., Schaber, M., Hartman, G. D., Huber, H. E., and Kuo, L. C. (2004). Inhibition of a mitotic motor protein: where, how, and conformational consequences. *J Mol Biol 335*, 547–554.
59. Kapoor, T. M., Mayer, T. U., Coughlin, M. L., and Mitchison, T. J. (2000). Probing spindle assembly mechanisms with monastrol, a small molecule inhibitor of the mitotic kinesin, Eg5. *J Cell Biol 150*, 975–988.
60. Ingber, D., Fujita, T., Kishimoto, S., Sudo, K., Kanamaru, T., Brem, H., and Folkman, J. (1990). Synthetic analogues of fumagillin that inhibit angiogenesis and suppress tumour growth. *Nature 348*, 555–557.
61. Sin, N., Meng, L., Wang, M.Q., Wen, J. J., Bornmann, W. G., and Crews, C. M. (1997). The anti-angiogenic agent fumagillin covalently binds and inhibits the methionine aminopeptidase, MetAP-2. *Proc Natl Acad Sci U S A 94*, 6099–6103.
62. Griffith, E. C., Su, Z., Turk, B. E., Chen, S., Chang, Y. H., Wu, Z., Biemann, K., and Liu, J. O. (1997). Methionine aminopeptidase (type 2) is the common target for angiogenesis inhibitors AGM-1470 and ovalicin. *Chem Biol 4*, 461–471.
63. Yeh, J. R., Mohan, R., and Crews, C.M. (2000). The antiangiogenic agent TNP-470 requires p53 and p21CIP/WAF for endothelial cell growth arrest. *Proc Natl Acad Sci U S A 97*, 12782–12787.
64. Zhang, Y., Yeh, J. R., Mara, A., Ju, R., Hines, J. F., Cirone, P., Griesbach, H. L., Schneider, I., Slusarski, D. C., Holley, S. A., and Crews, C. M. (2006). A chemical and genetic approach to the mode of action of fumagillin. *Chem Biol 13*, 1001–1009.
65. Cirone, P., Lin, S., Griesbach, H. L., Zhang, Y., Slusarski, D. C., and Crews, C. M. (2008). A role for planar cell polarity signaling in angiogenesis. *Angiogenesis 11*, 347–360.
66. Zheng, X. S., Chan, T. F., and Zhou, H. H. (2004). Genetic and genomic approaches to identify and study the targets of bioactive small molecules. *Chem Biol 11*, 609–618.
67. Heitman, J., Movva, N. R., and Hall, M. N. (1991). Targets for cell cycle arrest by the immunosuppressant rapamycin in yeast. *Science 253*, 905–909.
68. Miyamoto, Y., Machida, K., Mizunuma, M., Emoto, Y., Sato, N., Miyahara, K., Hirata, D., Usui, T., Takahashi, H., Osada, H., and Miyakawa, T. (2002). Identification of *Saccharomyces cerevisiae* isoleucyl-tRNA synthetase as a target of the G1-specific inhibitor reveromycin A. *J Biol Chem* 277, 28810–28814.
69. Sundberg, S. A. (2000). High-throughput and ultra-high-throughput screening: solution- and cell-based approaches. *Curr Opin Biotechnol 11*, 47–53.
70. Zhang, J. H., Chung, T. D., and Oldenburg, K. R. (1999). A simple statistical parameter for use in evaluation and validation of high throughput screening assays. *J Biomol Screen 4*, 67–73.
71. Auld, D. S., Thorne, N., Nguyen, D. T., and Inglese, J. (2008). A specific mechanism for nonspecific activation in reporter-gene assays. *ACS Chem Biol 3*, 463–470.
72. Kwok, T. C., Ricker, N., Fraser, R., Chan, A. W., Burns, A., Stanley, E. F., McCourt, P., Cutler, S. R., and Roy, P. J. (2006). A small-molecule screen in *C. elegans* yields a new calcium channel antagonist. *Nature 441*, 91–95.
73. MacRae, C. A., and Peterson, R. T. (2003). Zebrafish-based small molecule discovery. *Chem Biol 10*, 901–908.
74. Peterson, R. T., Link, B. A., Dowling, J. E., and Schreiber, S. L. (2000). Small molecule developmental screens reveal the logic and timing of vertebrate development. *Proc Natl Acad Sci U S A* 97, 12965–12969.
75. Khersonsky, S. M., Jung, D. W., Kang, T. W., Walsh, D. P., Moon, H. S., Jo, H., Jacobson, E. M., Shetty, V., Neubert, T. A., and Chang, Y. T. (2003). Facilitated forward chemical genetics using a tagged triazine library and zebrafish embryo screening. *J Am Chem Soc 125*, 11804–11805.
76. Mitsopoulos, G., Walsh, D. P., and Chang, Y. T. (2004). Tagged library approach to chemical genomics and proteomics. *Curr Opin Chem Biol 8*, 26–32.
77. MacBeath, G., Koehler, A. N., and Schreiber, S. L. (1999). Printing small molecules as microarrays and detecting

protein-ligand interactions en masse. *J Am Chem Soc 121*, 7967–7968.

78. Winssinger, N., Pianowski, Z., and Debaene, F. (2007). Probing biology with small molecule microarrays (SMM). *Top Curr Chem 278*, 311–342.
79. Hergenrother, P. J., Depew, K. M., and Schreiber, S. L. (2000). Small-molecule microarrays: covalent attachment and screening of alcohol-containing small molecules on glass slides. *J Am Chem Soc 122*, 7849–7850.
80. Bryan, M. C., Fazio, F., Lee, H. K., Huang, C. Y., Chang, A., Best, M. D., Calarese, D. A., Blixt, O., Paulson, J. C., Burton, D., Wilson, I. A., and Wong, C. H. (2004). Covalent display of oligosaccharide arrays in microtiter plates. *J Am Chem Soc 126*, 8640–8641.
81. Kanoh, N., Kumashiro, S., Simizu, S., Kondoh, Y., Hatakeyama, S., Tashiro, H., and Osada, H. (2003). Immobilization of natural products on glass slides by using a photoaffinity reaction and the detection of protein-small-molecule interactions. *Angew Chem Int Ed Engl 42*, 5584–5587.
82. Kanoh, N., Asami, A., Kawatani, M., Honda, K., Kumashiro, S., Takayama, H., Simizu, S., Amemiya, T., Kondoh, Y., Hatakeyama, S., Tsuganezawa, K., Utata, R., Tanaka, A., Yokoyama, S., Tashiro, H., and Osada, H. (2006). Photo-cross-linked small-molecule microarrays as chemical genomic tools for dissecting protein-ligand interactions. *Chem Asian J 1*, 789–797.
83. Dilly, S. J., Bell, M. J., Clark, A. J., Marsh, A., Napier, R. M., Sergeant, M. J., Thompson, A. J., and Taylor, P. C. (2007). A photoimmobilisation strategy that maximises exploration of chemical space in small molecule affinity selection and target discovery. *Chem Commun (Camb)*, 2808–2810.
84. Bradner, J. E., McPherson, O. M., Mazitschek, R., Barnes-Seeman, D., Shen, J. P., Dhaliwal, J., Stevenson, K. E., Duffner, J. L., Park, S. B., Neuberg, D. S., Nghiem, P., Schreiber, S. L., and Koehler, A. N. (2006). A robust small-molecule microarray platform for screening cell lysates. *Chem Biol 13*, 493–504.
85. Bradner, J. E., McPherson, O. M., and Koehler, A. N. (2006). A method for the covalent capture and screening of diverse small molecules in a microarray format. *Nat Protoc 1*, 2344–2352.
86. Vegas, A. J., Fuller, J. H., and Koehler, A. N. (2008). Small-molecule microarrays as tools in ligand discovery. *Chem Soc Rev 37*, 1385–1394.
87. Bailey, S. N., Sabatini, D. M., and Stockwell, B. R. (2004). Microarrays of small molecules embedded in biodegradable polymers for use in mammalian cell-based screens. *Proc Natl Acad Sci U S A 101*, 16144–16149.
88. Schmitz, K., Haggarty, S. J., McPherson, O. M., Clardy, J., and Koehler, A. N. (2007). Detecting binding interactions using microarrays of natural product extracts. *J Am Chem Soc 129*, 11346–11347.
89. Dove, A. (2007). High-throughput screening goes to school. *Nature Methods 4*, 523–529.
90. Kaiser, J. (2008). Industrial-style screening meets academic biology. *Science 321*, 764–766.

CHAPTER 3

Chemical Proteomics: A Global Study of Protein–Small Molecule Interactions

Akihisa Matsuyama
Yoko Yashiroda
Minoru Yoshida

1. WHAT IS CHEMICAL PROTEOMICS?

Proteins exert their functions through interaction with other proteins or ligands. Therefore, to understand protein function, it is important to analyze protein-protein and protein-ligand interactions. As a result of the completion of the genome sequencing, studies of protein-protein interactions can be conducted on a genome-wide scale. Similarly, biologically active small molecules exert their actions through interaction with their targets, mostly proteins. Thus, the same holds true for analyses of chemicals and proteins. Chemical proteomics is an approach to clarify the biological function of proteins or chemicals through analyses of protein-chemical interactions.

As with genetics, there are two types of chemical proteomics, viz. forward chemical proteomics and reverse chemical proteomics [1]. Although the strategies are different, the primary goals of these two approaches are the same. Both methodologies employ chemicals of interest, so-called bioprobes, at the beginning [2]. The bioprobes with unique biological activities not only lead to drug discovery and development, but they also can be used for investigating many aspects of proteins, such as expression and subcellular localization. If the mode of action of a compound is unknown, we seek to identify the specific cellular targets in cells from the proteome. In forward chemical proteomics, whole-cell lysates are generally screened for interacting proteins, whereas in reverse chemical proteomics, purified or recombinant proteins expressed from the cloned ORFeome (a whole set of open reading frames in an organism) are employed for the target screen. These approaches have both advantages and disadvantages based on the differences in experimental procedures. This section describes their features and typical examples.

2. FORWARD CHEMICAL PROTEOMICS

Forward chemical proteomics is a systematic effort to identify targets starting from a chemical with interesting biological activity. To fish the specific target from the ocean of the proteome, it is generally necessary to modify the compound for labeling. Forward chemical proteomics sometimes reveals an unexpected target of a compound. For example, a kinase inhibitor, paullone, was originally shown to inhibit several cyclin-dependent protein kinases (CDKs) and glycogen synthase kinase 3 (GSK-3) [3, 4]. However, when gwennpaullone (2-[4-aminobutoxy]-9-bromo-7,12-dihydroindolo[3,2-*d*][1]benzazepin-6[5*H*]-one hydrochloride: a derivative of paullone) was immobilized to an affinity matrix, mitochondrial malate dehydrogenase (mMDH) was identified from porcine brain lysates as a specific binding protein. Paullone was then shown by detailed analysis to specifically bind and inhibit mMDH. Such results provide deep insight into the mode of drug action as well as potentially adverse side effects. Thus, screens at the proteome-wide scale for the target of drugs or drug candidates will become increasingly important.

2.1. Affinity Chromatography

Affinity chromatography has been widely used for purification of proteins. This traditional technique is also the central one in forward chemical proteomics, used to fish out a given compound's target protein. In this approach, a small molecule ligand of interest is immobilized via a suitable functional group onto a solid matrix. Total cell lysates are then subjected to this column to allow binding of the target to the immobilized ligand. After several washing steps, the bound targets are eluted by competition with free ligand or denaturation (Figure 3.1). Eluted proteins are then identified by physicochemical and spectral techniques such as mass spectrometry. A typical example of the target identification of a biological active compound using affinity chromatography is the isolation of *cis–trans* peptidyl-prolyl isomerase (FKBP) as the binding protein of an immunosuppressant FK506. FK506 was fixed onto the polymer packed in the column, and total cell lysates of bovine thymus and human spleen were loaded onto the affinity column to allow

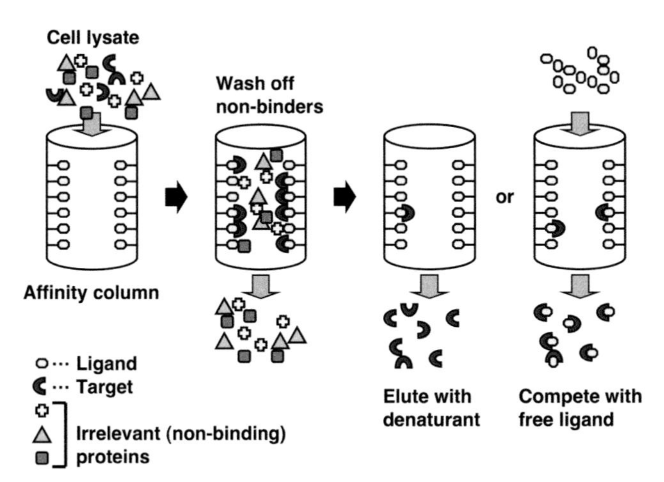

***FIGURE 3.1:** Affinity purification. At the first step for drug target identification, a small molecule is directly immobilized onto a solid matrix. The protein lysate is then loaded onto the resin to allow selective purification of proteins that bind to the molecule. After washing several times, targets are eluted by using denaturants (i.e., sodium dodecyl sulfate) or free ligand.*

binding of target proteins. After elution of the affinity-bound proteins followed by mass spectrometry, FKBP was identified as the specific protein whose binding was competitively inhibited by free FK506 as a competitor (Figure 3.2) [5]. Similarly, the first cloning of histone deacetylase was accomplished by using affinity beads conjugated with trapoxin [6], which had been reported as a specific, irreversible inhibitor of histone deacetylase [7].

Type II methionine aminopeptidase (MetAp-2) was isolated as a protein that specifically binds to fumagillin, an antiangiogenesis drug [8, 9]. In this case, fumagillin was biotinylated and mixed with the total cell lysate from bovine brain. After incubation, the mixture was subjected to an avidin column to capture the biotinylated compound complexed with its target protein. Indeed, biotinylated probes were widely used for the identification of many drug targets, including leptomycin B (LMB) [10], epolactaene [11], pateamine A [12], chromoceptin [13], and spliceostatin A [14]. This biotin-conjugate system has an advantage over the affinity-matrix method in that the modified chemicals can be used for isolating target proteins from not only the cell lysate but also living cultured cells. For example, LMB, an inhibitor of Crm1 nuclear export factor, was biotinylated (Figure 3.2) and directly added to the culture of HeLa cells. The proteins bound to biotinylated LMB in vivo were then isolated using streptavidin beads from the total cell lysates. This experiment clearly demonstrated that Crm1 is the only protein in the cell that binds covalently to LMB [15].

In all of these studies, the first step was to label or tag the biologically active compound without losing its activity by

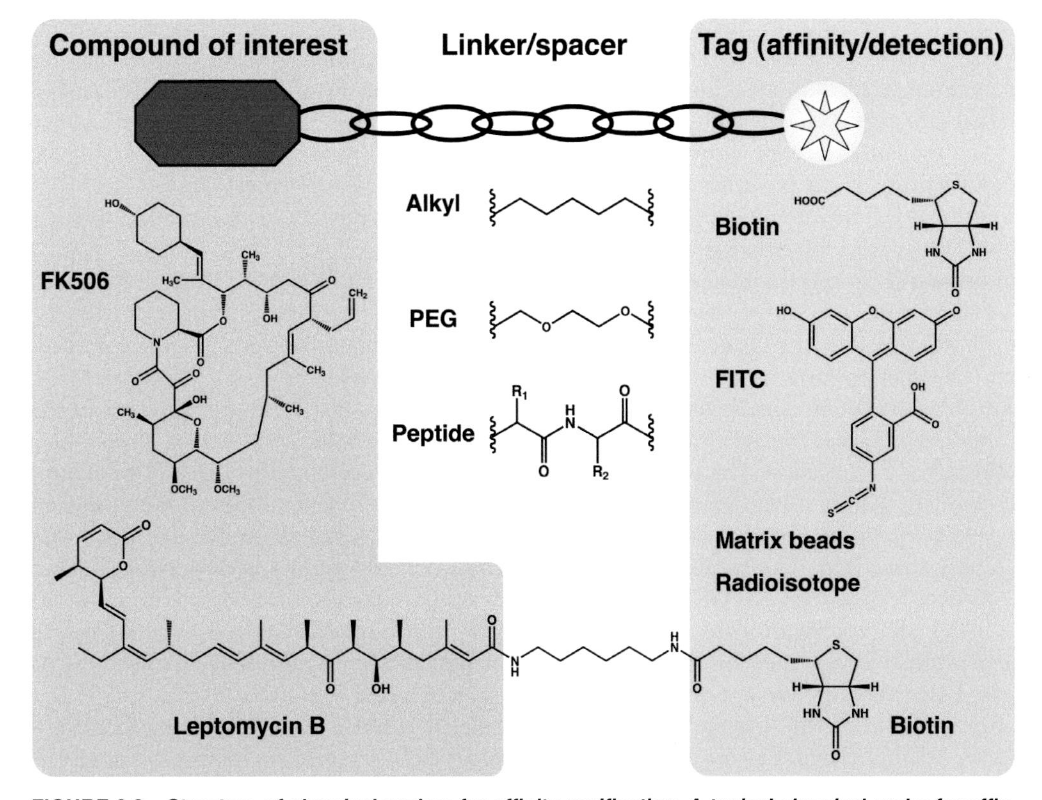

***FIGURE 3.2:** Structure of chemical probes for affinity purification. A typical chemical probe for affinity purification is composed of three components: label or tag, linker, and the active small molecule. Biotinylated LMB used for identification of exportin (Crm1) is shown at the bottom.*

chemical modification or total synthesis. As seen in Figure 3.2, the structure of those probes usually includes three elements. A tag for conjugating or labeling compounds should be attached to an appropriate group via a linker. Thus, structure-activity relationship (SAR) studies using a series of synthesized analogs are needed to determine which functional group should be modified for tagging. Recently, however, a unique method for immobilizing a variety of small molecules in a functional group–independent manner was established by Kanoh et al. using a photocrosslinking approach [16]. This method depends on the reactivity of the carbene species generated from trifluoromethylaryldiazirine upon UV irradiation. It was demonstrated in a number of model experiments that photogenerated carbenes were able to react with the small molecules, producing multiple conjugates in most cases. Although the compounds were randomly immobilized on the substrate, the linker-compound conjugates generated in this manner contained those with appropriate orientations to show biological activity. This method relieves researchers from the burden of SAR studies. Indeed, by taking advantage of this nonselective nature of the photocrosslinking process, chemicals can be directly spotted onto a solid surface such as glass slides followed by photocrosslinking to generate chemical microarrays. The immobilized small molecules can be demonstrated to retain their ability to interact with their binding proteins in on-array immobilization experiments.

Proteins or peptides displayed on phage particles as fusions with the coat protein can be used instead of the total cell lysate as the source of target proteins. A protein-protein interaction has been successfully elucidated by a phage display technique [17]. By addition of an appropriate signal sequence, such fusion proteins can be directed to the bacterial periplasm or inner cell membrane during the phage assembly process. The genetic information encoding the displayed fusion protein is packaged into the same phage particle. Phages that present a binding protein (or peptide) are enriched by affinity selection of a phage library on the immobilized bait. The bound phages are eluted and amplified by reinfection of *Escherichia coli* cells. By repeating these procedures, the binding target can be specifically enriched, leading to identification. For example, the phage display was successfully used to identify a peptide motif (CKGGRAKDC) that selectively targets the vascular endothelial cells of white adipose tissue [18]. The peptide motif of a proapoptotic sequence (producing CKGGRAKDC-GG-D-KLAKLAKKLAKLAK) was synthesized and injected into mice, which resulted in the loss of an impressive amount of weight. Essentially, the same technique can be used for identification of chemical-protein interactions. For example, using a biotinylated FK506, a full-length gene clone of the FK506-binding protein (FKBP12) was again specifically isolated from a human brain cDNA-based phage library [19]. Phage display screening revealed that HBC, a curcumin derivative with antitumor activity, directly binds and inhibits Ca^{2+}/calmodulin [20]. Similarly, the target of Bz-423, which specifically induces apoptosis in pathogenic lymphocytes in the systemic lupus erythematosus, was determined to be a component of the mitochondrial F1F0–adenosine triphosphate (ATP)-ase; Bz-423 was shown to inhibit ATPase activity [21].

2.2. Affinity Labeling

The affinity chromatography approach is based on the assumption that the interaction between a protein and a chemical is noncovalent binding. Therefore, a target showing a weak affinity would easily be dissociated from the compound-bound matrix and lost. Such weak or unstable small molecule–protein interactions represent one of the major difficulties encountered during affinity purification of a target. In such cases, affinity labeling is a powerful strategy for drug target identification.

A chemical probe for affinity labeling essentially consists of three functional elements that are similar to those used in affinity chromatography: tag, linker, and reactive group. Figure 3.3 indicates the moieties that are generally used for affinity labeling. The affinity or labeling tags are employed for capture, detection, and visualization of the target protein. Because of its compatibility with both purification and a gel-based detection technique, biotin is the most commonly used tag. The biotin-streptavidin bond is one of the strongest known noncovalent interactions, and exploiting this interaction allows effective purification of low-abundance targets. However, strong binding to endogenous cellular biotinylated proteins by the streptavidin beads sometimes prevents accurate detection of the target. Furthermore, strong binding to streptavidin interferes with effective elution from the beads under mild conditions.

The general role of the linker is to connect the tag and the reactive group and add enough distance between them to prevent steric hindrance. For these purposes, long-chain alkyls or polyethylene glycol (PEG) are often used. A long-chain alkyl is suited for chemical probes that are directly used in living cells, because its hydrophobicity is expected to allow permeability through cell membrane. In contrast, PEG confers hydrophilicity to the probe, making it more soluble in aqueous solutions. Recently, a cleavable linker responsive to light has been developed. Using this photocleavable linker connected to biotin, the difficulty in the elution of biotinylated probe-bound proteins captured in streptavidin beads, as discussed previously, could be overcome [22, 22a].

If the reactive group is an enzyme substrate analog that serves as an irreversible inhibitor, the molecule will covalently bind to the active site of the enzyme together with the tag. An example of this type of affinity labeling is the use of a probe bearing sulfonate esters in screens of mouse heart

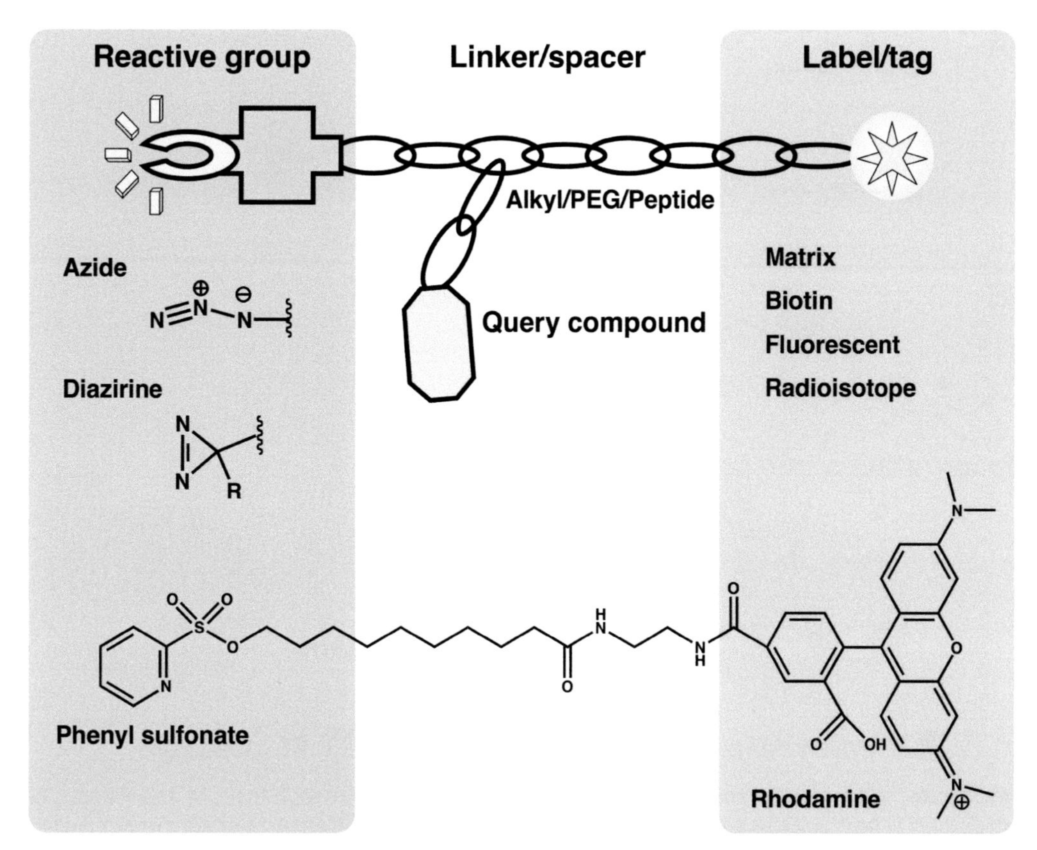

***FIGURE 3.3:** Structure of chemical probes for affinity labeling. A typical chemical probe for affinity labeling is made of three components: label or tag, linker, and a reactive group. A tag and a linker are similar to those for affinity chromatography. Rhodamine-tagged phenyl sulfonate (bottom) shows broad reactivity with many metabolic enzymes in the cell. Some linkers have an additional branch for affinity binding to the target.*

cell lysates (Figure 3.3) [23]. Using rhodamine-tagged sulfonate esters, several cellular proteins were visualized after incubation with the lysates. These proteins included acetyl-CoA acetyltransferease, enoyl-CoA hydratase, and aldehyde dehydrogenase, among others. Affinity labeling of these proteins depended on which reactive group was used for reaction. These observations suggest that activity-based proteomics can be conducted by selecting a reactive group that commonly interacts with a family of enzymes. In the case of the reactive groups that are highly reactive and nonspecific, however, their usefulness for target identification might be limited by their toxicity to cells.

If the small molecule binds noncovalently to the target but the interaction is not strong enough to allow affinity purification, it is necessary to support the specific interaction by inducing a secondary interaction that occurs in the vicinity of the target. Chemical crosslinking reagents have been used for this purpose. In particular, arylazide and aryldiazirine functionalities are frequently used as photosensitive linkers. Hatanaka et al. designed and synthesized a photoaffinity labeling reagent in which the aryldiazirine was coupled with biotin [24] and identified the substrate binding site of β1,4-galactosyltransferase using a photoreactive *N*-acetylglucosamine derivative [25]. Recently, a similar photoaffinity/biotin probe was used for identification of SAP130 of the SF3b subcomplex in the mammalian spliceosome complex as the target of novel antitumor agent pladienolide B [26].

2.3. Targeted Labeling using Click Chemistry

Click chemistry is an approach proposed in 2001 by Karl Barry Sharpless, who won the Nobel Prize in Chemistry in 2001 for his work on stereoselective oxidation reactions [27]. Although it was developed originally in the area of organic synthesis, it has recently been expanded to biological research. The method employs reactive molecular building blocks designed to "click" together selectively and covalently with a few practical and reliable reactions. It is defined to satisfy several characteristics as follows: simple reaction conditions (ideally, insensitive to oxygen and water), readily available starting materials and reagents, no solvent or a solvent that is benign (such as water) or easily removed, and products that can be easily isolated. Of the various click chemistry reactions, the most representative one is the Huisgen's 1,3-dipolar cycloaddition of

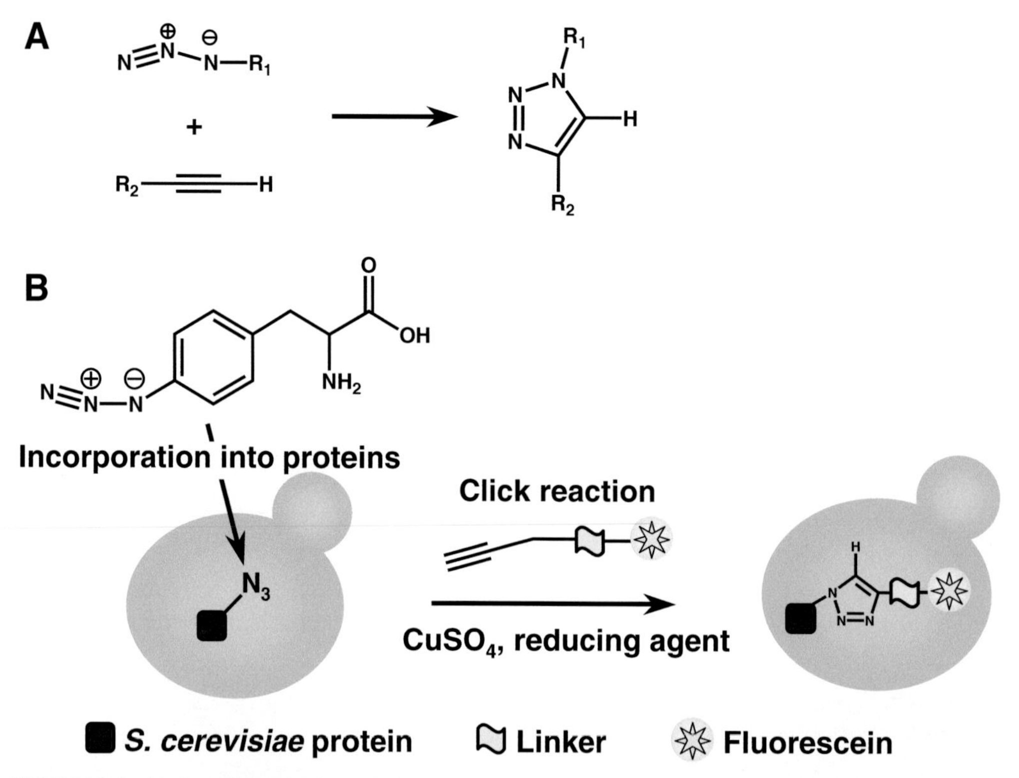

***FIGURE 3.4:** In vivo click chemistry. **A.** Click reactions between azides and acetylenes. This reaction is widely used in biology as well as in organic chemistry. **B.** Labeling of all proteins within a cell by a click reaction. By incorporation of azide tyrosine instead of tyrosine, budding yeast proteins are genetically labeled with azides. Using these reactive azides, fluorescein fused with acetylenes can be attached to the azide-containing proteins via Huisgen's 1,3-dipolar cycloaddition.*

alkynes to azides, which leads to the generation of triazoles (Figure 3.4A) [28]. In particular, the copper (I)-catalyzed 1,2,3-triazole formation from azides and terminal acetylenes is the premier example that is very effective for biological applications. For example, it has been reported that proteins could be genetically labeled with click chemistry–compatible functional groups by addition of the azide or acetylene tyrosine analogs into the culture of yeast (Figure 3.4B) [29]. Ideally, the structure of a chemical probe should be analogous to those of natural co-factors or intracellular biomolecules. In this sense, azides and alkynes are ideal because they are small, relatively easy to install, and stable in biological environments (i.e., reduced, aqueous conditions), despite their high energetic potentials as functional groups. Recent success in the design of a copper-free reagent for cyclooctyne derivatives provides great possibilities to expand the applicability of this method into many fields of biomedical research [30, 31]. For example, mucin-type *O*-linked glycosylation is an abundant post-translational modification involved in a variety of biological interactions in eukaryotic cells. The carbohydrate moiety in this modification is synthesized from *N*-acetylgalactosamine (GalNAc) by a family of polypeptide *N*-acetyl-α-galactosaminyltransferases (ppGalNAcTs). By feeding cells peracetylated *N*-azidoacetylgalactosamine (Ac4GalNAz), membrane-associated glycans were successfully visualized in developing zebrafish using the difluorinated cyclooctyne-Alexa Fluor 488 conjugate [32]. Thus, an alkyne click chemistry tag can be used in place of biotin for affinity labeling. MacKinnon et al. installed a photo-leucine residue and a propargyl substituent into HUN-7293, a cyclodepsipeptide inhibitor of cotranslational translocation of vascular cell adhesion molecule-1 (VCAM-1) to the endoplasmic reticulum (ER); they identified Sec61α as the protein photocrosslinked by the photo-leucine–containing click chemistry probe [33]. Thus, the click chemistry–based approach will become increasingly popular in the field of forward chemical proteomics.

2.4. Differential Display

The differential display technique was invented in 1992 to allow rapid and sensitive detection of altered gene expression [34]. Messenger RNA (mRNA) subpopulations from related sample cells are visualized by denaturing polyacrylamide electrophoresis, which allows direct side-by-side comparison of most of the mRNAs. Because of its simplicity and sensitivity, the mRNA differential display method has become a standard application in many types of biological research. A similar strategy is also possible for proteomics. In this case, all proteins from two samples are labeled and separated by electrophoresis, usually 2D-PAGE. By comparing images side by side, spots of up- or downregulated proteins are picked up and then identified by mass spectrometry. Recently, a new method has been developed called differential in-gel electrophoresis (DIGE); in this technique, two samples for comparison are labeled with different fluorescent dyes, mixed together, and electrophoresed in the same gel [35, 36]. In this system, up- or downregulated proteins can be easily discriminated according to the color of each spot. These systems allow proteome-wide screening for proteins that are influenced by addition of a drug of interest by comparing total cell lysates between cells treated with the drug and those treated with vehicle alone. Although such strategies do not reveal the target of a drug directly, they sometime provide insights into the mode of action.

The strategy of differential display technique is also applicable for more focused target identification. For example, deacetylation of microtubules by histone deacetylase 6 (HDAC6) was found using the differential display technique by taking advantage of the difference in the specificity of two HDAC inhibitors, trichostatin A (TSA) and trapoxin B (TPX-B) [37]. TSA is a broad inhibitor blocking activity of both classes I and II of HDACs, whereas TPX-B cannot efficiently inhibit HDAC6 [38, 39]. When cells are treated with these drugs, enhanced acetylation of some proteins should be observed in Western blotting using an anti-acetyllysine antibody as a result of inhibition of deacetylation. Acetylation of a protein with the molecular size of 54 kDa was enhanced in TSA-treated cells but not in TPX-B-treated cells, suggesting that this protein is deacetylated specifically by HDAC6. This protein was eventually determined to be the alpha subunit of tubulin.

3. REVERSE CHEMICAL PROTEOMICS

Recent success in cloning of entire sets of protein-coding open reading frames (ORFeome) in various organisms has enabled a variety of systematic proteomic studies using recombinant proteins (i.e., *reverse proteomics*) [40]. Some early efforts have been made to construct gene libraries, in which randomly cloned cDNAs are characterized by sequencing of both ends. In contrast, an ORFeome-based library can cover essentially all ORFs and can be used systematically in one assay after another, enabling one to set up a genome-wide experiment with a minimum of exertion. Thus, cloning of the ORFeome of an organism may give strong momentum to experimental efficiency once the challenging and daunting task of constructing the ORFeome collection has been completed. Reverse chemical proteomics employs these ORFeome-based protein sets to discover new drugs or drug targets.

3.1. ORFeome Libraries for Reverse Proteomics

Early efforts in large-scale reverse proteomics focused on the use of the ORFeome of the budding yeast *Saccharomyces cerevisiae*, as this organism is the first eukaryote for which the whole genome sequence was determined [41]. The ORFeome library was initially created by cloning each ORF directly into an expression vector by gap repair, although this system is disadvantageous in that the cloned ORF cannot be easily transferred into alternative expression vectors [42]. The other cloning method is to employ site-specific homologous recombination, such as the Gateway system [43]. This system can easily shuttle ORFs between different expression vectors; therefore, the ORFeome library based on the versatile Gateway system is an excellent platform for reverse chemical proteomics. To date, ORFeome libraries using the Gateway system have become available in budding yeast, fission yeast, bacteria, viruses, nematodes, and human [42, 44–49, 49a]. The reverse chemical proteomic studies using ORFeome libraries thus far reported include drug target identification using proteome microarrays and drug target pathway identification using the *localizome* (the set of all protein subcellular localizations) [47, 54] (Figure 3.5).

3.2. Proteome Microarray and Drug Target Identification

Comprehensive proteome microarrays are composed of individually cloned, expressed, and purified proteins and have been shown to be useful for identification of protein-protein, protein–nucleic acid, and protein-lipid interactions. For instance, yeast proteome microarray technology revealed proteins binding to calmodulin and phosphatidylinositide-containing liposomes [42, 50], DNA-binding proteins [51], and proteins subject to glycosylation [45], phosphorylation [52], or ubiquitination [53], and so forth. The proteome microarray is also powerful as a tool to analyze the ability of proteins to bind small molecules. Schreiber and colleagues screened for small molecule inhibitors of rapamycin (SMIRs), which suppress rapamycin's effect on the budding yeast and identified putative intracellular target proteins by employing the budding yeast proteome array [54]. The proteome array was probed with bioactive biotin-conjugated SMIR molecules. Binding of the biotin-conjugated SMIRs to the target proteins was detected using a fluorescence-labeled streptavidin conjugate (Figure 3.5). Proteome array–based target identification potentially may be more effective than affinity chromatography, which may be biased toward high-abundance proteins. Goshima et al. created a large collection of human ORFeome clones that cover approximately 70% of the 22,000 human genes [49]. They also established a high-throughput in vitro protein expression method using a wheat germ cell-free protein synthesis system in which all the proteins are produced

- PCR amplification of each ORF using genomic DNA or cDNA as templates
- Cloning to expression vectors using Gateway technology etc.

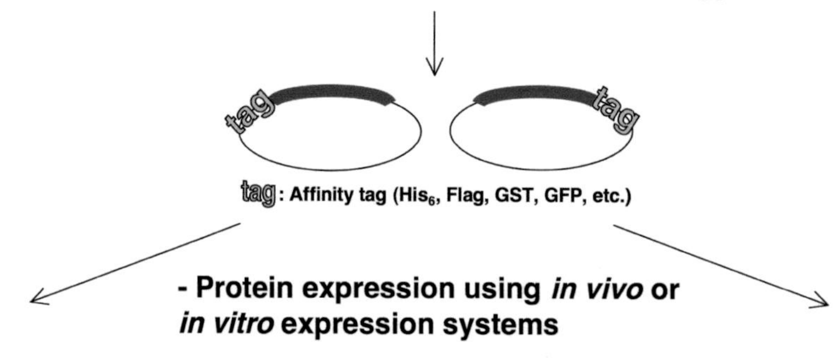

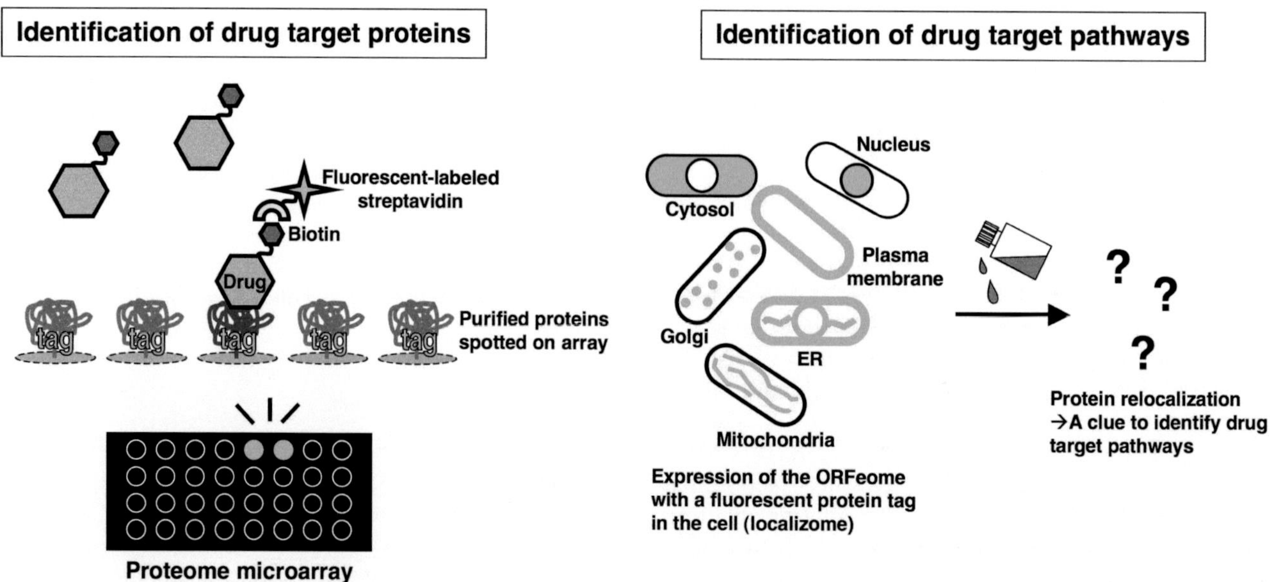

FIGURE 3.5: ORFeome-based reverse chemical proteomics. ORFeome libraries are constructed by creating expression vectors with the PCR-amplified ORFs and desired tags either at the N- or C-terminus. Each tagged protein is expressed in the test tube using either an in vitro translation system or live cells. The entire protein collection can be used for generation of proteome microarrays (left) or for identification of protein localization (right). Localization changes upon the drug treatment provide important information that can be used to identify drug target pathways .

theoretically independent of their intrinsic abundance in cells. They manufactured a proteome microarray of glutathione S-transferase (GST)-tagged human proteins synthesized by using the wheat germ system; this tool will allow proteome-scale functional analyses, including chemical proteomics.

3.3. Localizome and Drug Target Pathway Identification

ORFeome-based reverse proteomics provides another approach to the discovery of drug target pathways. In *Schizosaccharomyces pombe*, a data set of protein subcellular localization (*localizome*) covering 90% of the fission yeast proteome was determined by expressing the ORFeome fused to yellow fluorescent protein (YFP) [47]. Leptomycin B (LMB) is a specific inhibitor of Crm1, which transports proteins bearing the nuclear export signal from the nucleus to the cytoplasm. LMB has been established as a useful tool for testing whether a protein of interest is regulated by Crm1 [10, 55]. The fission yeast strain collection expressing each ORF-YFP fusion was treated with LMB, and fluorescent images were collected to allow determination of protein localization [47]. This global analysis identified 285 proteins as potential Crm1 cargoes in the fission yeast proteome. Not only cytoplasmic proteins but also proteins normally localized in the microtubules, cell tip, and septum were relocalized to the nucleus upon LMB treatment, suggesting that Crm1-dependent nuclear export plays a broader role than originally thought. Thus, the changes in the localizome profiles upon treatment with various small molecules will provide information regarding how cellular proteins respond to perturbation by small molecules at the systems level (Figure 3.5).

4. FUTURE PROSPECTS

Identification of the target(s) of a biologically active small molecule is a central challenge of chemical biology and therapeutic development. Recent advances in mass spectrometry and the complete sequencing of many genomes will greatly accelerate forward chemical proteomics. Because it ideally covers all proteins expressed within a cell, forward chemical proteomics often results in the discovery of unexpected protein species. However, it is still difficult to detect proteins whose amounts in the cell are very low or whose physical interactions with the molecule of interest are weak. To overcome the disadvantages of forward chemical proteomics, a number of new technologies such as clickable tags, photoaffinity linkers, and phage display have been established. In contrast, reverse proteomics based on the ORFeomes enables one to overproduce the naturally low abundant proteins. By tagging the ORFeome at the N- or C-terminus, the expressed proteins can be detected in the cell lysate in a highly sensitive manner. It should be noted, however, that affinity tagging at the N- or C-terminus of a protein sometimes interferes with the functions or correct localization of proteins. Nonetheless, ORFeome-based reverse chemical proteomics is a powerful method for revealing previously unknown protein-chemical interactions. We note that the proteome microarray assays have been optimized over the past few years, not only for protein-protein but also for protein–small molecule interactions. Establishment of the localizome data sets also has great potential for chemical proteomics [64]. If the cellular target of a small molecule can be visualized by modifying the compound with an appropriate tag such as fluorescent dyes, then the candidate clones can be retrieved from the data set by comparing the cellular distribution of the labeled compound with the database of images of subcellular localizations. Furthermore, comprehensive analysis of protein localization changed upon drug treatment will reveal what is happening in the cells in response to the compound.

One of the most important future directions for chemical proteomics will be a link to chemical genomics, a global study of chemical genetic interactions. Chemical genetic interactions can be detected when the mutated genes are involved in the pathway of the small molecule target, even in cases where it is difficult to observe physical interactions. In general, mutations or overexpression of the particular genes involved in the target pathway result in altered sensitivity of cells to the biologically active compound. Completion of genome sequencing enables one to do systematic observation of chemical genetic interactions between small molecules and genes [56–63, 65]. In particular, the budding yeast *Saccharomyces cerevisiae* has been widely used for drug target prediction and validation, mostly by haploinsufficiency or synthetic lethality analysis employing comprehensive collections of gene deletion mutants [56–63]. Screens using induced haploinsufficiency are based on the gene dosage effect on drug sensitivity, whereas synthetic lethality analysis is based on the hypothesis that the drug sensitivity profiles and genetic interaction profiles should match if the query mutation is the drug target–encoding gene. Both the characteristics of each hit gene that alters drug sensitivity and the genome-wide patterns of the hit genes (chemical genomic profile) are important for predicting the target. Indeed, two-dimensional hierarchical clustering analysis of chemical-genomic interaction profiles is a powerful method for grouping compounds with their known target pathways or proteins. However, the usefulness of chemical genomic profiling is not limited to the pattern-matching analysis with the data set of reference compounds whose targets are known. Combining chemical genomic profile information with chemical proteomics information, such as physical interactions with particular proteins or protein subcellular localization, will provide deep insights into the drug target identification. Thus, with the help of bioinformatics, integrating forward and reverse chemical proteomics will facilitate comprehensive understanding of the biological systems that are targeted by small molecules.

ACKNOWLEDGMENTS

We thank Takeo Usui, Tsukuba University, for valuable discussion of this manuscript.

REFERENCES

1. Palcy, S., and Chevet, E. (2006). Integrating forward and reverse proteomics to unravel protein function. *Proteomics 6*, 5467–5480.
2. Osada, H. (2000). Bioprobes (Tokyo: Springer-Verlag).
3. Schultz, C., Link, A., Leost, M., Zaharevitz, D. W., Gussio, R., Sausville, E. A., Meijer, L., and Kunick, C. (1999). Paullones, a series of cyclin-dependent kinase inhibitors: synthesis, evaluation of CDK1/cyclin B inhibition, and in vitro antitumor activity. *J Med Chem 42*, 2909–2919.
4. Leost, M., Schultz, C., Link, A., Wu, Y. Z., Biernat, J., Mandelkow, E. M., Bibb, J. A., Snyder, G. L., Greengard, P., Zaharevitz, D. W., Gussio, R., Senderowicz, A. M., Sausville, E. A., Kunick, C., and Meijer, L. (2000). Paullones are potent inhibitors of glycogen synthase kinase-3beta and cyclin-dependent kinase 5/p25. *Eur J Biochem 267*, 5983–5994.
5. Harding, M. W., Galat, A., Uehling, D. E., and Schreiber, S. L. (1989). A receptor for the immunosuppressant FK506 is a *cis–trans* peptidyl-prolyl isomerase. *Nature 341*, 758–760.
6. Taunton, J., Hassig, C. A., and Schreiber, S. L. (1996). A mammalian histone deacetylase related to the yeast transcriptional regulator Rpd3p. *Science 272*, 408–411.
7. Kijima, M., Yoshida, M., Sugita, K., Horinouchi, S., and Beppu, T. (1993). Trapoxin, an antitumor cyclic tetrapeptide, is an irreversible inhibitor of mammalian histone deacetylase. *J Biol Chem 268*, 22429–22435.

8. Sin, N., Meng, L., Wang, M. Q., Wen, J. J., Bornmann, W. G., and Crews, C. M. (1997). The anti-angiogenic agent fumagillin covalently binds and inhibits the methionine aminopeptidase, MetAP-2. *Proc Natl Acad Sci U S A 94*, 6099–6103.
9. Griffith, E. C., Su, Z., Turk, B. E., Chen, S., Chang, Y. H., Wu, Z., Biemann, K., and Liu, J. O. (1997). Methionine aminopeptidase (type 2) is the common target for angiogenesis inhibitors AGM-1470 and ovalicin. *Chem Biol 4*, 461–471.
10. Kudo, N., Wolff, B., Sekimoto, T., Schreiner, E. P., Yoneda, Y., Yanagida, M., Horinouchi, S., and Yoshida, M. (1998). Leptomycin B inhibition of signal-mediated nuclear export by direct binding to CRM1. *Exp Cell Res 242*, 540–547.
11. Nagumo, Y., Kakeya, H., Shoji, M., Hayashi, Y., Dohmae, N., and Osada, H. (2005). Epolactaene binds human Hsp60 Cys442 resulting in the inhibition of chaperone activity. *Biochem J 387*, 835–840.
12. Low, W. K., Dang, Y., Schneider-Poetsch, T., Shi, Z., Choi, N. S., Merrick, W. C., Romo, D., and Liu, J. O. (2005). Inhibition of eukaryotic translation initiation by the marine natural product pateamine A. *Mol Cell 20*, 709–722.
13. Choi, Y., Shimogawa, H., Murakami, K., Ramdas, L., Zhang, W., Qin, J., and Uesugi, M. (2006). Chemical genetic identification of the IGF-linked pathway that is mediated by STAT6 and MFP2. *Chem Biol 13*, 241–249.
14. Kaida, D., Motoyoshi, H., Tashiro, E., Nojima, T., Hagiwara, M., Ishigami, K., Watanabe, H., Kitahara, T., Yoshida, T., Nakajima, H., Tani, T., Horinouchi, S., and Yoshida, M. (2007). Spliceostatin A targets SF3b and inhibits both splicing and nuclear retention of pre-mRNA. *Nat Chem Biol 3*, 576–583.
15. Kudo, N., Matsumori, N., Taoka, H., Fujiwara, D., Schreiner, E. P., Wolff, B., Yoshida, M., and Horinouchi, S. (1999). Leptomycin B inactivates CRM1/exportin 1 by covalent modification at a cysteine residue in the central conserved region. *Proc Natl Acad Sci U S A 96*, 9112–9117.
16. Kanoh, N., Asami, A., Kawatani, M., Honda, K., Kumashiro, S., Takayama, H., Simizu, S., Amemiya, T., Kondoh, Y., Hatakeyama, S., Tsuganezawa, K., Utata, R., Tanaka, A., Yokoyama, S., Tashiro, H., and Osada, H. (2006). Photo-cross-linked small-molecule microarrays as chemical genomic tools for dissecting protein-ligand interactions. *Chem Asian J 1*, 789–797.
17. Paschke, M. (2006). Phage display systems and their applications. *Appl Microbiol Biotechnol 70*, 2–11.
18. Kolonin, M. G., Saha, P. K., Chan, L., Pasqualini, R., and Arap, W. (2004). Reversal of obesity by targeted ablation of adipose tissue. *Nat Med 10*, 625–632.
19. Sche, P. P., McKenzie, K. M., White, J. D., and Austin, D. J. (1999). Display cloning: functional identification of natural product receptors using cDNA-phage display. *Chem Biol 6*, 707–716.
20. Shim, J. S., Lee, J., Park, H. J., Park, S. J., and Kwon, H. J. (2004). A new curcumin derivative, HBC, interferes with the cell cycle progression of colon cancer cells via antagonization of the Ca2 + /calmodulin function. *Chem Biol 11*, 1455–1463.
21. Johnson, K. M., Chen, X., Boitano, A., Swenson, L., Opipari, A. W., Jr., and Glick, G. D. (2005). Identification and validation of the mitochondrial F1F0-ATPase as the molecular target of the immunomodulatory benzodiazepine Bz-423. *Chem Biol 12*, 485–496.
22. Olejnik, J., Sonar, S., Krzymanska-Olejnik, E., and Rothschild, K. J. (1995). Photocleavable biotin derivatives: a versatile approach for the isolation of biomolecules. *Proc Natl Acad Sci U S A 92*, 7590–7594.

22a. Heney, G., and Orr, G. A. (1981). The purification of avidin and its derivatives on 2-iminobiotin-6-aminohexyl-Sepharose 4B. *Anal Biochem 114*, 92–96.

23. Adam, G. C., Sorensen, E. J., and Cravatt, B. F. (2002). Proteomic profiling of mechanistically distinct enzyme classes using a common chemotype. *Nat Biotechnol 20*, 805–809.
24. Hatanaka, Y., Hashimoto, M., and Kanaoka, Y. (1994). A novel biotinylated heterobifunctional cross-linking reagent bearing an aromatic diazirine. *Bioorg Med Chem 2*, 1367–1373.
25. Hatanka, Y., Hashimoto, M., Nishihara, S., Narimatsu, H., and Kanaoka, Y. (1996). Synthesis and characterization of a carbene-generating biotinylated N-acetylglucosamine for photoaffinity labeling of beta-(1->4)-galactosyltransferase. *Carbohydr Res 294*, 95–108.
26. Kotake, Y., Sagane, K., Owa, T., Mimori-Kiyosue, Y., Shimizu, H., Uesugi, M., Ishihama, Y., Iwata, M., and Mizui, Y. (2007). Splicing factor SF3b as a target of the antitumor natural product pladienolide. *Nat Chem Biol 3*, 570–575.
27. Kolb, H. C., Finn, M. G., and Sharpless, K. B. (2001). Click chemistry: diverse chemical function from a few good reactions. *Angew Chem Int Ed Engl 40*, 2004–2021.
28. Huisgen, R. (1984). *1,3-Dipolar Cycloaddition Chemistry*, vol. 1 (New York: Wiley).
29. Deiters, A., Cropp, T. A., Mukherji, M., Chin, J. W., Anderson, J. C., and Schultz, P. G. (2003). Adding amino acids with novel reactivity to the genetic code of *Saccharomyces cerevisiae*. *J Am Chem Soc 125*, 11782–11783.
30. Baskin, J. M., Prescher, J. A., Laughlin, S. T., Agard, N. J., Chang, P. V., Miller, I. A., Lo, A., Codelli, J. A., and Bertozzi, C. R. (2007). Copper-free click chemistry for dynamic in vivo imaging. *Proc Natl Acad Sci U S A 104*, 16793–16797.
31. Codelli, J. A., Baskin, J. M., Agard, N. J., and Bertozzi, C. R. (2008). Second-generation difluorinated cyclooctynes for copper-free click chemistry. *J Am Chem Soc 130*, 11486–11493.
32. Laughlin, S. T., Baskin, J. M., Amacher, S. L., and Bertozzi, C. R. (2008). In vivo imaging of membrane-associated glycans in developing zebrafish. *Science 320*, 664–667.
33. MacKinnon, A. L., Garrison, J. L., Hegde, R. S., and Taunton, J. (2007). Photo-leucine incorporation reveals the target of a cyclodepsipeptide inhibitor of cotranslational translocation. *J Am Chem Soc 129*, 14560–14561.
34. Liang, P., and Pardee, A. B. (1992). Differential display of eukaryotic messenger RNA by means of the polymerase chain reaction. *Science 257*, 967–971.
35. Unlu, M., Morgan, M. E., and Minden, J. S. (1997). Difference gel electrophoresis: a single gel method for detecting changes in protein extracts. *Electrophoresis 18*, 2071–2077.
36. Tonge, R., Shaw, J., Middleton, B., Rowlinson, R., Rayner, S., Young, J., Pognan, F., Hawkins, E., Currie, I., and Davison, M. (2001). Validation and development of fluorescence two-dimensional differential gel electrophoresis proteomics technology. *Proteomics 1*, 377–396.

37. Matsuyama, A., Shimazu, T., Sumida, Y., Saito, A., Yoshimatsu, Y., Seigneurin-Berny, D., Osada, H., Komatsu, Y., Nishino, N., Khochbin, S., Horinouchi, S., and Yoshida, M. (2002). In vivo destabilization of dynamic microtubules by HDAC6-mediated deacetylation. *Embo J 21*, 6820–6831.
38. Yoshida, M., Kijima, M., Akita, M., and Beppu, T. (1990). Potent and specific inhibition of mammalian histone deacetylase both in vivo and in vitro by trichostatin A. *J Biol Chem 265*, 17174–17179.
39. Furumai, R., Komatsu, Y., Nishino, N., Khochbin, S., Yoshida, M., and Horinouchi, S. (2001). Potent histone deacetylase inhibitors built from trichostatin A and cyclic tetrapeptide antibiotics including trapoxin. *Proc Natl Acad Sci U S A 98*, 87–92.
40. Yashiroda, Y., Matsuyama, A., and Yoshida, M. (2008). New insights into chemical biology from ORFeome libraries. *Curr Opin Chem Biol 12*, 55–59.
41. Goffeau, A., Barrell, B. G., Bussey, H., Davis, R. W., Dujon, B., Feldmann, H., Galibert, F., Hoheisel, J. D., Jacq, C., Johnston, M., Louis, E. J., Mewes, H. W., Murakami, Y., Philippsen, P., Tettelin, H., and Oliver, S. G. (1996). Life with 6000 genes. *Science 274*, 546, 563–547.
42. Zhu, H., Bilgin, M., Bangham, R., Hall, D., Casamayor, A., Bertone, P., Lan, N., Jansen, R., Bidlingmaier, S., Houfek, T., Mitchell, T., Miller, P., Dean, R. A., Gerstein, M., and Snyder, M. (2001). Global analysis of protein activities using proteome chips. *Science 293*, 2101–2105.
43. Rual, J. F., Hirozane-Kishikawa, T., Hao, T., Bertin, N., Li, S., Dricot, A., Li, N., Rosenberg, J., Lamesch, P., Vidalain, P. O., Clingingsmith, T. R., Hartley, J. L., Esposito, D., Cheo, D., Moore, T., Simmons, B., Sequerra, R., Bosak, S., Doucette-Stamm, L., Le Peuch, C., Vandenhaute, J., Cusick, M.E., Albala, J. S., Hill, D. E., and Vidal, M. (2004). Human ORFeome version 1.1: a platform for reverse proteomics. *Genome Res 14*, 2128–2135.
44. Lamesch, P., Milstein, S., Hao, T., Rosenberg, J., Li, N., Sequerra, R., Bosak, S., Doucette-Stamm, L., Vandenhaute, J., Hill, D. E., and Vidal, M. (2004). *C. elegans* ORFeome version 3.1: increasing the coverage of ORFeome resources with improved gene predictions. *Genome Res 14*, 2064–2069.
45. Gelperin, D. M., White, M. A., Wilkinson, M. L., Kon, Y., Kung, L. A., Wise, K. J., Lopez-Hoyo, N., Jiang, L., Piccirillo, S., Yu, H., Gerstein, M., Dumont, M. E., Phizicky, E. M., Snyder, M., and Grayhack, E. J. (2005). Biochemical and genetic analysis of the yeast proteome with a movable ORF collection. *Genes Dev 19*, 2816–2826.
46. Kitagawa, M., Ara, T., Arifuzzaman, M., Ioka-Nakamichi, T., Inamoto, E., Toyonaga, H., and Mori, H. (2005). Complete set of ORF clones of *Escherichia coli* ASKA library (a complete set of *E. coli* K-12 ORF archive): unique resources for biological research. *DNA Res 12*, 291–299.
47. Matsuyama, A., Arai, R., Yashiroda, Y., Shirai, A., Kamata, A., Sekido, S., Kobayashi, Y., Hashimoto, A., Hamamoto, M., Hiraoka, Y., Horinouchi, S., and Yoshida, M. (2006). ORFeome cloning and global analysis of protein localization in the fission yeast *Schizosaccharomyces pombe*. *Nat Biotechnol 24*, 841–847.
48. Lamesch, P., Li, N., Milstein, S., Fan, C., Hao, T., Szabo, G., Hu, Z., Venkatesan, K., Bethel, G., Martin, P., Rogers, J., Lawlor, S., McLaren, S., Dricot, A., Borick, H., Cusick, M. E., Vandenhaute, J., Dunham, I., Hill, D. E., and Vidal, M. (2007). hORFeome v3.1: a resource of human open reading frames representing over 10,000 human genes. *Genomics 89*, 307–315.
49. Goshima, N., Kawamura, Y., Fukumoto, A., Miura, A., Honma, R., Satoh, R., Wakamatsu, A., Yamamoto, J., Kimura, K., Nishikawa, T., Andoh, T., Iida, Y., Ishikawa, K., Ito, E., Kagawa, N., Kaminaga, C., Kanehori, K., Kawakami, B., Kenmochi, K., Kimura, R., Kobayashi, M., Kuroita, T., Kuwayama, H., Maruyama, Y., Matsuo, K., Minami, K., Mitsubori, M., Mori, M., Morishita, R., Murase, A., Nishikawa, A., Nishikawa, S., Okamoto, T., Sakagami, N., Sakamoto, Y., Sasaki, Y., Seki, T., Sono, S., Sugiyama, A., Sumiya, T., Takayama, T., Takayama, Y., Takeda, H., Togashi, T., Yahata, K., Yamada, H., Yanagisawa, Y., Endo, Y., Imamoto, F., Kisu, Y., Tanaka, S., Isogai, T., Imai, J., Watanabe, S., and Nomura, N. (2008). Human protein factory for converting the transcriptome into an in vitro-expressed proteome. *Nat Methods 5*, 1011–1017.
49a. Pellet, J., Tafforeau, L., Lucas-Hourani, M., Navratil, V., Meyniel, L., Achaz, G., Guironnet-Paquet, A., Aublin-Gex, A., Caignard, G., Cassonnet, P., Chaboud, A., Chantier, T., Deloire, A., Demeret, C., Le Breton, M., Neveu, G., Jacotot, L., Vaglio, P., Delmotte, S., Gautier, C., Combet, C., Deleage, G., Favre, M., Tangy, F., Jacob, Y., Andre, P., Lotteau, V., Rabourdin-Combe, C., and Vidalain, P. O. (2010). ViralORFeome: an integrated database to generate a versatile collection of viral ORFs. *Nucleic Acids Res 38*, D371–378.
50. Popescu, S. C., Popescu, G. V., Bachan, S., Zhang, Z., Seay, M., Gerstein, M., Snyder, M., and Dinesh-Kumar, S. P. (2007). Differential binding of calmodulin-related proteins to their targets revealed through high-density *Arabidopsis* protein microarrays. *Proc Natl Acad Sci U S A 104*, 4730–4735.
51. Hall, D. A., Zhu, H., Zhu, X., Royce, T., Gerstein, M., and Snyder, M. (2004). Regulation of gene expression by a metabolic enzyme. *Science 306*, 482–484.
52. Ptacek, J., Devgan, G., Michaud, G., Zhu, H., Zhu, X., Fasolo, J., Guo, H., Jona, G., Breitkreutz, A., Sopko, R., McCartney, R. R., Schmidt, M. C., Rachidi, N., Lee, S. J., Mah, A. S., Meng, L., Stark, M. J., Stern, D. F., De Virgilio, C., Tyers, M., Andrews, B., Gerstein, M., Schweitzer, B., Predki, P. F., and Snyder, M. (2005). Global analysis of protein phosphorylation in yeast. *Nature 438*, 679–684.
53. Gupta, R., Kus, B., Fladd, C., Wasmuth, J., Tonikian, R., Sidhu, S., Krogan, N. J., Parkinson, J., and Rotin, D. (2007). Ubiquitination screen using protein microarrays for comprehensive identification of Rsp5 substrates in yeast. *Mol Syst Biol 3*, 116.
54. Huang, J., Zhu, H., Haggarty, S. J., Spring, D. R., Hwang, H., Jin, F., Snyder, M., and Schreiber, S. L. (2004). Finding new components of the target of rapamycin (TOR) signaling network through chemical genetics and proteome chips. *Proc Natl Acad Sci U S A 101*, 16594–16599.
55. Wolff, B., Sanglier, J. J., and Wang, Y. (1997). Leptomycin B is an inhibitor of nuclear export: inhibition of nucleocytoplasmic translocation of the human immunodeficiency virus type 1 (HIV-1) Rev protein and Rev-dependent mRNA. *Chem Biol 4*, 139–147.

56. Giaever, G., Shoemaker, D. D., Jones, T. W., Liang, H., Winzeler, E. A., Astromoff, A., and Davis, R. W. (1999). Genomic profiling of drug sensitivities via induced haploinsufficiency. *Nat Genet 21*, 278–283.
57. Lum, P. Y., Armour, C. D., Stepaniants, S. B., Cavet, G., Wolf, M. K., Butler, J. S., Hinshaw, J. C., Garnier, P., Prestwich, G. D., Leonardson, A., Garrett-Engele, P., Rush, C. M., Bard, M., Schimmack, G., Phillips, J. W., Roberts, C. J., and Shoemaker, D. D. (2004). Discovering modes of action for therapeutic compounds using a genome-wide screen of yeast heterozygotes. *Cell 116*, 121–137.
58. Parsons, A. B., Brost, R. L., Ding, H., Li, Z., Zhang, C., Sheikh, B., Brown, G. W., Kane, P. M., Hughes, T. R., and Boone, C. (2004). Integration of chemical-genetic and genetic interaction data links bioactive compounds to cellular target pathways. *Nat Biotechnol 22*, 62–69.
59. Giaever, G., Flaherty, P., Kumm, J., Proctor, M., Nislow, C., Jaramillo, D. F., Chu, A. M., Jordan, M. I., Arkin, A. P., and Davis, R. W. (2004). Chemogenomic profiling: identifying the functional interactions of small molecules in yeast. *Proc Natl Acad Sci U S A 101*, 793–798.
60. Parsons, A. B., Lopez, A., Givoni, I. E., Williams, D. E., Gray, C. A., Porter, J., Chua, G., Sopko, R., Brost, R. L., Ho, C. H., Wang, J., Ketela, T., Brenner, C., Brill, J. A., Fernandez, G. E., Lorenz, T. C., Payne, G. S., Ishihara, S., Ohya, Y., Andrews, B., Hughes, T. R., Frey, B. J., Graham, T. R., Andersen, R. J., and Boone, C. (2006). Exploring the mode-of-action of bioactive compounds by chemical-genetic profiling in yeast. *Cell 126*, 611–625.
61. Hoon, S., Smith, A. M., Wallace, I. M., Suresh, S., Miranda, M., Fung, E., Proctor, M., Shokat, K. M., Zhang, C., Davis, R. W., Giaever, G., St. Onge, R. P., and Nislow, C. (2008). An integrated platform of genomic assays reveals small-molecule bioactivities. *Nat Chem Biol 4*, 498–506.
62. Giaever, G., Chu, A. M., Ni, L., Connelly, C., Riles, L., Veronneau, S., Dow, S., Lucau-Danila, A., Anderson, K., Andre, B., Arkin, A. P., Astromoff, A., El-Bakkoury, M., Bangham, R., Benito, R., Brachat, S., Campanaro, S., Curtiss, M., Davis, K., Deutschbauer, A., Entian, K. D., Flaherty, P., Foury, F., Garfinkel, D. J., Gerstein, M., Gotte, D., Guldener, U., Hegemann, J. H., Hempel, S., Herman, Z., Jaramillo, D. F., Kelly, D. E., Kelly, S. L., Kotter, P., LaBonte, D., Lamb, D. C., Lan, N., Liang, H., Liao, H., Liu, L., Luo, C., Lussier, M., Mao, R., Menard, P., Ooi, S. L., Revuelta, J. L., Roberts, C. J., Rose, M., Ross-Macdonald, P., Scherens, B., Schimmack, G., Shafer, B., Shoemaker, D. D., Sookhai-Mahadeo, S., Storms, R. K., Strathern, J. N., Valle, G., Voet, M., Volckaert, G., Wang, C. Y., Ward, T. R., Wilhelmy, J., Winzeler, E. A., Yang, Y., Yen, G., Youngman, E., Yu, K., Bussey, H., Boeke, J. D., Snyder, M., Philippsen, P., Davis, R. W., and Johnston, M. (2002). Functional profiling of the *Saccharomyces cerevisiae* genome. *Nature 418*, 387–391.
63. Winzeler, E. A., Shoemaker, D. D., Astromoff, A., Liang, H., Anderson, K., Andre, B., Bangham, R., Benito, R., Boeke, J. D., Bussey, H., Chu, A. M., Connelly, C., Davis, K., Dietrich, F., Dow, S. W., El Bakkoury, M., Foury, F., Friend, S. H., Gentalen, E., Giaever, G., Hegemann, J. H., Jones, T., Laub, M., Liao, H., Liebundguth, N., Lockhart, D. J., Lucau-Danila, A., Lussier, M., M'Rabet, N., Menard, P., Mittmann, M., Pai, C., Rebischung, C., Revuelta, J. L., Riles, L., Roberts, C. J., Ross-MacDonald, P., Scherens, B., Snyder, M., Sookhai-Mahadeo, S., Storms, R. K., Veronneau, S., Voet, M., Volckaert, G., Ward, T. R., Wysocki, R., Yen, G. S., Yu, K., Zimmermann, K., Philippsen, P., Johnston, M., and Davis, R. W. (1999). Functional characterization of the *S. cerevisiae* genome by gene deletion and parallel analysis. *Science 285*, 901–906.
64. Nishimura, S., Arita, Y., Honda, M., Iwamoto, K., Matsuyama, A., Shirai, A., Kawasaki, H., Kakeya, H., Kobayashi, T., Matsunaga, S., and Yoshida, M. (2010). Marine antifungal theonellamides target 3b-hydroxysterol to activate Rho1 signaling. *Nat Chem Biol 6*, 519–526.
65. Arita, Y., Nishimura, S., Matsuyama, A., Yashiroda, Y., Usui, T., Boone, C., and Yoshida, M. (2011). Microarray-based target identification using drug hypersensitive fission yeast expressing ORFeome. *Mol Biosyst* 7, 1463–1472.

SECTION TWO

MOLECULES FOR CHEMICAL GENOMICS

CHAPTER 4

Diversity-Oriented Synthesis

Warren R. J. D. Galloway
Richard J. Spandl
Andreas Bender
Gemma L. Thomas
Monica Diaz-Gavilan
Kieron M. G. O'Connell
David R. Spring

1.1. INTRODUCTION

Chemical genetics describes the use of molecules as "chemical probes" to investigate biological systems [1–3]. In contrast with traditional genetics, in which gene knockouts on the level of the DNA are used, chemical genetics uses biologically active small molecules to directly attenuate the corresponding biological macromolecular (usually protein) product. Thus, the ready availability of bioactive small molecules is of crucial importance in chemical genetics studies. Such small molecules can be identified by screening compound collections (libraries) in suitably designed assays. This chapter describes the use of *diversity-oriented synthesis* (DOS) to prepare structurally diverse small molecule libraries. Structurally diverse libraries show a greater variety in not only their physiochemical properties but also, and of most relevance here, in their biological activities. Herein we describe some of the most effective strategies that have been used in DOS library design and preparation.

1.2. SMALL MOLECULES, CHEMICAL GENETICS, AND CHEMICAL GENOMICS

Chemical genetics experiments can be performed in either a forward or a reverse sense (Figure 4.1). The first step of both approaches requires the identification of a small molecule that either induces a desired phenotype (forward chemical genetics) or modulates the function of a specific protein of interest (reverse chemical genetics). Thus, in the former case, investigations proceed from phenotype to protein, whereas in the latter case, investigations progress from protein to phenotype.

Chemical genomics has been defined as the search for selective small molecule modulators of each function of all gene products (e.g., proteins) – that is, the application of reverse chemical genetics on a genome-wide scale [2]. Possession of such molecules would allow the systematic exploration and perturbation of biological systems with obvious potential for the development of improved chemotherapeutic treatments [2]. The challenges presented by a chemical genomics approach are daunting. It has been estimated that roughly 10% of the human genome (approximately thirty thousand genes) encode proteins that can bind to "drug-like" compounds [4]. However, small molecule partners for only approximately one thousand members of this total "druggable" proteome have been identified [5]. Fewer still can be considered as specific in their interaction. Clearly, there exists a significant need for the discovery of many more biologically active small molecules that are capable of selectively modulating gene product function.

One of the fundamental considerations of chemical biology is what type of compounds should be synthesized and employed in biochemical or biological screenings [6]. Ultimately, this is determined by the requirements placed on the compounds. For example, if the goal of a screening process is to identify a small molecule for application as an orally bioavailable drug, then several observations have been made regarding molecular characteristics that are desirable, such as molecular size, shape, and tolerable functional groups [5, 7]. However, these demands are considerably different from those placed on a compound required for cell-based or in vitro assays [5, 8].

Biologically active small molecules can be identified by screening libraries of compounds in either phenotypic or

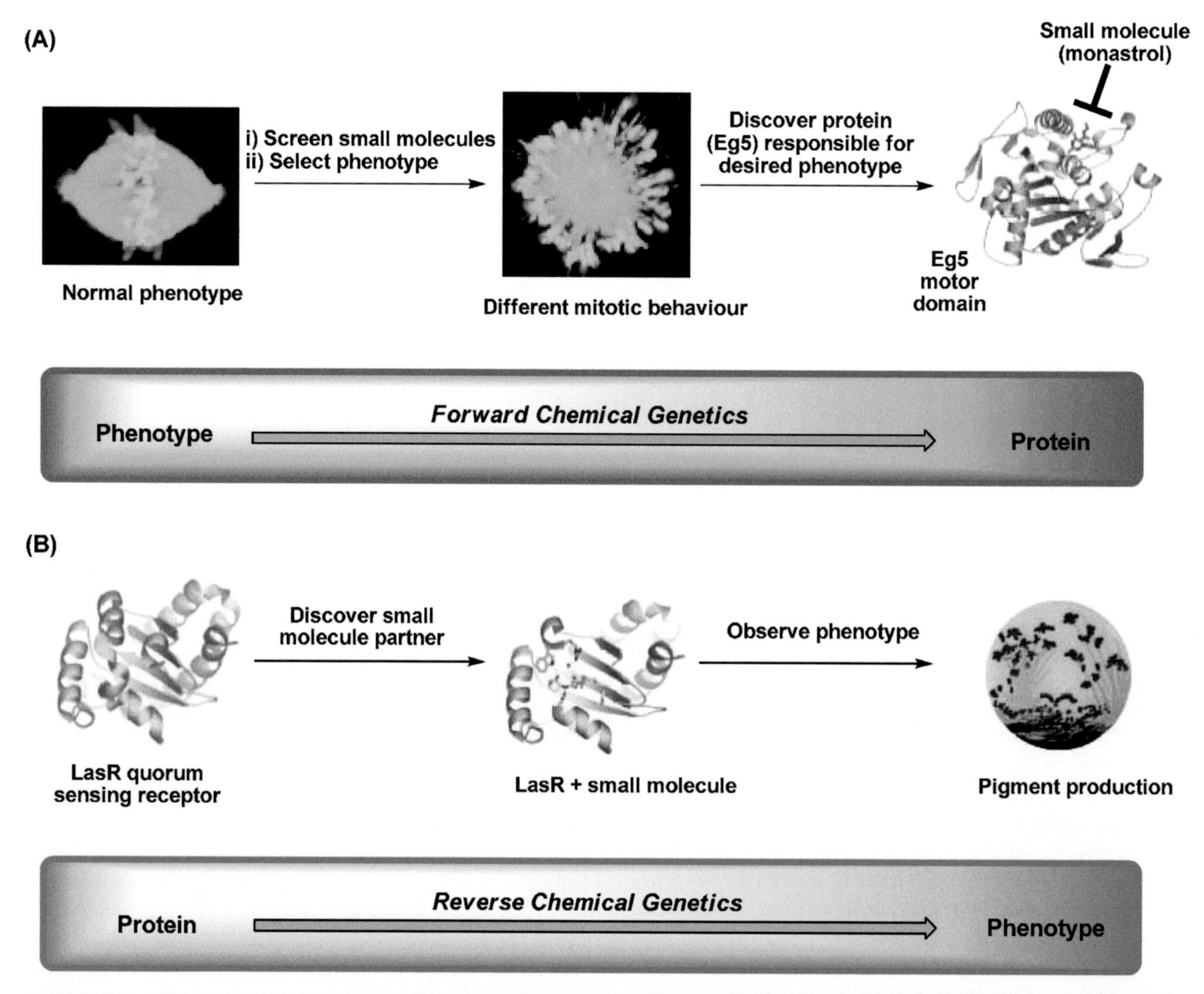

***FIGURE 4.1:* (A) In a forward chemical genetics approach, a library of small molecules is screened to identify those that that induce a desired phenotypic effect (e.g., different mitotic behavior) on the biological system under investigation. Once a suitable small molecule has been identified, further investigation allows the gene product (i.e., protein) with which the molecule interacts to be discovered. (B) Reverse chemical genetics involves the use of small molecules against a known protein of interest (e.g., the LasR quorum sensing receptor protein). A small molecule that binds to the protein is identified, and the phenotypic effect induced by the action of this small molecule helps to define the role of its protein partner.**

protein-binding assays. In the latter case, where the target protein is known (e.g., in a reverse chemical genetics experiment), libraries containing compounds synthesized in a "hypothesis-driven" fashion – that is, designed to interact with a specific target – are more appropriate. Such compounds are usually selected/designed based on knowledge of the target's structure or the structure of known natural ligands [9]. In phenotypic assays, in which the precise nature of the eventual biological target is unknown, the selection criterion for small molecules is complicated dramatically. In these situations, the successful identification of "useful" biologically active small molecules may be aided by screening functionally (biologically) diverse compound libraries, because it has been argued that a greater sample of the bioactive chemical universe (i.e., of all bioactive molecules) increases the chance of identifying a compound with the desired properties [10, 11]. Such an approach may also be advantageous in the case of a reverse chemical genetic experiment in which the precise structural features of the biological target of interest are unknown (e.g., the structure of the target protein has not yet been fully characterized), or known natural ligands are unselective and interact with more than one protein target, meaning that closely related analogs may not be suitable for use.

The functional (biological) diversity of a library of small molecules has been shown to be directly correlated with its structural diversity, which in turn is related to the amount of *chemical space* the library occupies [11, 12].

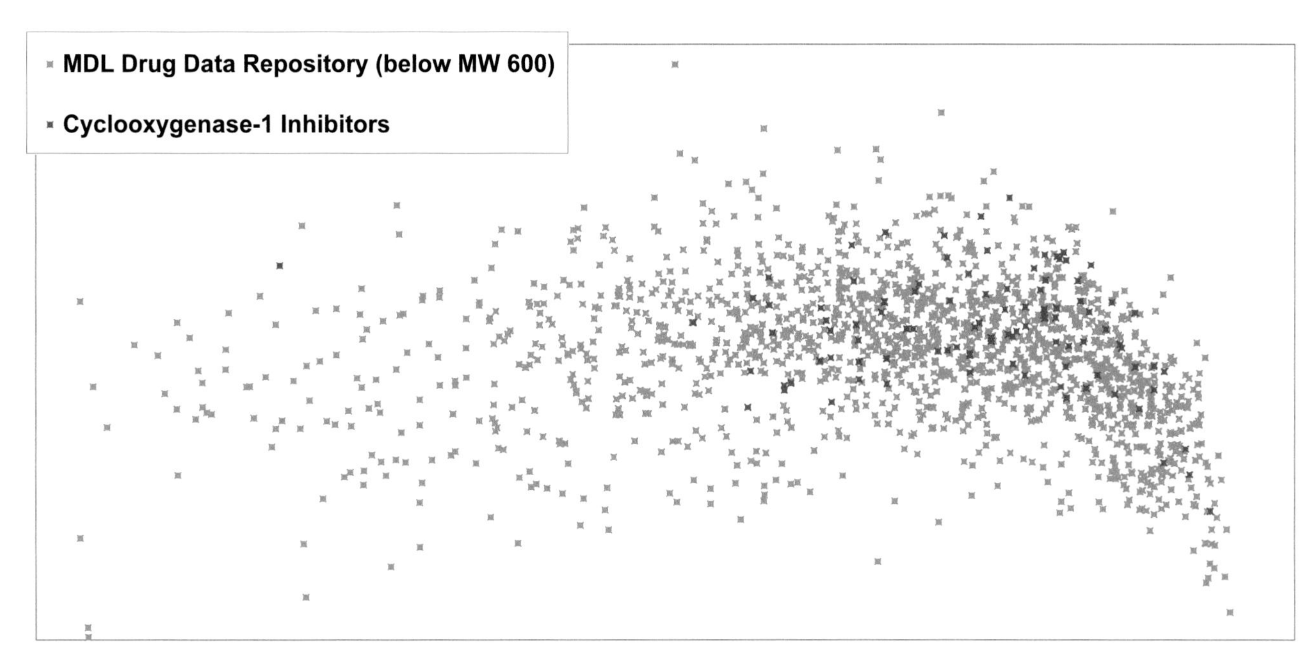

***FIGURE 4.2:** The distribution in chemical space of cyclooxygenase-1 inhibitors (red squares) and a subset of compounds from the MDL Drug Data Repository (MDDR) database (green squares). Once the chemical descriptors have been defined and calculated for each compound, this information can be condensed using a mathematical process known as principle component analysis (PCA). This allows for the construction of two-dimensional (2D) or three-dimensional (3D) displays that are accessible to human interpretation. This 2D visual representation shows that the cyclooxygenase-1 inhibitors populate a broad region of chemical space on the background of the MDDR compounds. In this diagram, each compound is plotted at a discrete point in chemical space. MW = molecular weight.*

1.3. STRUCTURAL DIVERSITY AND THE CONCEPT OF CHEMICAL SPACE

Chemicals can be characterized by a wide range of physiochemical and topological *descriptors* that contain information about either the bulk properties of the compound, such as molecular mass and lipophilicity, or its topological features, such as degree of branching [13–16]. *Chemical space* is a term often used in place of multidimensional descriptor space; it is a region defined by a particular choice of descriptors and the limits placed upon them [17]. In the context of small molecule libraries, chemical space can be defined as the "total descriptor space that encompasses all the small carbon-based molecules that could, in principle, be created" [17].

A specific molecule will reside at a discrete point in chemical space because of its unique combination of molecular descriptor values. The structural features of a collection of molecules will therefore influence the distribution of the molecules in chemical space (Figure 4.2). Therefore, it follows that the more structurally diverse the library, the more chemical space it interrogates. Maximizing the structural diversity and thus chemical space coverage of a library should, in turn, increase its overall functional (biological) diversity; as "molecular shape is intrinsically linked to biological activity, the greater the structural diversity in a library, the better the odds of identifying ligands for a broad range of targets" [12].

1.4. BIOLOGICALLY RELEVANT CHEMICAL SPACE

The degree of overlap between total chemical space and biologically relevant chemical space is somewhat of a contentious issue and subject to much debate in the literature [5, 6, 18]. The limits of biologically relevant chemical space are defined by the specific binding interactions that must occur between small molecules and the three-dimensional (3D) molecular recognition patterns on biological molecules such as proteins [5]. What it not known is the size of this region in comparison to total chemical space – that is, whether the biologically relevant region is "small" and most of the chemical universe is "empty" (containing no therapeutically interesting compounds) [5]. In other words, are the regions of chemical space defined by natural products and known drugs the best or most fertile regions for discovering small molecules, or is their scope for discovering biologically useful molecules, particularly those with novel modes of action, from "untapped" areas of chemical space? [5, 18]

In spite of this controversy, structural diversity (and thus total chemical space coverage), though rarely the "end game" in a synthesis project, is generally perceived to be an important consideration in small molecule library synthesis, particularly when the precise nature of the biological target molecule is unknown or the identification of a novel biologically active molecule is desired [19]. Indeed, the generation of libraries of pure, structurally diverse compounds is thought by some to be the "key to the discovery of new medicines and to the elucidation of biological pathways through chemical genetics" [20].

1.5. STRUCTURAL DIVERSITY AND THE IMPORTANCE OF STRUCTURAL COMPLEXITY

The synthesis of a collection of structurally diverse small molecules offers a unique challenge to the synthetic chemist [6, 21]. It is widely accepted that it is not synthetically feasible to produce all theoretically stable, small carbon-based molecules [5, 21]. Thus, selectivity in synthesis is an important consideration. This issue has spurred the development of a variety of different approaches that aim to efficiently interrogate wide regions of chemical space simultaneously or to identify regions of chemical space that have an enhanced probability of containing biologically active compounds [6]. Before evaluating the relative merits of some of the most commonly used methods, it is useful to consider what is meant by the term *structural diversity* in the context of library synthesis.

Though the word *diversity* is, to some degree, a subjective one, there are four principle components of structural diversity that have been consistently identified in the literature: [19, 21, 22]

1) *Appendage diversity* (or building-block diversity): Variation in different structural moieties around a common skeleton;
2) *Functional group diversity*: Variation in the functional groups present;
3) *Stereochemical diversity*: Variation in the orientation of potential macromolecule-interacting elements; and
4) *Skeletal diversity*: Presence of many distinct molecular skeletons (or frameworks/scaffolds)*.

Increasing the skeletal diversity in a small molecule library is widely regarded as one of the most effective ways of increasing the overall structural diversity of the library [12, 24, 25]. Furthermore, computational analyses have been carried out to support the notion that small multiple-scaffold libraries are superior to large single-scaffold libraries in terms of biorelevant diversity [11]. Libraries based around a single scaffold, regardless of their size, are restricted to a limited number of molecular shapes, as opposed to smaller libraries designed around multiple scaffolds [11, 12]. Libraries of "compounds that have a common molecular skeleton display chemical information similarly in 3D space, thus limiting the pool of potential binding partners to only those macromolecules with a complementary 3D binding surface" [18, 24]. Thus, variation in 3D structure, rather than the nature of the peripheral substituents, is key for interaction with a broad range of molecular targets [26]. Conversely, conservation of 3D structure (i.e., the nature of the core molecular skeleton) in a small molecule collection generally means that the molecules will bind a narrower range of molecular targets.

Despite the acknowledged correlation between skeletal diversity and overall structural and functional (biological) diversity, the need to incorporate skeletal diversity in small molecule libraries is a somewhat contentious issue and is very much application dependent. In instances where a specific protein is being targeted, small molecule libraries are generally based around a single molecular skeleton – for example, the skeleton of a known natural ligand that has demonstrated the ability to bind to the desired protein. Furthermore, the synthesis of small molecule libraries around scaffolds present in known biologically active compounds has been cited as a possible means of identifying small molecules with novel biological properties. The synthesis of such "biased" small molecule libraries is discussed later in this chapter.

In addition to structural diversity, *structural complexity* is another characteristic that is important in small molecule libraries. Although there is some debate in the literature, it has been argued that molecules that are structurally complex are likelier to interact with biology in a selective and specific manner [5, 27].

1.6. SOURCES OF SMALL MOLECULES

There are a number of potential sources of small molecules for use in biological screens, each of which addresses various aspects of the structural diversity and structural complexity criteria in a different manner and with varying degrees of success.

1.6.1. Natural Products

Traditionally, nature has served as a rich source of biologically active molecules [22, 28, 29]. Natural products exhibit enormous structural diversity [30], and though they may vary in terms of structural complexity, many exhibit a high degree of specificity for their biological target. Numerous natural products have proven to be useful as drugs or leads [31, 32] and are still a major source of innovative therapeutic agents for infectious diseases [28].

Unfortunately, there are several problems associated with using natural product compounds in screening experiments. These include difficulties with purification and bioactive component(s) identification. Additionally, chemical modification and analog synthesis, processes that are

* The term *molecular skeleton* has no strict definition. Within the context of this discussion, the description recently outlined by Schreiber is appropriate; "we use the term skeleton loosely to denote rigidifying elements in small molecules; these can be atom connectivities that yield either linked, fused, bridged or spiro rings, or acyclic conformational elements that provide substantial rigidification by avoiding non-bonding interactions" [23].

particularly pertinent in the drug development process, may be extremely challenging because of the highly complex nature of most natural products [19]. Furthermore, natural products occupy only a small proportion of chemical space [5, 17], which runs the risk of omitting a vast number of possibly biologically valuable small molecules from any screening process [18].

1.6.2. The Synthesis of Small Molecule Libraries

The crucial need for high-quality compound libraries in biological screening experiments has spurred the development of several *synthetic* approaches toward small molecule library synthesis. It is possible to analyze these approaches in terms of the various design concepts they employ. One of the most fundamental of these is the control in the nature of the molecular skeletons present in the small molecules. Based on this consideration, synthetic approaches to structurally diverse small molecule libraries can be divided into two distinct, broad groups:

1) *Biased approaches*: The synthetic route is designed with a pre-encoded structural bias, such that all of the resultant molecules are based around a similar molecular skeleton (i.e., a biased library).
2) *Nonbiased approaches*: The synthetic route is designed with no pre-encoded structural bias, such that there are a variety of different molecular skeletons present in the final products (i.e., a nonbiased library).

1.6.2.1. Biased Approaches: Combinatorial Chemistry

Commercially available combinatorial libraries and pharmaceutical proprietary compound collections are traditionally very important sources of small molecules [12, 30]. Combinatorial chemistry may be defined, in a very broad sense, as the rapid synthesis and screening of libraries of varied compounds to identify agents with desired functional properties [33]. Combined with established high-throughput screening (HTS) techniques, the development of combinatorial chemistry strategies in the early 1990s enabled the generation and testing of libraries of hundreds of thousands of different compounds at comparably low cost [30]. The method was quickly embraced by the pharmaceutical industry, with the hope that drug leads would be produced by sheer weight of numbers. However, the expected surge in productivity has not materialized. Indeed, as of the end of 2007 there was only one reported de novo new chemical entity **(1)** resulting from this method of chemical discovery that had been approved for drug use (Figure 4.3) [32].

This disappointing degree of productivity is generally attributed to defects in the nature of the libraries produced. Early combinatorial libraries have been described as being "intrinsically useless for drug discovery" [34] because the compounds were too similar to each other, having limited structural diversity [30] and consequently offering "only a narrow slice" of chemical space [34]. This limited degree of overall structural diversity may be chiefly attributed to a lack of skeletal diversity. Traditionally, combinatorial libraries were based on a "one synthesis/one skeleton" approach, which resulted in a high degree of appendage diversity (and possibly stereochemical diversity) but little variety in the nature of the core molecular scaffold [12, 18, 19].

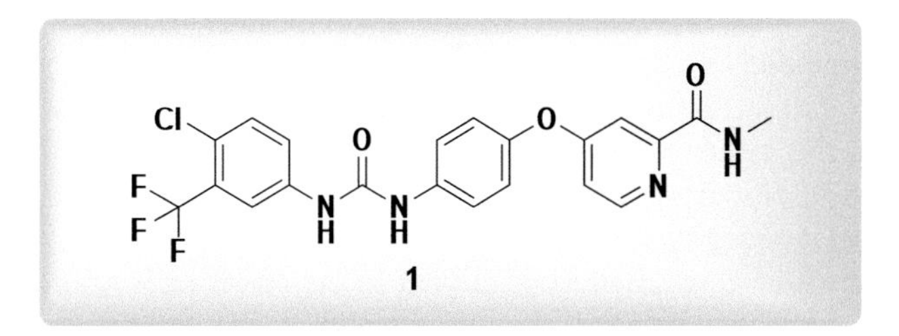

***FIGURE 4.3:** The antitumor compound 1 known as sorafenib (Nexavar) from Bayer, approved by the U.S. Food and Drug Administration (FDA) in 2005. It is a multikinase inhibitor and is in multiple clinical trials as both a combination and a single-agent therapy.*

In an attempt to combat this problem, a more considered approach has been taken to combinatorial chemistry in recent years to try to increase the structural diversity exhibited by combinatorial libraries [12]. Nevertheless, even these approaches are generally limited to known biologically active frameworks and, as such, have met with limited success in identifying novel biologically active small molecules.*

It is widely recognized that molecules that have a common molecular skeleton tend to display chemical information similarly in 3D space and, as a result, are predisposed to bind to certain molecular targets that possess a complementary 3D binding surface [11, 18, 24]. In the case of a reverse chemical genetics experiment, in which small molecule modulators of a specific protein of interest are desired, it therefore can be advantageous to have a library of compounds based around a specific 3D structure (molecular skeleton) that has demonstrated affinity for the protein. The presence of specific structural features (such as a certain molecular skeleton) in a small molecule should predispose it toward binding the specific protein of interest. These desired/prerequisite binding structural features may be deduced from knowledge of the 3D structural features of the desired protein itself (e.g., a crystal structure) or of a known natural ligand for the protein. In the context of a reverse chemical genetics, it could be argued that it is inherently wasteful to synthesize (and screen) a library of skeletally diverse molecules that are designed to bind to a

* In the context of this discussion, a *novel* biologically active molecule can be defined as a molecule with a previously unknown biological activity that exerts this biological effect through a unique mechanism/mode of action.

wide range of proteins when only a specific protein is being targeted.

Although the 3D structure of a small molecule – that is, the nature of its core molecular skeleton – is of central importance in determining the range of molecular targets it can interact with, variation in the functional groups (structure and stereochemistry) present in the molecule is often important in terms of the strength and specificity of these interactions. Thus, reverse chemical genetics experiments may benefit from the screening of small molecule collections that display appendage, stereochemical, and functional group diversity around a common molecular skeleton, as this may facilitate the identification of ligands that bind with increased selectivity (and/or affinity) for the protein of interest.

1.6.2.2. Biased Approaches: Synthesis around a Privileged Molecular Skeleton Although molecules based around a specific core skeleton are generally limited to a smaller range of biological partners than molecules based around a diverse range of skeletons, it has emerged that there is a subset of molecular skeletons whose presence in molecules confers upon them more flexible binding properties. The frameworks in question are so-called privileged frameworks that have been defined as "molecular scaffolds with versatile binding properties, such that a single scaffold is able to provide potent and selective ligands for a range of different biological targets through modification of functional groups" [35]. These are molecular scaffolds that are common to known bioactive molecules (usually natural products) and thus have proven biological relevance [6, 36].

In recent years, there has been an increase in the synthesis of small molecule libraries that are based around such privileged structures. The rationale behind using such an approach is based on two main hypotheses. The first is that evolutionary pressure over millions of years has "prevalidated" natural products, and thus compounds that are structurally similar, to be able to modulate protein function. Second, the chemical space explored by natural products and protein structure during evolution is strongly limited in size and highly conserved. That is, there is a concept of *evolutionary convergence of structures* in that natural products have evolved to interact with multiple proteins [37, 38]. Synthesis around a privileged scaffold, which has also been referred to as natural product–guided synthesis, has been described as being distinct from the process of (focused) combinatorial library synthesis because the ultimate goal is the identification of compounds with novel biological properties distinct from those of the original privileged compound [39]. It has been argued that privileged structure–based synthesis may permit such compounds to be found with enhanced probability and quality.

For example, Park et al. have recently reported the construction of a library of small molecules composed of twenty-two discrete and novel core skeletons embedded with a privileged benzopyran motif **(2)** through a branching DOS strategy [40]. The outline of the synthetic routes used is given in Scheme 1.

Starting from compounds **3a** and **3b**, intermediates **4** and **5** could be generated by reaction pathways A and B, respectively. These served as branch points for various chemical transformations such as Diels-Alder reactions (paths A1 and B1), click chemistry (A2 and B2), and palladium-mediated cross-coupling (A3 and B3). Using this strategy, Park and colleagues were able to synthesize twenty-two novel molecular architectures, each one containing the benzopyran substructure. The biological diversity of the library was demonstrated by the dramatic differences in the biological activities (IC_{50} values against a human cancer cell line) of compounds sharing the same appendices but having different 3D structures; that is, the variation in biological activity was reported to be a function of the core molecular skeleton rather than of the appendages present.

The use of a synthesis around a privileged structure approach again highlights the controversial issue of the degree of overlap between total chemical space and biologically relevant chemical space – that is, whether there is any point in exploring seemingly uncharted regions of chemical space (which nature may have indeed "sampled" through the process of evolution over the course of millions of years but ultimately ignored as a source of biologically useful molecules) when chemical space occupied by known natural products and medicinal compounds is enriched with bioactive structures. What is clear is that if we do not try to access such regions, we will never know! Clearly, library synthesis around a privileged structure is particularly relevant when a specific protein target is being considered. When a less focused approach is required – for example, if the protein target is unknown or we are hoping to find novel biologically active molecules – the use of nonbiased synthetic approaches that aim to access a wider range of chemical descriptor space may be more useful.

1.6.3. Nonbiased Approaches: DOS

The aim of a nonbiased approach toward library synthesis is to create a structurally diverse (including skeletal diversity) and functionally diverse small molecule collection with the potential to provide hits against a panel of biological targets, allowing the discovery of small molecules with previously unknown (and potentially novel) biological effects [41]. DOS has recently emerged as a new synthetic approach to achieving this objective.

The goal of a DOS is to efficiently interrogate wide areas of chemical space simultaneously; this may include known bioactive regions of chemical space *and* unexplored regions of chemical space [9, 18, 21, 42]. The hope is that by sampling a greater total area of chemical space, the functional

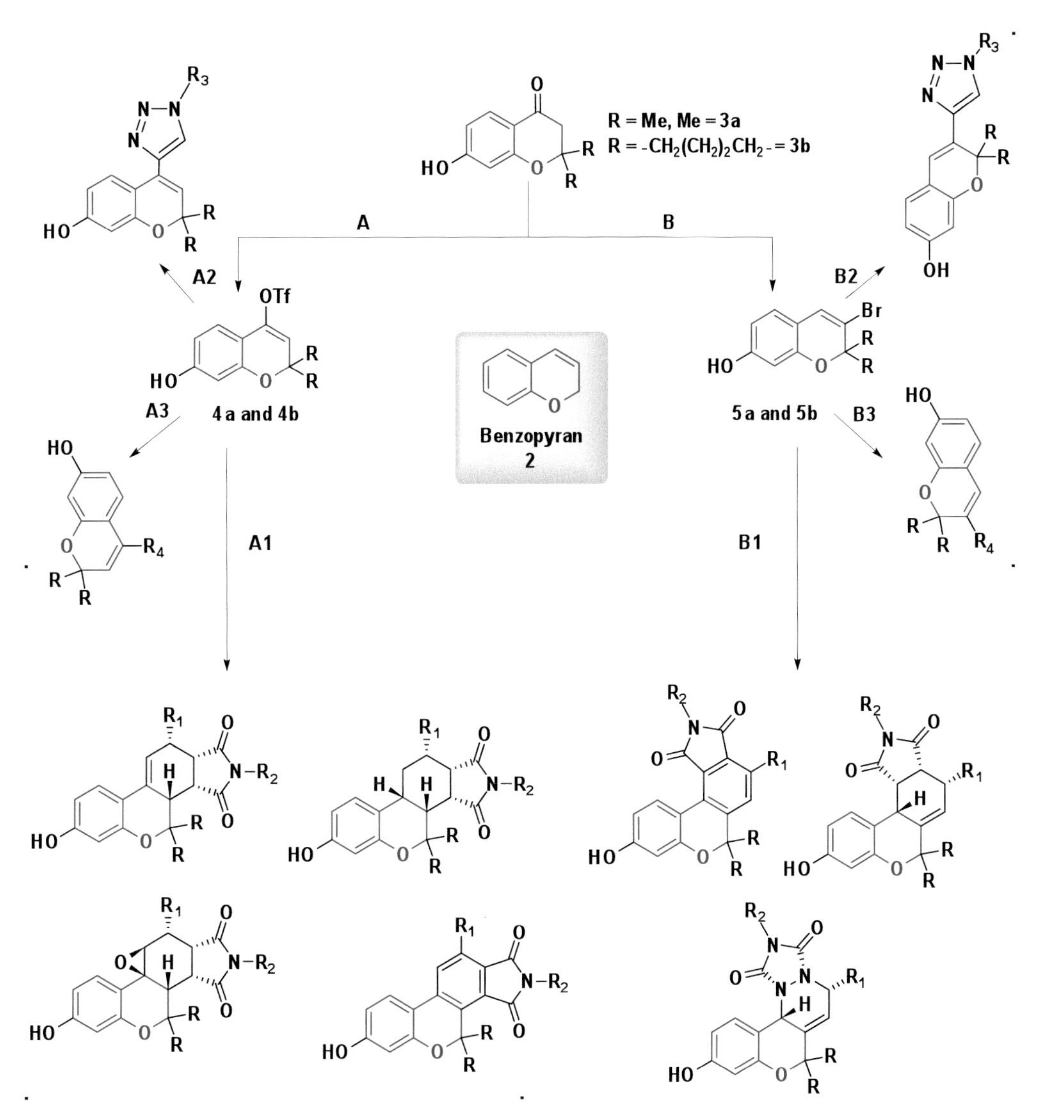

***SCHEME 1:** Outline of the synthetic strategy employed by Park and co-workers to generate a library based around twenty-two distinct core molecular skeletons embedded with a privileged benzopyran motif. The benzopyran motif common to all the members is highlighted in red. Tf = trifluoromethanesulfonate.*

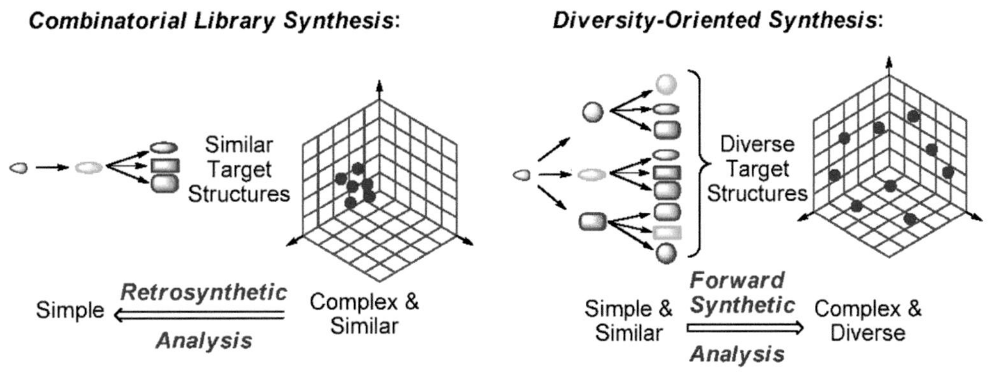

Combinatorial Library Synthesis:

Similar Target Structures

Simple ⟸ *Retrosynthetic Analysis* Complex & Similar

Diversity-Oriented Synthesis:

Diverse Target Structures

Simple & Similar ⟹ *Forward Synthetic Analysis* Complex & Diverse

***FIGURE 4.4:** A comparison of the synthetic planning strategies used in a traditional combinatorial synthesis with a DOS, together with a visual representation of the chemical space coverage achieved in both cases (i.e., focused around a specific point or diverse coverage).*

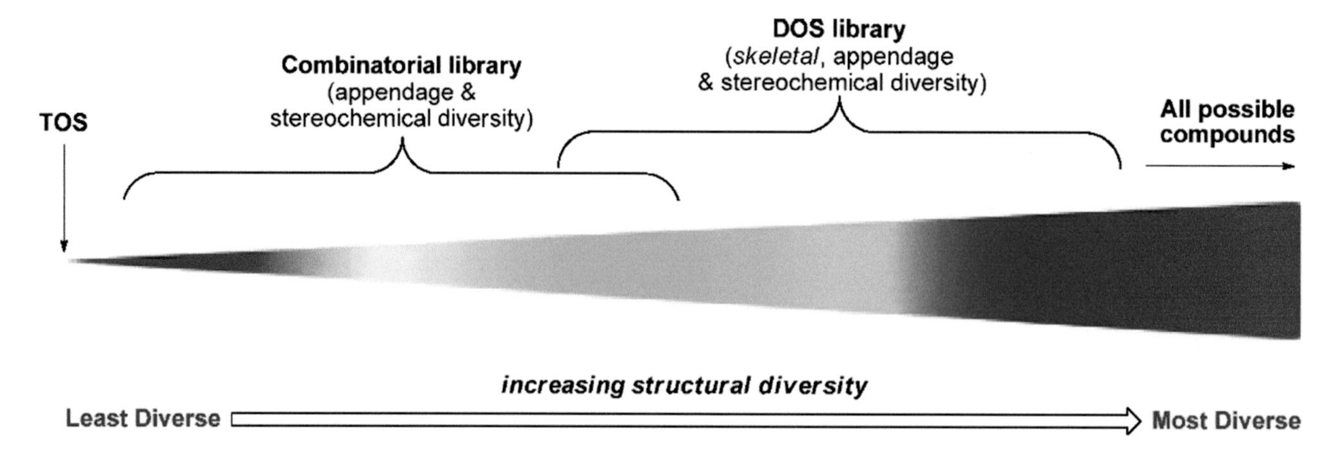

***FIGURE 4.5:** The "molecular diversity spectrum." "Diversity" can be viewed as a spectrum ranging from a target-oriented synthesis (TOS) of a specific molecule (i.e., a total synthesis) to the synthesis of all possible compounds (i.e., total chemical space coverage); compound collections arising from a traditional combinatorial approach and those arising from a DOS sit between these two extremes [19].*

(biological) diversity of the library will be greater, increasing the chances of identifying a compound with the desired properties [10–12].

An illustration of the synthetic strategy used in a DOS is given in Figure 4.4. A DOS synthetic pathway is analyzed in the *forward* sense; a single, simple starting material is converted into a collection of structurally diverse small molecules in no more then five synthetic steps (to maximize synthetic efficiency) [18, 21, 43]. The overall aim is the broad, nonfocused coverage of chemical space, which can be contrasted with the outcomes of more traditional combinatorial syntheses.

A successful DOS must address the four principle types of structural diversity mentioned previously (Section 1.5), that is, appendage, functional group, stereochemical, and skeletal* [6, 18, 19, 43, 44]. It is the ability of a DOS to incorporate skeletal diversity into a compound collection that is the most challenging facet of this method and of central importance to its success [18, 24, 42, 45]. The efficient generation of multiple molecular scaffolds is regarded as one of the most effective methods of increasing the overall structural diversity of a collection of molecules and has been reported to "increase the odds of addressing a broad range of biological targets" [11] (relative to a single-scaffold library) [12, 19, 24, 46].

There is a clear distinction between DOS and traditional combinatorial methods; DOS libraries are generally smaller but consist of molecules that are structurally more complex, have a greater variety of core structures (skeletons), and possess richer stereochemical variation [33]. However, the boundary between modern, more considered combinatorial approaches and DOS is less clear-cut, and the terms DOS and combinatorial chemistry are often used interchangeably in the literature. Indeed, because many of the principles of combinatorial chemistry are used in DOS, it is probably best to consider DOS as a more evolved version of traditional combinatorial methodologies.† Recently, the concept of the "molecular diversity spectrum" has been introduced as a useful qualitative means for comparing the structural diversity associated with a particular molecular collection (Figure 4.5). It should therefore "be the goal of a DOS to synthesize, in a qualitative sense, collections of small molecules which are as near as possible to the right hand side of this spectrum" [19].

It should be noted that reverse chemical genetics experiments may benefit from the use of nonbiased (i.e., skeletally as well as functionally diverse) small molecule collections. Access to a range of different molecular architectures may allow the discovery of a small molecule that is completely unrelated structurally to any natural ligand(s) for the protein of interest but is capable of modulating the activity of this protein in a much more useful manner (e.g., is more selective, binds with a higher affinity, is easier to access).

* The process of varying functional group, appendage, and stereochemical diversity around privileged scaffolds is occasionally referred to in the literature as *DOS around a privileged scaffold*. However, the true ethos of DOS is based around a diverse, nonfocused coverage of chemical space, which is most efficiently achieved though variation in all aspects of diversity, including skeletal. In this context, we believe that the term *DOS around a privileged scaffold* is somewhat of a contradiction in terms and that other descriptions are more appropriate for the process of library generation around a privileged scaffold, e.g., *natural-product-inspired synthesis*.

† It is widely accepted in the literature that the use of a traditional combinatorial approach (diversity around a single scaffold) as a means for structural optimization once a biologically active molecular skeleton has been identified is without par; the principle benefit of DOS is in the initial discovery of (potentially novel) biologically active skeletons [32].

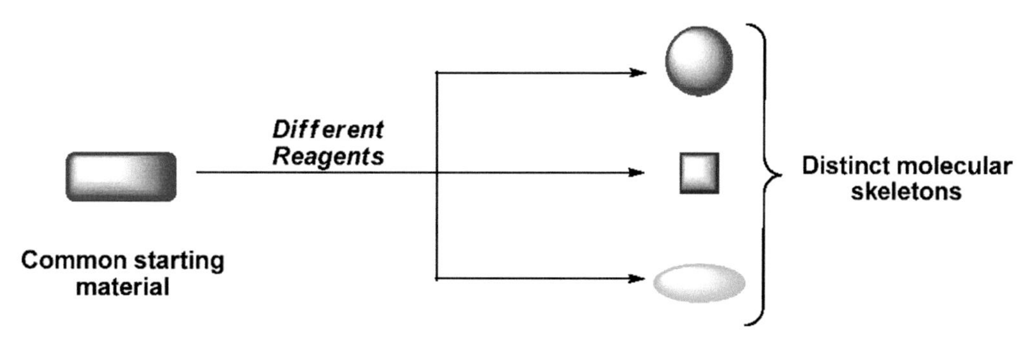

FIGURE 4.6: Achieving skeletal diversity using the reagent-based approach.

1.6.3.1. Achieving Skeletal Diversity via a DOS from Simple Starting Materials The efficient synthesis of a functionally, stereochemically, and *skeletally* diverse collection of small molecules from a common, simple starting material was the process described and developed by Schreiber in much of his pioneering work in the field of DOS [18, 24, 47]. The remainder of this chapter focuses on the general methods that have been developed to maximize skeletal diversity in small molecule libraries generated through a DOS approach.

There are two main approaches that have been developed to generate skeletal diversity in DOS libraries, based either upon the *reagent* (the reagent-based approach) or the *substrate* (the substrate-based approach) [18, 23, 48].

1.6.3.1.1. The reagent-based approach. The reagent-based approach is a branching synthetic strategy that involves a short series of divergent, complexity-generating reactions from a common starting material to generate a collection of compounds with distinct molecular skeletons (Figure 4.6) [18].

Critical to the success of this strategy is the choice of a synthetically versatile starting substrate that has the potential to be converted into two or more products with different molecular skeletons through the variation of reagents alone. These products in turn should be suitable for further diversification, preferably in further branching reactions (i.e., conversion into two or more skeletally diverse products again through choice of reagents).

In practice, reagent-based skeletal diversity is achieved via two main methods: [18, 19]

1) The use of a *pluripotent functionality* whereby exposure of a given molecule to different reagents results in different reactions occurring at the same part (functional group) of the molecule; and
2) The use of a *densely functionalized molecule* whereby different functionalities in the same molecule are transformed by different reagents.

1.6.3.1.2. The use of a pluripotent functionality. Thomas et al. have recently reported an example of the use of a pluripotent functional group strategy to generate a skeletally diverse compound collection (Scheme 2) [49]. This work employed the solid-supported phosphonate **6** as the starting unit. The immobilization of **6** on a silylpolystyrene support greatly simplified product purification during library synthesis.

In the first step of the DOS, **6** was reacted with a variety of aldehyde building blocks (building block diversity) to deliver twelve α,β-unsaturated acyl-imidazolidinones **7**. The second step of the DOS involved three catalytic enantioselective divergent reaction pathways (stereochemical diversity): 1) dihydroxylation (reaction b); 2) [2 + 3] cycloaddition (reaction c); and 3) [4 + 2] cycloaddition (reaction d) to yield a collection of molecules based on three molecular frameworks (skeletal diversity). The next step of the DOS (step 3) involved a series of branching reactions to diversify these key branch-point substrates further. For example, the pyrrolidine products **8** could be acylated or alkylated (reactions e and f, respectively) to yield **9** and **10** (appendage diversity). The norbornene derivatives **11** (formed in step d) served as suitable intermediates for a series of branching reactions (reactions l to o) to generate five different molecular scaffolds (skeletal diversity). For example, a tandem ring-closing-opening-closing metathesis reaction was carried out (reaction o) to give skeletally diverse tricyclic products **12a** (7-5-7) and **12b** (7-5-8).* Of particular note was the generation of the *cis-trans*-fused 7-5-7 scaffold of **12a**, which has no known representation in nature, highlighting the capability of this DOS approach to generate products that populate new, unexplored regions of chemical space. In the final step of the DOS (step 4), the compounds were cleaved off the solid support using a variety of reagents (appendage diversity).

Using the chemistry shown in Scheme 2 and a limited number of structurally diverse building blocks, a DOS of 242 small molecules that have eighteen molecular frameworks, among other unique structural features, was achieved.

* Spandl and co-workers have recently developed a related tandem metathesis process that allows the generation of complex polycyclic molecular architectures from substituted norbornene derivatives in a highly efficient and atom-economical manner [50].

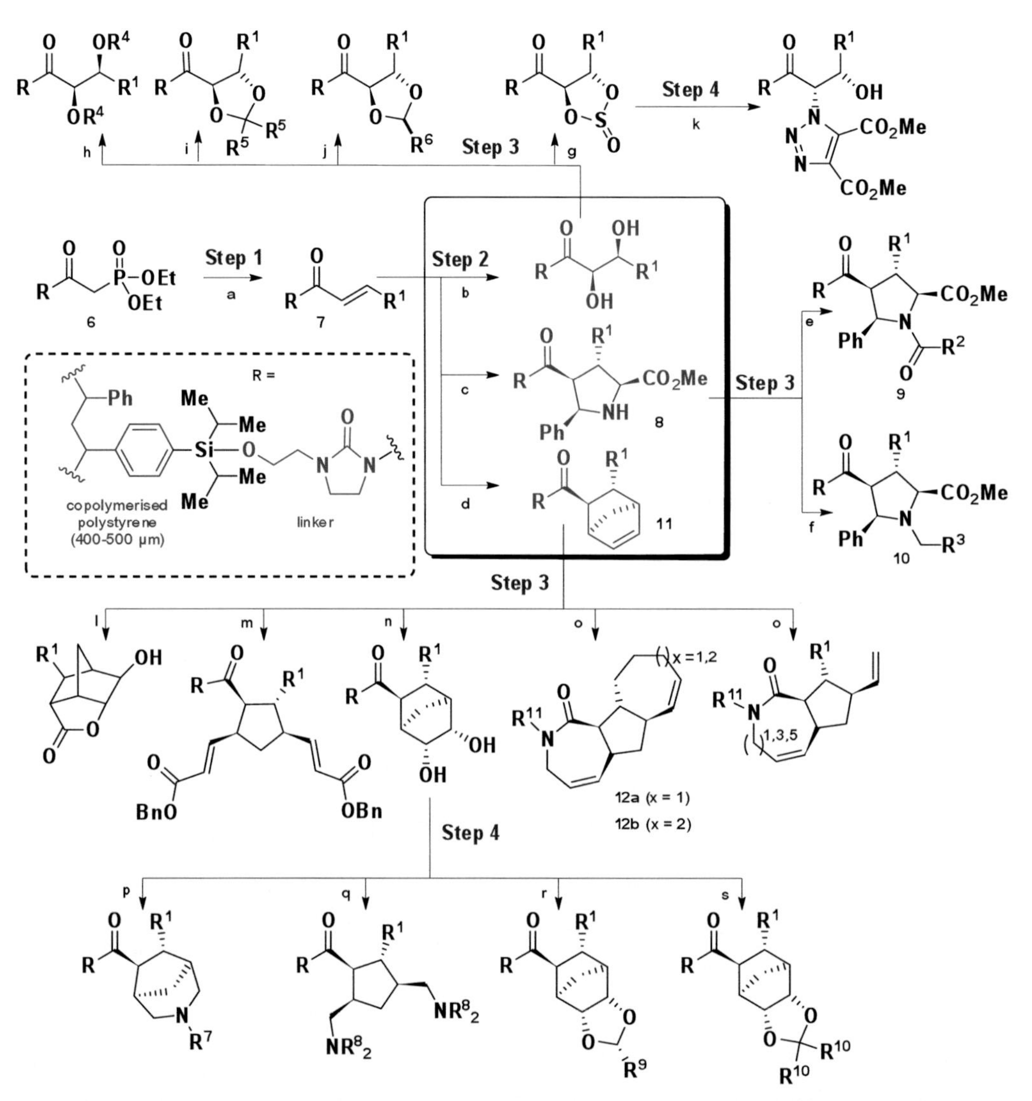

SCHEME 2: DOS of 242 compounds based of 18 discrete molecular frameworks. Conditions: (a) LiBr, 1,8-diazabicyclo[5.4.0]undec-7-ene, R^1CHO, MeCN; (b) AD-mix β , THF:H2O (1:1); (c) (R)-QUINAP, AgOAc, iPr2NEt, α -imino-ester, THF, $-78°C$ to 25°C; (d) chiral bis(oxazoline), Cu(OTf)2, 3Å MS, CH2Cl2, C5H6; (e) R^2COCl, DMAP, pyridine, CH2Cl2; (f) R^3CHO, BH3-pyridine, MeOH; (g) SOCl2, pyridine, CH2Cl2, 40°C; (h) R^4Br, Ag2O, CH2Cl2, 40°C; (i) $R^5C(O)R^5$, TsOH, DMF, 65°C; (j) R^6CHO, TsOH, DMF, 65°C; (k) NaN3, DMF, 100°C then DMAD, toluene, 65°C; (l) mCPBA, CH2Cl2 then MeOH, 65°C; (m) CH2=CHCO2Bn, Grubbs' second-generation catalyst, ethylene, toluene, 120°C; (n) OsO4, NMO, CH3C(O)CH3:H2O (10:1); (o) RNH2, Me2AlCl, toluene 120°C; then NaH, R^{11}X, DMF, THF; then toluene, 120°C, Grubbs' second-generation catalyst, ethylene; (p) NaIO4, THF:H2O (1:1); then R^7NH2, NaB(OAc)3H, CH2Cl2; (q) NaIO4, THF:H2O (1:1); then R^8NHR^8, NaB(OAc)3H, CH2Cl2; (r) R^9CHO, DMF, TsOH, 60°C; (s) $R^{10}C(O)R^{10}$, DMF, TsOH, 60°C. (DHQD)PHAL = hydroquinidine 1,4-phthalazinediyl diether; DMAD = dimethyl acetylenedicarboxylate; DMAP = N,N-dimethylaminopyridine; DMF = N,N-dimethylformamide; Grubbs II = 1,3-(bis(mesityl)-2-imidazolidinylidene) dichloro (phenylmethylene) (tricyclohexylphosphine) ruthenium; mCPBA = meta-chloroperbenzoic acid; NMO = 4-methylmorpholine-N-oxide; THF = tetrahydrofuran; Ts = para-toluenesulfonyl.

A branching DOS strategy was also utilized by Wyatt et al. for the synthesis of a skeletally diverse small molecule library (Scheme 3) [46]. The fluorous tagged diazoacetate compound **13** was identified as an attractive starting unit for two main reasons: 1) diazoacetate compounds exhibit enormous synthetic versatility, allowing a wide variety of different synthetic transformations to be carried out on the starting material; and 2) polyfluorocarbon tag technology allowed standard solution-phase parallel synthesis methods to be coupled with the benefits of fluorous-based purification protocols [51, 52], thus simplifying the isolation of the library compounds.

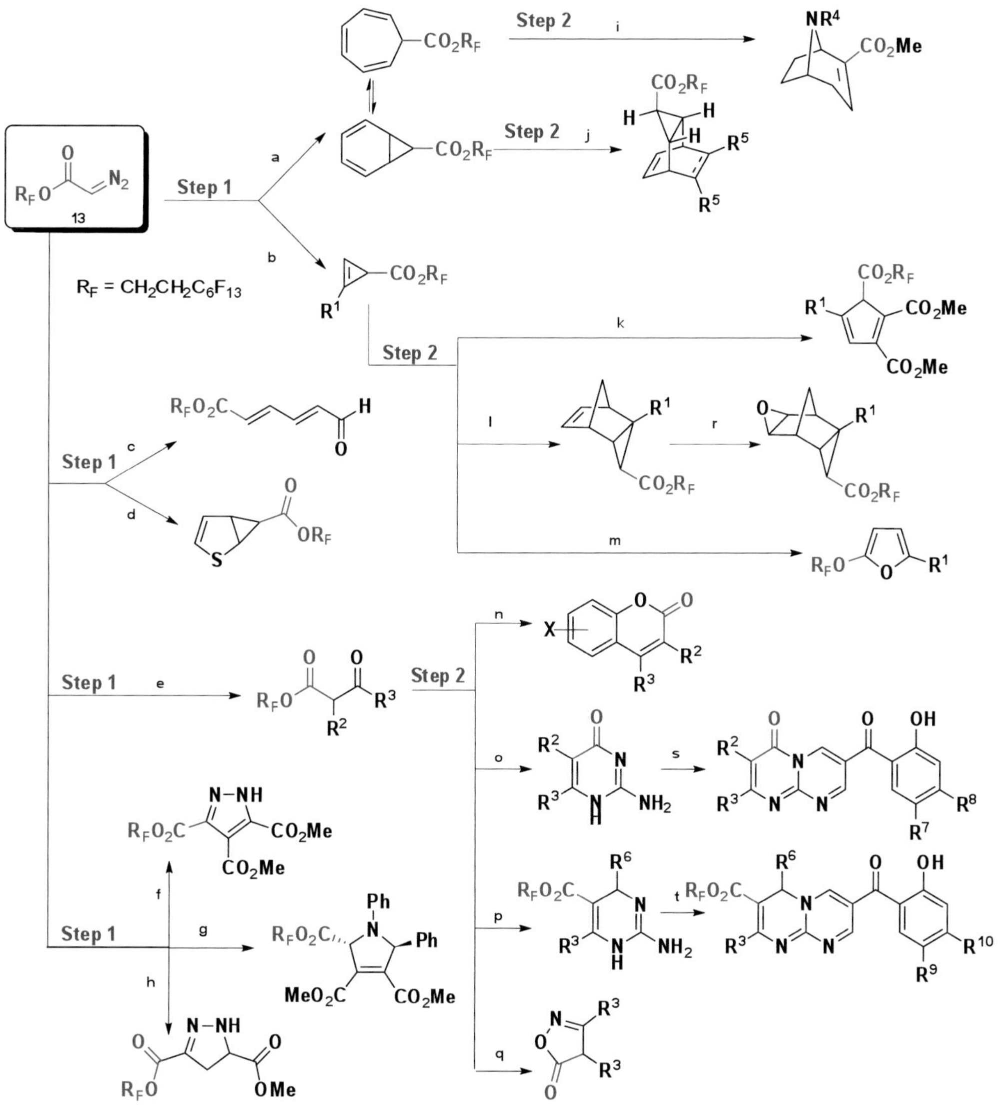

SCHEME 3: The synthetic plan for the DOS of a library of small molecules from a simple diazoacetate starting material 13. Step 1 refers to the first step of the DOS; Step 2 refers to the second step of the DOS. (a) C6H6, Rh2(OCOCF3)4; (b) R^1CCH, Rh2(OAc)4, CH2Cl2; (c) Furan, Rh2(OAc)4 then I2; (d) Thiophene, Rh2(OAc)4; (e) LDA – 78°C, then R^2COR^3, THF then Rh2(OAc)4, CH2Cl2; (f) DMAD; (g) PhCHO, PhNH2 then DMAD, Rh2(OAc)4 or toluene. [Cu(OTf)]2, CH2Cl2; (h) methyl acrylate; (i) R^4NH2, NaOH, H2O, 180°C then MeOH, H2SO4, 60°C; (j) dienophile, toluene, reflux; (k) DMAD, toluene, 100°C; (l) cyclopentadiene, CH2Cl2, 0°C to rt; (m) Grubbs' II, toluene, ethylene, reflux; (n) phenol derivative, conc. H2SO4; (o) guanidine, EtOH, reflux; (p) guanidine, R^6CHO, DMF, 75°C; (q) NH2OH, THF, reflux; (r) mCPBA, CH2Cl2, rt; (s) substituted 3-formyl chromone, EtOH, reflux; (t) substituted 3-formyl chromone, EtOH, reflux. DMAD = dimethyl acetylenedicarboxylate; Grubbs' II = 1,3-(bis(mesityl)-2-imidazolidinylidene) dichloro (phenylmethylene) (tricyclohexylphosphine) ruthenium; LDA = lithium diisopropylamide; mCPBA = meta-chloroperbenzoic acid; Tf = trifluoromethanesulfonate; THF = tetrahydrofuran.

The main pathways are summarized in Scheme 3. In the first stage of the DOS, the reactive diazoacetate functionality was exploited in four main branching reactions. Each branching reaction was chosen so as to generate at least one unique molecular skeleton (Step 1, Scheme 3):

1) Cycloaddition with benzene and alkynes (reactions a and b, respectively);
2) Cycloaddition with the heteroaromatic compounds furan and thiophene (reactions c and d, respectively);
3) 1,3-Dipolar cycloadditions with dimethyl acetylenedicarboxylate (DMAD) and methyl acrylate and a three-component ylide-mediated cycloaddition (reactions f, g, and h, respectively); and
4) α-Deprotonation followed by trapping of the resultant anion with an electrophile and subsequent

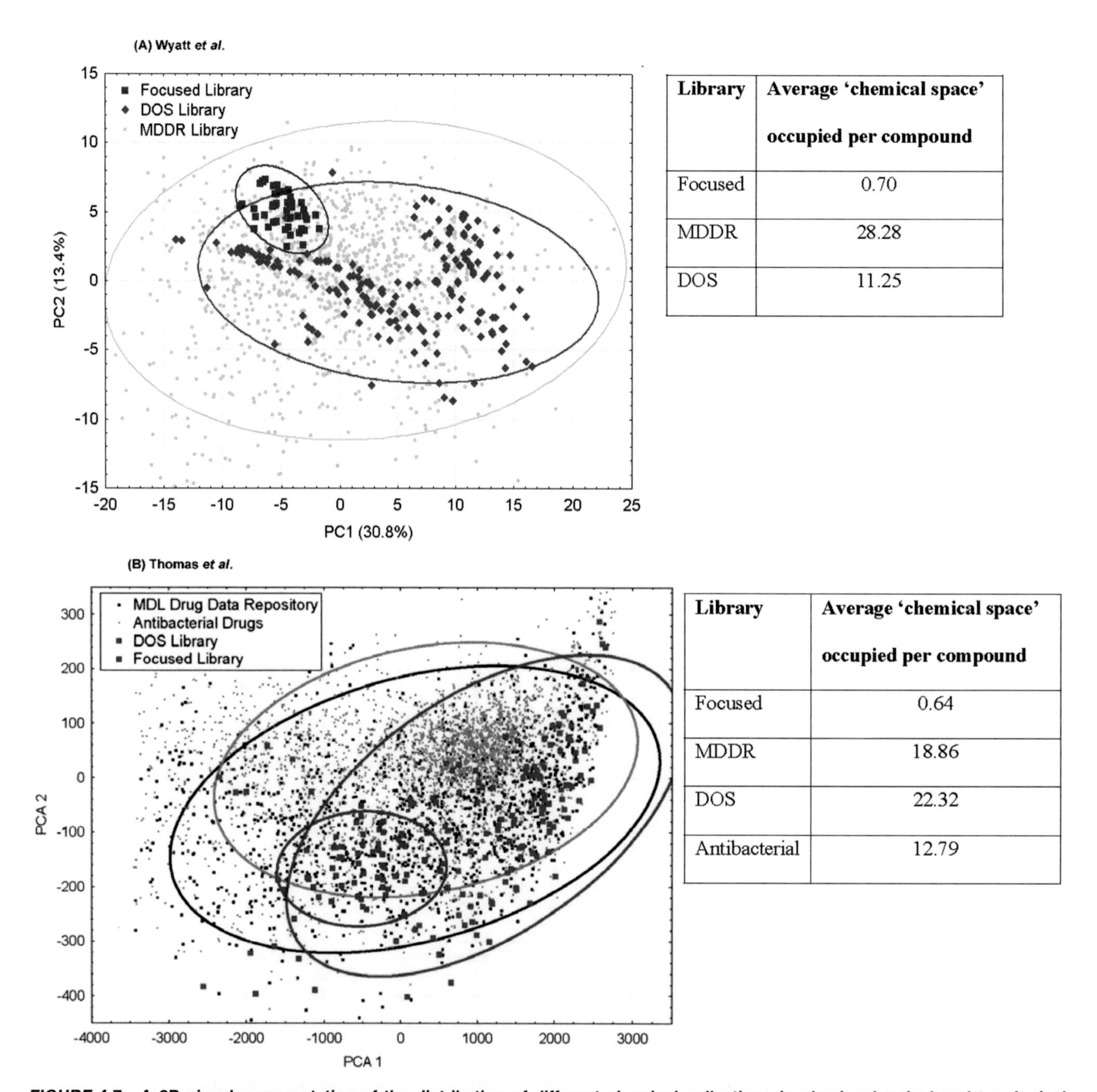

Library	Average 'chemical space' occupied per compound
Focused	0.70
MDDR	28.28
DOS	11.25

Library	Average 'chemical space' occupied per compound
Focused	0.64
MDDR	18.86
DOS	22.32
Antibacterial	12.79

FIGURE 4.7: A 2D visual representation of the distribution of different chemical collections in physicochemical and topological space derived using molecular operating environment (MOE) descriptors followed by principal component analysis (PCA). The DOS libraries synthesized are depicted by red squares (A: Wyatt et al. B: Thomas et al.). For comparison, a focused library (blue squares), the MDL Drug Data Repository (MDDR; black dots), and antibacterial drugs (gray dots) (B only) are depicted. Data for the average "chemical space" occupied per compound calculated in the context of each of the DOS libraries are shown in the tables on the right of the corresponding chemical space diagrams.

metal-catalyzed hydrogen or carbon migration to form β-dicarbonyl compounds (reaction e).

The second stage of the DOS involved a series of complexity-generating reactions to diversify these molecular frameworks further, leading to the generation of more skeletal diversity in the library (Step 2, Scheme 3, reactions i through q). In some cases, a third stage of reactions was carried out (reactions r, s, and t) to introduce additional complexity and diversity. Appendage and functional group diversity were introduced into the library through variation in the substrates used in these branching reactions (i.e., variation in R^{1-10}). In addition, additional appendage and functional group diversity were incorporated into the

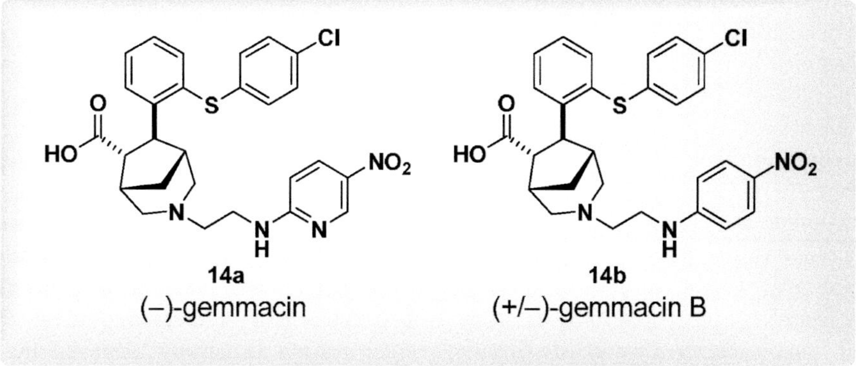

	MIC_{50} (μg ml^{-1})	
Compound	**EMRSA 15**	**EMRSA 16**
(-)-gemmacin	8	16
(+/-)-gemmacin B	8	8
erythromycin	>64	>64
oxacillin	>32	>32

***FIGURE 4.8:** Structure and activity of gemmacin (14a) and gemmacin b (14b) with growth inhibitory activity (MIC_{50}) against two strains of methicillin-resistant* Staphylococcus aureus*, EMRSA 15 and EMRSA 16. For comparison, the MIC_{50} values for erythromycin and oxacillin are also shown. ND = not determined. MSSA = methicillin-susceptible* S. aureus*. MIC_{50} = minimum inhibitory concentration required to inhibit the growth of 50% of organisms.*

products from these pathways via the use of different ester cleavage mechanisms (i.e., ester hydrolysis, transesterification, transamidation, and ester reduction; R replaced by R^{11}, R^{12}, etc.).

Various substrates (building blocks) were used in the branching synthetic routes outlined in Scheme 3 to synthesize a library of 223 compounds based on thirty different core molecular skeletons in no more than four linear synthetic steps from a simple diazoacetate starter unit.

A fundamental problem when attempting the synthesis of a diverse small molecule library is the subjective nature of diversity itself; that is, how does one determine how successful one library synthesis is compared with another in terms of diversity generation? Recent years have witnessed significant progress toward the development of computational methods that allow an assessment of the relative diversity present in different chemical collections in a more quantitative fashion. For example, for each compound in the DOS libraries synthesized by Wyatt et al. and Thomas et al., the values of 184 different physiochemical and topological chemical descriptor properties (e.g., molecular weight, degree of branching, pKa, charges) were calculated. The data sets produced for each compound were analyzed using principal component analysis (PCA) to generate a unique set of coordinates for each compound in chemical space.

Using this method, a two-dimensional (2D) visual representation of the distribution of the compounds of the DOS libraries in chemical space was derived (Figure 4.7A for the library of Wyatt et al., Figure 4.7B for the library of

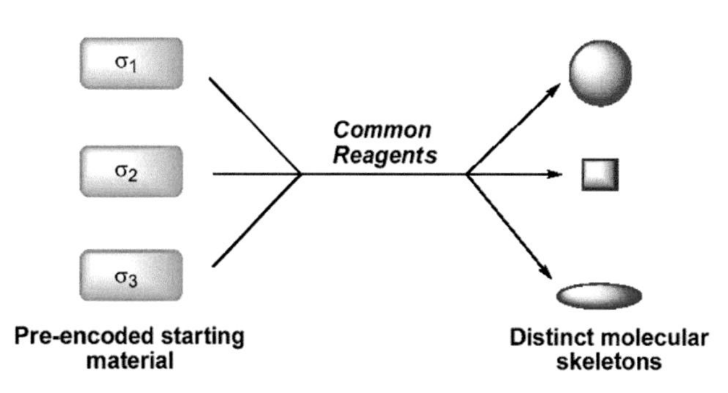

***FIGURE 4.9:** The substrate-based approach (folding process) to generating skeletal diversity.*

Thomas et al.). For comparison, the distribution in chemical space of two "benchmark" molecule collections was also computed and included in these diagrams. The first of these collections was a focused library (indicated by blue squares) that was produced via a more traditional combinatorial approach (whereby a common scaffold is decorated with different appendages) [53]. The second of these benchmark collections was a sample of all known drug molecules with a similar weight range to the compounds present in the DOS libraries (molecular weight cutoff of 650) taken from the MDL Drug Data Repository (MDDR) database (small gray dots) [54]. Finally, in the case of the library synthesized by Thomas et al. (Figure 4.7B), the chemical space coverage achieved by the 3,762 compounds marked as "antibacterial" in the MDDR database is also included. A quantitative estimation of the diversity achieved on a per-compound basis for each of these three chemical collections was also made (this can be interpreted as a measure of the average chemical space occupied per compound for each compound collection).

In both cases, the DOS libraries (high skeletal diversity) span a larger region of chemical space than that occupied by the focused library (low skeletal diversity). This supports the premise that maximizing skeletal diversity in a small molecule library is critical in terms of maximizing overall structural diversity and thus chemical space coverage. In the case of the library synthesized by Wyatt et al., the largest coverage of chemical space is achieved by the MDDR sample. However, the DOS library, despite having significantly fewer compounds, seems to occupy a relatively large region of chemical space, illustrating the value of this DOS approach to generate structurally diverse products that span a wide area of chemical space in an efficient manner. Perhaps of greater significance was the fact that the compound collection produced by Thomas et al. was shown to be even more diverse than the MDDR library in terms of (relative) diversity units – that is, 22 for the DOS library, 19 for MDDR, 13 for the antibacterials, and 0.6 for the focused library.

Computational analyses such as those described previously (Figure 4.7) can be performed to determine if a library synthesis has been successful in terms of achieving a high degree of diversity. However, it is important to remember that the ultimate success of any small molecule library is determined by the biological relevance of the compounds it contains; if the small molecule library does not yield hits in a chosen biological screening experiment, it will be deemed unsuccessful, no matter how structurally diverse it is. Phenotypic screening experiments performed using the DOS libraries of Wyatt et al. and Thomas et al. identified a number of structurally novel compounds that displayed antibacterial activity against pathogenic strains of methicillin-resistant *Staphylococcus aureus* (EMRSA 15 and EMRSA 16, the strains responsible for the majority of MRSA infections in the United Kingdom), demonstrating the utility of the DOS approach for the discovery of new antibacterial agents. The most active compound identified was **14a**, named gemmacin, which displayed a broad range of activity against Gram-positive bacteria and is believed to operate as a cell-membrane disruptor (Figure 4.8).

These examples clearly illustrate that biologically active molecules (so-called hits) can be identified through the screening of structurally diverse small molecule collections. However, optimization is usually required to transform these "hits" into "leads" that possess more useful properties (e.g., increased potency or specificity). Such optimization is usually achieved by the synthesis of a focused library around the original bioactive molecule through variation in appendage, functional group, and stereochemical diversity elements. Subsequent screening of these compounds allows structure-activity relationships (SARs) to be investigated. Recently, Thomas and coworkers have reported such a SAR investigation of the antibacterial compound gemmacin that was identified in the DOS campaign discussed previously (Scheme 2)[55]. Their studies identified compound **14b**, named gemmacin B, which demonstrated higher levels of bioactivity against EMRSA 16 (Figure 4.8). Interestingly, these SAR studies suggested that antibacterial activity was very dependent on the original structural features displayed by gemmacin; the authors surmised that "the gemmacin architecture appears to be situated on an 'isolated island of bioactivity' where little manoeuvrability (i.e., chemical diversification) is possible if antibacterial activity is to be retained."

Further examples of the use of a pluripotent functionality approach to DOS can be found in some reviews and recently published articles [18, 43, 45, 56].

1.6.3.1.3. The use of a densely functionalized molecule. Schreiber et al. have reported the synthesis of a skeletally and stereochemically diverse small molecule collection via a DOS approach based around the varied reactivity of densely functionalized β-amino alcohol derivatives [57].

A Petasis three-component coupling of **15**, **16**, and **17** generated compound **18**; subsequent amine propargylation furnished the highly functionalized β-amino alcohol derivative **19** (Scheme 4). The stereochemical outcome of the Petasis reaction was found to be controlled by the lactol **15**, suggesting that all four possible anti–amino alcohol stereoisomers could (in principle) be generated (stereochemical diversity).

Initially, a series of seven skeletal diversification reactions was carried out on **19**, yielding products **20** through **26** (Scheme 5). These reactions were based around reactivity at four of the different functionalities present in **19** – that is, the hydroxyl group, the alkene, the alkyne, and the cyclopropane moieties: 1) palladium-catalyzed cycloisomerization (route a); 2) ruthenium-catalyzed cycloisomerization (route b); 3) cobalt-mediated Pauson-Khand

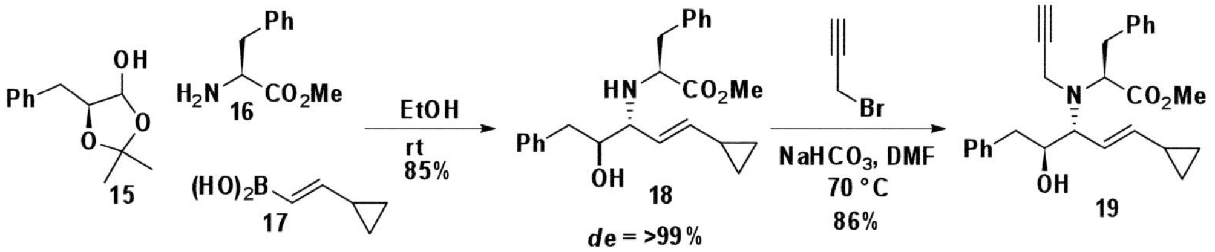

SCHEME 4: *Synthesis of β-amino alcohol 19. DMF = N,N-dimethylformamide; rt = room temperature.*

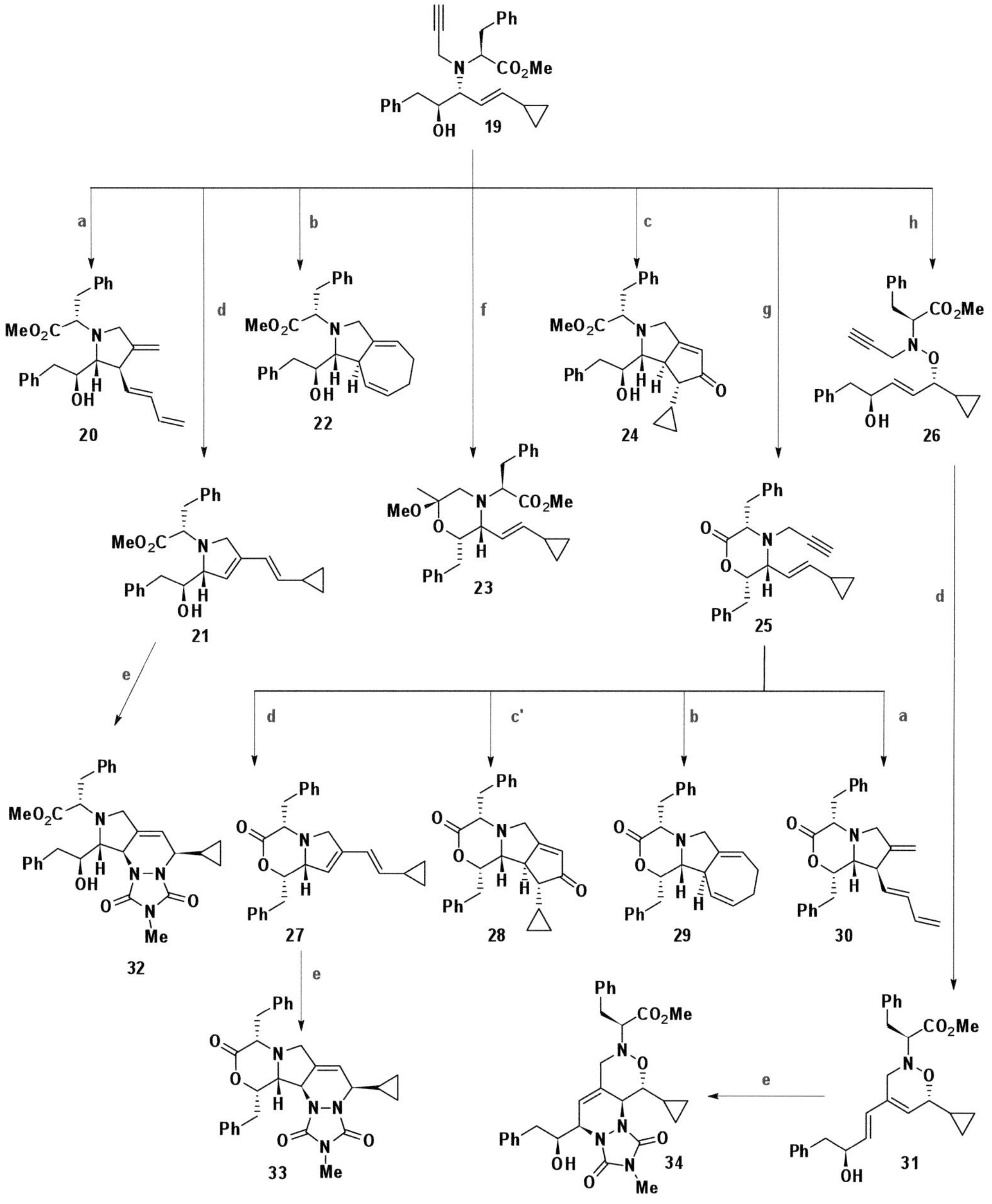

SCHEME 5: *Conditions: (a) [Pd(PPh3)2(OAc)2] (10 mol%), benzene, 80°C; (b) [CpRu(CH3CN)3PF6] (10 mol%), acetone, rt; (c) [Co2(CO)8], trimethylamine N-oxide, NH4Cl, benzene, rt; (c′) [Co2(CO)8], trimethylamine N-oxide, benzene, rt; (d) Hoveyda-Grubbs second-generation catalyst (10 mol%), CH2Cl2, reflux; (e) 4-methyl-1,2,4-triazoline-3,5-dione, CH2Cl2, rt; (f) NaAuCl4 (10 mol%), MeOH, rt; (g) NaH, toluene, rt; (h) mCPBA, THF, −78 to 0°C. rt = room temperature; mCPBA = meta-chloroperbenzoic acid.*

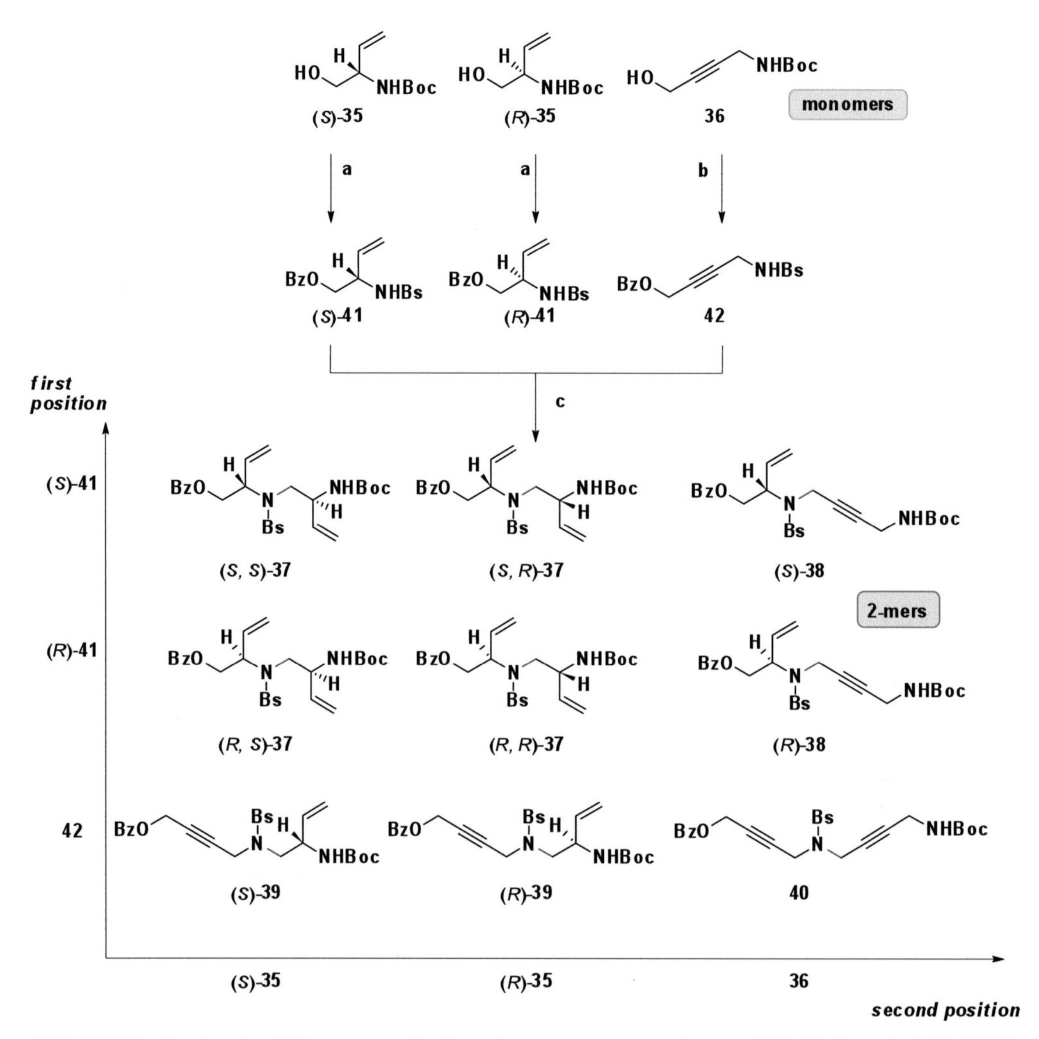

SCHEME 6: Synthesis of 2-mer combinations from monomer units; (a) (i) TFA then BsCl/NaHCO3, EtOAc, (ii) BzCl, pyridine; (b) (i) BzCl, pyridine, (ii) TFA then BsCl, NEt3, CH2Cl2; (c) 2-mers were synthesized from the corresponding alcohol and brosylate using PPh3, DEAD, THF, 0°C to rt. Newly formed bonds are shown in red. Boc = tert-butoxycarbonyl; Bz = benzoyl; Bs = brosylate (BrC6H4SO2); Ac = acetyl; TFA = trifluoroacetic acid; DEAD = diethyl azodicarboxylate; THF = tetrahydrofuran.

reaction (route c); 4) enyne metathesis (route d); 5) gold-catalyzed intramolecular cyclization (route f); 6) lactone formation (route g); and 7) *m*CPBA-mediated rearrangement (route h).

Subjecting lactone **25** to a similar set of conditions used in the diversification of **19** led to products **27** through **30**. Further diversification of compounds containing a 1,3-diene moiety (**21**, **27**, and **31**) was achieved via a Diels-Alder reaction with 4-methyl-1,2,4-triazolin-3,5-dione to yield products **32** through **34** (route e). The whole DOS pathway was repeated using alternative amine building blocks in the Petasis reaction (substitutional diversity). The net result was that this DOS strategy enabled the synthesis, in only three to five steps, of a diverse collection of single-isomer small molecules whose members displayed substitutional, stereochemical, *and* skeletal diversity, with fifteen different molecular skeletons being present among the molecules produced.

1.6.3.2. The Substrate-Based Approach The substrate-based approach to skeletal diversity is based around a *folding process*. It involves the conversion of a collection of substrates that contain appendages with suitable "pre-encoded" skeletal information (so-called σ-elements) into products have distinct molecular skeletons using a common set of conditions (Figure 4.9) [18, 42, 48].

Spiegel and co-workers have reported the synthesis of a library of skeletally diverse small molecules using such an approach, based around reactive "oligomers" that could be subjected to a chemical transformation that causes them to fold up into distinct 3D shapes (Figure 4.10) [58].

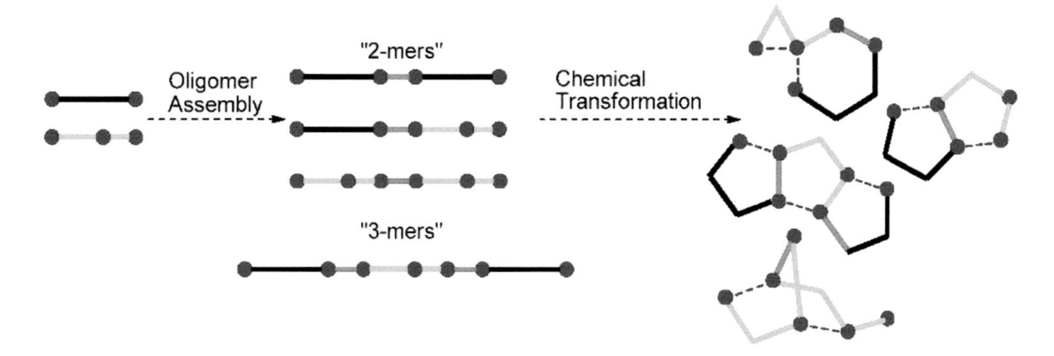

***FIGURE 4.10:** Schematic depiction of the oligomer-based approach used by Spiegel et al. [58] Red circles indicate reactive groups, green lines indicate monomer attachment sites, and dashed red lines indicate newly formed covalent bonds.*

The library synthesis was based around three monomer units, (*S*)-**35**, (*R*)-**35** and **36**, which were converted into all nine possible 2-mers (a combination of 2 monomers) **37** to **40** via intermediate sulfonamides **41** and **42** (Scheme 6).

Treatment of all stereoisomeric variants **37** to **40** with Grubbs' first-generation catalyst provided eight different products having three types of skeletons: disubstituted tetrahydopyridine **43**, vinyl tetrahydropyridine **44**, and dihydropyrrole **45** (three selected examples are illustrated in Scheme 7).

Various 3-mers (a combination of 3 monomers) (*S*, *R*)-**46**, (*S*, *S*)-**46**, and **47** were then accessed from the 2-mers

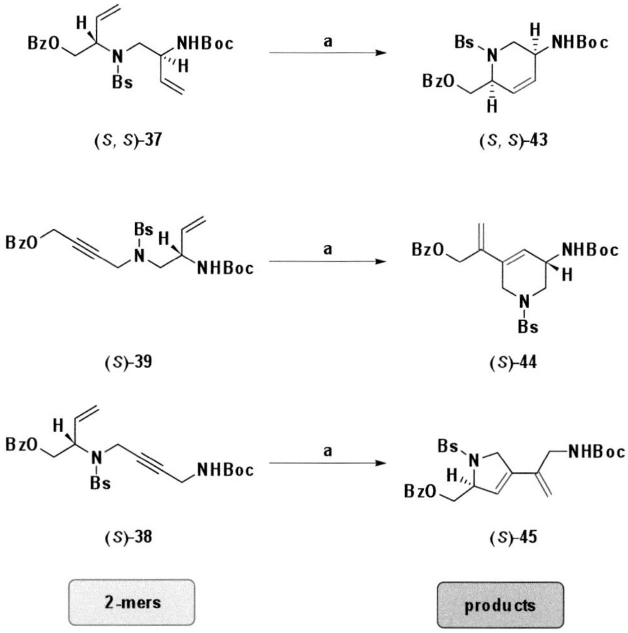

(*S, S*)-37 —a→ (*S, S*)-43

(*S*)-39 —a→ (*S*)-44

(*S*)-38 —a→ (*S*)-45

2-mers products

***SCHEME 7:** Three examples of the generation of skeletally diverse products by treatment of three different 2-mers with a common reagent; (a) Grubbs' I (5 mol%), toluene, ethylene, reflux, 16–30 h. Newly formed bonds are shown in red. Boc = tert-butoxycarbonyl; Bs = brosylate (BrC6H4SO2); Bz = benzoyl; Grubbs' I = benzylidene-bis(tricyclohexylphosphine)dichloroorruthenium.*

(*S*)-**38** and **40** using a Fukuyama-Mitsunobu coupling reaction (Scheme 8). Under the common metathesis conditions employed for the reaction of the 2-mers, polycyclic compounds (*S*, *R*)-**48**, (*S*, *S*)-**48**, and **49** were formed.

By the methods outlined above, a library of small molecules based on twelve distinct molecular skeletons was produced. The 1,3-diene products of the oligomerization-skeletalization sequence also served as substrates in a second skeletal diversification step using Diels-Alder reactions (not shown). Overall, starting from only three simple monomer units, a library of small molecules based around seventeen unique molecular skeletons was produced. Further examples of the use of a substrate-based approach to DOS can be found in some reviews and recently published articles [19, 23, 24, 48].

1.6.3.3. The Build-Couple-Pair Strategy Recent work by Schreiber has identified a common strategic feature present in many DOS pathways. This is the so-called build/couple/pair (B/C/P) three-phase strategy [23] (Figure 4.11).

The three phases involve can be defined in the following fashion: [23]

1) *Build:* Asymmetric syntheses of chiral building blocks;
2) *Couple:* Intermolecular coupling reactions that join the building blocks are performed; this process provides the basis for stereochemical diversity; and
3) *Pair:* Intramolecular coupling reactions that join pairwise combinations of functional groups incorporated in the "build" phase are performed; this process provides the basis for skeletal diversity.

The characteristics of a B/C/P approach toward skeletal diversity construction can be identified in both reagent- and substrate-based DOS pathways. For example, the DOS pathway outlined in Scheme 4 and Scheme 5 can be analyzed in terms of a B/C/P strategy; the "build" phase involved assembly of fragments **15**, **16**, and **17**; the Petasis

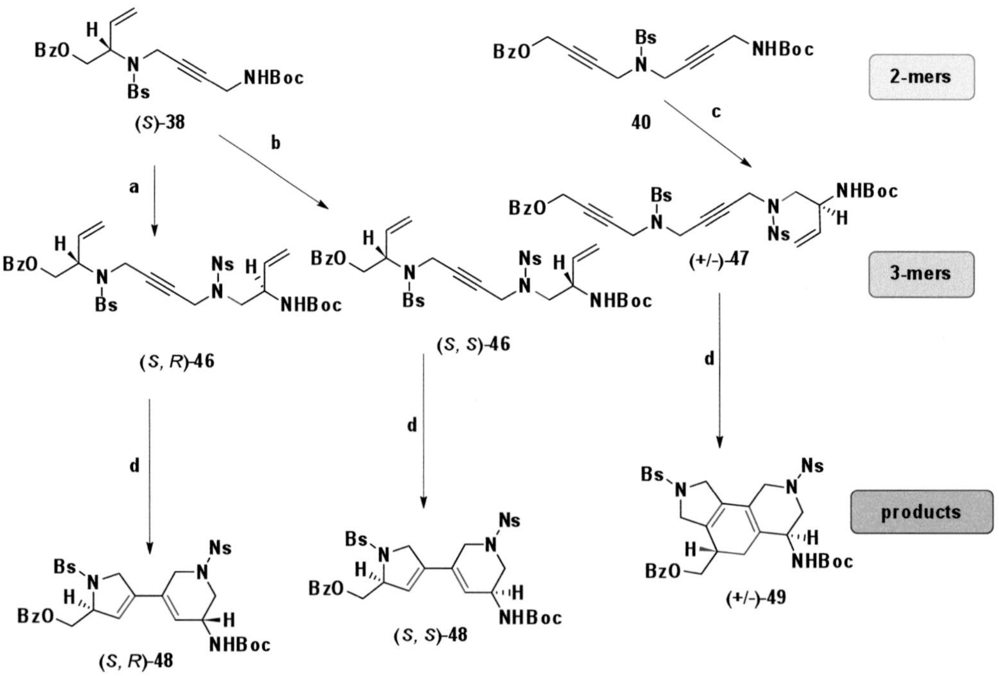

SCHEME 8: Formation of 3-mers and their subsequent diversification. (a) (i) TFA then NsCl, NaHCO3, EtOAc, (ii)(R)-29, PPh3, DEAD, THF, 0°C to rt; (b) (i) TFA then NsCl, NaHCO3, EtOAc, (ii)(S)-29, PPh3, DEAD, THF, 0°C to rt; (c) TFA then NsCl, NaHCO3, EtOAc, (ii)(rac)-29, PPh3, DEAD, THF, 0°C to rt; (d) Grubbs' I (5 mol%), toluene, ethylene, reflux, 16–30 hr. Newly formed bonds from each step are shown in red. Ac = acetyl; Boc = tert-butoxycarbonyl; Bs = brosylate (BrC6H4SO2); Bz = benzoyl; DEAD = diethyl azodicarboxylate; Grubbs' I = benzylidene-bis(tricyclohexylphosphine)dichlororuthenium; Ns = para-nitrophenylsulphonyl; TFA = trifluoroacetic acid; THF = tetrahydrofuran.

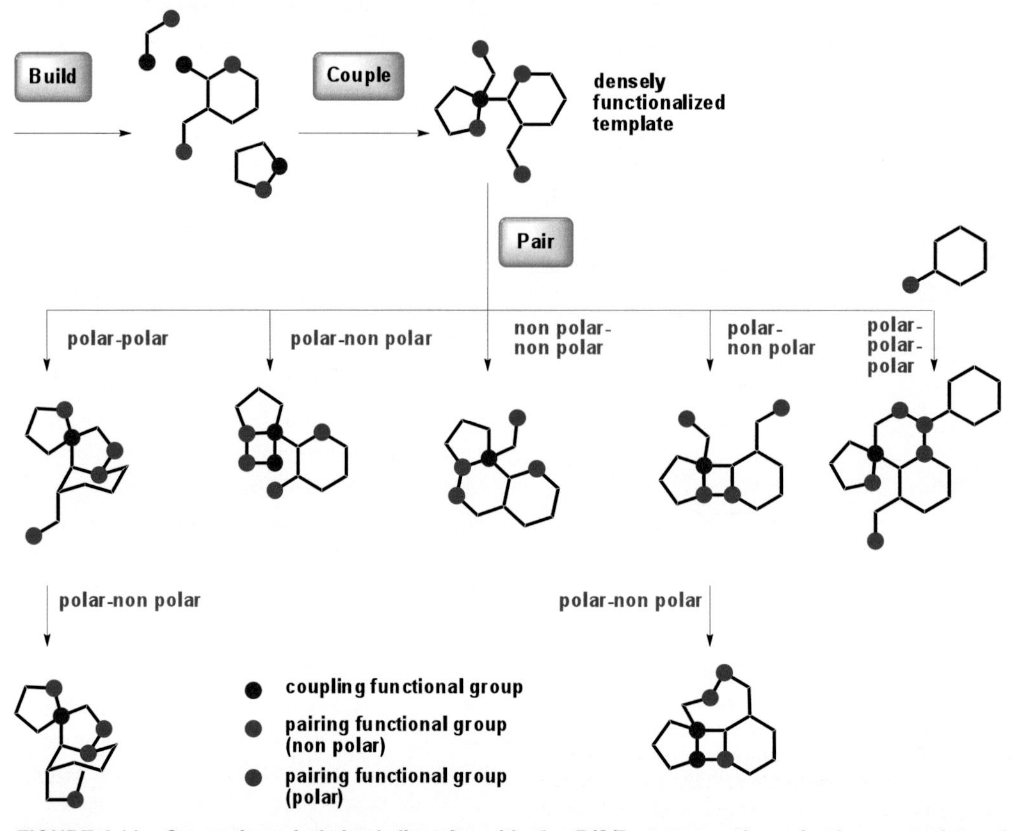

FIGURE 4.11: Generation of skeletal diversity with the B/C/P strategy: the pair phase consists of chemoselective and intramolecular joining of strategically placed polar (blue) and nonpolar (red) functional groups to afford diverse skeletons [23].

reaction followed by alkylation served as "couple" phases; and the subsequent reagent-controlled skeletal diversification reactions served as "pair" phases in which different combinations of the moieties of **19**, both polar and nonpolar, were "paired" in functional group specific reactions. For example, the palladium- and ruthenium-based catalysts selectively paired the nonpolar alkene, alkyne, and cyclopropane groups of **19**, enabling the cycloisomerization reactions leading to compounds **20** to **22**, whereas sodium hydride–mediated lactonization selectively paired the polar functional groups to form **25**.

The DOS pathway of Spiegel et al. outlined in Schemes 6, 7, and 8 can also be analyzed in a similar fashion. The "build" phase involved the synthesis of non-racemic monomers **35** and **36**. The combination of these monomers by reaction of their polar groups to form 2-mers and 3-mers comprised the "couple" phase, and the "pair" phase focused on joining the nonpolar groups of these 2-mers and 3-mers through the use of ruthenium-catalyzed metathesis reactions [23].

1.6.4. Conclusions

Small molecule libraries that display a high degree of structural and thus functional (biological) diversity have proven to be valuable in the discovery of molecules that can modulate the activities of biological macromolecules in a useful fashion. The synthesis of such diverse small molecule collections presents many challenges to the synthetic chemist. In recent years, many innovative DOS strategies have been developed in an attempt to increase the overall structural diversity of small molecule libraries in an efficient manner. The issue of maximizing skeletal diversity is widely recognized as the key to achieving this goal; "that the diversity of the (small molecule) library is defined by the diversity of the scaffolds (molecular skeletons) that make up the library is becoming an axiom" [25]. Many compound collections synthesized using a DOS approach have been successfully exploited in chemical genetics to identify useful modulators for biological systems [59, 60]. Nonetheless, there still remains a need for the development of new sources of diverse small molecules that can be exploited as potential therapeutic agents and research tools in biological systems. However, it is once again worth emphasizing that the ultimate success of any small molecule library is determined by the biological relevance of the compounds it contains. If the library does not yield hits in a chosen biological screening experiment, it will be deemed unsuccessful, no matter how structurally diverse it is.

REFERENCES

1. Walsh, D. P., and Chang, Y. T. (2006). Chemical genetics. *Chem Rev 106*, 2476–2530.
2. Spring, D. R. (2005). Chemical genetics to chemical genomics: small molecules offer big insights. *Chem Soc Rev 34*, 472–482.
3. Schreiber, S. L. (1998). Chemical genetics resulting from a passion for synthetic organic chemistry. *Bioorg Med Chem 6*, 1127–1152.
4. Hopkins, A. L., and Groom, C. R. (2002). The druggable genome. *Nat Rev Drug Discov 1*, 727–730.
5. Lipinski, C., and Hopkins, A. (2004). Navigating chemical space for biology and medicine. *Nature 432*, 855–861.
6. Kaiser, M., Wetzel, S., Kumar, K., and Waldmann, H. (2008). Biology-inspired synthesis of compound libraries. *Cell Mol Life Sci 65*, 1186–1201.
7. Lipinski, C. A., Lombardo, F., Dominy, B. W., and Feeney, P. J. (2001). Experimental and computational approaches to estimate solubility and permeability in drug discovery and development settings. *Adv Drug Deliv Rev 46*, 3–26.
8. Di, L., and Kerns, E. H. (2003). Profiling drug-like properties in discovery research. *Curr Opin Chem Biol 7*, 402–408.
9. Tan, D. S. (2005). Diversity-oriented synthesis: exploring the intersections between chemistry and biology. *Nat Chem Biol 1*, 74–84.
10. Haggarty, S. J. (2005). The principle of complementarity: chemical versus biological space. *Curr Opin Chem Biol 9*, 296–303.
11. Sauer, W. H. B., and Schwarz, M. K. (2003). Molecular shape diversity of combinatorial libraries: a prerequisite for broad bioactivity. *J Chem Inf Comput Sci 43*, 987–1003.
12. Kennedy, J. P., Williams, L., Bridges, T. M., Daniels, R. N., Weaver, D., and Lindsley, C. W. (2008). Application of combinatorial chemistry science on modern drug discovery. *J Comb Chem 10*, 345–354.
13. Estrada, E., and Uriarte, E. (2001). Recent advances on the role of topological indices in drug discovery research. *Curr Med Chem 8*, 1573–1588.
14. Fergus, S., Bender, A., and Spring, D. R. (2005). Assessment of structural diversity in combinatorial synthesis. *Curr Opin Chem Biol 9*, 304–309.
15. Oprea, T. I., and Gottfries, J. (2001). Chemography: the art of navigating in chemical space. *J Comb Chem 3*, 157–166.
16. Oprea, T. I. (2002). Chemical space navigation in lead discovery. *Curr Opin Chem Biol 6*, 384–389.
17. Dobson, C. M. (2004). Chemical space and biology. *Nature 432*, 824–828.
18. Burke, M. D., and Schreiber, S. L. (2004). A planning strategy for diversity-oriented synthesis. *Angew Chem Int Ed 43*, 46–58.
19. Spandl, R. J., and Spring, D. R. (2008). Diversity-oriented synthesis; a spectrum of approaches and results. *Org Biomol Chem 6*, 1149–1158.
20. Adriaenssens, L. V., Austin, C. A., Gibson, M., Smith, D., and Hartley, R. C. (2006). Stereodivergent diversity oriented synthesis of piperidine alkaloids. *Eur J Org Chem*, 4998–5001.
21. Spring, D. R. (2003). Diversity-oriented synthesis; a challenge for synthetic chemists. *Org Biomol Chem 1*, 3867–3870.
22. Thomas, G. L., Wyatt, E. E., and Spring, D. R. (2006). Enriching chemical space with diversity-oriented synthesis. *Curr Opin Drug Discov Devel 9*, 700–712.

23. Nielsen, T. E., and Schreiber, S. L. (2008). Diversity-oriented synthesis – towards the optimal screening collection: a synthesis strategy. *Angew Chem Int Ed 47*, 48–56.
24. Burke, M. D., Berger, E. M., and Schreiber, S. L. (2003). Generating diverse skeletons of small molecules combinatorially. *Science 302*, 613–618.
25. Shelat, A. A., and Guy, R. K. (2007). Scaffold composition and biological relevance of screening libraries. *Nat Chem Biol 3*, 442–446.
26. An, H., Eum, S., Koh, M., Lee, S., and Par, S. (2008). Diversity-oriented synthesis of privileged benzopyranyl heterocycles from *s-cis*-enones. *J Org Chem 73*, 1752–1761.
27. Hopkins, A. L., Mason, J. S., and Overington, J. P. (2006). Can we rationally design promiscuous drugs? *Curr Opin Struct Biol 16*, 127–136.
28. Clardy, J., and Walsh, C. (2004). Lessons from natural molecules. *Nature 432*, 829–837.
29. Pucheault, M. (2008). Natural products: chemical instruments to apprehend biological symphony. *Org Biomol Chem 6*, 424–432.
30. Schneider, G., and Grabowski, K. (2007). Properties and architecture of drugs and natural products revisited. *Curr Chemical Biol 1*, 115–127.
31. Butler, M. S. (2004). The role of natural product chemistry in drug discovery. *J Nat Prod 67*, 2141–2153.
32. Newman, D. J., and Cragg, G. M. (2007). Natural products as sources of new drugs over the last 25 years. *J Nat Prod 70*, 461–477.
33. Borman, S. (2004). Rescuing Combichem. *Chem Eng News: Sci Technol 82*, 32–40.
34. Rouhi, A. M. (2003). Rediscovering natural products. *Chem Eng News 81*, 77–78, 82–83, 86, 88–91.
35. DeSimone, R. W., Currie, K. S., Mitchell, S. A., Darrow, J. W., and Pippin, D. A. (2004). Privileged structures: applications in drug discovery. *Comb Chem High Throughput Screening 7*, 473–493.
36. Koch, M. A., and Waldmann, H. (2005). Protein structure similarity clustering and natural product structure as guiding principles in drug discovery. *Drug Discov Today 10*, 471–483.
37. Balamurugan, R., Dekker, F. J., and Waldmann, H. (2005). Design of compound libraries based on natural product scaffolds and protein structure similarity clustering (PSSC). *Mol Biosyst 1*, 36–45.
38. Breinbauer, R., Vetter, I. R., and Waldmann, H. (2002). From protein domains to drug candidates – natural products as guiding principles in the design and synthesis of compound libraries. *Angew Chem Int Ed 41*, 2879–2890.
39. Goess, B. C., Hannoush, R. N., Chan, L. K., Kirchhausen, T., and Shair, M. D. (2006). Synthesis of a 10,000-membered library of molecules resembling carpanone and discovery of vesicular traffic inhibitors. *J Am Chem Soc 128*, 5391–5403.
40. Ko, S. K., Jang, H. J., Kim, E., and Park, S. B. (2006). Concise and diversity-oriented synthesis of novel scaffolds embedded with privileged benzopyran motif. *Chem Commun*, 2962–2964.
41. Cordier, C., Morton, D., Murrison, S., Nelson, A., and O'Leary-Steele, C. (2008). Natural products as an inspiration in the diversity-oriented synthesis of bioactive compound libraries. *Nat Prod Rep 25*, 719–737.
42. Burke, M. D., Berger, E. M., and Schreiber, S. L. (2004). A synthesis strategy yielding skeletally diverse small molecules combinatorially. *J Am Chem Soc 126*, 14095–14104.
43. Burke, M. D., and Lalic, G. (2002). Teaching target-oriented and diversity-oriented organic synthesis at Harvard University. *Chem. Biol. 9*, 535–541.
44. Mishra, J. K., and Panda, G. (2007). Diversity-oriented synthetic approach to naturally abundant S-amino acid based benzannulated enantiomerically pure medium ring heterocyclic scaffolds employing inter- and intramolecular Mitsunobu reactions. *J Comb Chem 9*, 321–338.
45. Taylor, S. J., Taylor, A. M., and Schreiber, S. L. (2004). Synthetic strategy toward skeletal diversity via solid-supported, Otherwise unstable reactive intermediates. *Angew Chem Int Ed 43*, 1681–1685.
46. Wyatt, E. E., Fergus, S., Galloway, W. R., Bender, A., Fox, D. J., Plowright, A. T., Jessiman, A. S., Welch, M., and Spring, D. R. (2006). Skeletal diversity construction via a branching synthetic strategy. *Chem Commun*, 3296–3298.
47. Schreiber, S. L. (2000). Target-oriented and diversity-oriented organic synthesis in drug discovery. *Science 287*, 1964–1969.
48. Oguri, H., and Schreiber, S. L. (2005). Skeletal diversity via a folding pathway: synthesis of indole alkaloid-like skeletons. *Org Lett* 7, 47–50.
49. Thomas, G. L., Spandl, R. J., Glansdorp, F. G., Welch, M., Bender, A., Cockfield, J., Lindsay, J. A., Bryant, C., Brown, D. F. J., Loiseleur, O., Rudyk, H., Ladlow, M., and Spring, D. R. (2008). Anti-MRSA agent discovery using diversity-oriented synthesis. *Angew Chem Int Ed* 47, 2808–2812.
50. Spandl, R. J., Rudyk, H., and Spring, D. R. (2008). Exploiting domino enyne metathesis mechanisms for skeletal diversity generation. *Chem Commun*, 3001–3003.
51. Curran, D., and Luo, Z. Y. (2001). Fluorous techniques for the synthesis and separation of organic molecules. *Green Chem 3*, G3–G7.
52. Zhang, W. (2003). Fluorous technologies for solution-phase high-throughput organic synthesis. *Tetrahedron 59*, 4475–4489.
53. Faghih, R., Dwight, W., Pan, J. B., Fox, G. B., Krueger, K. M., Esbenshade, T. A., McVey, J. M., Marsh, K., Bennani, Y. L., and Hancock, A. A. (2003). Synthesis and SAR of aminoalkoxy-biaryl-4-carboxamides: novel and selective histamine H3 receptor antagonists. *Bioorg Med Chem Lett 13*, 1325–1328.
54. MDL Drug Data Report, http://www.symyx.com/products/pdfs/mddr_ds.pdf.
55. Robinson, A., Thomas, G. L., Spandl, R. J., Welch, M., and Spring, D. R. (2008). Gemmacin B: bringing diversity back into focus. *Org Biomol Chem 6*, 2978–2981.
56. Kumar, N., Kiuchi, M., Tallarico, J. A., and Schreiber, S. L. (2005). Small-molecule diversity using a skeletal transformation strategy. *Org Lett* 7, 2535–2538.
57. Kumagai, N., Muncipinto, G., and Schreiber, S. L. (2006). Short synthesis of skeletally and stereochemically diverse small molecules by coupling petasis condensation reactions to cyclization reactions. *Angew Chem Int Ed 45*, 3635–3638.

58. Spiegel, D. A., Schroeder, F. C., Duvall, J. R., and Schreiber, S. L. (2006). An oligomer-based approach to skeletal diversity in small-molecule synthesis. *J Am Chem Soc 128*, 14766–14767.
59. Koehler, A. N., Shamji, A. F., and Schreiber, S. L. (2003). Discovery of an inhibitor of a transcription factor using small molecule microarrays and diversity-oriented synthesis. *J Am Chem Soc 125*, 8420–8421.
60. Spring, D. R., Krishnan, S., Blackwell, H. E., and Schreiber, S. L. (2002). Diversity-oriented synthesis of biaryl-containing medium rings using a one bead/one stock solution platform. *J Am Chem Soc 124*, 1354–1363.

CHAPTER 5

Targeted Chemical Libraries

Gregory P. Tochtrop
Ryan E. Looper

1. TARGETED CHEMICAL LIBRARIES

This chapter addresses both technical and strategic issues germane to the synthesis of chemical libraries that are targeted in nature. What makes a library targeted versus not targeted has not been clearly delineated in the literature (and is somewhat subjective), but here we define three ideological types (hit to lead, natural product inspired, and protein class targeted) and address topics relevant to their design and realization. Unbiased or prospecting libraries are dealt with in other chapters of the section and are not addressed here. Organizationally, an initial section on technical aspects of library construction is followed by sections dedicated to each ideological library type. For each section, a brief introduction is given on general considerations for the given library type, but a focus is placed on illustrative examples. This is not meant to be a comprehensive review, but care has been taken to identify specific examples that clearly illustrate both the intellectual and technical challenges of such an endeavor. An emphasis is placed on the initial discovery/hypothesis and how that is subsequently translated into the synthetic design and technical execution of the library.

A major demarcation between targeted and nontargeted libraries is commonly the size (how many discrete molecules synthesized) and the scale (the target mass of each final product). Targeted libraries are generally smaller in size but greater in scale; therefore, the experimental techniques used are not completely translatable from those used for libraries of large size. Consequently, the first section of this chapter addresses experimental techniques specifically geared toward libraries of smaller size but greater scale.

2. METHODS FOR SMALL TO MEDIUM-SIZED LIBRARY SYNTHESIS

In 1985, Richard Houghten published a landmark paper describing the synthesis of 248 different 13-residue peptides to study the implications of hemagglutinin variability [1]. The method described in this paper utilized polypropylene mesh packets into which standard *t*-butoxycarbonyl (Boc) amino acid resin was placed and sealed. The bags were then subjected to standard amino acid synthesis conditions. Variability was introduced by immersing the individual bags in different solutions of appropriate activated amino acids, and deprotections and washings were carried out with all the bags pooled together. The bags were subsequently reseparated, and the process was repeated. A critical component of this method was the use of a code or tag to keep track of each packet and which conditions it had been subjected to. This tag was then used to deconvolute the amino acid sequence of the peptide contained in each bag. Houghten's tag was very simple (a number written in black marker on each bag) but remains a critical component of chemical libraries synthesized on the solid phase (discussed in more detail 2.3). This general method has come to be known as the *tea bag* method because of the polypropylene bags that were utilized, but it stands as a very important moment in the synthesis of chemical libraries, and the Houghten library is widely considered one of the first combinatorial chemical libraries. Central to the theme of this chapter, the Houghten library is likely the first example of a targeted chemical library. The peptides synthesized systematically studied the antigen-antibody interaction of a specific sequence within the influenza hemagglutinin molecule. Although this paper is mainly cited for the method of synthesizing chemical libraries, the information gleaned from the targeted nature of the work remains an important insight.

2.1. Solution-Phase Versus Solid-Phase Synthesis

Until recently, combinatorial chemistry had traditionally been closely linked with the solid phase. This is likely due in part to the field's origins in the combinatorial peptide field but also to the restrictions placed on library size using parallel solution-phase techniques. The landmark papers of a 2-million-member, natural product–like library from Tan et al. clearly illustrate this limitation [2, 3]. This type of library would have been impossible to synthesize using parallel solution-phase techniques, as the final library realization step would have necessitated the execution of more than 2 million individual reactions. Figure 5.1 illustrates this idea in pictorial form and shows the clear superiority that solid-phase techniques possess when considering the synthesis of large-sized

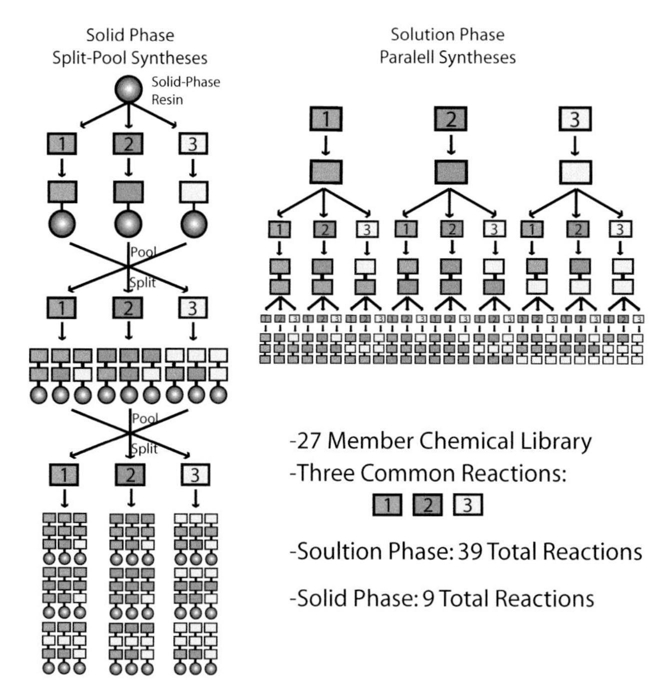

***FIGURE 5.1:* A diagrammatic representation of solution- versus solid-phase library synthesis.**

libraries. Even for the small, twenty-seven-member library depicted in the figure, using split-pool solid-phase techniques, the library can be made inclusive of nine total reactions. If the same library were made using solution-phase techniques, a total of 39 reactions would be necessary. If this example were taken one step further (not pictured) and the same three reactions could be used again to synthesize an 81-member library, then only three more reactions (12 total) would be added in the solid-phase example, whereas an additional 81 (120 total) would be necessary for the solution phase. It was this superiority with respect to library size that historically drove the majority of combinatorial chemistry and diversity-oriented synthesis (DOS) toward the solid phase. Recently, this idea has been reexamined, and the trend in combinatorial chemistry and DOS has been toward libraries that are smaller (in terms of size) but of greater scale (multimilligram quantities of final product). This transition has been emblematic of a rethinking of a fundamental principal that, given a large enough cadre of candidate molecules for any given biological target (regardless of chemical structure), the discovery of a lead would be inevitable. Furthermore, this is likely the cause of the pharmaceutical industry moving away from large-scale combinatorial chemistry as a source of lead drug molecules.

Before beginning the synthesis of a targeted library, an important (if not critical) decision is the format on which the library will be synthesized. Although not a thorough treatment of the various approaches available, sections **2.2** and **2.3** address the basic formats for synthesizing a small to medium-sized targeted chemical library in either the solution phase (**2.2**) or the solid phase (**2.3**).

2.2. Solution-Phase Techniques for Small to Medium-Sized Libraries

Although working in the solid phase is enabling in terms of library size, it is restrictive in terms of the reactions available to the synthetic chemist. The translation of an efficient, straightforward chemical transformation from the solution phase to the solid phase can take months (if not years) to accomplish. This difficulty is often a result of spatial separation of reactive intermediates and kinetic perturbations; however, it is partially mitigated by the fact that large excesses of reagents can be added to drive a reaction to completion and simply filtered away. Regardless, one of the main attractions of synthesizing libraries in the solution phase is the wealth of reactions that have been developed and optimized for this medium. If one chooses to synthesize a library in the solution phase, the key logistic hurdle that must be overcome is the final step of the library realization. This step necessitates the setup and execution of a number of reactions that is equal to the final size of the library. If the end goal is a modestly sized five-hundred-member library, careful planning and reaction development are necessary to set up, execute, work up, and purify five hundred reactions. Using the traditional round bottom flask and standard organic glassware, this would be a daunting endeavor for even the hardest-working and most dedicated chemist. This problem was recognized in the 1990s in conjunction with the explosion of combinatorial chemistry, and various efforts began to make solution-phase chemistry competitive with the solid phase by setting up several parallel reactions in one operation. One approach involved the development of instrumentation that would fully automate the parallel synthesis process. In 1994 and 1995, respectively, Argonaut Technologies [4] and Diversomer Technologies [5] developed instruments that promised to enable parallel solution phase to be competitive with the solid phase for synthesizing large-sized chemical libraries. Since these instruments were developed, several other companies have taken the same approach. However, the many nuances of chemical transformations and the vast difference in the physical properties of reaction end products have limited the success of fully automated parallel reactors. At the same time, in the 1990s a group of chemists at Bristol-Myers Squibb started developing a series of tools that were designed to be an iterative step from the traditional organic chemist's tool set to the manual parallelization of a reaction. In conjunction with Mettler-Toledo AutoChem, they developed the highly successful MiniBlocks [6]. Pictured in Figure 5.2, the MiniBlock allows the chemist to set up several reactions in parallel (up to ninety-six) during one operation and addresses many of the challenges of workup via

Methods for the Synthesis of Small to Medium Sized Libraries of Large Scale

Solution Phase — Solid Phase

A

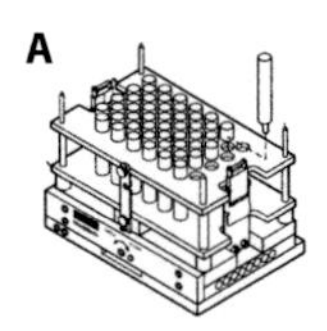

B

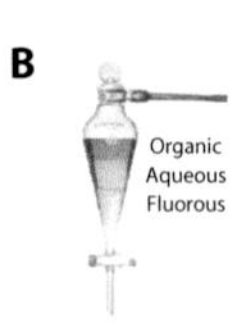

C

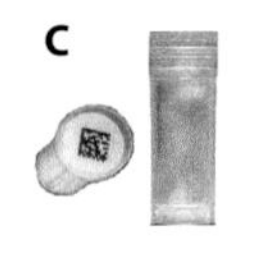

D

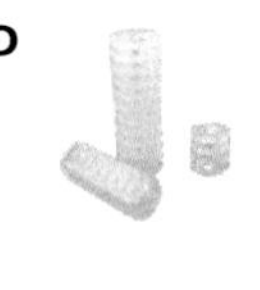

FIGURE 5.2: ***Several enabling technologies for library synthesis. A. Mettler Toledo AutoChem MiniBlock; B. Fluorous tagging utilizing a triphasic extraction; C. IRORI Kan technology illustrating the two-dimensional bar code; D. Mimotope lanterns.***

solid-phase extraction. Several companies make similar variants of the MiniBlock, and this is an appealing approach for initial entry into solution-phase library synthesis.

A recent development in the field of solution-phase library synthesis is the availability of fluorous tags that simplify workup and purification of reactions via a polyfluorinated alkyl chain [7–10]. Often called ponytails, these perfluorinated chains can contain thirty-nine or more fluorines and can help with synthetic workup and purification in two ways. First, if the ponytail is large, it will facilitate a triphasic liquid-liquid extraction in which the tag and pendant compound will partition selectively into fluorinated solvent (Figure 5.2). Second, small, fluorous tags will facilitate highly efficient solid-phase extraction with a fluorinated solid-phase extraction resin, in essence acting as affinity chromatography [9].

2.3. Solid-Phase Techniques for Libraries of Large Scale

The original 1963 report on solid-phase chemistry (which led to the 1984 Nobel Prize in Chemistry) detailed the synthesis of a tetrapeptide [11]. The requirements for synthesizing a library on the solid phase are still the same as for the synthesis of this initial peptide. First, a suitable polymer is necessary that is insoluble in any of the used solvents, but that can be appropriately derivatized with a linker that will allow the covalent coupling of starting material and cleavage of the final product. For comprehensive reviews of the various polymers and linkers, see Haag et al. [12] and Bräse et al. [13] Synthesizing a library of large scale historically has been difficult on the solid phase because of limitations in how much starting material can be loaded (and ultimately cleaved) per bead; however, several advances in available reagents and techniques have made these endeavors more feasible. For example, the 200–400 mesh polymeric resin originally described by Merrifield (with loading capacities of less than 1 nmol) has evolved into a family of resins in which as much as 110 nmol can be loaded on a single macrobead [14]. Furthermore, the original tea bag approach from Houghten has been modernized into the IRORI Kan system, in which the tea bag has been reengineered to accept either radiofrequency tags or two-dimensional (2D) bar codes for tracking and decoding the final product. Finally, the SynPhase lanterns use grafted polymers to create solid supports that are easy to handle and track and can be loaded at 75 μmol.

2.4. Chemical Planning for a Targeted Library

In 1997, Spaller et al. published an important review that clearly illustrated a key difference between focused and unbiased libraries (referred to in this report as "prospecting libraries") related to the emphasis during development of the chemistry. Intuitively, the focused chemical library will emphasize the structure and composition of the targeted end products. Multiple routes will be considered, but the end products remain consistent. For example, if a targeted library is being designed for taxol, one would consider how many ways there are to synthesize the taxane ring system but would not put as much thought into the starting materials that would be necessary for the transformations. This is in contrast to an unbiased library, in which an emphasis is placed on the input/starting materials and creativity toward what can ultimately be synthesized from them. Furthermore, unbiased libraries are usually much larger than their targeted counterparts and consequently may be limited by the commercial availability of various building blocks.

The remainder of this chapter is organized to better illustrate and expand on the idea shown in Figure 5.3. Several key examples of focused chemical libraries have been

Targeted Chemical Library

Emphasis is on the Target

Retrosynthetic Analysis

Non-targeted Chemical Library

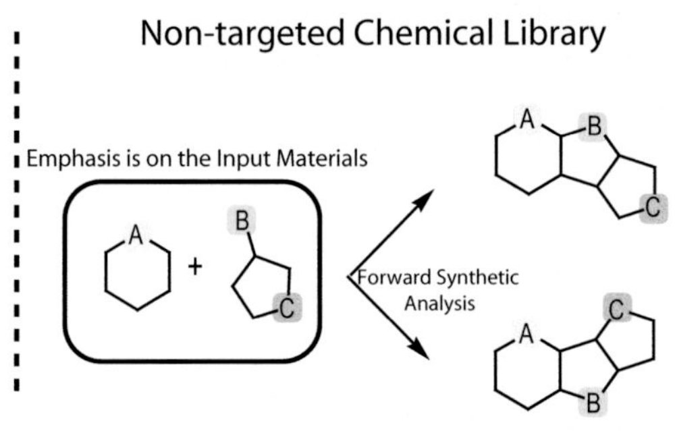

FIGURE 5.3: ***Diagrammatic representation of the differences in approach between targeted and non-targeted chemical libraries.***

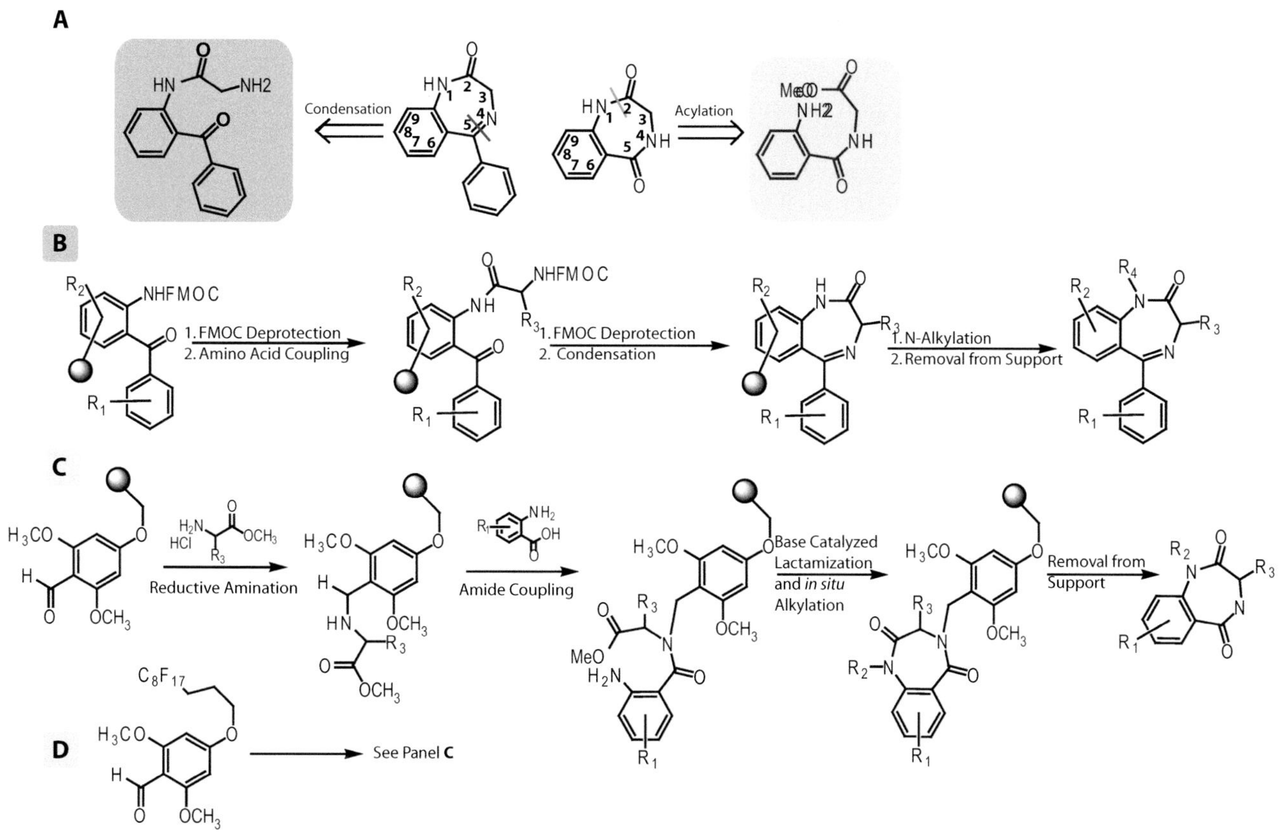

FIGURE 5.4: **A.** Two retrosynthetic analyses and **B, C.** two approaches to benzodiazepine-focused libraries. **D.** A fluorous tagging approach to the chemistry shown in **C.**

identified from the literature and are discussed from a pedagogical point of view. The examples by no means represent a thorough review of the field. Rather, care was taken to identify representative examples to illustrate the retrosynthetic logic and the logistics of constructing a targeted library.

3. HIT TO LEAD

Targeted or focused library synthesis has been most thoroughly developed by the pharmaceutical industry. This is not surprising given that the central scientific platform of most companies lends itself well to the practice. Many, if not most, of the small molecules screened today in industrial settings were originally the products of combinatorial chemistry efforts. Consequently, most of the synthetic planning and analysis for a targeted hit to lead effort has already been accomplished. The job of the chemist involved in making a focused library is then relegated to choosing different building blocks that cover new chemical space and then simply using the previously defined chemistry to assemble the molecules. This is also an emerging theme in academic screening centers, as commercially procured screening sets typically are at least partially populated with products of combinatorial chemistry.

Unfortunately, targeted libraries are rarely this simple, conceptually or logistically, and much more planning is typically necessary. Initially, a retrosynthetic analysis must be undertaken that not only devises a strategy for making the molecule, but also must allow for the incorporation of diversity in key areas to test specific hypotheses as they correlate with activity. Additionally, during this planning stage, a keen eye must be turned toward feasibility. The cost of reagents (which can easily spiral out of control during library realization) in addition to the reaction conditions must be considered.

3.1. Benzodiazepines

Since their serendipitous synthesis in 1954 as potential dyes and subsequent discovery of a unique pharmacologic profile in 1957, benzodiazepines are likely the best example of a lead molecule being reexamined and refined through the synthesis of derivatives and targeted libraries. Since 1954, many reports have devised new synthetic routes, proposed different biological targets, and evaluated a pleiotropic pharmacology, but it was not until 1992 that systematic methodology was developed to make benzodiazepines in a library format [15]. As seen in Figure 5.4, 1,4-benzodiazepines are defined by a bicyclic structure with nitrogen heteroatoms at positions 1 and 4 and a conserved amide functionality at carbon-2. The carbons at the fusion of the bicycle are not numbered, as they cannot be substituted. Two logical disconnections exist in these molecules and are illustrated Panel A. Shown in red, the linkage between positions 4 and 5 is achieved via a condensation to form the imine (or an amide coupling, depending on the structure of the benzodiazepine). Equally intuitive, shown in blue, a disconnection could be made between positions 1 and 2 via an amide coupling. These two disconnections were taken advantage of, and solid-phase library syntheses were reported in four papers from the Ellman Laboratory [15–18]. In 1992 and 1994 the imine disconnection between positions 4 and 5 was exploited in two publications from Bunin et al. [15, 18], and in 1997, an approach utilizing the amide disconnection was reported in two publications by Boojamra et al. [16, 17]. Not only were the reports from Bunin et al. important publications for the benzodiazepine field, but they also represent one of the first combinatorial chemistry reports of nonpeptidyl molecules synthesized on solid support.

In the first approach, shown in Panel B, the benzodiazepine was broken down into three principle components: substituted 2-aminobenzophenones, amino acids, and alkylating agents. The library was ultimately constructed using the Geysen pin apparatus, which was originally designed for peptide epitope mapping [19] but quickly gained popularity in the 1990s because of its inherent encoding and simplification of a parallel solid-phase approach (this is the original technology that eventually matured into the current-day lanterns described in Section 2.3. In this method, the solid support is spatially arrayed into formats so that they could then be dipped into multiwell (96-well in this case) microtiter plates containing reagents. The pins were pred-erivatized with aminoalkyl groups to provide attachment sites for the 4-hydroxymethylphenoxyacetic acid linker (depicted as the support), and they were subsequently able to attach this linker via a benzylic ether to the aryl moiety of the benzophenone. After fluorenylmethyloxycarbonyl (FMOC) protection of the aminobenzophenone, it was loaded onto the resin to begin the library construction. After deprotection of the FMOC group, an amino acid was coupled to the aminobenzophenone using an FMOC-protected amino acid fluoride. The most intuitive way to couple the amino acid to the benzophenone would have been to use standard peptide coupling reagents, as they had been developed to work well on the solid phase, but, as the authors note, these reagents did not work in this instance, and they were forced to develop the acid fluoride approach. The newly incorporated amino acid also incorporated an area of diversity (from the various α-amino acids used), depicted as R3. After deprotection and condensation/cyclization, the amide nitrogen was alkylated for the final diversity step. In total, the authors synthesized 192 total benzodiazepines (spanning two 96-well microtiter plates).

In 1995 and 1997, Boojamra et al. took the approach depicted in Panel C, where the major disconnection was between positions 1 and 2. A major reason for the further development of the methods described in the second set of two papers is that the substitution patterns on the aryl rings (which are critical for many of the pharmacologic properties of benzodiazepines) were limited by the access to derivatized 2-aminobenzophenones. In this example, the R1 functionality could be thoroughly examined through more than forty commercially available anthranillic acids. Furthermore, the chemistry described by Bunin et al. relegated the synthesis of aryl substitution at position 5, whereas this work allowed access to the 1,4-benzodiazipene-2,5-diones. In comparison to the modestly sized 192-member library above, more than 2,500 compounds were synthesized in this example. Also, this study illustrates several key examples of the types of issues that plague transitioning chemistry to the solid phase. Initially, the authors desired to proceed through library realization using the same parallel synthesis pin system described in the Bunin et al. papers but encountered significant difficulty in transitioning their chemistry from the Merrifield resin (with which they had developed the chemistry) to the polyethylene pins. No explanation was given for why this might have occurred, but nevertheless, it points out some of the major logistic hurdles when transitioning from the solution to solid phase. Not only do the reactions need to be optimized for the solid phase, but many times they also need to be individually optimized for different solid phases. Eventually, the authors utilized Merrifield resin and designed their own parallel synthesis apparatus (very similar to the MiniBlock previously described) by using the sawed-off top of a 96-well microtiter plate as a guide for arranging 7-mm glass tubing that had been fitted with a polyethylene frit. They then delivered the reagents to the reaction vessels by submerging them in a 96-well 2-mL microtiter plate. The chemistry is described in Figure 5.4C and begins with the loading of the solid support, which had been prefunctionalized with 4-hydroxy-2,6-dimethoxybenzaldehyde via a reductive amination of several α-amino esters with $NaBH(OAc)_3$. During this step, the authors had observed that they could control whether or not the enantiomerically pure α-amino esters would racemize under the reaction conditions. If the resultant imines were immediately reduced, they would retain the chirality at the α-position. However, if the imine was allowed to equilibriate for three hours, complete racemization was observed. For the library, they decided to racemize the α-amino esters to gain access to a larger number of molecules. Acylation of the resultant secondary amine with several commercially available anthranilic acids provided the tertiary amide. The authors reported that although this was anticipated to be a simple amide coupling of a secondary amine, it required extensive optimization. They were able to identify carbodiimides as the only reagents that could complete the transformation and further found that only EDC (1-ethyl-3-[3-(dimethylamino)-propyl]carbodiimide HCl) gave satisfactory results. For the lactamization/ring closure, they found in solution-state pilot studies that a base-catalyzed route provided the most general route to the benzodiazepine. They further reasoned that if they could perform the lactamization under sufficiently basic conditions, they would also be able to alkylate the 4-position nitrogen at the same time. During library realization, they were able to incorporate nineteen α-amino esters (nine racemized enantiomeric pairs plus the methyl ester of glycine), twelve substituted anthranilic acids, and eleven alkylating agents, totaling 2,508 compounds.

In 2007, work on benzodiazepine chemistry was published again, this time arising from a small company developing the fluorous tags discussed in section 2.2. In this example, instead of the 4-hydroxy-2,6-dimethoxybenzaldehyde being coupled to the solid phase, a fluorous tag was appended to the phenolic position. The authors utilized fluorous solid-phase extraction cartridges to work up and purify each reaction. Although not an extensive library realization example (nine benzodiazepines were synthesized), this highlights proof of principle for the fluorous tagging and illustrates the diversity of approaches available.

3.2. Styryl-Based Fluorophores

Whereas the benzodiazepine is clearly an example of a targeted library with the goal of optimizing the biological activity of a molecule, the second example deals with using a targeted library to optimize the activity of a lead molecule, but not toward any particular biological target. In a publication from the Chang laboratory, the authors decided to study the ability of several styryl-based molecules to act as organelle-specific fluorescent dyes [20, 21]. The logic behind the endeavor was the observation (or, in the context of this chapter, a "hit") that an entire class of commercially available styryl dyes specifically targeted the mitochondria [22]. Styryl dyes had been utilized based on the inherent electrochemical potential across the mitochondrial inner membrane and the prediction that lipophilic cations would accumulate there [23]. The hypothesis that the authors then put forth was that several classes of biological molecules would potentially have inherent affinity for these library memers based on the ionic attraction (such as DNA) and that a thorough examination of the different chemical structures could confer different physiochemical features and consequently different subcellular localization.

Library construction for this idea was relatively straightforward. Two types of building blocks were selected, and the library was constructed in one step using solution-phase parallel techniques. To illustrate different chemical approaches to making libraries, the authors utilized the emerging technology of microwave irradiation to accelerate reaction rates and drive the transformation to completion. The general synthetic scheme is shown in Figure 5.5 and depicts the condensation of a 2- or 4- substituted N-substituted pyridinium salts containing an aromatic aldehyde. In a 2003 publication, the authors describe the synthesis of a 574-member library (building blocks shown in Figure 5.5) [20] and then subsequently expanded on this work in a 2006 publication in which a 1,336-member library was constructed [21]. The reactions were all carried out in solution phase with pyrrolidine and microwave irradiation. The products were characterized for purity using liquid chromatography mass spectrometry (LCMS) and subsequently tested for inherent fluorescence in microtiter plates and then in live cell imaging (shown in Figure 5.5). In the 2003 report, several organelle-specific dyes were discovered, and in the 2006 report, they were able to further show that they could specifically target RNA and study RNA distribution with live cell imaging.

4. NATURAL PRODUCT–INSPIRED LIBRARIES

The appeal in the drug discovery arena of focused library development is obvious. The natural products field is abundant with examples of molecules that have either dissected a biological process or developed new classes of drugs. From colchicine being used to identify microtubules to the discovery of morphine and the opiate drugs, natural products have played a dominant role in the history (and likely the future) of chemical biology. This point can be well illustrated by carefully examining a review of natural products as drugs presented by Newman et al. in which the authors showed that the majority of drugs targeted to either cancer or infectious disease (70% and 80%, respectively) had their origins in natural products [24].

From a purely chemical point of view, natural products, or secondary metabolites, can well be regarded as the pinnacle of small molecule chemical library design. The synthesis of a single natural product through traditional solution-phase target-oriented synthesis using defined chemical transformations thoroughly developed over many years can be considered an incredible feat. Therefore, when considering the synthesis of a library of natural products, or natural product–like molecules, the chemistry dictates highly refined methods and ingenuity to be able to synthesize not just one natural product, but an entire library of them. For the organization of this section, two key examples have been identified that encompass the idea of natural products being the basis for a focused library; additionally, the first example represents the exploitation of the guanidine functional group. This represents an area that likely will be emerging in the field of focused library development: libraries designed to test structure-activity questions of a specific functional group.

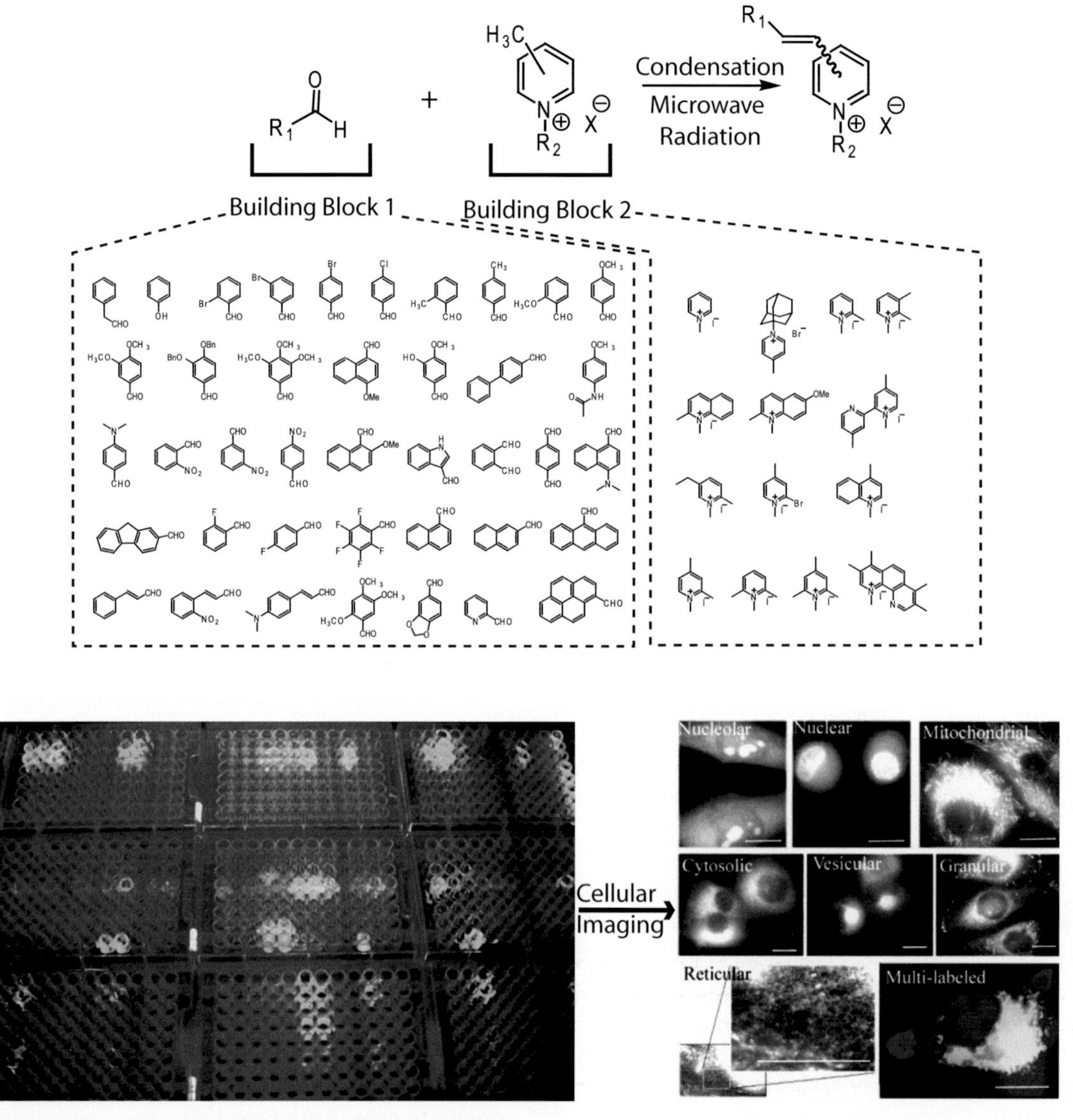

***FIGURE 5.5:** A library of styryl-based fluorescent dyes.*

4.1. Guanidine Recognition

In the context of small molecule–protein recognition, guanidine is important in the recognition of active-site aspartate and glutamate residues. As an example, this binding interaction is displayed for transition-state stabilization as seen in chorismate mutase, in which both Arg7 and Arg116 make two-point hydrogen bond interactions with the carboxylates of prephenic acid (Figure 5.6A) [25]. Conversely, the guanidinium ion of peptidylarginine interacts with protein active-site aspartate with four-point hydrogen bonding. This is seen in the recognition of histone arginines by peptidylarginine deiminase IV (PAD4) (Figure 5.6B) [26]. This is also an important recognition motif in the coagulation cascade and has been the basis for the extensive development of thrombin inhibitors [27]. Because of the importance and frequency of this enzyme-substrate recognition motif, it has been extensively mimicked by natural products and unnatural pharmacophores for medicinal applications [28]. Several complex, polycyclic guanidine-containing natural products have been shown to replicate this binding motif. For example, saxitoxin (STX, **2**) exerts its neurotoxic effects by binding to voltage-gated sodium channels [29]. Subsequent site-directed mutagenesis studies indicated that a single point mutation at Glu387 drastically decreased the sensitivity of these channels to STX [30]. Again, this suggests an important guanidinium hydrogen bond–donor interaction (Figure 5.6D).

The batzelladines (illustrated by batzelladine F[**1**]) inhibit human immunodeficiency virus (HIV)-1 envelope mediated cell-cell fusion by inhibiting gp120-CD4 binding, aided by hydrogen bonding of the polycyclic guanidine core to Asp368 on gp120 (Figure 5.6C) [31]. The structural uniqueness of the batzelladine alkaloids and the

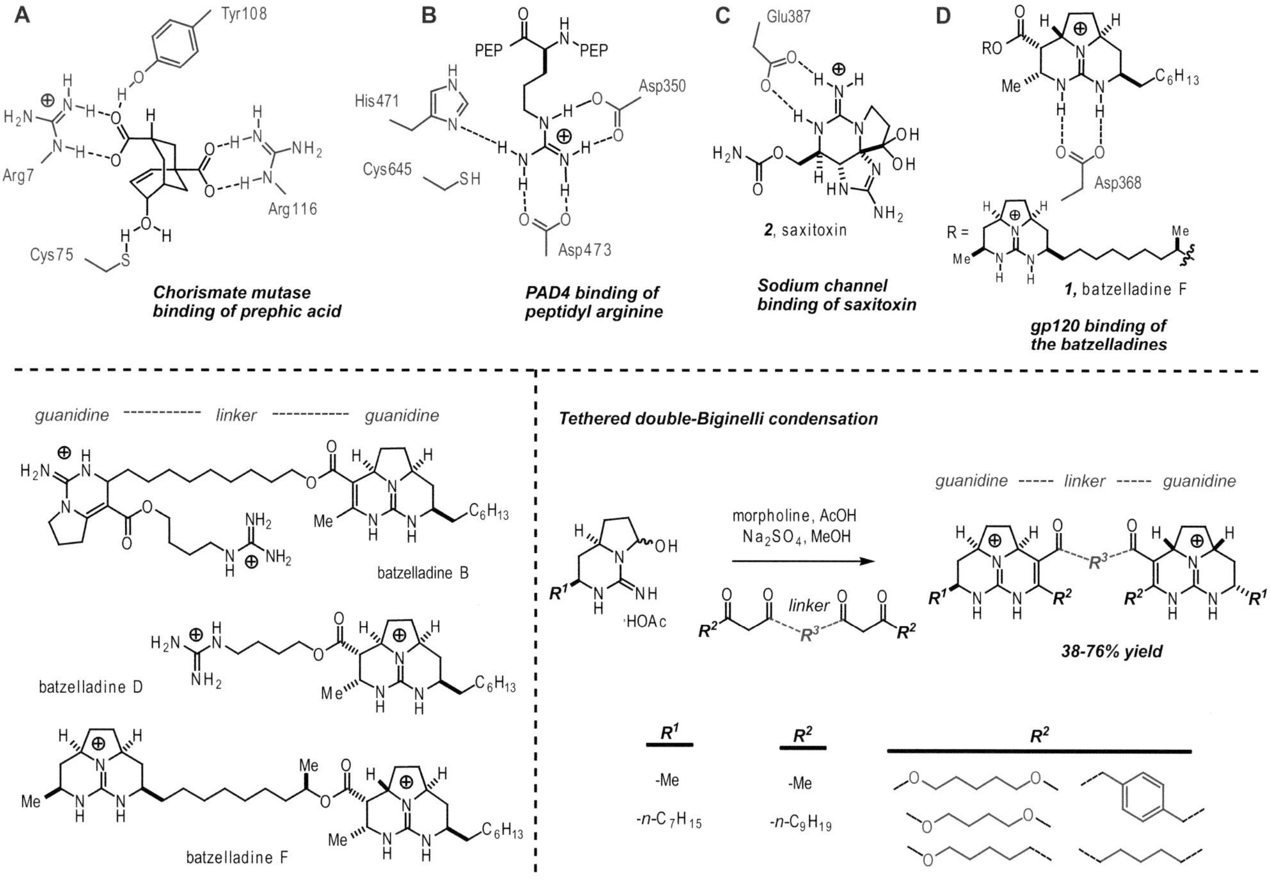

FIGURE 5.6: ***A-D.*** ***Guanidine recognition and a library of batzelladine-like molecules.***

realization that this alkaloid family has a propensity to inhibit protein-protein interactions, a rare activity for natural products, have made them targets for the construction of targeted libraries. Given the technical difficulties of handling guanidines, the fact that they likely will require purification by high-performance liquid chromatography (HPLC) methods, and the structural complexity of the target compounds, complex guanidine analogs are frequently pursued through solution-phase parallel synthesis. Overman has developed several approaches to the batzelladine cores using a tethered Biginelli condensation. This condensation relies on the addition of a β-ketoester or a *β* -diketone to a guanidine-iminium ion generated in situ from the hemiaminal. Depending on the exact nature of the guanidine precursor, access to both the *trans*- and *cis*-fused five-membered ring of the tricyclic guanidine cores can be realized (as present in batzelladine F) [32]. Noting that most of the batzelladine alkaloids are composed of a tricyclic guanidinium nucleus tethered to a second guanidine, which can be tri-, bi-, or acyclic, they reasoned that a simplified pharmacophore model might consist of a guanidine-linker-guanidine motif [32]. Using an extension of their methodology, the tethered double-Biginelli condensation, they were able to generate a series of *C*2-symmetric batzelladine analogs by performing the reaction on β-ketoesters linked by a series of alkyl- or aryl-containing spacers. This synthetic approach generated a series of approximately thirty complex analogs. Structure-activity relationships (SARs) for the ability of these analogs to inhibit HIV-1 envelope-mediated cell fusion were quite clear and resulted in the identification of structural variants able to inhibit HIV entry with IC_{50}s ranging from 0.8 to 3.0 μM.

4.2. Benzopyran-Inspired Libraries

To date, one of the greatest success stories in natural product–inspired libraries is derived from three back-to-back papers from the Nicolaou laboratory in 2000 describing the synthesis of a ten-thousand-member library of benzopyran natural product–like molecules [33–35]. In this report, the authors postulated that the benzopyran moiety should be considered *privileged* such that it would allow chemical libraries with this functionality the ability to modulate multiple biological pathways. This idea of a *privileged structure* was first put forth in 1988, when Evans et al. [36] (and later reviewed [24]) described how specific modifications of the benzodiazepine structure could expand the pharmacology from the known anxiolytic properties to new activities as cholecystokinin antagonists.

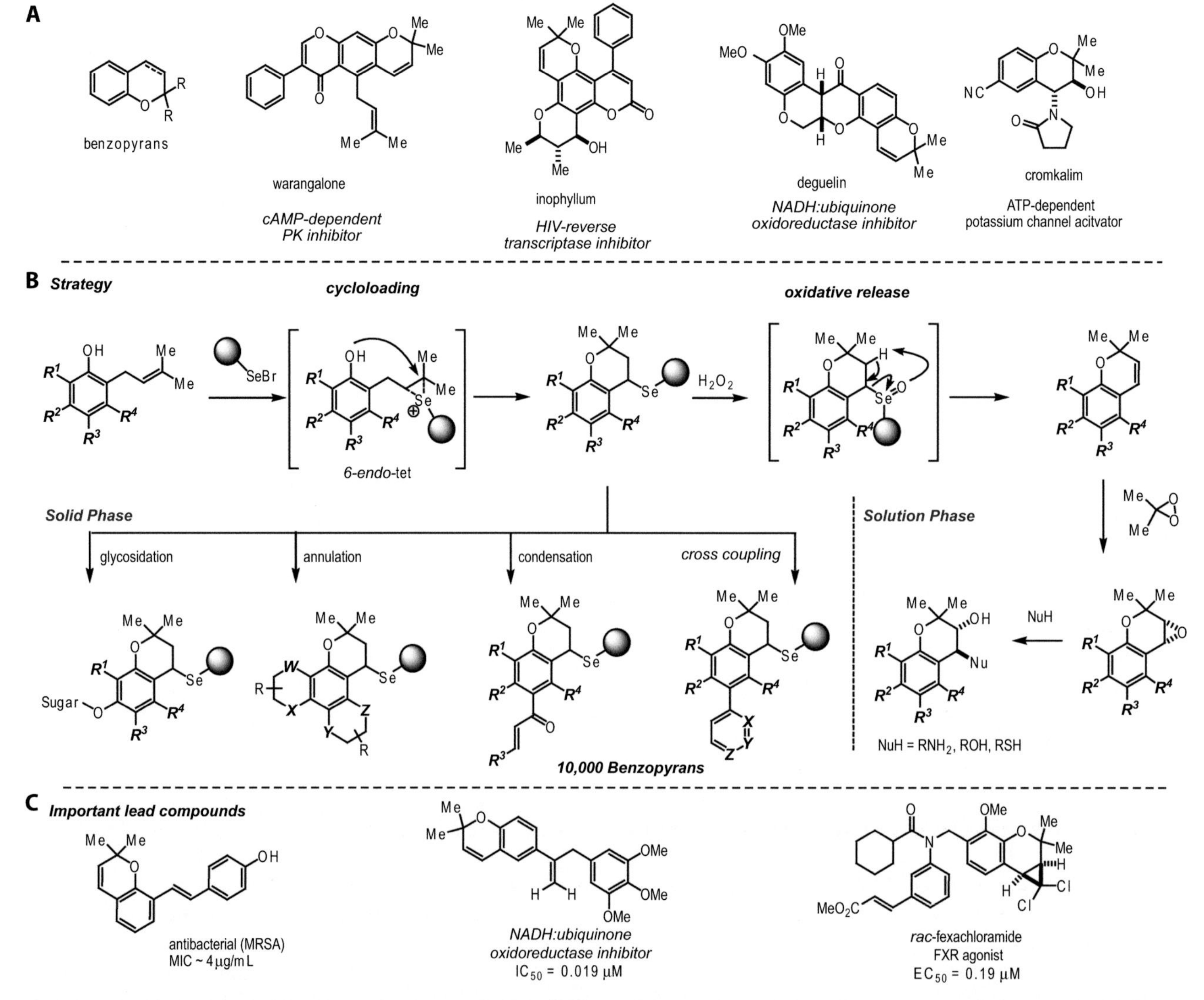

FIGURE 5.7: ***A.*** ***Key examples of the occurrence of the benzopyran functionality in natural products and pharmaceuticals.*** ***B.*** ***Strategy for the construction of benzopyran-containing targeted library.*** ***C.*** ***Biologically active molecules from the library efforts in B.***

The argument for benzopyrans being considered privileged was logical when considering the diverse biological activities of various benzopyran-containing natural products (Figure 5.7A). Four examples of natural products containing benzopyrans with diverse biological activities are shown, but the authors were able to identify more than four thousand such examples of natural products and pharmaceutical ligands with activity in a wide array of biological systems. The ultimate test of their hypothesis therefore was to synthesize a library of benzopyran-containing molecules and test their activity against a broad swath of biological targets.

To develop their chemistry, the authors wanted to devise chemistry whereby the loading and/or cleavage operations would further diversify their intermediates with new substituents. Their reasoning for this was intuitive, as it would ultimately minimize the number of extraneous steps that were not directly related to developing further diversity in the library. The chemistry began by loading a substituted *ortho*-prenylated phenol onto a polystyrene-based seleneyl bromide resin that had been previously reported [37]. The benzopyran would then be formed via a precedented intramolecular 6-*endo*-trig cyclization. At this point, the benzopyran would be tethered to the solid support via a selenoether. The library was then elaborated; in essence, four separate focused libraries were synthesized in which the aryl moiety was substituted using glycosidation, annulation, condensation, and cross-coupling reactions. This was not the entirety of the diversifying steps, however. The step of cleaving from the resin was via the oxidation of the selenoether to the selenoxide, which they found to spontaneously undergo *syn*-elimination at 25°C. Considering that one of the initial goals of the loading/cleavage steps would contribute to the target structure, the cleavage step was key because it provided not only the unsaturated benzopyran but also the opportunity for further diversification in the solution phase through manipulations of the carbon-carbon π-bond. In the third publication of the series, the authors discussed how the unsaturation could be exploited via an epoxidation and subsequent nucleophilic ring opening. In the end, a ten-thousand-member library was constructed using the IRORI Kan system.

Constructed on the premise that the benzopyran represents a privileged pharmacophore, it is not surprising that these synthetic activities led to the identification of several important small molecules, as highlighted in Figure 5.7C. As is often the case, these targeted libraries inspire the construction of second-generation targeted libraries to maximize the specific activity of leads identified after primary screening. This is particularly true in those examples in which follow-up chemistry has developed highly advanced medicinal candidates, which is unusual for academic research. The first compound was arrived at after identification of library members capable of inhibiting the growth of methicillin-resistant *Staphylococcus aureus* (minimum inhibitory concentration [MIC] <50 μg/mL) [38]. Structural refinement generated this significantly more active lead displaying an MIC of 4 μg/mL. The second compound was identified as a potent inhibitor of NADH:ubiquinone oxidoreductase complex with an IC_{50} of 0.019 μM [39]. Because this mitochondrial oxidoreductase complex is used to reduce molecular oxygen during oxidative phosphorylation, this agent is capable of halting cellular proliferation without nonspecific cytotoxic side effects, representing an advanced lead for chemotherapeutic/chemopreventive therapeutics. The third compound, *rac*-fexachloramide, was developed as a potent agonist of the farnesoid X receptor (FXR), a transcriptional sensor for bile acids ultimately controlling cholesterol metabolism. Initial screening of the benzopyran library identified several candidate agonists with EC_{50}s in the 5- to 10-μM range. These hits were systematically optimized employing both parallel solution-phase and solid-phase synthesis techniques to provide several lead compounds that potently activate FXR, including *rac*-fexachloramide (IC_{50} = 0.19 μM) [40, 41].

5. PROTEIN CLASS–TARGETED LIBRARIES

The rational design of an inhibitor or library of inhibitors for a given class of proteins is a nontrivial endeavor. Contemplating the synthesis of an entire library of molecules directed at a particular protein class must be approached carefully. Very few examples exist of entire protein classes that have conserved small molecule binding pockets. An illustrative example of this point is the nuclear receptor class of proteins. These proteins are ideal targets for drug development (and are key targets in the pharmaceutical industry), but the design of a library of molecules targeted generally at the nuclear receptor class of proteins would be extraordinarily difficult, if not impossible. The difficulty lies in the fact that, that although there is a binding pocket that upon ligand binding induces a transcriptional cascade specific for each family member, each individual protein displays exquisite affinity for specific small molecules. This is not to say that libraries could not be developed for each individual member. These endeavors have already been shown to be successful with the extensive pharmacology known for some individual family members. Taken together, when considering a focused library around a specific class of proteins, a key analysis must be made to identify a specific protein or class of proteins that will have conserved ligand-binding motifs.

Two general approaches have been taken to achieve protein class–targeted libraries. In the first, a conserved protein motif is exploited, and libraries of compounds are designed around this conserved motif. The first illustrative example we present shows how a conserved enzyme cofactor binding pocket (in this case, adenosine triphosphate [ATP]) can be exploited for the design and synthesis of a small molecule inhibitor. This is a useful model and does not always have to utilize a cofactor. In the case of histone deacetylase enzymes, a conserved active site can be targeted with a properly constructed hydroxamic acid–containing molecule using the elegant structural work with trichostatin A and suberoylanilide hydroxamic acid [42, 43]. The second general approach is to perform a post hoc analysis of all of the types of molecules that bind to a given class of proteins, provided a large enough data set is available. In the second illustrative example, an approach is briefly outlined to define a focused library of molecules using in silico analysis of the MDDR database.

5.1. Kinase-Targeted Libraries

ATP is a key cofactor in many enzymatic processes (Figure 5.8). Protein kinases are the largest family of proteins in humans. Their activity hinges on the utilization of ATP to transfer phosphate groups to other proteins. Typically, they catalyze the phosphorylation of threonine, serine, or tyrosine residues as part of an incredibly complex and dynamic network that propagates signal transduction within the cell. Among others, they play key roles in regulating metabolism, differentiation, and apoptosis. Dysregulation of kinase activity is commonly implicated in a number of disease states, including cancer, diabetes, and inflammation. Targeted inhibition of these protein kinases has thus become an attractive therapeutic strategy. The purine ring system incorporated as the heterocyclic base in ATP is a key structural element for ligand recognition and has served as the inspiration for the design of many targeted libraries. As one can imagine, the ATP biding site of enzymes is highly conserved and presents a daunting challenge to develop agents capable of *selectively* modulating an ATP-dependent

competetive ATP-binding kinase inhibitors

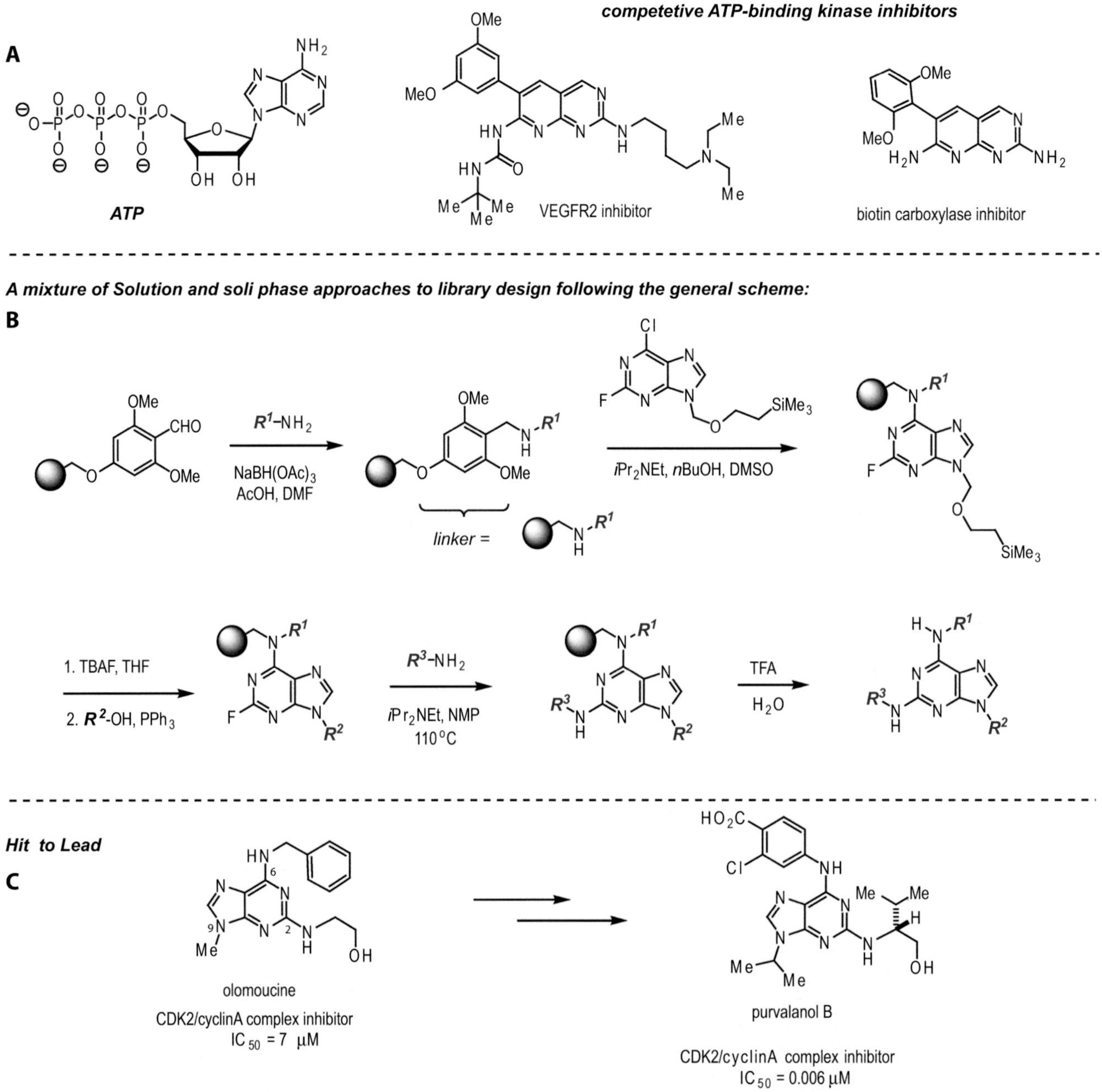

A

ATP

VEGFR2 inhibitor

biotin carboxylase inhibitor

A mixture of Solution and soli phase approaches to library design following the general scheme:

B

R^1-NH_2

$NaBH(OAc)_3$
AcOH, DMF

linker =

iPr_2NEt, $nBuOH$, DMSO

1. TBAF, THF
2. R^2-OH, PPh_3

R^3-NH_2
iPr_2NEt, NMP
110°C

TFA
H_2O

Hit to Lead

C

olomoucine
CDK2/cyclinA complex inhibitor
IC_{50} = 7 μM

purvalanol B

CDK2/cyclinA complex inhibitor
IC_{50} = 0.006 μM

***FIGURE 5.8:* A. The conserved structure of ATP and two competitive binders. B. A purine-based library of possible kinase inhibitors. C. Optimization of the lead natural product to a potent specific kinase inhibitor.**

process by competitively binding the ATP recognition site.

The natural metabolite olomoucine, shown in Figure 5.8, was found to inhibit several cyclin-dependent kinases (CDKs) at micromolar concentrations. For example, it is a competitive inhibitor of CDK2/cyclin A with an IC_{50} of 7 μM [44]. Because cyclin-dependent kinases, including CDK2, regulate important cell cycle checkpoints including DNA replication and cellular division, they have stimulated intense efforts to develop potent and selective inhibitors as chemotherapeutics. The structural basis for olomoucine-CDK2 binding was revealed by x-ray crystallography. [45] This structure revealed that the purine base occupies the adenine binding pocket of ATP; however, the ring is oriented in a plane orthogonal to that of ATP. Schultz and co-workers used this structure to plan the development of a targeted library that would explore diversification of olomoucine at C2, C6, and C9 with the aim of developing more potent and selective inhibitors of the CDK2/cyclin A complex.

They developed several approaches to these derivatives that relied on both solid- and solution-phase strategies to explore this three-point diversification [46, 47]. The general outline of their strategy is presented in Figure 5.8.

Their solid-phase strategy relied on the use of an acid labile rink linker resin arrayed in 96-well microtiter plates. The first point of diversification at C6 was realized by reductive amination to introduce R1. Nucleophilic aromatic substitution on the chloropurine, which was prepared in solution, gave the resin bound nucleoside. Deprotection of the silyl protecting group gave the free N9 intermediate. By virtue of its relative acidity, N9 is readily alkylated under Mitsunobu conditions, thus introducing a variety of substituents (R2) at this position. A spatially separated library of analogs was also prepared by solution-phase chemistry (not shown) that introduces substituents at this position, which cannot be introduced via the Mitsunobu protocol. A final nucleophilic aromatic substitution reaction on the 2-fluoro position with a primary amine at relatively high temperatures introduces the last point of diversity (R3). Release of the fully derivatized purine from the Rink linker was then accomplished using aqueous trifluoroacetic acid. This strategy enabled the development of a 348-member library, which was evaluated for anti-CDK2 activity using a microtiter-based assay for protein kinase activity. A hit was identified ($IC_{50} = 0.6\ \mu M$) in this library, which was an order of magnitude more potent than the parent natural product omoloucine.

Second-generation targeted libraries based on this new hit structure led to the discovery of purvalanol B, which was shown to be a highly potent inhibitor of the human CDK2-cyclin A with an IC_{50} of 0.006 μM. More importantly, purvalanol B was shown to be highly selective for the inhibition of CDK2 when profiled against twenty-two other highly related human kinases.

5.2. Focused Libraries from Database Searching

To this point, this chapter has focused on chemical approaches to synthesizing focused libraries. However, not all researchers are going to have access to chemists who are capable of library synthesis, nor will all researchers be able to garner enough enthusiasm to put in the effort required for a successful focused library realization. In the final example, we point to a paper from Schnur et al. [48] that describes the assembly of a targeted library from databases of available ligands. The scientific focus of this paper is to determine if previously defined privileged structures (for protein classes such as GPCRs, nuclear receptors, and serine proteases) were truly privileged or simply represented classes of molecules that were more bioavailable or possibly promiscuous protein binders. Regardless of the findings, the authors present nice methodology for searching the MDDR with computational tools such as ClassPharmer. Briefly, the authors systematically searched the MDDR individually to identify classes of molecules that seemed targeted to specific classes. This approach has been well developed and is discussed, but here the authors took their work a step further and subsequently compared the identified classes of molecules to each other. The approaches described could potentially be a viable alternative, enabling individuals to construct targeted libraries from commercially available compounds.

REFERENCES

1. Houghten, R. A. (1985). General method for the rapid solid-phase synthesis of large numbers of peptides: Specificity of antigen-antibody interaction at the level of individual amino acids. *Proc Natl Acad Sci U S A 82*, 5131–5135.
2. Tan, D. S., Foley, M. A., Stockwell, B. R., Shair, M. D., and Schreiber, S. L. (1999). Synthesis and preliminary evaluation of a library of polycyclic small molecules for use in chemical genetic assays. *J Am Chem Soc 121*, 9073–9087.
3. Tan, D. S., Foley, M. A., Shair, M. D., and Schreiber, S. L. (1998). Stereoselective synthesis of over two million compounds having structural features both reminiscent of natural products and compatible with miniaturized cell-based assays. *J Am Chem Soc 120*, 8565–8566.
4. Gooding, O., Hoeprich, P. D., Jr., Labadie, J. W., Porco, J. A., Jr., van Eikeren, P., and Wright, P. (1996). Boosting the productivity of medicinal chemistry through automation tools: novel technological developments enable a wide range of automated synthetic procedures. *Mol Diversity Comb Chem: Libr Drug Discovery*, 199–206.
5. DeWitt, S., Sherblom, J., and Bristol, J. (1996). Press release.
6. Weller, H. N., Rubin, A. E., Moshiri, B., Ruediger, W., Li, W.-J., Allen, J., Nolfo, J., Bertok, A., and Rosso, V. W. (2005). Development and commercialization of the MiniBlock synthesizer family: a historical case study. *JALA 10*, 59–71.
7. Studer, A., Hadida, S., Ferritto, R., Kim, S.-Y., Jeger, P., Wipf, P., and Curran, D. P. (1997). Fluorous synthesis: a fluorous-phase strategy for improving separation efficiency in organic synthesis. *Science 275*, 823–826.
8. Gladysz, J. A., and Curran, D. P. (2002). Fluorous chemistry: from biphasic catalysis to a parallel chemical universe and beyond. *Tetrahedron 58*, 3823–3825.
9. Zhang, W., and Curran, D. P. (2006). Synthetic applications of fluorous solid-phase extraction (F-SPE). *Tetrahedron 62*, 11837–11865.
10. Gladysz, J. A., Curran, D. P., Horvath, I. T., eds. (2004). *Handbook of Fluorous Chemistry* (Hoboken, N.J.: Wiley).
11. Merrifield, R. B. (1963). Solid phase peptide synthesis. I. The synthesis of a tetrapeptide. *J Am Chem Soc 85*, 2149–2154.
12. Haag, R., Hebel, A., and Stumbe, J.-F. (2002). Solid phase and soluble polymers for combinatorial synthesis. In *Handbook of Combinatorial Chemistry*, vol. 1, K. C. Nicolaou, R. Hanko and W. Hartwig, eds. (Weinheim: Wiley-VCH), 24–58.
13. Brase, S., and Dahmen, S. (2002). Linkers for solid-phase synthesis. In *Handbook of Combinatorial Chemistry*, vol. 1, K. C. Nicolaou, R. Hanko, and W. Hartwig, eds. (Weinheim: Wiley-VCH), 59–169.
14. Tallarico, J. A., Depew, K. M., Pelish, H. E., Westwood, N. J., Lindsley, C. W., Shair, M. D., Schreiber, S. L., and Foley, M. A. (2001). An alkylsilyl-tethered, high-capacity solid support amenable to diversity-oriented synthesis for one-bead, one-stock solution chemical genetics. *J Comb Chem 3*, 312–318.

15. Bunin, B. A., and Ellman, J. A. (1992). A general and expedient method for the solid-phase synthesis of 1,4-benzodiazepine derivatives. *J Am Chem Soc 114*, 10997–10998.
16. Boojamra, C. G., Burow, K. M., and Ellman, J. A. (1995). An expedient and high-yielding method for the solid-phase synthesis of diverse 1,4-benzodiazepine-2,5-diones. *J Org Chem 60*, 5742–5743.
17. Boojamra, C. G., Burow, K. M., Thompson, L. A., and Ellman, J. A. (1997). Solid-phase synthesis of 1,4-benzodiazepine-2,5-diones. Library preparation and demonstration of synthesis generality. *J Org Chem 62*, 1240–1256.
18. Bunin, B. A., Plunkett, M. J., and Ellman, J. A. (1994). The combinatorial synthesis and chemical and biological evaluation of a 1,4-benzodiazepine library. *Proc Natl Acad Sci U S A 91*, 4708–4712.
19. Geysen, H. M., Rodda, S. J., Mason, T. J., Tribbick, G., and Schoofs, P. G. (1987). Strategies for epitope analysis using peptide synthesis. *J Immunol Methods 102*, 259–274.
20. Rosania, G. R., Lee, J. W., Ding, L., Yoon, H.-S., and Chang, Y.-T. (2003). Combinatorial approach to organelle-targeted fluorescent library based on the styryl scaffold. *J Am Chem Soc 125*, 1130–1131.
21. Li, Q., Kim, Y., Namm, J., Kulkarni, A., Rosania, G. R., Ahn, Y.-H., and Chang, Y.-T. (2006). RNA-selective, live cell imaging probes for studying nuclear structure and function. *Chem Biol 13*, 615–623.
22. Haugland, R. (2001). *Handbook of Fluorescent Probes and Research Chemicals* (Eugene, Ore.: Molecular Probes).
23. Chen, L. B. (1988). Mitochondrial membrane potential in living cells. *Annu Rev Cell Biol 4*, 155–181.
24. Patchett, A. A., and Nargund, R. P. (2000). Privileged structures – an update. *Annu Rep Med Chem 35*, 289–298.
25. Lee, A. Y., Karplus, P. A., Ganem, B., and Clardy, J. (1995). Atomic structure of the buried catalytic pocket of Escherichia coli chorismate mutase. *J Am Chem Soc 117*, 3627–3628.
26. Arita, K., Hashimoto, H., Shimizu, T., Nakashima, K., Yamada, M., and Sato, M. (2004). Structural basis for Ca2+-induced activation of human PAD4. *Nat Struct Mol Biol 11*, 777–783.
27. Masic, L. P. (2006). Arginine mimetic structures in biologically active antagonists and inhibitors. *Curr Med Chem 13*, 3627–3648.
28. Sullivan, J. D., Giles, R. L., and Looper, R. E. (2009). 2-Aminoimidazoles from *Leucetta* sponges: synthesis and biology of an important pharmacophore. *Curr Bioact Compd 5*, 39–78.
29. Anger, T., Madge, D. J., Mulla, M., and Riddall, D. (2001). Medicinal chemistry of neuronal voltage-gated sodium channel blockers. *J Med Chem 44*, 115–137.
30. Noda, M., Suzuki, H., Numa, S., and Stuehmer, W. (1989). A single point mutation confers tetrodotoxin and saxitoxin insensitivity on the sodium channel II. *FEBS Lett 259*, 213–216.
31. Bewley, C. A., Ray, S., Cohen, F., Collins, S. K., and Overman, L. E. (2004). Inhibition of HIV-1 envelope-mediated fusion by synthetic batzelladine analogues. *J Nat Prod 67*, 1319–1324.
32. Cohen, F., Collins, S. K., and Overman, L. E. (2003). Assembling polycyclic bisguanidine motifs resembling batzelladine alkaloids by double tethered Biginelli condensations. *Org Lett 5*, 4485–4488.
33. Nicolaou, K. C., Pfefferkorn, J. A., Barluenga, S., Mitchell, H. J., Roecker, A. J., and Cao, G. Q. (2000). Natural product-like combinatorial libraries based on privileged structures. 3. The "libraries from libraries" principle for diversity enhancement of benzopyran libraries. *J Am Chem Soc 122*, 9968–9976.
34. Nicolaou, K. C., Pfefferkorn, J. A., Roecker, A. J., Cao, G. Q., Barluenga, S., and Mitchell, H. J. (2000). Natural product-like combinatorial libraries based on privileged structures. 1. General principles and solid-phase synthesis of benzopyrans. *J Am Chem Soc 122*, 9939–9953.
35. Nicolaou, K. C., Pfefferkorn, J. A., Mitchell, H. J., Roecker, A. J., Barluenga, S., Cao, G. Q., Affleck, R. L., and Lillig, J. E. (2000). Natural product-like combinatorial libraries based on privileged structures. 2. Construction of a 10 000-membered benzopyran library by directed split-and-pool chemistry using NanoKans and optical encoding. *J Am Chem Soc 122*, 9954–9967.
36. Evans, B. E., Rittle, K. E., Bock, M. G., DiPardo, R. M., Freidinger, R. M., Whitter, W. L., Lundell, G. F., Veber, D. F., Anderson, P. S., Chang, S. L., Lotti, V.J., Cerino, D. J., Chen, T. B., Kling, P. J., Kunkel, K. A., Springer, J. P., Hirshfield, J. (1988). Methods for drug discovery: development of potent, selective, orally effective cholecystokinin antagonists. *J Med Chem 31*, 2235–2246.
37. Nicolaou, K. C., Pastor, J., Barluenga, S., and Winssinger, N. (1998). Polymer-supported selenium reagents for organic synthesis. *Chem Commun*, 1947–1948.
38. Nicolaou, K. C., Roecker, A. J., Barluenga, S., Pfefferkorn, J. A., and Cao, G. Q. (2001). Discovery of novel antibacterial agents active against methicillin-resistant Staphylococcus aureus from combinatorial benzopyran libraries. *Chembiochem 2*, 460–465.
39. Nicolaou, K. C., Pfefferkorn, J. A., Schuler, F., Roecker, A. J., Cao, G. Q., and Casida, J. E. (2000). Combinatorial synthesis of novel and potent inhibitors of NADH:ubiquinone oxidoreductase. *Chem Biol* 7, 979–992.
40. Nicolaou, K. C., Evans, R. M., Roecker, A. J., Hughes, R., Downes, M., and Pfefferkorn, J. A. (2003). Discovery and optimization of non-steroidal FXR agonists from natural product-like libraries. *Org Biomol Chem 1*, 908–920.
41. Downes, M., Verdecia, M. A., Roecker, A. J., Hughes, R., Hogenesch, J. B., Kast-Woelbern, H. R., Bowman, M. E., Ferrer, J.-L., Anisfeld, A. M., Edwards, P. A., Rosenfeld, J. M., Alvarez, J. G. A., Noel, J. P., Nicolaou, K. C., and Evans, R. M. (2003). A chemical, genetic, and structural analysis of the nuclear bile acid receptor FXR. *Mol Cell 11*, 1079–1092.
42. Richon, V. M., Emiliani, S., Verdin, E., Webb, Y., Breslow, R., Rifkind, R. A., and Marks, P. A. (1998). A class of hybrid polar inducers of transformed cell differentiation inhibits histone deacetylases. *Proc Natl Acad Sci U S A 95*, 3003–3007.
43. Finnin, M. S., Donigian, J. R., Cohen, A., Richon, V. M., Rifkind, R. A., Marks, P. A., Breslow, R., and Pavletich, N. P. (1999). Structures of a histone deacetylase homologue bound to the TSA and SAHA inhibitors. *Nature 401*, 188–193.
44. Vesely, J., Havlicek, L., Strnad, M., Blow, J. J., Donella-Deana, A., Pinna, L., Letham, D. S., Kato, J.-Y., Detivaud, L., Leclerc, S., Meijer, L. (1994). Inhibition of cyclin-dependent kinases by purine analogs. *Eur J Biochem 224*, 771–786.

45. Schulze-Gahmen, U., Brandsen, J., Jones, H. D., Morgan, D. O., and Meijer, L. (1995). Multiple modes of ligand recognition: crystal structures of cyclin-dependent protein kinase 2 in complex with ATP and two inhibitors, olomoucine and isopentenyladenine. *Proteins Struct Funct Genet 22*, 378–391.
46. Norman, T. C., Gray, N. S., Koh, J. T., and Schultz, P.G. (1996). A structure-based library approach to kinase inhibitors. *J Am Chem Soc 118*, 7430–7431.
47. Gray, N. S., Wodicka, L., Thunnissen, A. M., Norman, T. C., Kwon, S., Espinoza, F. H., Morgan, D. O., Barnes, G., LeClerc, S., Meijer, L., Kim, S.-H., Lockhart, D. J., and Schultz, P. G. (1998). Exploiting chemical libraries, structure, and genomics in the search for kinase inhibitors. *Science 281*, 533–538.
48. Schnur, D. M., Hermsmeier, M. A., and Tebben, A. J. (2006). Are target-family-privileged substructures truly privileged? *J Med Chem 49*, 2000–2009.

CHAPTER 6

Fragment-Based Ligand Discovery

Sandra Bartoli
Antonella Squarcia
Daniela Fattori

"Reversible molecular interactions are at the heart of the dance of life." Lubert Stryer

Biochemical systems, their communication pathways, and the related transformations that take place are based on molecular interactions, so we may safely say that these regulate life at a molecular level. For basic science studies or for therapeutic purposes, we can perturb these systems with chemical manipulations. Because of the accumulated knowledge in the fields of organic chemistry and molecular recognition, our investigative instruments have become increasingly powerful. During the last fifty years, there has been a continuous evolution in chemical approaches to interact with biological systems, and fragment-based drug discovery can be considered one of the products of this evolution [1]. In short, fragment-based ligand discovery (FBLD) may be described as the search for a ligand for a macromolecule, which may constitute a lead for a medicinal chemistry program, through the use of very small molecules. This approach, as we will see, may provide a number of advantages over the classical approaches together with the logical consequence that the observed affinities for a good lead compound will fall in the micromolar-millimolar range rather than in the nanomolar range, necessitating that chemists and biologists leave behind the high-affinity paradigm.

This chapter discusses in detail the historical background, key concepts, and basis for the FBLD approach. An illustration of the technology involved will follow, together with a selection of practical and successful applications.

1. HISTORICAL BACKGROUND

Today in the pharmaceutical industry, the most commonly used drug discovery strategy is the target-based one. On the basis of previous studies, a molecular target is proposed for a certain disease, which is then validated by verifying that the interaction of a chemical entity (protein or small molecule) with the target causes the expected biological effect. Once these steps have been more or less accomplished, a screening program can start with the goal of identifying a lead compound, which, once selected, is submitted to a series of modifications to improve its pharmacodynamic and pharmacokinetic properties (Figure 6.1). During the 1980s, drug design and computational techniques emerged, thanks also to impressive improvements in computer science, which were followed in the 1990s by an explosion of high-throughput technologies. These developments have brought about a major revolution in the field of drug discovery from both the biological and the chemical perspectives; indeed, a large number of the molecules that are presently in clinical trials or have recently been registered and launched are from leads that emerged during a high-throughput screening (HTS) campaign.

After the initial excitement about and effort into HTS approaches, the methods reached an equilibrium state and became routine in most pharmaceutical companies. The need for further improvements in drug discovery productivity gradually shifted the attention of managers and researchers from the quantity of compounds produced to the quality of hits and leads; consequently, a new philosophy emerged in the field of drug discovery. In the mid-1990s, many research groups, almost exclusively from industry, began a process of lead- and drug-database mining with the goal of finding a more rational approach to the drug discovery process, which until then had been largely driven by intuition, experience, and serendipity. Some research groups, such as Vertex [2], analyzed the structure of approved drugs and advanced clinical candidates in an attempt to identify the most commonly occurring fragments. The general result was the finding that a restricted number of frameworks were common to a large number of drugs. The issue of whether these fragments had intrinsic characteristics that gave them drug-like properties or were only chosen on the basis of chemical versatility or market availability was recognized but not specifically addressed.

This finding was not completely unexpected; in 1988, Evans, after his experience with the alimentary cholecystokinin (CCK-A) antagonists, had already observed that certain substructural motifs (*privileged structures*) were

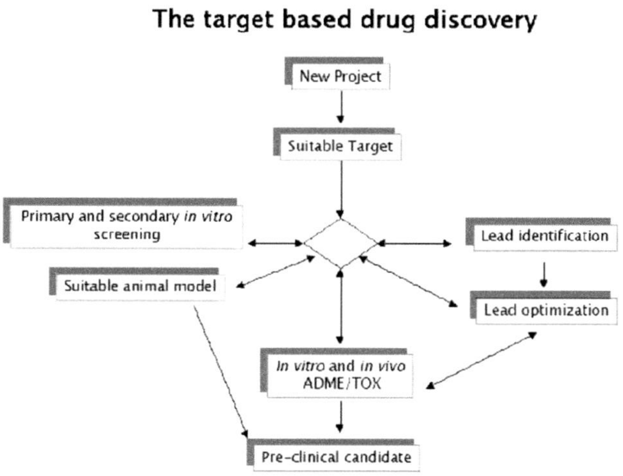

***FIGURE 6.1:** Generic flow chart of target-based drug discovery.*

capable of providing useful ligands for more than one receptor: "... these structures appear to contain common features which facilitate binding to various proteinaceous receptor surfaces... " [3]. The selection and derivatization of these privileged scaffolds became a working strategy pursued by Merck in the mid-1990s for the construction of combinatorial chemistry libraries, which have produced a number of interesting leads [4].

Other groups took a different approach, focusing on the analysis of physicochemical properties of compounds and taking advantage of the large amount of data accumulated by pharmaceutical companies [5]. These efforts resulted in a better understanding of the relationships between these properties and the optimal profile of a drug or a lead compound. The study of the characteristics of oral drugs resulted in Lipinsky's "Rule of Five" [6], and analysis of the evolution of physicochemical properties during the successful optimization process revealed additional interesting trends. In fact, it was quite clear that, from a statistical point of view, optimal leads have, on average, lower molecular weight and greater lipophilicity and structural complexity than their corresponding drug. These results suggested that, to prevent failure, a considerable amount of energy had to be devoted to the evaluation of lead-like properties in lead generation programs.

2. THE RATIONALE FOR SCREENING FRAGMENTS

On the basis of the data and observations accumulated from past experience, a strategy was devised to improve the lead discovery process – to initiate screening with molecules smaller than those usually considered for HTS, following a so-called minimalist approach. This was in strong contrast to the high-affinity paradigm followed by most medicinal chemists.

Before the idea of changing screening strategies was widely accepted, a number of issues had to be addressed. First, was such an approach compatible with the recognition process and the selectivity required by a successful medicinal chemistry project? In a paper now considered a classic, Hann and co-workers [7] demonstrated in theory and in practice that, in order to maximize the hit rate, there is an optimal molecular complexity that should be considered while screening a compound collection. As the system becomes more complex, the chance of observing a useful interaction for a randomly chosen ligand falls dramatically. The optimal complexity derives from a compromise between the probability of having a positive match between the ligand and the target (the so-called useful event) and the ability to detect their interaction (the screening method).

In general, the fewer the interaction features of the ligand, the higher the number of energetically similar and structurally different binding modes. This effect, which is called the "frustrated energy landscape," occurs often but fortunately not always and was illustrated by the work of Rejito and Verkhivker [8]. These two researchers, using FKB12-ligand complexes, streptavidine-peptide complexes, and lysozyme-ligand complexes as examples, proposed that the molecular recognition event is very likely fulfilled only by the interaction of a small portion of the ligand, called the anchoring fragment, with a relatively rigid portion of the protein target. The binding of these molecular anchors has a pivotal characteristic: It generally displays a pronounced gap between the free energy of the favorable binding mode and that of the other possible binding modes (Figure 6.2A) – that is, the formation of a single binding mode that is both thermodynamically stable, compared with possible alternatives, and kinetically accessible. Indeed, considering some classes of target, a molecular anchor motif can be easily identified by intuition, as, for instance, the hydroxamic acid for histone deacetylase inhibitors (HDACs) and other matrix metalloproteinases (MMP) inhibitors and the β-amino alcohol for aspartic proteases (Figure 6.2B). In the reported cases, the fragment shown in red seems to be essential for binding to the target and cannot be modified, even if it exhibits very poor binding or no binding itself. The remainder of the molecule contributes significantly to the affinity, generally is amenable to significant modifications, and, of note, does not bind in the absence of the anchor.

Abbott was the first company to demonstrate that FBLD was a practical opportunity with the pioneering work of Shuker *et al.*, published in *Science* in 1996 [9], which was followed in a relatively short time by many other publications. Today, it seems that at least twenty companies, both big pharma and small biotechs, have firmly established FBLD units [10]. A number of compounds in clinical trials derive from this approach.

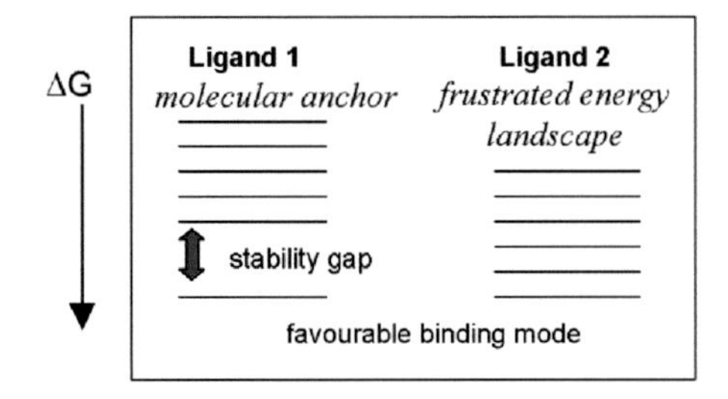

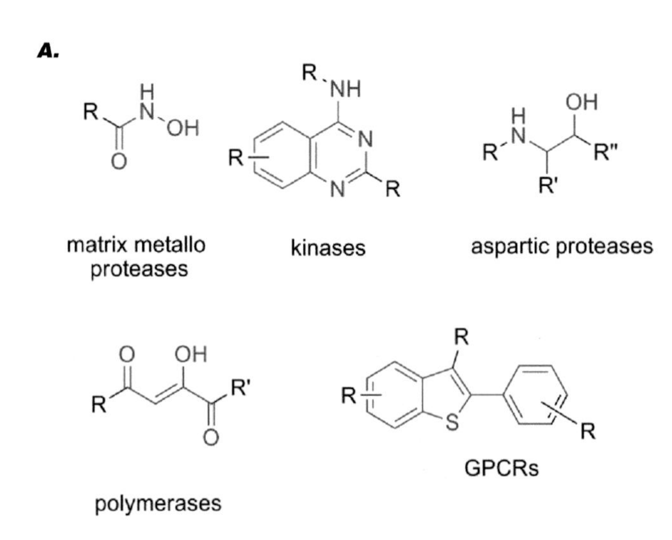

FIGURE 6.2:* A. *The concept of stability gap in binding energy for molecular anchors;* B. *examples of key fragments for selected targets (in red).

2.1. On the Target Side

When working on a target-based strategy, the chosen target must satisfy two conditions: 1) It is therapeutically relevant; and 2) it is *druggable*. A protein is druggable when it is possible to find a ligand that is able to interfere with the biological activity of the target, a condition that is not always satisfied. Proteins can be more or less druggable depending on their structure. An HTS campaign is quite expensive; thus, estimating the druggability of the new target in advance is both time and money saving. A common approach to establish the druggability of a target is to examine its genetic family. It is clear that it would be easier to inhibit an enzyme than a protein-protein interaction. Often, a possible starting lead is a molecule interacting with a protein belonging to the same family of the chosen target.

Proteins in living organisms are always surrounded by an enormous number and variety of molecules; life would be simply impossible if they interacted with all of these chemical entities. Under evolutionary pressure, proteins have developed well-defined recognition regions on their surfaces, called hot spots [11], surrounded by large, sometimes flat, normally well-solvated nonrecognition areas. The hot spots account for most of the binding energy involved in the ligand-protein interaction, with either another protein or a small molecule. Interactions of the ligand with additional regions of the protein surface serve mainly to increase specificity, with a minimal contribution to the overall binding energy.

By making use of the data accumulated during their nuclear magnetic resonance (NMR) fragment screening process, Abbott researchers have demonstrated that these data can be used to predict target druggability. Hajduk and co-workers screened a ten-thousand-fragment library on twenty-three different proteins and found that the experimental NMR hit rate correlated with the known capability of the proteins under investigation to bind ligands; the higher the fragment hit rate, the greater the druggability [12]. The advantage is that a fragment screening campaign is usually quicker and much less expensive than an HTS one. There is also a subtle additional advantage: through the use of fragment screening, protein hot spots not previously known may be detected, with the chance of disclosing unknown physiologic functions.

2.2. On the Ligand Side

Because of the reduced size of fragments, fragment screening generally results in weakly binding hits. This raises the question of what the hit selection criteria should be in the case of multiple hit compounds. The criterion most proposed, and now also widely accepted, is ligand efficiency (LE) [13]. LE is defined as the free energy of binding divided by the heavy atom count (HAC), which recalls somewhat the binding potential of molecules proposed by Andrew in 1984 [14]. The logarithmic nature of the relationship between the binding constant and the free energy means that for every -1.4 kcal/mol change of ΔG, there is a ten-fold increase in potency. A survey of experimental data on a large number of strong binding ligands, done by Kuntz et al. [15], indicates that the maximal free-energy contributions per non-hydrogen atom are approximately -1.5 kcal/mol for up to fifteen non-hydrogen atoms.

Because the free energy of binding is not always available, the LE is often approximated by the equation $LE = -RT\ln(IC_{50})/HAC$, where IC_{50} = concentration that produces 50% inhibition of the measured biological effect, R = gas constant (1.987 $calK^{-1}mol^{-1}$), and T = temperature (K). For relative comparison purposes, the calculated value results are as good as those calculated using K_D. The theory is to first optimize the hits having higher LE instead of being attracted by those having higher potency. Let us consider the following example: In a screening campaign, a 50-nM lead is found with a molecular weight (MW) of about 500 (an average of 36–37 HAC), and also a 0.5-μM molecule with an MW of 200 (12 HAC). The former has an LE of 0.27 (36 HAC), whereas the latter has an LE of 0.70 despite being ten times less potent. According to previously held logic, efforts would have been concentrated on the 50-nM lead. According to the fragment approach, the 500-nM lead is more promising in terms of optimization potential.

The analysis of lead quality has gone further, and an additional set of "efficiency" parameters has been proposed by Abbott researchers [16]. The first is the percentage efficiency index (PEI), which is the percentage of inhibition at a given concentration/MW. This parameter is mainly useful at the very beginning of a screening program, when, for economical and time reasons, the percent inhibition is evaluated instead of the IC_{50}. Clearly, only PEI values obtained at the same concentration can be compared. The second index is the binding efficiency index (BEI), which is the pKi, pK_D, or IC_{50}/MW; it is practically analogous to the LE but more rapidly evaluated. Last is the surface-binding efficiency index (SEI), which is pKi, pK_D, or IC_{50}/PSA (polar surface area) of the ligand, which introduces a new concept. It allows one to evaluate the activity of a molecule versus its PSA, a crucial parameter that takes into account a number of factors among which is the passive passage through biological membranes. These three indices have been applied to the analysis of in-house–obtained series of biological results and have identified two main issues. First, when high values of PEI are obtained at the early stages of lead validation, they are very likely due to artifacts. Second, the simultaneous monitoring of BEI and SEI (instead of only LE or BEI) during the optimization process may be useful to improve qualities of the molecule [17].

3. THE SCREENING TECHNOLOGY

One of the biggest challenges in fragment-based drug discovery is to identify and select fragments, because their affinities for the target protein are generally low (typically from high micromolar to low millimolar) compared with those found for drug-like molecules. Apart from the classical biological screening, nearly all of the biophysical methods commonly used in drug discovery, such as NMR, mass spectrometry (MS), x-ray crystallography, and surface plasmon resonance (SPR), with opportune modifications, have been applied to the analysis of fragment binding [18]. The choice of the technique depends on the nature of the target and the available facilities, and often a combination of two or more techniques has been successfully used.

3.1. High-Concentration (biological) Screening

Because of the weak interactions involved, the screening concentration of a fragment-detecting biological assay has to be much higher than that used in classical HTS (typically 250–1000 μM, compared with the 10–30 μM used in HTS). This kind of assay has the advantage of being faster and better related to physiologic conditions than physicochemical methods but presents a series of potential problems because of the high concentration of ligand required, which sometimes causes artifacts leading to false positives and false negatives. For these reasons, the value of biochemical screening for fragment-based drug discovery was questioned during the initial development of the technique. Nevertheless, a wide variety of assay types based mainly on fluorescence readout methods together with a robust statistical treatment of the results have been successfully applied [19]. The most common artifacts generally arise from nonspecific effects of the compounds, such as low solubility, tendency to aggregate, denaturation of the protein target, and interference with the assay signal, which can be minimized by careful selection of the fragment libraries together with technical improvements in the assay. Particularly important is the monitoring of compound solubility, which must be optimally 1 mM or greater; this will minimize not only precipitation, but also compound aggregation. Once the assay conditions have been optimized, a large number of compounds (>1,000) can be screened per day.

3.2. NMR-Based Screening

NMR-based fragment screening methodologies were first pioneered by Fesik and co-workers at Abbott in 1997 [20]. Since then, much work has been carried out to improve sensitivity, feasibility, and performance via the application of digital recording, cryogenic probes, autosamplers, and higher magnetic fields. The most widely used NMR techniques to date are summarized in Table 6.1. They can be divided into two categories: target based (when the target protein is monitored) and ligand based (when the ligand is monitored). The former involves two-dimensional (2D) experiments with an isotopically labeled protein, which is very useful for gaining structural information but is complex, expensive, and time-consuming; in addition, the molecular weight of the target protein is limited to 30 to 40 kDa because of signal broadening with increased molecular weight. For these reasons, when no structural information is required (at least in the hit-finding stage), many research groups replace these experiments with faster, less demanding one-dimensional (1D) ligand-based methods. The latter make use of unlabeled proteins, which are readily available and less costly; moreover, they do not suffer from molecular weight limitations and generally have a greater applicability in terms of throughput. Such 1D experiments are particularly suited for the very early stages of hit finding, when identifying weakly binding molecules within large mixtures is crucial and time consumption is relevant. An additional advantage of 1D NMR for screening purposes is that it can reliably detect binding using compound concentrations much lower than the K_D of the interaction.

Two ligand-based methods in particular, saturation transfer difference (STD) [21] and water-ligand optimized gradient spectroscopy (waterLOGSY) [22], have been broadly applied. STD is based on the Nuclear

TABLE 6.1: Key Methods in NMR-Based Fragment Screening

Observed Molecule	Method Description	Scientific Rationale	Advantages
Target Protein	^{1}H-^{15}N HSQC (2D) a ^{15}N-labeled protein is required	Comparison of the spectrum of the protein in the absence and presence of the ligand	Gives information on binding mode
Target Protein	^{13}C-^{1}H HSQC (2D), a ^{13}C-labeled protein, is required	Comparison of the spectrum of the protein in the absence and presence of the ligand	Gives information on binding mode
Ligand (weak binders)	1D NMR WaterLOGSY	Magnetization transfer from bulk water	Fast, can analyze mixtures
Ligand (weak binders)	1D NMR: saturation transfer diffusion (STD), mapping of the active regions of the ligand	Changes in the resonances of the ligand located in the protein binding site by saturation transfer	Fast, can analyze mixtures
Ligand (weak binders)	1D ^{19}F-NMR FAXS (fluorine chemical shift anisotropy and exchange for screening)	Changes in the ^{19}F resonances of the "spy molecule" (reporter ligand) upon displacement or modification	Fast, very sensitive
Ligand (weak binders)	1D ^{19}F-NMR 3-FABS (three fluorine atoms for biochemical screening), for enzymes	Changes in the ^{19}F resonances of a CF3 tagged substrate upon reaction	Fast, very sensitive

HSQC, Heteronuclear Single Quantum Coherence; WaterLOGSY, water-ligand optimized gradient spectroscopy.

Overhauser Effect (NOE) between the bound ligand protons and all of the target protein protons, whereas waterLOGSY exploits the magnetization transfer from bulk water. ^{19}F-NMR 1D ligand observation has also been used. Fluorine chemical shift anisotropy and exchange for screening (FAXS) and three fluorine atoms for biochemical screening (3-FABS), recently developed by Dalvit and coworkers [23], allow the determination of the K_D and the IC_{50}, respectively, of the compounds under study at low protein concentrations (150 nM for FAXS, 200 fM for 3-FABS). Because the lack of spectral interference allows the analysis of mixtures composed of a large number of compounds, a library of ten thousand compounds requires only a few days to be screened.

All of the techniques described previously [21, 22, 23] are applicable to soluble proteins. For insoluble proteins such as G protein-coupled receptors (GPCRs), a method called target-immobilized NMR screening (TINS), based on 1D experiments, has been developed at Leiden University in collaboration with Abbott [24]. In this method, the protein is immobilized on a solid support, which means that it does not necessarily have to be soluble. Binding is detected as a simple reduction in peak amplitude by recording a 1D spectrum of the ligand in the absence and in the presence of protein, and a library of ten thousand compounds can be screened in less than a week.

3.3 X-ray Crystallography

The application of x-ray crystallography as the initial screening technique has the advantage that it allows the immediate optimization of fragment hits based on the obtained structural information. The idea of co-crystallizing protein targets and small fragments was introduced by researchers at Abbott in 2000 [25]. Since then, several companies have used this technique.

It is quite intuitive that, for x-ray to be applicable, the protein target must form suitable crystals for x-ray analysis; thus, for example, GPCRs or other membrane proteins cannot be studied in this way at the moment. Although it initially was considered a rather low-throughput method, at present a relatively high throughput is possible because of the advances and improvements in both instrumentation and related software, allowing the screening of up to one hundred protein structures per day.

The actual process consists of soaking a mixture of fragments (100–300 Da) into preformed protein crystals for up to twenty-four hours; fragments that have an affinity for a protein will bind at its active site, whereas those that do not will remain in the bulk solvent. Clearly, solubility in the selected solvent is a must, and the fragment library choice must be made accordingly. An additional characteristic of libraries directed toward x-ray screening is the often strong presence of aromatic bromine to facilitate determination of bound hits by collecting x-ray data at the bromine anomalous dispersion wavelength. After soaking, the x-ray structure is acquired and compared with that of the original protein. Hits are then identified by differences in the electron density map of the active site. The upper affinity limit can be considered 100 μM to 10 mM.

3.4. Surface Plasmon Resonance

Surface plasmon resonance (SPR) is another powerful tool for the study of biomolecular interactions. It relies on the change in refractive index (because of the mass change) at

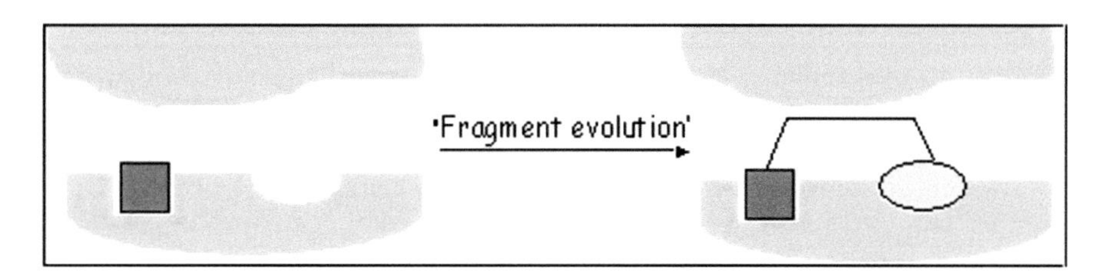

***FIGURE 6.3:** Schematic representation of the fragment evolution approach.*

a metal (gold)/liquid interface. At first, the technique was essentially applied to detect protein-protein interactions, but hardware and software improvements have enabled the determination of binding affinities of smaller and weaker ligands. In the field of fragment screening, the technique has been pioneered by a German biotech company, Graffinity [26]: small molecules are immobilized on a carrier surface and provided in a standardized format, ready for screening; incubation of the immobilized compounds with the purified and soluble target protein yields comprehensive affinity fingerprints in a label- and assay-free procedure. One such chip contains as many as 9,216 microarrays of various fragments organized in predetermined locations. More recently, researchers at Biacore have developed a series of SPR-based screening systems for primary and secondary screening of fragments in which the protein is fixed on the microarray. Their current throughput is 1,400 fragments per day per 10 μg of protein, and their methods have been recognized as an important primary screening technique for use as the first step in a fragment screening campaign. Array-based biosensors can confidently measure the binding of four fragments to four proteins in parallel and provide information on the selectivity of binding and target site specificity, which are important criteria in fragment selection. In addition to measuring affinities in the range from millimolar to picomolar, the approach provides values for the residence time (off rate).

3.5. Mass Spectrometry

MS has also been used to detect fragment binding, although to a lesser extent. Some biotech firms have set up their own patented screening technology to detect fragment binding to targets. In MS, fragments can be identified by measuring the quantity of "free" ligand (that not involved in the binding event) in the absence and in the presence of the target, or in some cases by actually analyzing the noncovalent target–fragment complex in the gas phase.

Electrospray ionization MS (ESI-MS) experiments are relatively quick to run via direct infusion of the fragments and the target, provided that the target is soluble. New techniques, such as frontal affinity chromatography, are being developed that use immobilized targets in conjunction with MS detection [27], and the system can be readily automated to improve throughput. The sensitivity of the method means that less protein is required compared with other techniques, and there is no need to label the target or the ligand because the mass of the complex identifies the fragment. If the masses of the fragments are sufficiently different (ca. 10 atomic mass units), mixtures can be also analyzed.

4. PRACTICAL APPLICATIONS

Because the potency of the initial fragment hit (FRIT) is generally very low, as discussed previously [18], further manipulations are required to improve the pharmacodynamic and pharmacokinetic properties of the selected molecules. Many successful strategies have been developed and applied over the past years to optimize fragments, ranging from very simple to quite complex and creative approaches. This section focuses on the description of the most representative examples of each published strategy.

4.1. Fragment Evolution

The fragment evolution approach is by far the most popular and successful method of fragment optimization. A gain in potency is achieved by adding functionalities in iterative cycles to the initial fragment hit that are able to make additional interactions with the target; this requires the initial fragment to act as an "anchor" and not to change its binding mode during the optimization process. This can be particularly useful not only when no structural information on the binding mode is available but also, impressively, when such information is available to guide the choice of the functional groups to be introduced (Figure 6.3).

A noteworthy application of this approach has been the discovery of novel, nonpeptide inhibitors of the aspartyl protease β-secretase (BACE-1, Table 6.2, entry 1), an important target for the treatment of Alzheimer's disease (AD) that is responsible for the regulation of the release of precursors of α-amyloids (Aα). The initial hit with an isocytosine core (**1**) was identified via NMR screening as a moiety interacting with the catalytic aspartates Asp228/Asp32 through the formation of hydrogen bonds [28]. Evolution of this fragment under the guidance of x-ray crystallography, along with structure-based design (SBD), led to the identification of evolved fragments bearing features able to interact with the hydrophobic S1 subsite of the

TABLE 6.2: Examples of Lead Molecules Developed by a Fragment Evolution Approach

Entry [ref.]	Target/Method	Fragment	Evolved Fragment	Lead Molecule
1[28]	BACE-1/NMR	Insert unnumbered figure **1**, 28% at 1 mM	Insert unnumbered figure **2**, $IC_{50} = 130$ μM (FRET assay)	Insert unnumbered figure **3**, $IC_{50} = 0.08$ μM (FRET assay)
2[29]	DNA-Gyrase/VS and SBD	Insert unnumbered figure MNEC > 250 μg/mL	Insert unnumbered figure MNEC = 8 μg/mL	Insert unnumbered figure MNEC = 30 ng/mL
3[30]	ErmAM/NMR	Insert unnumbered figure $K_d = 1$ mM	Insert unnumbered figure $K_d = 75$ μM	Insert unnumbered figure $K_i = 7.5$ μM
4[31]	IMPDH/VS	Insert unnumbered figure $IC_{50} = 21$ μM	Insert unnumbered figure $IC_{50} = 1.2$ μM	Insert unnumbered figure $IC_{50} = 0.076$ μM
5[32]	DHNA/x-ray	Insert unnumbered figure $IC_{50} = 28$ μM	Insert unnumbered figure $IC_{50} = 1.5$ μM	Insert unnumbered figure $IC_{50} = 0.068$ μM

BACE, aspartyl protease β-secretase; DHNA, dihydroneopterin aldolase; ErmAM, erythromycin-resistance methylase AM; FRET, fluorescence resonance energy transfer; IMPDH, inosine monophosphate dehydrogenase; MNEC, maximal noneffective concentration, a measure of activity; SBD, structure-based design; VS, virtual screening.

enzyme (**2**), with a consequent increase in affinity. Finally, a lead molecule with submicromolar potency emerged (**3**), in which the methoxy group (forming a hydrogen bond with Ser-229) and the 6-methyl group were identified as essential features that fixed the molecule in the correct conformation. Other recent interesting examples are the inhibitors of DNA-girase [29], of erythromycin-resistance methylase (ErmAM) [30], of inosine monophosphate dehydrogenase (IMPDH) [31], and of dihydronepterin aldolase (DHNA) [32], reported in Table 6.2.

4.2. Fragment Linking

When two or more fragments are detected to bind separately to nonoverlapping binding sites of the target protein, they can be linked together directly or by a suitable linker to give a larger molecule with a net increase in binding affinity (Figure 6.4). For this to be an efficient approach, the selected linker must allow each fragment to maintain its original binding mode; otherwise, ligand efficiency will drop sharply. Clearly, proteins with additional binding pockets near the active site are particularly suited targets for a fragment-linking strategy. This approach can also be used to identify new inhibitors of protein targets for which several potent ligands are already known and novelty is an issue. From a practical point of view, the first step to undertake is to block a "hot spot" of the protein target by binding one fragment and then to block the resulting complex with a second binding event.

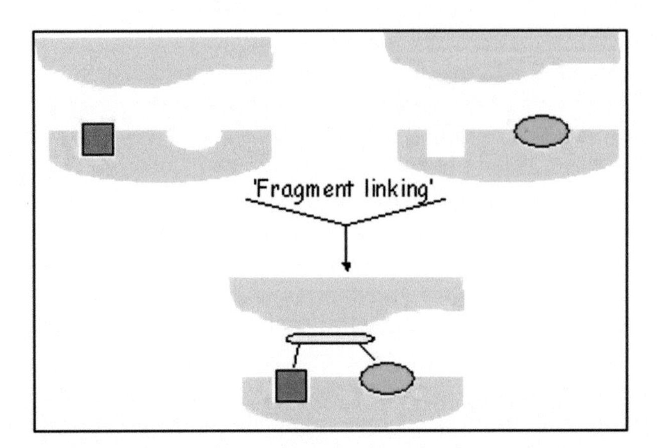

***FIGURE 6.4:** Schematic representation of the fragment linking approach.*

This method has been used to find inhibitors for Hsp90, a molecular chaperone abundantly expressed in cells that has been shown to promote cancer survival [33]. By using 2D NMR-based fragment screening against the *N*-terminal adenosine triphosphate (ATP)-ase domain, a first ligand (**4**) was found that forms four hydrogen bonds, one with the carboxylic side chain of Asp93 and the other three with well-conserved water molecules (Table 6.3, entry 1). A further 2D-NMR screening on a set of 3,360 fragments, in the presence of saturating amounts of **4**, led to the identification of second-site ligand **5**. The structure of the ternary complex revealed a π-stacking interaction between the phenyl of compound **5** and the pyrimidine ring of **4**, so a suitable linker, able to ensure the same orientation, was searched for and found, leading to compound **6** [34]. It is worth mentioning that LE for **6** is much lower than that of the initial fragments (0.31 vs. 0.53), suggesting that the linking strategy is not yet optimal. Additional recent examples are the discovery of inhibitors of MMP3 [35], Bcl-2 [36], thrombin [37], and FKBP (Table 6.3) [38].

4.3. In Situ Fragment Self-Assembly

This approach involves the use of chemically reactive fragments that bind separately to adjacent regions of the target protein and self link in the presence of the protein, which

TABLE 6.3: Examples of Lead Molecules Developed by Fragment Linking

Entry [ref.]	Target/Method	Fragment 1	Fragment 2	Lead molecule
1 [33–34]	Hsp90/NMR	Insert unnumbered figure **4**, $K_d = 20\ \mu M$, LE = 0.53	Insert unnumbered figure **5**, $K_d > 5$ mM (without **4**), $K_d = 150\ \mu M$ (with **4**)	Insert unnumbered figure **6**, $K_i = 1.9\ \mu M$, LE = 0.31
2 [35]	MMP/NMR and SBD	Insert unnumbered figure $K_d = 17$ mM	Insert unnumbered figure $K_d = 0.02$ mM	Insert unnumbered figure $IC_{50} = 15$ nM (MMP3)
3 [36]	Bcl-2/NMR	Insert unnumbered figure $K_d = 300\ \mu M$	Insert unnumbered figure $K_d = 6000\ \mu M$	Insert unnumbered figure $K_i = 0.036\ \mu M$
4 [37]	Thrombin/x-ray	Insert unnumbered figure $IC_{50}50 = 100\ \mu M$	Insert unnumbered figure $IC_{50} = 330\ \mu M$	Insert unnumbered figure $IC_{50}50 = 3.7$ nM
5 [38]	FKBP/NMR	Insert unnumbered figure $K_d = 2\ \mu M$	Insert unnumbered figure $K_d = 100\ \mu M$	Insert unnumbered figure $K_d = 19$ nM

Bcl, B-cell lymphoma; Hsp90, heat shock protein 90; MMP, matrix metalloproteinases; FKBP, FK506-binding protein; SBD, structure-based design.

TABLE 6.4: Examples of Lead Molecules Developed by In Situ Fragment Self-Assembly

Entry [ref.]	Target	Fragment 1	Fragment 2	Lead Molecule
1 [39]	CA	Insert unnumbered figure **7**, $K_d = 37$ nM	Insert unnumbered figure **8**	Insert unnumbered figure **9**, $K_d = 0.2$–0.5 nM
2 [40]	HIV-1 protease	Insert unnumbered figure $IC_{50} > 100$	Insert unnumbered figure $IC_{50} = 4.2\ \mu M$	Insert unnumbered figure $IC_{50} = 6$ nM
3 [41]	AChE	Insert unnumbered figure $K_d = 18\ \mu M$	Insert unnumbered figure $K_d = 10$–100 nM	Insert unnumbered figure $K_d = 33$ fM
4 [42]	CDK2	Insert unnumbered figure $IC_{50} > 1$ nM	Insert unnumbered figure $IC_{50} > 1$ nM	Insert unnumbered figure $IC_{50} = 30$ nM
5 [43]	Neuro-aminidase	Insert unnumbered figure K_i not reported	Insert unnumbered figure $IC_{50} = 31\ \mu M$	Insert unnumbered figure $K_i = 85$ nM

AChE, acetylcholinesterase; CA, carbonic anhydrase; CDK2, cyclin-dependent kinase; HIV, human immunodeficiency virus.

acts as a template. Only those reagents able to react at room temperature and in an aqueous environment will link in situ to produce a more potent inhibitor (Figure 6.5).

A noteworthy application is the synthesis of the novel carbonic anhydrase (CA) inhibitor **9** (Table 6.4, entry 1) [39]. CA is a zinc-containing enzyme that catalyzes the interconversion of HCO_3^- and CO_2 and is involved in many key biological processes. Aromatic acetylenic sulfonamide **7** was selected as the first reactive fragment able to strongly coordinate with the Zn^{+2} cation of the enzyme to form an inhibitor in situ by a 1,3-dipolar cycloaddition reaction (click chemistry). An array of twenty-four azides was used as complementary reagents, and the reactions were run at 37°C in a buffer solution and in the presence of bovine CA II (bCAII). Only twelve azides reacted to form the corresponding triazole, and of these, eleven did not form the corresponding products when the experiments were done in the presence of another enzyme (bovine serum albumin [BSA]) or a known inhibitor of bCAII, proving the involvement of the enzyme active site in the cycloaddition reaction. Click chemistry was also used for the in situ assembly of HIV-1 protease [40], AchE [41], CDK2 [42], and neuroaminidase inhibitors (Table 6.4) [43].

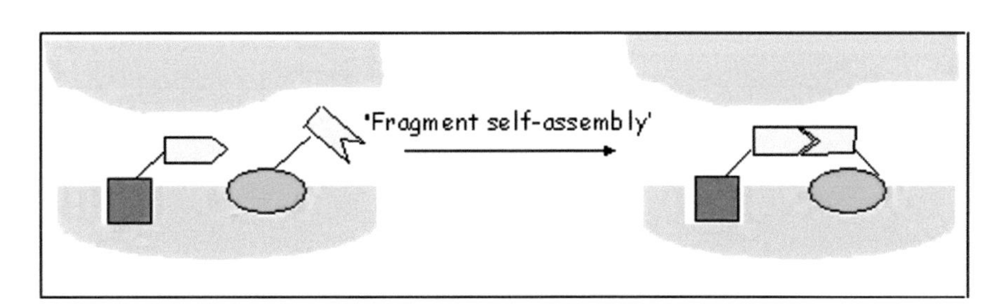

***FIGURE 6.5:** Schematic representation of the in situ fragment self-assembly approach.*

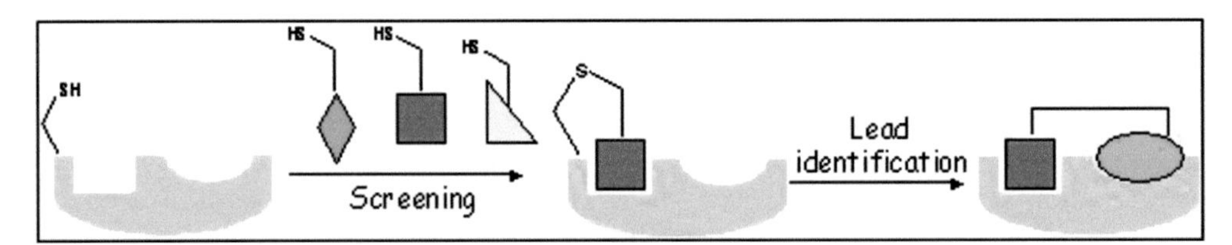

FIGURE 6.6: Schematic representation of the tethering approach.

4.4. Fragment Tethering

This approach involves the formation of a disulfide bond between a cysteine residue on the target protein and a chemically reactive fragment (Figure 6.6). The greater the affinity of the fragment, the more stable the disulfide bond, which will be detected by MS.

Researchers at Sunesis Pharmaceutical found a potent inhibitor (**12**) of caspases (cysteinyl aspartate specific proteases), a family of endopeptidases involved in the inflammatory processes occurring in various diseases (Table 6.5, entry 1) [44]. At first, an aspartyl residue containing a thiol group was linked through a thioether linkage to the cysteine in the active site of the protein (**10**); next, fragments capable of forming a disulfide bond with the modified protein were screened and identified by mass spectrometry. The selected fragment **11** was further

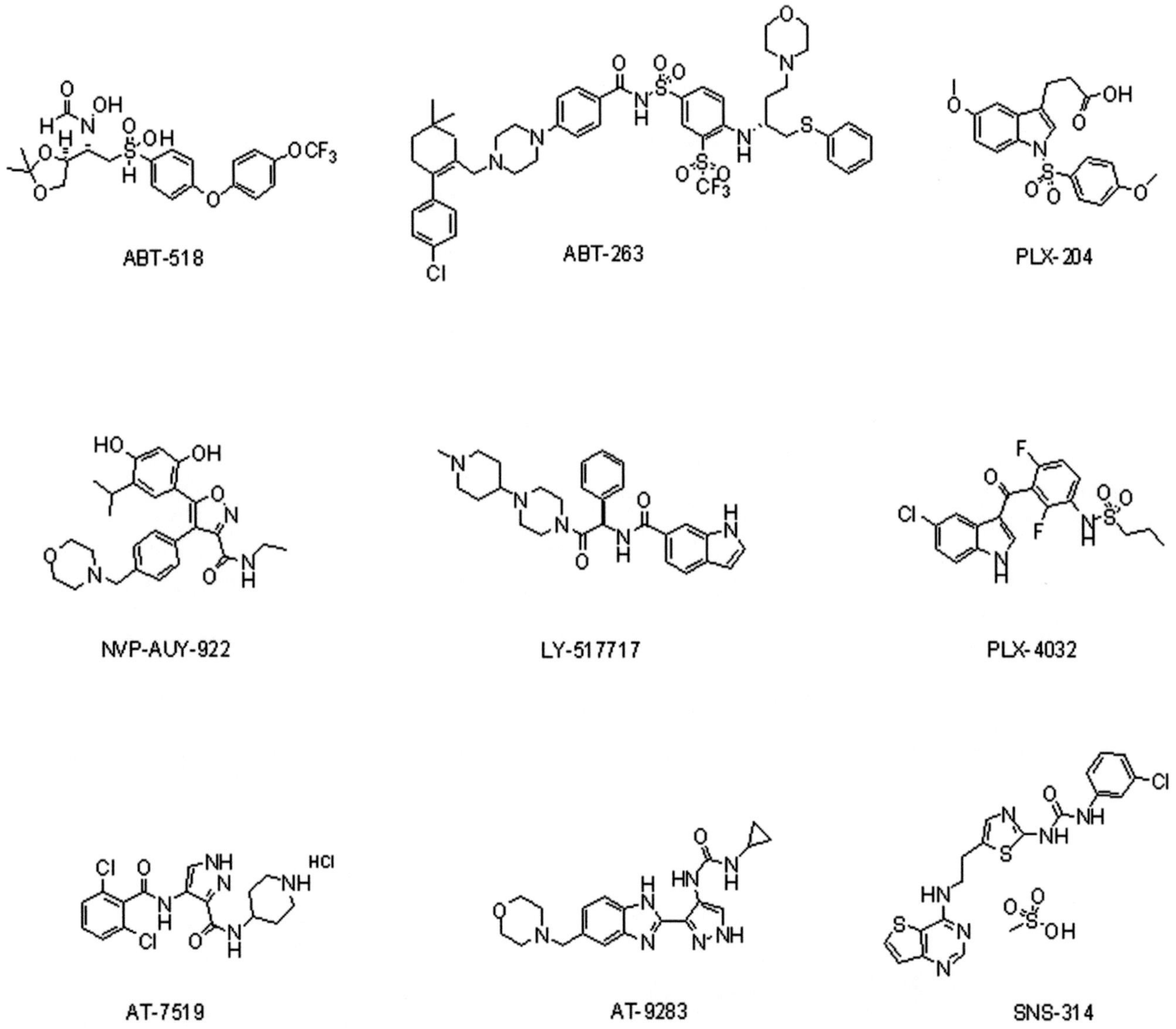

ABT-518 ABT-263 PLX-204

NVP-AUY-922 LY-517717 PLX-4032

AT-7519 AT-9283 SNS-314

FIGURE 6.7: Structure of some clinical candidates derived from fragment screening.

TABLE 6.5: Examples of Fragment-Based Lead Discovery by the Tethering Approach

Entry [ref.]	Target	Fragment 1	Fragment 2	Lead Molecule
1 [44]	Caspase	Insert unnumbered figure **10**	Insert unnumbered figure **11**	Insert unnumbered figure **12**, $IC_{50} = 5$ nM
2 [45]	IL-2	Insert unnumbered figure	Insert unnumbered figure $IC_{50} = 3\ \mu M$	Insert unnumbered figure $IC_{50} = 60$ nM

IL-2, interleukin 2.

modified by substituting the disulfide bond with a linker unit and the modified aspartyl residue with a reversibly bound one, affording a nanomolar inhibitor (**12**). Another great application has been in the field of protein-protein interactions with the discovery of a nanomolar inhibitor of interleukin-2 [45].

5. CLINICAL CANDIDATES DERIVED FROM FRAGMENT SCREENING

The industry's answer to the newly emerging philosophy and the potential of fragment screening technologies has been quite rapid. Some compounds deriving from fragment-based approaches have already entered preclinical development and even clinical trials (Figure 6.7, Table 6.6), and, optimistically, the first registration is expected for 2011, given the typical fifteen-year pipeline for drug discovery. Unfortunately, the exact structures of compounds are not always available, and the corresponding stories have been disclosed for only a restricted number of molecules.

6. CONCLUSIONS

Initially regarded with some amount of skepticism, mainly because of the low affinities of the resulting hits, fragment-based screening is becoming increasingly popular among medicinal chemists and in other scientific communities; indeed, the number of sessions in congresses or entire symposia dedicated to the subject is still increasing. The presence of compounds derived from an initial FBLD in clinical trials clearly has proven that the strategy works. Statistical work published by Abbott suggests that FBLD gives higher hit rates than classical HTS [1], and it is now a commonly held opinion that, when starting with fragments, optimization through iterative chemical manipulations is easier: the smaller, the better. In addition, some indications about how to proceed in a systematic and rational way along the optimization path are emerging with the work published on ligand efficiency metrics and the relative statistics derived from successful and unsuccessful projects.

There is also an economic component. HTS is expensive, mainly because of the size of the libraries: In a normal HTS campaign, million of compounds are screened. These numbers are necessary to cover the chemical space in a reasonable way, because the compounds typically have a molecular weight of about 500 Da, which has led to an estimation of 10^{62} possible different molecular scaffolds [10]. In the fragments world, the average molecular size is about 200 Da, and the number of possible molecular scaffolds is calculated to be about 10^{10}, which means that a good chemical diversity can be achieved with a smaller number of compounds. Indeed, the larger reported fragment library is about ten thousand units, but routine libraries can be sized one thousand or even less. This makes the technique affordable not only to large pharmaceutical companies, but also to small biotechs and nonprofit research institutes and universities [46].

The exact role of fragment-based screening in the drug discovery world remains to be seen once the initial excitement wanes, as often happens with new techniques and as happened previously in the drug discovery world with the development of computational and the combinatorial chemistry approaches.

TABLE 6.6: Clinical Candidates Reported to have Evolved from FRITs

Company	Target	Compound (highest development status)[a]
Abbott	Bcl-2/Bcl-xL	**ABT-263**[b, c] (Phase II)
Abbott	MMP	**ABT-518**[b, d] (Phase I)
Astex	CDK	**AT-7519**[b] (Phase I)
Astex	Hsp90	**AT13387**[e] (Phase I)
Astex	Aurora-kinase	**AT-9283**[b] (Phase II)
Lilly/Protherics	Factor Xa	**LY-517717**[b] (Phase II)
Plexxikon	PPAR	**PLX-204**[b] (Phase II)
Plexxikon	B-Raf	**PLX-4032**[b] (Phase I)
Sunesis	Aurora-kinase	**SNS-314**[b] (Phase I)
Vernalis/Novartis	Hsp 90	**NVP-AUY-922**[b] (Phase I)

[a]Data from Thomson Pharma™.
[b]Figure 6.8.
[c]Table 6.3, entry 3.
[d]Table 6.3, entry 2.
[e]Structure not disclosed.

7. GLOSSARY

BEI	Binding efficiency index
ESI-MS	Electron spray ionization mass spectrometry
FBLD	Fragment-based ligand design

FAXS	Fluorine chemical shift anisotropy and exchange for screening
FRITs	Fragment hits (hits derived from fragment screening)
HAC	Non-hydrogen atom count
HTS	High-throughput screening
LE	Ligand efficiency
MS	Mass spectrometry
NOE	Nuclear Overhauser Effect
NMR	Nuclear magnetic resonance
PSA	Polar surface area
PEI	Percentage efficiency index
SBD	Structure-based design
SEI	Surface-binding efficiency index
SPR	Surface plasmon resonance
STD	Saturation transfer difference
TINS	Target-immobilized NNR screening

REFERENCES

1. Hajduk, P. J., and Greer, J. A. (2007). A decade of fragment-based drug design: strategic advances and lessons learned. *Nature Rev Drug Discov 6*, 211–219.
2. Bemis, J. W., and Murcko, M. A. (1996). The properties of known drugs. 1. Molecular frameworks. *J Med Chem 39*, 2887–2893.
3. Evans, B. E., and Bock, M. G. (1993). Promiscuity in receptor ligand research: benzodiazepine-based cholecystokinin antagonists. *Adv Med Chem 2*, 111–152.
4. Patchett, A. A. (2002). *2002* Alfred Burger Award address in medicinal chemistry. Natural products and design: interrelated approaches in drug discovery. *J Med Chem 45*, 5609–5616.
5. Teague, S. J., Davis, A. M., Leeson, P. D., and Oprea, T. (1999). The design of lead-like combinatorial libraries. *Angew Chem IEE 38*, 3743–3748.
6. Lipinski, C., Lombardo, F., Dominy, B., and Feeney, P. (1997). Experimental and computational approaches to estimate solubility and permeability in drug discovery and development settings. *Adv Drug Delivery Rev 23*, 3–25.
7. Hann, M. M., Leach, A. R., and Harper, G. (2001). Molecular complexity and its impact on the probability of finding leads for drug discovery. *J Chem Inf Comput Sci 41*, 856–864.
8. Rejito, P. A., and Verkhivker, G. M. (1996). Unraveling principles of lead discovery: from unfrustrated energy landscapes to novel molecular anchors. *Proc Natl Acad Sci USA 93*, 8945–8950.
9. Shuker, S. B., Hajduk, P. J., Meadows, R. P., and Fesik, S. W. (1996). Discovering high-affinity ligands for proteins: SAR by NMR. *Science 274*, 1531–1534.
10. Everts, S. (2008). Piece by piece. More and more companies are using fragment-based lead design as a drug discovery strategy. *Chem & Eng News 86*, July 21, 15–23.
11. Clackson, T., and Wells, J. A. (1995). A hot spot of binding energy in a hormone-receptor interface. *Science 267*, 383–386.
12. Hajduk, P. J., Huth, J. R., and Fesik, S. W. (2005). Druggability indices for protein targets derived from NMR-based screening data. *J Med Chem 48*, 2518–2525.
13. Hopkins, A. L., Groom, C. R., and Alex, A. (2004). Ligand efficiency: a useful metric for drug selection. *Drug Discov Today 9*, 430–431.
14. Andrews, P. R., Craik, D. J., and Martin, J. L. (1984). Functional group contributions to drug-receptor interactions. *J Med Chem 27*, 1648–1657.
15. Kunz, I. D., Chen, K., Sharp, K. A., and Kollman, P. A. (1999). The maximal affinity of ligands. *Proc Natl Acad Sci U S A 96*, 9997–10002.
16. Abad-Zapatero, C., and Metz, J. T. (2005). Ligand efficiency indeces as guidepost for drug discovery. *Drug Discov Today 10*, 464–469.
17. Hajduk, P. J. (2006). Fragment-based drug design. How big is too big ? *J Med Chem 49*, 6972–6976.
18.(a) Bartoli, S., Fincham, C. I., and Fattori, D. (2006). The fragment approach: an update. *Drug Discov Today Technol 3*, 425–431; (b) Fattori, D., Squarcia, A. and Bartoli, S. (2008). Fragment-based approach to drug lead discovery. Overview and advances in various techniques. *Drugs R. D. 9*(4), 217–227.
19. Barker, J., Courtney, S., Hesterkamp, T., Ullmann, D., and Whittaker, M. (2006). Fragment screening by biochemical assay. *Expert Opin Drug Discov 1*, 225–236.
20. Hajduk, P. J., Sheppard, G., Nettesheim, D. G., Olejniczak, E. T., Shuker, S. B., Meadows, R. P., Steinman, D. H., Carrera, G. M., Jr., Marcotte, P. A., Severin, J., Walter, K., Smith, H., Gubbins, E., Simmer, R., Holzman, T. F., Morgan, D. W., Davidsen, S. K., Summers, J. B., and Fesik, S. W. (1997) Discovery of potent nonpeptide inhibitors of stromelysin using SAR by NMR. *J Am Chem Soc 119*: 5818–5827.
21. Meyer, M., and Meyer, B. (1999). Characterization of ligand binding by saturation transfer difference NMR spectroscopy. *Angew Chem Int Ed 38*, 1784–1788.
22. Dalvit, C., Fogliatto, G. Steward, A., Veronesi, M. and Stockman, B. (2001). WaterLOGSY as a method for primary NMR screening: practical aspects and range of applicability. *J Biomol NMR 21*, 349–359.
23. Dalvit, C., Mongelli, N., Paper, G., Giordano, P., Veronesi, M, Moskau, D, Kümmerle, R. (2005). Sensitivity improvement in 19F-based screening experiments: theoretical considerations and experimental applications. *J Am Chem Soc 127*, 13380–13385.
24. Vanwetswinkel, S., Heetebrij, R. J., van Duynhoven, J., Hollander, J. G., Filippov, D. V., Hajduk, P. J., and Siegal, G. (2005). TINS, target immobilized NMR screening: an efficient and sensitive method for ligand discovery. *Chem Biol 12*, 207–216.
25. Nienaber, V. L., Richardson, P. L., Klighofer, V., Bouska, J. J., Giranda, V. L., and Greer, J. (2000). Discovering novel ligands for macromolecules using X-ray crystallographic screening. *Nat Biotechnol 18*, 1105–1108.
26. Vetter, D. (2002) Chemical microarrays, fragment diversity, label-free imaging by plasmon resonance – a chemical genomic approach. *J Cell Biochem Suppl 39*, 79–84.
27. Slon-Usakiewicz, J. J., Ng, W., Dai, J.-R., Pasternak, A., and Redden, P. R. (2005). Frontal affinity chromatography with MS detection (FAC-MS) in drug discovery. *Drug Discov Today, 30*, 409–416.

28.(a) Murray, C. W., Callaghan, O., Chessari, G., Cleasby, A., Congreve, M., Frederickson, M., Hartshorn, M. J., McMenamin, R., Patel, S., and Wallis, N. (2007). Application of fragment screening by X-ray crystallography to beta-secretase. *J Med Chem 50*, 1116–1123; (b) Geschwinder, S., Olsson, L. L., Albert, J. S., Deinum, J., Edwards, P. D., Beer, and Folmer, R. H. (2007). Discovery of a novel warhead against beta-secretase through fragment-based lead generation. *J Med Chem 50*, 5903–5911; (c) Edwards, P. D., Albert, J. S., Sylvester, M., Aharony, D., Andisik, D., Callaghan, O.,Campbell, J. B., Carr, R. A., Chessari, G., Congreve, M., Frederickson, M., Folmer, R. H., Geschwindner, S., Koether, G., Kolmodin, K., Krumrine, J., Mauger, R. C., Murray, C. W., Olsson, L. L., Patel, S., Spear, N., and Tian, G. (2007). Application of fragment-based lead generation to the discovery of novel cyclic amidine beta-secretase inhibitors with nanomolar potency, cellular activity, and high ligand efficiency. *J Med Chem 50*, 5912–5925.

29. Boehm H. J., Boehringer M., Bur, D., Gmuender H., Huber, W., Klaus, W., Kostrewa, D., Kuehne, H., Luebbers, T., Meunier-Keller, N., and Mueller, F. (2000). Novel inhibitors of DNA gyrase: 3D structure based biased needle screening, hit validation by biophysical methods, and 3D guided optimization. A promising alternative to random screening. *J Med Chem 43*, 2664–2674.

30. Hajduk P. J., Dinges, J., Schkeryantz, J. M., Janowick, D., Kaminski, M., Tufano, M., Augeri, D. J., Petros, A., Nienaber, V., Zhong, P., Hammond, R., Coen, M., Beutel, B., Katz, L., and Fesik, S. W. (1999). Novel inhibitors of Erm methyltransferases from NMR and parallel synthesis. *J Med Chem 42*, 3852–3859.

31.(a) Beevers, R. E., Buckley, G. M., Davies, N., Fraser, J. L., Galvin, F. C., Hannah, D. R., Haughan, A. F., Jenkins, K., Mack, S. R., Pitt, W. R., Ratcliffe, A. J., Richard, M. D., Sabin, V., Sharpe, A., and Williams, S. C. (2006). Low molecular weight indole fragments as IMPDH inhibitors. *Bioorg Med Chem Lett 16*, 2535–2538; (b) Beevers, R. E., Buckley, G. M., Davies, N., Fraser, J. L., Galvin, F. C., Hannah, D. R., Haughan, A. F., Jenkins, K., Mack, S. R., Pitt, W. R., Ratcliffe, A. J., Richard, M. D., Sabin, V., Sharpe, A., and Williams, S. C. (2006). Novel indole inhibitors of IMPDH from fragments: synthesis and initial structure-activity relationships. *Bioorg Med Chem Lett, 16*, 2539–2542; (c) Pickett, S. D., Sherborne, B. S., Wilkinson, T., Bennett, J., Borkakoti, N., Broadhurst, M., Hurst, D., Kilford, I., McKinnell, M., and Jones, P. S. (2003). Discovery of novel low molecular weight inhibitors of IMPDH via virtual needle screening. *Bioorg Med Chem Lett 13*, 1691–1694.

32. Sanders, W. J., Nienaber, V. L., Lerner, C. G., McCall, J. O., Merrick, S. M., Swanson, S. J., Harlan, J. E., Stoll, V. S., Stamper, G. F., Betz, S. F., Condroski, K. R., Meadows, R. P., Severin, J. M., Walter, K. A., Magdalinos, P., Jakob, C. G., Wagner, R., and Beutel, B. A. (2004). Discovery of potent inhibitors of dihydroneopterin aldolase using CrystaLEAD high-throughput X-ray crystallographic screening and structure-directed lead optimization. *J Med Chem 47*, 1709–1718.

33. Solit, D. B., and Rosen, N. (2006). Hsp90: a novel target for cancer therapy. *Curr Top Med Chem 6*, 1204–1214.

34. Huth, J. R., Park, C., Petros, A. M., Kunzer, A. R., Wendt, M. D., Wang, X., Lynch, C. L., Mack, J. C., Swift, K. M., Judge, R. A., Chen, J., Richardson, P. L., Jin, S., Tahir, S. K., Matayoshi, E. D., Dorwin, S. A., Ladror, U. S., Severin, J. M., Walter, K. A., Bartley, D. M., Fesik, S. W., Elmore, S. W., and Hajduk, P. J. (2007). Discovery and design of novel HSP90 inhibitors using multiple fragment-based design strategies. *Chem Biol Drug Des 70*, 1–12.

35. Hajduk, P. J., Sheppard, G., Nettesheim, D. G., Olejniczak, E. T., Shuker, S. B., Meadows, R. P., Steinman, D. H., Carrera, G. M., Jr., Marcotte, P. A., Severin, J., Walter, K., Smith, H., Gubbins, E., Simmer, R., Holzman, T. F., Morgan, D. W., Davidsen, S. K., Summers, J. B., and Fesik, S. W. (1997). Discovery of potent nonpeptide inhibitors of stromelysin using SAR by NMR. *J Am Chem Soc 119*, 5818–5827.

36.(a) Oltersdorf, T., Elmore, S. W., Shoemaker, A. R., Armstrong, R. C., Augeri, D. J., Belli, B. A., Bruncko, M., Deckwerth, T. L., Dinges, J., Hajduk, P. J., Joseph, M. K., Kitada, S., Korsmeyer, S. J., Kunzer, A. R., Letai, A., Li, C., Mitten, M. J., Nettesheim, D. G., Ng, S., Nimmer, P. M., O'Connor, J. M., Oleksijew, A., Petros, A. M., Reed, J. C., Shen, W., Tahir, S. K., Thompson, C. B., Tomaselli, K. J., Wang, B., Wendt, M. D., Zhang, H., Fesik, S. W., and Rosenberg, S. H. (2005). An inhibitor of Bcl-2 family proteins induces regression of solid tumours. *Nature 435*, 677–681; (b) Petros, A. M. Dinges, J., Augeri, D. J., Baumeister, S. A., Betebenner, D. A., Bures, M. G., Elmore, S. W., Hajduk, P. J., Joseph, M. K., Landis, S. K., Nettesheim, D. G., Rosenberg, S. H., Shen, W., Thomas, S., Wang, X., Zanze, I., Zhang, H., and Fesik, S. W. (2006). Discovery of a potent inhibitor of the antiapoptotic protein Bcl-xL from NMR and parallel synthesis. *J Med Chem 49*, 656–663.

37. Howard, N., Abell, C., Blakemore, W., Chessari, G., Congreve, M., Howard, S., Jhoti, H., Murray, C.W., Seavers, L. C., and Van Montfort, R. L. (2006). Application of fragment screening and fragment linking to the discovery of novel thrombin inhibitors. *J Med Chem 49*, 1346–1355.

38. Shuker, S. B., Hajduk, P. J., Meadows, R. P., and Fesik, S. W. (1996). Discovering high-affinity ligands for proteins: SAR by NMR. *Science 274*, 1531–1534.

39. Mocharia, V. P., Colasson, B., Lee, L. V., Roper, S. Sharpless, K. B., Wong, C. H., and Kolb, H. C. (2005). In situ click chemistry: enzyme-generated inhibitors of carbonic anhydrase II. *Angew Chem Int Ed 44*, 116–120.

40. Whiting, M., Muldoon, J., Lin, Y. C., Silverman, S. M., Lindstrom, W., Olson, A. J., Kolb, H. C., Finn, M. G., Sharpless, K. B., Elder, J. H., and Fokin, V. V. (2006). Inhibitors of HIV-1 protease by using in situ click chemistry. *Angew Chem Int Ed 45*, 1435–1439.

41. Krasinski, A., Radic, Z., Manetsch, R., Raushel, J., Taylor, P., Sharpless, K. B., and Kolb, H. C. (2005). In situ selection of lead compounds by click chemistry: target-guided optimization of acetylcholinesterase inhibitors. *J Am Chem Soc 127*, 6686–6692.

42. Congreve, M. S., Davis, D. J., Devine, L., Granata, C., O'Reilly, M., Wyatt, P. G., and Jhoti, H. (2003). Detection of ligands from a dynamic combinatorial library by X-ray crystallography. *Angew Chem Int Ed 42*, 4479–4482.

43. Hochgurtel, M., Biesinger, R., Kroth, H., Piecha, D., Hofmann, M. W., Krause, S., Schaaf, O., Nicolau, C., and Eliseev, A. V. (2003). Ketones as building blocks for dynamic

combinatorial libraries: highly active neuraminidase inhibitors generated via selection pressure of the biological target. *J Med Chem 46*, 356–358.

44.(a) O'Brien, T., Fahr, B. T., Sopko, M. M., Lam, J. W., Waal, N. D., Raimundo, B. C., Purkey, H. E., Pham, P., and Romanowski, M. J. (2005). Structural analysis of caspase-1 inhibitors derived from Tethering. *Acta Crystallogr Sect F Struct Biol Cryst Commun 61*, 451–458; (b) Fahr, B. T., O'Brien, T., Pham, P., Waal, N. D., Baskaran, S., Raimundo, B. C., Lam, J. W., Sopko, M. M., Purkey, H. E., and Romanowski, M. J. (2006). Tethering identifies fragments that yield potent inhibitors of human capsase-1. *Bioorg Med Chem Lett 16*, 559–562.

45. Braisted, A. C., Oslob, J. D., Delano, W. L., Hyde, J., McDowell, R. S., Waal, N., Yu, C., Arkin, M. R., and Raimundo, B. C. (2003). Discovery of a potent small molecule IL-2 inhibitor through fragment assembly. *J Am Chem Soc 125*, 3714–3715.

46. Murray, C. W., and Rees, D. C. (2009). The rise of fragment-based drug discovery. *Nature Chemistry 1*, 187–192.

CHAPTER 7

Basics and Principles for Building Natural Product–based Libraries for HTS

Ronald J. Quinn

Natural product libraries require collection of samples of biota and extraction of compounds for screening against a biological target. Although this is a simple task, the establishment of a quality natural product–based library requires a good understanding of the modern drug discovery paradigm. A quality natural product library will allow both the discovery of natural products that may be able to be developed into therapeutic agents as they occur in nature and the identification of starting points for medicinal chemistry optimization. The process of establishing a natural product screening library for use in high-throughput screening (HTS) is illustrated in Figure 7.1. The collected sample may be plant material, marine organisms, or terrestrial or marine microorganisms. Once collected, the samples need to be processed to provide the compounds contained in the samples in a soluble form to allow HTS. Extraction with organic and aqueous solvents provides the soluble constituents that may be processed further into fractions or pure compounds.

1. REQUIREMENTS PRIOR TO COLLECTING NATURAL PRODUCTS

In recent years, it has been recognized that genetic material is a sovereign right of the country of origin. Collection of biota samples must be undertaken according to the United Nations Convention on Biological Diversity (CBD). The CBD opened for signature on June 5, 1992, at the United Nations Conference on Environment and Development (UNCED), also known as the Earth Summit, in Rio de Janeiro, Brazil. The text of the CBD is available on the Web site of the Convention of Biological Diversity at www.cbd.int/convention/convention.shtml.

The objectives (Figure 7.2) of the Convention are

1) the conservation of biological diversity;
2) the sustainable use of its components; and
3) the fair and equitable sharing of the benefits arising out of the utilization of genetic resources.

Figure 7.3 highlights important aspects of the CBD for those interested in biodiscovery, the use of biological diversity as a discovery resource for new potential therapeutic agents. Article 15 of the CBD recognizes the sovereign rights of States over their natural resources and that the authority to determine access to genetic resources rests with national governments and is subject to national legislation. Article 15 also specifies that access shall be on mutually agreed terms (Article 15. 4), subject to the prior informed consent of the Contracting Party (Article 15.5) and the benefits arising from the commercial and other utilization of genetic resources shared on mutually agreed terms (Article 15.7).

All collections should adopt "best practices" in access and benefit sharing. It is necessary to have formal contracts between the parties and approval of the government of the country that has the sovereign rights to the genetic resources. Negotiations can be time consuming and require a good knowledge of the commercialization pathway, risks, and rewards from all sides involved. It may realistically take several years to negotiate appropriate benefit sharing and access agreements. The chance of finding a new therapeutic agent are low; there is significant attrition in the pathway to market; and large numbers of samples will need to be investigated in HTS campaigns. Although rewards for a successful drug may be large, not all drugs introduced to the market repay the investment in their discovery. The holder of genetic resources needs to have an excellent understanding of the commercialization pathway in order to understand the value of the resource and enact the mutually agreed terms of the CBD. Investors need a guarantee that they can acquire intellectual property rights to specific compounds, and the CBD provides the mechanism to ensure certainty that commercialization can take place.

2. COLLECTION

The next step in the discovery process is the collection of samples. A biota library may consist of microbial cultures or macro biota such as plants and marine and terrestrial invertebrates. There are a number of approaches taken in collecting samples. It is possible to rely on prior knowledge, such as that generated from traditional use of the biota, or to undertake random collections. Irrespective of the

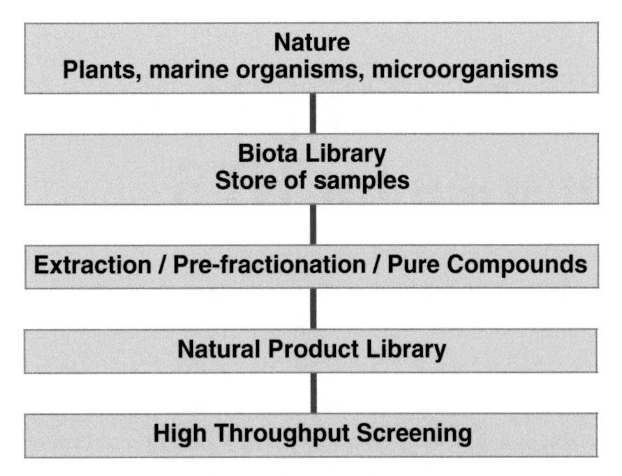

***FIGURE 7.1:** Building a natural product screening library.*

philosophy underpinning the strategy, it is essential to document and curate the collection. Getting this aspect right will be important for downstream activities (see section 3 on recollection). Two approaches are generally adopted with respect to collection. Collection may be undertaken by the biodiscovery team that brings the screening and isolation expertise or by botanists and biologists who bring expertise in the taxonomy and environmental areas. Regardless of the approach, there is no doubt that biota samples collected for biodiscovery purposes can play a vital role in conservation knowledge and biodiversity research, and placing samples into national collections will facilitate research into conservation and understanding of genetic resources. Taxonomic identification of the species is vital for both increasing the chances of finding novel species containing novel constituents and avoiding known compounds (see section 5 on dereplication). Current strategies for collection of biota samples require a great deal less material than was previously required by screening programs. [1] Advances in screening technology, in particular 384-well and 1536-well assays, mean that a sample of 200 mg can provide sufficient extract for screening in multiple HTS campaigns. Advances in spectroscopy for structure elucidation mean that 1 mg of compound can, in most cases, provide enough material for both structure elucidation and biological profiling in the primary HTS assay as well as several selectivity assays. Initial collection needs to be undertaken bearing in mind conservation of the species and the habitat from which it was collected. Development of a compound of interest will

Objectives

Conservation of biological diversity

Sustainable use of biological diversity

Fair and equitable sharing of the benefits arising out of the utilization of genetic resources

***FIGURE 7.2:** Objectives of the Convention on Biological Diversity.*

Article 15

Affirms the sovereign right of countries to control the use of their own resources

15.4 Access shall be on mutually agreed terms

15.5 Access shall be subject to the prior informed consent of the Contracting Party

15.7 Benefits arising from commercial and other utilization of genetic resources shared on mutually agreed terms

***FIGURE 7.3:** Article 15 of the CBD recognizes the sovereign rights of states over their natural resources and that the authority to determine access to genetic resources rests with national governments and is subject to national legislation.*

require progressively larger amounts of material. Collection of smaller quantities at the beginning followed by recollection of larger quantities only when results indicate the presence of a compound that is of continuing interest has the least environmental impact. In undertaking biodiscovery, it needs to be recognized that recollection of larger quantity may not be possible, and supply of compound by synthesis may be required.

The eventual objective of the natural product library is to obtain a source of diverse compounds for HTS evaluation. Compound diversity correlates with biota diversity. Therefore, it is worth bearing in mind that the seventeen megadiverse countries of the world (Australia, Brazil, China, Colombia, Democratic Republic of Congo, Ecuador, India, Indonesia, Madagascar, Malaysia, Mexico, Papua New Guinea, Peru, Philippines, South Africa, the United States, and Venezuela) account for more than 70% of all biodiversity [2]. Figure 7.4 shows the seventeen megadiverse countries of the world. Figure 7.5 shows the number of endemic plant species for the seventeen megadiverse countries as one indicator of biodiversity.

The establishment of a database, at the stage of collection, is vital to track the obligations under the CBD and also to track samples through the HTS campaigns. Data such as taxonomy, collection date and location, collecting institution and individual collector, and species abundance should be captured in the database. This assists with both tracking and monitoring samples throughout the research process for access and benefit-sharing purposes, recollection (including any concerns associated with sustainability), and the identification of factors that contribute to bioactivity, such as season, location, and stage in reproductive cycle.

A choice can be made to extract the whole biota sample, to extract a subsample of the biota sample, or to extract a fresh sample prior to each screening campaign. Extraction of the whole biota sample ensures that one extract is available for the screening, isolation, and structure elucidation stage. Sequoia, for example, extracted 150 g of dried plant

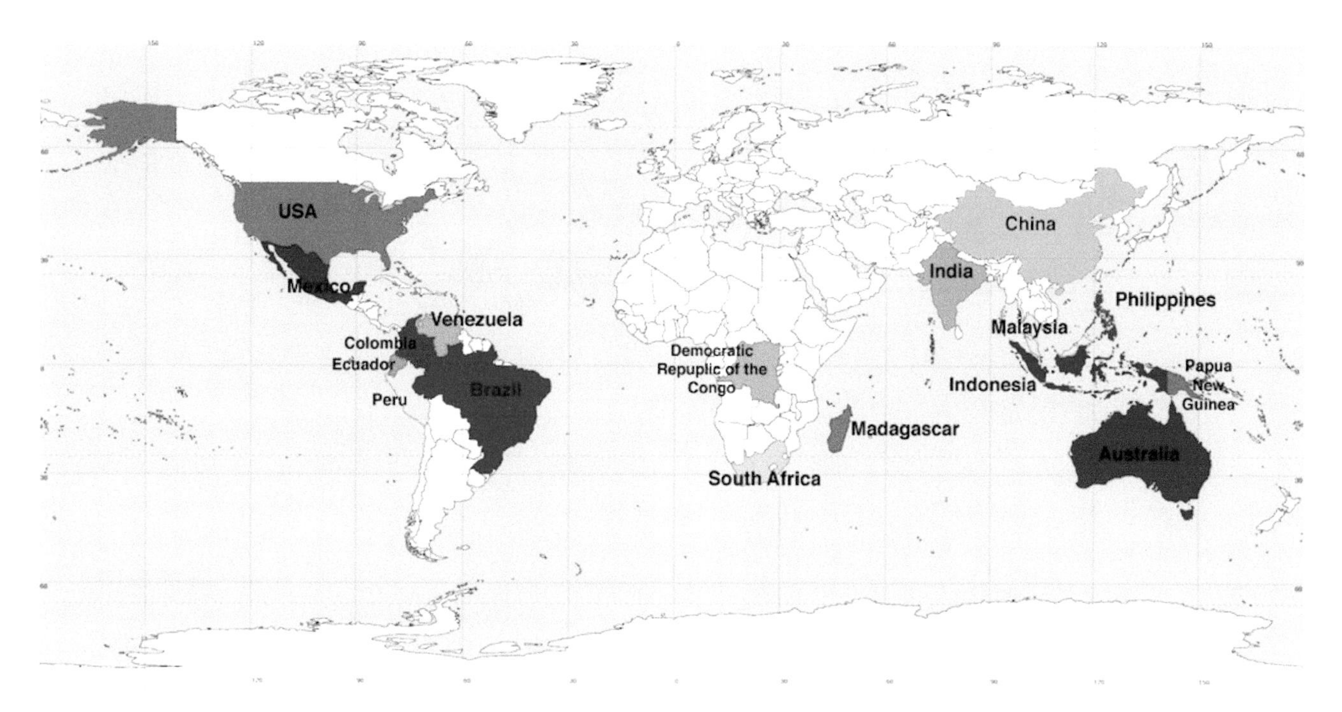

FIGURE 7.4: The seventeen megadiverse countries of the world [2].

material in generating its fractionated library [3]. Compound degradation may occur in extracts that are held for a long time. The cost in time and solvent will be greater than those for a procedure that extracts a subsample of the biota sample. In our experience, plant samples that are air-dried and ground to a powder and marine samples that are freeze-dried and ground retain compound integrity. Extraction prior to each assay is not efficient. Figure 7.6 illustrates aspects of a biota store. The humidity needs to be controlled to prevent moisture from entering the samples; high humidity may result in fungal or other microorganism spoilage. Samples have been stored in individual containers that are bar coded. Individual samples are stored in larger containers that are bar coded. The database system tracks the location of individual containers in each of the larger containers.

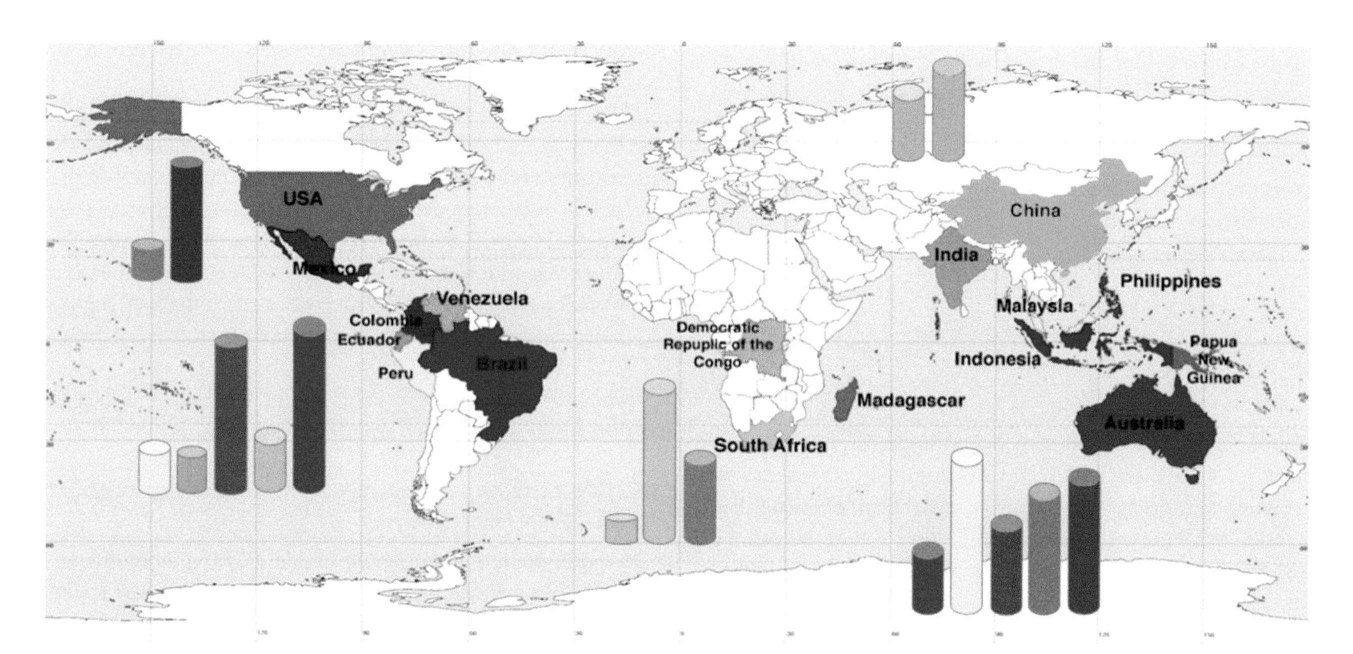

FIGURE 7.5: Number of endemic vascular plant species in seventeen megadiverse countries, adapted from Williams using Conservation International (2000) data [2].

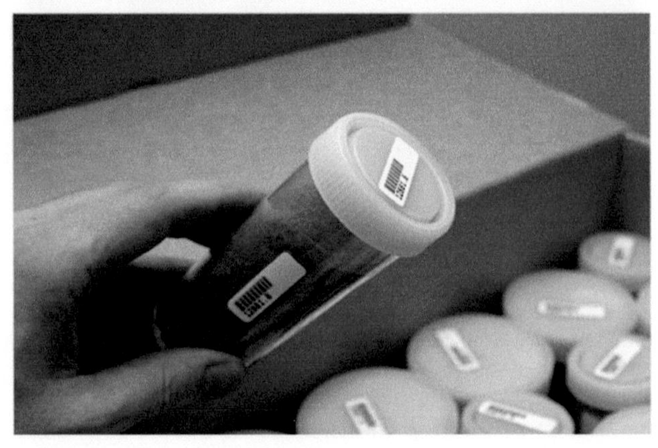

***FIGURE 7.6:* Characteristics of a biota store. Individual samples are stored in larger containers that are bar coded. The database system tracks the location of individual containers in each of the larger containers. Thanks to David Camp for logistics and Peter Walve for the software. Photo: Stuart Newman, Griffith University.**

3. RECOLLECTION

Larger quantities of compound will be required to progress any interesting compounds. The probability of recollection will be enhanced if the initial collections are well documented. The ability to have the original collector return to the same site armed with the GPS location, photographs (underwater and above water for marine samples), and taxonomic knowledge cannot be underestimated. Again, conservation requirements must be observed when larger quantities are sought.

4. EXTRACTION

The methodology for natural product screening should conform to well-established screening practice. HTS campaigns are usually conducted on compounds dissolved in dimethyl sulfoxide (DMSO). Compound libraries can be screened for many years, and preparation of a natural product screening library should ensure that it can be used for the same time span. To avoid continual extraction of the entire biota library for every HTS campaign, it is much more efficient to store sufficient amount of the biota extract to last five years (>100 screens). Figure 7.7 shows a multiparallel extraction apparatus designed to extract ninety-six solid biota samples simultaneously. An extraction cycle comprising a hexane wash and two extraction solvents with overnight drying using nitrogen and vacuum dryers allowed a twenty-four-hour cycle.

The two most popular ways of achieving long-term storage that maintains the integrity of the samples in DMSO are microtubes and minitubes held anywhere between $-20^\circ C$ and $20^\circ C$ under an atmosphere of nitrogen or low relative humidity. Figure 7.8 shows typical microtubes. Figure 7.9 shows a store that has a capacity of 300,000 microtubes and 1,500,000 wells in 384-well microtiter plate format. Figure 7.10 shows a formatting robot for conversion from tubes to microtiter plates.

There are a number of ways in which biota samples can be processed into a form suitable for screening:

- Crude extracts – an extract using organic or organic/aqueous solvent mixtures;
- Prefractionated libraries of crude extracts – crude extracts that are fractionated using solid-phase extraction (SPE), conventional liquid chromatography techniques, or a combination of both; and
- Pure natural products.

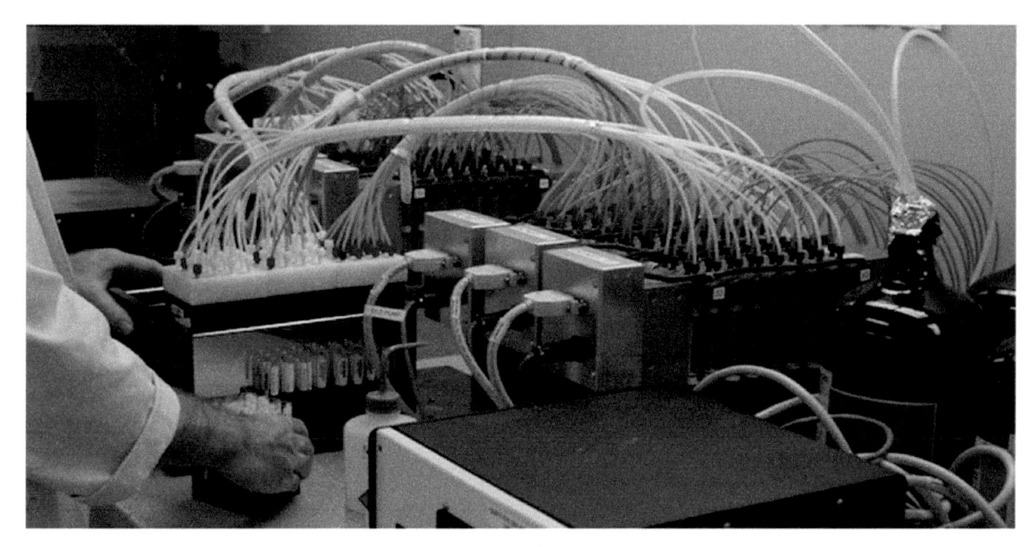

***FIGURE 7.7:* Parallel extraction – Six banks of sixteen syringes for solid phase extraction. Acknowledgement to Stephen Toms (Eskitis Institute) and J-Kem Scientific for design and construction of the 96-way parallel extraction unit. Photo: Stuart Newman, Griffith University.**

5. HANDLING DEREPLICATION

Dereplication, the use of biological profiling and chromatographic/spectroscopic analysis to recognize known substances present in an extract, has been used widely [4–6]. Dereplication aims to identify known compounds prior to investment in isolation. Another important outcome of dereplication is the identification of multiple extracts or fractions that contain the same active component or biological profile. Dereplication has been particularly necessary for those groups working on microbial extracts [7]. Differential analysis of two-dimensional (2D) nuclear magnetic resonance (NMR) spectra has been used to identify new compounds from fungal extracts [8].

***FIGURE 7.8:* Bar-coded microtubes that hold solubilized compound. Compounds are stored in individual tubes. Photo: Chris Stacey, Griffith University.**

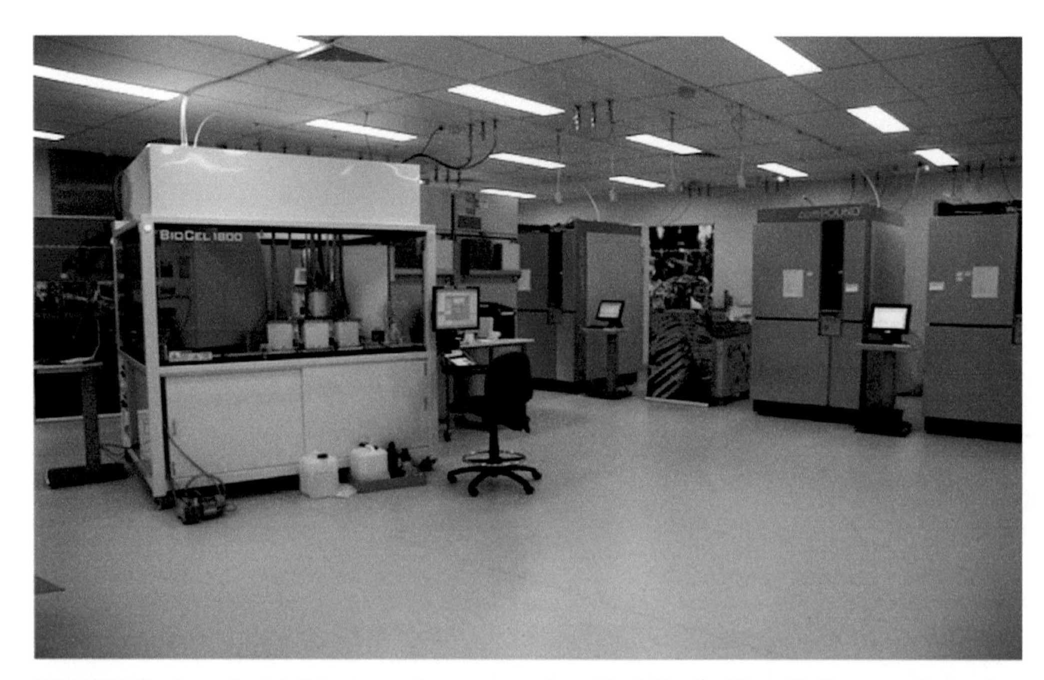

FIGURE 7.9: An automated compound management and logistics facility with the capacity to store 300,000 microtubes. Tubes can be selected via software control to provide any combination (subset) for formatting into the required high-throughput screening plate. This facility can also hold 1,500,000 wells in 384-well microtiter plates prior to commencement of screening campaigns. Photo: Ron Quinn, Griffith University.

6. PRO AND CONS OF SCREENING CRUDE EXTRACTS/FRACTIONS/PURE COMPOUNDS

A quality natural product–based library requires that the compounds within the screening set be able to be developed into therapeutic agents. A good understanding of the modern drug discovery paradigm is therefore necessary. Analysis of the 126,000 unique entries in the *Dictionary of Natural Products* [9] against Lipinski's "Rule of Five" indicated that 60% of natural products comply with Lipinski's rules. [10] Only about 10% of NPs exhibited two or more violations, the same as for trade drugs [11].

Compound diversity is greatest in crude extracts, but many components are commonly occurring or redundant. Much of the chemical space in crude extracts consists of

FIGURE 7.10: Formatting robot for generating a screening set of microtiter plates from individual tubes. Photo: Chris Stacey, Griffith University.

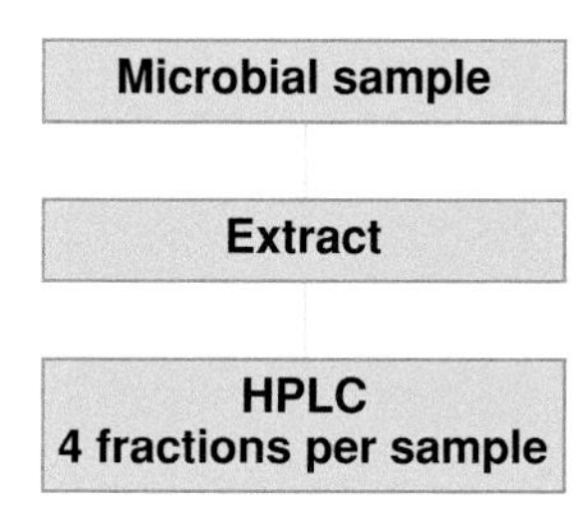

***FIGURE 7.11:** Prefractionation reported by MerLion using reverse-phase C18 HPLC and collection of four fractions per extract [15].*

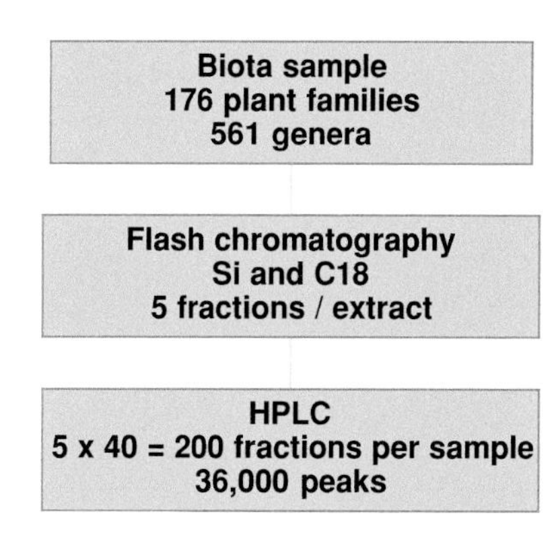

***FIGURE 7.12:** Prefractionation reported by Sequioa using flash chromatography followed by HPLC. Forty fractions were collected per each of five flash fractions from the biota sample. This process led to a library of 36,000 fractions containing one to five compounds per well (Figure 7.12) [3].*

compounds that are frequent hitters, that is, have promiscuous pharmacology, do not conform to the physicochemical parameters required of drugs, or indeed may not be druggable in the sense that they are not easily synthesized and it is difficult to develop chemical libraries or structure-activity relationships (SARs) around the compounds [12].

In addition to the screening of crude extracts, two approaches have been utilized. One has been to purify as many components of an extract as possible and determine the structures of the compounds [10, 13, 14]. This approach has several obvious advantages: 1) The compounds isolated can be treated the same as any other compound in a library; and 2) immediate assessment of the compound's potential can be made during hit evaluation, thus eliminating the time delay between hit extract identification and isolation of the hit natural product. With a structure in hand, it is possible to exclude compounds from the screening set that do not conform to the physicochemical parameters of choice or have chemical alerts associated with the structures. Two major disadvantages of this approach are 1) minor components may be overlooked and thus never reach a screening set and 2) limited throughput that results in a limited diversity.

A second approach has been to generate libraries of semipurified fractions for screening. Prefractionation involves the separation of crude extracts by SPE, column chromatography, high-performance liquid chromatography (HPLC), liquid-liquid partitioning, or some combination of the above to obtain fractions containing simpler mixtures of compounds. Prefractionation can eliminate highly hydrophilic and hydrophobic fractions, and compounds found within each fraction can be tested at a higher dose than would be possible in a crude extract. Even though active fractions require purification to isolate the active constituent, the use of MS and NMR data as well as bioassay data can guide further purification.

Numerous prefractionation strategies have been reported [3, 14–19]. Approaches are wide-ranging, from the preparation of four to two hundred fractions per sample. MerLion has taken the approach of using reverse-phase C18 HPLC and collecting four fractions with collection commencing after the solvent front peak (Figure 7.11) [15]. Sequioa performed an organic extraction followed by silica flash chromatography and an aqueous extract followed by C18 flash chromatography. Subsequent HPLC collected forty fractions per each of four flash fractions from the organic extract. The aqueous extract was passed through C18, polyamide, and a 3,000 molecular cutoff filter, and the single aqueous flash fraction was subjected to HPLC, again collecting forty fractions. The HPLC fractions that contained quantifiable compounds, approximately 60% of the total, constituted the library. This process led to a library of 36,000 fractions containing one to five compounds per well (Figure 7.12) [3]. Wyeth has used prefractionation to generate ten fractions plus crude from the combined extract from at least two different culture conditions (Figure 7.13) [18]. The Ireland group has used SPE followed by HPLC to generate eighty fractions per sample (Figure 7.14) [19].

Prefractionation together with taxonomic identification, GPS location, photographs, and voucher specimens

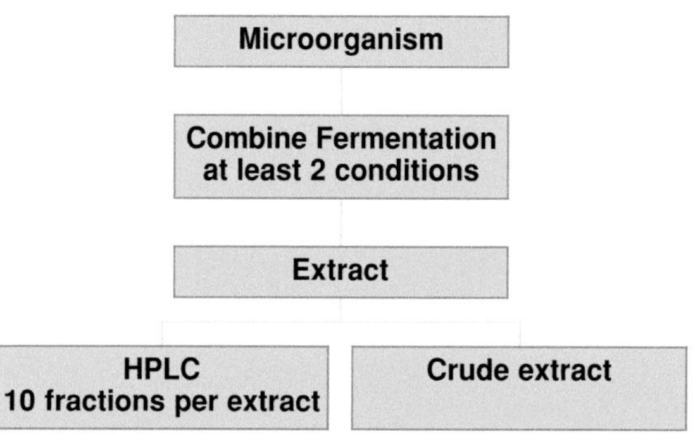

***FIGURE 7.13:** Prefractionation reported by Wyeth generating ten fractions plus crude from the combined extract form at least two different culture conditions [18].*

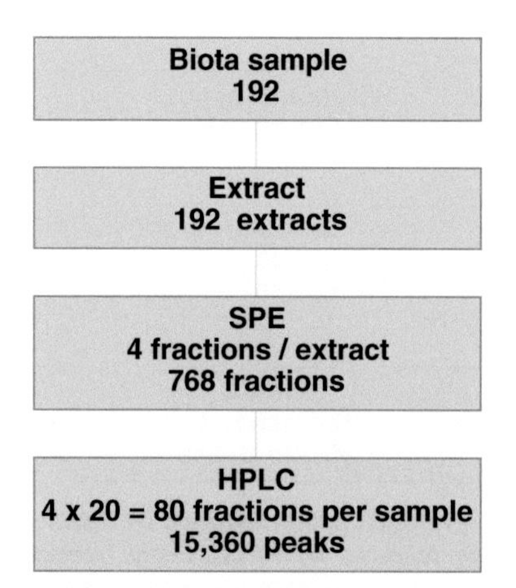

FIGURE 7.14: Prefractionation by the Ireland group used SPE followed by HPLC to generate eighty fractions per sample [19].

provide a systematic documentation of a biota collection (Figure 7.15). A number of questions can be answered with this data set. On the one hand, Figure 7.16 shows five chromatograms of the same genus that clearly demonstrated chemical diversity at the species level. On the other hand, Figures 7.17 and 7.18 shows a cluster of four and two biota, respectively, displaying similar chromatographic profiles. If biological screening results were to indicate activity in the same fraction, a useful dereplication would be achieved.

HTS data have been used to evaluate prefractionation strategies. Of the 1,882 active cultures following nine HTS campaigns on a microbial natural product library, 79.9% of the activity was found only in the fractions, 12.5% of the activity was detected only in the crude extract, and 7.6% was found in both fractions and crude extract (Figure 7.19) [18]. A very similar result was achieved using the simpler four fractions per extract approach. For 80% of the 1,700 active fractions from eleven HTS screens, the associated crude extract did not show activity (Figure 7.19) [15].

In considering the approach to developing the natural product library, the throughput of the HTS technology as well as the cost need to be taken into account. The Wyeth publication quotes assay costs at $0.25 to $0.50 per well and the screening of one million wells at between $250,000 and $500,000 per HTS campaign. If ten thousand biota samples were extracted to give ten thousand extracts, they would produce twenty-nine 384-well microtiter plates in which the plate has two columns empty for controls. If four fractions were produced per biota sample, the resultant forty thousand fractions would produce one hundred

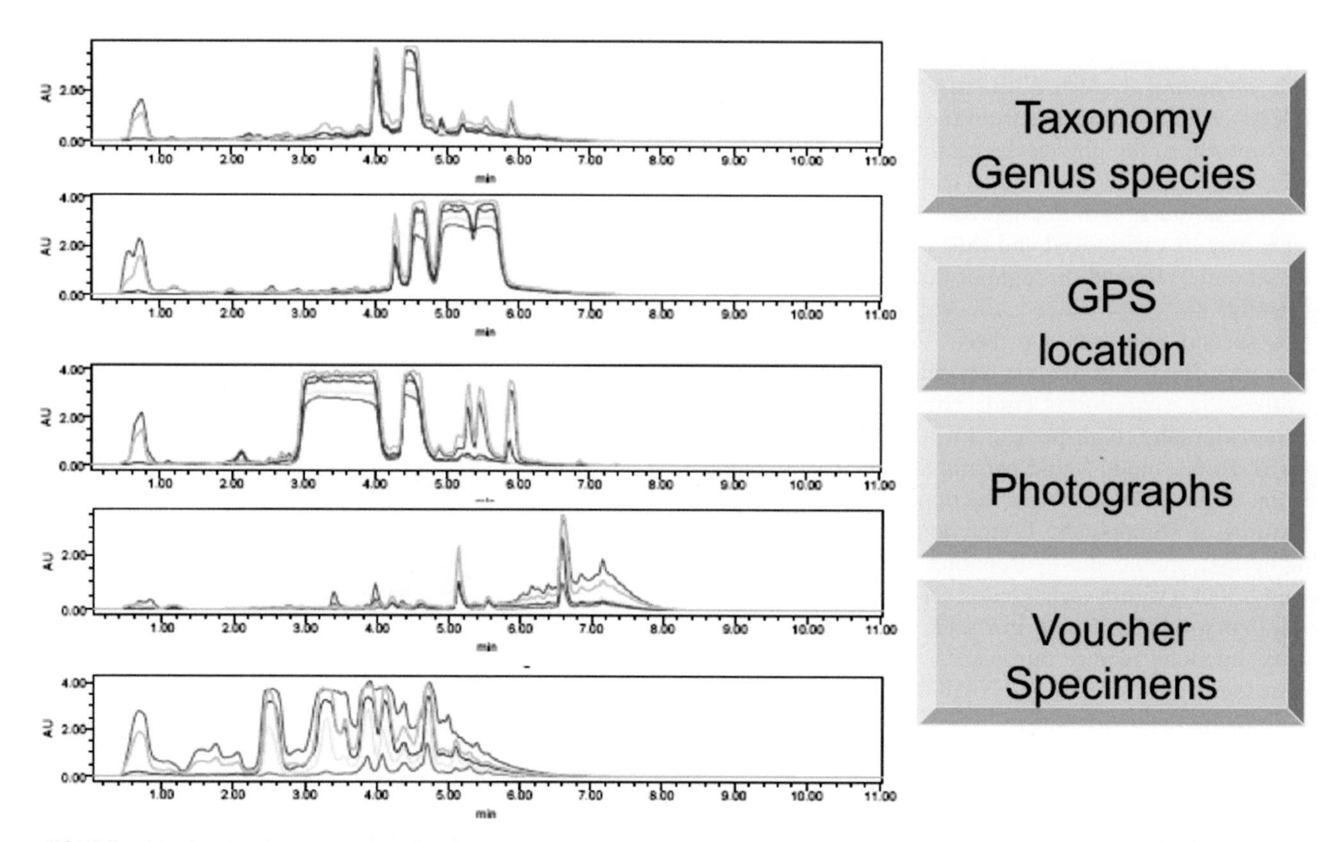

FIGURE 7.15: Prefractionation together with taxonomic identification, GPS location, photographs, and voucher specimens provide a systematic documentation of a biota collection.

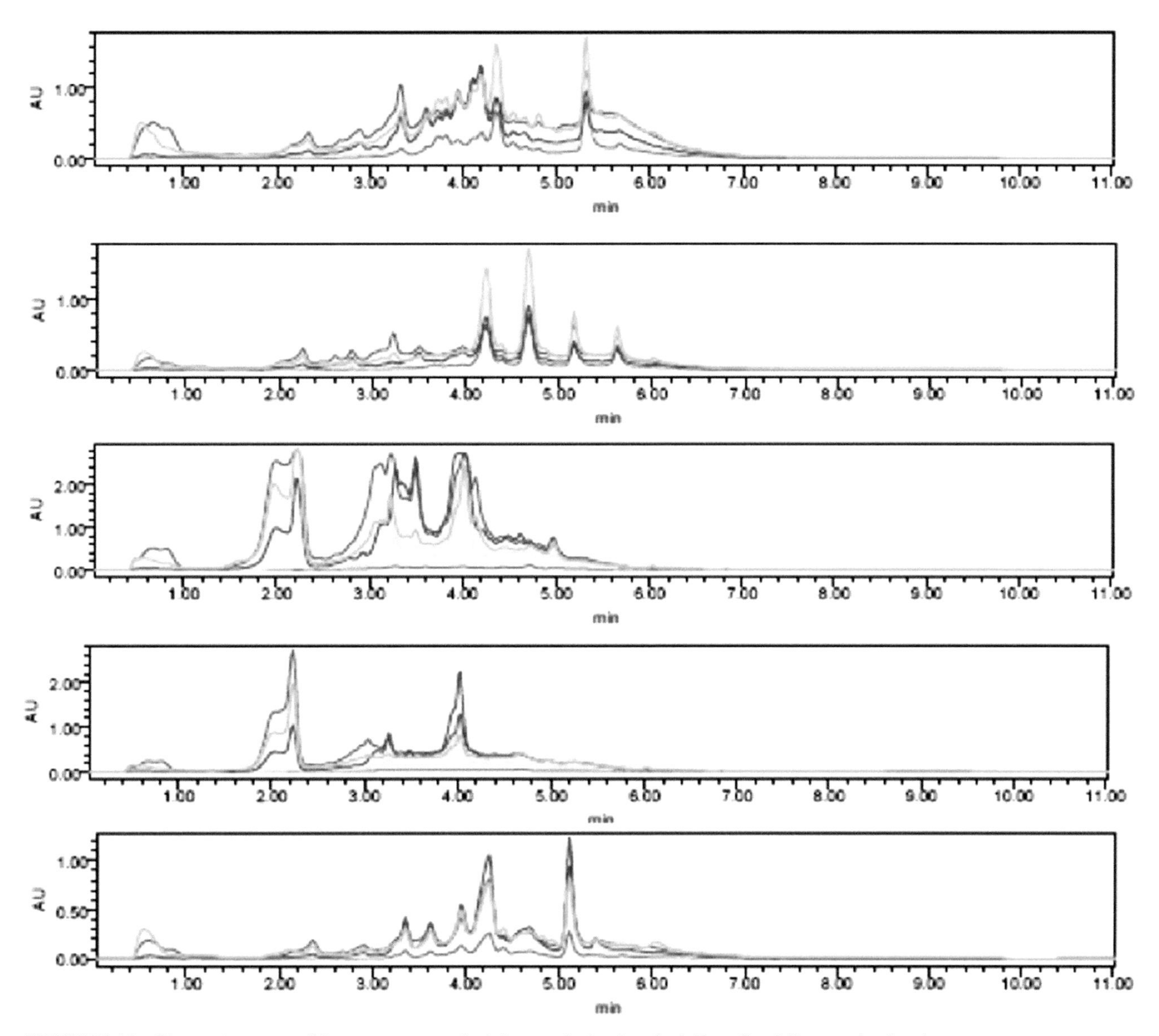

FIGURE 7.16: Chromatograms of the same genus that demonstrate chemical diversity at the species level.

and fourteen 384-well microtiter plates. Increasing the prefractionation to give twenty fractions per sample would require 569 microtiter plates to house the two hundred thousand fractions. Isolation of pure compounds from ten thousand biota samples, which may contain approximately two hundred thousand unique compounds, will result in a smaller number of compounds because of the time and effort involved in isolation and structure elucidation. A set of twenty thousand pure compounds would be accommodated in six microtiter plates. Figure 7.20 shows the cost for the resultant HTS campaigns at a cost of $0.10 per well. A cost-benefit analysis (Figure 7.21) indicates that prefractionation provides a significant benefit in screening relative to the investment in enrichment of the screening set. The cost associated with preparation of a pure natural product library is significantly greater than the effort to prefractionate. The arrow indicates a balance along this continuum where the cost-benefit ratio is optimal.

CONCLUSION

Natural products interact with a variety of proteins during their biosynthesis and can target specific therapeutic target proteins. [20] If modern drug discovery principles are applied in the preparation of the natural product screening library using the general approaches discussed, the resultant high-quality libraries will be more efficient in harnessing the natural product world in the modern drug discovery paradigm.

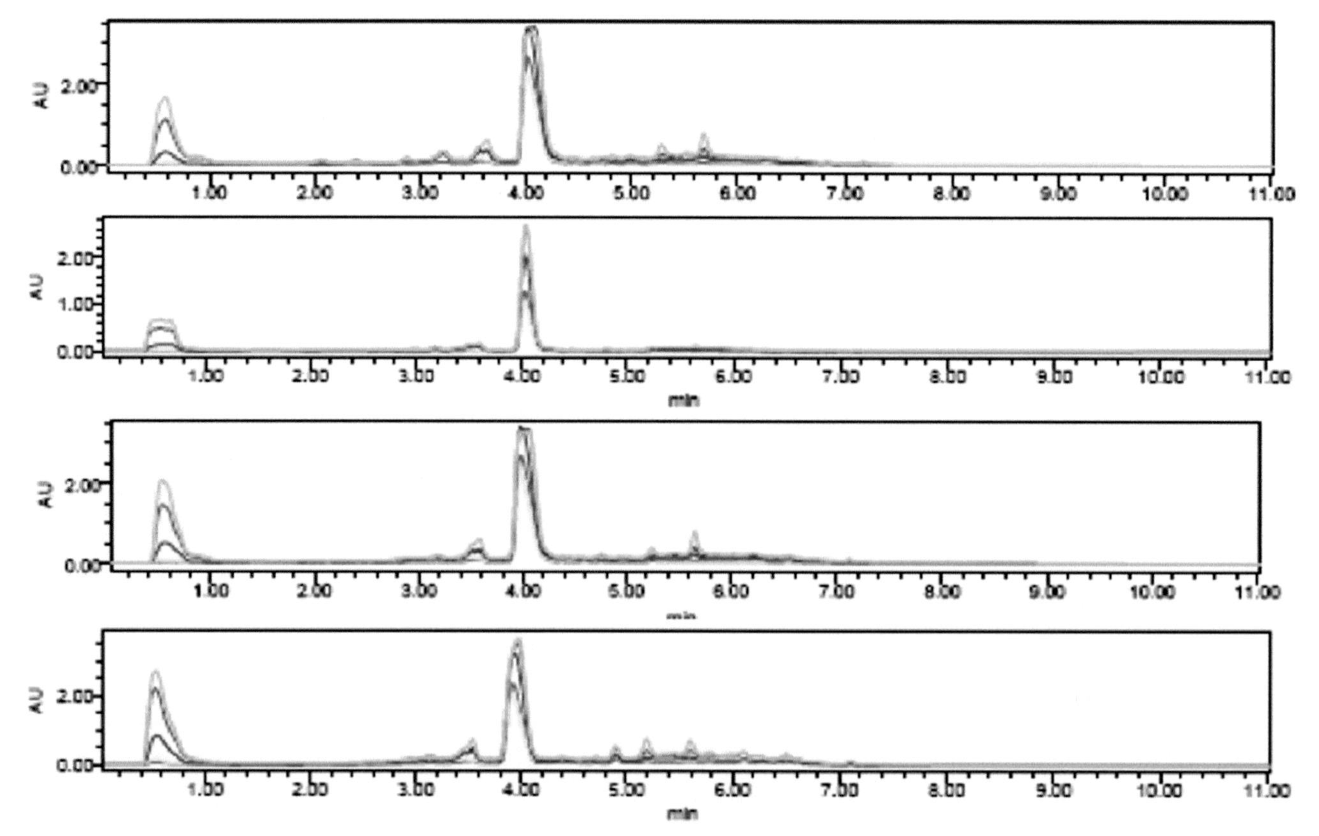

FIGURE 7.17: Prefractionation aids dereplication.

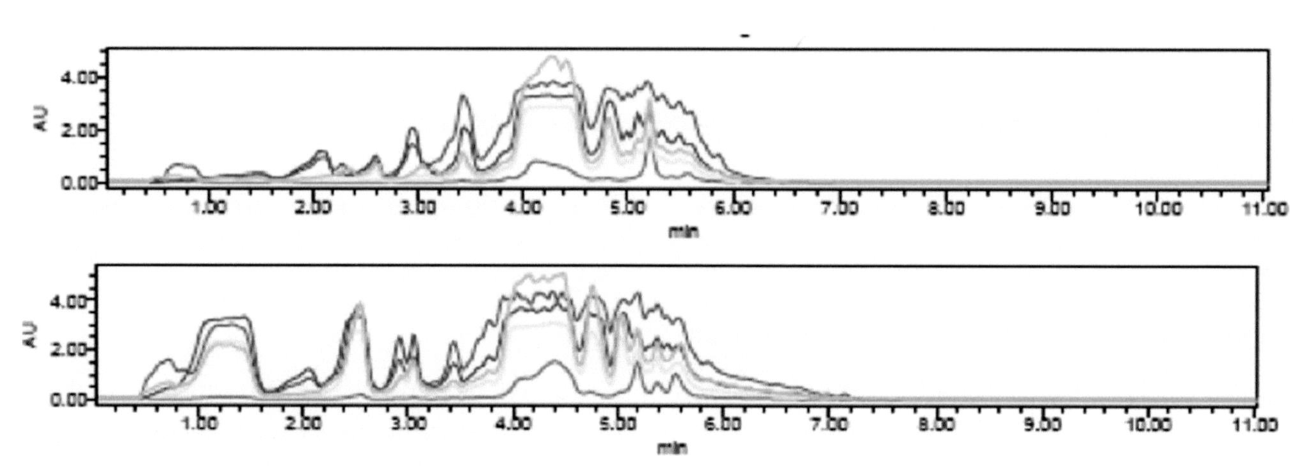

FIGURE 7.18: Prefractionation aids dereplication.

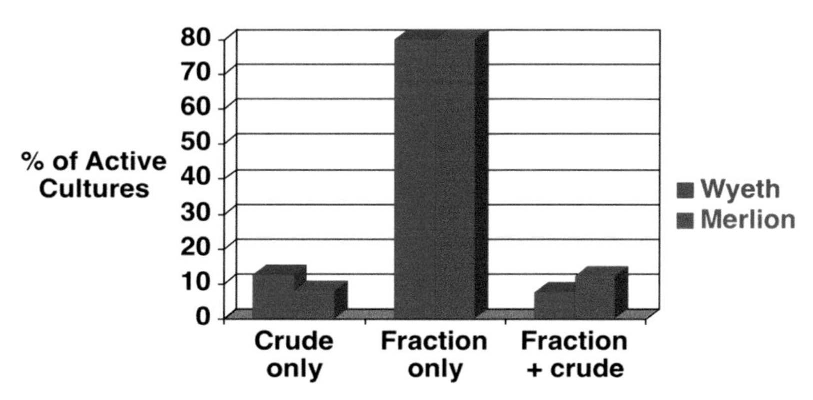

FIGURE 7.19: Distribution of Hits from 1,882 active cultures across nine HTS campaigns and from 1,700 active fractions from eleven HTS campaigns. Adapted from Wagenaar, M. M. (2008). Molecules 13*, 1406–1426, and Appleton, D. R., Buss, A. D., Butler, M.S. (2007).* Chimia 61*, 327–331. [15, 18]*

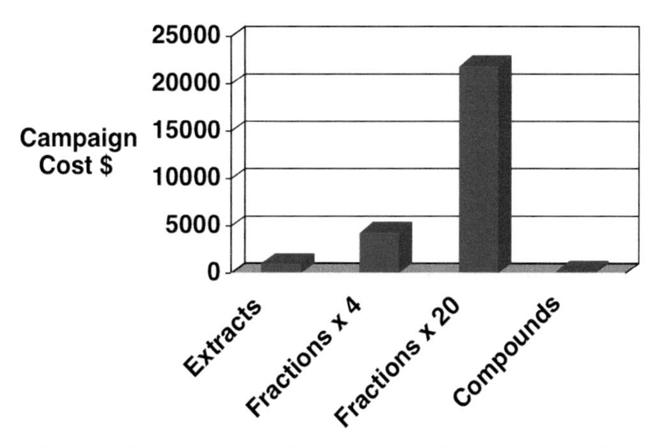

FIGURE 7.20: The greater the prefractionation, the greater the resultant cost of the HTS campaign.

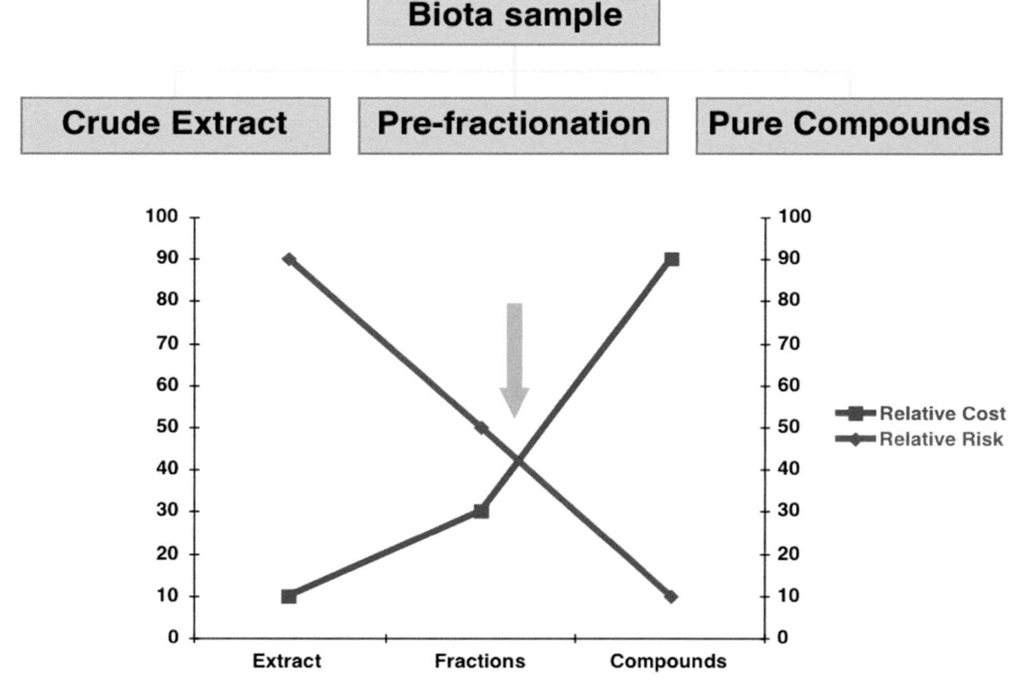

FIGURE 7.21: Cost-benefit analysis of screening sets.

REFERENCES

1. Camp, D., and Quinn, R. J. (2007). The evolution of compound libraries for applied and basic research. *Chem Aust 74*, 14–16.
2. Williams, J., Read, C., Norton, A., Burgman, M., Proctor, W., and Andeerson, H. (2001). *Biodiversity, Australia State Of The Environment Report 2001 (Theme Report)* (Canberra: CSIRO Publishing on behalf of the Department of the Environment and Heritage).
3. Eldridge, G. R., Veroort, H. C., Lee, C. M., Cremin, P. A., Williams, C. T., Hart, S. M., Goering, M. G., O'Neil-Johnson, M., and Zeng, L. (2002). High-throughput method for the production and analysis of large natural product libraries for drug discovery. *Anal Chem 74*, 3963–3971.
4. Strege, M. A. (1999). High-performance liquid chromatographic-electrospray ionization mass spectrometric analyses for the integration of natural products with modern high-throughput screening. *J Chromatogr B 725*, 67–68.
5. Wolf, D., and Siems, K. (2007). Burning the hay to find the needle – data mining strategies in natural product dereplication. *Chimia 61*, 339–345.
6. Julian R. K., Jr., Higgs, R. E., Gygi, J. D., and Hilton, M. D. (1998). A method for quantitatively differentiating crude natural extracts using high-performance liquid chromatography – electrospray mass spectrometry. *Anal Chem 70*, 3249–3254.
7. Bitzer, J., Köpcke, B., Stadler, M., Helwig, V., Ju, Y.-M., Seip, S., and Henkel, T. (2007). Accelerated dereplication of natural products, supported by reference libraries. *Chimia 61*, 332–338.
8. Schroeder, F. C., Gibson, D. M., Churchill, A. C. L., Sojikul, P., Wursthorn, E. J., Krasnoff, S. B., and Clardy, J. (2007). Differential analysis of 2D NMR spectra: new natural products from a pilot-scale fungal extract library. *Angew Chem Int Ed 46*, 901–904.
9. *Dictionary of Natural Products on CD-Rom* (2005). (London:. Chapman and Hall/CRC Press). www.crcpress.com.
10. Quinn, R. J., Carroll, A. R., Pham, N. B., Baron, P., Palframan, M. E., Suraweera, L., Pierens, G. K., and Muresan, S. (2008). Developing a drug-like natural product library. *J Nat Prod 71*, 464–468.
11. Lee, M. L., and Schneider, G. (2001). Scaffold architecture and pharmacophoric properties of natural products and trade drugs: application in the design of natural product-based combinatorial libraries. *J Comb Chem 3*, 284–289.
12. Harrigan, G. G., and Goetz, G. H. (2005). Chemical and biological integrity in natural products screening. *Comb Chem High Throughput Screen 8*, 529–534.
13. Bindseil, K. U., Jakupovic, J., Wolf, D., Lavayre, J., Leboul, J., and van der Pyl, D. (2001). Pure compound libraries; a new perspective for natural product based drug discovery. *Drug Discov Today 6*, 840–847.
14. Abel, U., Koch, C., Speitling, M., and Hansske, F. G. (2002). Modern methods to produce natural-product libraries. *Curr Opin Chem Biol 6*, 453–458.
15. Appleton, D. R., Buss, A. D., and Butler, M. S. (2007). A simple method for high-throughput extract prefractionation for biological screening. *Chimia 61*, 327–331.
16. Schmid, I., Sattler, I., Grabley, S., and Thiericke, R. (1999). Natural products in high throughput screening: automated high-quality sample preparation. *J Biomol Screen 4*, 15–25.
17. Jia, Q. (2003). Generating and screening a natural product library for cyclooxygenase and lipoxygenase dual inhibitors. *Stud Nat Prod Chem 29*, 643–718.
18. Wagenaar, M. M. (2008). Pre-fractionated microbial samples – the second generation natural products library at Wyeth. *Molecules 13*, 1406–1426.
19. Bugni, T. S., Richards, B., Bhoite, L., Cimbora, C., Harper, M. K., and Ireland, C. M. (2008). Marine natural product libraries for high-throughput screening and rapid drug discovery. *J Nat Prod.71*, 1095–1098.
20. McArdle, B. M., Campitelli, M. R., and Quinn, R. J. (2006). A common protein fold topology shared by flavonoid biosynthetic enzymes and therapeutic targets. *J Nat Prod 69*, 14–17.

SECTION THREE

BASICS OF HIGH-THROUGHPUT SCREENING

CHAPTER 8

Essentials for High-Throughput Screening Operations

Stewart P. Rudnicki
James V. Follen
Nicola J. Tolliday
Caroline E. Shamu

As the field of chemical genomics grows, many universities and other research institutions are establishing high-throughput screening (HTS) facilities for screening of small molecule libraries. The authors of this chapter are current and former staff members at the Harvard Medical School HTS facility, which was originally established as part of the Institute for Chemistry and Cell Biology (ICCB). The ICCB screening facility was one of the first HTS facilities in academe. Over the years, its staff members have advised many colleagues at other institutions as they build their own HTS facilities.

1. PLANNING AND DESIGN OF THE FACILITY WORKSPACE

Space planning is the first concern in the design of a screening facility. Adequate space is required for bench work, screening equipment, compound storage, staff office space, data analysis workstations, and IT infrastructure. An area of approximately 400 square feet is enough to accommodate a small office for facility staff as well as two to three plate readers or automated screening microscopes, several small liquid handling devices, freezer(s) for compound storage, and a tissue culture area.

The design of facility workspace is wholly dependent upon the types of assays being performed and the equipment required to process the assays. For example, mammalian cell-based assays require access to a tissue culture area with water-jacketed CO_2 incubators, whereas yeast or bacterial assays require only a standard 37°C incubator. Separation of these types of screen workflows should be considered in the design of laboratory areas. All assays require an adequate amount of bench workspace for individual researchers to prepare and carry out their screens. Flexibility can be introduced into a facility from the outset by purchasing carts designed for laboratory equipment. These carts can be used separately or connected to provide modular, integrated workstations. In lieu of this, however, standard laboratory benches and a reasonable amount of open floor space for a small number of freestanding machines is sufficient.

For tissue culture, 6-foot, laminar-flow tissue culture hoods are recommended because they have room for an automated plate filler, which is required to dispense cells into assay plates. To accommodate a wide variety of assay conditions, incubators may be equipped with circulating coolers to allow a temperature range between 20°C and 42°C. Clinical centrifuges used for tissue culture can be equipped with microplate carriers so that they can be used as necessary for compound stock plates and assay plates. It is helpful to have a refrigerator in the screening facility or access to a cold room nearby for temporary storage of cell media and assay reagents.

Consideration should be given to ensuring that adequate and dependable vacuum and gas (e.g., CO_2 and air) services are available for the facility, as some instruments, such as high-end liquid handling instruments, require them for their operation. In many academic laboratories, because of age of infrastructure or intermittent heavy usage, one or both of these services may be unreliable. One solution to this problem is to employ small, dedicated vacuum pumps and compressors where they are needed. Soundproof cabinets or enclosures can keep noise at a reasonable level.

Computer network connections (data jacks) are essential for the assay detection and data capture phases of screens. Large amounts of data are generated during HTS, and their final destination (e.g., file server and/or database), as well as high-speed network connections, should be planned during the design of the facility. It is helpful to maintain close working relationships with the local computing support group and the data management group to ensure a smooth

TABLE 8.1: Examples of Academic HTS Facility Staff Positions

Assay Developer	Software Engineer
Lab Automation Specialist	Microscopy Specialist
Data Analyst	Image Analyst
Database Administrator	Medicinal Chemist

transfer of data from the screening facility to institutional file servers and databases.

2. STAFF

A wide variety of skills are necessary to run an HTS facility (examples of staff positions dedicated to particular areas of academic screening are listed in Table 8.1). Depending on the size of the facility, one or more people may be dedicated to particular areas of screening. For example, there might be one Ph.D.-level assay developer, one laboratory robotics specialist, and one data analyst/statistician on staff. Alternatively, the responsibilities of these areas might be the combined responsibility of just one or two individuals. An aptitude for troubleshooting the computer and mechanical problems that invariably arise with complex instrumentation is essential. At least one staff member should be proficient in computer programming languages such as Visual Basic, C++, or JAVA, which are used to integrate the operation of individual screening robots with each other. In addition to operating and maintaining laboratory equipment, staff members likely will also be asked to organize compound collections and assist investigators in performing screens. Thus, some formal training in the biological sciences or chemistry is desirable. Experienced laboratory automation specialists can be hired from industry to the academic setting, but many academic facilities have also successfully trained recent college graduates for their staff positions. Organization, formatting, and tracking of library and screening data can be large responsibilities, as can be data analysis and database support.

3. COMPOUND ACQUISITION

There are many compound libraries available for purchase from commercial suppliers. These fall into two broad classes: libraries assembled from compounds that were collected as discretes from chemistry laboratories worldwide (especially from Russia and Ukraine) and combinatorial chemistry libraries that are synthesized by laboratories affiliated directly with vendors. Generally, the collected compounds are cheaper than the vendor-synthesized compounds, but vendors often provide more follow-up options (e.g., guaranteed resupply/resynthesis) for their own compounds.

In choosing a compound supplier, one needs to consider not only the cost per library and the quality/purity of the compounds sold, but also the resupply cost and availability of individual compounds for follow-up experiments. As time passes and vendors add new compounds to their collections, availability of older compounds generally falls. It is important to ask a potential supplier how long resupply of its compounds is guaranteed (one year from the purchase date is typical) and what options are available once compounds are out of stock. Another area of consideration is whether the vendor retains any intellectual property (IP) rights on the compounds it sells. Academic screening facilities are probably most interested in compounds without IP entanglements that might complicate publication and follow-up of screening results.

Once a supplier is selected, the next step is to choose which compounds to purchase. The degree of choice offered and the level of assistance provided by the vendors during this process vary greatly. Some suppliers sell pre-assembled libraries for a set cost; others allow the buyer to choose individual compounds from a larger collection and charge per compound. If possible, it is very helpful to get advice from colleagues who are medicinal chemists or who have experience with small molecule screening.

Some groups recommend collections that are enriched for complex heterocyclic compounds and compounds of higher molecular weight (MW) (average MW of ~350–400) because they feel that these are likelier to provide interesting hits in screens. However, low molecular weight compounds are preferred for fragment-based ligand discovery (see Chapter 6). Eliminating those with groups that might make them unstable or toxic is another strategy to minimize the number of potentially "bad" compounds. Unstable imines, compounds with free carboxyl groups, and compounds with building block elements that might chelate metals are examples of potentially undesirable compounds.

When placing an order, one must specify how the compounds should be shipped, including the type of plate and the number of wells left empty per plate. Generally, 0.5 to 1.0 mg of each compound is enough to make several copies of each screening plate. Depending on compound transfer method, a purchase of 1 mg of each compound should support at least one hundred to two hundred screens of the library in 384-well format. Sometimes it is easier to request that the compounds arrive already dissolved in dimethyl sulfoxide (DMSO), as it is quite time-consuming to resuspend more than ten thousand dry compounds. A good concentration at which to store commercial compound stocks is 5 mg/mL (see the Compound Storage and Handling section below for more details). Note that deep-well liquid handling capabilities may be required for compound reformatting into 384-well screening stock plates, as compounds are almost always shipped from the supplier in 96-deep well plates or in 96-tube racks. Consider asking for at least one column to be left empty in each 96-well plate. This results in two empty columns per plate for screening controls once the compounds have been formatted into 384-well plates.

4. COMPOUND STORAGE AND HANDLING

Because the characteristics of individual compounds within screening collections vary greatly, there is no single ideal storage solution for compound libraries. Typically, compound stocks are dissolved in DMSO and stored frozen, either at 4°C or – 20°C. Because DMSO is hygroscopic and because many compounds used in screening are not soluble at high concentration in water, compound stock plates should be stored in a desiccated environment. To promote compound stability, it is recommended that the number of freeze/thaw cycles experienced by compounds be limited to as few as reasonably possible (fewer than ten to fifteen cycles is preferred).

Expensive plate storage devices and servers are available to organize and store compound stock plates under controlled atmosphere at controlled temperatures. These generally require bar coding of stock plates and integration with other screening instruments via robotic arms to retrieve library plates. Such storage systems are wonderful if money and space allow but are probably not practical purchases for starting screening facilities in academe.

One alternative to an automated storage system is to store microplates in a – 20°C freezer in an airtight container containing desiccant (e.g., calcium carbonate). Such airtight containers can be custom-made to any size or shape, allowing for storage of microplates in many formats such as racks. Using this type of system, as many as 1,056 microplates can fit inside a standard Revco freezer. The library microplates themselves are made from polypropylene and are sealed with foil seals that adhere either by heating (applied using an automated heat sealer) or by adhesive. Both polypropylene plates and foil seals resist DMSO well and are suitable for long-term storage.

Reformatting libraries from deep-well 96-well blocks to screening format is a process that depends on many factors. Plate format, volume of compound available, and available freezer space usually determine how many copies of screening, stock, and cherry-pick plates to make. At the ICCB-Longwood Screening Facility at Harvard Medical School, 1 mg of each compound is purchased, and seven copies of each compound library plate are made when the library is formatted initially: four copies for screening (pin transfer into assay plates), one for cherry picking, and two reserve stock plates (for subsequent formatting into screening or cherry-pick copies). The screening and cherry-pick copies contain 10 μL each, whereas the stock copies contain 75 μL each. Only one copy of each screening plate is active at a time; the others are held in reserve. The concentration of the compounds in DMSO is 5 mg/mL (corresponding to ~10 mM for a compound of MW 500). After a screening copy has undergone ten to fifteen freeze/thaw cycles, it is retired and a new screening copy is activated.

Plates to be screened are typically thawed at room temperature in desiccators before they are used for screening. Library plates are ideally stored in a dry/inert atmosphere at all times, but short of this, open air exposure should be as brief as reasonably possible to prevent water absorption. After screening, the library plates are sealed and returned to the – 20°C freezer.

5. INSTRUMENTATION

Prior to the purchase of any equipment, it is important to consider the particular requirements of the users. For example, fairly inexpensive plate fillers are sufficient to perform all of the liquid handling steps of some screens, whereas more accurate, but very expensive, low-volume, automated pipettors are required for other screens. For assay detection, some instruments provide multimode assay detection and are therefore useful for multiple assay types, whereas others are specialized for individual assays. An additional consideration is whether radioactive assays will be performed in the screening facility.

Instrument calibration and maintenance are important. Liquid handling machines usually arrive from the factory with calibration certification, but the conditions under which this verification was performed do not necessarily correspond with the conditions that exist in the screening facility. It is good practice, after delivery, to validate the accuracy and precision of a machine for each specific assay run in the laboratory and to monitor the findings at regular intervals (e.g., weekly, monthly, or quarterly). The initial in-facility calibration may involve leveling the machine itself, as bench tops and floors may not be level. Because of the small size of dispense needles or pin arrays in some instruments, a surface that is only slightly off level can produce significant inaccuracies in the volumes of liquid transferred. Finally, one common option when purchasing a new machine is an extended service contract. These can offer substantial discounts on of the cost of standard service contracts, which can be very expensive, usually 10% of the purchase price per year. For non–mission critical instrumentation, it is often a good policy to track the maintenance costs for the machines throughout their warranty period (usually one to two years) and to purchase a contract only if the yearly costs exceed the price of the contract.

The following sections discuss considerations to take into account when choosing instrumentation for HTS.

5.1. Liquid Handling

There are a wide variety of options available for high-throughput liquid handling. Frequently, improvements in accuracy at low volumes correspond with a large increase in instrument price. There are several different types of liquid handlers in use in HTS today.

Plate fillers (also known as bulk reagent dispensers) are designed to quickly dispense one reagent at a time to many different plates. They pump assay reagents or cells through a manifold (e.g., 8- or 16-channel) into microplates. Plate fillers tend to be straightforward to operate and can normally be used independently by screeners after only a short training session. Some instruments employ a dispensing cartridge that can be removed and sterilized in an autoclave – a convenient feature when performing operations that require sterile conditions. In some instruments, it is not possible to completely sterilize the path from reagent bottle to microplate. Basic plate fillers are generally only accurate to volumes greater than 2 μL per well. However, recent advances in technology now allow plate fillers to dispense accurately in the nanoliter volume range. Some fillers are even capable of dispensing different volumes and different reagents to individual wells. Plate fillers with these advanced capabilities tend to be more expensive, however.

Pipettors aspirate and dispense liquid and can make copies of whole microplates in one step or transfer the contents of specific wells to new plates. Most high-accuracy ("high-end") automated pipettors (liquid handling instruments) are multifunctional and are used for library reformatting, compound transfer, assay reagent addition, and cherry picking (see the Hit Picking section below for more information about cherry picking). Significant training is generally required to program, run, and maintain high-end liquid handling instruments. Thus, these instruments might be operated exclusively by screening facility staff. Most platforms can be described as having a deck with positions for microplates, pipet tips, and so forth, and one or more robotic arms with pipettors, pin tools, or grippers attached. Accessories are nearly unlimited and include items such as stackers, shakers, plate hotels, plate readers, tip loaders, and waste bins to allow the instrument to be tailored to a specific task or tasks.

A *pin tool* is an array of carefully machined stainless steel pins capable of transferring very low volumes of liquid from one plate to another. When transferring volumes less than 1 μL, most liquid handling systems require expensive specialized equipment to pipette accurately. Pin tools are a good alternative to pipettors in this volume range. Pins can be obtained in arrays of nearly any size (24, 48, 96, 384, or 1536 pins) that transfer volumes from 5 to 4,000 nL. When calibrated regularly, stainless steel pins reliably transfer sub-1 μL volumes into microplate wells already filled with assay reagents or cells. A pin array can be rapidly washed, dried with compressed air, and reused with undetectable levels of carryover between stock plates. Pin arrays do wear over time, however, and must be sent back to the manufacturer approximately once a year for refurbishment. One limitation of pin arrays is that, for accurate transfer, they require a minimum volume of liquid in the source plate (in addition to the transfer volume) and must transfer into destination wells containing liquid.

Plate washers aspirate liquid from microplate wells and dispense wash buffers. Fully automated plate washers are relatively inexpensive and can carry out a wash step (aspiration followed by buffer addition) on a 384-well assay plate in thirty to sixty seconds. One drawback of these instruments is that the needles that perform these tasks can clog easily, despite regular cleaning. Sonication of the needles, a new feature on some plate washers, has proven effective at reducing or eliminating these problems. Nevertheless, plate washers tend to require more care and maintenance than other liquid handling machines. Another problem is that vigorous plate washing will often wash cells off the microplates used in cell-based assays. Specially designed plate washers with gentle washing capabilities should be selected when used for this application. Washing 384-well plates can also be accomplished in a semiautomated fashion using commercially available, handheld 24-channel "wands" (or aspirating manifolds) attached to a vacuum line for the aspiration step, followed by use of a plate filler to add wash reagents.

Acoustic transfer is a new technology that has proven useful for compound transfer and hit picking: A sound wave is bounced off the bottom of a well in a source microplate to transfer liquid into a specific well of a destination microplate that is inverted above the source plate. Acoustic transfer offers the advantage of no contact between the instrument and the liquid being transferred and can be set to transfer a range of volumes by adjusting the size/number of acoustic pulses. The technology can also transfer from any well to any well, which makes it useful as a hit picker. A few drawbacks of the current acoustic technology prevent it from being more widely adopted. Because it can only be used where the surface tension of the liquid in the destination plate allows the plate to be inverted over the top of the source plate, it is really suitable only for 384-well and higher-density microplates. Also, only one well at a time can be transferred from plate to plate, and a specific flat-bottom source plate is required. Finally, the technology is still very expensive, preventing its adoption by many academic laboratories.

5.2. Compound Transfer

Every small molecule screen requires transfer of compounds from library stock plates to assay plates. Compounds can be transferred into microplate wells already containing cells and assay reagents, or compounds can be transferred into empty assay plates as the first step of screening. The pros and cons of each approach are detailed in Table 8.2. When transferring compound into filled wells, pin tools are often used, allowing for lower-volume transfers (in the nanoliter range). Using pin arrays to transfer library compounds conserves library reagents and reduces operating costs because the only consumable is the methanol used for washing the pins. Aliquoting compounds into empty assay plates in advance of screening requires a method other than pin transfer. Because of the challenges in pipetting volumes less than 1 μL using automation, this usually means transferring volumes in the microliter range to the empty wells.

TABLE 8.2: Transferring Compound to Empty versus Filled Assay Plates: Advantages and Disadvantages

Transfer Compounds to Empty Wells		Transfer Compounds to Filled Wells	
Advantages	**Disadvantages**	**Advantages**	**Disadvantages**
Assay plates containing compound only can be made ahead and stored	More freezer space required to store preprepared plates	Can add dynamic volumes at screen time	Potential for higher number of freeze/thaw cycles for library stock plates prior to screening
Plates can be taken off-site for assays	Less flexibility for volume changes in assays – need to plan ahead	Can use pin tools – inexpensive way to deliver lower volumes	Compound transfer must occur at facility
Potentially limits number of compound freeze/thaw cycles prior to screening	More difficult to deliver lower volumes, may result in higher DMSO concentrations in assay wells	Less freezer space required to store plates – only library plates stored, not assay plates	Because of requirement for minimum volume in the source plate, some library reagent waste occurs

Compound libraries are stored in DMSO. Ideally, the amount of DMSO transferred to an assay well should be less than 5% of the final well volume. In most cases, especially for cell-based assays, less than 1% DMSO is preferred in the final assay volume. Because assay volumes typically range from 25 to 50 μL per well (in 384-well plates), many groups transfer approximately 100 nL of compound stock solution to maintain the DMSO concentration in a desirable range. In low-volume 384-well plates, sub-100 nL compound volumes (e.g., 30 nL) can be added to 10 μL assay volumes.

5.3. Hit Picking

Compounds that score as positive in the primary screening assay can be validated by *cherry picking* them from library stocks and rescreening them to confirm their activity. A cherry-pick plate that consists solely of screening positives (and possibly control compounds as well) is created using a pipetting platform that is capable of transferring liquid to and from individual microplate wells using pipette tips that move independently. Throughput is limited by the number of individual pipette tips that can operate simultaneously, and eight tips are recommended on a liquid handling platform to maximize the number of wells that can be transferred in one cycle of picking.

5.4. Robotic Integration

Although screening instruments tend to be purchased for their stand-alone capabilities, they can often be integrated with each other to automate sequential steps in screening protocols. This generally requires detailed discussion with knowledgeable salespeople or technicians so that appropriate software and hardware components can be purchased. Integration of screening instruments can be accomplished either by contracting the vendor for the task or by employing an on-staff robotics programmer. Creating custom robotic integrations in-house using such tools as Visual Basic 6.0 and vendor-provided ActiveX controls can be an effective strategy because the ongoing support for the effort remains local. For example, a robotic arm can be integrated with a liquid handling device and a plate reader. Usually, the robotic arm will ship with scheduling software, but it is necessary for the staff to write software drivers to enable communication between the arm and each instrument. The result of this integration is an extremely flexible environment in which the robotic arm can serve microplates to any or all instruments. If this integration were purchased commercially, it would likely require future expenditures even if equipment configurations changed only slightly.

Many fully automated screening platforms are available that combine most or all of the elements of the screening process into a system that is capable of running a screen from start to finish with little or no human intervention. These systems tie together automated compound storage, incubators, robotic arms, liquid handlers, and readout devices that are coordinated by sophisticated scheduling software that tracks plate movement through the entire system. Fully automated screening systems offer advantages such as hands-off screening, higher throughput, and less day-to-day assay variability because of precise timing between each step of the assay. However, there are some disadvantages. The systems are extremely complicated and require highly trained specialists to install, operate, and maintain them. They also take up very large amounts of space (measured in rooms) and are generally quite costly, with prices in the millions of dollars.

5.5. Assay Detection

The results of high-throughput assays are typically detected using uniform well readout methods with a plate reader or by imaging at the level of individual cells with an automated microscope. Several recently developed *laser scanning cytometer* technologies offer the sensitivity of a microscope with the speed of a plate reader. Fluorescence

TABLE 8.3: Assays Using Uniform Well Readout Detection Methods

Detection Method	Assay Examples
Absorbance	Growth/nongrowth of bacteria or yeast; colorimetric assays for enzyme activity
Luminescence	Luciferase production to monitor gene expression or protein stability Luciferase activity to measure ATP levels (cell viability)
Fluorescence Intensity (FI)	Growth/non-growth of cells expressing GFP Fluorescent products generated by substrate cleavage (e.g., proteasome or RNAase activity) or by polymerization (e.g., actin polymerization) Indicator dyes to measure calcium levels
Fluorescence Polarization (FP)	Monitoring peptide/protein binding or small molecule/protein binding
Fluorescence Resonance Energy Transfer (FRET)	Monitoring peptide/protein binding or protein/protein binding

activated cell sorting (FACS) can also be used as a readout for screening, but it currently has some limitations that make it not ideally suited for high-throughput work.

Uniform well readout assays are quantitative assays that monitor only one or two parameters at a time from each well, often the increase or decrease in the amount of light emitted. Multimode plate readers measure absorbance or emitted light from several assay types, including colorimetric, luminescence, fluorescence intensity (FI), fluorescence polarization (FP), time-resolved fluorescence (TRF), and fluorescence resonance energy transfer (FRET) assays. Examples of applications using these assay detection techniques are shown in Table 8.3.

Most plate readers can read a microplate in a short amount of time – often one to two minutes per 384-well plate. However, plate readers also provide less information for each well – for example, for cell-based assays, phenotypic information for the population of cells in each well is averaged across the well and provided as a single value. Also, plate readers are generally not as sensitive at detecting fluorescent signals as microscopes.

In *high-content screening* (HCS), automated screening microscopes are used to monitor changes at the level of individual cells within an assay well. Using immunofluorescence or other cell staining methods, high-throughput imaging assays might monitor changes in protein levels, protein localization, cell morphology, or cell motility. Screening microscopes acquire images of each well of a microplate using either laser- or image-based autofocusing. The images are then analyzed using image analysis software. Most screening microscopes are packaged with image analysis software. All image analysis software requires training to use effectively, but some programs are more user friendly than others. For sophisticated image analysis applications, it might be necessary to collaborate with image analysis experts who can write custom algorithms.

More information is captured for each well in an imaging assay than in an assay read out on a plate reader – multiple parameters in individual cells can be tracked. However reading times using a screening microscope are an order of magnitude longer (e.g., twenty to ninety minutes per 384-well plate via automated microscopy versus one to two minutes via plate reader). The number of wavelengths acquired, the number of sites per well imaged, and the exposure length all contribute to the overall time to read a plate. For live cell imaging, additional consideration must be given to temperature and CO_2 requirements of the cells while they are being imaged. Also, automated analysis of live cell movie or time-lapse data is significantly more challenging than analysis of images of fixed cells.

For some assays, a plate reader is simply not sensitive enough to detect the fluorescence present in the assay well (e.g., for many assays monitoring green fluorescent protein [GFP] in whole cells), and a microscope, although certainly sensitive enough, is too slow to make screening of large numbers of compounds practical. This niche between plate readers and automated microscopes has recently been filled by a new class of detectors called laser scanning cytometers that use lasers to excite fluorophores in individual wells and then capture a low-resolution image of each well. These instruments can read 384-well plates in less than ten minutes and have preprogrammed analysis modules to quantitate simple parameters in each image.

5.6. Data Capture

Capture and storage of raw data generated by high-throughput assays is not a trivial issue. Plate readers generate data in the form of text files that are generally small in size (measured in kilobytes [KB]). In contrast, image files are large in size (measured in megabytes [MB] for a single image). It is not unusual for an imaging screen in which two or more wavelengths and multiple sites per well are imaged to generate more than one terabyte (10^{12} bytes of data)! Certainly for image files, it is advisable to use a file server for data storage. Server storage with backup is also recommended for any mission-critical data, including screening data, library inventory management data, and analysis results.

5.7. Databases and Software Tools

Databases are required to store information about the contents of library wells and to match that information with screening data obtained for each well. To put the results into context, the databases must also store information about the screening protocol as well as the investigators who carried out the screen, the dates screening took place, and so forth. Software tools are also needed to analyze screening data.

The ideal screening database is capable of storing large amounts of data from many screens and allows comparison of data across multiple assays and integrated access to chemistry information such as substructure searching. Several commercial database packages are available that integrate data storage and data analysis functions. These packages are generally quite expensive and should be considered only if there is adequate IT staff to support them. In addition, these packages require much thought and work to configure correctly so that their features are utilized appropriately. Some screening facilities have hired software engineers to develop custom database solutions (e.g. the Screensaver laboratory information management system; Tolopko et al. (2010) *BMC Bioinformatics* 11:260). This strategy is probably not less costly than purchasing commercial database packages but can produce solutions perfectly suited to the needs of the individual facility.

In many cases, commercial database and data analysis solutions are not feasible because of cost, personnel, or simply because the number of assays to be run is small. It is possible to work with assay data using much simpler and cheaper tools. For data analysis, most raw numeric data from plate readers can be easily handled within Microsoft Excel or in any other good data analysis/statistics software package. It is quite common for researchers to analyze the data from their own assays within Excel and then provide the results for entry in a centralized database for comparison with other screens or for access to compound structures. Very simple custom databases can be constructed to maintain analyzed assay data, but care should be taken to design a common format for data entry that is adhered to by all users.

Some very simple and cost-effective tools exist for dealing with chemistry data, the most common being ChemOffice from CambridgeSoft and ISIS from MDL. These tools provide basic databases for cataloging and maintaining chemical compound collections. Most compound collections are provided from the supplier with an associated SD format file, which contains the compound structure and reorder information. These files can easily be imported into either ISIS or ChemOffice ChemFinder so that structures can be browsed and searched. Plate and well information can also be added for the formatted compounds if this information does not exist already.

6. CONCLUDING THOUGHTS

In this chapter, we have outlined the essential infrastructure needed to support HTS operations in an academic environment. Over the past ten years, many academic screening centers have been established, ranging from small departmental centers to large networks of screening centers whose complexity rivals that of screening operations in the pharmaceutical industry. Whereas the details and scope of each varies, we feel that the fundamental principles outlined here – namely, design of space, staffing considerations, acquisition and storage of compounds, choice of instrumentation, and data analysis tools – can serve as a model for anyone considering establishing an academic HTS center.

ACKNOWLEDGMENT

The authors thank Dr. Su Chiang (HMS) for her helpful comments on the manuscript.

CHAPTER 9

High-Content Analysis and Screening: Basics, Instrumentation, and Applications

Paul A. Johnston

1. BASICS OF HIGH-CONTENT ANALYSIS AND SCREENING

Drug discovery is now moving toward the implementation of cell- and whole organism–based assays in which the target is screened in a more physiologic context than in biochemical assays of isolated targets. Automated high-content analysis and screening (HCA/HCS) platforms are ideally suited to such chemical genomics approaches [1–4]. HCA/HCS encompasses an integrated process involving the use of fluorescent labeling techniques combined with automated multiwavelength fluorescent light microscopy and image analysis algorithms to extract multiparameter quantitative and qualitative data and information on cellular macromolecular structures and the localization of cellular components and to define the temporal dynamics of cellular functions [2, 5–17]. In automated imaging platforms, the acquisition of multiwavelength fluorescence images is integrated with image analysis algorithms and informatics tools to automate the unbiased capture and analysis of fluorescent images from millions of cells arrayed in the wells of microtiter plates [5–11]. What distinguishes HCA/HCS assays from the more typical single-parameter high-throughput screening (HTS) assay formats is their ability to acquire images in multiple fluorescent channels and by image analysis output multiparameter data from a variety of fluorescent measurements and features. These include fluorescence intensity and intensity ratios, texture within regions, cellular and subcellular morphometrics, and a total count of features [6, 9, 12–17]. HCA/HCS platforms have provided sufficient throughput and capacity to generate multiparametric cellular data at a scale that could be applied to drug discovery and high-throughput cell biology approaches, genome-wide RNA interference and overexpression strategies, phenotypic chemical biology screens, and cellular systems biology [2, 4–6, 11, 13–27]. Automated imaging platforms are being deployed in many phases of the drug discovery, basic research, and the chemical genomics process for target identification/validation, primary screening and lead generation, hit characterization, lead optimization, toxicology, biomarker development, diagnostic histopathology, and other clinical applications [2, 4–6, 11, 13–27].

There are four basic components of HCA/HCS systems: sample preparation, image acquisition, image analysis, and data analysis (Figure 9.1). The development of HCA/HCS assays involves the optimization of sample preparation, image acquisition procedures, and image analysis algorithms [15, 16, 27]. In recent years, three excellent books have been published on the field of HCA/HCS, edited by and with contributions from the pioneers and early practitioners of HCA/HCS [28–30], and I would highly recommend these texts to both beginner and experienced operator alike.

1.1. Sample Preparation

It will not matter how good your image acquisition and image analysis capabilities are if your sample preparation is deficient. The source of the cells selected for the imaging assay will have a profound impact on both the biology under investigation and the implementation of the HCA/HCS assay. Primary cells are arguably the most physiologically relevant model system, and selected primary cells have been utilized for HCA/HCS [31–33]. However, primary cells may not always be available at the scale required for HCA/HCS, and there may be significant issues with both scale up and the maintenance of the cells in their primary state. In addition, the morphology and growth characteristics of some primary cells can present significant challenges for image analysis. Stem cells have been utilized in HCA/HCS assays to screen for modulators of stem cell growth and differentiation [34, 35] and may ultimately lead to methods to derive reproducible cell lineages for chemical genomics screens. Transformed cell lines are the most commonly used cell-based platform for HCA/HCS and HTS [3, 15, 16, 18, 20, 22, 24–27, 36]. Transformed cell lines can be screened in their native state for targets endogenously expressed in the cells or in cell lines that have been engineered to express or overexpress the target(s) of interest through transfection with expression plasmids or by

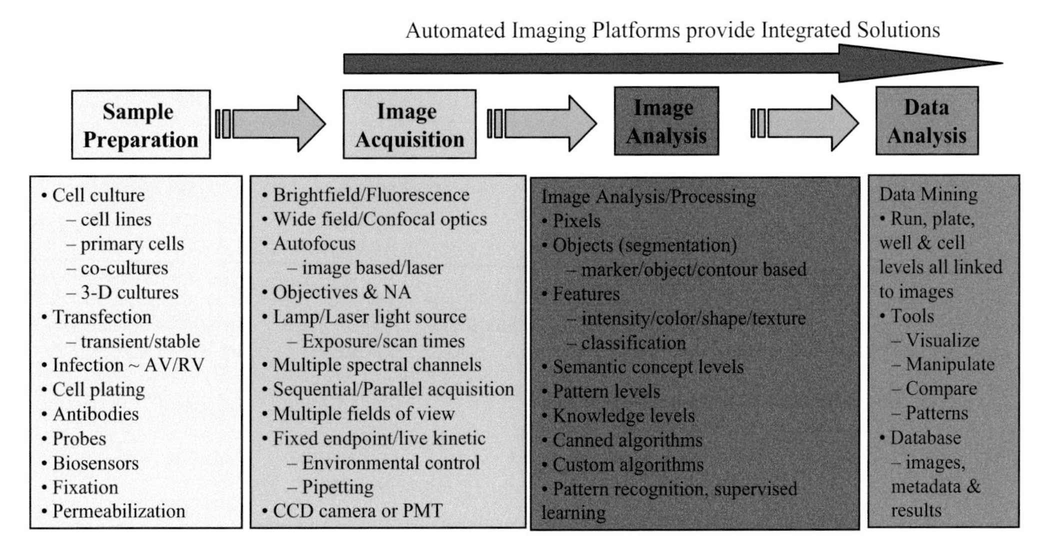

***FIGURE 9.1:** Components of automated high-content screening. There are four basic components of HCA/HCS systems: sample preparation, image acquisition, image analysis, and data analysis. The development of HCA/HCS assays involves the optimization of sample preparation, image acquisition procedures, and image analysis algorithms, and some of the major components of these processes are listed. Also listed are some of the challenges associated with HCA/HCS databases and data mining.*

infection with recombinant retrovirus or adenovirus constructs [3, 15, 22, 27, 37].

Through the selection of the appropriate fluorescent probes (antibodies, protein fusion partners, biosensors, and stains), HCA/HCS assays can be applied to many target protein classes, including G protein-coupled receptors (GPCRs), kinases, phosphatases, nuclear hormone receptors, transporters, ion channels, proteases, and others [2, 5–7, 13–19, 21–27, 38]. HCS sample preparation can be a complex, multicomponent process that includes selection and optimization of the cell line, microtiter plate, fixative, permeabilization buffer, blocking buffer, wash buffer, primary and secondary antibodies, and fluorescent probes (Figure 9.1) [15, 27, 37, 39, 40]. The number and types of fluorescent probes will determine how many separate channels are to be collected and which excitation and emission filters are required [6, 7, 9, 15, 17]. The sample quality and optical resolution required will influence the selection of the objective and exposure times required for image acquisition [6, 7, 9]. The cell seeding density and/or the relative frequency of the biological response will determine the number of image fields that need to be acquired [15, 16, 27]. The selection of a specific microtiter plate can influence both the biological response and several of the image acquisition parameters, including autofocus options, auto-exposure times, and whether large numerical aperture (NA) objectives can be utilized to acquire images in all of the wells of the plate. The choices made on automating the process of cell plating, compound treatment, and sample preparation will also have a significant impact on the biology and the consistency of HCA/HCS assays [41].

1.2. Image Acquisition

A critical component of any automated HCA/HCS platform is the software that serves to control and set up the instrument for image capture and provides tools for image analysis and data visualization [6, 7, 9]. The instrument settings for image acquisition are either controlled by the software or have to be selected in the software, along with other experimental details such as the plate type, plate density, and number of wells to be imaged. Typically, the image acquisition for each HCA/HCS assay will require input on the objective, the number and types of fluorescent probes and therefore the number of channels to be collected, the excitation and emission filters, focal offsets required relative to the autofocus point, exposure times, and the number of image fields that need to be captured (Figure 9.1).

Multiwavelength fluorescence imaging is the key technology for HCA/HCS platforms. It provides the ability to detect and quantify multiple cellular components or responses that have been specifically labeled in a single preparation [5–7, 9, 13, 14, 17, 40]. HCS platforms acquire wavelength channels either sequentially or in parallel. In the sequential acquisition mode, each fluorophore is separately excited and detected on a single monochrome charged-coupled digital (CCD) camera or photomultiplier tube (PMT). Channels are typically selected using a fast excitation filter wheel combined with a multiband emission filter, although single-band emission filters can be used to improve selectivity. Most HCA/HCS platforms come equipped with filter sets for the most commonly used fluorescent probes and can distinguish up to four labels in a single preparation with minimal cross talk between channels [6, 7, 9]. Cross talk arises from overlap in the spectra

TABLE 9.1: High-Content Analysis/Screening Platforms

Vendor	HCA/HCS Platform	Optical Design	Light Source	Autofocus Method	Live well Capability
ThermoFisher	ArrayScan V[TI]	Wide field	Lamp	Image based	Yes
Applied Precision	CellwoRx	Wide field	Lamp	Image based	No
GE Healthcare	In cell 1000	Wide field	Lamp	Image based and laser	Yes
GE Healthcare	In cell 3000	Confocal Line scanning	3x Lasers	Laser	Yes
BD	Pathway 415	Wide field	Lamp	Image based and laser	No
BD	Pathway 435	Wide field or confocal spinning disk	Lamp	Image based and laser	No
BD	Pathway 855	Wide field or confocal spinning disk	Lamp	Image based and laser	Yes
MDS Analytical Technologies	ImageXpress Micro	Wide field	Lamp	Image based and laser	Yes
MDS Analytical Technologies	ImageXpressUltra	Confocal point scanning	2–4x Lasers	Laser	No
Perkin Elmer	Opera	Confocal spinning disk	3x Lasers	Laser	Yes
CompuCyte	iCyte	Wide field	3x Lasers	Laser	No

of the fluorophores and the bandwidth and nonideal performance of filters [6, 7, 9]. For bright fluorescent samples, sequential channel acquisition can be reasonably fast. Some automated HCA/HCS systems have been designed with multiple excitation lasers together with multiple CCD cameras or PMTs that allow simultaneous parallel imaging in three to four emission channels. Through the appropriate selection of fluorescent probes that avoid spectral overlap, parallel channel detection can be significantly quicker than sequential detection. However, commonly utilized near-UV excited fluorophores like DAPI and Hoechst DNA labels can be a problem because they have broad blue emission bands that overlap the green fluorescein (or green fluorescent protein [GFP]) emission channel. The selection of fluorophores, filter performance, and mode of image acquisition all impact the sensitivity, signal-to-background (S/B), signal-to-noise (S/N), and throughput of the HCA/HCS system.

The choice of magnification (objective) profoundly affects HCA/HCS assay performance and throughput by impacting the resolution, field of view, detection sensitivity, and output from the image analysis algorithm [6, 7, 9]. The inverse relationship between resolution and field of view means that larger-magnification, higher-resolution objectives provide a smaller field of view with fewer cells per image and longer scan times [6, 7, 9]. The majority of HCA/HCS platforms may be configured with multiple, standard, long working distance, low NA air/dry objectives (5x, 10x, 20x, and 40x) that can effectively acquire images of cells grown on the thick (0.5–1.1 mm) clear-bottom microtiter plates. The low NA objectives can be exchanged for larger NA, shorter working distance objectives to capture brighter and sharper images, but this may require a switch to more expensive thin-glass microtiter plates or may sometimes prevent the imaging of the wells on the outer edges of plates because their larger dimensions are physically blocked by the side edges of plates [6].

Fluorescence imaging HCA/HCS systems require high-intensity illumination sources such as arc lamps or lasers to appropriately excite fluorophores (Table 9.1) [6, 7, 9]. There are three common types of broad-spectrum arc lamps – mercury, xenon, and metal halide – which all provide good intensity and, when coupled with the wide array of interference filters available, provide the flexibility to utilize nearly any fluorescent probe. However, the downside is that, depending on the sample quality, fluorophore, and objective, broad-spectrum arc lamps may require longer exposure times [6, 7, 9]. The high power available from laser illumination typically results in faster scan times, and an additional benefit may be a reduction in the amount of fluorescent reagents required for some applications. However, because lasers produce sharp emission lines that may limit the choice of probes that can be utilized, most HCA/HCS platforms using laser excitation require multiple lasers to provide the wavelengths needed for multiplexed HCS assays (Table 9.1), thereby increasing their cost [6, 7, 9].

1.3. Image Analysis and Data Handling

In addition to controlling the image acquisition process, the integrated software for most automated HCA/HCS platforms typically includes a number of image analysis algorithms that address specific biological applications [6, 7, 9, 12, 42, 43]. Image analysis or processing can be achieved at several levels – pixels, objects, and semantic concepts – and at the pattern and knowledge levels (Figure 9.1)[6, 12, 42]. Digital images are composed of pixels, or squares of uniform gray values sometimes referred to as intensity levels/values captured by a CCD camera or PMT, that are assigned to objects established through segmentation. Segmentation subdivides an image into several regions that might represent several objects using a number of approaches, including marker-based, object-based or contour-based segmentation. Information about objects in an image is condensed into features such as color, shape, and texture. Objects and regions can be classified into user-defined categories based on their features or properties, known or extracted from the image. Image analysis algorithms derive quantitative and qualitative measures of features that are calculated on a per-cell basis and/or as a well average [12, 42, 43]. Such features include counting objects and calculating ratios, width, length, spatial distribution, texture, motion, and behavior over time [12, 42, 43]. The user defines the objects and features to be extracted automatically from every image prior to the analysis procedure. The selection and optimization of the final image analysis parameters typically involve the iterative use of a training set of images, most commonly the assay controls for the top and bottom of the signal window [6, 7, 9, 12, 42, 43]. The morphology and growth behavior of the cells being imaged will have a profound impact on the image analysis algorithm. Images of large cells that grow as monolayers of well-spaced individual cells are typically easier to analyze than smaller cells that grow as three-dimensional colonies. Similarly, optimization of the analysis algorithm is usually easier with images of cells that adhere well to the microtiter plate substrate and assume a well-spread, flat morphology with large, well-defined cytoplasm and nuclear regions. It sometimes may be challenging to adapt an image analysis algorithm that has been optimized for cell lines that have desirable imaging characteristics (e.g., U2OS, HeLa, A549) to primary cells that can have significantly different growth behavior and/or morphology.

The integrated software of the automated HCA/HCS platform also needs to provide an environment that supports data mining at several levels (Figure 9.1) [6, 7, 12, 42]. During image acquisition, the software should provide images of the fields being captured in real time, together with plate views of the data analysis readout for wells already imaged. These outputs demonstrate that the instrument is acquiring quality images, that the image analysis algorithm has been appropriately optimized, and that the plate controls are behaving as expected. Postacquisition, the software must provide methods to inspect and interpret the multiparameter analysis results in the context of the images, raw data, experimental conditions, and procedures utilized. The interactive software should provide data visualization and analysis tools to assess the quality of the experiment and programmed analysis [6, 7, 12, 42]. The software should allow the user to mine (visualize, extract, and analyze) all of the multiparameter data and features that the algorithm provides, both at the well-based and the cell-based levels. The ability to toggle between the images (fields and individual cells) and the extracted data is critical, and software tools to visualize, manipulate, and compare the multiparameter data help the investigator to recognize high-level patterns and relationships (Figure 9.1) [6, 7, 12, 42].

Automated HCA/HCS platforms generate large volumes of images, metadata, and image processing data that all need to be securely stored, organized, and effectively managed in a database [6, 7, 12]. Metadata include the nature of the samples, the experimental conditions, and the procedures used to acquire and analyze the images. The image-derived information includes the objects, features, classifications, and calculated data. The data model needs to capture and integrate the raw images, metadata, and image analysis/processing data together with any associated data (e.g., compound IDs and structures) to provide an effective data-mining environment [12]. The raw images, metadata, and derived data should be archived and stored in an untouched form for scientific and regulatory reasons. The integrated HCA/HCS database and software should provide efficient methods to query and retrieve images and data for review and potentially for reanalysis (Figure 9.1).

2. INSTRUMENTATION

At the time of this writing, there are more than ten commercially available automated imaging platforms that have adopted a diverse array of imaging technologies to meet the requirements and the performance of an HCA/HCS system (Table 9.1) [7, 9, 44]. The major differences in the configurations of the available HCA/HCS platforms are in the choices made for their optical design, light source (see above), autofocus procedures, and whether an environmentally controlled chamber with on-board liquid handling for live cell assays is or can be incorporated. In recent years, some platforms have also been modified to incorporate brightfield imaging to complement their fluorescent capabilities.

2.1. Optical Design

Wide-field optical systems illuminate a "large" area of the specimen, directly image that area all at once, and typically exhibit a high S/N ratio with thin specimens such as cell monolayers [7, 9, 44]. In contrast, confocal optical systems work by illuminating the specimen in one or more small regions (spots or lines) and building up an image by scanning the illumination through the specimen while measuring the emission in synchrony with the scanning [7, 9, 44]. Confocal HCA/HCS systems utilize either point scanning, line scanning, or multipoint scanning (spinning disk) illumination designs (Table 9.1). The pinhole, slit, or spinning disk barrier blocks out-of-focus light from entering the detector, thereby rejecting background fluorescence from material outside the plane of focus [7, 9, 10, 44]. Out-of-focus light may arise from sample preparations that are significantly thicker than the depth of field or from excess fluorescent label in the surrounding media. Confocal systems will likely perform better with tissue sections and multilayer cell sample preparations [7, 9, 10, 44].

2.2. Autofocus

Most HCA/HCS platforms utilize either image analysis software or specular reflectance to provide quick, reliable autofocus, and some have both capabilities (Table 9.1). Image analysis–based autofocus systems image a target fluorescent probe in cells, commonly stained nuclei although any feature can be used, and then an imaging software algorithm is used to measure relative sharpness in the image [7, 9, 44]. Sophisticated image-based autofocus can handle plate irregularities, unevenness, and heterogeneous sample preparations in the wells of microtiter plates because the focus is based on the specific contents of each well. Specular reflectance systems project a small spot of illumination, typically an infrared (IR) laser, onto the substrate and maximize the intensity of the reflected light [7, 9, 44]. Specular reflectance autofocus systems typically identify two reflected peaks, one corresponding to the bottom of the plate and the other from the interface between the buffer in the well and the well bottom. Laser autofocus systems are very fast, but the focus quality relies on a consistent position of the feature of interest relative to the substrate to which the cells are attached. Typically, operators will collect Z-stacks of images to determine the offset(s) for the features of interest relative to the interface at the bottom of the well.

2.3. Live Cell Imaging

A number of HCA/HCS platforms either come equipped with an environmentally controlled chamber and liquid handling capability for live cell kinetic imaging, or these features can be added on in a modular fashion (Table 9.1). Live cell HCA/HCS systems are significantly more complex than fixed end point platforms because they require more sophisticated acquisition software with provision for scheduling various sequences of image acquisition and liquid additions to accommodate a wide range of biological timing, as well as sophisticated software applications for analyzing the time course of the response [6, 7, 10]. High-throughput time-lapse microscopy in living cells broadens the scope of phenotypic imaging assays to dynamic functional assays, including cell movement, cell spreading, wound healing, phagocytosis, GFP-fusion protein redistribution, cell division, Ca^{2+} mobilization, and membrane depolarization [6, 7, 10]. Live cell imaging capability can sometimes be an important assay development tool for optimizing the time course of fixed end point HCS assays.

2.4. Benchmarking HCA/HCS Platforms

To conduct an extensive comparison of every HCA/HCS system available today is quite a daunting exercise because of the array of features and options that are available (Table 9.1). Furthermore, each platform has strengths and weaknesses that complicate the interpretation of the matrix of features. A pragmatic approach to benchmarking the various HCA/HCS platforms is to narrow the choice on the basis of preferred options and to then directly compare the performance of specific bioassays on the candidate platforms. We recently had an opportunity to compare the performance of a Glucocorticoid nuclear hormone translocation HCS assay on four automated imaging platforms: the ArrayScan V^{TI} (Thermofisher), the Opera (Perkin Elmer), the ImageXpress Ultra (MDS Analytical Technologies), and the Pathway 855 (Becton Dickenson).

The ArrayScan V^{Ti} houses a Zeiss 200M inverted microscope outfitted with 5 × /0.25 NA, 10 × /0.3NA, 20 × /0.4 NA, and 40 × /0.5 NA Zeiss objectives in a four-position automated objective turret, and the specific objective can be selected in the software. One of the objective positions is available for exchange with different objectives such as larger NA versions if desired. Illumination is provided by a full-spectrum (300–2,000 nM) Hg-halide arc lamp source (EXFO, Quebec, Canada), and fluorescence is detected with a high-sensitivity cooled Orca CCD Camera (Photometrics Quantix). The ArrayScan V^{Ti} uses an image-based autofocus system and has the capability of imaging multiwavelength fluorescence with up to six excitation and emission channels excited and acquired sequentially. Channel selection is accomplished using a fast-excitation filter wheel combined with a multiband emission filter.

The Opera is a fully automated, spinning disk, point-confocal imaging platform with four diode laser–based excitation sources (405, 488, 561, and 640) with up to four independent cooled CCD detectors, and it has the ability to acquire multiwavelength fluorescence in up to four channels in parallel. The optical system is a modified Yokogawa Nipkow CSU10 spinning disk confocal system with approximately one thousand concurrent spots, a pinhole diameter of 50 μm, axial resolution of 1.5 to 10 μm, and lateral resolution down to 0.25 μm. The Opera can be outfitted with 10 × , 20 × , 40 × , and 60 × water or air objectives that can be selected in the software and uses an IR laser autofocus.

The ImageXpress Ultra is a fully integrated point-scanning confocal system with two to four solid-state lasers providing up to four simultaneous excitation wavelengths (405, 488, 532 or 561, and 635 nm), each with a dedicated photomultiplier tube for detection. The size of the detection pinhole diameter of the confocal optics is configurable using the software. The ImageXpress Ultra has a dedicated, high-speed laser autofocus system and a four-position automated objective changer that is compatible with air objectives ranging between 4 × and 100 × and oil objectives ranging between 40 × and 100 × that can be selected in the software.

The Pathway 855 uses dual HBO 100W mercury arc full spectrum UV-NIR (near infrared) lamp units with sixteen excitation and eight emission filter positions, each with five dichroic mirror positions. It has a single

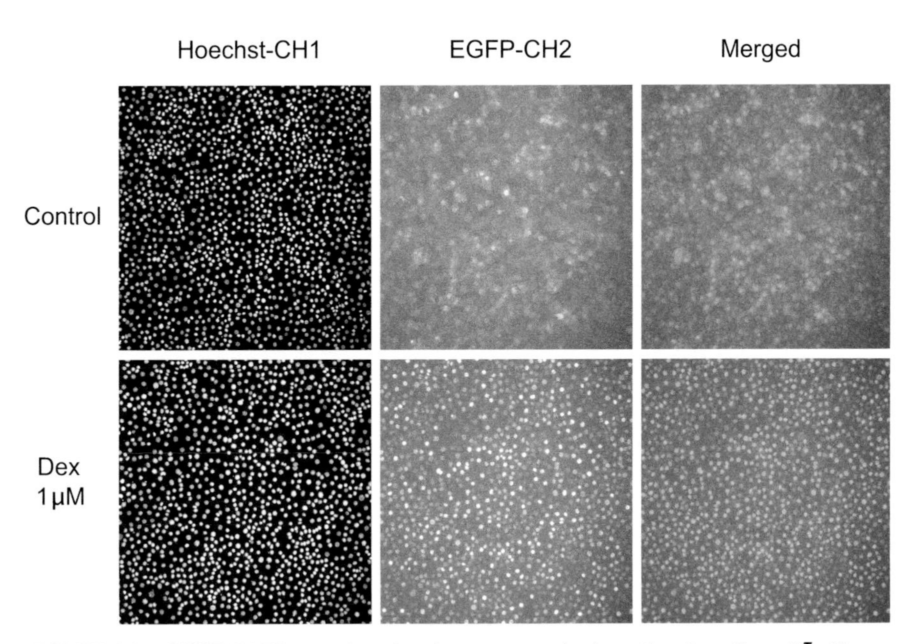

***FIGURE 9.2:** GNHR-EGFP translocation images acquired on the ArrayScan V^T. Mouse mammary adenocarcinoma cells (3617.4) stably expressing a rat glucocorticoid nuclear hormone receptor–enhanced green fluorescent protein (GNHR-EGFP) under the control of a tetracycline (tet)-regulated promoter were seeded at 2,500 cells per well in Greiner black-walled clear-bottomed 384-well plates and cultured for 48 hours at 37°C, 5% CO_2 and 95% humidity in media without tetracycline. After 48 hours, cells were either treated with 1 μM Dexamethasone (Dex) or left untreated for 30 minutes at 37°C, 5% CO_2 and 95%. After the 30-minute Dex treatment, cells were fixed for 10 minutes at ambient temperature by the addition of prewarmed (37°C) paraformaldehyde (3.7% final) containing 2 μg/mL Hoechst 33342. Cells were then washed twice with 50 μL of PBS and then plates were sealed. Wide-field images of a single field of view of the Hoechst (nucleus) and GNHR-EGFP fluorescent channels of untreated and 1 μM Dex-treated 3617.4 cells were then acquired on the ArrayScan V^{TI} using a 10× objective and XF100 filters.*

high-resolution Hamamatsu ORCA ER CCD camera, and multiwavelength fluorescence channels are acquired sequentially. The Pathway 855 has a Nipkow spinning disk confocal system with a pinhole diameter of 70 μm that can be removed from the optical path so that the imager can be operated in wide-field mode. The Pathway 855 has both laser and image-based autofocus and comes with a single-position objective mount that can hold a single 4×, 10×, 20×, 40×, or 60× objective.

The mouse mammary adenocarcinoma cell line (3617.4), which stably expresses a rat glucocorticoid nuclear hormone receptor–enhanced GFP (GNHR-EGFP) under the control of a tetracycline (tet)-regulated promoter, has been utilized as a model system to investigate hormone-dependent translocation of the GNHR [45–47]. The following protocol was developed and optimized for the GNHR-EGFP translocation HCS assay conducted on the ArrayScan V^{TI} platform: 1) 2,500 mouse 3617.4 cells per well are seeded in Greiner black-walled, clear-bottomed 384-well plates and cultured for forty-eight hours at 37°C, 5% CO_2, and 95% humidity in media without tetracycline; 2) after forty-eight hours, cells are treated with the indicated concentrations of Dexamethasone (Dex) for thirty minutes at 37°C, 5% CO_2, and 95%; 3) after the thirty-minute Dex treatment, cells are fixed for ten minutes at ambient temperature by the addition of prewarmed (37°C) paraformaldehyde (3.7% final) containing 2 μg/mL Hoechst 33342; 4) fixed and stained cells are then washed twice with 50 μL of phosphate buffered saline (PBS) and plates containing 50 μL of PBS in the wells are sealed; 5) fluorescent images are acquired on the indicated automated imaging platform, and the Dex-induced translocation of GNHR-EGFP translocation is quantified using the image analysis algorithms provided by the manufacturer (Figures 9.2–9.6, Table 9.2).

Representative images of the Hoechst (nucleus) and GNHR-EGFP fluorescent channels of untreated and 1 μM of Dex-treated 3617.4 cells acquired on the ArrayScan V^{TI} (Figure 9.2), the Opera (Figure 9.3), the ImageXpress Ultra (Figure 9.4) and the Pathway 855 (Figure 9.5) using a 10× objective are presented. The ArrayScan V^{TI} and Pathway 855 platforms were operated in the wide field mode, whereas the Opera and ImageXpress Ultra were operated in confocal mode. In the absence of Dex treatment, the GNHR-EGFP appears to be diffusely distributed throughout the cytoplasm and nuclear compartments of the cell.

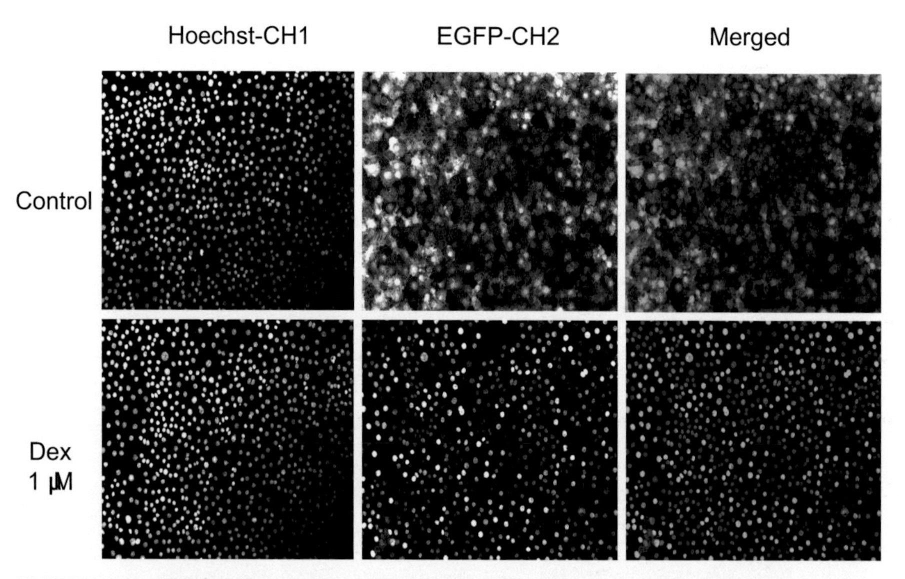

FIGURE 9.3: GNHR-EGFP translocation images acquired on the Opera. 3617.4 cells stably expressing GNHR-EGFP under the control of a tetracycline (tet)-regulated promoter were seeded at 2,500 cells per well in Greiner black-walled clear-bottomed 384-well plates and cultured and treated as described in legend for Figure 9.2. Confocal mages of a single field of view of the Hoechst (nucleus) and EGFP (GNHR) fluorescent channels of untreated, and 1 µM Dex-treated 3617.4 cells were then acquired on the Opera using a 10× objective.

However, after thirty minutes of exposure to 1 µM of Dex, the GNHR-EGFP is predominantly colocalized within the nucleus. Existing image analysis algorithms integrated with the four platforms were then used to quantify the relative distribution of the GNHR-EGFP between the cytoplasm and the nuclear compartments (Figures 9.6 and 9.7). All four image analysis algorithms utilized the same basic segmentation strategy to quantify the relative subcellular distribution of the GNHR-EGFP (Figure 9.6). The nucleic acid dye Hoechst 33342 was used to stain and identify the nucleus, and this fluorescent signal was used to define a nuclear mask (Figure 9.6). The mask was eroded to reduce cytoplasmic contamination within the nuclear area, and the reduced mask was used to quantify the amount of target channel GNHR-EGFP fluorescence within the nuclear region (Figure 9.6). The nuclear mask was then dilated to cover as much of the cytoplasmic region as possible without going outside the cell boundary. Removal of the

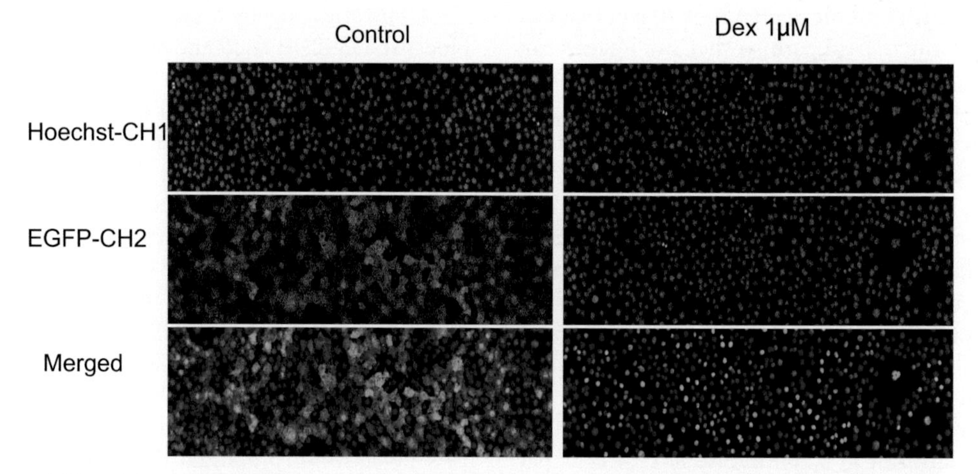

FIGURE 9.4: GNHR-EGFP translocation images acquired on the ImageXpress Ultra. 3617.4 cells stably expressing GNHR-EGFP under the control of a tet-regulated promoter were seeded at 2,500 cells per well in Greiner black-walled clear-bottomed 384-well plates and cultured and treated as described in legend for Figure 9.2. Confocal images of a single field of view of the Hoechst (nucleus) and EGFP (GNHR) fluorescent channels of untreated and 1 µM Dex-treated 3617.4 cells were then acquired on the ImageXpress Ultra using a 10x objective.

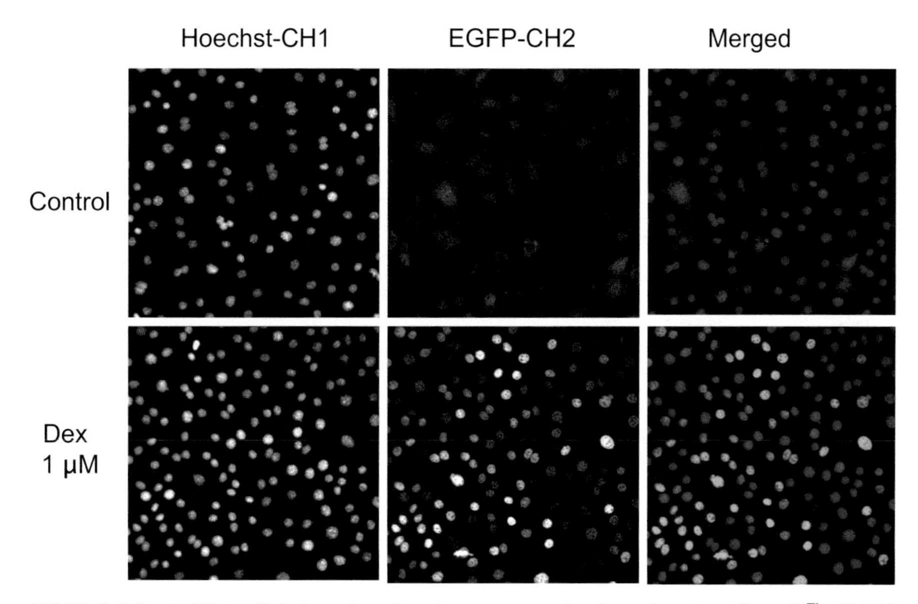

***FIGURE 9.5:** GNHR-EGFP translocation images acquired on the ArrayScan V^{TI}. 3617.4 cells stably expressing GNHR-EGFP under the control of a tet-regulated promoter were seeded at 2,500 cells per well in Greiner black-walled clear-bottomed 384-well plates and cultured and treated as described in legend to Figure 9.2. Wide-field images of a single field of view of the Hoechst (nucleus) and EGFP (GNHR) fluorescent channels of untreated and 1 μM Dex-treated 3617.4 cells were then acquired on the Pathway 855 using a 10x objective.*

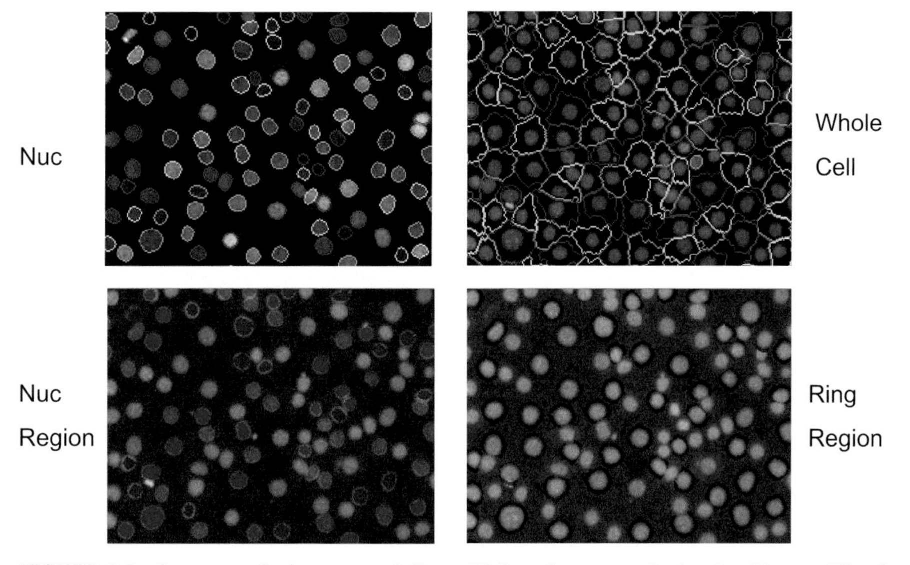

***FIGURE 9.6:** Image analysis segmentation. All four image analysis algorithms utilized the same basic segmentation strategy to quantify the relative subcellular distribution of the GNHR-EGFP. The nucleic acid dye Hoechst 33342 was used to stain and identify the nucleus, and this fluorescent signal was used to define a nuclear mask (Nuc). The mask was eroded to reduce cytoplasmic contamination within the nuclear area, and the reduced mask was used to quantify the amount of target channel GNHR-EGFP fluorescence within the nuclear region (Nuc region). The nuclear mask was then dilated to cover as much of the cytoplasmic region as possible without going outside the cell boundary (whole cell). Removal of the original nuclear region from this dilated mask creates a ring mask that covers the cytoplasmic region outside the nuclear envelope (ring region). The image analysis algorithm outputs quantitative data such as the total or average fluorescent intensities of the GNHR-EGFP signal in the nucleus or cytoplasm on a per-cell basis that may also be reported as an overall well average value. We utilized the same mean average intensity difference, nucleus region–cytoplasm region, to directly compare the outputs of the different algorithms.*

TABLE 9.2: Benchmarking the Automated Imaging Platforms

GNHR-EGFP Translocation HCS Assay: 384-Well Greiner Plate				
Instrument	**ArrayScan V^{TI}**	**Opera**	**ImageX-Ultra**	**Pathway 855**
Objective and NA	10x 0.3NA	10x air 0.4NA	10x 0.3NA	10x 0.4NA
Number of wells	96	84	128	128
Number of images	192	168	256	256
CH-1 fluorophore	Hoechst	Hoechst	Hoechst	Hoechst
CH-1 exposure sec	0.0281	0.28	1x averaging	200 ms
CH-2 fluorophore	EGFP	EGFP	EGFP	EGFP
CH-2 exposure sec	0.0974	0.28	1x averaging	100 ms
Start scan	2:42:27	5.04.32	16:02:13:912	2:39PM
End scan	2:49:49	5.09.54	16:06:42:662	2:50PM
Total scan time	7 min 22 sec	5 min 22 sec	4 min 29 sec	11 min
Scan 384-wells	29 min 28 sec	24 min 32 sec	13 min 27 sec	33 min
Analysis algorithm	Mol. translocation	Nuc – Trans	MWTL + journal	Nuc-Cyt
Mean # of FOV	1	1	1	1
Mean cell #/FOV	1,239	707	512	400
Mean EC_{50} μM	2.21	1.999	1.172	1.263
EC_{50} 95% confid.	1.924 to 2.538	1.095 to 3.647	0.9514 to 1.445	0.929 to 1.715
Hill slope	1.625	1.206	1.614	1.208
R^2	0.98	0.782	0.95	0.938
Bottom	19.32	70.02	217.8	48.76
Top	171.5	436.5	2151	504.6
Max:min ratio	8.9	6.2	9.9	10.3
Analysis time	× to scan time	× to scan time	6.4 min	× to scan time

FOV, field of view; MWTL, multiwavelength translocation; Mol, molecular translocation.

original nuclear region from this dilated mask creates a ring mask that covers the cytoplasmic region outside the nuclear envelope (Figure 9.6). The image analysis algorithm outputs quantitative data such as the total or average fluorescent intensities of the GNHR-EGFP signal in the nucleus (Nuc region) or cytoplasm (Ring region) on a per-cell basis that may also be reported as an overall well average value (Figure 9.7). Although each algorithm provided a variety of alternative methods to quantify the GNHR-EGFP translocation, we utilized the same mean average intensity difference, nucleus region–cytoplasm region, to directly compare their outputs (Figure 9.7, Table 9.2).

A 10 × objective was used to acquire images of a single field of view for both fluorescent channels on each platform because 1) it provided sufficient resolution to enable the image analysis algorithm to segment the images appropriately; 2) a single field of view captured enough cells for the image analysis to generate statistically significant data; and 3) it reduced the number of images acquired, which in turn shortened plate scan times and significantly decreased the size of the data and image files produced. Although the GNHR-EGFP images of untreated cells acquired on the Opera and ImageXpress Ultra appear marginally sharper (Figures 9.2–9.5), all four platforms were able to adequately quantify cell counts and the Dex-induced GNHR-EGFP translocation concentration response (Figure 9.7, Table 9.2). Table 9.2 lists the performance data for each of the imaging platforms used to acquire images of 3617.4 cells exposed to a 10-point series of Dex agonist concentrations, together with the analyzed output of their respective image analysis algorithms. The average cell count per field of view ranged from 400 cells with the Pathway 855 to 1,239 cells with the ArrayScan V^{TI} and was not significantly modulated by treatment with the indicated concentrations of Dex (Figure 9.7). It should be noted that on the ImageXpress Ultra, the length of the scan and therefore the size of the field of view can be adjusted in the software. In general, the total scan times were marginally shorter on the confocal platforms with laser excitation than on the wide field imagers with lamp excitation (Table 9.2). Images from both channels were acquired in parallel on the ImageXpress Ultra. Even though the Opera also has parallel acquisition capability, the two channels were acquired sequentially, just as they were on the ArrayScan V^{TI} and the Pathway 855. Not surprisingly, the shortest total scan time was observed on the ImageXpress Ultra platform, but when the additional time required for postacquisition image processing and analysis was also considered, the apparent faster throughput was eliminated (Table 9.2). Unlike the ImageXpress Ultra, the ArrayScan V^{TI}, Opera, and Pathway 855 platforms can all perform image analysis in real time as the images are acquired. If we had selected Draq 5 to stain the nuclei of the GNHR-EGFP expressing 3617.4 cells, we likely could have reduced scan times further on the confocal platforms by exciting both fluorophores with a single pass of one laser and simultaneously collecting the

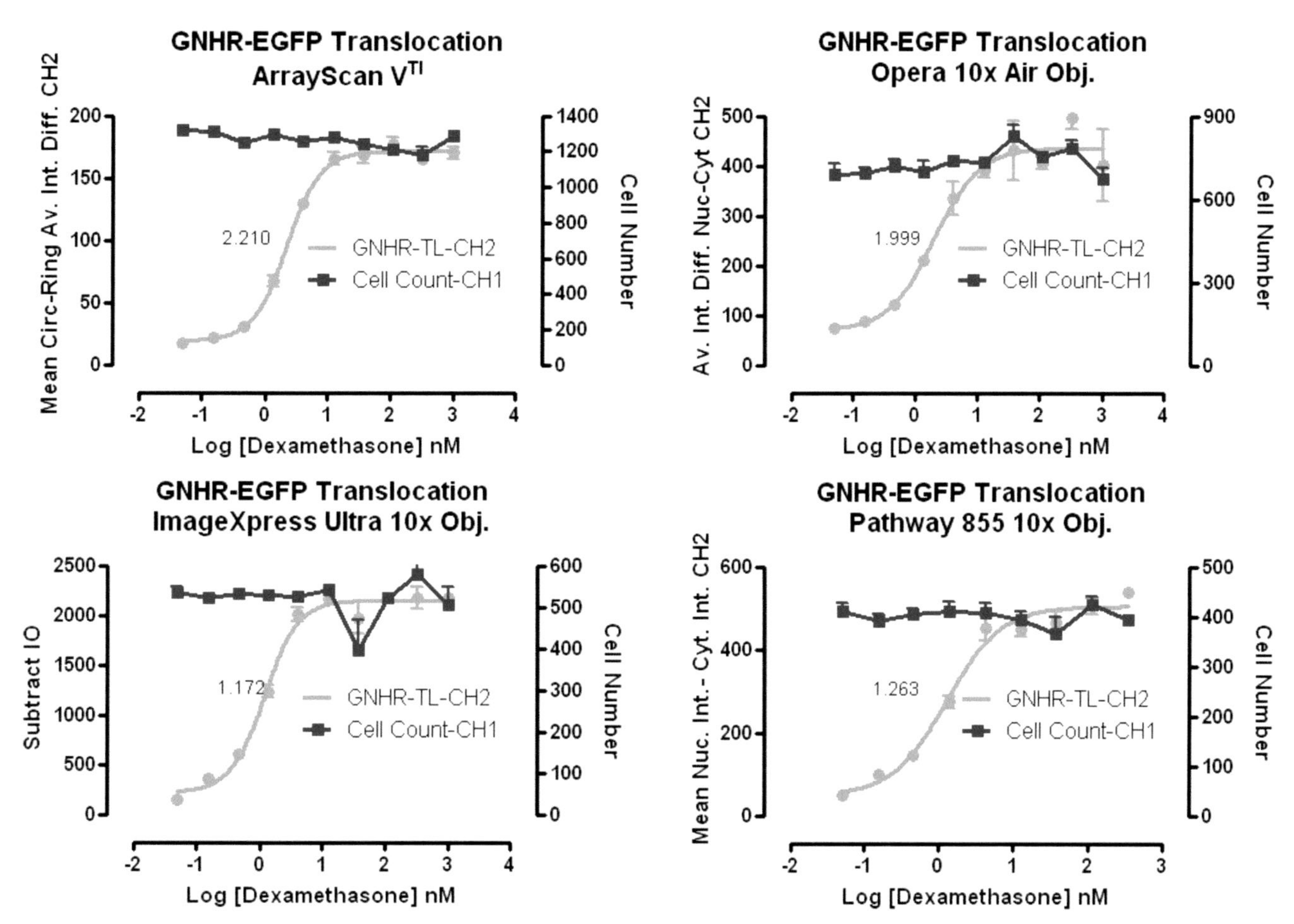

***FIGURE 9.7:** Comparison of dexamathasone-induced GNHR-EGFP translocation. The four automated imaging platforms were used to acquire images of 3617.4 cells exposed to the indicated concentrations of Dex agonist in replicate wells (n x 6 or n x 7), and the analyzed output of their respective image analysis algorithms are presented; total cell counts per field of view (cell count-CH1), and mean average intensity difference nucleus region–cytoplasm region (GNHR-TL-CH2). **A.** ArrayScan V^{TI}, **B.** Opera, **C.** ImageXpress Ultra, and **D.** Pathway 855. These data were fit to curves using the sigmoidal dose response variable slope equation Y = Bottom + (Top-Bottom)/(1 + 10^((LogEC50-X)*HillSlope)) using Graphpad Prism software 4.03.*

emissions on separate red and green CCD cameras (Opera) or PMTs (ImageXpress Ultra). In terms of the performance of the image analysis algorithms, all four platforms generated very similar data with respect to EC_{50}s, hill slopes, r^2 correlation coefficients, and S/B ratios (Figure 9.7, Table 9.2). With the GNHR-EGFP translocation assay configured as described here, only relatively minor differences were apparent among the four imaging platforms evaluated.

3. APPLICATIONS

HCS assays can be configured to measure many target protein classes, including GPCRs, kinases, phosphatases, nuclear hormone receptors, transporters, ion channels, proteases, and others [2, 4, 5, 15, 16, 18, 21–27, 48–52]. In addition to the target readout, multiwavelength fluorescent microscopy and imaging software can extract qualitative and quantitative data on fluorescent parameters in a single preparation that provides information on cell morphology and other cellular features that may reveal both on- and off-target compound effects [2, 4, 5, 15, 16, 18, 21–27, 48–52]. Cell-based imaging assays provide a mechanism to screen against protein targets that are expressed in the context of their endogenous protein substrates and scaffolding proteins in a complex microenvironment involving cross talk between signaling pathways, positive and negative regulatory control, and intracellular concentrations of substrates and co-factors [15, 16, 27]. Multifluorescence channel multiparameter HCS assays are also ideally suited to systems biology approaches to signaling pathway and network mapping because they can be configured to simultaneously detect multiple targets (multiplexing) [13, 14, 16, 18, 24–27, 30–32, 51, 53]. The analysis of complex cellular pathways and cell functions requires large-scale experimental approaches that take advantage of HCS systems in conjunction with powerful reagent tools [13, 14, 30, 38, 53, 54]. For example, HCS is particularly compatible with small interfering RNA (siRNA) technology to investigate and understand the consequences of target protein knockdown in a cell, and investigators are applying these to genome-wide screens to identify targets that contribute to specific cell phenotypes [38]. The ability to perform subpopulation analyses makes HCS ideally suited to stem cell research and assays involving mixed cell populations [1, 34, 35]. HCS technology can also be configured to measure complex cellular functions in phenotypic assays. Although many phenotypic outputs, such as cell proliferation and cell death (necrosis or apoptosis), can also be addressed in other formats, HCS is an enabling technology for more complex assays and challenging target classes such as cell cycle progression, mitosis, differentiation, and senescence; cell adherence, spreading, motility, and migration; wound healing; vascular tube formation; microtubule and actin cytoskeleton reorganization; protein-protein interaction; protein translocation; and neurite extension and synaptogenesis [1, 4, 10, 17, 20, 21, 23, 35, 36, 38, 48, 55–64]. Kinetic live well imaging will also significantly expand the number and complexity of the available types of HCS assays [4, 5, 10, 37, 60, 64]. The versatility of applications that automated HCA/HCS platforms provide for the implementation of cell- and whole organism–based assays to investigate protein targets, to map signaling pathways and networks, and to screen complex phenotypic assays makes the technology an invaluable tool for chemical genomics research.

REFERENCES

1. Ding, S. and P. G. Schultz, *A role for chemistry in stem cell biology*. *Nat Biotechnol*, 2004. 22: p. 833–40.
2. Haney, S. A., et al., *High-content screening moves to the front of the line*. *Drug Discov Today*, 2006. 11: p. 889–94.
3. Johnston, P. A. and P. A. Johnston, *Cellular platforms for HTS: three case studies*. *Drug Discov Today*, 2002. 7: p. 353–63.
4. Lang, P., et al., *Cellular imaging in drug discovery*. *Nat Rev Drug Discov*, 2006. 5: p. 343–56.
5. Giuliano, K. A., et al., *High-Content Screening: A New Approach to Easing Key Bottlenecks in the Drug Discovery Process*. *J Biomol Screen*, 1997. 2: p. 249–259.
6. Gough, A. H. and P. A. Johnston, *Requirements, features and performance of high content screening platforms, in high content screening: a powerful approach to systems cell biology and drug discovery*. *Methods in Molecular Biology*, 2006. 356: p. 41–61.
7. Johnston, P., *Automated High Content Screening Microscopy, in High Content Screening*, S. Haney, Editor. 2008, John Wiley and Sons Inc.: Hoboken, NJ. p. 25–42.
8. Keefer, S.a.Z., J., *Approaching High Content Screening and Analysis: Practical Advice for Users, in High Content Screening*, S. Haney, Editor. 2008, John Wiley & Sons Inc.: Hoboken, NJ. p. 3–24.
9. Lee, S., and Howell, B. J., *High Content Screening: Emerging Hardware and Software Technologies, in Measuring Biological Responses with Automated Microscopy*, J. Inglese, Editor. 2006, Academic Press, Elsevier: San Diego, CA. p. 468–483.
10. Stephens, D.a.A., V. J., *Light Microscopy Techniques for Live Cell Imaging*. *Science*, 2003. 300(5616): p. 82–86.
11. Taylor, D., *Introduction to High Content Screening, in High Content Screening*, D. Taylor, Haskins J. R., and Giuliano, K. A., Editor. 2007, Humana Press: Totowa, NJ. p. 3–18.
12. Berlage, T., *Analyzing and mining image databases*. *Drug Discov Today*, 2005. 10: p. 795–802.
13. Giuliano, K. A., et al., *Systems cell biology knowledge created from high content screening*. *Assay Drug Dev Technol*, 2005. 3: p. 501–14.
14. Giuliano K. A., J. P., Gough A., Taylor D. L., *Systems cell biology based on high-content screening*. *Methods Enzymol.*, 2006. 414: p. 601–619.
15. Nickischer, D., et al., *Development and implementation of three mitogen-activated protein kinase (MAPK) signaling pathway imaging assays to provide MAPK module selectivity profiling for kinase inhibitors: MK2-EGFP translocation, c-Jun, and ERK activation*. *Methods Enzymol*, 2006. 414: p. 389–418.
16. Trask, O. J., Jr., et al., *Assay development and case history of a 32K-biased library high-content MK2-EGFP translocation screen to identify p38 mitogen-activated protein kinase inhibitors on the ArrayScan 3.1 imaging platform*. *Methods Enzymol*, 2006. 414: p. 419–39.
17. DeBiasio, R., et al., *Five-parameter fluorescence imaging: wound healing of living Swiss 3T3 cells*. *J Cell Biol*, 1987. 105: p. 1613–22.
18. Almholt, D. L., et al., *Nuclear export inhibitors and kinase inhibitors identified using a MAPK-activated protein kinase 2 redistribution screen*. *Assay Drug Dev Technol*, 2004. 2: p. 7–20.
19. Arnold, D. M., et al., *Synthesis and biological activity of a focused library of mitogen-activated protein kinase phosphatase inhibitors*. *Chem Biol Drug Des*, 2007. 69: p. 23–30.
20. Mayer, T. U., et al., *Small molecule inhibitor of mitotic spindle bipolarity identified in a phenotype-based screen*. *Science*, 1999. 286: p. 971–4.
21. Mitchison, T. J., *Small-molecule screening and profiling by using automated microscopy*. *Chembiochem*, 2005. 6: p. 33–9.
22. Oakley, R. H., et al., *The cellular distribution of fluorescently labeled arrestins provides a robust, sensitive, and universal assay for screening G protein-coupled receptors*. *Assay Drug Dev Technol*, 2002. 1: p. 21–30.
23. Ramm, P., et al., *Automated screening of neurite outgrowth*. *J Biomol Screen*, 2003. 8: p. 7–18.
24. Vogt, A., et al., *Cell-active dual specificity phosphatase inhibitors identified by high-content screening*. *Chem Biol*, 2003. 10: p. 733–42.
25. Vogt, A. and J. S. Lazo, *Chemical complementation: a definitive phenotypic strategy for identifying small molecule inhibitors of elusive cellular targets*. *Pharmacol Ther*, 2005. 107: p. 212–21.
26. Vogt, A. and J. S. Lazo, *Implementation of high-content assay for inhibitors of mitogen-activated protein kinase phosphatases*. *Methods*, 2007. 42: p. 268–77.
27. Williams, R. G., et al., *Generation and characterization of a stable MK2-EGFP cell line and subsequent development of a high-content imaging assay on the Cellomics ArrayScan platform to screen for p38 mitogen-activated protein kinase inhibitors*. *Methods Enzymol*, 2006. 414: p. 364–89.
28. Haney, S. A., ed. *High Content Screening, Science, Techniques and Applications*. 2008, John Wiley & Sons, Inc.: Hoboken, NJ. 389.

29. Inglese, J., ed. *Measuring Biological Responses with Automated Microscopy*. Methods in Enzymology, ed. J.a.S. Abelson, MI. Vol. 414. 2006, Academic Press, Elsevier: San Diego, CA. 679.
30. Taylor, D., Haskins J. R. and Giuliano, K. A., ed. *High Content Screening, A Powerful Approach to Systems Cell Biology and Drug Discovery*. Methods in Molecular Biology, ed. J. Walker. Vol. 356. 2007, Humana press: Totowa, NJ. 435.
31. Borchert K. M., G. R., Frolik C. A., Hale L. V., Halladay D. L., Gonyier R. J., Trask O. J., Nickischer D. R., Houck K. A.., *High-content screening assay for activators of the Wnt/Fzd pathway in primary human cells*. *Assay Drug Dev Technol.*, 2005. 3(2): p. 131–141.
32. Borchert K. M., S. G. R., Hale L. V., Trask O. J., Nickischer D. R., Houck K. A., Screening for activators of the wingless type/Frizzled pathway by automated fluorescent microscopy, in *Measuring Biological Responses with Automated Microscopy*, J. Inglese, Editor. 2006, Academic Press, Elsevier: San Diego, CA. p. 140–150.
33. Mayer T., J. B., Wyler M. R., Kelly P. D., Aulner N., Beard M., Barger G., Többen U., Smith D. H., Brandén L., Rothman J. E.., Cell-based assays using primary endothelial cells to study multiple steps in inflammation., in *Measuring Biological Responses with Automated Microscopy*, J. Inglese, Editor. 2006, Academic Press, Elsevier: San Diego, CA. p. 266–283.
34. Bushway P. J., M. M., High-throughput screening for modulators of stem cell differentiation., in *Measuring Biological Responses with Automated Microscopy*, J. Inglese, Editor. 2006, Academic press, Elsevier: San Diego, CA. p. 300–316.
35. Sammak, P., Abraham, V., Ghosh, R., Haskins, J., Jane, E., Petrosko, P., Erb, T. M., Kinney, T. N., Jefferys, C., Desai, M. and Mangoubi, R., high content analysis of stem cell growth and differentiation, in *High Content Screening, Science, Techniques and Applications*, S. A. Haney, Editor. 2008, John Wiley & Sons Inc.: Hoboken, NJ. p. 205–224.
36. Vogt, A., Kalb E. N., and Lazo J. S., *A scalable high-content cytotoxicity assay insensitive to changes in mitochondrial metabolic activity*. *Oncol Res*, 2004. 14: p. 305–14.
37. Giuliano K. A., T. D., Waggoner A. S., *Reagents to measure and manipulate cell functions*. *Methods Mol Biol.*, 2007. 356: p. 141–63.
38. Moffat, J. and et. al., *A Lentiviral RNAi library of human and mouse genes applied to an arrayed viral high content screen.*. *Cell*, 2004. 124: p. 1283–1298.
39. Giuliano, *Optimizing the integration of immunoreagents and fluorescent probes for multiplexed high content screening assays*. *Methods Mol Biol.*, 2007. 356: p. 189–93.
40. Howell, B., Lee, S, and Sepp-Lorenzino, L., *Development and implementation of Multiplexed Cell-based Imaging Assays*, in *Measuring Biological Responses with Automated Microscopy*, J. Inglese, Editor. 2006, Academic Press (Elsevier): San Diego, CA. p. 284–300.
41. Richards, G., Kerbu, J. E., Chan, G. K. Y., and Simpson, P. B., automating cell plating and sample treatments for fixed cells in high content assays, in *High Content Screening*, D. L. Taylor, Haskins J. R., and Giuliano, K. A., Editor. 2007, Humana: Totowa, NJ. p. 109–119.
42. Ghosh, R., Lapets, O., and Haskins, J. R., *Characteristics and Value of Directed Algorithms in High Content Screening, in High Content Screening, A Powerful Approach to Systems Cell Biology and Drug Discovery*, D. Taylor, Haskins, J. R., and Giuliano, K. A., Editor. 2007, Humana Press Inc.: Totowa, NJ. p. 63–81.
43. Zhou, X.a.W., T. C., *A Primer on Image Informatics of High Content Screening, in High Content Screening, Science, Techniques and Applications*, S. A. Haney, Editor. 2008, John Wiley & Sons, Inc.: Hoboken, NJ. p. 43–84.
44. Gough, A.a.J., P. A., *Requirements, Features and Performance of High Content Screening Platforms, in High Content Screening*, D. Taylor, Haskins, J. R., and Giuliano, K. A., Editor. 2007, Humana Press: Totowa, NJ. p. 41–61.
45. Elbi C., W. D., Lewis M., Romero G., Sullivan W. P., Toft D. O., Hager G. L., DeFranco D. B., *A novel in situ assay for the identification and characterization of soluble nuclear mobility factors*. *Sci STKE.*, 2004. 238: p. pl10.
46. Elbi C, W. D., Romero G, Sullivan W. P., Toft D. O., Hager G. L., DeFranco D. B., *Molecular chaperones function as steroid receptor nuclear mobility factors*. *Proc Natl Acad Sci U S A.*, 2004. 101(9): p. 2876–2881.
47. Johnston, *An interview with Paul A. Johnston, Ph.D., by Vicki Glaser*. *Assay Drug Dev Technol.*, 2007. 5(3): p. 289–97.
48. Lundholt, B. K., et al., *Identification of Akt pathway inhibitors using redistribution screening on the FLIPR and the IN Cell 3000 analyzer*. *J Biomol Screen*, 2005. 10: p. 20–9.
49. St. Croix, C. M., B. R. Pitt, and S. C. Watkins, *The use of contemporary fluorescent imaging technologies in biomedical research Medicine and Science*, 2005. 10: p. 16–29.
50. Vogt, A., et al., *Spatial analysis of key signaling proteins by high-content solid-phase cytometry in Hep3B cells treated with an inhibitor of Cdc25 dual-specificity phosphatases*. *J Biol Chem*, 2001. 276: p. 20544–50.
51. Vogt, A. and J. S. Lazo, *Discovery of protein kinase phosphatase inhibitors*. *Methods Mol Biol*, 2007. 356: p. 389–400.
52. Vogt, A., et al., *The benzo[c]phenanthridine alkaloid, sanguinarine, is a selective, cell-active inhibitor of mitogen-activated protein kinase phosphatase-1*. *J Biol Chem*, 2005. 280: p. 19078–86.
53. Butcher, E. C., *Can cell systems biology rescue drug discovery? Nat Rev Drug Discov*, 2005. 4: p. 461–7.
54. Butcher, R. A. and S. L. Schreiber, *Using genome-wide transcriptional profiling to elucidate small-molecule mechanism*. *Curr Opin Chem Biol*, 2005. 9: p. 25–30.
55. Clemons, P. A., *Complex phenotypic assays in high-throughput screening*. *Curr Opin Chem Biol*, 2004. 8: p. 334–8.
56. Ghosh, R. N., et al., *Cell-based, high-content screen for receptor internalization, recycling and intracellular trafficking*. *Biotechniques*, 2000. 29: p. 170–5.
57. Ghosh, R. N., et al., *Quantitative cell-based high-content screening for vasopressin receptor agonists using transfluor technology*. *J Biomol Screen*, 2005. 10(5): p. 476–84.
58. Keyel, P. A., Watkins S. C., Traub. L. M., *Endocytic adaptor molecules reveal an endosomal population of clathrin by total internal reflection fluorescence microscopy*. *J Biol Chem*, 2004. 279: p. 13190–204.
59. Lidov, H. G., et al., *Localization of dystrophin to postsynaptic regions of central nervous system cortical neurons*. *Nature*, 1990. 348: p. 725–8.
60. Salter, R. D., et al., *Rapid and extensive membrane reorganization by dendritic cells following exposure to bacteria revealed by high-resolution imaging*. *J Leukoc Biol*, 2004. 75: p. 240–3.

61. Sorkina, T., et al., *Oligomerization of dopamine transporters visualized in living cells by fluorescence resonance energy transfer microscopy*. *J Biol Chem*, 2003. 278(30): p. 28274–83.
62. Trask, O., DeMarco, C. T., Dunn, D, Gainer, T. G., Eudailey, J., Kalyenbach, L. and Lo, D. C., *Live Brain Slice Imaging for Ultra High Content Screening: Automated Fluorescent Microscopy to Study Neurodegenerative Diseases, in High Content Screening, Science, Techniques and Applications*, S. A. Haney, Editor. 2008, John Wiley & Sons Inc.: Hoboken, NJ. p. 189–203.
63. Watkins, S., et al., *Imaging secretory vesicles by fluorescent protein insertion in propeptide rather than mature secreted peptide*. *Traffic*, 2002. 3: p. 461–71.
64. Watkins, S. C., Salter R. D., *Functional connectivity between immune cells mediated by tunneling nanotubules*. *Immunity*, 2005. 23: p. 309–18.

CHAPTER 10

Phenotypic Screens with Model Organisms

Grant N. Wheeler
Robert A. Field
Matthew L. Tomlinson

Chemical genetics (see this book and [1]) for an extensive review) offers a complementary approach to loss-of-function mutations in the analysis of complex, multicomponent biological processes. Large-scale mutagenesis screens using genetic model systems have been used very successfully for many years in identifying genes involved in developmental and physiological events. However, such screens are expensive and time consuming. In addition, with respect to later development and organogenesis, these screens can be limited in scope. During embryonic development, many patterning and signaling systems are used multiple times [2–4]. Thus, if an embryo is disrupted at an early time point because of a mutation, later events using the same pathway become difficult to study. Such mutagenesis screens therefore do not allow for the finer temporal control of protein function. The ability to have temporal control over compound addition and thus the modulation of protein function provides a more focused approach to phenotypic assays. This also means that chemical genomic screens are applicable to maternal proteins, which many traditional mutagenesis screens are not, significantly extending the opportunity to identify key endogenous players in biological processes.

In this chapter, we focus on some of the multicellular organisms commonly used for chemical genomic screens, including plants, worm (*Caenorhabditis elegans*), fruit fly (*Drosophila melanogaster*), fish (zebrafish), and frog (*Xenopus*). Table 10.1 shows some of the general advantages and disadvantages of commonly used model organisms. With respect to chemical genetics, which requires high-throughput screening (HTS), organisms producing large numbers of embryos are essential, so chick and mouse are not suitable for such screens. However, because of the conservation of protein sequence and especially protein structure between species, small molecules identified in screens in one species can be tested in another species, with the reasonable expectation that the compound will interact with a similar protein. Thus, molecules identified in screens can be tested in higher organisms such as mouse and chick as well as in other systems such as cell-based assays using human cells. In the case of plants, compounds identified in *Arabidopsis* spp. can then be tested in other species such as crops. Another advantage is that, in many cases, compounds that are toxic will have been screened out before testing in higher organisms.

Of course, the closer an organism is in evolutionary terms to the target organism, such as human or a particular plant, the better. Figure 10.1 shows a phylogenetic tree highlighting the various animal model organisms used for chemical genomic screens and their evolutionary distance from man. Figure 10.2 outlines the general steps involved in a chemical genomic screen. The initial step is to design an assay or identify a phenotype that will be the target for the screen and to choose the relevant model organism. The screen could be to identify a morphological or behavioral change in the organism or to characterize an alteration in gene expression either by variations in green fluorescent protein (GFP) reporter levels, in situ hybridization, or microarray technology. Once the assay has been designed, a library of compounds is chosen. There are many sources for such libraries [5]. Once the screen has been carried out and hits obtained, these hits need to be validated. This can be done in two ways, either by determining the mechanism of action (MOA) and validating the target and/or by validating the compound effects in secondary assays or by showing an effect in higher organisms.

There are two purposes for carrying out phenotypic chemical genomic screens in model organisms. One is to promote research in a given area by obtaining small molecules that can be used to investigate fundamental questions, and the other is as a screen for pharmaceutical reagents that potentially can be used for clinical purposes.

1. *ARABIDOPSIS THALIANA* AND OTHER PLANTS

To complement classical biochemical and genetic methods, the development of small molecule inhibitors that induce phenotypes of interest is now emerging as a powerful approach in plant biology. The screening of small molecules against plants has a long history in the agrochemical sector. Such experiments often have been aimed at identifying potential new herbicides or growth modifiers – in other words, often the phenotype being screened for

TABLE 10.1: Advantages and Disadvantages to Using the Common Model Organisms for Chemical Genomic Phenotypic Screens

	C. elegans	*D. melanogaster*	Zebrafish	Xenopus	Chick	Mouse	Arabidopsis
Numbers of embryos	High	High	High	High	Low	Low	High
Cost	Low	Med	Med	Med	Low	High	Med
Access	Good	Good	Good	Good	Good	Poor	Good
micromanipulation	Limited	Limited	Fair	Good	Good	Limited	Limited
Genetics	Good	Good	Good	None/fair[a]	None	Good	Good
Gene inventory	Known	Known	Known	Known[a]	Known	Known	Known
Chemical genomics	Good	Good	Good	Good	None	None	Good

[a] *Xenopus tropicalis.*

was "dead"! Reflecting on just what widespread use plants make of small molecules in their natural metabolism, the realization that organic molecules might be used to manipulate plants rather than to kill them was perhaps overdue. A number of recent review articles highlight the variety of uses to which chemical genetics approaches have been used recently in plants [6–9], both to understand the fundamentals of plant biology per se, but also for commercial exploitation [10]. Examples include the generation/identification of chemical tools with which to study germination, membrane trafficking, gravitropism, cell wall biosynthesis, plant immunity, hormone biosynthesis and signaling, and stress responses. Herein, the use of chemical approaches to study aspects of membrane trafficking, cellulose biosynthesis and deposition, and auxin and brassinosteroid (BR) signaling are highlighted.

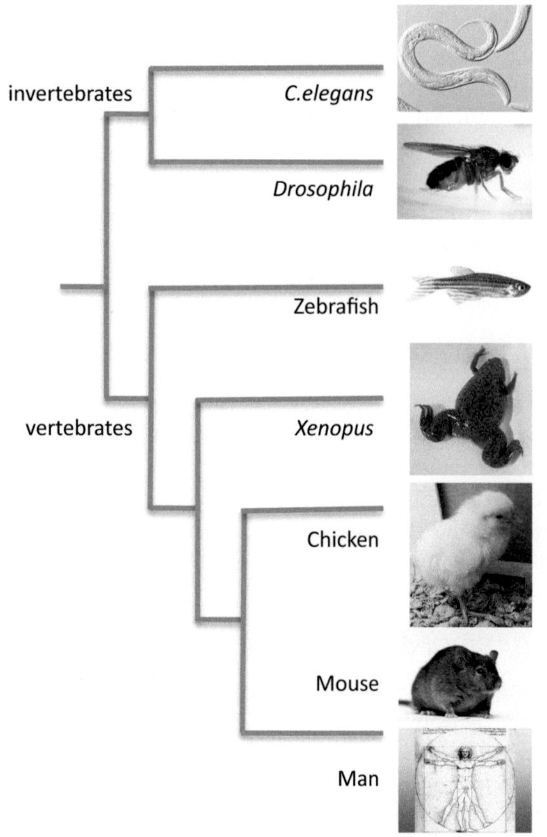

FIGURE 10.1: ***Phylogenetic tree showing the main animal organisms commonly used in research and their evolutionary relationship.*** **Source: *adapted from [5].***

The attraction of manipulating plants by chemical means rather than genetically is numerous. First, one does not become embroiled in the genetic modification (GM) debate. In addition, the ability to move between species is (relatively) straightforward. There is also potential value where genetic resources are suboptimal, which at this stage includes some commercially important cereals. As with many plant studies, the use of *A. thaliana* as a model organism is attractive, given that its complete genome sequence is available (*Arabidopsis* Genome Initiative, [11]), along with a variety of genetics and informatics resources [12] that enable broad-scale systems biology approaches to be adopted (e.g., screening large numbers of homozygous T-DNA lines with parallel morphological, physiological, and chemical phenotypic assays [13]). Although the obvious use of *Arabidopsis* is as a model plant, in principle there is scope to use it in relation to biomedical programs: A comparison of the list of human disease genes with the complete *Arabidopsis* gene set demonstrated that, of 289 human disease genes, 139 (48%) had hits in *Arabidopsis*. Interestingly, there are at least seventeen human disease genes that are more similar to *Arabidopsis* genes than to those found in yeast, *Drosophila*, or *C. elegans* (*Arabidopsis* Genome Initiative, [11]).

1.1. Membrane Trafficking

The evolutionary conservation of the endomembrane system between plants, yeast (*Saccharomyces cerevisiae*), and other eukaryotes has been exploited to identify small molecules that induce the secretion of carboxypeptidase Y, which is normally targeted to the vacuole [14]. A screen in yeast of 4,800 compounds identified 14 compounds termed sorting inhibitors (sortins). Application of some of these compounds to *Arabidopsis* seedlings led to reversible defects in vacuole biogenesis and root development [14]. These

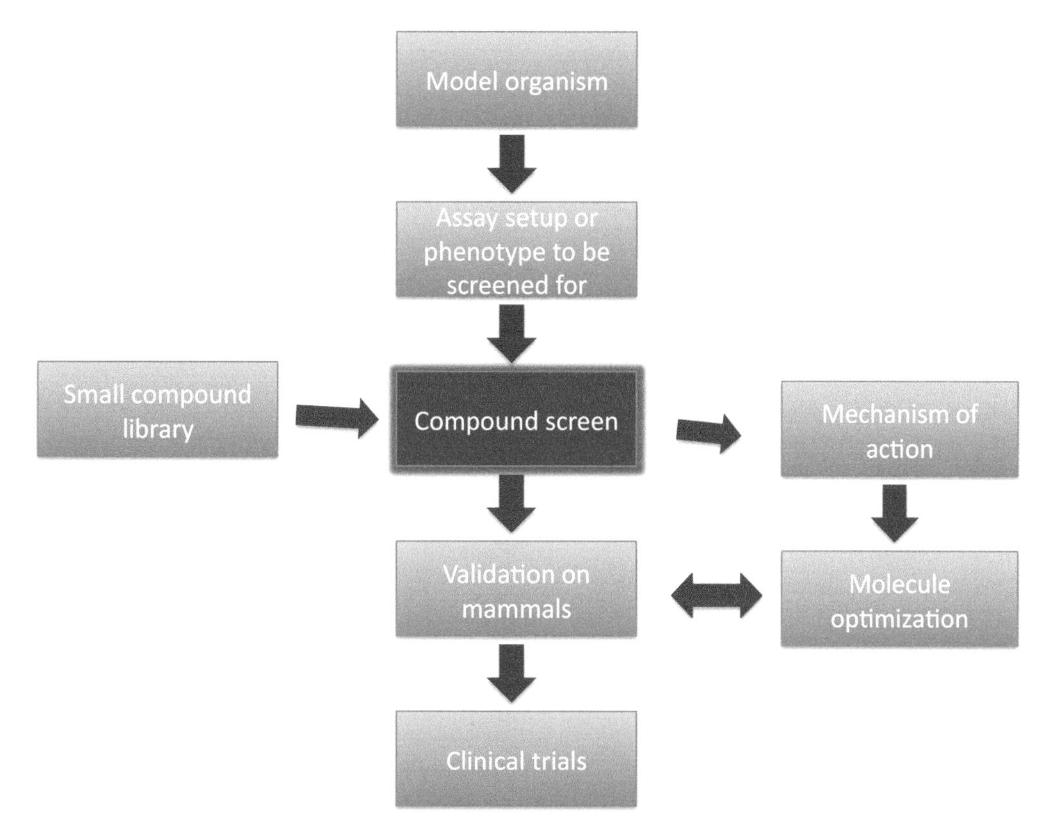

FIGURE 10.2:** Animal or plant models in the drug discovery process. The model shows the readouts and procedures that enable the rapid and reliable detection of hits during a screen. Once hits are identified, processes of validation and target identification can proceed.* **Source: adapted from [43].*

studies, which employed a sublethal dose of inhibitors, demonstrate the power of using temporally controlled, chemically induced phenotypes in a manner that would be difficult to achieve using conventional genetic methods.

Chemical genetics approaches have also been used to explore the link between endomembrane trafficking and gravitropism [15–17]. A library of ten thousand diverse compounds was screened for molecules that affected (positively or negatively) the gravitropic response of *Arabidopsis* seedlings (Figure 10.3), which led to thirty four confirmed actives [15]. Four of these compounds were found to cause aberrant endomembrane morphologies, affecting both gravitropism and vacuole morphology in a tissue-specific manner.

1.2. Control of Cellulose Biosynthesis and Deposition

The inhibition of cellulose biosynthesis has received much attention because of its potential as an herbicide target. In addition, we still don't fully understand the composition, arrangement, and regulation of the array of protein components required for exquisitely controlled cellulose microfibril deposition in the plant cell wall. The combination of genetic mutations and established inhibitors of cellulose biosynthesis, such as the herbicides isoxaben and thiazolidinone, has proved useful in establishing that although the *IXR1* gene is expressed in the same cells as the structurally related *RSW1* (*AtCESA1*) cellulose synthase gene, these two cellulose synthase (*CesA*) genes are not functionally redundant [18].

Cortical microtubules control the direction of cellulose microfibril deposition in the cell wall, which in turn determines directional cell expansion and plant cell morphogenesis. Chemical genomic screening has been used to identify compounds that induce a spherical swelling (SS) phenotype

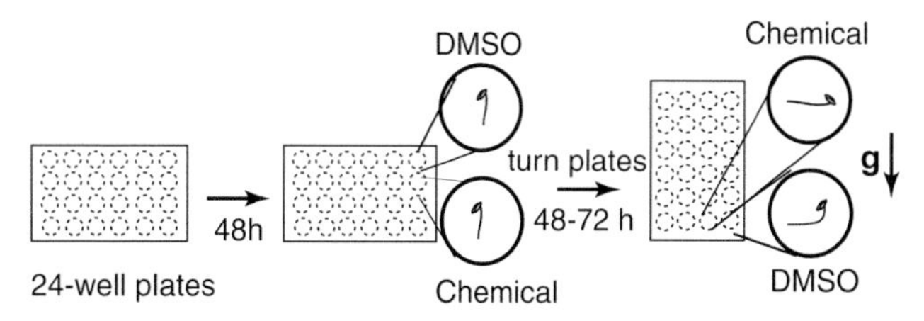

FIGURE 10.3:** Membrane trafficking and gravitropism. Screen for chemicals that affected gravitropism. The chemical library was screened in a 24-well format, and seedlings were scored for gravitropic response after reorientation. Chemicals dissolved in 20% DMSO were added to wells. Control wells contained an equivalent concentration of the solvent. The gravity vector is indicated by an arrow next to the g (on the right).* **Source: Reproduced from [15].*

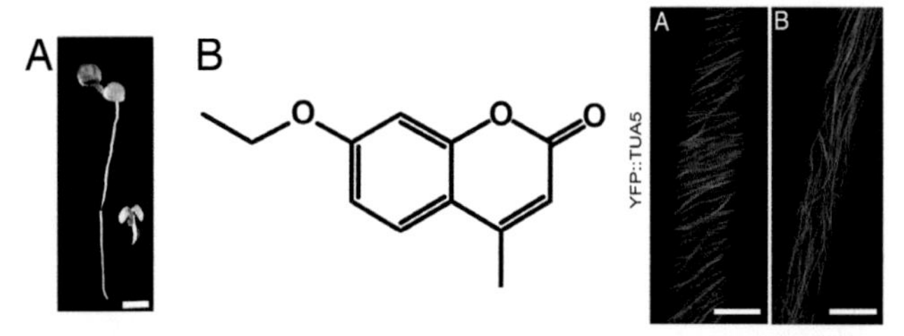

FIGURE 10.4: ***Control of cellulose biosynthesis and deposition. Morlin inhibits plant growth. A. Arabidopsis seedling grown under continuous light for five days on agar-containing morlin (5 μM) (right) showed reduced growth compared with control plants (left). B. Morlin, 7-ethoxy-4-methyl chromen-2-one. RIGHT. Effect of morlin on microtubule organization. Confocal images of cortical microtubules labeled with YFP::TUA5 in hypocotyl cells of 3-d etiolated*** **Arabidopsis** ***seedlings. A. DMSO control. B. Morlin (50 μM; 2 h).*** **Source:** ***Reproduced from [20].***

in tobacco BY-2 cells [19]. Among the most potent SS-inducing compounds, a chemical designated cobtorin perturbed the parallel alignment of preexisting cortical microtubules and nascent cellulose microfibrils. Hence, cobtorin potentially represents a novel tool with which to dissect how cortical microtubules guide the deposition of cellulose microfibrils.

Recently, a forward chemical genetics screen in *Arabidopsis* with twenty thousand small molecules was used to identify compounds capable of generating altered cell morphology (Figure 10.4). Swollen root phenotypes were analyzed, leading to the identification of 7-ethoxy-4-methyl chromen-2-one (morlin) [20]. Live-cell imaging of fluorescently labeled *CesA* and microtubules showed that morlin acts on the cortical microtubules and alters the movement of *CesA*. Thus, morlin represents a useful new probe of the mechanisms that regulate microtubule cortical array organization and its functional interaction with *CesA*.

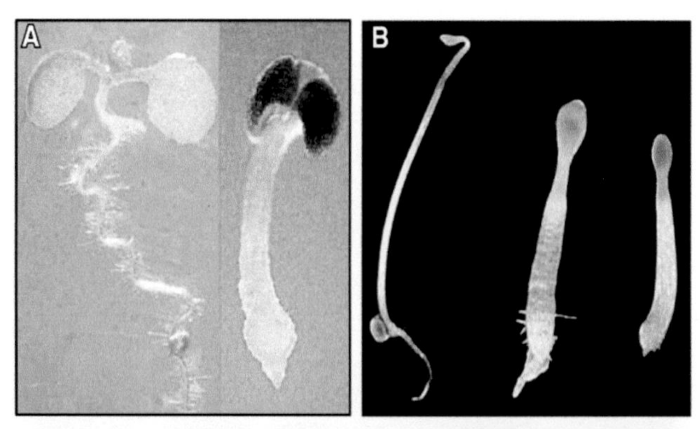

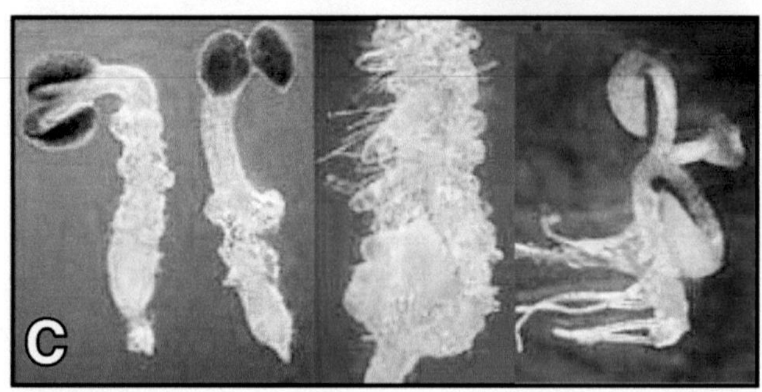

FIGURE 10.5: ***Auxin and brassinosteroid signaling. The effects of sirtinol treatment on*** **Arabidopsis** ***growth and development. A. WT*** **Arabidopsis** ***grown on 0.5 × MS media (left) and 25 μM sirtinol (right) for 5 days. B. Arabidopsis seedlings grown in total darkness on MS media (left), 10 μM sirtinol (right) for three days. C. Stimulation of adventitious root growth by sirtinol. Left two seedlings: WT grown on 25 μM sirtinol for eight days. The third plant from the left: an*** **Arabidopsis** ***seedling grown on 25 μM sirtinol for five days and then transferred to 0.5 × MS media for one day (only the hypocotyl structure is shown). Right seedling: WT grown on 25 μM sirtinol for five days and then grown on 0.5 × MS for three days.*** **Source:** ***Reproduced from [21].***

1.3. Auxin and BR Signaling

Plant hormones are involved in a wide array of processes, including cell division, cell elongation, tissue patterning, phototropism, gravitropism, and root development. Despite recent descriptions of the machinery involved in auxin transport and perception, additional regulatory mechanisms and components remain to be identified and characterized. Chemical genetics has proven to be a valuable means by which to investigate hormone response pathways in *A. thaliana* and other plant species.

A regulator of many auxin-inducible genes, *SIR1*, was identified using a chemical genetics approach [21]. A *SIR1* mutant was shown to be resistant to sirtinol, an NAD-dependent deacetylase inhibitor originally identified in yeast phenotypic screens and also shown to interfere with body axis formation in *Arabidopsis* [22]. Sirtinol activates many auxin-inducible genes and promotes auxin-related developmental phenotypes (Figure 10.5). Subsequent studies employing genetic and chemical analyses have demonstrated the MOA of sirtinol in *Arabidopsis* [23]. Analysis of sirtinol derivatives indicated that sirtinol itself is not the active compound and that it serves as a *pro-drug* – that is, it is metabolized *in planta* to 2-hydroxy-1-naphthaldehyde (HNA) and subsequently to 2-hydroxy-1-naphthoic acid (HNC), the latter being the auxin-signaling activator. The utility of sirtinol in dissecting auxin signaling is further demonstrated by the fact that it is readily taken up by *Arabidopsis* seedlings, whereas the active HNC is not.

Chemical genetics approaches have also been used to understand the MOA of BR hormones [24]. Exploration of small molecules that induce the BR deficient–like phenotype in *Arabidopsis* identified brassinazole as the first candidate BR biosynthesis inhibitor. This compound binds directly to DWF4, a cytochrome P450 monooxygenase that catalyzes side chain hydroxylation of BRs and behaves in a similar manner to a conditional mutation in BR biosynthesis. As a practical alternative to mutant screens using

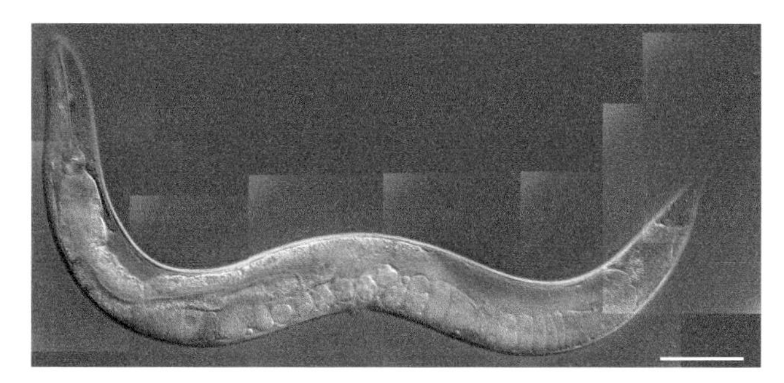

FIGURE 10.6: The model organism* C. elegans. *Hermaphrodite adult worm showing the translucent nature of the model and its relative complexity.* Source*: Taken from www.chin-sang.ca.

BR-deficient mutants as a background, application of BR biosynthesis inhibitors to a standard genetic screen was used to identify mutants that confer resistance to these inhibitors, resulting in the identification of new components of the BR signal transduction pathway. The ubiquitous nature of brassinosteroids in plants and the ease of transferring between species suggest potential for similar chemical genetic investigations of BR function in other plants.

2. *C. ELEGANS*

The round worm *C. elegans* is a soil-dwelling organism found in most countries that feeds on bacteria and grows to only 1 mm in length (Figure 10.6) [25]. The life cycle is completed in as few as three days at 25°C [25]. *C. elegans* was first used as a model organism in the 1960s for developmental and genetics-based research and has been extensively used for these purposes since. For chemical genomic screens, it is the simplest multicellular model used to date. Although there are only 959 cells in the adult hermaphrodite, it contains complex structures such as a digestive tract, nervous system, and muscles and has intricate behavior patterns [26]. *C. elegans* has an established cell lineage map (the origin of each adult cell can be tracked from an early embryo blastomere), and the sequencing of its genome was essentially completed in 1998.

The transparency of the worm (Figure 10.6) at all stages along with its small size and ability to generate large, synchronous populations facilitates HTS projects. The worms are easy to culture in liquid media and in 96-well format, making them highly amenable to a chemical genomic approach [27] and ideal for high-throughput assays [28]. The development of the Union Biometrica COPAS Biosort to robotically analyze and sort individuals using size, optical density, and up to two channels of fluorescence further increases the screening capacity of this model organism [29, 30].

Parallel technologies such as RNAi and the ease with which transgenics can be generated along with an array of mapped mutants mean that an extensive number of chemical genomic screening combinations are possible (Figure 10.9). This enables the researcher to focus the biological interest of the screen on a particular gene (via enhancer/suppressor screens or RNAi knockdown) or cell type (through a reporter gene such as GFP) (Figure 10.7). Although the adults have fewer than a thousand cells, almost one-third are specified to be neurons (302) whose connectivity map has been completed [31], making *C. elegans* attractive to chemical genomic researchers interested in neural development.

The use of enhancer/suppressor screens has also been applied to *C. elegans*–based chemical genomic screening. The enhancer/suppressor screen uses a mutant that has a mutated allele giving rise to a weak phenotype – for instance, a partially developed organ in the homozygous state. It would be possible to screen for compounds able to enhance the phenotype and give a fully developed organ or suppress the phenotype where the organ would not develop at all. The use of fluorescently labeled organs could greatly assist in this type of phenotype-based screen (Figure 10.7).

An excellent example of a forward chemical genomic screen using genetic mutants as the screening model was aimed at rescuing a mutation (enhancer screen) in the *C. elegans* insulin-signaling pathway, which gives a doubled life span (called *daf-2* [32]). The screen involved identifying chemicals that rescue the phenotype of increased life span (these would therefore be acting in the insulin-signaling pathway). The small molecule GAPDS was identified both as increasing glucose uptake (insulin levels control glucose uptake) and by its ability to rescue *daf-2* mutants (restore normal life span). GAPDS was generated as part of a tagged library. This means that all of the molecules in the library were easily tagged with an appropriate affinity tag to facilitate proteomic approaches aimed at identifying their targets. Proteomics, with embryo lysates and affinity matrices, identified fourteen candidate targets. RNAi was then used to identify which of these candidate genes are responsible for the phenotype [33]. From this, the insulin pathway enzyme glyceraldehyde-3-phosphate dehydrogenase was identified as the target of GAPDS [33]. The amenability of *C. elegans* to RNAi and the availability of a genetic mutant made this phenotype-based chemical genomic screen possible.

The genetics of the worm has proven valuable in its use as a chemical genomic model. For instance, to identify the target/pathway being modulated by a small molecule compound either at the screening or the target identification stage, a genetic candidate approach involving the use of genetic mutants can be taken (an example is shown in Figure 10.7) [34]. In a search for novel calcium channel inhibitors, the small molecule compound nemadipine was identified by a genetic suppressor screen [34]. The mutant in this case was a calcium channel mutant (*egl-19*) [34].

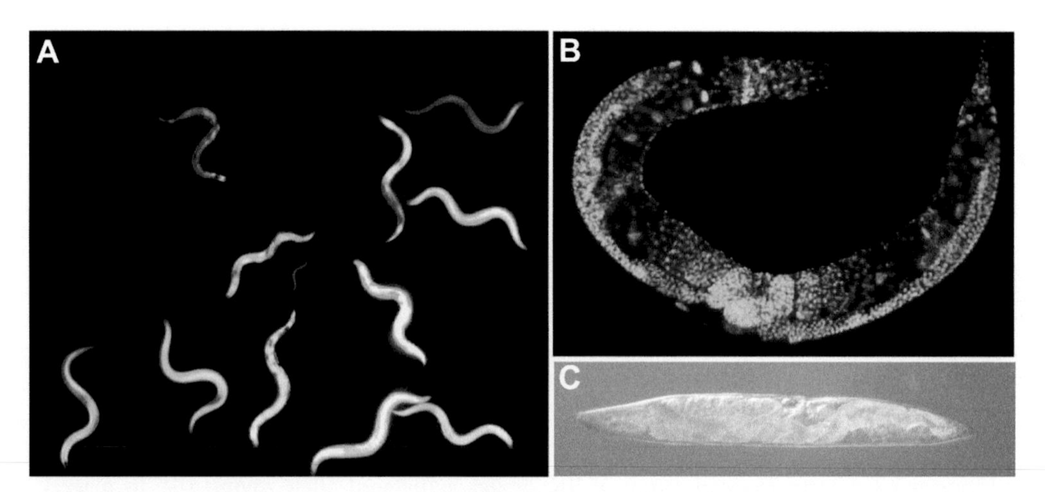

FIGURE 10.7: ***A.*** *Worms expressing a constitutively fluorescent active reporter gene (dsred:*col 12*) in the hypodermis and a GFP gene fusion (nlp 29), which is activated upon infection with a fungal pathogen. The resulting worms appear yellow.* Source: *J. Ewbank.* ***B.*** *Nuclei (Dapi) stained adult hermaphrodite worm.* Source: *S. Shaham.* ***C.*** *Mutant worm* (dpy-13 *[e458]) with a malformed cuticle leading to a dumpy phenotype.* Source: *Taken from http://www.wormbook.org/chapters/www_cuticle/cuticle.html.*

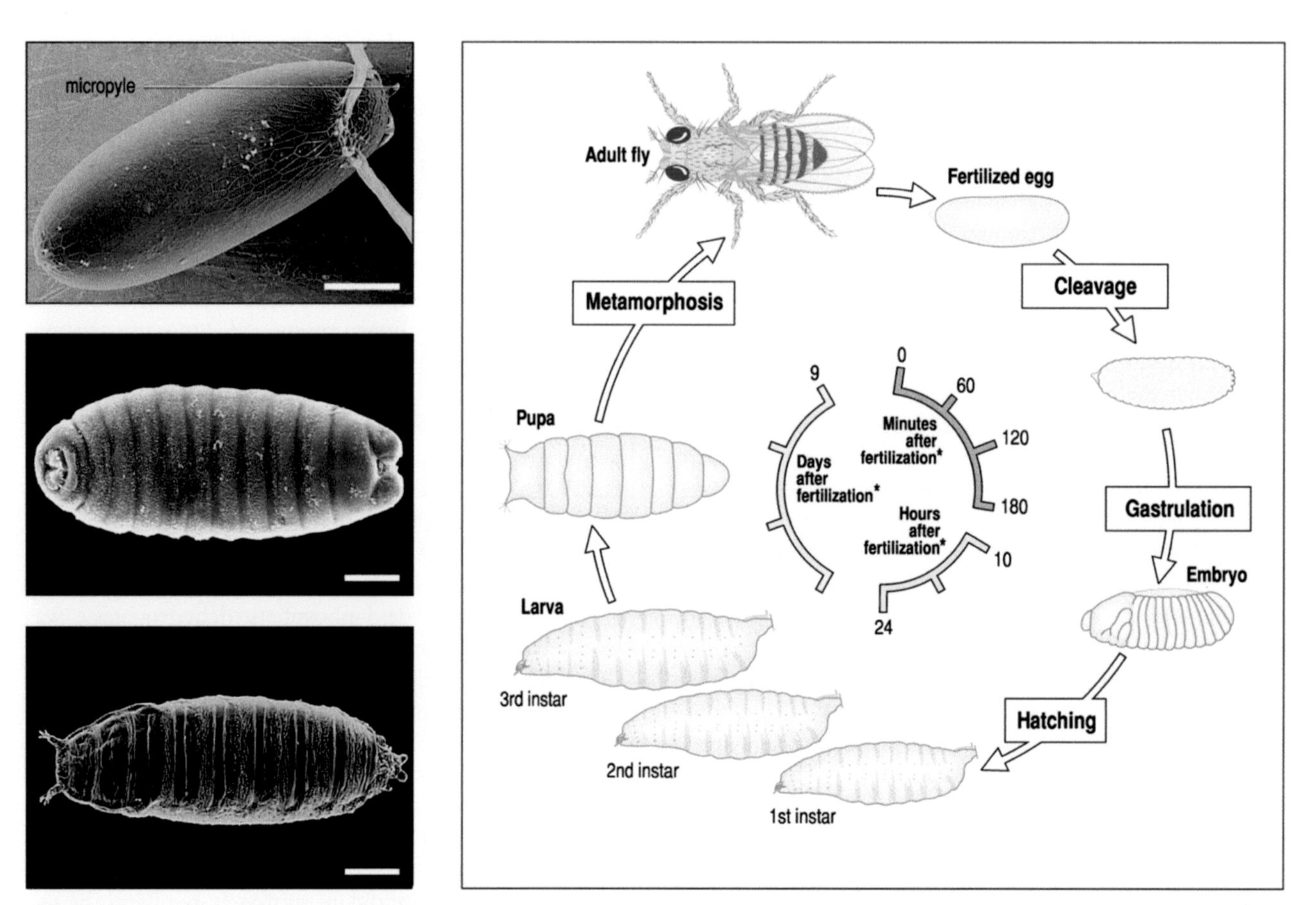

FIGURE 10.8: Life cycle of Drosophila melanogaster. *After cleavage and gastrulation, the embryo becomes segmented and hatches out as a feeding larva. It eventually forms a pupa and metamorphoses into the adult fly. The photographs show scanning electron micrographs of a Drosophila egg (top), a Drosophila second instar larva (middle), and a Drosophila pupa (bottom). Scale bars = 0.1 mm. Taken from Principles of Development by Wolpert, by permission of Oxford University Press.*

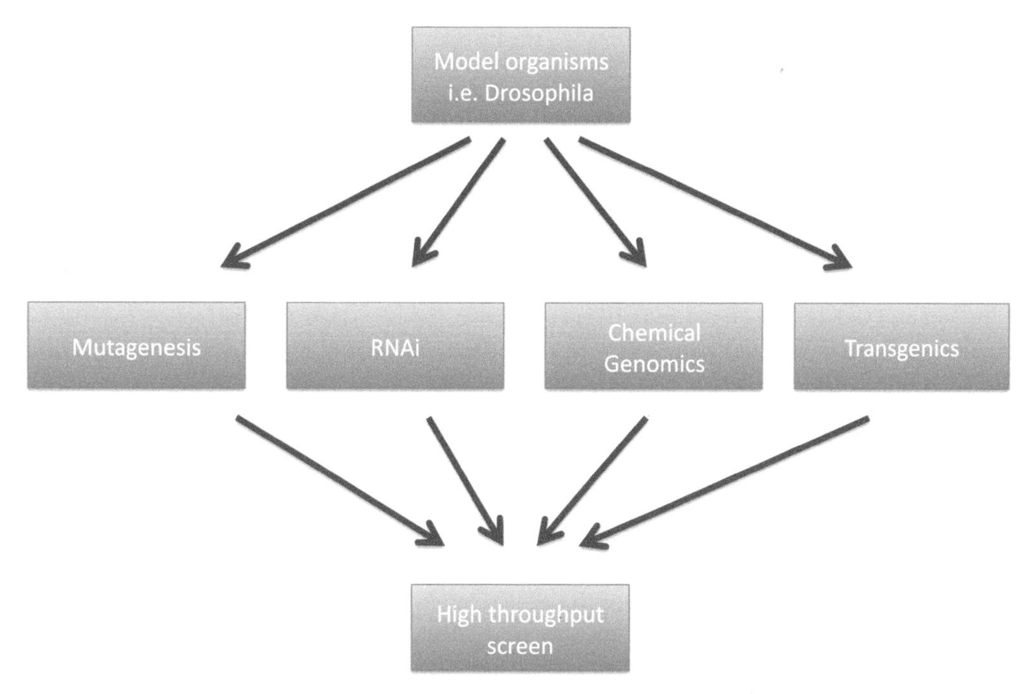

FIGURE 10.9: ***High-throughput phenotypic screens can be carried out using a number of different methods. These include inactivation of genes by mutation or RNAi, introduction of a pathogenic version of a gene (transgenesis), or exposure to chemicals.***

The target of nemadipine was determined from the following evidence. Worms treated with nemadipine phenocopy *egl-19* mutants. *egl-19* hypomorphs (mutation causing a partial loss-of-gene function) are hypersensitive to nemadipine, whereas *egl-19* hypermorphs (an increase in otherwise normal gene function) are resistant. This strongly suggests that nemadipine acts as an L-type calcium channel inhibitor [36].

An interesting and recent exploitation of *C. elegans* ability to be cultured in a liquid media makes use of a pathogenic fungal infection to the worm to perform live-animal infection screens (also termed pathogenicity assays) [35, 36]. The screens involve co-culturing the worms along with the pathogens to look for compounds that block infection and therefore lethality. Infection and progression of the fungal infection can be followed with GFP fluorescent variants of the fungal pathogen (Figure 10.7A) [36]. This type of chemical genomic screen allows a greater amount of biological information to be gained when screening the compounds as opposed to screening the fungal pathogen on its own. It is possible to block infection and host pathogen interactions rather than simply affecting the free-living form of the pathogen. These types of chemical genomic screens not only look for compounds capable of killing pathogens, but they also identify the in vivo concentration – that is, the compound efficacy (which in some cases has been found to be lower than the in vitro concentration) – and can give preliminary toxicology data [35, 36]. The assays are also capable of finding compounds that do not exhibit an effect on the pathogen in vitro – for instance, pro-drugs that may need to be processed by the organism [35].

C. elegans has attracted interest from researchers in both academia and industry looking to use chemical genomic screens to identify novel drug-like compounds [37–39]. An example of the worm's potential utility in drug discovery comes from the development of transgenic worms, which express GFP driven by cytochrome P450 promoters [40]. Cytochrome proteins cause problems by rapidly removing hydrophobic substances (such as drugs) from the body [40]. Drugs that induce cytochrome proteins are a major cause of adverse drug interactions because they potentially reduce the half-life of the drug [40]. The transgenic worms have been screened with a set of known cytochrome inducers, and these compounds were found to behave in the same way in worms, inducing green fluorescence [40]. A primary screen of drug libraries for those inducing cytochrome expression would eliminate compounds at an early stage, reducing time and costs of further drug development.

C. elegans also has potential to be useful in phenotype screens looking for small molecules, which modulate cell division genes. Misregulation of the rate of dividing cells in the body is one of the major causes of cancer. Adult hermaphrodite worms essentially function as egg-laying machines, with two large gonad arms continuously producing oocytes (Figure 10.6). This makes them suitable for studying the cell cycle, as different stages of the cell cycle can be tracked as the oocyte progresses down the transparent arm of the gonad [41].

Although *C. elegans*, with its large screening capacity, makes an impressive chemical genetics model that acts as an important bridge between unicellular and higher multicellular model organisms, it does have limitations. Its evolutionary distance from man has to be taken into account (Figure 10.1), signaling pathways are often simpler, and organogenesis is limited. Also, the lower degree of conservation between human and *C. elegans* proteins could mean that small molecule compounds discovered in the worm may not inhibit the human orthologous protein or that they inhibit at a much higher concentration.

3. DROSOPHILA

For nearly a hundred years, *Drosophila melanogaster* (the common fruit fly) genetics has been a central contributor to research on many fundamental biological mechanisms (Figure 10.8). *Drosophila* is particularly well known for research into comparative genomics to mammalian systems, genetic tools, and genetic screens. However, *Drosophila* chemical genetics is still an emerging field [27, 42, 43].

Drosophila has provided a rich treasure of mutant phenotypes that disrupt developmental, physiological, and behavioral processes. Genes identified as responsible for mutant phenotypes revealed the conserved nature of genes, proteins, and cellular networks in evolution. Other methods used in *Drosophila* to complement mutagenesis screens include transgenesis to overexpress a mutant version of a particular gene, RNAi screens to degrade specific mRNAs, and chemical genomic screens in which embryos or adults are exposed to disease-inducing chemicals (Figure 10.9). *Drosophila* mutants are often mutated in genes known to cause health problems in humans, such as cancer, diabetes, and inflammation [44, 45]. In addition, *Drosophila* is amenable to research into insecticide resistance. For instance, a chloride ion channel gene was found to be mutated in *Drosophila* populations resistant to the insecticide cyclodiene. This ion channel was subsequently shown to be a direct target of cyclodiene [46].

The mechanism of a compound action can also be explored with genomic technologies such as microarrays. Feeding drugs to flies and testing global gene changes at discrete time points can be informative. In aging studies, groups have combined drug treatment of aging flies with transcriptional profiling. For example, 4-phenylbutyrate, a histone deacetylase, was shown to increase life span and to increase global histone acetylation when fed to *Drosophila* [47].

Recently, two studies have highlighted the potential of chemical genetics in *Drosophila*. In the first, chemical modifiers of a CUG triplet repeat toxicity model in *Drosophila* were identified that could lead to potential treatments for myotonic dystrophy I (DMI), a chronic, slowly progressing disease characterized mainly by wasting of the muscles [44]. Noncoding CUG nucleotide repeat expansions interfere with the activity of human muscleblind-like (MBNL) proteins, contributing to DMI. The molecules that rescue this phenotype included nonsteroidal anti-inflammatory agents; muscarinic, cholinergic, and histamine receptor inhibitors; and drugs that can affect sodium and calcium metabolism. Thus, the screen has suggested novel candidates for DMI treatments.

In the second study, a chemical genomic screen was carried out to identify small molecules that rescue fragile X syndrome phenotypes in *Drosophila*. Fragile X syndrome is a syndrome of X-linked mental retardation. The screen revealed that muscarinic cholinergic receptors may have a role in fragile X syndrome in parallel to the gamma-aminobutyric acid (GABA)-ergic pathway, pointing to potential therapeutic approaches for treating fragile X syndrome [45].

Drosophila does have drawbacks that limit its use. The major one is that there are many diseases that cannot be modeled, such as cardiovascular diseases, acquired immunity diseases, and nondegenerative behavioral disorders. The second disadvantage is the thick cuticle that surrounds the fly and is a physical barrier to the penetration of molecules. Thus, *Drosophila* may not be as popular as *C. elegans*, but as outlined previously, it does have its advantages and therefore a place in the field of chemical genetics.

4. ZEBRAFISH

Zebrafish, along with *C. elegans*, is probably the most widely used model organism in chemical genetics studies (Figure 10.10). Its advantages include a short reproductive cycle combined with a high fecundity. For example, a single pair of adults can produce several hundred eggs in one week. Embryos are transparent, develop outside the mother, and mature into adults in three to six months (Table 10.1) (Figure 10.10A) [48]. In addition, zebrafish embryos have been shown to absorb small molecule compounds from the surrounding water [49]. A major advantage over the invertebrate models is primarily its evolutionary closeness to mammals (Figure 10.1), with the discrete development of an array of complex organs such as the eye, brain, and heart [50].

As with other model systems, the utility of the fish as a model organism has increased greatly with the near completion of the sequenced genome, an annotated collection of mapped mutants and associated databases (Zfin) [51]. This provides an exciting mutant resource for enhancer/suppressor-based chemical genomic screening, in which a mutagenized population has an allele of a gene that leads to a weak mutant phenotype in the biological process of interest [34]. A variety of transgenic lines are available to perform phenotype-focused chemical genomic

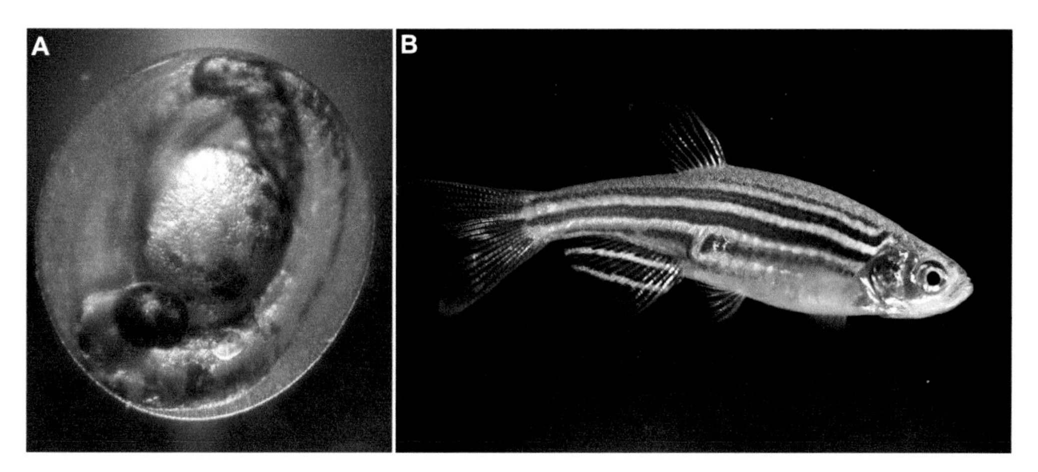

FIGURE 10.10: ***Wild type zebrafish embryo (A) and adult (B). The embryo illustrates the model's opacity. The developing eye, somites, and pigmentation can clearly be seen.*** **Source:** ***Taken from http://www.ufz.de/index.php?en=6556.***

screens on the development of specific organs, such as the eye and liver (Figure 10.11). The optical clarity of the developing embryo is important when combined with fluorescent markers to highlight the location and activity of specific cell populations [50], an example of which is shown for blood vessels (Figure 10.11).

Another recent development is the use of zebrafish to create disease models akin to those normally restricted to the mouse (Figure 10.12). This opens up the possibility of using chemical genetics to phenotypically screen for small molecule compounds able to perturb disease processes [52]. The potential high throughput of the zebrafish means that it is in a unique position between in vitro disease assays and mouse models. One example is cancer research. The aquatic nature of the zebrafish life cycle means that embryos can be exposed to chemical carcinogens and will develop a range of benign and malignant tumors in many different organs [53]. These tumors are comparable to human tumors on many levels, for instance, histological appearance, rate of cell division, and the role of conserved cancer-related genes [53]. Transgenic zebrafish with T-cell lymphoma can be generated by introducing the cancer-causing gene *m-myc* to lymphoid cells, driving its expression from a lymphoid-specific promoter (*rag2*) [54, 55]. A number of cancer models exist, ranging from leukemia to melanoma (Figure 10.12) [56]. The zebrafish model of melanoma (skin cancer) offers the potential to phenotypically screen for inhibitors of this highly metastatic cancer [57]. The melanoma transgenic fish were created by overexpressing the oncogene *BRAF*, an activated serine/threonine kinase in human melanoma, in melanocytes (pigment cells) of zebrafish via the melanocyte-specific promoter *mitf* [57]. These fish were then crossed into fish lacking the tumor suppressor gene *p53*, which resulted in zebrafish with large, spontaneously arising melanomas (Figure 10.12B) [57]. An extension of this mutant's utility is with cell transplantation assays. The tumors can be dissected and dissociated to a cell suspension, which can then be injected into a host embryo [56]. More detailed analysis of cancer progression can be followed in vivo when melanoma cells (from the *BRAF* mutant) are injected into the colorless casper mutant fish. [58, 59]. Xenotransplant assays are also possible with developing zebrafish into which cancer cells derived from

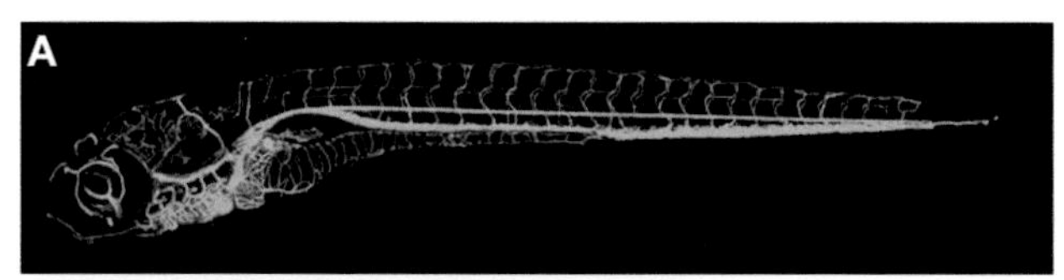

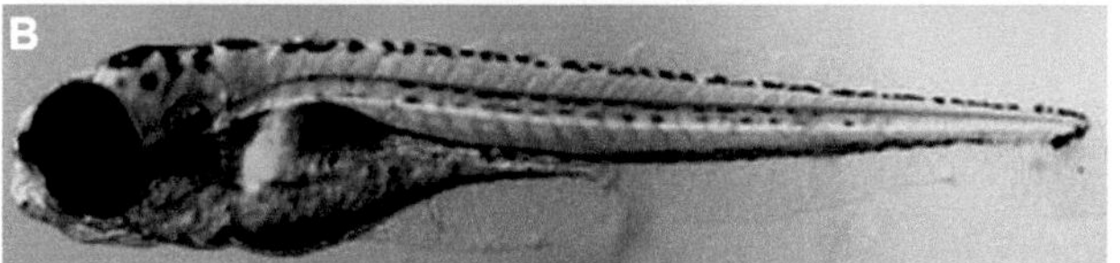

FIGURE 10.11: ***Zebrafish transgenics can be used to visualize tissues and cell populations. A. Blood vessels of the developing zebrafish embryo.*** **Source:** ***http://gpp.nih.gov/Faculty/Mentors/NICHD/BrantWeinstein.htm. B. Distinct organs such as the liver shown here can be easily visualized in the developing embryo with the use of fluorescent reporter genes.*** **Source:** ***Taken from http://www.nimr.mrc.ac.uk/research/elke-ober. Courtesy of S. Isogai and B. M. Weinstein.***

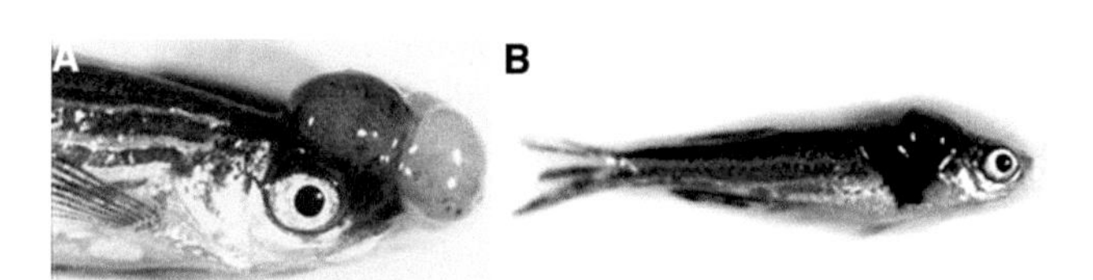

FIGURE 10.12: ***Zebrafish based cancer models. A. Adult fish with pronounced tumours*** **(crash&burn** ***mutant) taken from http://www.jyi.org/news/nb.php?id=474. B. BRAF zebrafish mutant showing spontaneous melanoma development.*** **Source:** ***Taken from http://zon.tchlab.org/.***

different species can be introduced by injection and the progression of these cancer cells tracked, such as with a fluorescent reporter gene [56, 60]. Recently the use of these zebrafish models have led to exciting developments in melanoma cancer research [61–62]. An excellent current review of cancer models in zebrafish is given in [56].

An example of a forward chemical genomic screen using a zebrafish mutant as a model for another human disease (aortic contraction) was performed using the mutant gridlock (*grl*) [63]. Gridlock has a disrupted blood flow in a similar fashion to human aortic contraction [63]. The screen discovered small molecules that restore the wild type without targeting the affected gene directly, something that would have been technically challenging without the use of chemical genetics, because few other genes in this disease pathway are known [63]. Developmental processes can also be used to model human disease in zebrafish. An example of this is inner ear sensory hair death, a cause of hearing and balance disorders in the human population [64]. To model this in the fish, the hair cells of the lateral line (sensory organ of the fish) known as neuroblasts are used. These small aggregates of cells are present on the outer surface of the head and midline of the embryo and provide a visual assay for a screen [64]. Lateral line hair death can be specifically induced by the application of the compound neomycin and the hair cells labeled by specific dyes. A total of 10,960 compounds were screened in 96-well format, and two compounds were discovered that offered protection against neomycin-induced hair cell death [64].

Another strategy to look for compounds able to affect cell division in the developing zebrafish embryos is to use an antibody stain, phospho-histone H3 (pH3), after the embryos are exposed to the chemical library [65]. The pH3 antibody is a specific known marker of actively dividing cells [65]. Along with pH3, staining incorporation of the fluorescent stain propidium iodide also allows DNA content of treated embryos to be quantified by flow cytometry [65], further confirming the number of cells in the process of dividing within the developing embryo. Not only did this study identify compounds able to affect the rate of cell division in vivo, but it also allowed a structure-activity–based relationship assay to be performed to measure potency and, importantly, toxicity of the compounds [52, 65].

As a chemical genetics model for assaying toxicology, zebrafish as well as *Xenopus* is proving valuable in providing a non-rodent vertebrate model [66–70]. An emerging area of interest related to chemical genetics is the use of developing zebrafish embryos in the drug discovery pipeline. Applications range from target identification to lead selection to assessing a drug's potential toxicology in vivo before more standard mammalian experiments are undertaken [52, 71]. This has the possibility of reducing the time and expense of producing new drugs, as toxic side effects can be identified at the early stage of the drug discovery pipeline [71]. Several phenotype-based assays have been successfully undertaken in these areas. The potential of zebrafish to act as a model in drug safety has been assayed by phenotype using cardiac (QT prolongation), visual (optometer response assay), locomoter, and gut contraction assays [71, 72]. Prolongation of the heart QT interval (a measurement of the heart electrical cycle) indicates a side effect on heart rate and is a frequent cause of drug leads failing to progress to clinical trials. Automated video microscopy was used to assay the embryo's heartbeat in a high-throughput fashion. This was possible because of the embryo's transparency [73]. A transgenic line carrying a heart-specific fluorescent GFP also exists. This made it easier to measure embryo heart rate using automated video microscopy [74]. This type of zebrafish-based phenotypic assay offers great potential for chemical genetics and will have an impact in areas where rodents are normally the models of choice. However, this area of research has to be more fully developed, especially where the presence of drug targets and conserved function needs to be ascertained [71]. As an extension of fish utility in assessing the toxicology of early-stage drugs, there has been some use with respect to environmental toxicology, that is, assessment of aquatic contaminants. Fish have been used to assess developmental toxicity to dioxins, a known environmental pollutant. Dioxin toxicity was evident through a number of phenotypes, including edema, anemia, hemorrhage, and ischemia [75].

Chemical genetics using zebrafish to investigate vertebrate developmental processes was pioneered by the Schreiber group [76]. They carried out a forward chemical genomic screen using a structurally diverse library of compounds that revealed several developmental phenotypes. The developing embryos were simply scored by eye for developmental abnormalities or malformations in a similar fashion to forward genetic screens using mutagens. Importantly, these effects were not only specific, with the embryos otherwise seemingly unaffected, but also diverse in the developmental processes they affected, such as central nervous system, cardiovascular, ear, and skin development [76]. Their results were obtained from phenotypically screening a library of only 1,100 compounds, giving a hit rate of approximately 1% [76].

From this initial establishment of the zebrafish as a viable model to study development using chemical genetics, several other screens have gone on to investigate a number of developmental processes. One example was a screen to study zebrafish fin regeneration. The caudal fin of embryos was removed and the embryos then arrayed in 96-well plates and exposed to two thousand small molecule compounds [77]. At three days post-amputation, the embryos were microscopically imaged to assess regenerative progression [77]. Seventeen small molecules (0.8% of the library) inhibited tissue regeneration [77].

Pigmentation, commonly targeted in forward genetic screens because of its ease of visual assay, has also been used in chemical genomic screens with developing zebrafish embryos. In a screen looking for compounds modulating pigmentation, one compound (PPA) was discovered that increased levels of pigmentation [78]. The work suggested that the target of PPA, a mitochondrial adenosine triphosphate (ATP)-ase, was responsible for modulating pigmentation in melanocytes [78]. A chemical genomic screen looking for changes in normal zebrafish brain and eye morphologies discovered one compound that retarded the normal development of these organs in a dose-dependent fashion [79]. Another recent chemical genomic screen looking for small molecules affecting embryonic development discovered phenotypes in eleven different categories (ranging from touch responsiveness to defects in the blood system of the developing embryos) [80]. From the 5,760 compounds screened, 3% gave a developmental effect [80]. The same screen also identified the retinoid-like compound DTAB, which causes anterior-posterior axis shortening in the zebrafish embryo [80]. DTAB has selective activity toward retinoic acid receptors c and b [80]. Blood vessel formation (angiogenesis) in zebrafish has been used as an assay to screen for compounds that inhibit the process [81] (also see Chapter 16). Zebrafish were assayed either by in situ hybridization with known markers or by injecting a fluorescent dye into the embryo's heart and observing blood flow in live embryos (microangiography) (Figure 10.11) [81]. The use of transgenic zebrafish embryos in chemical genomic screens was utilized in a screen to look for compounds able to modulate fibroblast growth factor (FGF) signaling [82]. The FGF-signaling pathway is used in many developmental processes, including limb and bone development. Zebrafish expressing GFP driven by a promoter (*dusp6*) known to be controlled by FGF signaling were screened for compounds that suppress GFP, indicating that the compound is acting as an FGF inhibitor [82].

Zebrafish offers many advantages in chemical genomic screens. These include its relative complexity, the available molecular tool set, genetics, and disease models. The history of chemical genomic screening with zebrafish means that it is one of the most well-established model organisms. It does, however, have some drawbacks: The setup costs of a fish facility may deter some investigators. For this reason, several companies are now offering screening services with zebrafish embryos. The zebrafish has yet to be fully tested as a safety pharmacology model, and it is limited in some biological processes, such as regeneration (only the fin regenerate) and processes such as limb development, which cannot be studied.

5. *XENOPUS LAEVIS*

X. laevis (the African clawed frog) has been extensively used in biological research for more than seventy years. It initially began to be used in the 1930s and 1940s in the first pregnancy tests. Many hospitals therefore kept colonies, which facilitated its use in biomedical research. *Xenopus* has a long history in the fields of teratology and toxicology. Frog Embryo Teratogenesis Assay–*Xenopus* (FETAX) is an assay used for many years to screen for effects of known chemicals on early developmental stages [70, 83]. However, until recently, chemical genomic screens had not been carried out in *Xenopus*, and screens using zebrafish have led the way with respect to genetic and chemical genomic screens in vertebrates. Recently, the use of genetics in *Xenopus tropicalis* [84] and work by the authors and others in chemical genetic phenotypic screens have shown *Xenopus* to be an excellent model system for such screens [70, 85–90, 61–62]. In fact, from an evolutionary point of view, *Xenopus* is the only tetrapod vertebrate to have free-living embryos in which embryonic development is neither in utero nor in ovo and thus is the highest order in which high-throughput screens can be carried out (Figure 10.1).

Xenopus has many of the advantages that are necessary to carry out phenotypic screens (Table 10.1). Its eggs can be easily obtained in large numbers at any time during the year by simple hormone injection, and it can then be synchronously fertilized. This facilitates biochemical, pharmacologic, and statistical analyses. It develops in the petri dish in simple salt solutions at room temperature. *X. laevis* embryos are bigger than zebrafish but can still be screened in 96-well plates (Figure 10.13) [86, 89, 62]. *X. tropicalis*, which is being developed as a genetically useful organism [84], is half the size of *X. laevis* embryos and thus would be even better with respect to assaying in 96-well plates. Compounds can be added to the media, and the vitelline membrane around the embryo is highly porous, so accessibility of compounds to the embryo is usually good. The detailed fate map for *Xenopus* facilitates injection of compounds, as they can be easily targeted to specific areas of the embryo [91]. This is not the case with zebrafish, in which cell migration and mixing during gastrulation prevent targeted injections.

Also unique to *Xenopus* among model organisms is the use of *Xenopus* oocytes as "laboratories" for the study of ion translocators, neurotransmitter receptors, second messenger cascades, calcium-dependent events, and cytoskeletal rearrangements. In addition, *Xenopus* oocyte and egg extracts are used extensively to study DNA damage and cell cycle progression [92, 93].

We have shown *Xenopus* to be a useful model for developmental chemical genomic screens [86]. We initially showed that compounds that affect zebrafish development also affected *Xenopus* development, giving rise to comparable phenotypes.

In all screens, it is important to target a specific phenotype or organ to identify compounds with defined, specific effects or modes of action. Mutagenesis screens often examine a phenotype of the whole embryo. For instance, in *Drosophila*, phenotypic changes in the outer cuticle pattern led to insight into important patterning and morphogenesis pathways [94]. In our initial screens, we have concentrated

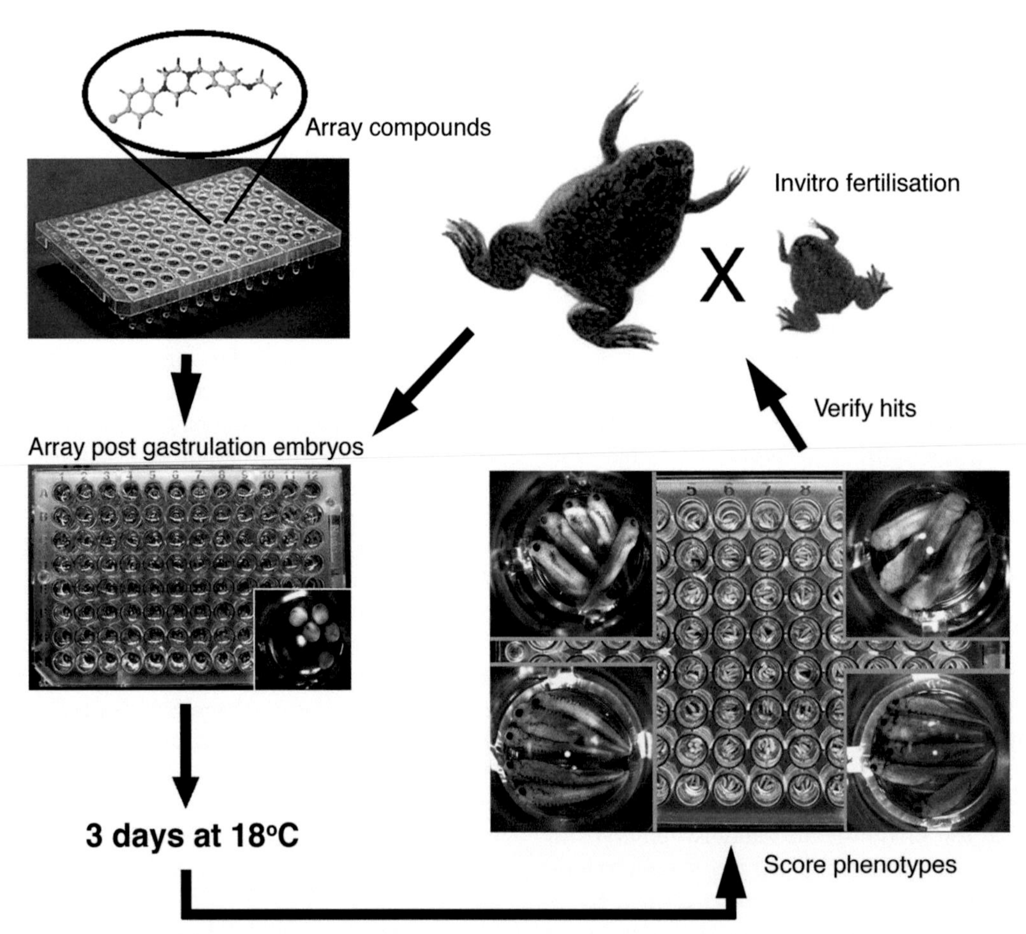

FIGURE 10.13: Flow diagram of a Xenopus laevis chemical genomic screen. Compounds are arrayed in 96-well plates and five embryos at stage 15 are added per well. These are incubated for three days and then scored. Examples of some of the possible pigmentation phenotypes observed are shown. Reproduced by permission of the Royal Society of Chemistry [89].

on the pigmentation of the embryo as easily scoreable. Pigment cells are derived from pluripotent neural crest cells and can give rise to melanoma cancer. So effects on pigment can relate to issues of neural crest induction, cell morphology, cell migration, cell biochemistry as well as melanoma cancer biology [89, 90, 62].

Figure 10.13 outlines how we do a large-scale screen with *Xenopus* embryos. Unless interested in specific early developmental effects or gastrulation phenotypes, we apply compounds post-gastrulation. Our previous studies have shown that the degree of toxicity of compounds is much lower if applied at this time [86]. Embryos are viewed under a low-magnification dissecting microscope. The example shown in Figure 10.13 highlights the ease with which pigment patterns can be observed. Other phenotypes for which we visually score include effects on the eye (i.e., small eye) and effects on shape of the head, body, and fin.

An initial screen of three thousand compounds has identified forty-one compounds with varying effects on the embryo [89]. As part of this screen, we identified a number of compounds that affect pigment cell migration. Further analysis of one of these showed it to be a novel matrix metalloproteinase (MMP) inhibitor [90]. MMPs are a family of proteins known to have important roles in cell migration, inflammation, angiogenesis, and cancer [95]. Another compound (NSC210627) was characterized as an inhibitor of Dihydroorotate dehydrogenase (DHODH). This led to the identification of Leflunomide, a structurally distinct DHODH inhibitor, which phenocopied NSC210627 in *Xenopus* and zebrafish. Subsequent tests on melanoma cell lines and in vivo mouse models showed leflunomide to be a potent inhibitor of melanoma growth [62]. We have also identified compounds that affect pigment cell morphology and the production of pigment and have begun to determine their cellular targets [89]

Levin and co-workers have developed an interesting chemical genomic strategy using *Xenopus* [87]. They use known pharmacologic compounds to correlate specific families of proteins with a chosen biological phenotype. They have taken advantage of a hierarchical structure that can be shown for drug reagents in a number of fields, such as ion transport, neurotransmitter function, metabolism, and the cytoskeleton. They call this an *inverse drug screen* [87]. The advantage is that it is more efficient than carrying out an exhaustive screen with large numbers of drugs and instead quickly reveals a manageable number of specific molecular candidates that can then be validated and targeted. They have used this method to determine whether the ion flow was important for embryonic left/right asymmetry [96] and to uncover novel pre-nervous roles for the neurotransmitter serotonin [97, 98]. They first test for a specific phenotype using known global inhibitors of a cellular function, that is, ion transport. If this causes a phenotype, they then test more specific drugs that target subset pathways of this global inhibitor. Thus, they use a general ion channel inhibitor to show a phenotype and then narrow this down to a specific ion channel or pump [87].

Future screens that can be or are being done will make use of *Xenopus*' evolutionary closeness to mammals. Therefore, screens looking at aspects of limb development and regeneration (in the eye, tail, or limb) can be envisaged.

6. CONCLUSION

We have described some of the varied model organisms that can be utilized for chemical genomic phenotypic screens. We have discussed some of the advantages and disadvantages of these model systems with examples. They are an increasingly important complement to in vitro and/or targeted screens.

ACKNOWLEDGMENTS

M. L. T. was supported by a BBSRC CASE studentship with Pfizer. R. A. F. acknowledges support from the BBSRC. G. N. W. was supported by a BBSRC new investigator award (Grant no. G15793). G. N. W. and R. A. F. are grateful for the support of an MRC Discipline Hopper Award at the start of their collaboration (Grant no. G0100722).

REFERENCES

1. Walsh, D. P., and Chang, Y. T. (2006). Chemical genetics. *Chem Rev 106*, 2476–2530.
2. Reya, T., and Clevers, H. (2005). Wnt signalling in stem cells and cancer. *Nature 434*, 843–850.
3. Van Raay, T. J., and Vetter, M. L. (2004). Wnt/frizzled signaling during vertebrate retinal development. *Dev Neurosci 26*, 352–358.
4. Davidson, E. H., Rast, J. P., Oliveri, P., Ransick, A., Calestani, C., Yuh, C. H., Minokawa, T., Amore, G., Hinman, V., Arenas-Mena, C., Otim, O., Brown, C. T., Livi, C. B., Lee, P. Y., Revilla, R., Rust, A. G., Pan, Z., Schilstra, M. J., Clarke, P. J., Arnone, M. I., Rowen, L., Cameron, R. A., McClay, D. R., Hood, L., and Bolouri, H. (2002). A genomic regulatory network for development. *Science 295*, 1669–1678.
5. Wheeler, G. N., and Brandlii, A. (2009) Simple vertebrate models for chemical genetics and drug discovery screens: lessons from zebrafish and *Xenopus*. *Dev Dyn 238*, 1287–1308.
6. Blackwell, H. E., and Zhao, Y. (2003). Chemical genetic approaches to plant biology. *Plant Physiol 133*, 448–455.
7. Kaschani, F., and Van Der Hoorn, R. (2007). Small molecule approaches in plants. *Curr Opin Chem Biol 11*, 88–98.
8. Walsh, T. A. (2007). The emerging field of chemical genetics: potential applications for pesticide discovery. *Pest Manag Sci 63*, 1165–1171.
9. Armstrong, J. L. (2007). Chemical genetics: catalysing pathway exploration and new target discovery. *J Sci Food Agric 87*, 1985–1990.
10. Wolfson, W. (2006). Spray-on special effects. Mendel Biotechnology uses chemical genetics to help plants cope with stress. *Chem Biol 13*, 919–921.
11. Arabidopsis, G. I. (2000). Analysis of the genome sequence of the flowering plant Arabidopsis thaliana. *Nature 408*, 796–815.
12. The Arabidopsis Information Resource. http://www.arabidopsis.org.
13. Lu, Y., Savage, L. J., Ajjawi, I., Imre, K. M., Yoder, D. W., Benning, C., Dellapenna, D., Ohlrogge, J. B., Osteryoung, K. W., Weber, A. P., Wilkerson, C. G., and Last, R. L. (2008). New connections across pathways and cellular processes: industrialized mutant screening reveals novel associations between diverse phenotypes in Arabidopsis. *Plant Physiol 146*, 1482–1500.
14. Zouhar, J., Hicks, G. R., and Raikhel, N. V. (2004). Sorting inhibitors (sortins): chemical compounds to study vacuolar sorting in Arabidopsis. *Proc Natl Acad Sci U S A 101*, 9497–9501.
15. Surpin, M., Rojas-Pierce, M., Carter, C., Hicks, G. R., Vasquez, J., and Raikhel, N. V. (2005). The power of chemical genomics to study the link between endomembrane system components and the gravitropic response. *Proc Natl Acad Sci U S A 102*, 4902–4907.
16. Carter, C. J., Bednarek, S. Y., and Raikhel, N. V. (2004). Membrane trafficking in plants: new discoveries and approaches. *Curr Opin Plant Biol* 7, 701–707.
17. Surpin, M., and Raikhel, N. (2004). Traffic jams affect plant development and signal transduction. *Nat Rev Mol Cell Biol 5*, 100–109.
18. Scheible, W. R., Eshed, R., Richmond, T., Delmer, D., and Somerville, C. (2001). Modifications of cellulose synthase confer resistance to isoxaben and thiazolidinone herbicides in Arabidopsis Ixr1 mutants. *Proc Natl Acad Sci U S A 98*, 10079–10084.
19. Yoneda, A., Higaki, T., Kutsuna, N., Kondo, Y., Osada, H., Hasezawa, S., and Matsui, M. (2007). Chemical genetic screening identifies a novel inhibitor of parallel alignment of cortical microtubules and cellulose microfibrils. *Plant Cell Physiol 48*, 1393–1403.
20. DeBolt, S., Gutierrez, R., Ehrhardt, D. W., Melo, C. V., Ross, L., Cutler, S. R., Somerville, C., and Bonetta, D. (2007). Morlin, an inhibitor of cortical microtubule dynamics and cellulose synthase movement. *Proc Natl Acad Sci U S A 104*, 5854–5859.

21. Zhao, Y., Dai, X., Blackwell, H. E., Schreiber, S. L., and Chory, J. (2003). SIR1, an upstream component in auxin signaling identified by chemical genetics. *Science 301*, 1107–1110.
22. Grozinger, C. M., Chao, E. D., Blackwell, H. E., Moazed, D., and Schreiber, S. L. (2001). Identification of a class of small molecule inhibitors of the sirtuin family of NAD-dependent deacetylases by phenotypic screening. *J Biol Chem 276*, 38837–38843.
23. Dai, X., Hayashi, K., Nozaki, H., Cheng, Y., and Zhao, Y. (2005). Genetic and chemical analyses of the action mechanisms of sirtinol in Arabidopsis. *Proc Natl Acad Sci U S A 102*, 3129–3134.
24. Asami, T., Nakano, T., Nakashita, H., Sekimata, K., Shimada, Y., and Yoshida, S. (2003). The influence of chemical genetics on plant science: shedding light on functions and mechanism of action of brassinosteroids using biosynthesis inhibitors. *J Plant Growth Regul 22*, 336–349.
25. Kaletta, T., and Hengartner, M. O. (2006). Finding function in novel targets: C. elegans as a model organism. *Nat Rev Drug Discov 5*, 387–398.
26. Li, C., and Chalfie, M. (1990). Organogenesis in C. elegans: positioning of neurons and muscles in the egg-laying system. *Neuron 4*, 681–695.
27. Carroll, P. M., Dougherty, B., Ross-Macdonald, P., Browman, K., and FitzGerald, K. (2003). Model systems in drug discovery: chemical genetics meets genomics. *Pharmacol Ther 99*, 183–220.
28. Kamath, R. S., Fraser, A. G., Dong, Y., Poulin, G., Durbin, R., Gotta, M., Kanapin, A., Le Bot, N., Moreno, S., Sohrmann, M., Welchman, D. P., Zipperlen, P., and Ahringer, J. (2003). Systematic functional analysis of the Caenorhabditis elegans genome using RNAi. *Nature 421*, 231–237.
29. Pulak, R. (2006). Techniques for analysis, sorting, and dispensing of C. elegans on the COPAS flow-sorting system. *Methods Mol Biol 351*, 275–286.
30. Boyd, W. A., McBride, S. J., and Freedman, J. H. (2007). Effects of genetic mutations and chemical exposures on Caenorhabditis elegans feeding: evaluation of a novel, high-throughput screening assay. *PLoS ONE 2*, e1259.
31. Schafer, W. R. (2005). Deciphering the neural and molecular mechanisms of C. elegans behavior. *Curr Biol 15*, R723–R729.
32. Taguchi, A., and White, M. F. (2008). Insulin-like signaling, nutrient homeostasis, and life span. *Annu Rev Physiol 70*, 191–212.
33. Min, J., Kyung Kim, Y., Cipriani, P. G., Kang, M., Khersonsky, S. M., Walsh, D. P., Lee, J. Y., Niessen, S., Yates, J. R., III, Gunsalus, K., Piano, F., and Chang, Y. T. (2007). Forward chemical genetic approach identifies new role for GAPDH in insulin signaling. *Nat Chem Biol 3*, 55–59.
34. Kwok, T. C., Ricker, N., Fraser, R., Chan, A. W., Burns, A., Stanley, E. F., McCourt, P., Cutler, S. R., and Roy, P. J. (2006). A small-molecule screen in C. elegans yields a new calcium channel antagonist. *Nature 441*, 91–95.
35. Moy, T. I., Ball, A. R., Anklesaria, Z., Casadei, G., Lewis, K., and Ausubel, F. M. (2006). Identification of novel antimicrobials using a live-animal infection model. *Proc Natl Acad Sci U S A 103*, 10414–10419.
36. Breger, J., Fuchs, B. B., Aperis, G., Moy, T. I., Ausubel, F. M., and Mylonakis, E. (2007). Antifungal chemical compounds identified using a C. elegans pathogenicity assay. *PLoS Pathog 3*, e18.
37. Jones, A. K., Buckingham, S. D., and Sattelle, D. B. (2005). Chemistry-to-gene screens in Caenorhabditis elegans. *Nat Rev Drug Discov 4*, 321–330.
38. Segalat, L. (2006). Drug discovery: here comes the worm. *ACS Chem Biol 1*, 277–278.
39. Bhavsar, A. P., and Brown, E. D. (2006). The worm turns for antimicrobial discovery. *Nat Biotechnol 24*, 1098–1100.
40. Chakrapani, B. P., Kumar, S., and Subramaniam, J. R. (2008). Development and evaluation of an in vivo assay in Caenorhabditis elegans for screening of compounds for their effect on cytochrome P450 expression. *J Biosci 33*, 269–277.
41. Van Den Heuvel, S. (2005). The C. elegans cell cycle: overview of molecules and mechanisms. *Methods Mol Biol 296*, 51–67.
42. Chan, H. Y. (2002). Fly-ing from genes to drugs. *Trends Mol Med 8*, 99–101.
43. Segalat, L. (2007). Invertebrate animal models of diseases as screening tools in drug discovery. *ACS Chem Biol 2*, 231–236.
44. Garcia-Lopez, A., Monferrer, L., Garcia-Alcover, I., Vicente-Crespo, M., Alvarez-Abril, M. C., and Artero, R. D. (2008). Genetic and chemical modifiers of a CUG toxicity model in Drosophila. *PLoS ONE 3*, e1595.
45. Chang, S., Bray, S. M., Li, Z., Zarnescu, D. C., He, C., Jin, P., and Warren, S. T. (2008). Identification of small molecules rescuing fragile X syndrome phenotypes in Drosophila. *Nat Chem Biol 4*, 256–263.
46. ffrench-Constant, R. H., Steichen, J. C., Rocheleau, T. A., Aronstein, K., and Roush, R. T. (1993). A single-amino acid substitution in a gamma-aminobutyric acid subtype A receptor locus is associated with cyclodiene insecticide resistance in Drosophila populations. *Proc Natl Acad Sci U S A 90*, 1957–1961.
47. Kang, H. L., Benzer, S., and Min, K. T. (2002). Life extension in Drosophila by feeding a drug. *Proc Natl Acad Sci U S A 99*, 838–843.
48. Chen, E., and Ekker, S. C. (2004). Zebrafish as a genomics research model. *Curr Pharm Biotechnol 5*, 409–413.
49. Berger, J., and Currie, P. (2007). The role of zebrafish in chemical genetics. *Curr Med Chem 14*, 2413–2420.
50. MacRae, C. A., and Peterson, R. T. (2003). Zebrafish-based small molecule discovery. *Chem Biol 10*, 901–908.
51. Sprague, J., Clements, D., Conlin, T., Edwards, P., Frazer, K., Schaper, K., Segerdell, E., Song, P., Sprunger, B., and Westerfield, M. (2003). The Zebrafish Information Network (ZFIN): the zebrafish model organism database. *Nucleic Acids Res 31*, 241–243.
52. Zon, L. I., and Peterson, R. T. (2005). In vivo drug discovery in the zebrafish. *Nat Rev Drug Discov 4*, 35–44.
53. Goessling, W., North, T. E., and Zon, L. I. (2007). New waves of discovery: modeling cancer in zebrafish. *J Clin Oncol 25*, 2473–2479.
54. Langenau, D. M., Traver, D., Ferrando, A. A., Kutok, J. L., Aster, J. C., Kanki, J. P., Lin, S., Prochownik, E., Trede, N. S., Zon, L. I., and Look, A. T. (2003). Myc-induced T cell leukemia in transgenic zebrafish. *Science 299*, 887–890.
55. Langenau, D. M., and Zon, L. I. (2005). The zebrafish: a new model of T-cell and thymic development. *Nat Rev Immunol 5*, 307–317.

56. Stoletov, K., and Klemke, R. (2008). Catch of the day: zebrafish as a human cancer model. *Oncogene 27*, 4509–4520.
57. Patton, E. E., and Zon, L. I. (2005). Taking human cancer genes to the fish: a transgenic model of melanoma in zebrafish. *Zebrafish 1*, 363–368.
58. White, R. M., Sessa, A., Burke, C., Bowman, T., LeBlanc, J., Ceol, C., Bourque, C., Dovey, M., Goessling, W., Burns, C. E., and Zon, L. I. (2008). Transparent adult zebrafish as a tool for in vivo transplantation analysis. *Cell Stem Cell 2*, 183–189.
59. Stoletov, K., Montel, V., Lester, R. D., Gonias, S. L., and Klemke, R. (2007). High-resolution imaging of the dynamic tumor cell vascular interface in transparent zebrafish. *Proc Natl Acad Sci U S A 104*, 17406–17411.
60. Haldi, M., Ton, C., Seng, W. L., and McGrath, P. (2006). Human melanoma cells transplanted into zebrafish proliferate, migrate, produce melanin, form masses and stimulate angiogenesis in zebrafish. *Angiogenesis 9*, 139–151.
61. Ceol, C. J., Houvras, Y., Jane-Valbuena, J., Bilodeau, S., Orlando, D. A., Battisti, V., Fritsch, L., Lin, W. M., Hollmann, T. J., Ferre, F. et al. (2011). The histone methyltransferase SETDB1 is recurrently amplified in melanoma and accelerates its onset. Nature 471, 513–517.
62. White, R. M., Cech, J., Ratanasirintrawoot, S., Lin, C. Y., Rahl, P. B., Burke, C. J., Langdon, E., Tomlinson, M. L., Mosher, J., Kaufman, C., Chen, F., Long, H. K., Kramer, M., Datta, S., Neuberg, D., Granter, S., Young, R. A., Morrison, S., Wheeler, G. N., Zon, L. I. (2011) DHODH modulates transcriptional elongation in the neural crest and melanoma. Nature 471, 518–522.
63. Peterson, R. T., Shaw, S. Y., Peterson, T. A., Milan, D. J., Zhong, T. P., Schreiber, S. L., MacRae, C. A., and Fishman, M. C. (2004). Chemical suppression of a genetic mutation in a zebrafish model of aortic coarctation. *Nat Biotechnol 22*, 595–599.
64. Owens, K. N., Santos, F., Roberts, B., Linbo, T., Coffin, A. B., Knisely, A. J., Simon, J. A., Rubel, E. W., and Raible, D. W. (2008). Identification of genetic and chemical modulators of zebrafish mechanosensory hair cell death. *PLoS Genet 4*, e1000020.
65. Murphey, R. D., Stern, H. M., Straub, C. T., and Zon, L. I. (2006). A chemical genetic screen for cell cycle inhibitors in zebrafish embryos. *Chem Biol Drug Des 68*, 213–219.
66. Hill, A. J., Teraoka, H., Heideman, W., and Peterson, R. E. (2005). Zebrafish as a model vertebrate for investigating chemical toxicity. *Toxicol Sci 86*, 6–19.
67. Barros, T. P., Alderton, W. K., Reynolds, H. M., Roach, A. G., and Berghmans, S. (2008). Zebrafish: an emerging technology for in vivo pharmacological assessment to identify potential safety liabilities in early drug discovery. *Br J Pharmacol 154*, 1400–1413.
68. Spitsbergen, J. M., and Kent, M. L. (2003). The state of the art of the zebrafish model for toxicology and toxicologic pathology research – advantages and current limitations. *Toxicol Pathol 31 Suppl*, 62–87.
69. Redfern, W. S., Waldron, G., Winter, M. J., Butler, P., Holbrook, M., Wallis, R., and Valentin, J. P. (2008). Zebrafish assays as early safety pharmacology screens: paradigm shift or red herring? *J Pharmacol Toxicol Methods*.
70. Richards, S. M., and Cole, S. E. (2006). A toxicity and hazard assessment of fourteen pharmaceuticals to Xenopus laevis larvae. *Ecotoxicology 15*, 647–656.
71. Berghmans, S., Butler, P., Goldsmith, P., Waldron, G., Gardner, I., Golder, Z., Richards, F. M., Kimber, G., Roach, A., Alderton, W., and Fleming, A. (2008). Zebrafish based assays for the assessment of cardiac, visual and gut function – potential safety screens for early drug discovery. *J Pharmacol Toxicol Methods 58*, 59–68.
72. Winter, M. J., Redfern, W. S., Hayfield, A. J., Owen, S. F., Valentin, J. P., and Hutchinson, T. H. (2008). Validation of a larval zebrafish locomotor assay for assessing the seizure liability of early-stage development drugs. *J Pharmacol Toxicol Methods 57*, 176–187.
73. Milan, D. J., Peterson, T. A., Ruskin, J. N., Peterson, R. T., and MacRae, C. A. (2003). Drugs that induce repolarization abnormalities cause bradycardia in zebrafish. *Circulation 107*, 1355–1358.
74. Burns, C. G., Milan, D. J., Grande, E. J., Rottbauer, W., MacRae, C. A., and Fishman, M. C. (2005). High-throughput assay for small molecules that modulate zebrafish embryonic heart rate. *Nat Chem Biol 1*, 263–264.
75. Carney, S. A., Prasch, A. L., Heideman, W., and Peterson, R. E. (2006). Understanding dioxin developmental toxicity using the zebrafish model. *Birth Defects Res A Clin Mol Teratol 76*, 7–18.
76. Peterson, R. T., Link, B. A., Dowling, J. E., and Schreiber, S. L. (2000). Small molecule developmental screens reveal the logic and timing of vertebrate development. *Proc Natl Acad Sci U S A 97*, 12965–12969.
77. Mathew, L. K., Sengupta, S., Kawakami, A., Andreasen, E. A., Lohr, C. V., Loynes, C. A., Renshaw, S. A., Peterson, R. T., and Tanguay, R. L. (2007). Unraveling tissue regeneration pathways using chemical genetics. *J Biol Chem 282*, 35202–35210.
78. Jung, D. W., Williams, D., Khersonsky, S. M., Kang, T. W., Heidary, N., Chang, Y. T., and Orlow, S. J. (2005). Identification of the F1F0 mitochondrial ATPase as a target for modulating skin pigmentation by screening a tagged triazine library in zebrafish. *Mol Biosyst 1*, 85–92.
79. Khersonsky, S. M., Jung, D. W., Kang, T. W., Walsh, D. P., Moon, H. S., Jo, H., Jacobson, E. M., Shetty, V., Neubert, T. A., and Chang, Y. T. (2003). Facilitated forward chemical genetics using a tagged triazine library and zebrafish embryo screening. *J Am Chem Soc 125*, 11804–11805.
80. Sachidanandan, C., Yeh, J. R., Peterson, Q. P., and Peterson, R. T. (2008). Identification of a novel retinoid by small molecule screening with zebrafish embryos. *PLoS ONE 3*, e1947.
81. Chan, J., Bayliss, P. E., Wood, J. M., and Roberts, T. M. (2002). Dissection of angiogenic signaling in zebrafish using a chemical genetic approach. *Cancer Cell 1*, 257–267.
82. Molina, G. A., Watkins, S. C., and Tsang, M. (2007). Generation of FGF reporter transgenic zebrafish and their utility in chemical screens. *BMC Dev Biol* 7, 62.
83. Longo, M., Zanoncelli, S., Della Torre, P., Rosa, F., Giusti, A., Colombo, P., Brughera, M., Mazue, G., and Olliaro, P. (2008). Investigations of the effects of the antimalarial drug dihydroartemisinin (DHA) using the Frog Embryo Teratogenesis Assay-Xenopus (FETAX). *Reprod Toxicol 25*, 433–441.

84. Goda T, Abu-Daya A, Carruthers S, Clark MD, Stemple DL, Zimmerman LB. (2006). Genetic screens for mutations affecting development of Xenopus tropicalis. PLoS Genet. Jun;2(6):e91.
85. Brändli, A. (2004). Prospects for the XenopusEmbryo model in therapeutics technologies. *Chimia 58*, 694–702.
86. Tomlinson, M., Field, R. A., and Wheeler, G. N. (2005). Xenopus as a model organism in developmental chemical genetic screens. *Molecular BioSystems 1*, 223–228.
87. Adams, D. S., and Levin, M. (2006). Inverse drug screens: a rapid and inexpensive method for implicating molecular targets. *Genesis 44*, 530–540.
88. Fini, J. B., Le Mevel, S., Turque, N., Palmier, K., Zalko, D., Cravedi, J. P., and Demeneix, B. A. (2007). An in vivo multiwell-based fluorescent screen for monitoring vertebrate thyroid hormone disruption. *Environ Sci Technol 41*, 5908–5914.
89. Matthew L. Tomlinson, Martin Rejzek, Mark Fidock, Robert A. Field and Grant N. Wheeler. (2009). Chemical genomics identifies compounds affecting Xenopus laevis pigment cell development. *Molecular BioSystems*, *5*, 376–384.
90. Tomlinson, M. L., Guan, P., Morris, R. J., Fidock, M. D., Rejzek, M., Garcia-Morales, C., Field, R. A., and Wheeler, G. N. (2009). A chemical genomic approach identifies matrix metalloproteinases as playing an essential and specific role in Xenopus melanophore migration. *Chem Biol 16*, 93–104.
91. Dale, L., and Slack, J. M. (1987). Fate map for the 32-cell stage of Xenopus laevis. *Development 99*, 527–551.
92. Dupre, A., Boyer-Chatenet, L., Sattler, R. M., Modi, A. P., Lee, J. H., Nicolette, M. L., Kopelovich, L., Jasin, M., Baer, R., Paull, T. T., and Gautier, J. (2008). A forward chemical genetic screen reveals an inhibitor of the Mre11-Rad50-Nbs1 complex. *Nat Chem Biol 4*, 119–125.
93. Peterson, J. R., Lebensohn, A. M., Pelish, H. E., and Kirschner, M. W. (2006). Biochemical suppression of small-molecule inhibitors: a strategy to identify inhibitor targets and signaling pathway components. *Chem Biol 13*, 443–452.
94. Nusslein-Volhard, C., and Wieschaus, E. (1980). Mutations affecting segment number and polarity in Drosophila. *Nature 287*, 795–801.
95. Page-McCaw, A., Ewald, A. J., and Werb, Z. (2007). Matrix metalloproteinases and the regulation of tissue remodelling. *Nat Rev Mol Cell Biol 8*, 221–233.
96. Adams, D. S., Robinson, K. R., Fukumoto, T., Yuan, S., Albertson, R. C., Yelick, P., Kuo, L., McSweeney, M., and Levin, M. (2006). Early, H+-V-ATPase-dependent proton flux is necessary for consistent left-right patterning of non-mammalian vertebrates. *Development 133*, 1657–1671.
97. Fukumoto, T., Kema, I. P., and Levin, M. (2005). Serotonin signaling is a very early step in patterning of the left-right axis in chick and frog embryos. *Curr Biol 15*, 794–803.
98. Levin, M., Buznikov, G. A., and Lauder, J. M. (2006). Of minds and embryos: left-right asymmetry and the serotonergic controls of pre-neural morphogenesis. *Dev Neurosci 28*, 171–185.

CHAPTER 11

Screening Informatics and Cheminformatics

Melinda I. Sosa
Clinton Maddox
Iestyn Lewis
Cheryl L. Meyerkord
Pahk Thepchatri

Chemical genomics projects generate enormous amounts of data in a relatively short period of time. A screening campaign against a single target can produce millions of data points before data refinement is performed [1]. Generated data must then be stored and crosslinked with data from other projects. Informatics plays a critical role in all high-throughput screening (HTS) campaigns because of the sheer volume of data that need to be interpreted and the speed at which data are generated.

Informatics activities during a screening campaign are traditionally split into two primary functions, screening informatics and cheminformatics [2–5], which are usually treated as two separate disciplines because of the different training and experience required. In brief, screening informatics concentrates efforts on data organization, quality control, accessibility, and visualization, whereas cheminformatics focuses on identification of dominant chemical structural scaffolds that demonstrate a biological advantage in an HTS campaign and recommendation of functional changes for structural optimization. Although there are stages during a screening campaign when screening informatics and cheminformatics work independently, for most of a campaign, close collaboration is required to provide the highest-quality information for decision-making processes. Both types of informatics utilize databases and specialized software to support experimental biologists and chemists. Importantly, the integration of chemical structure data with assay results must be maintained by both screening informatics and cheminformatics teams in order to be useful for future work. This chapter introduces basic applications of screening informatics and cheminformatics for HTS.

1. INFORMATICS

1.1. Screening Informatics

Screening informatics involves the acquisition, storage, and management of HTS data generated through screening and biological assays. This includes computationally scriptable work flows such as:

- Managing compound libraries and related compound structures;
- Tracking plate and well locations of compounds during multiple replications and cherry-pick transfers;
- Gathering raw, unprocessed data from plate readers or file servers;
- Normalizing data from raw reader values to percentages of control or other relative values;
- Generating dose-response curves and making them accessible for analysis by investigators;
- Transforming data into formats required for downstream analyses such as structural clustering and subsequent medicinal chemistry;
- Making raw data easily accessible to other members of an organization for alternative types of data processing;
- Providing reporting and visualization tools to help investigators navigate and manipulate data in the system; and
- Archiving data for possible future analysis and intellectual property protection.

TABLE 11.1: Assay Stage and Corresponding Informatics Support

Stage	Screening Informatics	Cheminformatics
Compound Selection	Store compound structures, plate and well locations in appropriate formats	Make recommendations regarding diversity or class of compound selection
Assay Development	Provide support to investigators	Research properties of target
Assay Validation	Create templates for data reduction software; document and visualize QC parameters	
Primary Screening	Process data, provide reports and visualization tools for primary data analysis; continue observation of QC parameters	Cluster compounds for preliminary analysis
Dose-Response Analysis	Process data, provide dose-response curves; continue observation of QC parameters	Cluster hit compounds
Hit Confirmation	Maintain primary, confirmatory, and secondary data on hit compounds as screens are completed	Assist chemists with QC work
Series Expansion	Provide SAR database to allow grouping of related compounds	Perform series expansion and identify compounds from suppliers
Biologically Relevant Assays	Populate SAR database with secondary assay results and store source documents	
SAR – Series Expansion		Perform docking and modeling studies to rationalize existing SAR or suggest new modifications

1.2. Cheminformatics

Cheminformatics identifies the structural chemical features responsible for an observed biological phenotype. Thus, a cheminformaticist requires a chemical structure database annotated with assay data for all screened compounds that can be utilized later by the investigator for triaging hit candidates for future confirmatory assays. Responsibilities of a cheminformaticist include:

- Grouping primary hits by two-dimensional (2D) structural clustering;
- 2D similarity searching of vendor databases for compounds similar to the investigated hit compounds;
- Three-dimensional (3D) virtual screening of vendor databases using ligand- or receptor-based methodologies to further explore potential chemical space; and
- Interacting with medicinal chemists to identify potential synthetic routes to improve potency of lead compounds through computational methods such as 3D protein-ligand modeling.

1.3. Informatics Collaborations

Work flows designed to manage data for each step of the screening process can be combined to create a beginning-to-end pipeline that data management can follow through a screening campaign. Because of the cyclic nature of the compound refinement process, these work flows must include a constant exchange of information between screening informatics and cheminformatics. Thus, it is imperative that communications are effective and that the two disciplines function smoothly in an iterative manner.

2. INFORMATICS SUPPORT THROUGH STAGES OF A SCREENING CAMPAIGN

2.1. Stages of the Screening Process

A screening campaign progresses through several distinct stages (Table 11.1). Screening informatics and cheminformatics activities at each stage prepare the campaign for subsequent stages. It is important to note that hit-to-lead projects must start with hits for which the activity of a particular structure has been confirmed; otherwise, no chemistry, computational or experimental, can be performed in an investigative manner.

2.1.1. Compound Selection The onset of a screening campaign involves a significant amount of planning. Key considerations include the size and the chemical diversity of the library. Screening libraries may range from tens of thousands to more than one million compounds. Defined targets are often screened against a focused library. This approach is used frequently with screens against specific biochemical targets, in which case the compounds included in the screening set are modeled after known agonists, antagonists, or substrates. A compound library maintained specifically to screen against kinase targets would be an example of this type of "targeted library" (see Chapter 5). When screening large libraries against phenotypic targets or new biochemical targets, compounds included in a library are generally chosen for maximal chemical diversity. Cheminformatics support as well as recommendations from medicinal chemists should be taken into account and coupled with logistics regarding assay complexity and availability of funding to determine if a large library of diverse small molecules is feasible or if a subset of available compounds would be more practical. Following such selections, a screening informatics database would be populated with the chemical structural data as well as plate and well locations of all compounds intended for screening.

2.1.2. Assay Development During the assay development stage, biological scientists determine the optimal conditions for an assay through experimentation with assay components such as buffers, target concentrations, and incubation times to produce a model system that reflects either the biological system under investigation or the disease state targeted for therapeutic discovery. Assay development is characterized by multiple single-stage experiments with a need for immediate feedback on experimental results. The beginning of the development process usually involves bench-scale experiments, with the goal of miniaturizing assay volumes to suit the format of at least 384-well and often 1536-well or greater microtiter plates. As the well density of the plates used in development experiments increases, the need for screening informatics tools concurrently increases. These full-plate experiments usually generate thousands of data points, the analysis of which would be tedious and error prone if performed manually. Tasks of this nature can generally be performed by a computer program that can easily execute repetitive data analysis tasks in a time-efficient manner.

Scientists have a number of commercial software options for performing data analysis, such as Microsoft Excel (2007), TIBCO Spotfire [6, 7], and GraphPad Prism [8]. Because of the relatively small volume of data associated with each optimization run, specialized data reduction software is not an absolute requirement. It is important, however, to capture the information obtained as a result of these experiments. This can be done electronically either through an electronic lab notebook system or a document management system.

2.1.3. Assay Validation In the validation stage, an assay is run with controls over the course of many days and many replicates in order to predict the stability of an assay when performed in high-throughput mode. Some objectives of this phase are to achieve a repeatable and acceptable control value and Z' factor for the assay [9]. Generally, assays are considered sufficiently robust to be used in an HTS campaign if the Z' factors are between 0.5 to 1 and coefficient of variation (%CVs) for the control wells are less than 10. Once derived, these values establish a baseline to be used as a quality control reference through the completion of the screening campaign. Other incidental but critical information obtained relates to the batch size, which is the number of plates that can feasibly and reliably be screened at one time based on a variety of contingencies such as reagent stability, assay complexity, staff, and equipment availability.

Statistical and visualization packages, including Spotfire [6, 7], SAS (Statistical Analysis System) [10], visualization tools available in the ActivityBase suite (ID Business Solutions), and Statistica [11], may be used to analyze quality control parameters and statistical variation. It is also useful to visualize plates side by side to determine if there are patterns in the data either within plates or across multiple validation plates (Figure 11.1). These patterns usually indicate a systematic variation in the liquid handler, plate reader, or environment in which the plates are processed or incubated. The types of screening informatics tools discussed in this chapter are essential for viewing these sets of validation data to identify and eliminate causes of systematic problems, as they can result in a costly derailment of a screen if not resolved during the validation stage.

The assay validation stage is also an opportunity for a screening informatics team to establish work flow templates that will be necessary to efficiently manage data during the screening stage. A work flow template is a generalized set of instructions to streamline the movement of large amounts of raw experimental data from plate readers to various programs for annotation, organization, and predictive model generation. Reports and files are then distributed to the research group to assist in various decision-making processes for future experiments [12, 13]. These templates allow informaticists to consolidate large amounts of screening data automatically, with allowances for future customizations, to bring about queryable access to the resulting biological data.

2.1.4. Primary Screening Although the primary screening stage produces the largest amount of data relative to the other stages of the HTS process, biologists generally spend less time generating these data compared with the development and validation steps. If assay development and validation have been adequately performed, the actual screening may be completed in as little as a few days or a few months, depending on the nature of the assay and the size of the library. Screening data files may be imported in real time or saved from plate readers in a format immediately accessible for informatics processes. Both numeric and visual means are used to assess the data for intraplate (Figure 11.2) and interplate (Figure 11.3) quality control, a process that includes comparing each plate's control values with the other plates in the screen and with the values established in the assay validation process. Compound results are normalized [14], most often as a percentage or ratio of one or more of the controls on the same plate; however,

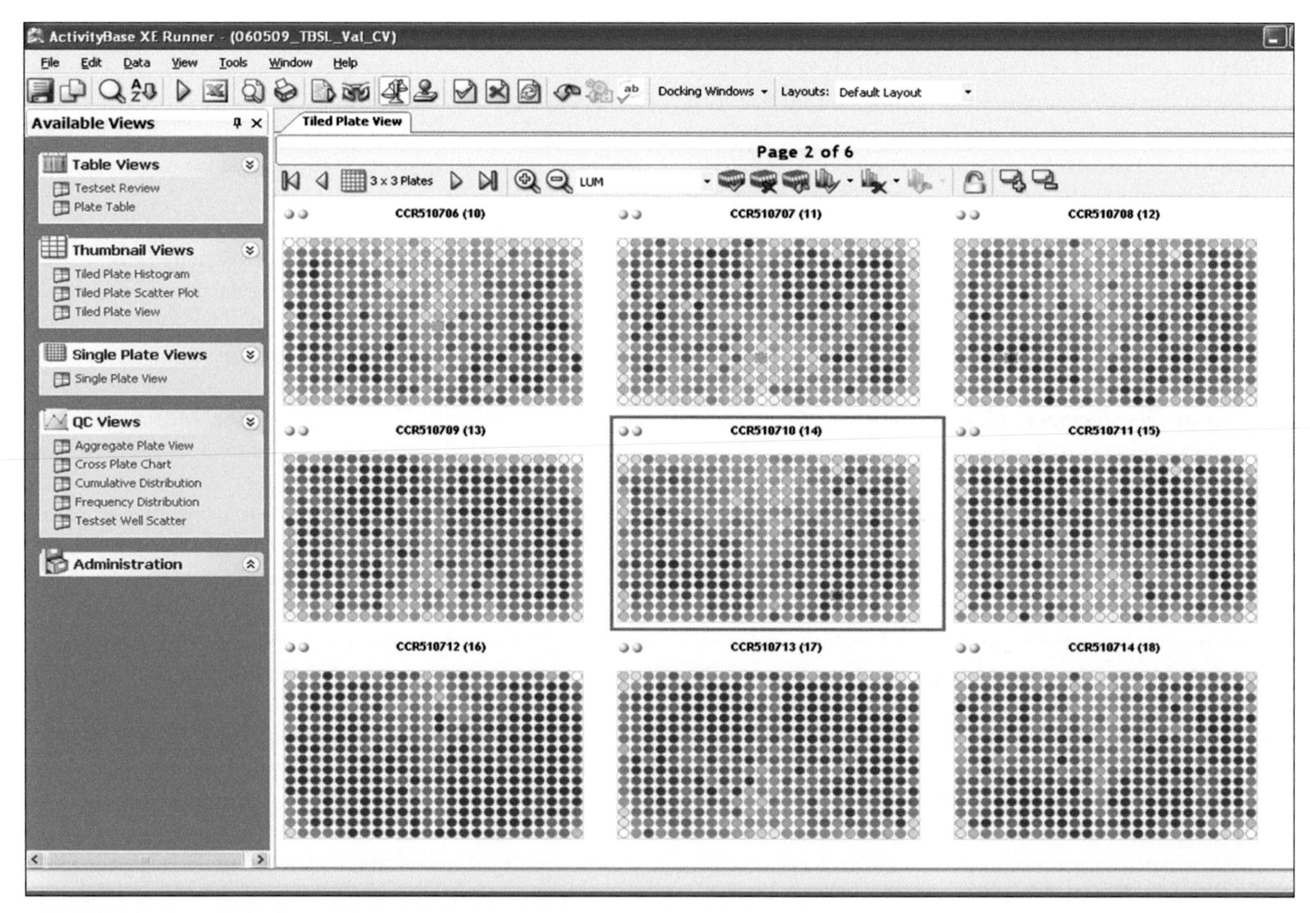

FIGURE 11.1: ***Cross-plate comparison. This single view of multiple plates enables visual comparison. Each well of these plates contains identical reagents and should produce similar numeric values. This figure demonstrates the presence of lower values at the edges of the plates (edge effects) and in the midsection. These observations should be addressed and corrected before proceeding with the primary HTS.***

other relative calculations such as a Z-score [15] may be preferred. These results are generally compared and used to determine whether the screening project will continue to the next phase.

2.1.5. Dose-response Analysis Compounds that are identified as positive hits, that is, those that produce the desired inhibition or activation in the primary screen, will be tested in the dose-response phase. Data reduction for dose-response assays usually proceeds through a similar informatics routine as single concentration screens, with the exception of altering the informatics templates to accommodate multiple concentrations of each compound tested. Typically, compounds will be tested in eight to ten concentrations in two to three replicates, with the highest concentration being two- to four-fold higher than the single-screening concentration used.

After individual values have been obtained for each concentration and replicates, a Levenburg-Marquardt 4-parameter (or another applicable) algorithm is used to fit the points on a dose response curve and to determine EC_{50}[1] values $\mathrm{Min}\,y + (\mathrm{Max}\,y - \mathrm{Min}\,y)/[1 + (EC50/x)^{\mathrm{slope}}]$. The validity of each compound's biological effect is then determined based on the dose-response curve, which is expected to be sigmoidal in shape. Statistical values are calculated for each of the compound curves including chi^2 and EC_{50}Stdev, both of which are attributes describing the quality of the fit, or how well the data comply with the curve algorithm applied. Lower values for these statistics indicate greater confidence in the calculated EC_{50} values. Compounds typically are then ranked by EC_{50} values, with the most potent compounds being allowed to enter the next phase of the process.

[1] Depending on assay types, an EC_{50} (compound effective concentration at which 50% of its maximal effect is observed) value might be referenced using a number of identifiers such as IC_{50} (inhibitory concentration: compound concentration at which 50% inhibition is achieved).

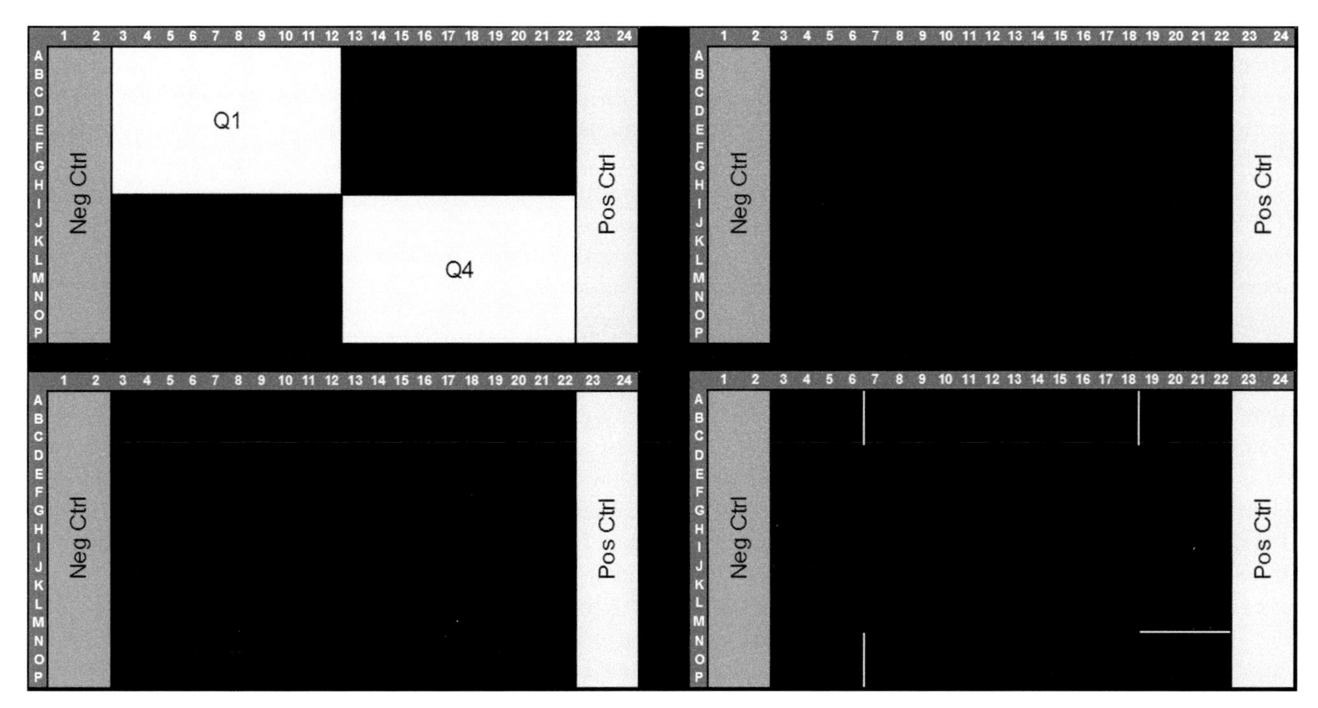

FIGURE 11.2: Intraplate QC. Divisions are made within each plate of a primary screen, and median values are compared. This type of analysis can only be useful if compounds are randomly distributed throughout the compound library. It is helpful to determine whether liquid handling or plate reading errors are weighting a certain region of a plate. Ratios of the different sections, Q1/Q4 or inner/outer, for example, should be close to 1. If a disproportionate number of active compounds appear in a specific region, further investigation should be conducted to confirm the results.

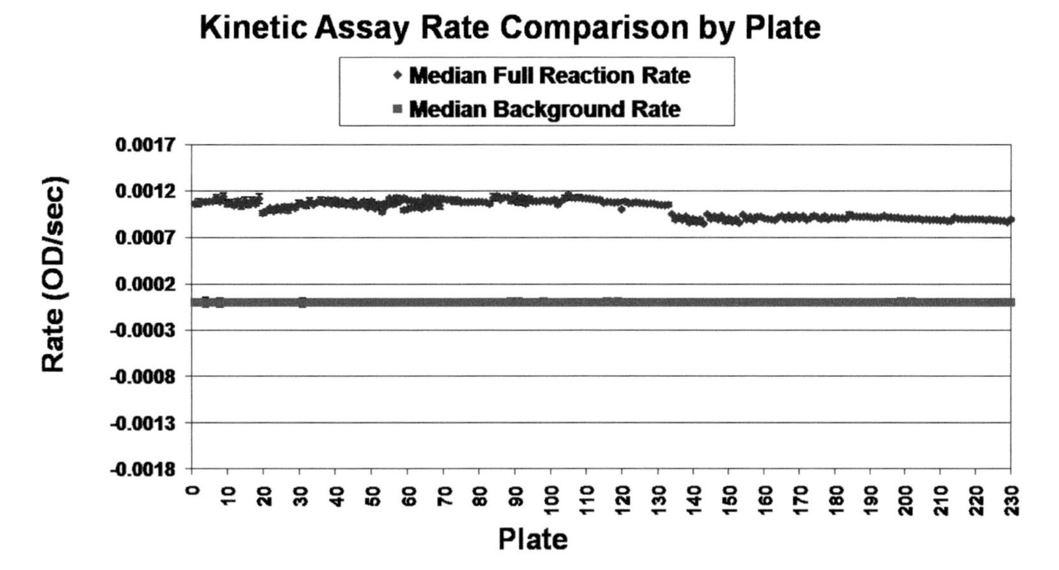

FIGURE 11.3: Interplate QC. Median values for each plate's control wells for the screening campaign are plotted. This chart illustrates relative consistency for each control and is maintained for the duration of a screening campaign. Here, the median background rate is very consistent, and the median full reaction rate shows only a slight decline near the hundredth plate. A significant shift would indicate a change in one or more of the assay conditions that needs to be identified and corrected before continuing the campaign.

2.1.6. Hit Confirmation Prior to lead optimization through medicinal chemistry, validation of a hit compound's chemical structure is critical. During lead optimization, a number of analogs are synthesized with small but significant changes to their functional groups, with the goal of creating analogs with improved potency or physiochemical properties such as solubility. This process is hampered by the fact that a large portion of compound libraries undergo chemical degradation more readily in situations such as being dissolved in dimethyl sulfoxide (DMSO) and when subjected to numerous freeze/thaw cycles [16–18]. Therefore, the biological modulation observed in a screen might have occurred because of a resultant degradation product of the original substance, not the chemical entity that is recorded in the structural database. Degradation of library compounds results in usage of unreliably annotated computer databases for latter stages of the project [16–18]. Structural confirmation and sample purity are therefore critical when considering how compounds are stored prior to screening. Although groups have performed various storage studies to identify the most effective conditions to extend compound library usage, it is still important that investigators structurally validate lead compounds following primary screens.

Validation provides the basis on which cheminformatics predictions are performed, as the calculations are dependent on knowing the chemical structure. Criteria for validation should include, but are not limited to:

- Confirmation that the initial hits from a primary screen are not artifacts of the experiment;
- Validation of the chemical structures of candidate lead compounds through methods such as nuclear magnetic resonance (NMR), liquid chromatography–mass spectrometry (LC-MS), or elemental analysis and demonstration of a high level of purity; and
- Demonstration that the hit compounds have a clear dose-response relationship in related confirmatory and secondary assays.

Impure or structurally undefined lead compounds require a considerable amount of time and effort to identify the true nature of the hit. One documented example is the follow-up of a polo-like kinase screen hit, in which a synthetically attractive lead was identified after screening [19]. However, biological testing of the resynthesized hit and subsequent synthesized analogs failed to reproduce the biological activity of the original sample. It was later concluded that an unidentified impurity was actually responsible for the observed biological activity. Thus, only after the chemical structures of lead compounds have been validated should structural optimization by medicinal chemistry or cheminformatics tools be performed to improve biological activity.

2.1.7. Series Expansion During the early stages of screening, the focus is usually on structural classes rather than individual compounds. A set of compounds that share a common chemical scaffold can be referred to as a structural cluster series. In many cases, scientists wish to expand a series by evaluating modifications of various functional groups that are attached to a core scaffold. Cheminformatics staff can then perform structural searches of chemical vendor catalogs to identify related compounds for series expansion.

2.1.8. Biologically Relevant Assays HTS protocols commonly utilize a readout that is quick and easy to measure, such as luminescence, fluorescence polarization, fluorescence resonance energy transfer (FRET) signal, or kinetic readouts [20]. Successful modification of an assay to be run in an HTS campaign generally requires bench-level experiments to be miniaturized and condensed. This is true not only for biochemical screens, also for but cell, viral, and bacterial screens. Although considerable effort is made to have HTS assays resemble in vivo processes as closely as possible, the practical approach will necessarily fall short, resulting in a tendency for false positives and negatives [14]. Secondary screening utilizes readouts and techniques that are more physiologically relevant. Western blots, whole-cell assays, or high-performance liquid chromatography (HPLC) substrate cleavage techniques may be used for secondary screening. These protocols cannot usually be adapted to HTS, so they are reserved for later stages, when the number of compounds has been reduced to a few dozen or hundred potential leads.

2.1.9. Structure-Activity Relationship Series Expansion The structure-activity relationship (SAR) series expansion stage falls particularly in the domain of cheminformaticists who work with medicinal chemists and biologists to locate or synthesize chemical analogs based on lead compounds but that are predicted to have greater potency or improved specificity or pharmacokinetic properties. Chemists and biologists will exchange compounds and biological results iteratively until these goals are achieved. In the meantime, cheminformaticists can also generate predictive molecular models to assist in the identification of compounds with desired biological properties based on structural and biological data generated during these exchanges.

2.2. Informatics Requirements for High-Content Screening

High-content screening (HCS) uses an automated imaging system coupled with image processing software to obtain a digital composite and numerous quantitative parameters for each well [21]. More broadly, it may refer to any assay and reading system that generates multiple readings per well. Thus, image storage and data manipulation require

TABLE 11.2: Examples of Currently Available HCS Software

Product	Vendor Web site	Platform
ImageXpress Ultra	http://www.moleculardevices.com/	Confocal
Pathway 855 and 435	http://www.bdbiosciences.com/	Confocal
PerkinElmer Opera	http://www.cellularimaging.com/	Confocal
IN Cell 3000	http://www.gelifesciences.com/	Confocal
Arrayscan VI	http://www.cellomics.com/	Wide Field
IN Cell 2000	http://www.gelifesciences.com/	Wide Field
Acumen eX3	http://www.ttplabtech.com/	Wide Field
Scanalyzer	http://www.lemnatec.com/	Wide Field
ImageXpress MICRO	http://www.moleculardevices.com/	Wide Field
Pathway HT	http://www.bdbiosciences.com/	Wide Field

more disk space and CPU usage than a conventional screen, which produces only numerical data. Each well may have multiple images associated with it, and these images must be stored in such a way that facilitates later access. In addition, the image processing software examines each image and assigns numerical values based on the content of the image. Data reduction for high-content screens usually resembles single-point data reduction. Often, multiple readings are combined to arrive at a single score for each well.

Analysis of high-content screens, specifically the analysis of images, is usually handled by software provided by the instrument vendor unless the screen is a novel or proprietary screen for which specific image analysis algorithms have been developed. Traditional HTS software can generally be used to manage the numeric output. Some examples of currently available software for HCS data analysis are shown in Table 11.2.

2.3. Screening Informatics Software

A substantial amount of specialized software supports screening informatics activities. The purchase and maintenance of this software is an additional responsibility of a screening informatics team. The software must be configured to meet the needs of each campaign performed by a screening center. Because of the numerous commercially available software packages and the constant evolution of features and capabilities, only a few examples are provided here (Table 11.3). Although most available packages are able to perform the same basic functions, significant variations exist in areas including user interface, flexibility or rigidity of system settings, ease of use, and whether the software code is proprietary or an open source. Some systems allow end users to make their own work flow modifications, whereas others require a vendor consultant to modify the software code. Detailed evaluation of a center's needs is advisable in order to determine the most appropriate solution, or combination of solutions, necessary to facilitate a successful informatics program. Common alternatives to commercial products include open-source or in-house proprietary programs. Two examples of HTS analysis software system are given in sections 2.3.1 and 2.3.2.

2.3.1. *The ActivityBase Suite*

2.3.1.1. General Description. IDBS's ActivityBase Suite is a commercially available software solution designed for screening informatics that relies on Oracle for its database. Various modules are marketed separately and can be purchased depending on the level of production and needs of a specific center. In this chapter, we focus on

TABLE 11.3: Examples of Currently Available HTS Software

Product	Vendor Web site	Licensing
Pipeline Pilot	http://www.accelrys.com	Commercial
ACD/Labs	http://www.acdlabs.com	Commercial
Agilent	http://www.home.agilent.com	Commercial
ChemBioOffice	http://www.cambridgesoft.com	Commercial
Symyx/MDL	http://www.accelrys.com	Commercial
ID-BS	http://www.id-bs.com	Commercial
BioRails	http://www.edgesoftwareconsultancy.com/	Open and Commercial
OpenHTS	http://www.openhts.ceuticalsoft.com	Open Source
Spotfire	http://spotfire.tibco.com	Commercial
StarLIMS	http://www.starlims.com	Commercial

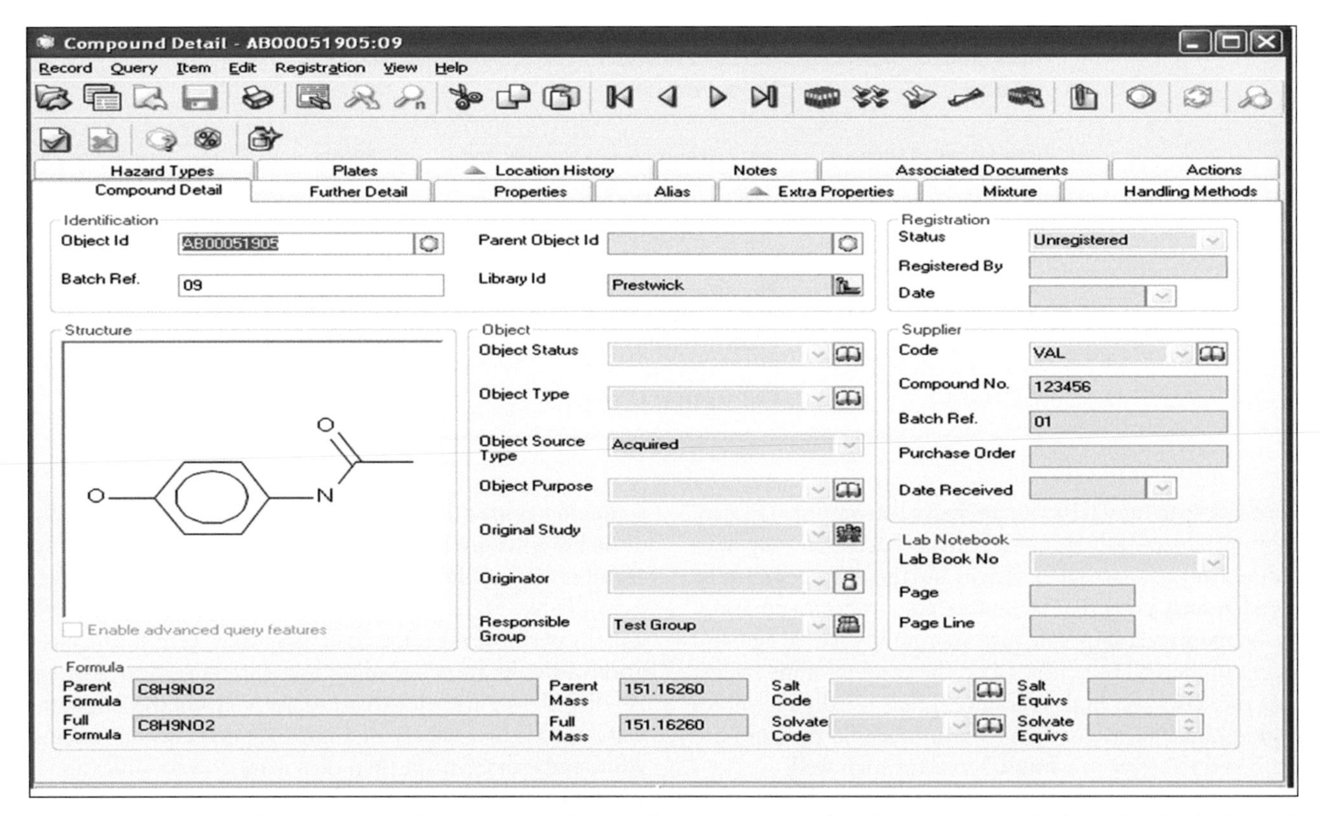

FIGURE 11.4:* *ActivityBase chemistry. Compound detail view allows the user to view the structure and other related data for each compound.

the basic features necessary for an HTS campaign. One necessity is the ability to link, store, and view compound-specific data with each compound's plate and well location. The Chemistry module of ActivityBase allows for tabbed viewing and manual entry of structures and structure-related data (Figure 11.4). Peripheral tools for importing large-structure data files in an automated fashion are provided as both graphical user interface (GUI) (Selection Assistant or Object Manager) and command line (StructureLoad) interfaces. Compound structures integrate with the Object Manager tools to track the plate and well location of each compound (Figure 11.5) through multiple replications, which can be visualized using the genealogy browser feature (Figure 11.6). Once test plates with the designated compounds have been created and made available through the database, processing of the biological data can begin.

The Biology module uses a "Protocol" to define raw data output, conditions, and calculated results that will be associated with a specific assay. The XE Designer module can be used to program various calculations for both validation and screening batches. Data can be imported in real time using the automation feature or in batches using the manual import capabilities of XE Runner. XE not only calculates data based on user defined criteria, but it also contains a variety of visualization tools that can be used by biologists to assess data quality. After biologists review the data, IDBS's query and reporting tools can be used to configure the data into the desired report format for distribution to appropriate team members.

2.3.1.2. An Application Example. During a screening campaign, dose-response analyses are the most common assay type used to evaluate lead compounds. An example of how a screening informatics team would plan and execute this type of study using commercially available tools is outlined here. The information in Table 11.4 is used as the foundation for an informatics work flow.

TABLE 11.4: Documentation of the Desired Data Fields Guides the Template Creation Process

Criteria for a Simple Dose Response Assay	
Plate Format	See Figure 11.6
Compound Conditions	Compound Concentration Control Drug Concentration
Well Level Data	Raw Data Value % Cell Viability
Compound Level Data	EC_{50} with Statistics
Plate Level Data	Control Median and/or Average Control CVs Z' factor

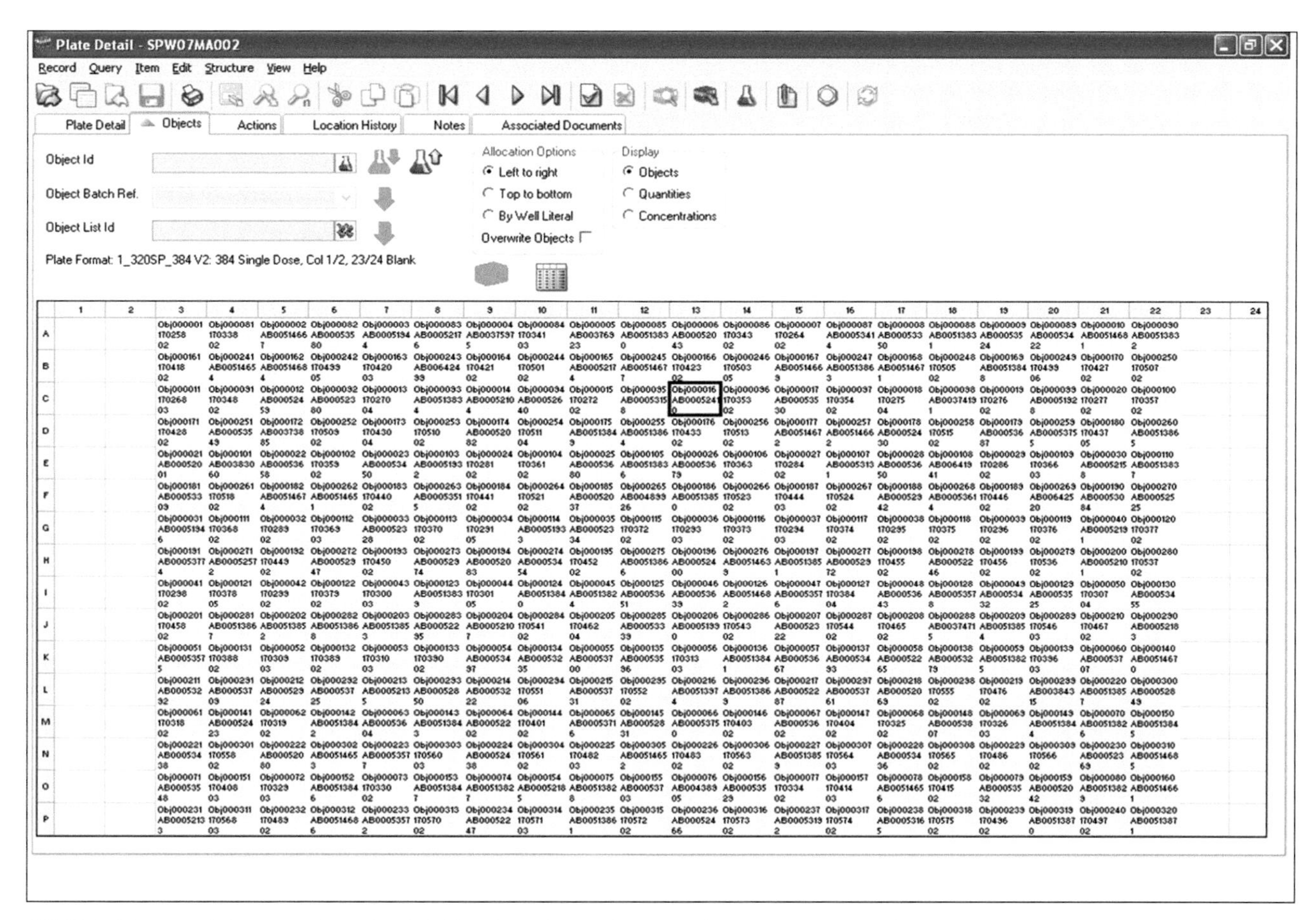

FIGURE 11.5: ActivityBase plate detail. This view shows the plate format and allows the user to visualize the well locations of compounds and control wells in a plate.

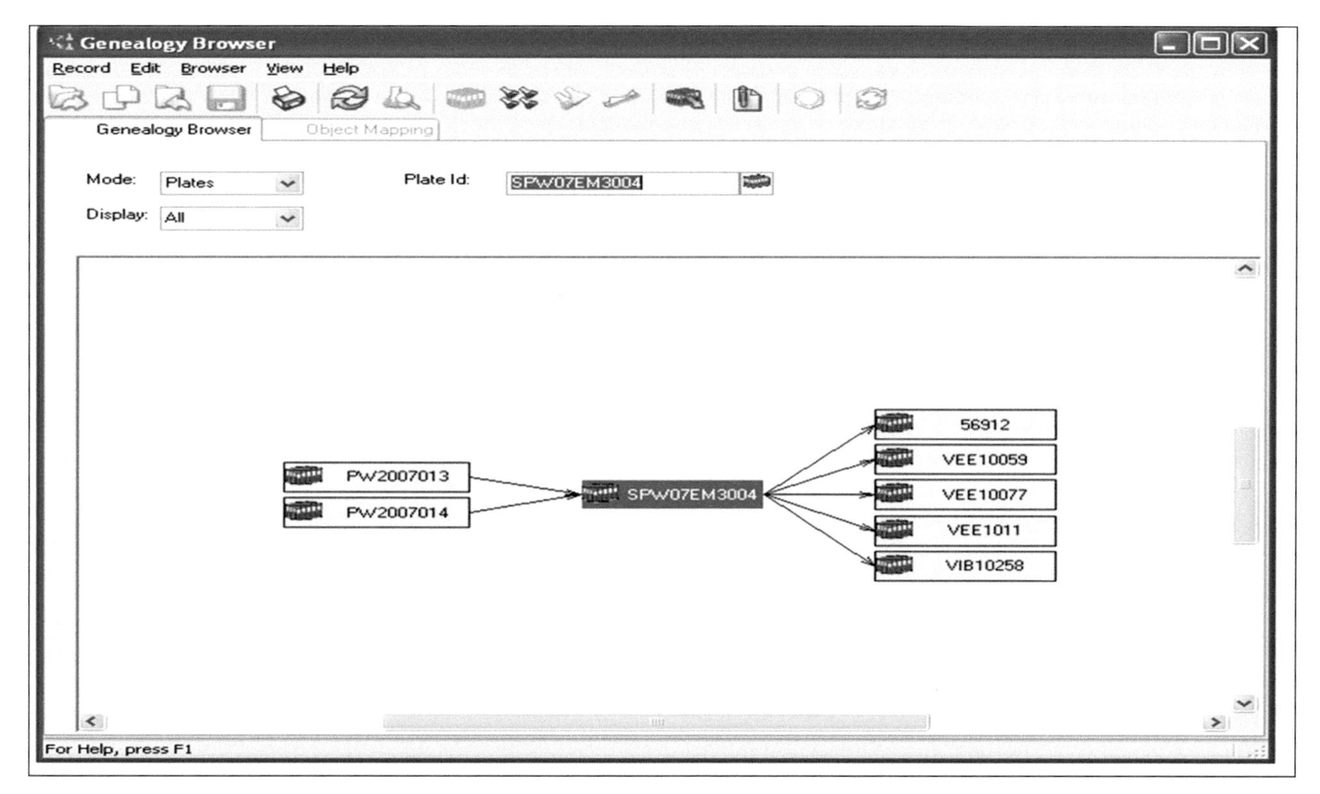

FIGURE 11.6: Genealogy browser. A visual representation of the relationship between plates through various replications required to create an assay plate for screening from the original library's master plate.

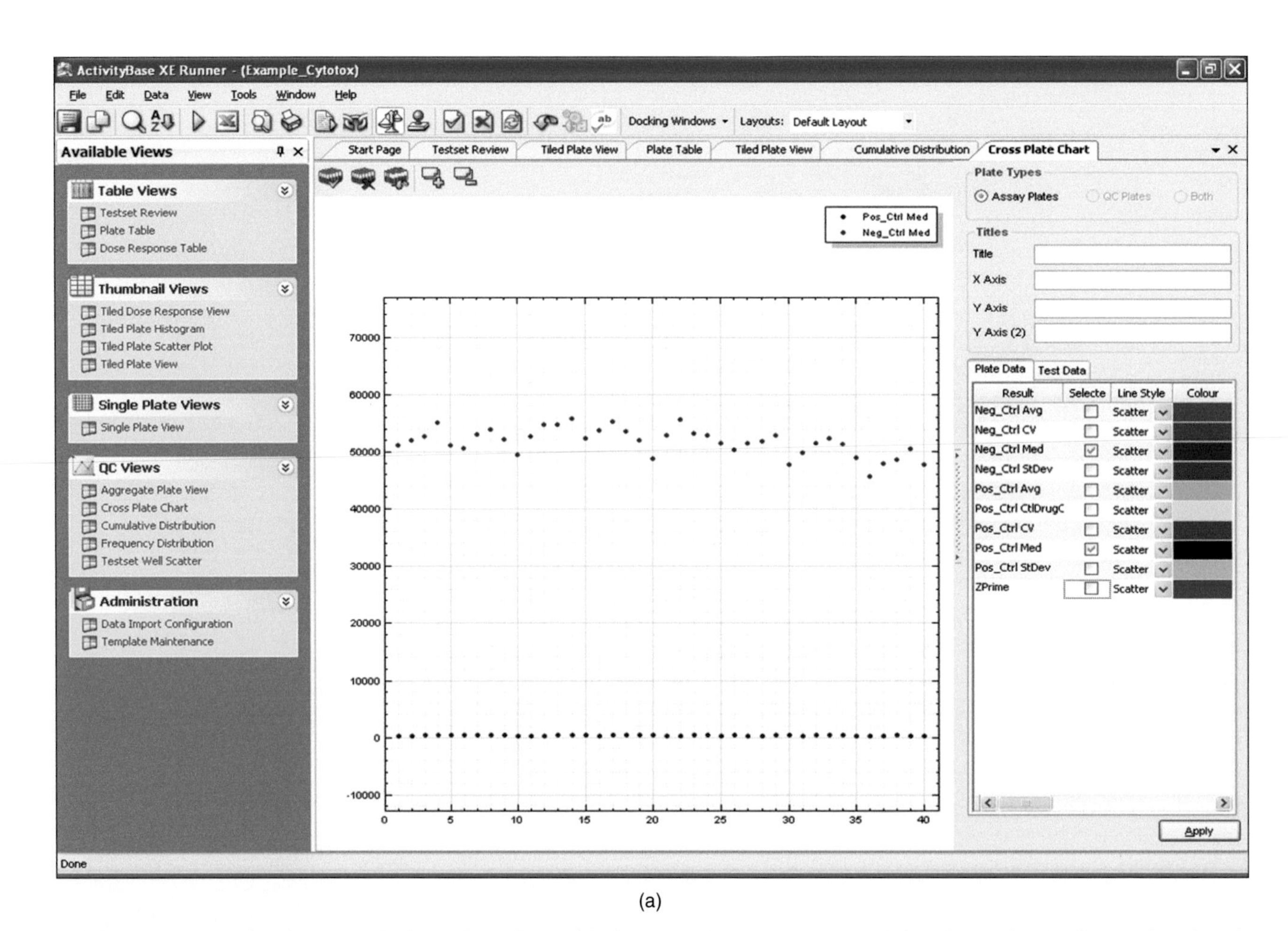

(a)

FIGURE 11.7: Visualization tools. A. Cross Plate Chart. The user can view selected control data from a batch of screening data. In this example, recurring low points can be observed for the negative control approximately every ten plates. Although the data are likely deemed acceptable because of the large signal window, the cause of the trend should be identified. B. Dose-Response Table. This view allows the user to compare curves and to sort data by criteria such as the EC_{50} values. C. Curve Explorer. Systematic or manual outlier exclusion can be performed by examining data points and statistics for individual curves. D. Tiled Plate Scatter. The distribution of values from multiple plates can be compared. In this example, the plates on the upper row return more hits than those on the lower rows. This might reflect a true result based on the distribution of compounds in the library but could also indicate false or skewed data due to a potential liquid handler problem. The background controls can also be seen in the lower right corner of each graph.

Programmatic instructions are provided to the ActivityBase XE module, which uses mapping tools and barcodes to accurately correlate numbers in raw data files with the correct compound structure for each plate and well. Following data import, biologists use the visualization tools in XE (Figure 11.7A–D) to determine whether to accept or reject data. A biologist should assess each curve manually, even though statistical limits can be set to automatically exclude outliers (Figure 11.7D). Following careful review, verified data are ready to be queried, reported, and distributed.

As with most software, query tools are available so that a scientist does not need to know the structured query language (SQL) and the database structure to retrieve data and graphics stored from test batches. SARgen and SARview are IDBS's tools for querying data and formatting reports. Numeric data and curve fit graphs from multiple data sets can be viewed, and structures can be incorporated. Reports can be exported in a variety of file formats ranging from delimited text or Excel format to structure data (SD) files. These tools can be used to generate the SAR databases that cheminformatics will need to identify and pursue hit compounds.

2.3.2. CambridgeSoft

2.3.2.1. General description CambridgeSoft's ChemBioOffice Enterprise software suite (which also uses an Oracle database) is another system designed to accommodate data from HTS. ChemBioOffice contains a series of applications that can be selected individually depending on a center's needs. Modules used for a similar exercise as described in section 2.3.1. are outlined in this section (Figure 11.8). ChemDraw is CambridgeSoft's structural drawing, analysis, and query tool that works

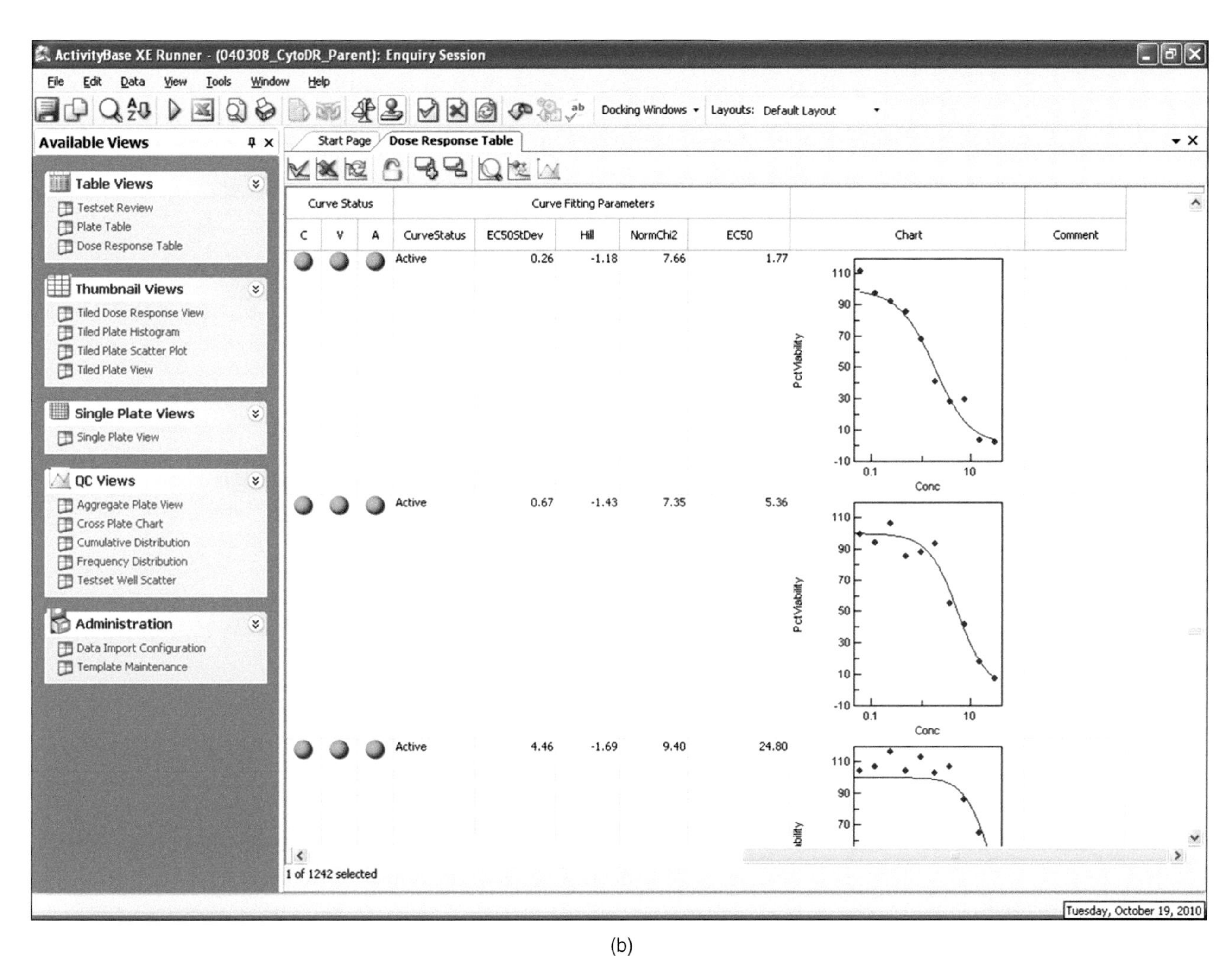

(b)

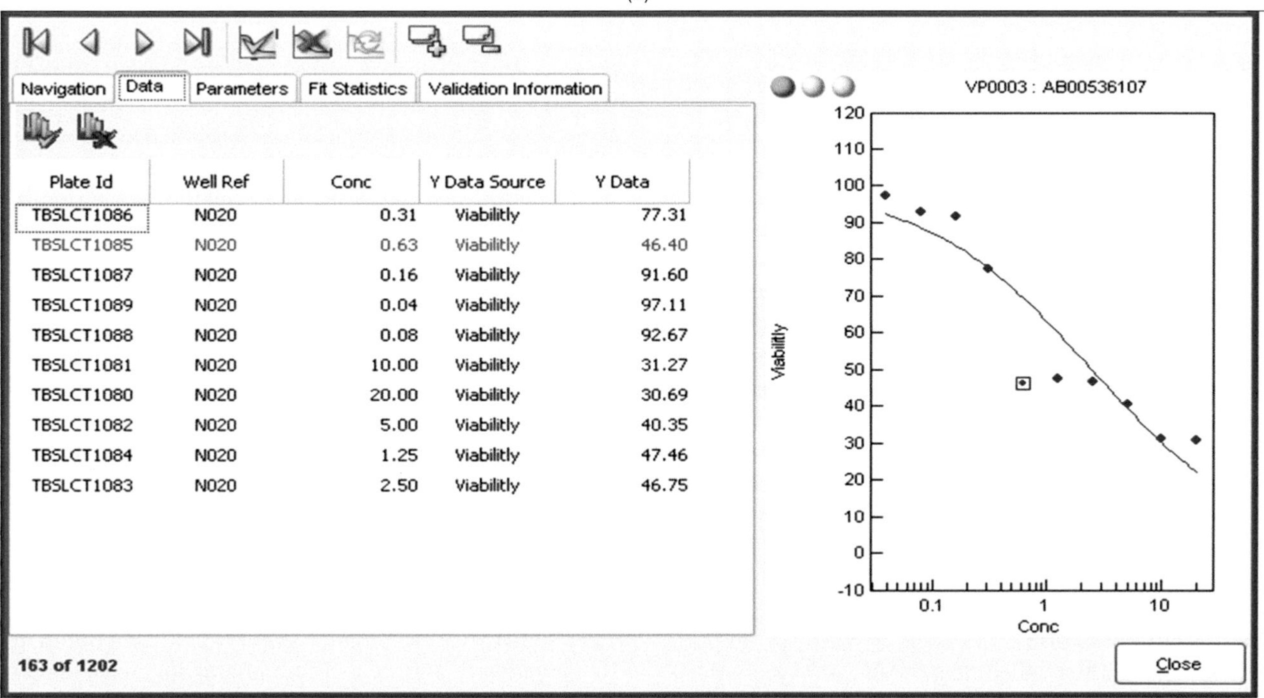

(c)

***FIGURE 11.7** (continued)*

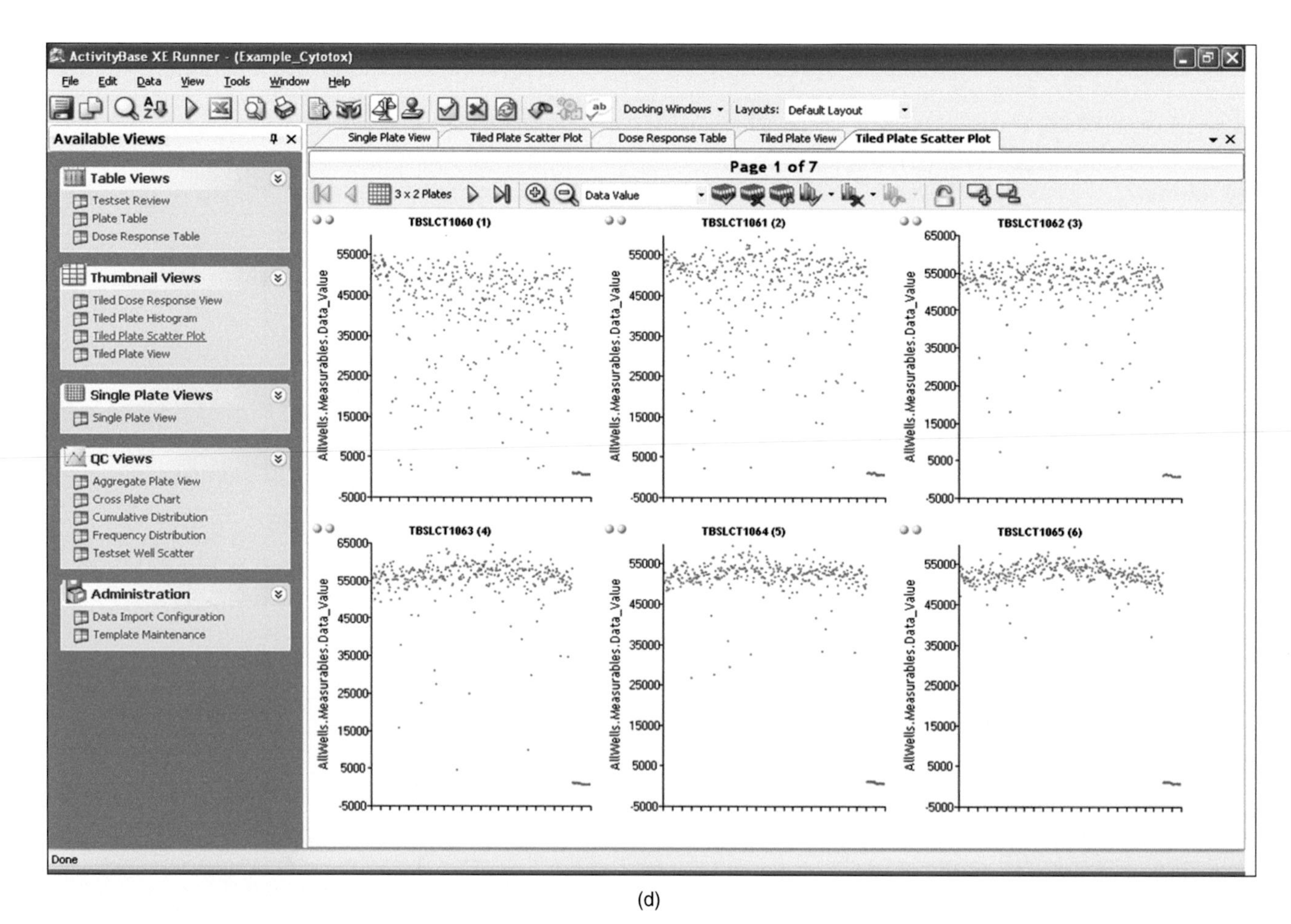

(d)

FIGURE 11.7 **(continued)**

in conjunction with the Registration module to enable compound structure entry. Compounds can be entered through a form or in bulk using batch loader or Inventory modules. Inventory is the chemical and biological reagent management system. This system keeps track of the location of all samples through material tracking, plate management, and substance control.

The BioAssay module is used for processing assay data and data reduction. The algorithms and formulas used for data processing are defined for each assay. Data are imported directly to BioAssay, where calculations are applied. The program provides a graphical user interface that allows for quick setup and definition of biological models by means of Excel templates and provides visualization tools to facilitate assessment. Approved data sets are queried and formatted for further visualization and analyses using the BioSAR and ChemFinder modules.

2.3.2.2. Applications. BioAssay is used to import, manage, and analyze experimental data. BioAssay's flexibility allows it to adapt to any assay, making it a scalable application. Once a biologist approves data upload, BioSAR serves as the data mining and analysis application. BioSAR can output data in many preferred formats, including SD files and Excel spreadsheets. ChemBioViz allows the user to perform more advanced statistical analysis and curve fitting and to generate customized SARs. ChemFinder is another application to search and view particular sets of data and also allows the user to produce desktop databases necessary for SAR studies.

3. APPLICATIONS OF CHEMINFORMATICS

Cheminformatics is the application of computer software to assist in the identification, optimization, and diversification of molecular entities being investigated in an HTS campaign. The basis for cheminformatics is the knowledge of a molecular entity's chemical structure, because its structural makeup confers its biological activity. This field encompasses methods that range from 2D neighborhood similarity searching of primary hit compounds to 3D modeling of a protein-ligand interaction. This section gives a brief overview of 2D and 3D methods employed following an HTS campaign.

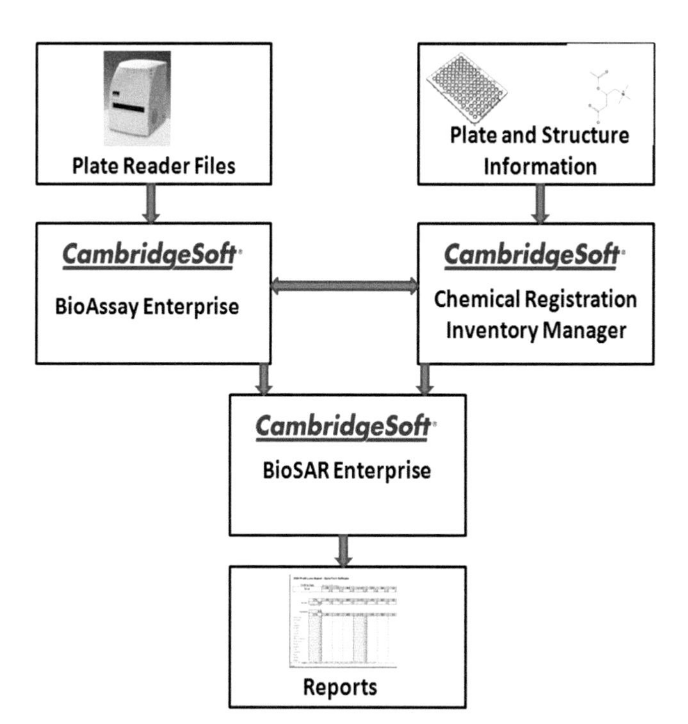

FIGURE 11.8: CambridgeSoft ChemBioOffice Enterprise HTS work flow. Plate reader files are imported into the database by BioAssay and combined with structural and plate information from Registration and Inventory. BioSAR Enterprise is used to generate reports and data files for review or further processing.

3.1. Identification of Lead Compounds through 2D Structural Clustering

Cheminformatics methods are often applied following primary HTS and secondary assays to identify hits with observed high biological potency that are structurally amenable through synthetic methodology. Relatively few of the potentially thousands of hits from a series of secondary assays are selected for optimization. These hits are referred to as lead compounds. One strategy to assist in the selection of lead compounds is to categorize them into groups that share a common chemical structure that theoretically confers the desired biological activity.

3.1.1. Structural Clustering Example in Lead Identification: Primary Screen for EP2 Modulators Structural clustering is a cheminformatics method that assists in identification of potential leads from a list of primary hits by grouping hits according to how similar they are in terms of their Lewis structure (atom connectivity). Clustering identifies compounds that possess a common scaffold in their structural makeup. As a compound's structure confers its biological activity, clustering should reveal the classes of molecules providing true hits.

The following is an example of structural clustering performed after a primary screening campaign. The prostaglandin EP2 receptor is a Gs-coupled protein with potential therapeutic value in the treatment of strokes [22]. Prior to the screen, EP2 had no known selective modulators; thus, HTS was employed to identify a first-in-class selective EP2 potentiator from a library of 255,000 compounds. Clustering was performed on the 2002 primary screen hits using LeadScope software [23], which employs the Jarvis-Patrick method [24, 25], with a threshold value of 40% similarity to generate clusters. Two groups of compounds sharing two different scaffolds were identified to confer an advantage in the assay (Figure 11.9). Both clusters shared a common motif of an amide functional group connecting two aromatic rings. The atom connectivity that distinguished the members of Cluster 1 is characterized by a bicyclic thiophene with an ester substituent, whereas members of Cluster 2 share a structural motif of two aryl units and a piperidine ring. Medicinal chemists in this study elected to synthesize analogs around Cluster 1 [22].

With increasing advances in and accessibility of high-speed computer hardware, structural clustering has become a computationally inexpensive and easily applicable routine for HTS groups in post-primary screening stages. As the previous example demonstrates, this method allows for the rapid categorization of structurally similar analogs with slightly varying biological activity due to differences in their functionality along a common scaffold core (Figure 11.9). In addition to quick triaging, clustering also offers researchers a preliminary understanding of the chemical space that can be explored around a scaffold. In drug development, this understanding is referred to as a SAR in which the small structural changes of a class of molecules that share the same scaffold are plotted against their varying biological activity to guide the design of future molecular analogs [26].

3.1.2. 2D Structural Clustering for Identification of False Positives In addition to lead identification, structural clustering can also be applied to eliminate potential false positives. Triaging hits can begin by eliminating compounds with chemically reactive or undesirable functionality due to known toxicity in cellular or whole organism evaluations. Structural clustering can then be used to examine the remaining compounds and identify compounds that showed chemical functionality but are actually false positives. Particularly in fluorescence polarization assays, a compound with fluorescent-sensitive functionality can mistakenly be classified as having potent biological activity. In fact, the compound may be reacting to the wavelength exposure required for the experiment rather than the modulation of a biological event. Examples of compounds with auto-fluorescent properties that share a coumarin substructure are shown in Figure 11.10. Identification of compounds with fluorescent functionality is useful to avoid losing time and effort in the chemical optimization of artifact leads following an HTS screen. Characterization of compounds in this regard is so important that groups have started cataloging known fluorescent compounds [27, 28].

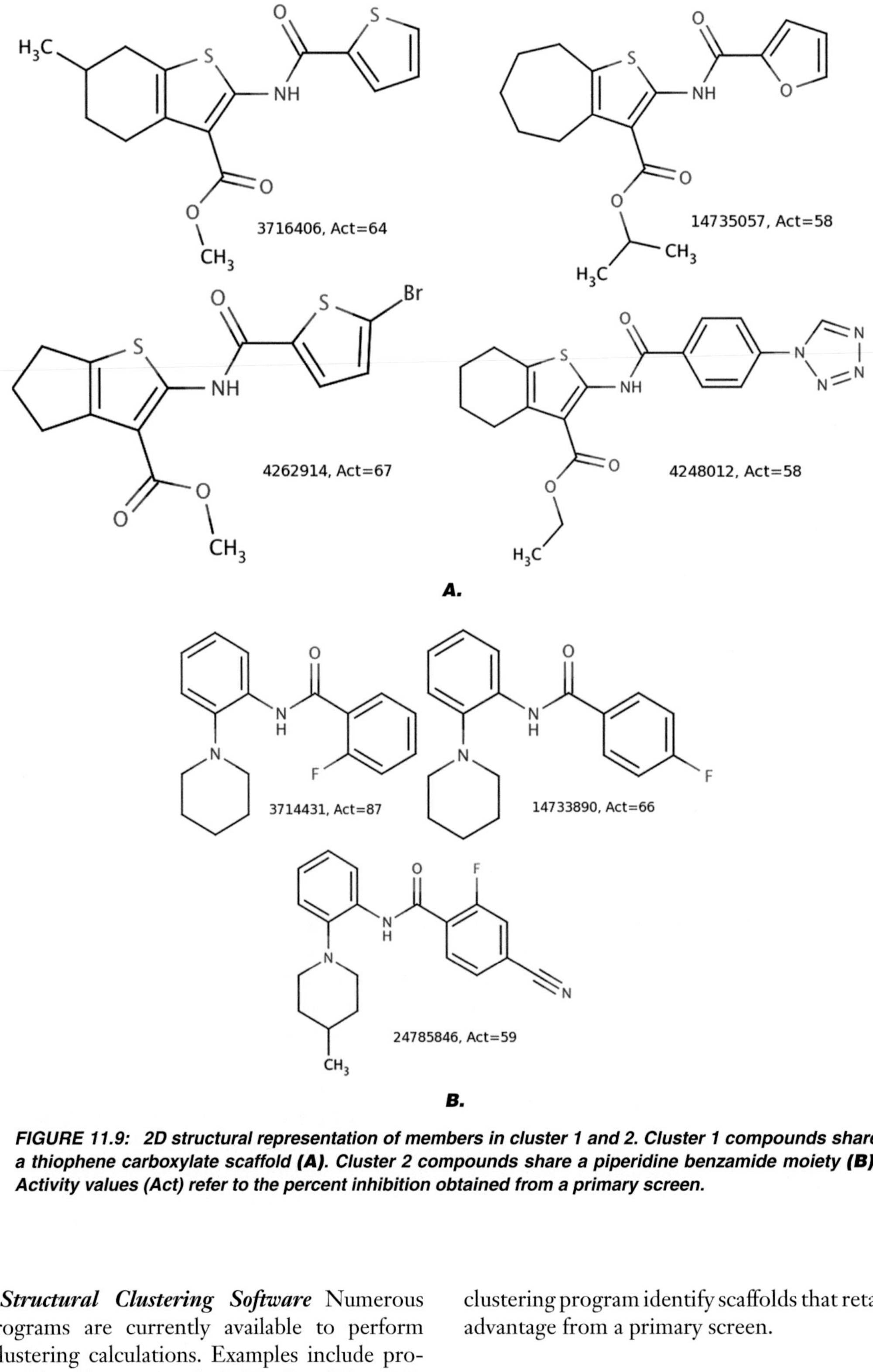

FIGURE 11.9: 2D structural representation of members in cluster 1 and 2. Cluster 1 compounds share a thiophene carboxylate scaffold (A). Cluster 2 compounds share a piperidine benzamide moiety (B). Activity values (Act) refer to the percent inhibition obtained from a primary screen.

*3.1.3. 2D **Structural Clustering Software*** Numerous software programs are currently available to perform structural clustering calculations. Examples include programs such as LeadScope, Accelry's Pipeline Pilot [29], Schrodinger's Canvas [30, 31], Substructure Enrichment Analysis (SEA), [32] and ChemAxon's LibraryMCS [33]. Many more programs exist that are available publicly (Table 11.5) and in-house. It is essential that any such 2D clustering program identify scaffolds that retain a biological advantage from a primary screen.

3.2. Data Mining and Virtual Screening in Lead Expansion

Once structural and biological verification have been established from a primary screen, investigators can begin

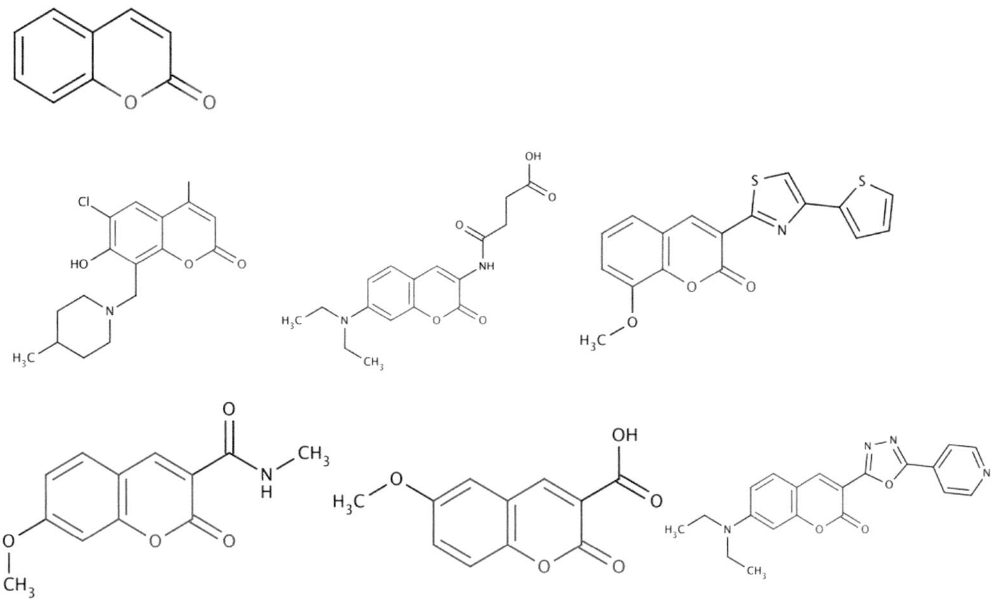

FIGURE 11.10: Compounds that share coumarin as a common structural motif. The coumarin functional group is depicted in the top left-hand corner. The coumarin substructure shared by each compound is highlighted in blue. This figure is derived from NIH Chemical Genomics Center qHTS profiling screening results, AID 593, PubChem [27].

commercial purchasing and in-house synthesis of leads and their structural analogs. This serves two purposes: verifying that leads are indeed biological modulators of the intended target or phenotype and expanding the explored chemical space around the lead's scaffold.

Exploring chemical space through commercial sample purchasing is convenient, as samples tend to be readily available for shipment by most chemical suppliers. This method also maps out SAR space that would be useful for synthetic chemistry optimization. The investigator in this instance is hampered because of the number of molecular choices available that otherwise might be accessible exclusively through synthetic methods. Cheminformatics software has been developed to easily organize choices for more rapid lead optimization.

3.2.1. Lead Expansion through 2D Similarity Searching

2D fingerprints, a numerical abstraction of the compound's atomic makeup and connectivity, are assigned to each compound in a database and compared with the fingerprint of the query compound. Examples of fingerprints for similarity searches include Daylight [39] and the extended connectivity fingerprints [40]. A percent similarity is then assigned to each compound through statistical methods. Tanimoto similarity searching is a 2D searching tool most often applied in structural database mining that utilizes these fingerprints. An example of 2D similarity searching involves the analoging of quinocide, an inhibitor for Hsp90 identified through primary screening (Figure 11.11). Prior to synthetic optimization of quinocide, 2D similarity searching was performed across several commercial databases to

TABLE 11.5: 2D Cheminformatics Software For Primary Screening Triaging

Software	Vendor	Web Site
Pipeline Pilot [29]	Accerlys, Inc	http://accelrys.com/products/pipeline-pilot/
LeadScope [23] [34, 35]	Leadscope, Inc	http://www.leadscope.com/
Canvas [30, 31]	Schrodinger, Inc	www.schrodinger.com/
LibraryMCS [33]	ChemAxon	www.chemaxon.com/
Distill [36]	Tripos	www.tripos.com/
JP Clustering [24]	NCGC	http://ncgc.nih.gov/pub/openhts/
KNIME [37]	University of Konstanz	www.knime.org/
FILTER [38]	Openeye Scientific Software, Inc	www.eyesopen.com/
Substructure Enrichment Analysis (SEA) [32]	Emory Chemical Biology Discovery Center	http://www.emory.edu/chemical-biology/

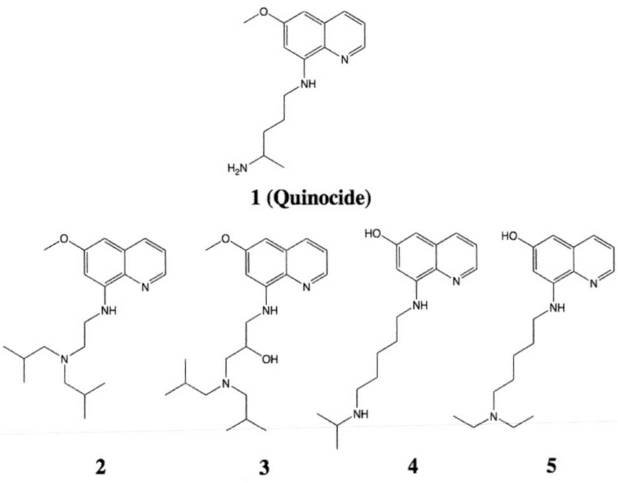

1 (Quinocide)

2 3 4 5

***FIGURE 11.11:* Quinocide and related structural analogs. Compounds 2 through 5 demonstrated consistent activities in the primary fluorescence polarization assay and the following cell-based Western blot assays.**

identify compounds that had similar activity to quinocide but that varied slightly in their structure. The results of a Tanimoto similarity search of the National Cancer Institute's (NCI) database (250,000 compounds) are shown in Figure 11.11. Thirty-five compounds were obtained based on the results of the data mining, but four (compounds 2–5) showed repeatable activity in primary and confirmatory assays and were the basis for medicinal chemistry optimization.

3.2.2. 3D Methods for Lead Expansion To explore additional chemical space, 3D ligand-based search methods are employed. 3D methods take into account a ligand's conformation, volume, spatial distribution of functional groups, and electrostatic character to identify similar compounds. The added advantage of 3D similarity searching over 2D searching is the ability to identify compounds with similar biological activity but different structural makeup, also known as scaffold hopping. Methods such as these are useful when no further chemical space can be explored around a current lead or if alternative scaffolds need to be identified for intellectual property reasons.

The compounds sildenafil (Viagra) [41, 42] and tadalafil (Cialis) [43–46] are examples of how different scaffolds can be used to bring about a similar biological result (Figure 11.12). Despite their widely divergent structures, both are able to bind to the same region of cyclic guanosine monophosphate (CGMP)-specific phosphodiesterase type 5, the primary target of erectile dysfunction therapeutics [47].

3.2.3. 33D Virtual Screening One successful usage of 3D screening was described in the publication by Rush et al. in which they describe the identification of novel structural inhibitors for the ZipA-FtsZ protein-protein interaction,

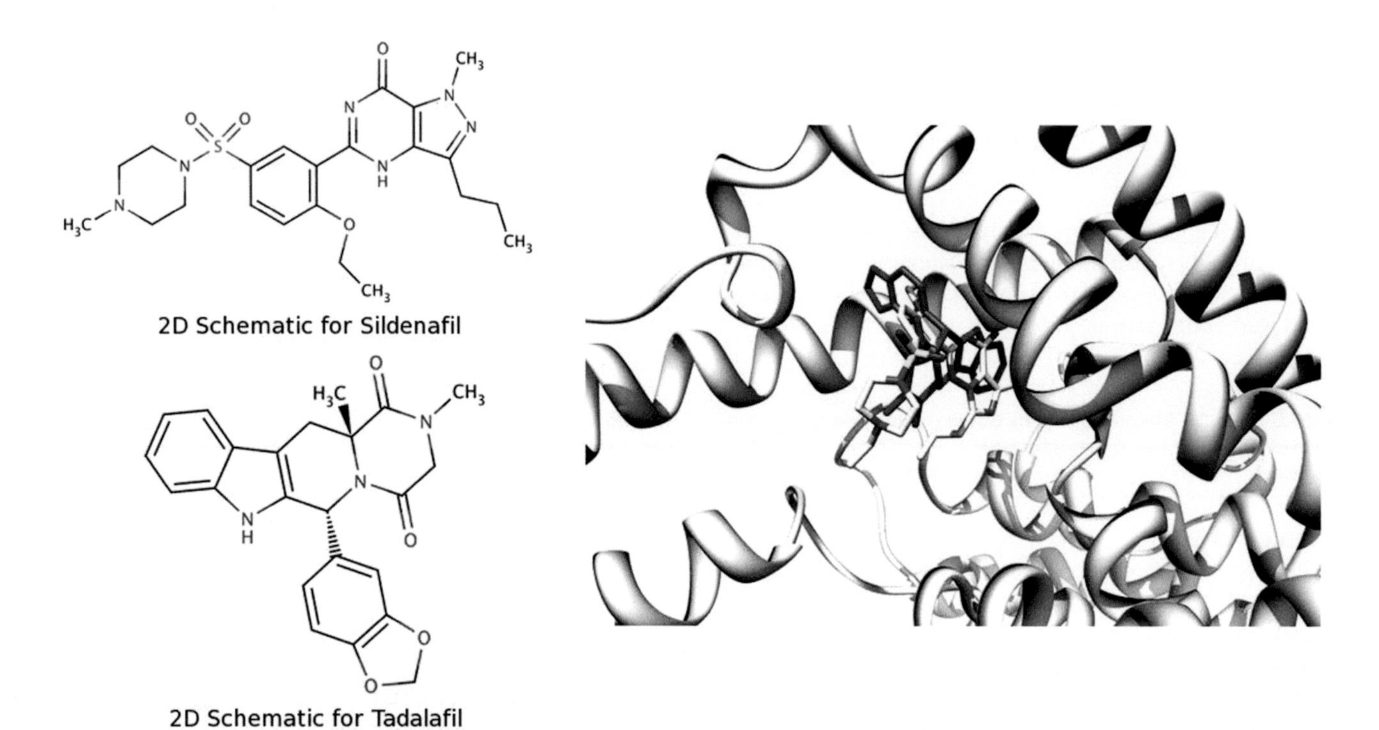

***FIGURE 11.12:* Overlap of sildenafil (white) with tadalafil (magenta). Despite having significantly different Lewis structures, both compounds bring about similar biological phenotypes. Coordinates for protein-ligand models retrieved from Sung et al. (2003) [47] and rendered in the molecular visualizer, Chimera [63]**

TABLE 11.6: 3D Similarity Searching Packages for Lead Expansion

Software	Vendor	Web Site
ROCS [49] [48]	Openeye Scientific Software	www.eyesopen.com
Topomers [50, 51]	Tripos, Inc	www.tripos.com
ShaEP [52]	Åbo Akademi University	http://users.abo.fi/mivainio/shaep/
PHASE [53–55]	Schrodinger, Inc	www.schrodinger.com
LigandFit [56]	Accerlys, Inc	www.accelrys.com
Molecular Operating Environment [57]	Chemical Computing Group, Inc	www.chemcomp.com

a potential target for overcoming bacterial resistance [48]. The investigators of this work illustrated the ability of 3D methods to identify novel weakly active inhibitors that were not selected by 2D similarity searching. These investigators used the OpenEye Scientific software, ROCS (Rapid Overlay of Chemical Structures) [49], for the identification of protein-protein interaction inhibitors based on a lead compound identified by HTS. An expanded listing of software packages available for 3D similarity screening is provided (Table 11.6). Virtual screening can also be performed in the absence of a lead, where the residues of a known or theorized protein binding site generated by x-ray or homology modeling is used to identify compounds that complement the shape, hydrophobicity, and electrostatic character of the receptor. Virtual screening methodologies are detailed in Chapter 19.

3.3. Molecular Modeling: Understanding Protein-ligand Interactions

A myriad of 3D molecular modeling tools are available to evaluate interactions between a ligand and a protein receptor (Table 11.7). Models are derived from experimental methods such as x-ray crystallography, which provides structural information on the target of interest as well as the location of the binding site, if co-crystallized with a ligand. By examining a model on a computer, chemists can identify ways of improving protein-ligand contacts, such as additional hydrogen bonds or increased hydrophobic interactions. In the absence of a ligand of interest bound to a receptor, docking calculations can be performed to hypothesize ligand binding to a desired receptor. Experimental methods such as single-point mutations can augment docking to identify residues that are essential to ligand activity and thus are assumed to play an important role in protein-ligand interactions. Examples of available docking software are given in Table 11.8.

Competition studies, in which a known ligand with a known binding site is used as a control, are useful in determining activity. Such a case is described in the identification of quinocide as a novel Hsp90 inhibitor (Figure 11.11). It was assumed that compounds showing activity in the Hsp90 primary assay dislodged control geldanamycin, a known N-terminal adenosine triphosphate (ATP) Hsp90 inhibitor. Several Hsp90 crystal structures were isolated; therefore, the binding position could be hypothesized through docking analysis of quinocide into geldanamycin's binding space. The work by Thepchatri et al. illustrates the proposed docking position of quinocide in the geldanamycin binding pocket and suggests that the compound makes interactions with Hsp90 residues similar to those of other known Hsp90 inhibitors [32].

TABLE 11.7: Molecular Visualization Software

Viewer	Distributor	Web Site
Maestro [58]	Schrodinger, Inc	www.schrodinger.com
PyMOL [59]	Schrodinger, Inc	www.schrodinger.com
Sybyl X [60]	Tripos, Inc	www.tripos.com
VIDA [61]	OpenEye Scientific Software	www.eyesopen.com
Discovery Studio Visualizer 2.5.5 [62]	Accerlys, Inc	www.accelrys.com
Chimera [63]	University of California, San Francisco	http://www.cgl.ucsf.edu/chimera/
VMD [64]	University of Illinois, Urbana-Champaign	http://www.ks.uiuc.edu/Research/vmd/
DeepView v4.0 [65]	Swiss Institute of Bioinformatics	http://spdbv.vital-it.ch/
Molecular Operating Environment [57]	Chemical Computing Group, Inc	www.chemcomp.com
Yasara [66]	Yasara Biosciences, GmbH	www.yasara.org

TABLE 11.8: Molecular Docking Software

Docking Software	Distributor	Web Site
Glide [67–70]	Schrodinger, Inc	www.schrodinger.com
FRED [71–73]	OpenEye Scientific Software	www.eyesopen.com
Surflex [74]	Jain Lab (UCLA)/Tripos, Inc	www.jainlab.org/www.tripos.com
AutoDock 4.0 [75]	Scripps Research Institute	http://autodock.scripps.edu/
DOCK 6.4 [75–78]	University of California, San Francisco	http://dock.compbio.ucsf.edu
Gold 4.1 [79–85]	Cambridge Crystallographic Data Centre	http://www.ccdc.cam.ac.uk/
RosettaLigand [86, 87]	Rosetta Commons	www.rosettacommons.org

4. FINAL REMARKS

Screening informatics and cheminformatics software packages offer users the ability to rapidly filter through large numbers of HTS data points to triage primary hits to lead optimization, to scan virtual databases to identify compounds for biological testing in an effort to augment SAR data, and to provide insights into the important structural characteristics of a compound that confer a desired biological effect. Although these are powerful tools to assess large amounts of data quickly, results should not be taken as is and should still be assessed for biological and chemical reasonability in relation to the biological target and the chemical space occupied by lead compounds. Given the number of informatics tools that are freely available for download and run on basic laptops or desktops, one is always encouraged to experiment with software to enrich group discussions on HTS lead identification and optimization projects.

ACKNOWLEDGMENTS

Our thanks to Dr. Andrew Prussia of the Emory Drug Discovery Institute for assistance in the molecular modeling concepts of the cheminformatics portion of this chapter.

ABBREVIATION

SAR
structure-activity relationship; QC
quality control.

REFERENCES

1. Ling, X. B. (2008). High throughput screening informatics. *Comb Chem High Throughput Screen 11*, 249–257.
2. Brown, F. K. (1998). Chemoinformatics: what is it and how does it impact drug discovery. *Ann Rep Med Chem 33*, 375–384.
3. Wishart, D. S. (2007). Introduction to cheminformatics. *Curr Protoc Bioinformatics*, chapter 14, unit 14.1.
4. Sukumar, N., Krein, M., and Breneman, C. M. (2008). Bioinformatics and cheminformatics: where do the twain meet? *Curr Opin Drug Discov Devel 11*, 311–319.
5. Oprea, T. I., Tropsha, A., Faulon, J. L., and Rintoul, M. D. (2007). Systems chemical biology. *Nat Chem Biol 3*, 447–450.
6. Kaushal, D., and Naeve, C. W. (2004). Analyzing and visualizing expression data with Spotfire. *Curr Protoc Bioinformatics*, chapter 7, unit 7.*9*.
7. Kaushal, D., and Naeve, C. W. (2004). Loading and preparing data for analysis in spotfire. *Curr Protoc Bioinformatics*, chapter 7, unit 7.*8*.
8. Motulsky, H., and Christopoulos, A. (2004). Fitting models to biological data using linear and nonlinear regression. *A Practical Guide to Curve Fitting* (New York: Oxford University Press).
9. Zhang, J. H., Chung, T. D., and Oldenburg, K. R. (1999). A simple statistical parameter for use in evaluation and validation of high throughput screening assays. *J Biomol Screen 4*, 67–73.
10. Slaughter, S., and Delwiche, L. (2010). The Little SAS Book for Enterprise Guide 4.2 (Cary, N.C.: SAS Institute Inc.).
11. Hill, T., and Lewicki, P. (2007). Statistics: Methods and Applications (Tulsa, Okla.: StatSoft, Inc.).
12. Hassan, M., Brown, R. D., Varma-O'Brien, S., and Rogers, D. (2006). Cheminformatics analysis and learning in a data pipelining environment. *Mol Divers 10*, 283–299.
13. Tiwari, A., and Sekhar, A. K. (2007). Workflow based framework for life science informatics. *Comput Biol Chem 31*, 305–319.
14. Assay Guidance Manual Version 5.0 (2008). Eli Lilly and Company and NIH Chemical Genomics Center. http://www.ncgc.nih.gov/guidance/manual_toc.html.
15. Larson, R. J., and Marx, M. L. (2000). An Introduction to Mathematical Statistics and Its Applications (Upper Saddle River, N. J.: Prentice Hall).
16. Kozikowski, B. A., Burt, T. M., Tirey, D. A., Williams, L. E., Kuzmak, B. R., Stanton, D. T., Morand, K. L., and Nelson, S. L. (2003). The effect of freeze/thaw cycles on the stability of compounds in DMSO. *J Biomol Screen 8*, 210–215.
17. Blaxill, Z., Holland-Crimmin, S., and Lifely, R. (2009). Stability through the ages: the GSK experience. *J Biomol Screen 14*, 547–556.
18. MacArthur, R., Leister, W., Veith, H., Shinn, P., Southall, N., Austin, C. P., Inglese, J., and Auld, D. S. (2009). Monitoring

compound integrity with cytochrome P450 assays and qHTS. *J Biomol Screen 14*, 538–546.

19. Wipf, P., Arnold, D., Carter, K., Dong, S., Johnston, P. A., Sharlow, E., Lazo, J. S., and Huryn, D. (2009). A case study from the chemistry core of the Pittsburgh Molecular Library Screening Center: the Polo-like kinase polo-box domain (Plk1-PBD). *Curr Top Med Chem 9*, 1194–1205.
20. Razvi, E. (2003). Emerging themes and quantitative metrics in high-throughput screening space. *Innovaro Pharmlicensing*. http://pharmalicensing.com/public/articles/view/1060857367_3f3b6617073e2/emerging-themes-and-quantitative-metrics-in-high-throughput-screening-space.
21. Abraham, V. C., Taylor, D. L., and Haskins, J. R. (2003). High-content screening applied to large-scale cell biology. *Trends Biotechnol. 1*, 15–22.
22. Jiang, J., Ganesh, T., Du, Y., Thepchatri, P., Rojas, A., Lewis, I., Kurtkaya, S., Li, L., Qui, M., Serrano, G., et al. (2010). Neuroprotection by selective allosteric potentiators of the EP2 prostaglandin receptor. *Proc Natl Acad Sci U S A*.
23. Richon, A. (2000). LeadScope: data visualization for large volumes of chemical and biological screening data. *J Mol Graph Model 18*, 76–79.
24. Jarvis, R. A., and Patrick, E. A. (1973). Clustering using a similarity measure based on shared nearest neighbors. IEEE Trans. *Comput C-22*, 1025–1034.
25. Willett, P., Winterman, V., and Bawden, D. (1986). Implementation of nonhierarchical cluster analysis methods in chemical information systems: selection of compounds for biological testing and clustering of substructure search output. *J Chem Inf Comput Sci 26*, 109–118.
26. Burger, A. (1994). Medical chemistry – the first century. *Med Chem Res 4*, 3–15.
27. Simeonov, A., Jadhav, A., Thomas, C. J., Wang, Y., Huang, R., Southall, N. T., Shinn, P., Smith, J., Austin, C. P., Auld, D. S., et al. (2008). Fluorescence spectroscopic profiling of compound libraries. *J Med Chem 51*, 2363–2371.
28. Haugland, R. P. (2006). The Handbook – A Guide to Fluorescent Probes and Labeling Technologies (Eugene, Ore.: Molecular Probes, Inc.).
29. *Pipeline Pilot*, Accelyrs: San Diego, Calif; 2007.
30. Sastry, M., Lowrie, J. F., Dixon, S. L., and Sherman, W. (2010). Large-scale systematic analysis of 2D fingerprint methods and parameters to improve virtual screening enrichments. *J Chem Inf Model 50*, 771–784.
31. *Canvas, version 1.2, Schrodinger*, LLC, New York, NY, 2009.
32. Thepchatri, P., Min, J., Ganesh, T., Du, Y., Lewis, I., Kurtkaya, S., Prussia, A., Li, L., Sneed, B., Plemper, R. K., et al. (2009). Cancer and virus leads by HTS, chemical design and SEA data mining. *Curr Top Med Chem 9*, 1159–1171.
33. JKlustor was used for clustering and diversity analysis of chemical sets, JChem 2.5.3, 2009, ChemAxon, http://www.chemaxon.com.
34. Roberts, G., Myatt, G. J., Johnson, W. P., Cross, K. P., Blower, P. E., Jr. (2000). LeadScope: Software for exploring large sets of screening data. *J. Chem*. Inf. Comput. Sci. *40*, 1302–1314.
35. Matthews, E. J., Kruhlak, N. L., Benz, R. D., Contrera, J. F., Marchant, C. A., and Yang, C. (2008). Combined use of MC4PC, MDL-QSAR, BioEpisteme, Leadscope PDM, and Derek for Windows software to achieve high-performance, high-confidence, mode of action-based predictions of chemical carcinogenesis in rodents. *Toxicol Mech Methods 18*, 189–206.
36. Distill, Tripos International, 1699 South Hanley Rd., St. Louis, Mo.
37. Berthold, M. R., Cebron, N., Dill, F., Gabriel, T. R., Kotter, T., Meinl, T., Ohl, P., Sieb, C., Thiel, K., and Wiswedel, B. (2008). KNIME: The Konstanz Information Miner. In Data Analysis, Machine Learning and Applications, C. Preisach, and L. Schmidt-Thieme, eds. (Berlin: Springer-Verlag), 319–325.
38. FILTER 2.02, OpenEye Scientific Software, Inc. Santa Fe, N.M., 2010.
39. Leo, A., and Weininger, A. (1995). *Daylight Chemical Information Systems, version 3* (Aliso Viejo, Calif.: Daylight Chemical Information Systems, Inc.).
40. Rogers, D., Brown, R. D., and Hahn, M. (2005). Using extended-connectivity fingerprints with Laplacian-modified Bayesian analysis in high-throughput screening follow-up. *J Biomol Screen 10*, 682–686.
41. Boolell, M., Gepi-Attee, S., Gingell, J. C., and Allen, M. J. (1996). Sildenafil, a novel effective oral therapy for male erectile dysfunction. *Br J Urol 78*, 257–261.
42. Padma-Nathan, H. (2006). Sildenafil citrate (Viagra) treatment for erectile dysfunction: An updated profile of response and effectiveness. *Int J Impot Res 18*, 423–431.
43. Frajese, G. V., and Pozzi, F. (2005). Phosphodiesterase type 5 inhibitors in older males. *J Endocrinol Invest 28*, 112–115.
44. Porst, H., Giuliano, F., Glina, S., Ralph, D., Casabe, A. R., Elion-Mboussa, A., Shen, W., and Whitaker, J. S. (2006). Evaluation of the efficacy and safety of once-a-day dosing of tadalafil 5mg and 10mg in the treatment of erectile dysfunction: results of a multicenter, randomized, double-blind, placebo-controlled trial. *Eur Urol 50*, 351–359.
45. Rajfer, J., Aliotta, P. J., Steidle, C. P., Fitch, W. P., III, Zhao, Y., and Yu, A. (2007). Tadalafil dosed once a day in men with erectile dysfunction: a randomized, double-blind, placebo-controlled study in the US. *Int J Impot Res 19*, 95–103.
46. Hatzichristou, D., Gambla, M., Rubio-Aurioles, E., Buvat, J., Brock, G. B., Spera, G., Rose, L., Lording, D., and Liang, S. (2008). Efficacy of tadalafil once daily in men with diabetes mellitus and erectile dysfunction. *Diabet Med 25*, 138–146.
47. Sung, B. J., Hwang, K. Y., Jeon, Y. H., Lee, J. I., Heo, Y. S., Kim, J. H., Moon, J., Yoon, J. M., Hyun, Y. L., Kim, E., et al. (2003). Structure of the catalytic domain of human phosphodiesterase 5 with bound drug molecules. *Nature 425*, 98–102.
48. Rush, T. S., III, Grant, J. A., Mosyak, L., and Nicholls, A. (2005). A shape-based 3-D scaffold hopping method and its application to a bacterial protein-protein interaction. *J Med Chem 48*, 1489–1495.
49. ROCS, 2.3.1; OpenEye Scientific Software: Santa Fe, N.M., 2007.
50. Jilek, R. J., and Cramer, R. D. (2004). Topomers: a validated protocol for their self-consistent generation. *J Chem Inf Comput Sci 44*, 1221–1227.

51. Cramer, R. D., Jilek, R. J., Guessregen, S., Clark, S. J., Wendt, B., and Clark, R. D. (2004). "Lead hopping." Validation of topomer similarity as a superior predictor of similar biological activities. *J Med Chem 47*, 6777–6791.
52. Vainio, M. J., Puranen, J. S., and Johnson, M. S. (2009). ShaEP: molecular overlay based on shape and electrostatic potential. *J Chem Inf Model 49*, 492–502.
53. Phase, version 3.1, Schrodinger, LLC, New York, N.Y., 2009.
54. Dixon, S. L., Smondyrev, A. M., Knoll, E. H., Rao, S. N., Shaw, D. E., and Friesner, R. A. (2006). PHASE: a new engine for pharmacophore perception, 3D QSAR model development, and 3D database screening: 1. Methodology and preliminary results. *J Comput Aided Mol Des 20*, 647–671.
55. Dixon, S. L., Smondyrev, A. M., and Rao, S. N. (2006). PHASE: a novel approach to pharmacophore modeling and 3D database searching. *Chem Biol Drug Des 67*, 370–372.
56. Venkatachalam, C. M., Jiang, X., Oldfield, T., and Waldman, M. (2003). LigandFit: a novel method for the shape-directed rapid docking of ligands to protein active sites. *J Mol Graph Model 21*, 289–307.
57. Molecular Operating Environment MOE 2009.10; C.C.G.I.M.; Quebec, Canada, 2008.
58. Maestro, version 9.0, Schro¨dinger, LLC, New York, N.Y., 2009.
59. DeLano, W. L. The PyMol Graphics Systems (2002), http://www.pymol.org.
60. Sybyl X, *Tripos International, 1699 South Hanley Rd.*, St. Louis, Mo.
61. VIDA 4.0, Openeye Scientific Software, Inc. Santa Fe, N.M., 2010.
62. *Discovery Studio Visualizer*, Accelyrs: San Diego, Calif., 2010.
63. Pettersen, E. F., Goddard, T. D., Huang, C. C., Couch, G. S., Greenblatt, D. M., Meng, E. C., and Ferrin, T. E. (2004). UCSF Chimera – a visualization system for exploratory research and analysis. *J Comput Chem 25*, 1605–1612.
64. Humphrey, W., Dalke, A., and Schulten, K. (1996). VMD: visual molecular dynamics. *J Mol Graph 14*, 33–38, 27–38.
65. Schwede, T., Kopp, J., Guex, N., and Peitsch, M. C. (2003). SWISS-MODEL: an automated protein homology-modeling server. *Nucleic Acids Res 31*, 3381–3385.
66. Krieger, E., Koraimann, G., and Vriend, G. (2002). Increasing the precision of comparative models with YASARA NOVA – a self-parameterizing force field. *Proteins 47*, 393–402.
67. Glide, version 5.5, Schrodinger, LLC, New York, NY, 2009.
68. Friesner, R. A., Banks, J. L., Murphy, R. B., Halgren, T. A., Klicic, J. J., Mainz, D. T., Repasky, M. P., Knoll, E. H., Shelley, M., Perry, J. K., et al. (2004). Glide: a new approach for rapid, accurate docking and scoring. 1. Method and assessment of docking accuracy. *J Med Chem 47*, 1739–1749.
69. Friesner, R. A., Murphy, R. B., Repasky, M. P., Frye, L. L., Greenwood, J. R., Halgren, T. A., Sanschagrin, P. C., and Mainz, D. T. (2006). Extra precision glide: docking and scoring incorporating a model of hydrophobic enclosure for protein-ligand complexes. *J Med Chem 49*, 6177–6196.
70. Halgren, T. A., Murphy, R. B., Friesner, R. A., Beard, H. S., Frye, L. L., Pollard, W. T., and Banks, J. L. (2004). Glide: a new approach for rapid, accurate docking and scoring. 2. Enrichment factors in database screening. *J Med Chem 47*, 1750–1759.
71. Fred, 2.2.3; OpenEye Scientific Software: Santa Fe, NM 2007.
72. McGann, M. R., Almond, H. R., Nicholls, A., Grant, J. A., and Brown, F. K. (2003). Gaussian docking functions. *Biopolymers 68*, 76–90.
73. Kellenberger, E., Rodrigo, J., Muller, P., and Rognan, D. (2004). Comparative evaluation of eight docking tools for docking and virtual screening accuracy. *Proteins 57*, 225–242.
74. Jain, A. N. (2003). Surflex: fully automatic flexible molecular docking using a molecular similarity-based search engine. *J Med Chem 46*, 499–511.
75. Ewing, T. J., Makino, S., Skillman, A. G., and Kuntz, I. D. (2001). DOCK 4.0: search strategies for automated molecular docking of flexible molecule databases. *J Comput Aided Mol Des 15*, 411–428.
76. Kuntz, I. D., Blaney, J. M., Oatley, S. J., Langridge, R., and Ferrin, T. E. (1982). A geometric approach to macromolecule-ligand interactions. *J Mol Biol 161*, 269–288.
77. Moustakas, D. T., Lang, P. T., Pegg, S., Pettersen, E., Kuntz, I. D., Brooijmans, N., and Rizzo, R. C. (2006). Development and validation of a modular, extensible docking program: DOCK 5. *J Comput Aided Mol Des 20*, 601–619.
78. Lang, P. T., Brozell, S. R., Mukherjee, S., Pettersen, E. F., Meng, E. C., Thomas, V., Rizzo, R. C., Case, D. A., James, T. L., and Kuntz, I. D. (2009). DOCK 6: combining techniques to model RNA-small molecule complexes. *RNA 15*, 1219–1230.
79. Jones, G., Willett, P., and Glen, R. C. (1995). Molecular recognition of receptor sites using a genetic algorithm with a description of desolvation. *J Mol Biol 245*, 43–53.
80. Jones, G., Willett, P., Glen, R. C., Leach, A. R., and Taylor, R. (1997). Development and validation of a genetic algorithm for flexible docking. *J Mol Biol 267*, 727–748.
81. Nissink, J. W., Murray, C., Hartshorn, M., Verdonk, M. L., Cole, J. C., and Taylor, R. (2002). A new test set for validating predictions of protein-ligand interaction. *Proteins 49*, 457–471.
82. Verdonk, M. L., Chessari, G., Cole, J. C., Hartshorn, M. J., Murray, C. W., Nissink, J. W., Taylor, R. D., and Taylor, R. (2005). Modeling water molecules in protein-ligand docking using GOLD. *J Med Chem 48*, 6504–6515.
83. Verdonk, M. L., Cole, J. C., Hartshorn, M. J., Murray, C. W., and Taylor, R. D. (2003). Improved protein-ligand docking using GOLD. *Proteins 52*, 609–623.
84. Hartshorn, M. J., Verdonk, M. L., Chessari, G., Brewerton, S. C., Mooij, W. T., Mortenson, P. N., and Murray, C. W. (2007). Diverse, high-quality test set for the validation of protein-ligand docking performance. *J Med Chem 50*, 726–741.
85. Cole, J. C., Nissink, J. W. N., and Taylor, R. (2005). Protein-ligand docking and virtual screening with GOLD. In Virtual Screening in Drug Discovery, B. K. Shoichet and J. Alvarez, eds. (Boca Raton, Fla.: CRC Press).
86. Meiler, J., and Baker, D. (2006). ROSETTALIGAND: protein-small molecule docking with full side-chain flexibility. *Proteins 65*, 538–548.
87. Davis, I. W., and Baker, D. (2009). RosettaLigand docking with full ligand and receptor flexibility. *J Mol Biol 385*, 381–392.

SECTION FOUR

CHEMICAL GENOMICS ASSAYS AND SCREENS

CHAPTER 12

Basics of HTS Assay Design and Optimization

Eduard A. Sergienko

High-throughput screening (HTS) is a production process aimed at testing tens of thousands of compounds per hour for the identification of a small subset of compounds that display a desired property, such as affecting a biological target of interest. As with any production process, the output of HTS depends profoundly on the quality of input materials, assay reagents and chemical compound libraries, as well as on the reliability of the process itself, a major component of which is a biological assay employed for testing compounds.

The utility of HTS assays depends on the correlation between biological properties of model systems, macromolecules or cellular pathways, and measurable biophysical properties gauging the effect of screened compounds on the HTS target of interest. Compounds with desired properties, identified as *actives* or *HTS hits*, need to be unequivocally distinguished from "inactive" compounds. Because of the number of compounds simultaneously tested in HTS, screening assays must be extremely reproducible and robust to provide statistically reliable results. Furthermore, HTS assays must be fine-tuned to enhance their ability to distinguish hits with desired pharmacologic properties apart from all other compounds. Achieving these generic and specific traits of HTS assays is described in detail in this chapter.

1. BASIC CONSIDERATIONS FOR HTS ASSAYS

The process of HTS embodies parallel testing of hundreds of thousands of samples wherein each compound is evaluated in a single replicate at a single concentration. A test for each compound can be considered as a separate biological experiment performed alongside a large number of similar experiments. By definition, experimentally determined values consist of two components, an actual value of the measured property and an experimental error associated with the measurement method, both of which are unknown a priori. Although a statistical approach for the determination of experimental errors from replicate measurements is not available for each sample in HTS, the degree of uncertainty in experimental measurements is readily extrapolated from control wells present in each plate.

The response typical to unaffected targets is exemplified by negative controls corresponding to the wells containing all assay components, with the exception of test compounds. Positive control wells exemplify the effect expected from the hits, which is usually achieved by adding a reference compound or omitting a critical assay component, such as an enzyme or substrate. In the majority of HTS projects, more than 99% of screened compounds are inactive in the assay, providing additional confirmation of assay performance. Dispersions of the values in compound and control wells are used to assess reliability of the HTS process and potential uncertainty in the screening results.

1.1. Assay Performance Measures

A number of parameters (commonly referred to as *assay performance measures*) are routinely used to ascertain assay performance. They are particularly helpful during the assay development stage, enabling a quantitative comparison of different screening approaches and specific screening conditions. During the HTS stage, these parameters provide quality control of the screening process and its products.

Signal-to-background ratio (S/B) determines the strength of a specific readout signal compared to its nonspecific component representing background (Figure 12.1). Background values provide an appropriate measure of the signal detected in the presence of a fully inhibited target. Assays with a high S/B allow more reliable culling of active compounds among their inactive counterparts. S/B is also referred to as the dynamic range of an assay. For most types of assays, an S/B of 10 or greater is desired in order to provide a reliable readout. A common exception to this rule is ratiometric assays, which can perform properly with an S/B of 5 or less because of their internal signal normalization, resulting in lower noise associated with the signal.

Another useful indicator of assay performance is *signal-to-noise ratio (S/N)*, which is defined as a signal normalized to the dispersion of the measured value. Although the shape of the dispersion distribution depends on the specific properties of the utilized experimental approach, Gaussian distribution provides an adequate approximation in most situations. It postulates that 99.8% of the repetitive

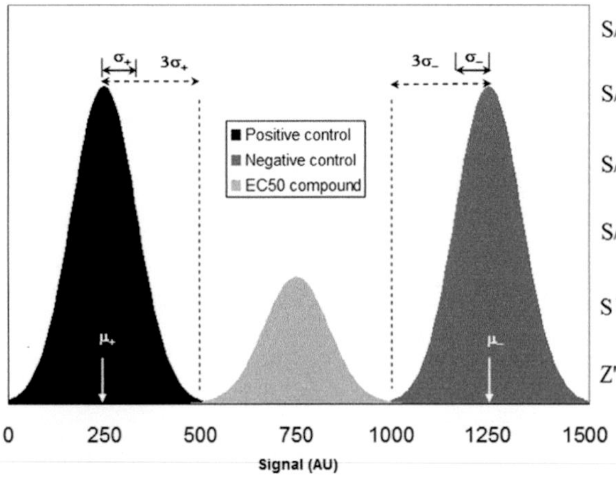

$$S/B = \frac{\mu_-}{\mu_+} = \frac{1250}{250} = 5$$

$$S/N_{(PC)} = \frac{\mu_+}{\sigma_+} = \frac{250}{83.3} = 3$$

$$S/N_{(NC)} = \frac{\mu_-}{\sigma_-} = \frac{1250}{83.3} = 15$$

$$S/N_{(assay)} = \frac{|\mu_+ - \mu_-|}{\sqrt{\sigma_+^2 + \sigma_-^2}} = 8.5$$

$$SW = \frac{\mu_- - \mu_+ - 3(\sigma_- - \sigma_+)}{\sigma_-} = 6$$

$$Z' = 1 - \frac{3 \cdot \sigma_+ + 3 \cdot \sigma_-}{|\mu_+ - \mu_-|} = 0.5$$

FIGURE 12.1 Comparison of assay performance measures.

measurements fall within a range defined by 3 standard deviations (σ) on either side of the mean value (μ) determined from the experiment. The S/N of an assay is calculated as the distance between positive and negative controls divided by the Pythagorean sum of their standard deviations. Assays with higher S/N facilitate the discrimination of the effects of active compounds from inherent experimental noise. In most types of assays, an S/N of 10 or greater is preferred. The inverse value of S/N is another valuable and commonly used assay descriptor, the *coefficient of variation (CV)*, which expresses the statistically uncertain percentage of the measured signal.

Both S/B and S/N describe distinct yet complementary properties of HTS assays. As with ratiometric assays, a low S/B can be compensated for, to some degree, by a high S/N value and vice versa. The S/B parameter provides an adequate forecast only for assays with low signal dispersion. In contrast, assays with high measurement noise are better assessed using the S/N parameter. The drawback of these parameters is that neither one alone provides the full scope of information regarding assay performance. In most cases, the knowledge of one without the other is insufficient to judge if an assay is ready for screening.

To address the necessity of a universal measure of HTS assay performance independent of assay specifics, a *signal window (SW)* parameter was proposed by a group from Eli Lilly in 1997 [1]. This parameter expressed the window between the signal and background in units of signal dispersion. Two years later, a team from DuPont introduced the term *Z′ factor* [2], a different parameter with a similar meaning that became the most commonly used measurement of HTS assay performance. SW and Z′ factor are composite parameters and are easily expressed in terms of S/B and S/N, combining the benefits of both parameters. All of the above-mentioned parameters are illustrated in Figure 12.1.

The more prevalent use of Z′ factor can probably be attributed to a more intuitive interpretation of the range of its values [3]. Z′ factor can take on any value below 1. Its upper limit corresponds to an assay with either no noise or an infinite assay window. Although this value is unattainable, it sets up a clear reference point. The higher the Z′-factor value and the closer it is to 1, the more robust the assay. Well-behaved assays are expected to have a Z′ factor of at least 0.5. This value corresponds to the situation illustrated in Figure 12.1, where positive and negative controls as well as an inhibitor with 50% inhibition give clearly separated measurement dispersion ranges with less than 0.1% overlap.

1.2. Selection of Active Compounds

Assay *actives*, also called HTS positives or hits, are selected using cutoff criteria formulated according to specific goals of a particular HTS project. These criteria take into account a compound's efficacy and other assessable properties acquired in the screening; the latter usually helps to establish that a compound is not a false positive that interferes with the detection system employed in an assay. If positive and negative controls of an assay are available, a common way of analyzing the HTS data is to express the response to tested compounds as a percentage of the response observed in control wells. In this case, the primary cutoff for hit selection is based on a subjectively selected response percentage.

In some cases, a positive control is not available, as frequently encountered in assays aimed at finding activators. In these special cases, one may express a compound's response in units of dispersion observed within negative control or compound wells. This numerical representation of the data is known as a *Z-score*; it provides an appraisal of the reliability of the results with respect to the noise of an assay [4]. Hits are selected using a certain Z-score value cutoff. For example, the EC_{50} compound shown in Figure 12.1 would have a Z-score of greater than 3 in 99.9% of independent tests.

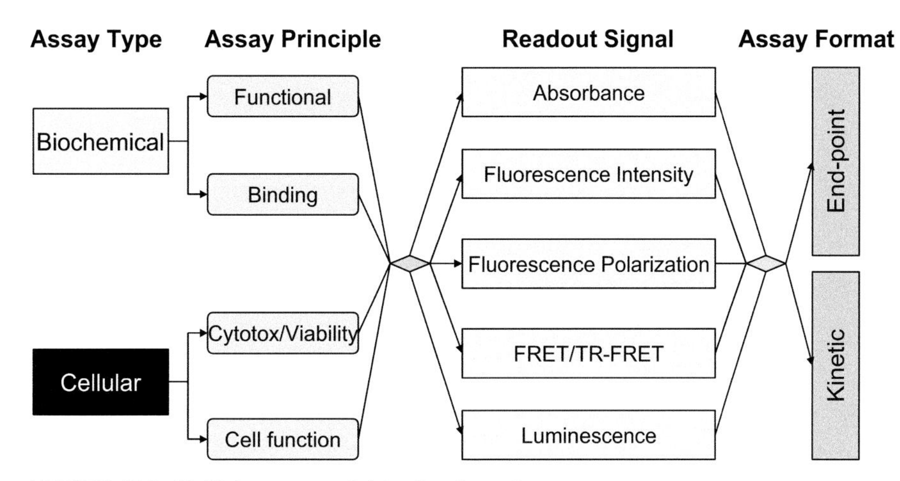

***FIGURE 12.2** Multiple assay and detection formats.*

During the assay development stage, the aforementioned assay performance parameters are used to judge if an assay is ready for HTS – in other words, to predict and evaluate how reliable or precise the screening data would be. Precision in determining the effect of each individual compound and ranking its effect relative to other compounds and controls is imperative because during HTS, each compound is tested in a single well without replicates. However, these performance parameters do not substantiate an assay's sensitivity, that is, its ability to detect HTS hits and provide an accurate quantification of their biological effect. The main principles of assay development that help to optimize an assay's sensitivity are described in the following sections.

2. LINKING BIOLOGY AND ASSAY READOUT

The first and paramount principle of each assay development effort is to ensure that the assay's results, which are acquired in the form of biophysical measurements, are directly correlated to the biology of interest. This principle is especially prone to being overlooked during the process of selecting an appropriate detection method for a biological event. The majority of biological events can be interrogated through multiple readout points and detection approaches. A search for an easily measured output occasionally leads to an unintentional disengagement between the biology and the signal, particularly for newly discovered and not-fully-understood biological targets and phenomena.

Potential pitfalls are illustrated in the example of selecting an assay readout for a phosphorylation event. Available approaches include using antibodies for detecting a phosphorylated substrate in biochemical or cellular reactions or employing a reporter gene assay to monitor a physiologic downstream event such as protein expression. Each of these approaches has its own strengths and weaknesses. Aberrant results may arise if the biology-to-signal correlation is not closely guarded. For example, detection antibodies could be unspecific and interact with other cellular proteins, especially if these are present at a higher concentration. Furthermore, a single post-translational modification site on the substrate could be modified by multiple upstream proteins. Reporter genes could also give misleading results because protein expression may be controlled by multiple pathways or regulated differently in cells with dissimilar backgrounds, especially if an overexpressing reporter gene is utilized.

To complicate matters further, some enzymes exert part of their cellular functions through interactions with other proteins rather than through catalytic activity. Thus, RNAi or gene knockout experiments for enzymes may be misinterpreted as proving the significance of enzyme activity rather than their protein expression levels. If the protein's primary involvement in a biological process is to serve as a binding scaffold for intracellular partners, targeting its catalytic activity through a biochemical assay would lead to a discovery of compounds ineffective in characterizing and targeting the primary binding function. Although one cannot predict all of these problems, a reasonable effort should be made to recognize and counteract them whenever possible. Development and application of multiple assays with independent readouts provide a good strategy for confirming the desired effect of HTS hits.

3. CLASSIFICATION OF HTS ASSAYS

The two main types of HTS assays are biochemical and cell-based assays. Other methods of classifying assays are based on 1) whether the functional or binding properties of a specific target are interrogated, 2) selected detection approaches, and 3) whether the signal development is monitored over a period of time or measured once (Figure 12.2).

3.1. Biochemical and Cell-Based Assays

In biochemical assays, biological macromolecules are tested outside their intracellular environment, whereas in cell-based assays, the effect of compounds is tested on live cells. Both of these approaches have clear advantages and disadvantages that should be considered when selecting an appropriate assay for a specific HTS campaign.

Biochemical assays that employ purified proteins provide a direct readout of a single selected function of the protein target. However, a low concentration of purified proteins makes them susceptible to denaturation and oxidation, leading to a loss of their native state and, thus, their biological function. Nonspecific interactions with plate materials and other assay components further facilitate these aberrant processes.

In cell-based assays, proteins are tested in a physiologic environment that is composed of intricate networks of macromolecular interactions and metabolic pathways. Cellular membranes provide a natural barrier for compounds, thus enabling the specific identification of physiologically relevant compounds. Unfortunately, membrane impermeability complicates the interpretation of screening data because affected compounds appear to be inactive in the assay, irrespective of their actual efficacy on the target. A further complication arises because cellular pathways represent a highly integrated network of interwoven processes, so multiple targets may result in similar cellular responses.

3.2. Alternative Assays

In the majority of cases, each biological event can be gauged via multiple detection approaches. For example, the phosphorylation of a protein visualized through an enzyme-linked immunosorbent assay (ELISA) can be based on absorbance, fluorescence, or luminescence. Major assay types and readouts are summarized in Figure 12.2. As mentioned in the previous section, all approaches have their advantages and disadvantages. Careful consideration of the approach in the context of the goals for a specific HTS project is an important part of selecting the appropriate detection for a particular target and compound library of interest. Examples in the next two paragraphs illustrate this concept.

Protein-peptide binding can be detected using homogeneous assays such as fluorescence polarization (FP) and time-resolved fluorescence resonance energy transfer (TR-FRET) or heterogeneous approaches such as ELISA. Homogeneous assays offer classical equilibrium-binding conditions but are affected by a compound's physicochemical properties, such as absorbance, fluorescence, and solubility. Heterogeneous assays are governed by surface adsorption phenomena that are disparate from equilibrium binding. On the other hand, heterogeneous assays with washing steps are immune to the optical properties of compounds that need to be taken in consideration when selecting an assay.

Fluorescence intensity (FI) or fluorescence resonance energy transfer (FRET) assays that are based on variants of green fluorescent protein (GFP) provide inexpensive and robust HTS platforms. Unfortunately, low wavelengths of fluorescent excitation and emission, broad spectra, and low extinction coefficients of GFP proteins lead to significant optical interference from fluorescent and colored compounds, or components of cell culture media. Specifics of the assay and compound library collection employed for screening may render a GFP-based assay inefficient, requiring laborious assay protocols and multiple follow-up assays to weed out artifacts. Substituting GFP with firefly luciferase in reporter gene assays improves sensitivity and decreases optical interference, as is expected with a luminescent readout. Unfortunately, this substitution also confers sensitivity to luciferase inhibitors, which comprise up to 3 percent of some small molecule libraries [5], and appear as false positives with nM and low-μM potencies. Intriguingly, some additional compounds exert their effect on firefly luciferase by affecting its stability or folding, which will invariably influence the HTS results of reporter gene assays employing this protein [6].

4. OPTIMIZATION OF ASSAY COMPONENTS

Most assays contain multiple components that affect their sensitivity, the degree of which is contingent on the exact role each component plays in the assay. Among assay components, those that participate in the interactions and/or enzymatic reaction should be distinguished from those that create the necessary environment for interactions, and therefore, that affect the interactions indirectly and nonspecifically. The first group is called *principal components*, whereas the second is called *supporting components*. Principal components include target proteins and their various ligands, such as substrates, co-factors, and essential ions. Supporting components are buffer ions, salts, and other additives, such as detergents and nonspecific proteins.

4.1. Interactions of Principal Components

Irrespective of their type or readout, all assays share common fundamental principles. For one, they all are based on detecting binding events between tested compounds and their intended target. The majority of HTS assays conform to dynamic equilibrium binding. Figure 12.3 A illustrates interactions between components of a reversible binary complex. It is common practice to designate a macromolecule as a *receptor (R)* and to name its binding counterpart as a *ligand (L)*. Their total concentrations in an assay are designated as $[R]_t$ and $[L]_t$, respectively. Both are distributed between free (R or L) and bound (RL) states. In equilibrium, the rates of forward and reverse reactions are equal. The law of mass action leads to the Langmuir equation [7], shown in Figure 12.3A. It is worth noting that this equation describes the interaction of a receptor with any single binding partner: ligand, substrate, small molecule inhibitor, or activator. Normally, the concentration of the

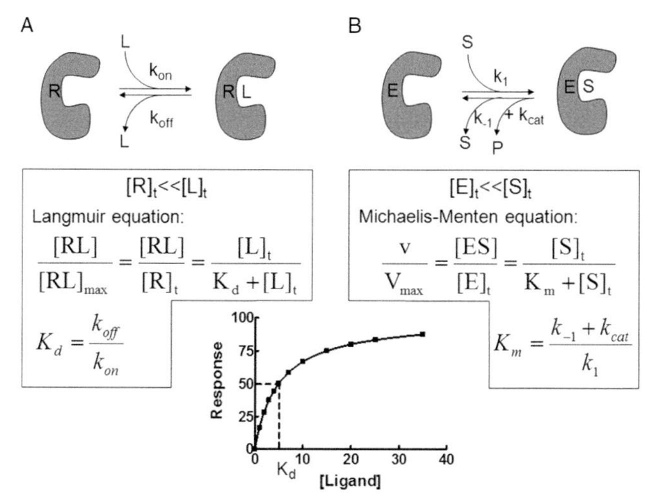

FIGURE 12.3 Phenomenological comparison of binding and enzyme assays.

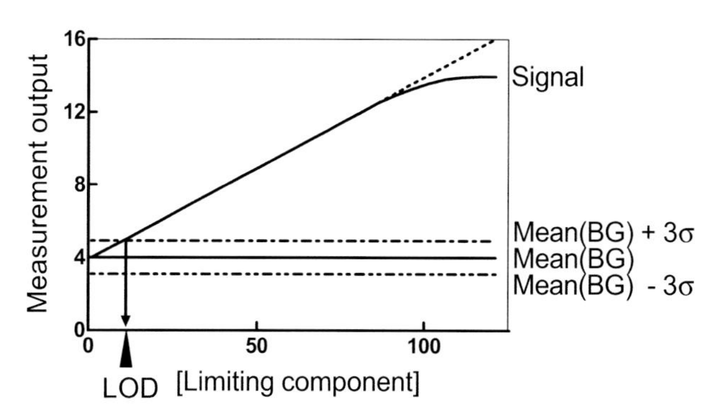

FIGURE 12.4 Effect of limiting component on assay signal.

receptor, and thus the receptor-ligand (RL) complex as well, is much lower than its ligand concentration, $[R]_t \ll [L]_t$. This condition permits the performance of the following reduction in the ligand law of mass action equation, $[L]_t = [L] + [RL] \approx [L]$, and the substitution of unbound ligand concentrations [L] with its total concentration $[L]_t$. A deviation from this scenario and its consequences in the HTS applications are discussed in section 5.3.

The readout of an HTS assay is selected to correlate directly with the degree of the target's occupancy with the tested compounds. As seen in the text box in Figure 12.3A, the ratio of [RL] over $[R]_t$, known as *fractional saturation*, is constant at a fixed $[L]_t$, leading to linear [RL] versus $[R]_t$ calibration curves. Any measurable property of [RL] that distinguishes it from [R] provides an adequate assay readout for monitoring fractional saturation. On some occasions, the assay consists of an unbound receptor and is based on changes in receptor properties upon the binding of small molecules. In this case, the equation in Figure 12.3A describes the small molecule binding with [RL] corresponding to the receptor–small molecule complex concentration. Most commonly, the assay involves an additional ligand that interacts with the receptor. The compound's binding to the receptor is then detected through the changes in receptor-ligand complex concentration [RL]. An interaction within the receptor-ligand complex is described in Figure 12.3.A. The presence of a small molecule compound does not change the equation describing receptor and ligand binding, although it does affect the values of the equation's parameters, as demonstrated in section 4.3.1.

In enzymatic reactions, the measured property is enzyme activity. Similar to binding processes, most enzymatic assays utilized in HTS have the characteristics of a dynamic quasi-equilibrium, commonly known as steady-state, for which the concentrations of each receptor-enzyme species are practically unchanged with time. One major prerequisite for this is that an enzyme's concentration must be much lower than the concentration of its substrates. The simplest steady-state enzymatic reaction with a one-substrate-binding step is illustrated in Figure 12.3B and is described by the Michaelis-Menten equation. Similar to receptor-ligand complex concentrations in binding assays, the rate of an enzymatic reaction is proportional to the concentration of enzyme in the assay. Interactions between the enzyme and small molecule effectors are measured indirectly through changes incurred in the concentration of ES complex, directly proportional to the enzymatic activity in the assay. Thus, both enzymatic and binding assays are described by almost identical equations and consequently are optimized in a very similar manner, as described in section 4.2.

4.2. Limiting Assay Components

The principal component that is present at lower concentrations than other reagents represents the limiting component of the assay. The signal measured in the assay is directly proportional to the concentration of the limiting component which determines an assay's sensitivity and performance. In most biochemical assays, the limiting component is either a receptor or an enzyme. In many cell-based assays, the readout is proportional to the number of cells, such as in the case of detecting cellular adenosine triphosphate (ATP) or expression of a reporter protein, which makes cell density the assay's limiting component.

4.2.1. Limits of Detection and Quantification As with any quantification approach, the outcome of a biological assay is constrained, both on the lower and upper boundaries of the detection range (Figure 12.4). At the lower end, its signal is limited by the background value and the precision of its assessment. The background value is estimated from replicate samples in the absence of the limiting component. Any signal that is within the dispersion of the background measurements is quantitatively indistinguishable from the background. It is customary to use three standard deviations (3σ) to set the threshold of a reliable separation between the signal and the background of the assay. In Figure 12.4, the range of 3σ above and below the background mean-value line is indicated by the dashed lines. The *limit of detection (LOD)* is defined as the lowest concentration of the limiting component that can be detected in an assay and is calculated as 3σ divided by the slope of the linear portion of the curve. One should note that the reliability of the standard deviation of the background should be comparable to that of the slope of the linear curve; thus, roughly the same number of data points should be utilized for the determination of background values and concentration-dependent titration curves. For optimum results, the concentration of the limiting component selected for screening should be at least ten-fold above the LOD value. This would ensure that at least 90% of inhibition is quantifiable and directly correlates with the detected signal.

The shape of the calibration curve at low concentrations of the limiting component can be used as an indicator of potential artifacts. Lag phases observed in the curves may be a result of nonspecific protein binding to the plate surfaces or the dissociation of functional oligomers into inactive monomers. In biochemical assays, supplementing a buffer with stabilizing reagents, nonspecific protein, or detergent additives often ameliorates artifacts and helps to "straighten" the curves. Lag phases observed in cell-based assays may indicate a requirement for cell-to-cell interactions, thus maintaining a certain level of confluency. Adjusting the serum level or adding other components to the media may reduce dependence on intercellular interactions.

Above a certain concentration of the limiting component, the upper portion of the calibration curve often deviates from linearity, as illustrated in Figure 12.4. This concentration defines the *upper limit of quantification (ULOQ)*. Nonlinearity above the ULOQ is usually caused by limitations in either assay components or an instrument's settings. In binding assays, this observation could be indicative of the limiting component being present at concentrations equal to or higher than some other principal reagents, leading to a change of the limiting component. It may also result from changes in protein properties due to protein aggregation at higher concentrations. The most common limitation in an enzymatic assay is caused by an overabundance of the enzyme, which results in the consumption of a significant amount of substrate and concomitant accumulation of product, thus deviating from steady-state postulates. The utilization of limiting component concentrations outside the linearity range significantly lowers the sensitivity of an assay and, if possible, should be avoided. The ratio between the ULOQ and LOD represents an assay's maximal *dynamic range*. In practice, the full dynamic range is rarely utilized in assays, as some portion of it is reserved to cushion day-to-day variations in the reagents' concentrations or other assay conditions, such as ambient temperature.

4.2.2. Assays for Identification of Activators and Inhibitors An assay that is intended for the identification of both inhibitors and activators in a single HTS test deserves special consideration. This type of assay must be able to measure an activity (or binding) as low as 5% to 10% and as high as 200% to 400% and, therefore, requires a dynamic range of greater than twenty-fold. Fluorescent and luminescent assays are characterized by extended dynamic ranges and, thus, are appropriate for the simultaneous identification of inhibitors and activators. One of the screens performed by the *Conrad Prebys Center for Chemical Genomics (CPCCG*; formerly known as the Burnham Center for Chemical Genomics, BCCG) represents an example of the successful application of an assay used to identify both inhibitors and activators. A luminescent assay developed for alkaline phosphatases was characterized by dynamic range of greater than 100-fold. Both inhibitors and activators were identified in a single HTS campaign. This campaign is utilized as a case study in this chapter and is described in more detail in section 6. These data can be found in the NCBI PubChem database (http://pubchem.ncbi.nlm.nih.gov/) under *PubChem Assay IDs (AIDs)* 518 and 813. In contrast, the colorimetric assay commonly utilized for alkaline phosphatases has less than ten-fold dynamic range, which is incompatible with the simultaneous detection of inhibitors and activators. When the development of a dual-purpose assay is not feasible, two separate assay conditions are employed for performing inhibition and activation studies.

4.2.3. Time as a Component of Enzymatic and Binding Assays One major difference between enzymatic and binding assays is their sensitivity to time. In enzymatic assays, incubation time defines the concentration of product formed in the reaction. Because the product concentration of a steady-state kinetic reaction is proportional to the length of incubation, the time of the reaction can be considered an auxiliary limiting component of enzyme assays, and is characterized with a similar quantification range.

In earlier stages of assay development, enzymatic assays are best studied using a kinetic mode in which progress of the reaction is monitored over time. This approach allows for the reliable determination of initial rates of progress curves and the detection of any aberrant behavior. Deviation from linearity at the initial stages of a reaction frequently results from either enzyme activation or an insufficient capacity of coupled assays utilized for detection. Both conditions are easily confirmed and corrected. Near-ULOQ nonlinearity most frequently is a result of the depletion of the substrate pool and product accumulation, which violates steady-state conditions and decreases the rate of further product formation. The linearity on the time scale can be extended by manipulating enzyme and/or substrate concentrations. Once kinetic assays are optimized, they are easily converted into an end-point format, which is preferred for HTS, by fixing the time component to an optimal value.

Binding assays are much less sensitive to time. In fast equilibrium processes, the length of incubation is irrelevant, because no changes in the distribution of components between binding states are expected or observed. Some biological interactions demonstrate relatively slow kinetics of binding and/or dissociation and would require an extended incubation to reach equilibrium. Because a lengthy incubation can potentially promote the denaturation and aggregation of proteins, which leads to a deviation from equilibrium, the acquisition of kinetic data at the earlier stages of binding assay development is beneficial in allowing the range of optimal incubation time to be established.

4.3. Nonlimiting Assay Components

In most assays, the interaction of a ligand (L) with a receptor (R) and the formation of their complex (RL) provides a basis for the visualization and quantification of small molecule effector (*I*) binding. Any detectable property that distinguishes the complex RL from unbound R or L species can be utilized to monitor the extent of compound binding, provided that the interaction between R and I affects the equilibrium between RL and its unbound species. The binding partner whose property is used for detection is utilized as the limiting component of the assay. As described previously, a receptor or enzyme most frequently serves as the limiting component of an assay. Their ligands are present at higher concentrations and participate as nonlimiting components. Optimization of the concentration of nonlimiting components and considerations that govern this process are described in section 4.3.1.

4.3.1. Modes of Inhibition and Their Effect on Nonlimiting Component For simplicity, this section focuses on optimizing assays for the identification of inhibitors, in which either an enzyme or a receptor serves as the limiting component. It is also assumed that both the ligand (L) and the inhibitor (I) directly interact with the enzyme or the receptor (R). A rectangular hyperbolic equation (Equation 12.1) describes the generic inhibitory effect of compound (I) exercised over receptor-ligand (RL) complex concentration:

$$[RL] = \frac{[R][L]_t}{K_d} = \frac{[RL]_{\max} EC_{50}}{EC_{50} + [I]_t} \qquad (12.1)$$

where $[L]_t$ and K_d are the ligand L total concentration and dissociation constant, respectively. [R], [RL], and $[RL]_{max}$ corresponds to the concentration of free receptor and receptor-ligand complex in the presence and absence of I, respectively. $[I]_t$ is the total inhibitor concentration in the assay. EC_{50} is an apparent dissociation constant of the receptor-inhibitor complex.

The effect of two ligands (L and I) on each other's affinity is defined by their interactions with the receptor and one another. There are three major modes in which two ligands can influence each other's binding. They are described by three models that are illustrated in Figure 12.5 and are collectively referred to as the *mechanism of action (MOA)*, also known as the mode of inhibition (MOI) in the case of inhibitors. Equation 12.1 is valid for all of the models, although the equations defining the EC_{50} parameter and its correlation with the inhibitor's (I) *equilibrium dissociation constant (K_i)* are MOA-specific.

The most frequent biligand system is characterized by both ligands binding in a mutually exclusive manner, competing with one another for a binding site (Figure 12.5A) or a binding conformation. These competitive systems are described by Equation 12.2:

$$EC_{50} = K_i \times \left(1 + \frac{[L]_t}{K_d}\right) \qquad (12.2)$$

which corresponds to the Cheng-Prussoff equation [8]. A noncompetitive system (Figure 12.5B) in which two ligands bind independently of one another is described by Equation 12.3:

$$EC_{50} = K_i \qquad (12.3)$$

Finally, an uncompetitive system (Figure 12.5C), in which binding of one ligand serves as a prerequisite for the binding of the second ligand, is illustrated by Equation 12.4:

$$EC_{50} = K_i \times \left(1 + \frac{K_d}{[L]_t}\right) \qquad (12.4)$$

A system in which two ligands affect each other's binding but do not have exclusivity is called a mixed inhibition system. Its properties are intermediate between competitive and uncompetitive systems, with one prevailing over the other depending on the values of the ligand *interaction coefficient*, α (Figure 12.5B). Thus, the conclusions obtained herein may be easily extended into mixed-inhibition systems as well. As mentioned in Section 4.1, the models and equations depicted in Figure 12.5 are applicable to enzymatic reactions after replacing K_d with K_m and $[L]_t$ with $[S]_t$. They are also consistent with activators and partial inhibitors. Minor modifications to Equation 12.1 will ensure its universality as well.

4.3.2. Nonlimiting Component Concentrations Allow for the Selection of MOA As demonstrated in section 4.3.1, the EC_{50} for all of the biligand systems is directly proportional to the dissociation constant K_i of the ligand I. Its correlation with the concentration of ligand L and its dissociation constant K_d (or K_m for enzymatic reactions) depends on the MOA. Thus, selection of the ligand L concentration in an assay can directly affect the MOA of hits identified in the assay. It is easy to deduce

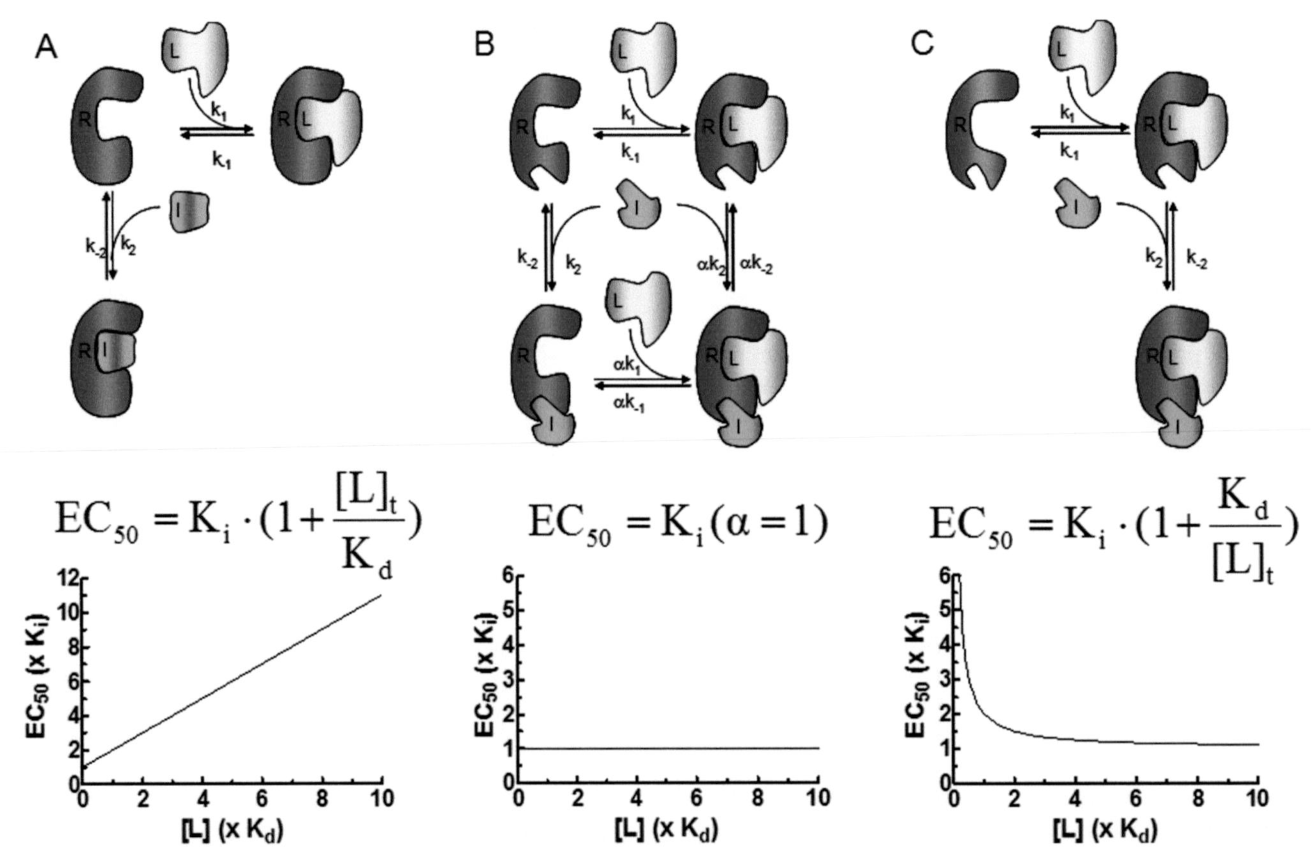

FIGURE 12.5 Three major biligand systems corresponding to the following mechanisms of action: A. Competitive. B. Noncompetitive. C. Uncompetitive.

that an assay aimed at detecting inhibitors with diverse MOA should be based on the concentration of nonlimiting component, ligand L, equal to its K_d or K_m, which is the most common way of performing HTS assays. In this case, the EC_{50} is expected to be within one- to two-fold of the inhibitor's K_i values for any type of inhibitor. In an alkaline phosphatase screening performed at CPCCG that was mentioned in section 4.2.2, concentration of its substrates was kept at their K_m values, leading to the identification of several scaffolds with different MOAs. In contrast, a colorimetric assay for alkaline phosphatases routinely performed in clinical labs employed the substrate at ten times its K_m value and was able to identify only one scaffold with a noncompetitive MOA [9].

The optimization of the principal ligand or substrate concentration allows the tailoring of assays to be more selective for a specific MOA. For example, if the preferred method of targeting a receptor or enzyme is uncompetitive, performing the assay at the concentration of ligand equal to K_m (or apparent K_d) value may not be the best approach, as the assay will also have the same sensitivity to competitive inhibitors. To ensure that competitive inhibitors are not detected in the assay, the concentration of the ligand L needs to be raised to the highest concentration that is practically acceptable. Similarly, in more complex systems in which agonists, antagonists, and potentiators are feasible, performing the binding assay in the absence and presence of a subsaturating concentration of principal ligand (for example, EC_{20}) allows for the identification of agonists and potentiators, respectively. Although utilization of concentrations closer to saturation (for example, EC_{80}) provides a reasonable platform for the observation of antagonists, oversaturation with the principal ligand further tailors the assay for the identification of allosteric modulators.

4.3.3. Mechanistic HTS Approaches Performing parallel HTS campaigns in the presence of different concentrations of principal ligand, which is referred to as *mechanistic HTS (mHTS)*, is especially beneficial and cost-effective if a very high HTS hit rate is expected and the desired mode of action is known. In addition to gaining information on the MOA of HTS hits, single-concentration confirmation is also achieved. An mHTS method facilitates the early selection of hit classes with required properties and thus saves effort and resources that otherwise would have been wasted on working with undesired hit classes. If impractical at the primary HTS stage, this approach can be applied at the single-concentration or dose-dependent stage of hit confirmation.

An example of this method was successfully utilized for the biochemical screening of human phosphomannose isomerase (PMI) at the CPCCG. The PMI enzyme serves as a link between glycolysis and glycosylation, interconverting mannose-6-phosphate and fructose-6-phosphate. In congenital disorder of glycosylation (GDG) Ib patients, this enzyme is compromised, leading to a deficiency in glycosylation and an insufficient mannose-6-phosphate supply. High mannose dietary supplement allows for an improvement of the condition through a glycolysis-independent substrate source. Patients with another glycosylation disorder, GDG Ia, have insufficient activity of a downstream enzyme, phosphomannomutase 2, which is responsible for converting mannose-6-phosphate into mannose-1-phosphate. This condition leads to deficiency in glycosylation, because PMI shunts the accumulated mannose-6-phosphate toward glycolysis, even in the presence of a high-mannose diet. Inhibitors of PMI can be useful in conjunction with a high-mannose diet. Under these conditions, PMI is expected to operate in the presence of saturating concentrations of its in vivo substrate mannose-6-phosphate; thus, any useful modulator of enzyme activity would need to be non- or uncompetitive with respect to the substrate. To help in the identification of a scaffold with the specific MOA, a primary HTS was performed both in the presence of near-K_m (PubChem AID 1209) and near-saturating (PubChem AID 1220) concentrations of mannose-6-phosphate substrate. Several chemical series with desired properties were identified and optimized; this work ultimately led to the identification of two small molecule classes active in cell-based disease models (PubChem AID 1545).

4.4. Supporting Assay Components

In biochemical assays, supporting components, such as buffer ions, salts, detergents, and nonspecific proteins, are also important for optimal assay performance. It is well known that a high salt concentration in a buffer may interfere with assay performance via nonspecific electrostatic competition with the principal components. Alternatively, if hydrophobic interactions provide large contributions to the binding affinity of a ligand, the presence of moderate to high concentrations of salt in the buffer may improve ligand binding.

Buffer ions also affect assay performance through either ionic strength or direct binding to assay components. For instance, citric acid–based buffers are infamous for their high ionic strength and metal-complexing properties. Replacing citrate with zwitterionic buffers can help to avert both of these undesired effects. Phosphate-based buffers are known to compete for binding with phosphate-containing ligands or substrates. These buffers often decrease the apparent affinity of such ligands, including the test compounds targeting their binding sites. In addition, phosphates are known to form complexes or precipitates with polyvalent metal ions, which decreases the effective concentration of the latter.

Some proteins display optimal ligand binding or activity in nonphysiologic conditions, such as alkaline or acidic pH. Although it is preferred to keep a screening assay as close to physiologic conditions as possible, this may not always be feasible or practical. An acceptable approach would be performing HTS in conditions that are optimal for enzyme activity or ligand binding if these conditions ensure that minimal protein or enzyme concentrations are maintained, thus allowing them to serve as the limiting component of the assay. The physiologic relevance of identified HTS hits must be confirmed in appropriate follow-up assays.

Testing several buffers of diverse nature, but with the same pH and ionic strength, allows a preferred buffer composition to be defined. If the optimal pH of the target is unknown, it can be determined in the presence of an appropriate buffer mixture with a constant composition and ionic strength [10]. Because V_{max} and K_m of enzymatic reactions frequently have different pH-optimum ranges, the best approach may be to perform Michaelis-Menten studies at different pH values to identify desired conditions.

The optimization of media in cell-based assays is not as intuitive as buffer optimization in biochemical assays, but nevertheless, is just as significant. For example, fluorescence-based detection in cell-based assays benefits from using media without phenol red indicator, the dye known to absorb light and produce fluorescence. In addition, lowering the concentration of serum and riboflavin can also help to decrease fluorescence background. This could be a valuable improvement for an assay and may help in eliminating wash steps.

Selecting plates that are consistent with the readout signal also ensures optimal assay sensitivity. Black plates decrease the amount of reflected and scattered light and are crucial for decreasing background in fluorescent assays. Most luminescent assays require white plates that help to gather the light, a major part of which is initially directed away from a light detector. A luminescent assay with a high signal may benefit from using black plates because of a significant decrease in background and, more importantly, from eliminating interwell cross talk. Though it may seem counterintuitive, the utilization of white plates in time-resolved fluorescence assays actually confers high sensitivity. This is because these assays utilize time-gated data acquisition. Background fluorescence decays prior to signal acquisition and, thus, is not a major concern in these assays, whereas white plates accumulate the relevant light signal and increase assay sensitivity.

Supporting assay components can significantly change assay behavior; therefore, after buffer or media optimization, it is recommended to confirm and, if necessary, adjust other assay parameters.

4.5. Considerations for Signal and Reagent Stability

Signal stability is an important factor for HTS performance. It defines how easily the assay can be automated and scheduled on screening robotic systems. Signal stability must be assessed and, if possible, improved during the assay development stage. Common reasons for signal deterioration include a reagent's light sensitivity, hydrolysis, or denaturation.

Steady-state enzymatic reactions by nature have signals that change over time in a linear fashion. Thus, the instability of their signal is exhibited as a deviation from the linearity of progress curves. In some assays, stopping the progress of an enzymatic reaction is strongly desired in order to help with HTS scheduling and to ensure uniform incubation time for all screened samples. This can easily be achieved by employing a pH shift or by adding an inhibitor. However, it is important to confirm that a utilized approach does not undermine the measured signal. For instance, ethylenediaminetetraacetic acid (EDTA) is frequently used to stop a kinase reaction by chelating Mg^{2+}, which is essential for the reaction. Yet EDTA is also known to form complexes with lanthanide cations, adversely affecting some TR-FRET detection reagents.

Another determinant of an assay is the stability of the reagents used for HTS. During the assay development stage, a reagent's stability should be tested, and if necessary, the conditions should be altered to ensure the extended stability of reagents. This may be achieved by protecting the reagent from light with foil or from ambient temperature by keeping it on ice. In addition, the composition of each reagent in multicomponent assays should be optimized to achieve the maximal compatibility of the components. Substrates of enzymatic reactions should be kept in the presence of other components that enhance their stability and prevent their conversion into products. Exposing principal limiting components such as enzymes or proteins to other components that could slowly change their activity or stability should also be avoided.

5. STRUCTURE-ACTIVITY RELATIONSHIP STUDIES

HTS hits identified in primary screening rarely possess all of the desired properties of a probe or investigative new drug. In most projects, they need further optimization, which is achieved through a series of iterative structural alterations and biological tests aimed at obtaining a compound with superior properties. This work is cumulatively called *structure-activity relationship (SAR)* studies. This stage of lead validation and optimization is a highly collaborative and integrated process, epitomizing the quintessence of chemical biology. Representatives of three disciplines, biology, medicinal chemistry, and cheminformatics, fulfill different yet equally important functions within the SAR process. Medicinal chemists synthesize compounds relying on guidance provided by biologists in the form of SAR data and on assistance from cheminformaticians in the interpretation of these data.

The most common criteria utilized in SAR studies are a compound's potency, selectivity, specificity, and mode of action. Potency is tested in specially optimized or designed HTS assays in which the effect of a compound is assessed as a function of its concentration. Information on selectivity and specificity is usually obtained through testing compounds in a panel of diverse dose-response assays. These panel assays are composed of a matrix combining multiple detection systems and targets aimed at separating artifacts from true hits, while quantifying the degree of a compound's specificity. Needless to say, the accuracy of the biological data and concomitant ability to distinguish and quantitatively rank compounds with close potencies are paramount to these activities. This section considers factors affecting the accuracy of SAR data.

5.1. Differences Between SAR and HTS Studies

To a major degree, the accuracy of SAR data is defined by biological assays used to generate the data. There are subtle differences in the development of assays tailored for HTS and SAR stages that reflect the dissimilarities of primary HTS and SAR approaches. During primary HTS, only a single concentration of a compound is tested (Figure 12.6A); normally, compounds are present in the micromolar range. Thus, a compound's effect needs to be quantifiable and reasonably accurate only at a single, sufficiently high concentration value. In SAR studies, compounds are tested at multiple concentrations, resulting in apparent binding curves, as shown in Figure 12.6B. Dose-response curves are fitted to equations describing the binding phenomena to arrive at the set of parameters describing them. The most common parameter utilized for the purposes of guiding the SAR is the EC_{50} described in section 4.3.1.

As discussed in the section 4.3.1, EC_{50} values rarely correspond to the thermodynamic dissociation constant, K_i. Nonetheless, correlation between EC_{50} and K_i values is easily deduced. Dose-response inhibition curves obtained in the presence of various substrate or ligand concentrations provide an array of EC_{50} values. Analysis of the data against the substrate or ligand concentration provides an indication of MOA and an extrapolated value of K_i.

5.2. Utilization of the Hill Equation

Many compounds exert multiple simultaneous and contradicting effects on an assay, resulting in a behavior

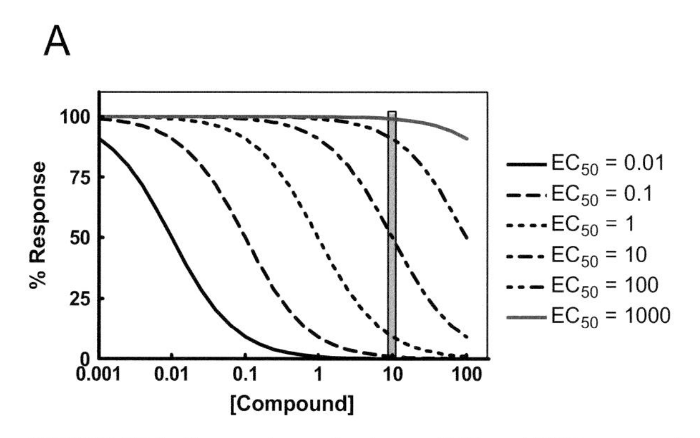

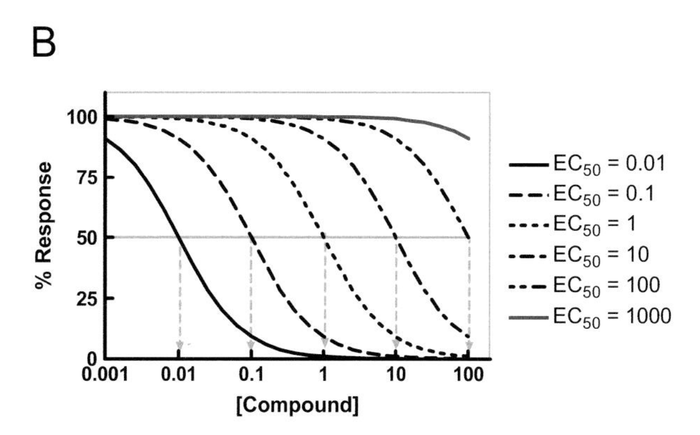

FIGURE 12.6** **Comparison of primary HTS and SAR assays. A. Primary HTS of compounds with different potencies. Blue box represents an arbitrary concentration selected for primary screening. B. SAR studies of compounds with different potencies. The values of EC_{50} are indicated with arrows.

distant from a canonical binary equilibrium complex. In addition to the intended effect on the target of an assay, compounds can form precipitates, aggregates, or micelles, shift pH, and/or have optical or chemical interference with an assay. As a result, many compounds exhibit nonhyperbolic curves.

As a general solution, a *Hill-type equation* (Equation 12.5), originally devised for describing cooperativity in hemoglobin-oxygen binding [11], is utilized for fitting the concentration-dependent inhibition:

$$[RI] = \frac{[R]_t [I]_t^{n_H}}{EC_{50}^{n_H} + [I]_t^{n_H}}, \tag{12.5}$$

where the EC_{50} is the apparent value of the dissociation constant and n_H is the *Hill coefficient*. This equation provides a foundation for the phenomenological quantitative description of diverse binding systems, specifically those deviating from ideal equimolar equilibrium-binding constraints. A value of the Hill coefficient (n_H) obtained from the equation provides a diagnostic tool and a parameter for quantifying the degree of deviation from the rectangular hyperbolic model. According to the Hill equation, perfect hyperbolic curves render n_H of 1. Taking into account random noise of experimental data, Hill coefficient values in the range between 0.8 and 1.3 are considered normal. The farther away the parameter value is from this range, the stronger the influences of exogenous processes are on the apparent affinity constant of the compound. The effect of the Hill coefficient parameter value on the appearance of curves with the same EC_{50} value is illustrated in Figure 12.7A.

5.3. Special Considerations for Systems with Tight Binding

The range of compound concentrations in SAR studies spans hundreds to hundred thousands fold. Successful lead optimization efforts result in the identification of highly potent compounds. Characterization of these compounds requires testing them in low-nM or even pM ranges. This frequently brings the compound concentration down to or below the concentration of limiting reagents in the assay, leading to the violation of a simplified Langmuir equation, shown in Figure 12.3A. In these circumstances, the total

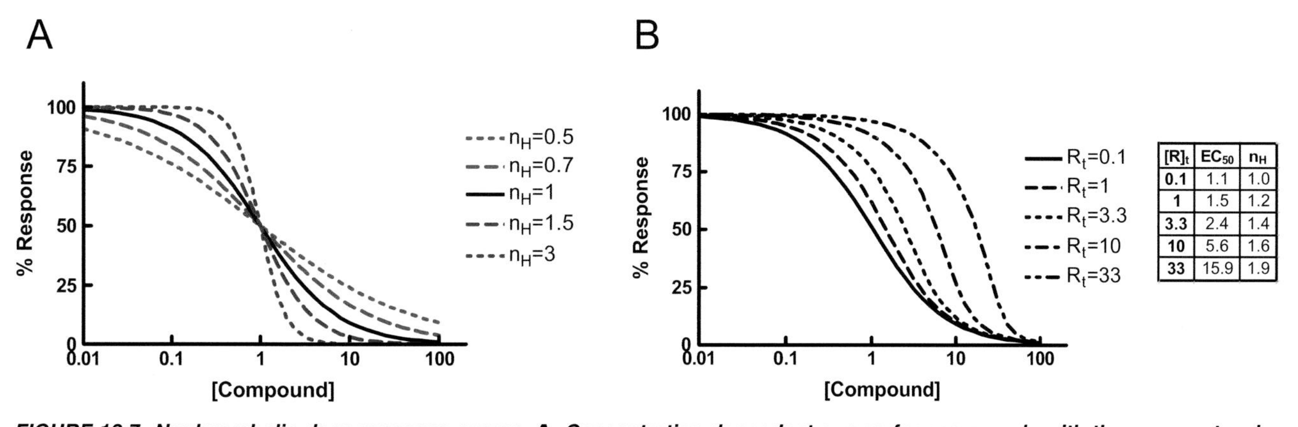

[R]t	EC50	nH
0.1	1.1	1.0
1	1.5	1.2
3.3	2.4	1.4
10	5.6	1.6
33	15.9	1.9

FIGURE 12.7** **Nonhyperbolic dose-response curves. A. Concentration-dependent curves for compounds with the same potencies (EC_{50} = 1), but different values of the Hill coefficient. B. Effect of receptor concentration (R_t) on the appearance of concentration-dependent curves of a potent compound (K_i = 1). The curves were generated using an explicit quadratic solution for ligand binding equation. Curve-fitting parameters in the table were obtained from nonlinear regression according to the Hill equation.

compound concentration no longer provides an appropriate representation of its free concentration. As a consequence, the shape of the dose-response curve does not correspond to the rectangular hyperbola equation meant for correlating the concentrations of unbound and bound ligand species. As seen in Figure 12.7B, the concentration-dependent curves for high-potency compounds that have dissociation constants below receptor concentrations ($K_i \ll [R]_t$) appear "angular" and are characterized with elevated Hill coefficients. These changes occur because a large portion of the compound is bound even at very low concentrations.

By definition, EC_{50}^{app} values correspond to compound concentrations, resulting in a half-maximal effect. Therefore, if these values are obtained in assays with a receptor as the limiting component, they cannot be lower than half of the receptor concentration and are described by the following semiempirical equation:

$$EC_{50}^{app} \approx EC_{50} + \frac{[R]_t}{2}, \qquad (12.6)$$

where EC_{50} is the $[R]_t$-independent component of EC_{50}^{app} that correlates with thermodynamic K_i through Equations 12.2, 12.3, or 12.4 described in section 4.3.1. These equations, along with equation 12.6, provide the basis for the assessment of K_i values of potent inhibitors. The EC_{50} values determined at different concentrations of receptors and fitted to Equation 12.6 would provide an extrapolated estimate of EC_{50} and subsequently of K_i, assuming that the concentration of nonlimiting ligand, its K_d, and MOA are known. The utility of this approach is limited to the characterization of compounds with K_i values of the same order of magnitude as receptor concentration. For an accurate determination of K_i of more potent compounds, it is imperative to develop and use a more sensitive approach, ensuring that the concentration of the receptor is less than or equal to the K_i value in an assay.

An assay that is utilized in primary HTS can tolerate much higher concentrations of receptor, because compounds are tested at a single high concentration. As seen in Figure 12.7B, assays with $[R]_t$ of 10 uM or less demonstrate almost identical compound effects if an employed compound concentration is 33 uM or greater. In contrast, their apparent EC_{50} values cover a range between 1 and 5.6 uM. Assays intended for SAR studies must employ the lowest possible concentration of a receptor or enzyme, because these concentrations define the lower limit of quantification for SAR studies.

6. A CASE STUDY: ASSAYS FOR HUMAN ALKALINE PHOSPHATASES

Clinical assays for detection of *alkaline phosphatases (APs)* employ hydrolysis of the chromogenic substrate p-nitrophenyl phosphate, which upon hydrolysis produces yellow-colored ($\lambda_{max} = 405$ nm) p-nitrophenol in the presence of a high concentration of amino alcohol–containing buffers, such as diethanolamine (DEA). In addition to serving as buffers, amino alcohols significantly accelerate the rate of phosphate release, replacing a water molecule in its role as a phosphate acceptor substrate. The assays provide a robust system for AP screening; however, they are sensitive to optical interference and lack dynamic range and sensitivity.

This assay was utilized for screening *tissue-nonspecific alkaline phosphatase (TNAP)* (PubChem AID 615). However, no hits were detected among the 20,500 compounds that were tested. In response to this observation, a luminescent assay employing a luminogenic substrate, CDP-Star, which was previously developed and optimized for the detection of alkaline phosphatases in blotting techniques [12], was adapted. The luminescent assay is about three orders of magnitude more sensitive than the colorimetric assay and allowed for a 400-fold decrease in the concentration of TNAP. The concentration of DEA in the assay was decreased to correspond to its K_m. The signal of the luminescent assay provided for the instantaneous measure of TNAP steady-state rate, which was stable for up to five hours, and the reagents remained stable for up to three days. This assay required no quenching, therefore was easily automatable.

A detailed description of the development and utilization of the novel luminescent HTS assay for TNAP is described elsewhere [13, 9]. Briefly, the assay was optimized to enable screening of TNAP in the presence of phosphate-donor and phosphate-acceptor substrates present at their respective K_m values. TNAP was screened against 64,394 compounds in the *Molecular Libraries Small Molecule Repository (MLSMR)* collection. The average Z′ factor for the full screen was equal to 0.82 (ranging between 0.75 and 0.89). From the primary screen, seventy-three hits were identified, and fifty-three of them were confirmed to have IC_{50} values below 20 uM (the concentration utilized in primary screening). In contrast, only four of these hits were active in the original colorimetric assay. It was later determined that the rest of the hits were inactive because of their competition with the DEA buffer-substrate.

From the luminescent assay-based screening, three major chemical series with diverse MOAs were identified [8,13]. Similar luminescent assays were developed for other AP: *placental (PLAP)* and *interstinal (IAP)*. One of the chemical series demonstrated extreme selectivity within and outside the AP enzyme family, with good pharmacokinetic properties and promising effects in rodent models [15]. Another scaffold inhibited several members of AP family. A subseries of compounds that selectively inhibit IAP and PLAP family members were identified through further SAR efforts [16]. In addition to these inhibitors, several chemical series of efficacious activators of TNAP with potencies ranging from 0.2 to 5 uM were fortuitously identified. This outcome would not have been possible in the colorimetric assay, which is constrained by its narrow dynamic range.

All of these projects were performed within the *Molecular Library Screening Center Network*, a part of the *Molecular Libraries Program*, a *National Institutes of Health (NIH) Roadmap Initiative*. The assay protocols and other experimental data are available through the PubChem database, a part of the NCBI resources.

7. COMMONLY AVAILABLE ASSAY DEVELOPMENT RESOURCES

There are a large number of resources available for assay development, both in printed formats and on the Internet. Most major assay kit vendors provide comprehensive booklets describing their technologies and products. An excellent source of additional information on specific well-established screening approaches for diverse types of targets can be found in *Assay Guidance Manual Version 5.0*, developed in collaboration between Eli Lilly and the NIH Center for Chemical Genomics [17]. It offers a wealth of useful information and cites a number of primary sources with further detailed information.

8. CONCLUDING REMARKS

Biological targets can be screened in a number of different ways. Each approach has specific advantages and disadvantages that are important and should be considered when selecting an assay for screening. A proper assay should highlight the HTS hits with desired properties while downplaying interference from the rest of the screening library. Because the main cost of an HTS-based lead discovery campaign is associated with post-HTS compound confirmation and characterization, selection of a suitable assay is critical for containing HTS project costs.

Assays utilized for HTS must be precise, sensitive, and robust. These properties are addressed during assay optimization. Multiple factors define assay precision and the ability to reliably distinguish active and inactive compounds. The concentration of the limiting component sets the S/B and the S/N of the assay, which are cumulatively described by the Z′-factor. Nonlimiting principal components define an assay's sensitivity and should be tailored to emphasize specific desired properties of the HTS hits, such as their MOA. Although supporting components participate in the assay indirectly, they are critical for assay performance, providing optimal conditions and helping to avert nonspecific phenomena and interactions. Of note, optimization of supporting reagents allows the limiting component concentration to be decreased, thereby enhancing the sensitivity of HTS and SAR assays.

Improving reagents and signal stability makes it possible to increase screening throughput and reduce assay failure rate. Assay automation and their implementation on robotic screening systems are necessary in large-scale screening campaigns to ensure consistency of data, yet on their own, they are unable to salvage an inferior assay. Assay optimization, which was discussed in this chapter, lays the foundation for assay performance and the ultimate success of a HTS campaign in the identification of hits with tractable properties.

REFERENCES

1. Sittampalam, G. S., Iversen, P. W., Boadt, J. A., Kahl, S. D., Bright, S., Zock, J. M., Janzen, W. P., and Lister, M. D. (1997). Design of signal window in high throughput screening assays for drug discovery. *J Biomol Screen 2*, 159–169.
2. Zhang, J.-H., Chung, T. D. Y., and Oldenburg, K. R. (1999). A simple statistical parameter for use in evaluation and validation of high throughput screening assays. *J. Biomol Screen 4*, 67–73.
3. Iversen, P. W., Eastwood, B. J., Sittampalam, G. S., and Cox, K. L. (2006). A comparison of assay performance measures in screening assays: signal window, Z′-factor, and assay variability ratio. *J. Biomol Screen 11*, 247–252.
4. Brideau, C., Gunter, B., Pikounis, B., and Liaw, A. (2003). Improved statistical methods for hit selection in high-throughput screening. *J Biomol Screen 8*, 634–647.
5. Auld, D. S., Southall, N. T., Jadhav, A., Johnson, R. L., Diller, D. J., Simeonov, A., Austin C. P., and Inglese, J. (2008). Characterization of chemical libraries for luciferase inhibitory activity. *J Med Chem 51*, 2372–2386.
6. Auld, D. S., Thorne, N., Nguyen, D. T., and Inglese, J. (2008). A specific mechanism for nonspecific activation in reporter-gene assays. *ACS Chem Biol 3*, 463–470.
7. Langmuir, I. (1916). The constitution and fundamental properties of solids and liquids. Part I. Solids. *J Am Chem Soc 38*, 2221–2295.
8. Cheng, Y.-C., and Prusoff, W. H. (1972). Relationship between the inhibition constant (Ki) and the concentration of inhibitor which causes 50 percent inhibition (IC50) of an enzymatic reaction. *Biochem Pharmacol 22*, 3099–3108.
9. Sergienko, E., Su, Y., Chan, X., Brock, B., Hurder, A., Narisawa, S., and Millán, J. L. (2009). Identification and characterization of novel tissue-nonspecific alkaline phosphatase inhibitors with diverse modes of action. *J Biomol Screen 14*, 824–837.
10. Ellis, K. J., and Morrison, J. F. (1982). Buffers of constant ionic strength for studying pH-dependent processes. *Methods Enzymol 87*, 405–426.
11. Hill, A. V. (1910). The possible effects of the aggregation of the molecules of hæmoglobin on its dissociation curves. *J Physiol (Lond.) 40*, IV–VII.
12. Schapp, A. P., Sandison, M. D., and Handley, R. S. (1987). Chemical and enzymatic triggering of 1,2-dioxetanes. 3: alkaline phosphatase-catalyzed chemiluminescence from an aryl phosphate-substituted dioxetane. *Tetrahedron Lett 28*, 1159–1162.
13. Sergienko, E. A., and Millán, J. L. (2010). High-throughput screening of tissue-nonspecific alkaline phosphatase for

identification of effectors with diverse modes of action. *Nat Protocols 5*, 1431–1439.
14. Sidique, S., Ardecky, R., Su, Y., Narisawa, S., Brown, B., Millán, J. L., Sergienko, E., and Cosford, N. D. (2009). Design and synthesis of pyrazole derivatives as potent and selective inhibitors of tissue-nonspecific alkaline phosphatase (TNAP). *Bioorg Med Chem Lett 19*, 222–225.
15. Dahl, R., Sergienko, E. A., Su, Y., Mostofi, Y. S., Yang, L., Simao, A. M., Narisawa, S., Brown, B., Mangravita-Novo, A., Vicchiarelli, M., Smith, L. H., O'Neill, W. C., Millán, J. L., and Cosford, N. D. (2009). Discovery and validation of a series of aryl sulfonamides as selective inhibitors of tissue-nonspecific alkaline phosphatase (TNAP). *J Med Chem 52*, 6919–6925.
16. Lanier, M., Sergienko, E., Simao, A. M., Su, Y., Chung, T., Millán, J. L., and Cashman, J. R. (2010). Design and synthesis of selective inhibitors of Placental Alkaline Phosphatase. *Bioorg Med Chem 18*, 573–579.
17. *Assay Guidance Manual Version 5.0* (2008). Eli Lilly and Company and NIH Chemical Genomics Center. http://www.ncgc.nih.gov/guidance/manual_toc.html.

CHAPTER 13

Molecular Sensors for Transcriptional and Post-Transcriptional Assays

Douglas S. Auld
Natasha Thorne

Cell-based assays allow researchers to probe the regulation of cell signaling, communication, and regulation within a system that is more physiologically relevant than assays that focus on isolated molecular components or cell extracts (biochemical assays). The application of cell-based assays in high-throughput screening (HTS) has increased in recent years. It is estimated that cell-based assays comprise more than half of HTS assays used in pharmaceutical industries [1].

Cell-based molecular sensor assays have been developed to measure either transcriptional or post-transcriptional events (Figure 13.1). A typical transcriptional reporter gene assay involves the placement of a *cis*-regulatory element in front of a gene that produces a measurable signal in response to modulation of transcriptional complexes that are regulated by ligands or complex networks. Assays favored in HTS are usually constructed to generate fluorescence or luminescence by expressing either an enzyme or a fluorescent protein reporter. Post-transcriptional assays have been developed by engineering the reporter to respond to events such as mRNA splicing, protein complex formation, translocation of a receptor between cellular compartments, or protein stability.

Choosing between the various molecular sensors that are available in cell-based assays requires careful consideration of several questions. For example, what is the cell type required? What is the time scale of the response, and what type of microtiter plate detector will be required? Will the assay require the measurement of signal from live cells, or can an end-point assay be used in which cells are lysed during the detection step? What are the potential artifacts associated with the assay? As we outline in this chapter, current assay technologies provide a large number of methods to assess the activity of signaling pathways and the regulation of proteins in a cell-based system. The purpose of this chapter is to describe transcriptional and post-transcriptional assays that are commonly used in HTS, as well as to highlight their applications and limitations. It is our hope that this information will help the reader choose the most suitable assay for the scientific question at hand.

1. TRANSCRIPTIONAL REPORTER GENE ASSAYS

There is a wide variety of transcriptional reporter gene assay designs that can be built to cater to the particular pathway, transcription factor, and readout desired by the scientist, giving this type of assay flexibility in targeting many different biological signaling pathways (Figure 13.2). Targets of transcriptional reporter gene assays are typically signaling pathways or the transcription factors themselves. Promoters used to control reporter gene transcription can be complex or simple and endogenous or synthetic (Figure 13.2 and discussed in more detail later). In addition, reporter gene constructs can be transiently or stably transfected. In this section, we focus on the type of reporter genes that are commonly used in HTS.

1.1. Reporters

There are a few key considerations in determining which reporter gene to use. First and most importantly, the type of reporter gene employed largely determines the sensitivity of the assay, as measured by the signal to background (S/B) and response to known modulators as well as the time scale of the response, which is a function of the reporter protein's half-life [2]. Second, one needs to consider the equipment (i.e., microtiter plate reader) required to measure the reporter. Currently, assay technologies that rely on radioisotopes and scintillates to generate light are falling out of favor and largely have been replaced by reporters that generate light, such as enzymes that are luminescent or generate fluorescent products. These types of reporters have become the most popular choice for HTS assays [3], as they provide the greatest sensitivity and are amenable to automation.

There are a quite a number of different reporter genes to choose from (Table 13.1), and it is worth pointing out that some of these reporters, such as the green fluorescent protein (GFP), are directly quantifiable, whereas enzymatic reporters require the reaction products to be detected and measured. These enzymatic reporters, such as luciferase or β-lactamase reporters, require the addition of substrate to

Transcriptional

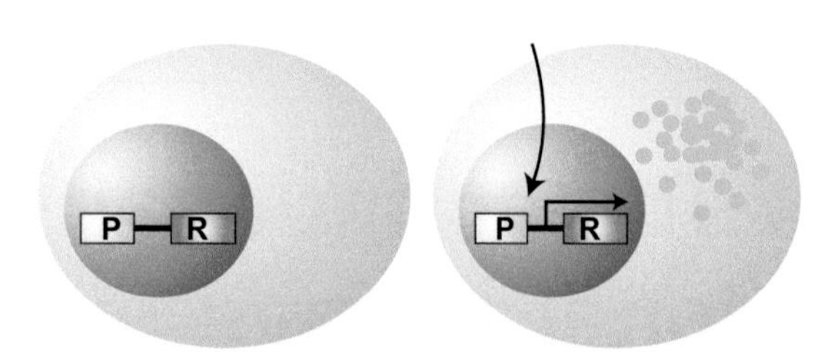

Post-Transcriptional

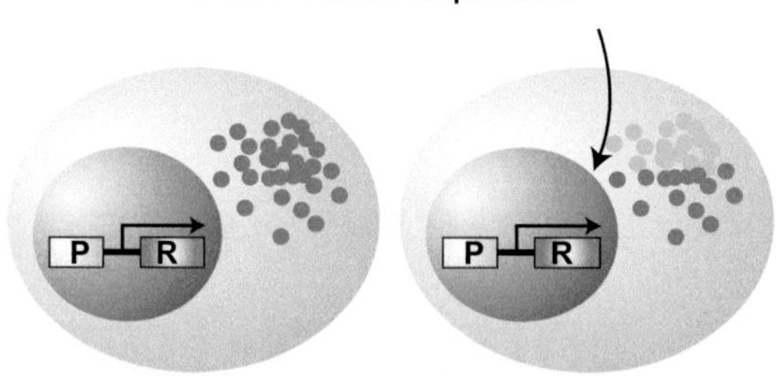

FIGURE 13.1: ***Transcriptional and post-transcriptional reporter assays. Left, transcriptional based reporter gene assay where a*** cis***-acting promoter (P) is used to drive the expression of a reporter (R) in response to an exogenously added ligand or test compound (arrow) resulting in the expression of a reporter with detectable activity (green circles). Right, a post-transcriptional assay where a constitutively active promoter drives expression of one or more reporters that undergo a measurable change in signal (blue vs. green circles) during post-translational processing.***

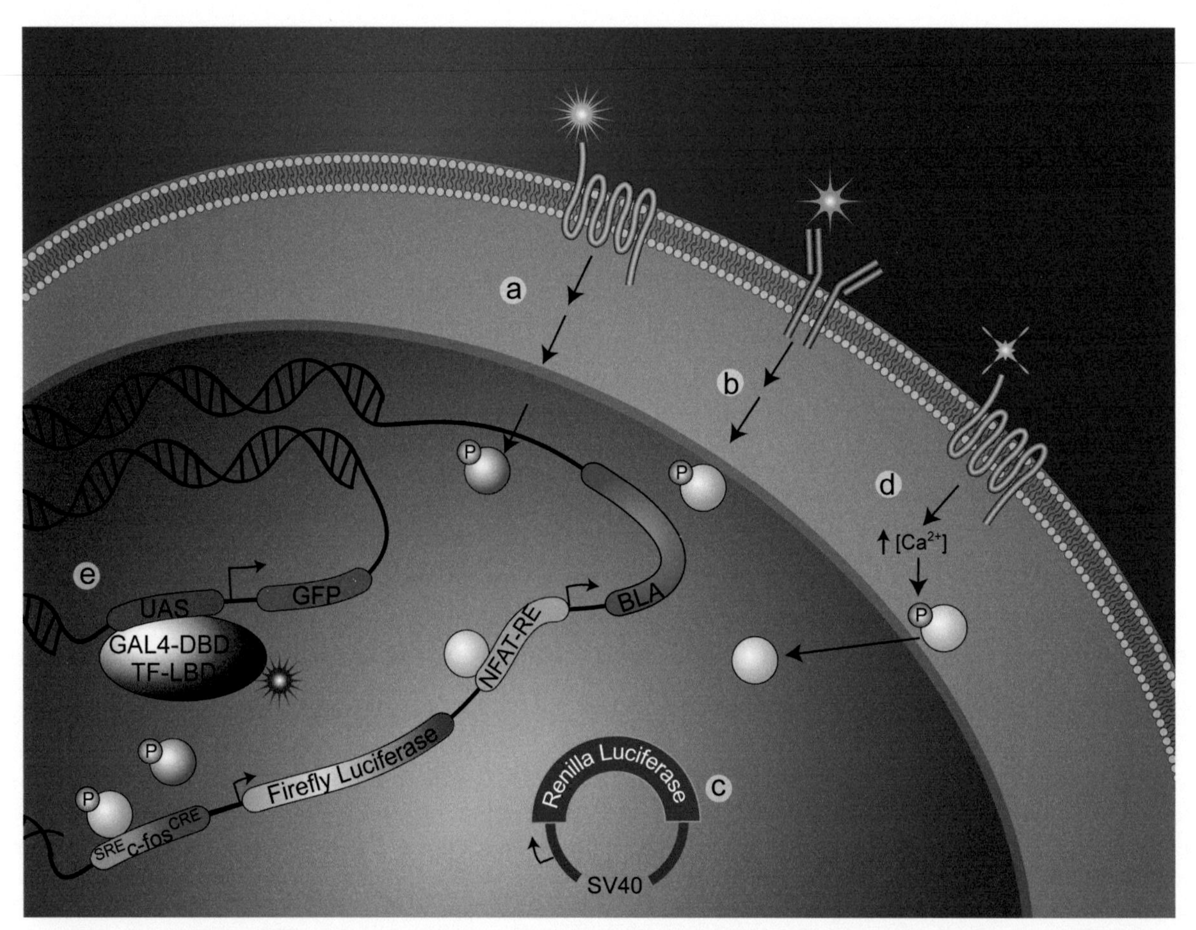

FIGURE 13.2: ***Representative examples of different transcriptional reporter gene assays in a cell: an introduction to different promoters and reporters. Activation of cell surface receptors initiate signaling cascades that eventually lead to gene expression. Modulation of these signaling pathways by small molecules, as occurs in HTS assays, can be monitored by reporter gene expression. Shown here are commonly used promoter elements and reporter genes that could be used to study signaling pathways. (a) Activation of a GPCR leads to phosphorylation of CREB, which binds to the CREB response element (CRE) in the*** c-fos ***promoter, leading to expression of FLuc. (b) Activation of the MAPK pathway leads to binding of a complex at the serum response element (SRE) in the*** c-fos ***promoter, which also results in expression of FLuc. A complex promoter,*** c-fos***, can be used to monitor the synergy between different signaling pathways. (c) Transient expression of*** Renilla ***luciferase, which is constitutively expressed under the control of the SV-40 promoter, is used as a control to normalize expression in FLuc reporter gene assays in which the FLuc reporter is also expressed transiently (not shown). The*** Renilla ***luciferase construct shown here represents a reporter that was introduced into the cells by transient transfection, whereas other reporter gene constructs shown here represent reporter genes that have been stably integrated into the DNA. (d) Activation of a different GPCR leads to an increase in intracellular calcium that allows the translocation of NFAT into the nucleus and expression of β-lactamase (BLA), which is under control of the synthetic, simple promoter containing only an NFAT response element. (e) Modulation of nuclear receptor transcription by small molecules (purple asterisk) can be examined by constructing a chimeric transcription factor made up of the DNA binding domain of the yeast GAL4 protein and the ligand-binding domain of the nuclear receptor. Activation of the nuclear receptor via its LBD leads to translocation to the nucleus and interaction at the upstream activation sequence (UAS) recognized by the GAL4 portion of the chimera. In the case shown, GFP is expressed.***

TABLE 13.1: Reporters for Cell-Based Molecular Sensor Assays

Molecular Sensor	Description	Ref
β-lactamase	Enzyme reporter used with FRET-based substrates such as CCF2/AM, CCF4/AM (λex = 409 nm and both 450 nm and 520 nm λem are monitored). Ratiometric signal processing allows flagging of nonspecific effects. Applied to live cell assays using conventional microplate readers or flow cytometry/laser-scanning imaging. Dynamic response enabled by short protein half-life (4 h).	[7]
FLuc (*Photinus pyralis*)	FLuc. Enzyme reporter (λem = 550–570 nm) used with cognate luminescent substrates D-luciferin and ATP. Dynamic response enabled by short protein half-life (4 h). Red-shifted variants have been designed, but wild type luciferase with yellow-green luminescence is widely applied in reporter gene assays. Detection reagents are formulated to promote glow luminescence response (4–5 h signal stability).	[53–55, 69, 79]
Click beetle luciferase (*Pyrophorus plagiophthalamus*)	CBLuc. Enzyme reporter, wild type enzyme shows yellow luminescence when D-luciferin and ATP substrates are supplied. Variants engineered with emission maximums showing green (537 nm) and red (613 nm) bioluminescence. If used together, spectral overlap of light from the two variant luciferases requires nontrivial corrective calculations of the data. Detection reagents are formulated to promote glow luminescence response (4–5 h signal stability). Used to construct two-color transcriptional assays. Dynamic responses enabled although half-life in mammalian cells of 7 h.[a]	[75, 182]
Sea pansy luciferase (*Renilla reniformas*)	RLuc. Enzyme reporter. Coelenterazine substrate shows blue luminescence (480 nm). Dynamic response enabled by short protein half-life (4.5 h).[a] Autoluminescence of coelenterazine leads to higher background. Weaker light output relative to FLuc. Often used in dual luciferase assays. Glow response with up to 2 h signal stability.	[66, 69]
Aequorin luciferase (*Aequorin victoria*)	Photoprotein showing blue luminescence (469 nm). Used as a Ca^{2+} sensor as luminescence intensity is proportional to the $[Ca^{2+}]$. Flash luminescence emitted, occurring within five seconds of exposure to Ca^{2+}. Low levels of light emitted (1 photon/aequorin). Requires high-sensitivity luminescent readers with auto-injection capabilities.	[83, 84]
β-galactosidase	Enzyme-based reporter where colorimetric, fluorescent, and luminescent substrates are available. Typical fluorescent substrate is methylumbelliferyl-β-D-galactoside (MUG; λex = 360 nm/λem = 410 nm); luminescent substrates include AMPGD[b] (λem = 470 nm) and 6-O-β-galactopyranosyl-luciferin (Lugal; λem = 550–570 nm). Widely used as cell-based molecular sensor using EFC.	[95, 183]
GFP and variants	Green fluorescent protein. Used in both reporter gene assays and post-transcriptional assays. Long half-life ~26 h. Popular in imaging assays but has also been used in HTS.	[126]
CAT	Chloramphenicol acetyl transferase. A bacterial enzyme that catalyzes the transfer of an acetyl group from acetyl-coenzyme A to chloramphenicol. Substrates are radiolabeled or fluorogenically labeled. An example substrate is BODIPY 1-deoxychloramphenicol (green and yellow substrates available; λex = 504 or 545 nm/ λem = 510 or 570 nm). Long protein half-life (50 h) does not make it dynamically responsive.	[96, 97, 184]
SEAP	Secreted placental alkaline phosphatase (AP). Enzymatic reporter. Endogenous AP expression in mammalian cells leads to high background; partially circumvented with secreted form. Output can be either colorimetric or chemiluminescent; coupled-enzyme assay with FLuc allows bioluminescent readout. Long protein half-life (72 h) does not make it dynamically responsive. Fluorescent substrates include 4-methylumbelliferyl phosphate (MUP; λex = 360 nm/λem = 410 nm), and a chemiluminescent substrate is CSPD (λem = 470 nm).[c]	[96, 183]
BRET	Bioluminescence resonance energy transfer; useful for protein-protein interactions; RLuc luciferase donor luminescence (480 nm) stimulates fluorescence of GFP/YFP acceptor; follows dipole-dipole interaction $1/r^6$ (~10 nm maximum).	[131, 132]
EFC	Enzyme fragment complementation. Based on β-galactosidase α-complementation; high affinity complementation. Multiple assay applications for nuclear receptors, proteases, kinases, GPCRs, and post-transcriptional events have been developed.	[99]
PCA	Protein complementation assay; recombinant split reporters (e.g., GFP, luciferase, β-lactamase) fused to interacting proteins; post-transcriptional assays; low affinity complementation.	[125]

[a] Promega, personal communication.
[b] 3-(4-methoxyspiro(1,2-dioxetane-3,2′-tricyclo(3.3.1.1(3,7))decan)-4-yl)phenylgalactopyranoside.
[c] 3-[2′-(spiro-5-chloroadamantane)]-4-methoxy-4-(3′-phosphoryloxy)phenyl-1,2-dioxetane.

produce a quantifiable signal, which may or may not be amenable to some experimental designs.

Some reporters are more commonly used than others for HTS applications for reasons that are discussed in greater detail throughout this chapter. As such, many of the traditional reporters are covered at least briefly here, but greater emphasis is paid to those commonly used in HTS, such as the luminescence-based reporter luciferase and the fluorescence-based reporter β-lactamase.

***1.1.1. Fluorescence-Based Catalytic Reporter – β-lactamase* (bla)** The enzyme β-lactamase is exclusively expressed in bacteria and is thought to have evolved for microbial defense against antibiotics containing a β-lactam ring, such as penicillins and cephalosporins [4, 5]. Currently, the most commonly used isoform of β-lactamase is derived from the TEM-1 β-lactamase, a 29-kDa protein encoded by the *Escherichia coli* Amp^R plasmid gene [5–7].

One aspect of catalytic reporters like β-lactamase is that substrate must be added to the assay. This substrate is often not cell permeable, requiring the cells to be lysed or, at the very least, the cell membranes to be permeabilized, a process that prevents live-cell analysis. However, Zlorarnik and colleagues [7] were able to overcome this step with the development of CCF2/AM – a β-lactamase substrate that is rendered membrane permeable by capping charged groups with esters, which upon entrance into the cell are hydrolyzed by cellular esterases, thus trapping the substrate (CCF2) in the cell.

The β-lactamase substrate CCF2 has been synthesized to contain the fluorophores 7-hydroxycoumarin and fluorescein connected by a cephalosporin core [7]. The basis of this assay is fluorescence resonance energy transfer (FRET) between the two fluorophores. In FRET, a donor fluorophore of short excitation and emission wavelengths transfers energy through a dipole-dipole interaction to an acceptor fluorophore of longer excitation and emission wavelengths in close proximity (~10 nm). For the β-lactamase substrate CCF2, excitation at 409 nm results in FRET from the coumarin moiety to fluorescein, with a resultant fluorescent emission at 520 nm (Figure 13.3A). Cleavage of the β-lactam ring of CCF2 by β-lactamase disrupts FRET, allowing detection of coumarin fluorescence emission at 450 nm. In this way, the level of β-lactamase expression in cells can be quantified by calculating the 450 nm/520 nm fluorescence ratio [5].

The ability to collect data at both the blue and green wavelengths as well as calculate the fluorescence ratio helps to alleviate problems intrinsic to cell-based assays. First, the ratiometric output of two emission wavelengths reduces the influence of well-to-well variability in, for example, cell number, as differences in cell number would affect fluorescence at both emission wavelengths equally [5]. Furthermore, collecting and examining emission output at both wavelengths separately can prove valuable in detecting cytotoxicity, which is characterized by a decrease in both blue and green fluorescence [8] .

The assay incubation time and the half-life of the reporter are always important to consider. Most reporter gene assays require sufficient incubation time after activation of a signaling pathway to allow for reporter gene transcription and reporter protein accumulation significant enough to provide a high S/B. For this reason, these assays are typically performed over the course of five hours or longer [8, 9]. The half-life of the TEM-1 β-lactamase is approximately two hundred minutes, allowing for a relatively dynamic response [7]. In addition, the kinetics of CCF2 cleavage by β-lactamase has been reported [5, 7], and analysis of β-lactamase activity indicates that maximum CCF2 substrate cleavage is reached approximately one hour after its addition. It has been noted, and we have also observed, that unstimulated cells will eventually demonstrate emission at 450 nm if the assay is allowed to proceed for extended periods of time. Zlokarnik et al. [7] attribute this to constitutive activity at the upstream promoter but find this only becomes significant at times of sixteen hours or longer after substrate addition.

One of the significant advantages of this reporter gene system is the sensitivity of detection possible, and it has been demonstrated using flow cytometry that a single β-lactamase–expressing cell can be identified in a population of 10^6 cells [10]. By comparison, a β-galactosidase reporter (*lacZ*) demonstrated sensitivity on the order of two magnitudes less than that of β-lactamase under identical conditions [10]. This sensitivity also allows β-lactamase reporter assays to be applied to a unique microtiter plate reader used in HTS – a laser-scanning fluorescence microplate cytometer [11] – which allows visualization and data collection of single cells within microtiter plate wells. We have successfully used such a detector in an assay to confirm NFκB inhibitors (using an NFκB-*bla* reporter construct) as well as agonists and antagonists of the NFAT signaling pathway (with an NFAT-*bla* reporter construct; Thorne and Auld, unpublished results). It should also be noted that because the β-lactamase assay is amenable to analysis by flow cytometry [10], it can be used as a cell sorter to select cells that optimally express β-lactamase in the process of producing stable β-lactamase reporter gene cell lines [7, 9, 12, 13].

There are significant commercial resources available to the researcher interested in implementing β-lactamase reporter gene assays to explore signaling pathways. Invitrogen offers not only constructs that allow the user to clone promoter/regulatory elements upstream of the β-*l*actamase reporter gene (pGeneBlAzer-TOPO and CellSensor vectors), but also numerous cell lines that contain a stably integrated *bla* reporter gene construct downstream of a response element specific for a signaling pathway (CellSensor cell lines). The β-lactamase substrates – CCF2/AM or CCF4/AM – are also commercially available (Invitrogen).

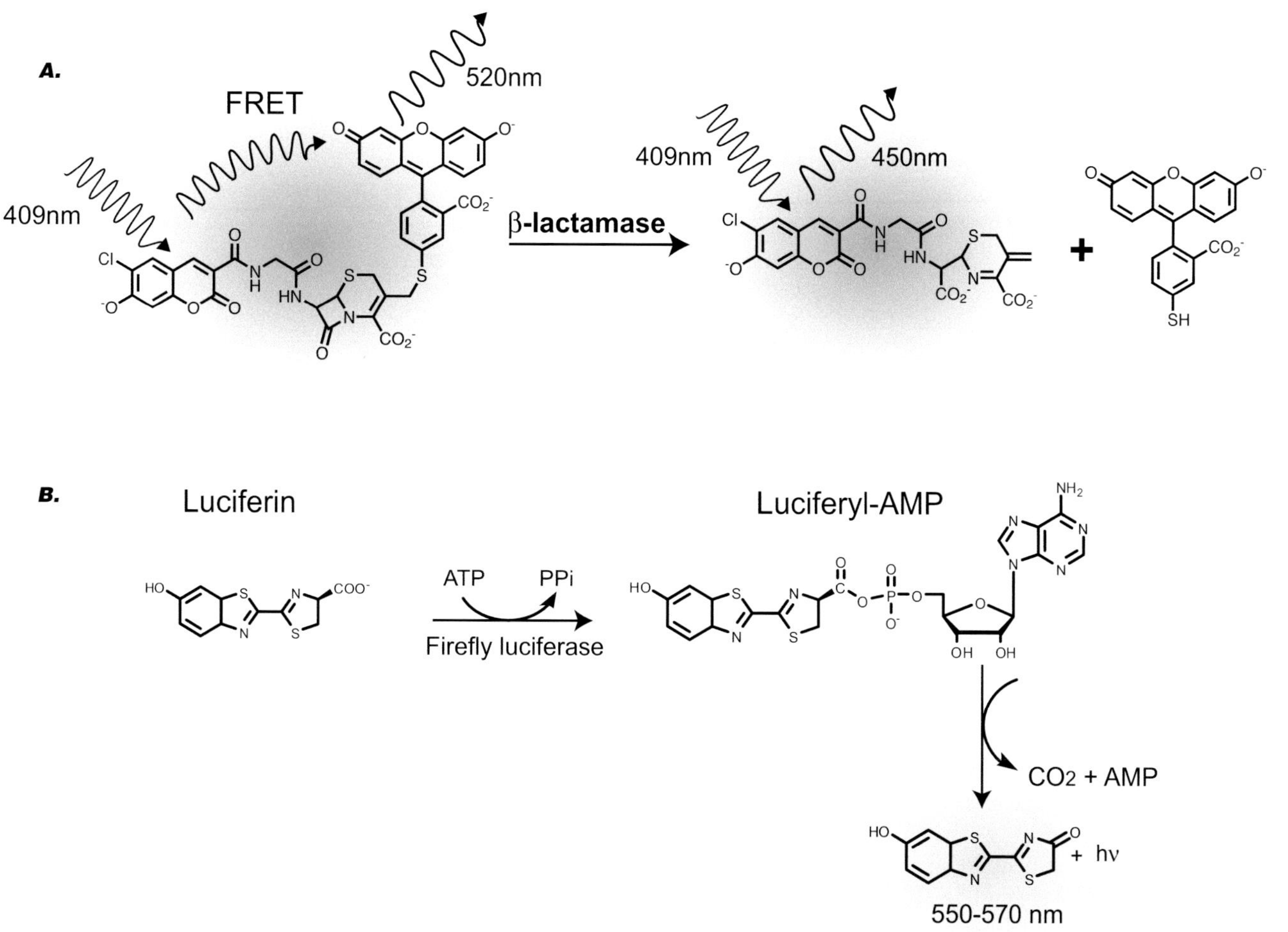

FIGURE 13.3: ***Common enzyme reporter substrates and products. (A) Disruption of FRET CCF2 signal upon cleavage by β-lactamase. CCF2 (shown at far left), once loaded into cells, fluoresces at 520 nm upon excitation at 409 nm because of FRET between the coumarin and fluorescein moiety. If β-lactamase has been expressed by the cell, the enzyme cleaves CCF2 at the cephalosporin core that links the two fluorophores, disrupting FRET, and allowing coumarin fluorescence at 450 nm. (B) The enzymatic reaction of FLuc. Luciferin and adenosine 5-triphosphate (ATP) substrates are used by the enzyme to form a luciferyl-adenosine 5′-monophosphate (AMP) intermediate that is then oxidized in the presence of oxygen to generate oxyluciferin, AMP, and CO_2 with the production of light (hν).***

CCF4-AM is generally used for HTS, as it is more stable in solution than CCF2-AM (Invitrogen). Additionally, a cytotoxicity indicator (ToxBlazer) that exhibits red fluorescence (λex 600 nm, λex 650 nm) is available that can be multiplexed with β-lactamase assays [14].

Assays used in HTS must be highly precise (variability <15%), produce an adequate S/B ratio (two-fold or greater), and be highly sensitive to the desired biology [3]. A measure of assay performance is the Z′ factor, whose value is determined by both signal variability and S/B. Generally, assays exhibiting Z′ factors greater than 0.5 are considered acceptable [15, 16]. Reporter gene assays employing β-lactamase, once optimized for HTS, clearly have these qualities and have been successfully implemented in HTS.

An example of an HTS protocol for a *bla* reporter gene assay is shown in Table 13.2, and of note, it requires only four reagent addition steps to complete the assay. Numerous examples exist where β-lactamase reporter gene assays have been implemented in HTS to target a variety of signaling pathways, including those of G protein-coupled receptors (GPCRs) and cytokine receptors [5, 8, 9, 17]. Oosterom et al. [8] performed a screen against 245,000 compounds for antagonists of the human gonadotropin-releasing hormone receptor (GnRH-R, a GPCR) using Chinese hamster ovary (CHO) cells that expressed the receptor and contained a reporter construct with β-lactamase downstream of an NFAT/Activator Protein-1 promoter. This assay was performed in 384-well microtiter plate format, and fluorescence was detected using a conventional fluorescent plate reader. The assay demonstrated high sensitivity with low variation between replicates (Z′ factor of 0.85) and resulted in the identification of seven

TABLE 13.2: *β* **-lactamase Reporter Gene Assay Protocol**

Sequence	Parameter	Description
1	Reagent	Seed cells at appropriate density in black, clear-bottom TC-treated microtiter plates
2	Time	For adherent cells –O/N incubation at 37°C, 5% CO_2
3	Reagent	Addition of compounds
4	Reagent	If antagonist assay – addition of stimulator (e.g., TNFα for NFκB or carbachol for NFAT)
5	Time	5-h incubation at 37° C, 5% CO_2
6	Time	15 min at RT to equilibriate plates
7	Reagent	Addition of CCF4/AM loading dye
8	Time	1-h incubation at RT in dark to allow CCF4 cleavage
9	Detector	Fluorescent plate reader or laser-scanning fluorescent microplate cytometer EX: 409 nm; EM: 450 nm and 520 nm Analyze 450/520 nm ratio

GnRH-R inhibitor compounds that were selected for follow-up assays [8]. In addition, this group found that the sensitivity of the β-lactamase assay was comparable to a Ca^{2+} assay for peptide ligands of GnRH-R, though both types of assays were less sensitive than a radioligand binding assay [8]. Efforts to optimize the reporter assay to identify agonists and antagonists of the GPCR human bradykinin B1 receptor in HTS (CHO cells; NFAT promoter; 384-well microplate; conventional fluorescent plate reader) have also been reported [9]. Orphan GPCRs have been expressed in cell lines containing either NFAT-*bla* or CRE-*bla* reporter constructs to measure Ca^{2+} or cAMP signaling, respectively. These constructs have been created in an effort to de-orphan GPCRs in high-throughput format [18]. Our group has also successfully implemented cell-based *bla* assays for internalization of anthrax lethal toxin (Pubchem [19] Assay Identifier [AID: 912]), JAK/STAT (AID: 446), NFAT (AID: 444), API (AID: 357), and hypoxia response element signaling (AIDs: 914, 915) using quantitative HTS (qHTS) [20] on libraries with twelve thousand to seventy thousand compounds in 1536-well microtiter plate format. An ultra-HTS (3456-well microtiter plate format) for agonists of the estrogen receptor β has also been performed using an upstream activation sequence (UAS) promoter element upstream of the β-lactamase reporter gene in CHO cells that expresses a chimeric Gal4-ERβ fusion protein, ultimately resulting in the identification of ten compounds with ERβ-specific activity at potencies less than 1 μM [13].

As with any assay, there are artifacts of which the user should be aware. For the fluorescence-based *bla* reporter gene assay, one of the main concerns is autofluorescence of small molecules [21, 22] in the chemical library; this is discussed in more detail later (section **3.2**) in this chapter. Users should also be aware of competitive substrates or inhibitors of β-lactamase. These compounds (such as clavulanic acid and cephalosporin) contain a β-lactam motif and can compete with binding of the CCF2/CCF4 substrate and inhibit the activity of β-lactamase. In the case of β-lactamase inhibition, the 450 nm/520 nm ratio is decreased, giving the appearance that the signaling pathway under investigation has been affected, which leads to false positives in a screen for antagonists [9]. A screen of a diverse library containing approximately seventy thousand compounds available within PubChem [19] (AID: 584) indicated that only a very small percentage of the compounds in the library – about 0.05% – were inhibitors of β-lactamase. The majority of these false positives were competitive inhibitors containing a β-lactam ring that covalently modified the enzyme [23]. However, this low percentage of false positives is comparable to the true positive hit rate found in HTS that is on the order of 0.001% to 0.1% [24].

Another concern when employing the β-lactamase assay is false positives in inhibitory screens due to cytotoxic compounds. In this case, however, the ratiometric output, as well as collection of data at both the 450-nm and 520-nm wavelengths, helps flag these instances. If the 450 nm/520 nm ratio decreases, true inhibition of signaling will show low fluorescence intensity at 450 nm but no decrease in signal at 520 nm. Conversely, if a cytotoxic effect is the cause of a decreased 450 nm/520 nm ratio, then there will be decreased fluorescence at both 450 nm and 520 nm.

1.1.2. Fluorescence-Based Reporter – GFP The GFP commonly used in cell biology is derived from a naturally occurring protein from the jellyfish *Aequorea victoria* [25, 26](Table 13.1). Each GFP contains one chromophore – a

TABLE 13.3: GFP Reporter Gene Assay Protocol

Sequence	Parameter	Description
1	Reagent	Seed cells with GFP reporter at appropriate density
2	Time	Incubate at 37°C 5% CO_2
3	Reagent	Addition of compounds
4	Reagent	Addition of ligands (e.g., stimulator if antagonist assay)
5	Detector	Fluorescent plate reader or laser-scanning fluorescent microtiter plate cytometer EX: 488 nm; EM: 500–530 nm

p-hydroxybenzylideneimidazolinone – formed by residues of the native protein that undergo modification during maturation of the protein [26, 27]. Excitation of this chromophore at 395 nm results in emission of fluorescence with a peak of 508 nm. There is also minor excitation of GFP at 475 nm, which allows excitation of the protein using a 488-nm laser [28] and produces peak emission at 503 nm [26, 29]. One of the main advantages of using GFP as a reporter is that it does not require addition of exogenous substrate to produce light, and this has been heavily exploited to monitor gene expression in live cells and transgenic animals [30, 31]. GFP is also commonly used in molecular biology to create tagged/fusion proteins [32] in an effort to monitor the subcellular localization of a protein of interest.

Screens of mutant GFP proteins have identified proteins with enhanced fluorescence with 488-nm excitation (enhanced green fluorescent protein [EGFP]) as well as fluorescent proteins with varying excitation and emission properties [28, 32–35]. Mutational variants of *Aequorea* GFP as well as the discovery of fluorescing proteins from other organisms such as coral (*Discosoma*) have broadened the fluorescence spectrum to cyan/blue, yellow, and red [34–39] and have introduced variants with increased fluorescent protein photostability [40], with the search for new fluorescent protein variants ongoing (for reviews, see references [41, 42]). Fluorescent proteins of different colors have supplemented the toolbox for exploring protein-protein interactions using FRET between fluorescent proteins with different emission spectra [26, 35, 41], and these have also been used to construct post-translational modification and cell-based protease assays (see section **2.5**). Also of note, photoactivatable and photoconvertable fluorescent proteins have been developed for use in analyzing live-cell dynamic processes [41, 42].

Several properties of fluorescent protein reporters distinguish them from enzyme-based reporters. For example, unlike β-lactamase, which has the advantage of catalytically turning over substrate to generate multiple fluorophores, each GFP molecule presents a single fluorophore. This, in combination with autofluorescence in the blue-green spectral region that is characteristic of mammalian cells, can lead to unacceptably low S/B [43] in an assay. Optimization techniques are possible, however, and include taking advantage of the 488-nm excitation and choosing filter sets tailored to GFP emission, which can reduce the collection of light due to autofluorescence [43]. Enhanced S/B can be achieved by expressing a GFP that is targeted to a cellular compartment [26, 43], such as the nucleus (nuclear GFP). In addition, to ensure robust GFP detection, strong, nonendogenous reporters (cytomegalovirus [CMV], simian virus 40 [SV-40], UAS; Figure 13.2) are most often used [26]. However, one must keep in mind that dependence on such a promoter puts into question the suitability of a GFP-based reporter gene assay for assessing transcriptional activation by a specific signaling pathway because of the high levels of reporter expression required. A very common use, however, for constitutively expressed EGFP or red fluorescent protein (RFP) is as a reporter of compound cytotoxicity.

GFP has a β-barrel structure that makes it a relatively stable protein [33] with a half-life of approximately twenty-six hours [44]. This allows GFP to build up in a cell, increasing signal output, which is desirable for any assay. However, such stability does not permit a signal that is responsive to down-regulation of reporter gene expression [2], because once GFP protein has accumulated to levels that are detectable above background, it takes considerable time to observe a decrease in GFP expression. For this reason, different versions of destabilized GFP (dEGFP) have been created, containing a PEST sequence [45] that targets the protein for degradation, thus reducing the half-life of GFP to as short as three hours [44, 46].

Although GFP reporter gene assays are not as commonly used in primary HTS because of their low sensitivity, assays that employ GFP as a reporter do offer several advantages: They are easy to implement because of a relatively simplified protocol (Table 13.3) compared to other assays used in HTS, and because detection reagents are not required, the cost of these assays can be considerably lower than those of enzyme-based reporter systems. Often, the detection of GFP signal may not be discernable over background for many microtiter plate readers used in HTS, particularly when miniaturized assay volumes ($<$10 μL/well;

1536-well microtiter plates) are employed, although GFP and RFP have been successfully implemented as reporters in 96-well and 384-well HTS assays [47, 48]. For example, EGFP was used to measure reactivation of latent HIV-1 by compounds in a human T-cell line in 384-well microtiter plate format [48, 49]. Broom and colleagues [47] developed an assay targeting superoxide dismutase gene (*SOD1*)-mediated amyotrophic lateral sclerosis (ALS) by cloning the promoter element of the Cu/Zn *SOD1* upstream of an EGFP reporter to screen for compounds that could inhibit *SOD1* expression. Not surprisingly, they found that they could not miniaturize the assay beyond a 96-well microtiter plate because of signal strength limitations. One group tried to offset the inherent low S/B by cloning four copies of a promoter response element (for NFκB) upstream of a dEGFP or EGFP reporter [50] but found because of the low accumulation of dEGFP (as it is destabilized), the assay could not be performed in microtiter plates, and thus was not suitable for further HTS development. The EGFP reporter did provide enough signal for microtiter plate-based assays; however, because of its long half-life, the GFP reporter was best suited for agonist assays in which upregulation of reporter gene expression was measured [50].

Although confocal microscope-based optical microtiter plate readers [51] are ideal for use with GFP-based assays, these readers generally possess low throughput. A sensitive microtiter plate reader that allows for miniaturized GFP assays in HTS is the Acumen laser-scanning microplate cytometer [11, 52], which offers cellular resolution and may provide optimal data for GFP-based reporter gene assays performed in miniaturized assay volumes.

A common artifact of GFP-based assays is optical interference by compounds in the library being screened. Some heterocyclic compounds fluoresce in the blue-green region of the spectrum and thus can potentially cause false positives in GFP-based agonist assays [3, 21](see following section). In addition, absorbance of the emitted light by colored compounds can produce false positives in antagonist assays.

1.1.3.* Bioluminescence – luciferases from firefly *(Photinus pyralis)*, sea pansy *(Renilla reniformas)*, and click beetles *(Pyrophorus plagiophthalamus) Like fluorescence, bioluminescence involves the generation of photons, which are detectable by a variety of microtiter plate readers. Unlike fluorescence, however, the emission of light does not require the input of excitation energy for its production and is instead the result of either a chemically or biologically catalyzed reaction [53]. Not surprisingly, the enzyme luciferase has been exploited as a transcriptional reporter, with the adenosine triphosphate (ATP)-dependent enzyme from firefly (*Photinus pyralis*) luciferase (FLuc) being the most commonly used. Applications using luciferases from sea pansy (*Renilla reniformas*; RLuc) and from the luminous Caribbean click beetle (*Pyrophorus plagiophthalamus*; CBLuc) have also become increasingly popular (Table 13.1).

The wild type *P. pyralis* luciferase catalyzes a two-step oxidation reaction. The first step involves activation of the substrate, D-luciferin (luciferyl carboxylate), using ATP to produce a luciferyl-adenylate intermediate (Figure 13.3B). The second step involves the reaction of this intermediate with oxygen to create a high-energy state oxyluciferin product, which, upon transition to ground state, emits yellow-green light (550–570 nm) [53, 54]. Light production is initially seen as a short, fifteen-second flash ("flash" type kinetics) followed by a low, steady level of luminescence ("glow"-type kinetics) [55]. Following the initial cloning of FLuc [55, 56] and its expression in mammalian cells [57], extensive work has been performed to optimize the commercially available enzyme for efficient expression and activity in mammalian cells, including removal of a peptide that targeted it to peroxisomes in cells of the firefly light organ [55]. In addition, commercially available substrate formulations have been optimized to extend the signal half-life to as long as four hours, depending on the formulation (PerkinElmer SteadyLite and Promega Steady-Glo). An extended-signal half-life is ideal for HTS, allowing larger batches of plates to be processed. Alternatively, substrate formulations that provide brighter light output, albeit with a shorter signal half-life of thirty minutes, are also available (PerkinElmer BriteLite and Promega Bright-Glo). Promega also now offers a substrate formulation called ONE-Glo that contains an alternative luciferase substrate (5′-fluoroluciferin) and demonstrates increased reagent stability compared to other formulations as well as enhanced light emission relative to Steady-Glo.

Luminescence-based reporter gene assays are considered one of the more dynamically responsive assay types – that is, responsive to changes in reporter transcription – for two main reasons. First, luciferase requires no post-translational modifications and is enzymatically active directly after protein synthesis [55]. Second, the half-life of the luciferase protein is short relative to other reporter gene products [2]. Whereas reporters with long half-lives potentially allow for increased signal because of accumulation of reporter gene product, dynamic changes in reporter gene expression, once accumulated in the cell, may be difficult to detect [53, 58]. It has been shown that wild type *P. pyralis* luciferase has a half-life of approximately four hours in mammalian cells [59]. Firefly luciferase variants have been engineered (and are commercially available) that contain a PEST sequence to reduce their half-life even further, resulting in a more responsive luciferase reporter [53, 58, 60].

Luminescence reporter gene assays are also generally one of the most sensitive assay types [2, 55]. Although light produced in a bioluminescence reaction is of lower intensity than that produced during fluorescence emission [53], this is offset by little to no background light inherent to luminescent assays. Thus, the luciferase reporter can be expressed at much lower levels than fluorescent proteins while still providing a robust signal. This ultimately results in an assay with high sensitivity, especially compared with fluorescence – which requires an input of excitation energy that can bleed into the detected emission signal as background [53]. Fluorescence, although ideal for imaging and microscopic techniques (small fields of view and high resolution) because of its higher-intensity emission, does not perform as well in HTS, where background levels must be low and data are collected at the resolution of a plate well – conditions in which luminescence-based assays often tend to perform better [53]. Microtiter plate readers commonly used for bioluminescence assays have charge-coupled device (CCD) cameras installed, in which light emitted from the entire microtiter plate is collected over a short period of time (typically less than one minute).

Live-cell analysis of FLuc expression and activity is possible and can provide an advantage for assays because a time-course of the FLuc signal is possible [61, 62]. ATP is a required substrate for FLuc–catalyzed luminescence [63] and the concentration of ATP in the cell is sufficient to allow for detection of FLuc activity by addition of D-luciferin alone. D-luciferin is charged at physiologic pHs and is somewhat cell membrane impermeable but luciferin esters have been synthesized to aid in cell uptake [63]. However, many protocols suggest cell lysis is necessary to detect a robust signal [57, 64]. Commercial formulations to detect luciferase activity generally contain luciferase substrate (e.g., luciferin) and the necessary substrates and co-factors (ATP, Mg^{2+}, coenzyme A) in a cell lysis buffer. The choice to use either D-luciferin alone or cell lysis with a formulation will depend on the strength of the luminescent signal, concerns around cell-health with prolonged luciferin exposure, and a consideration of assay artifacts associated with either mode of detection.

Beetle (both firefly and click variants) and RLuc do not share amino acid sequence similarity to one another [65], and although they catalyze an oxidation reaction to produce light, the substrates for these enzymes, as well as reaction pathways and necessary co-factors, are different (Figure 13.4A). RLuc is enzymatically active as a monomer and in vitro catalyzes the oxidative decarboxylation of the luciferin coelenterazine to form a dioxetane intermediate that ultimately produces coelenteramide and blue light with a spectral maximum of 480 nm [66]. In vivo in the sea pansy Renilla, the energy produced from coelenterazine degradation is transmitted through bioluminescence resonance energy transfer (BRET; Table 13.1) to a GFP that emits light at 505 nm [67]. This system has also been exploited for development of in vitro HTS assays (see next section).

RLuc offers the same advantages as FLuc as a reporter gene, but autoluminescence from the coelenterazine substrate – due to luciferase-independent oxidation in aqueous solutions – can increase background signal in an assay [55]. Formulations of cell lysis buffer with coelenterazine substrate are commercially available in which the components have been optimized to minimize autoluminescence (Dual-Luciferase Assay System/Dual-Glo Luciferase Assay System from Promega; Figure 13.4A). In addition, coelenterazine substrates that can be added to cell culture for live cell RLuc assays have been developed (ViviRen and EnduRen by Promega). Like FLuc, there are RLuc reporter gene constructs commercially available in which the luciferase coding region has been optimized for efficient transcription and translation in mammalian cell lines, in addition to constructs that code for destabilized luciferase protein. The use of RLuc has evolved significantly from its relatively simple use as a genetic reporter and red-shifted variants of RLuc have been engineered [62] as well as mutants such as RLuc8 which provide improved signal strength for use with the blue-shifted coelenterazine - DeepBlueC [68]. RLuc8 in combination with DeepBlueC is particularly enabling to BRET applications for the study of protein-protein interactions in cells (see following section).

A convenient and commercially available reporter gene assay exists to normalize bioluminescence assays and takes advantage of both the firefly and *Renilla* luciferases – the dual luciferase assay (offered by Promega as the Dual-Luciferase and Dual-Glo Luciferase Reporter Assays and by PerkinElmer as FireLite). In this system, the experimental reporter gene is generally FLuc, the control reporter is RLuc, and measurement of their activity is sequential (Figure 13.4A), capitalizing on the fact that the two luciferases have different substrate dependencies [69, 70]. After cell lysis, D-luciferin and ATP are added to the wells, and activity is measured by a detector. Subsequently, the reaction is stopped and the signal quenched by addition of another reagent, which also contains the RLuc substrate coelenterazine. Light production from this enzyme is then measured. The Dual-Luciferase Assay is designed for immediate detection of FLuc and RLuc activity after addition of their substrates. Use of this detection mix in microtiter plate format requires the use of detectors with auto-injector capabilities. However, more amenable to HTS is the Dual-Glo Luciferase Assay system, which is formulated for signal stability of up to two hours post–substrate addition [71]. Also of note, noncommercial detection formulations for FLuc and RLuc activity have also been described elsewhere [72, 73].

An alternative method of reporter gene assay normalization takes advantage of luminescence from two luciferase mutants derived from the yellow-emitting CBLuc. These variants share more than 95% of amino acid identity and catalyze the oxidation of the same substrate, D-luciferin, but were engineered to emit either green (537 nm) or red light (613 nm) [71, 74, 75]. An obvious advantage of this assay (offered by Promega as Chroma-Glo Luciferase Assay System) over the dual luciferase assay is that only

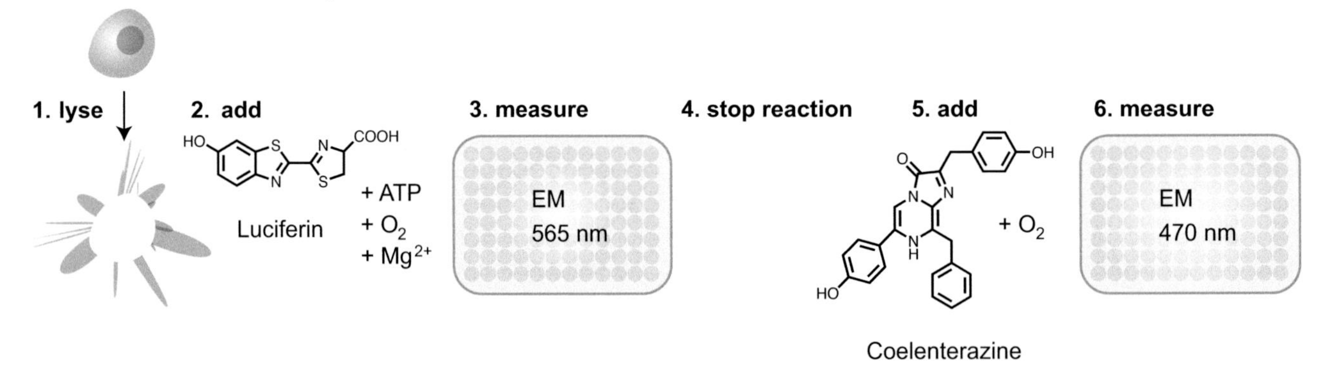

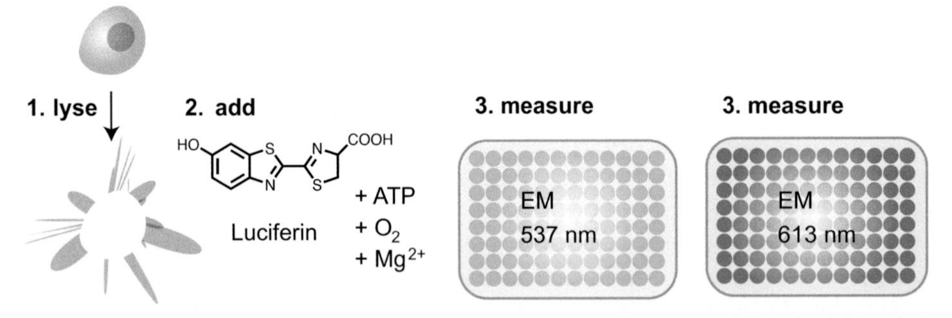

***FIGURE 13.4:** Methods of luciferase reporter gene assay normalization. Both the dual luciferase assay and two-color luciferase assay allow normalization of reporter gene expression by co-expression with a luciferase that responds to the biology of interest along with a different luciferase that is under the control of a constitutively active promoter, for example, SV40. **(A)** In the case of the dual luciferase assay, firefly and a* Renilla *luciferase are co-expressed. Each luciferase requires a different substrate for light production. To differentiate between the luminescent readout between the two luciferases, the appropriate substrate is added in separate addition steps and luminescence detection is sequential. Usually, a clear filter is used to collect all visible light. **(B)** In the two-color luciferase assay, luminescence signals from two highly similar click beetle luciferases are measured. Both require the same luciferin substrate but emit light at different wavelengths. One produces a green-shifted luminescence (537 nm) and the other produces a red-shifted luminescence (613 nm) that allow their activity, and subsequent expression, to be quantified separately. Appropriate filters and correction factors are used to allow collection of green and red luminescence in one detection step.*

a single addition of substrate is necessary, and light from the two variant luciferases can be differentiated based on the different bioluminescent emission wavelengths (Figure 13.4B). In addition, the fact that these two luciferase proteins are nearly identical means that they are expected to have similar responses to assay artifacts related to luciferase inhibitors, although one recent study has suggested that this may not be the case (unpublished observation). Many luciferase inhibitors may bind to the D-luciferin pocket and the mutations creating green or red color emissions are located in this pocket. One disadvantage to using this method of normalization is that because there is some spectral overlap of the light emitted by these luciferases, cell lines expressing a single CBLuc variant must be used as a control [74, 76].

Like all assays, a luciferase-based reporter gene assay has inherent artifacts. In HTS, awareness of such artifacts can prevent investing significant amounts of time, energy, and resources either following up off-target actives or, conversely, discarding potentially interesting leads. One potential source of off-target activity is inhibitors of the luciferase enzyme, which may appear as false positives in inhibitory screens. Our group screened a library of 72,000 small molecules for FLuc inhibitors and found that approximately 3% of these compounds were inhibitory (a percentage comparable to that found for blue fluorescent compound interference [21]), with 0.9% showing IC_{50} of less than 10 μM [77]. Not surprisingly, many of these small molecules identified as luciferase inhibitors were also identified as inhibitors of other FLuc-based reporter gene assays performed by our group. Major chemical series identified included likely mimics of the luciferin substrate, such as benzothiazoles, benzimidazoles, and benzoxazoles, which all acted as competitive inhibitors of FLuc, as did a series of quinolines that could be modeled into the adenylate pocket. Potent inhibitors ($IC_{50} < 1\ \mu M$) were also identified and included small molecules with a 3,5-diaryl-oxadiazole core or a benzyl amide core.

Small molecules that act as FLuc inhibitors can also increase the luminescent signal in luciferase reporter gene assays. The explanation for this counterintuitive observation is that the luciferase inhibitor binds to the enzyme to form a stable complex within the cells that stabilizes the luciferase enzyme and increases the protein half-life [59]. This leads to increased accumulation of luciferase in the cell, which upon detection mimics an upregulation of reporter gene expression, thus possibly leading to false positives. Small molecules found to have this effect include benzothiazoles, indolines, quinolines, and 3,5-diaryl-oxadiazoles [59, 77–79], and this effect is seen at concentrations commonly used in HTS [22, 79]. Retroactive analysis of numerous HTS experiments available within PubChem that used a cell-based FLuc reporter gene assay indicated that a significant proportion of actives identified in these screens were actually luciferase inhibitors (between 30% and 60%) [78]. FLuc undergoes a large conformational change from an open to a closed state upon substrate binding [80, 81], and it is possible that inhibitors may act similar to the substrates and promote the formation of a more stable closed state. The development of protease-resistant variants has been investigated, although these variants show altered light emission properties [82]. Regardless of these concerns, luciferase-based reporter gene assays still remain one of the most popular assays used in HTS because of their sensitivity and responsiveness, with the continued development of applications that take advantage of luciferase biology (see also section **3.3** for further discussion).

1.1.4. Bioluminescent Calcium Sensor – Aequorin The bioluminescent protein aequorin from the jellyfish *Aequorin victoria* exists as a photoprotein complex with coelenterazine and, upon exposure to calcium, leads to coelenterazine oxidation and the subsequent emission of light with peak emission at 469 nm [83, 84] (Table 13.1). Its relative specificity for calcium ions and its sensitivity to a wide range of biologically relevant calcium concentrations (10 nM to 10 μM and greater) has led to its development as a intracellular calcium sensor [85, 86]. Although this is not a transcriptional genetic reporter in the strictest sense, as its expression is not the variable measured in the assay, its activity can be used as a molecular sensor for the interaction of small molecules on signaling pathways that influence calcium signaling (e.g., GPCR targets), and it is briefly discussed here for this reason.

Apoaequorin protein needs to exist as a complex with coelenterazine to be a functional calcium sensor (called aequorin). Aequorin can be purchased as a lyophilized powder (AquaLite from Invitrogen) that is microinjected into cells. Alternatively, apoaequorin cDNA can be expressed in mammalian cells for use as a genetic reporter [86–90], requiring cells to be loaded with the membrane-permeable coelenterazine [84]. Cell lines stably expressing apoaequorin, a variety of G protein-coupled receptors (GPCRs), and a promiscuous Gαq to couple GPCR activation to calcium signaling have been developed and are commercially available (AequoScreen by PerkinElmer) [91]. Additionally, calcium changes in intracellular compartments can be measured by targeting apoaequorin [84, 87]. An advantage to using aequorin as a calcium sensor over fluorescent calcium indicators is that, like all bioluminescent reactions, there is no excitation energy required; therefore, there is less background inherent to the assay [92].

One complication of using aequorin as a reporter gene in HTS is that the luminescence reaction takes place extremely quickly – on the order of five seconds ("flash"-type kinetics) [83] – requiring immediate detection by a luminometer, which is not ideal for HTS assays using standard equipment [91, 93]. To aid in HTS of cell-based calcium assays with flash kinetic reads, detectors with multiple injection capabilities have been developed [91].

Signal generated by aequorin is characterized as relatively weak compared with other reporters. Like GFP, aequorin is not an enzymatic reporter. As such, aequorin does not have the advantage of signal amplification inherent to enzyme reporters, in which substrate is continually converted to product. Instead, one aequorin is capable of producing one photon, and at any given time, only a small percentage of the aequorin capable of producing a photon does so [87]. In addition, "spent" apoaequorin takes time to regenerate and is inhibited by Ca^{2+} [94]. To address this, efforts have been made to develop a new calcium-sensitive photoprotein for use in HTS. The photoprotein, derived from *Obelia longissima*, is reported to produce a stronger luminescent signal than aequorin [92].

1.1.5. β -Galactosidase (β -gal) β-galactosidase is an enzymatic reporter that traditionally is used to generate a colorimetric readout. However, depending on the substrate used, different readouts are possible, such as fluorescence and chemi- or bioluminescence [95] (Galacton chemiluminescent substrates, available from Applied Biosystems). Chemiluminescent reactions have fast kinetics, however, and may require plate readers with automatic injectors. Bioluminescence is achieved by providing a β-gal substrate, which upon hydrolysis becomes a substrate for FLuc, thus allowing indirect measurement of β-gal activity. The luminescence assays have the added benefit of producing significantly greater S/B and enhanced sensitivity compared with the colorimetric assay [96].

Regardless of the substrate used, the β-gal assay inherently has background signal issues, as mammalian cells have endogenous β-gal activity [96, 97]. Commercially available assay formulations often come with buffers that maintain conditions at a pH that minimizes endogenous β-gal activity. As previously mentioned, the β-gal reporter gene, like the RLuc reporter, can also be used as a control vector for normalization (offered as the Dual-Light System by Applied Biosystems) [97, 98]. Although not as commonly used in transcriptional reporter gene assays in HTS, β-gal has currently found its niche within HTS as a post-transcriptional reporter in enzyme fragment complementation (EFC) assays (Table 13.1) [99] through exploitation of α-complementation (e.g., DiscoveRx PathHunter; see following section).

1.1.6. Reporters Less Commonly Used in HTS: Chloramphenicol acetyl transferase (CAT) and Secreted Placental Alkaline Phosphatase (SEAP) CAT and SEAP (Table 13.1), although still commonly used in bench-science and which traditionally generate colorimetric or radiometric outputs, largely have been replaced in HTS applications with reporter enzymes that produce luminescent or fluorogenic products, such as luciferase or β-lactamase, respectively.

Chloramphenicol acetyl transferase was one of the first extremely popular reporter genes [97, 100]. As a bacterial enzyme, it catalyzes the transfer of an acetyl group from acetyl coenzyme A to chloramphenicol. Traditionally, CAT expression and activity were measured using radiolabeled substrates that were converted into products that could be separated by thin-layer chromatography and detected using autoradiography [97]. CAT has a relatively long protein half-life of fifty hours [59], making it a less dynamic reporter than others that are currently available. CAT assay formulations have been developed that produce a fluorogenic product but still require thin-layer chromatography to separate substrate from product (FAST CAT assay from Invitrogen). Thus, although highly sensitive as a reporter, it has failed to successfully transition for use in HTS, hindered by the methods necessary for product detection [96].

Secreted placental alkaline phosphatase (AP) is a modified form of alkaline phosphatase that is secreted from the cell, allowing aliquots of media to be removed from stimulated cells and SEAP activity to be measured. Traditionally, hydrolysis of *p*-nitrophenylphosphate resulted in a colorimetric product with light absorbance at 405 nm [100]. Like β-gal, this enzyme can be coupled to a variety of different substrates, enabling chemiluminescence or bioluminescence detection, which have increased sensitivity over the colorimetric assay (e.g., Clontech's Great EscAPe SEAP Chemiluminescent Kit). In contrast to enzymes such as luciferase or β-lactamase, endogenous AP activity exists in almost all mammalian cells, potentially resulting in high levels of background for this assay. Using a secreted form of alkaline phosphatase helps to alleviate this problem somewhat, as endogenous AP activity is intracellularly restricted [97]. The protein half-life of APs is on the order of seventy-two hours [97], rendering this assay less sensitive to changes in gene expression compared with other formats. As this assay generally must be coupled to a luminescence reaction to achieve high levels of sensitivity, it is not surprising that the SEAP assay has not translated well into assays developed for, and commonly used in, HTS.

1.2. Normalization of Reporter Expression/Activity

A problem intrinsic to reporter gene assays that can be a central issue when these assays are used in HTS is variability between assay wells that is due not to the experimental conditions being tested but, for example, to liquid handling variations that lead to differences in cell density or detection reagent concentrations [71]. In addition, differences in transfection efficiency between batches of cells used for a screen can also be an added source of variability. Ideally, the answer to this problem is a form of internal normalization. If a transiently transfected vector is used for the reporter gene assay, an additional vector – which serves as an internal control – can be co-transfected along with the experimental reporter gene vector. This control vector contains a reporter different from that found on the experimental reporter vector, driven by a constitutively active promoter such as a CMV or SV40 promoter [71](Figure 13.2C). Normalized data are reported simply by dividing the experimental reporter signal by the control reporter signal [71]. There are numerous examples of such normalized assays available. The most popular include the dual luciferase assay (Figure 13.4), which uses RLuc as a control for FLuc expression and activity; the two-color luciferase assay, which uses red and green light-emitting CBLuc; and dual reporter gene assays that involve the detection of FLuc and use β-galactosidase expression activity as a control [98] (e.g., Dual-Light Combined Reporter Gene Assay System from Applied Biosystems).

Along with the advantages of normalizing reporter gene assays, it should be noted that the use of two vectors can introduce additional variables. One issue is the possible *trans* effect of regulatory elements on reporter gene expression; that is, regulatory elements found on one vector can influence the expression of the reporter on the co-transfected vector, which can complicate results [101]. For this reason, it is recommended that promoters of similar strength be used to drive expression of each reporter gene and that the ratio of experimental to control vector be optimized when employing transiently transfected systems [102].

1.3. Targeting a Signaling Pathway or a Transcription Factor

As already briefly mentioned above, a transcriptional reporter gene assay can be used to measure the activity of a signaling pathway or to specifically target the activity of the transcription factor. It is not uncommon for cDNAs to be ectopically expressed so that a signaling pathway of interest can be coupled to reporter gene expression. One example of this is depicted in Figure 13.2D, in which modulation of intracellular calcium signaling and nuclear factor of activated T-cells (NFAT) transcriptional activation is studied by ectopically expressing a Gq protein-coupled GPCR that can be stimulated by known ligands [103–105].

When activation or inhibition of a signaling pathway is monitored by way of a reporter gene assay, often the specific target of the small molecule is not known. Although probing the activity of an entire signaling pathway may generate many interesting leads, one disadvantage to such an assay is the potentially time-consuming undertaking of determining the target of a compound of interest [3].

In some instances, only the activity of the transcription factor itself needs to be probed, not the activity of an entire signaling pathway. This type of screen is common for nuclear receptors (NRs), which are ligand-dependent transcription factors [106] whose activity is directly influenced and regulated by hormones, lipids, and fatty acids. Nuclear receptors have a DNA-binding domain (DBD) and a regulatory ligand-binding domain (LBD). Activation of the protein via its LBD leads to the transcription of genes whose promoter regions contain a recognized response element [107]. The modular nature of NRs allow for chimeric transcription factors to be designed [106, 107], which are essentially fusion proteins that contain the LBD of the transcription factor of interest and the DBD of another NR, such as the DBD from the yeast transcriptional element Gal4. In this case, a reporter gene downstream of the upstream activating sequence (UAS) recognized by Gal4 is transcribed upon activation of the chimeric transcription factor via its LBD (Figure 13.2E). One advantage to this assay design is that the promoter sequence recognized by the transcription factor need not be known or subcloned into a reporter gene construct, as one can rely on using a generic UAS-reporter construct. This type of reporter assay design has been widely used in HTS to screen for ligands of nuclear receptors such as the farnesoid X-activated receptor (FXR) and the peroxisome proliferator-activated receptor delta (PPARδ) [108], among others.

1.4. Promoter Choice

Transcriptional reporter gene assays can be used to measure the activity of a single signaling pathway or a few signaling pathways, depending on the type of promoter placed upstream of the reporter gene. If one chooses to focus on a single signaling pathway, a reporter gene construct containing a simple synthetic promoter with just one response element can be used. This is demonstrated in Figure 13.2D, in which the regulatory element recognized by NFAT is the promoter used to control the expression of the reporter gene β-lactamase. Whereas endogenous gene expression by NFAT can be more complex than this, with cooperative enhancement of gene expression by the activation of other signaling pathways and transcription [103, 105], this reduced promoter provides a simplified readout of NFAT activation [103–105]. Alternatively, one may be interested in the regulation and activity of a number of different signaling pathways whose output intersects in gene transcription. In this case, endogenous or complex promoters could be used. Such an example would be the use of the promoter for *c-fos*, which contains regulatory elements for a number of different transcription factors, such as the serum response element (SRE) and cAMP response element (CRE) [109–112], and thus provides a measure of activity for a number of different signaling pathways (Figure 13.2B).

1.5. Cell Lines: Stable or Transient

Development of a transcriptional reporter gene assay requires the production of a recombinant cell line containing at least one construct with the reporter gene downstream of the regulatory promoter element. This construct can be stably integrated into the host cell's DNA (as seen in Figure 13.2A, B, D, and E) or exist as a transient construct (Figure 13.2C). Additional constructs may be required for expression of a chimeric transcription factor, adding required but exogenous signaling pathway components, or addition of a control reporter construct that allows the normalization of experimental reporter gene expression to changes in transfection efficiencies or cell densities (such as in the dual luciferase assay; Figure 13.2C).

If one is considering the use of cell lines in which the reporter gene construct is stably maintained, it is worth mentioning that the expression of the reporter gene may be influenced by the environment in which it has integrated. Insertion near regulatory elements for endogenous genes can enhance or suppress transcription of the reporter gene. It is also worth mentioning again that co-transfected transient constructs can affect gene expression *in trans* [102].

2. POST-TRANSCRIPTIONAL REPORTER SYSTEMS

Post-transcriptional events include processing of mRNA transcripts, post-translational modifications of proteins, and protein interactions or redistribution. In some cases, the properties of the reporter can be exploited to create novel assays. For example, an assay for viral particle budding has been described in which cell lines co-expressing FLuc and the Ebola virus coat proteins were made [113]. The authors found that simple co-expression of the viral genes and luciferase leads to trapping of the luciferase enzyme within the virus bud particles that were secreted by the cells, allowing measurement of luminescent activity in the supernatant to report on viral activity. However, molecular sensors of post-transcriptional events more typically involve engineered reporters that employ luminescent or fluorescent constructs. These enable the measurement of

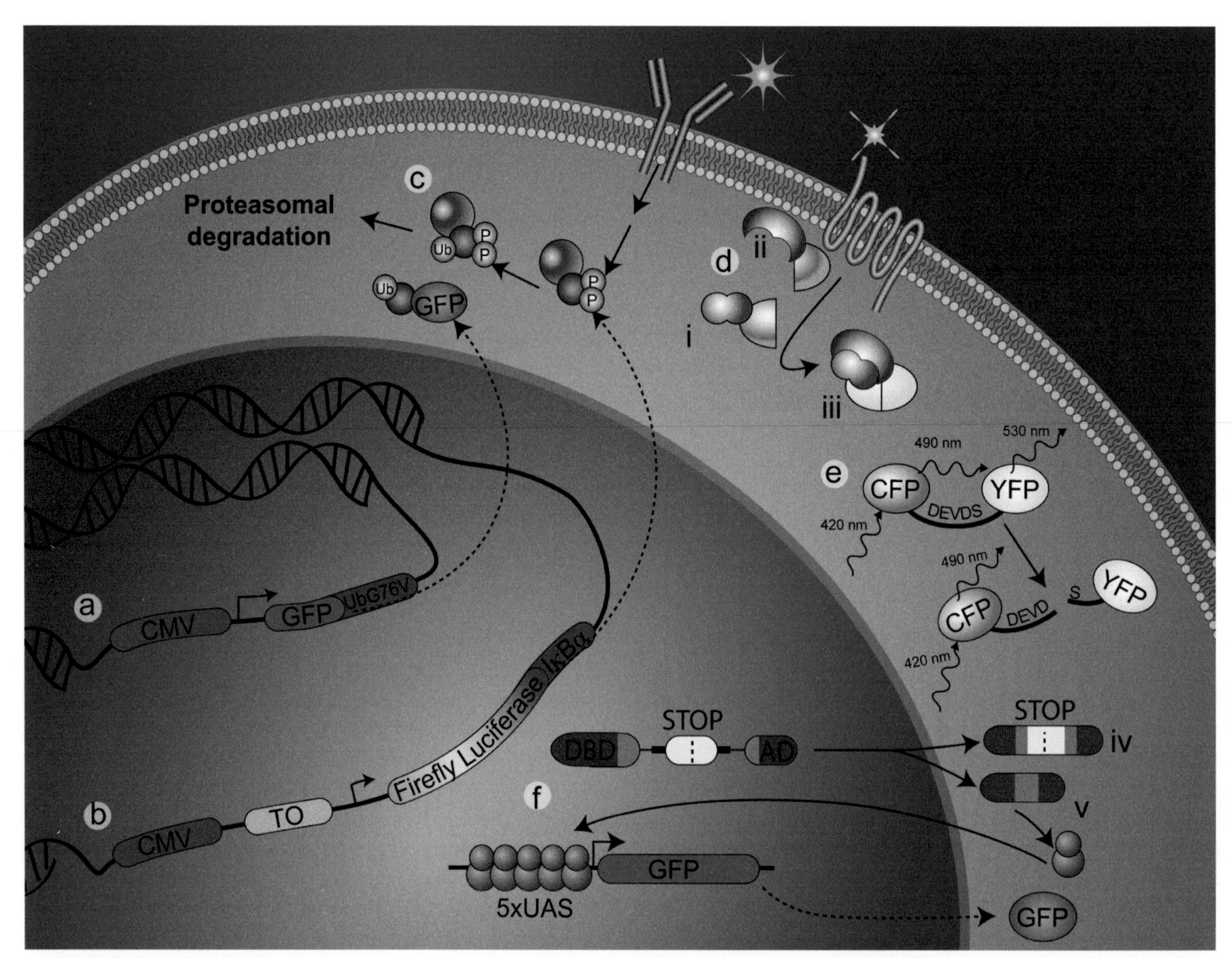

FIGURE 13.5: ***Examples of different post-transcriptional reporter assays. Regulation of signaling networks involve many different types of post-transcriptional events, such as mRNA processing, post-translational modification, and the formation of protein complexes. (a) An example where a reporter is constitutively expressed but is mutated to carry a destabilizing sequence, as in the case shown, where GFP is fused to the ubiquitin mutant G76V leading to rapid degradation of the reporter by the proteasome system. (b) An inducible system where a Tet-On (TO) regulatory element is placed downstream of a promoter such as CMV and used to express a genetically fused Iκ Bα -FLuc reporter. (c) Activation of the NFκ B pathway leads to phosphorylation and ubiquitination of the Iβ κBα -luciferase fusion, resulting in degradation of the reporter. (d) Example of a PCA system using split reporter proteins. i) An interacting protein (burgundy spheres) is fused to a nonfluorescent fragment of a YFP (gray half-spheres). ii) another interacting protein that shows a stimulus-dependent interaction with (i) is fused to a complementary nonfluorescent YFP fragment. iii) stimulation of a pathway promotes interaction between (i) and (ii) that reconstitutes the YFP fluorescence. (e) FRET-based sensor where fusion between YFP and a cyan mutant of GFP (CFP) is engineered with an intervening caspase-3 recognition site. The 530 nm/490 nm FRET ratio is recorded to measure intracellular caspase-3 activity. (f) An assay for mRNA splicing using a tandem reporter system [168], where a transcription factor composed of a DNA-binding domain (DBD) and a transcriptional activation domain (AD) is employed in which the coding sequence for each domain is flanked by exon sequences from a gene of interest containing splice sites to create a "mini-gene" capable of undergoing splicing. In the example shown, an exon containing stop codons is placed in the middle of the mini-gene to report on splicing-mediated exon skipping. Retention of the exon prevents the formation of an active transcription factor (iv), whereas exon removal results in the formation of a functional transcription factor (v) that then binds to a set of tandem upstream-activation sequences (UAS) to produce the reporter (GFP in the case shown) in an amplified manner.***

either post-translational modifications, such as ubiquitination and thus protein degradation, or protein-protein interactions through the reconstitution of split reporters. There currently is a wide range of HTS assays that address these events (Figure 13.5). In the following section, we illustrate specific examples of assay systems that have been developed to explore these processes, as well as provide a perspective on newly designed assays that expand their biological application into areas such as mRNA splicing.

2.1. Measurement of Protein Intermediates and Stability

Post-translational modification of proteins – such as phosphorylation or ubiquitination – that act to attenuate or augment cell signaling pathways have been studied using a range of assay types. Historically, assays for phosphorylated or ubiquitinated protein intermediates have utilized immunoassays in low-throughput formats such as Western blots, enzyme-linked immunosorbent assays (ELISAs), or microscope-based imaging systems. However, recently, many assays compatible with HTS systems have been developed that take advantage of fluorescent proteins (e.g., GFP) or enzymes such as FLuc.

A prototypical pathway for which several post-translational assays have been developed is the NFκB pathway. Activation of NFκB occurs when cytokines such as TNFα activate the pathway leading to phosphorylation of IκBα, a protein that sequesters the transcription factor NFκB to the cytoplasm by masking the nuclear translocation of the transcription factor. Phosphorylation of IκBα at two critical serine residues leads to disassociation of IκBα from NFκB and subsequent ubiquitination and degradation of IκBα by the proteasome [114]. The IκBα protein has been used to construct post-translational NFκB reporter assays by fusing this protein to reporters such as GFP [115] or FLuc [116](Figure 13.5A, B). For example, the Chroma-Luc system (Promega) was used to construct a two-color luciferase assay [76] for IκBα degradation using the red and green CBLuc variants mentioned in the previous section. The two-color luciferase assay fused the green beetle luciferase to IκBα, whereas the red beetle luciferase (CBR) was expressed in its native state and used for correction of nonspecific effects (Figure 13.5B, C). Importantly, this assay employed an inducible promoter system in which the same promoter was used for both luciferases. The reporter constructs were introduced into a human lymphoma cell line with constitutive IκB kinase (IKK) activity so that acute inhibition of IKK or the proteasome resulted in an increase in the amount of the IκBα-luciferase reporter.

Several issues should be considered when developing these types of reporter cell lines, including whether the signal rises or falls upon modulation of the pathway and the rapidness of the response. For post-transcriptional–based assays using dual-reporter systems, the use of a common promoter is often possible and should be employed whenever possible to normalize for transcriptional-based effects. Generally, constructing the assay so that a rise in signal is obtained for active compounds is desirable, as this allows for facile discrimination from nonspecific effects that can decrease the assay signal, such as cytotoxcity. An example of such an assay is one in which destabilization domains containing mutant forms of ubiquitin (UbG76V) that cannot be cleaved by ubiquitin hydrolases are genetically fused to either a GFP or β-lactamase reporter [117]. This leads to rapid destabilization of the reporters and provides a robust increase in the assay signal upon addition of proteasome pathway inhibitors (Figure 13.5A).

A technology that has been widely employed in HTS is EFC, which is based on α-complementation of the enzyme β-galactosidase (Table 13.1) [99]. The EFC assay contains a β-gal that has been split into a small peptide α fragment (enzyme donor [ED] fragment) and an enzyme acceptor (EA) fragment that interact with high affinity ($K_D < 1$ nM). This system has been widely used in cell-free assays such as those measuring protein kinase activity [118] but has also been useful for constructing cell-based assays, including those that measure protein expression and stability. The NFκB pathway has been explored with this technology by fusing the ED fragment to IκBα and expressing this along with the EA fragment. Upon activation of the NFκB pathway, the ED fragment was selectively degraded, resulting in a decrease in β-gal activity [119]. Fragments of β-lactamase have been also successfully implemented as EFC assays for HTS [120].

A set of assays targeted to probe intermediates of the NFκB pathway has been described that allows one to determine if inhibitors are working either at the level of IKK activation or further downstream in the ubiquitin pathway [121]. In this system, a time-resolved FRET (TR-FRET) assay was developed. In TR-FRET, the donor fluorophore is a lanthanide (such as europium [Eu] or terbium [Tb]) that demonstrates an extended fluorescence lifetime (~500 μsec), resulting in an assay signal in which an increase in S/B can be achieved by delaying the detection of the signal until background fluorescence (which has a much shorter fluorescent lifetime) has decayed. For the IKK assay, a TR-FRET assay was employed in which GFP (the acceptor) was genetically linked to IκBα and Tb-labeled antibodies ($\lambda_{ex} = 340$ nm, and the 490 nm Tb emission maximum was used to excite GFP) specific for either phosphorylated or ubiquitinated IκBα to allow inhibitors of phosphorylation acting at the level of IKK activation to be distinguished from those that stabilize the ubiquitinated form that likely act further downstream in the pathway. The TR-FRET signal in these assays is particularly attractive for HTS, as fluorescent interference from the cell matrix or test compounds is reduced.

2.2. Measurement of Protein Complexes

Cell signaling pathways involve the formation of multiple protein complexes that act as regulators within intricate signaling networks. Active compounds identified in transcriptional reporter assays (as described in the previous section) may operate in one or more interconnected or redundant signaling pathways, making it difficult to determine the mechanism of action. Post-transcriptional reporter assays provide direct measurement of protein complexes known to operate within particular pathways, allowing one to pinpoint the mechanism of active compounds to a specific area within multifaceted protein networks [122].

Several systems have been engineered to provide assays that report on the formation of protein complexes. The two most widely used methods utilize either a protein complementation assay (PCA) [123–125], in which a split reporter protein is genetically fused to interacting protein partners (Figure 13.5D), or a system that relies on FRET [126] or BRET to generate a signal upon protein complex formation (Table 13.1). A third system employs a specific protease from the tobacco etch virus (TEV) which is fused to one protein partner and a transcription factor that is fused to the other partner via a TEV cleavage sequence. Association of the two proteins releases the transcription factor which then drives expression of a reporter-gene such as FLuc (so-called "Tango assay";[127]).

The most common split reporters for PCA have included fluorescent proteins (such as GFP [128]) or luciferases [129]. To construct these PCA systems, the reporter must be split into two stable halves that are capable of associating and reconstituting activity. The reporter halves are kept separate by fusing each half to interacting proteins that associate only in response to a biological event such as a ligand-induced conformational change, translocation between cellular compartments, or protein phosphorylation. The split reporter fragments have relatively low affinity to each other compared with the split enzyme fragments used in α complementation systems such as EFC but are capable of efficiently associating upon interaction of the fusion proteins. The manner in which these two reporter fragments associate needs to be carefully considered to minimize the artificial influences of reporter reconstruction on the relevant biology. Ultimately, the sensitivity of the split reporter depends on how well the two reporter fragments are sequestered and prevented from nonspecific interactions as well as how efficiently the reporter is reconstructed upon interaction of the fusion protein partners.

One PCA-based study has been described in which forty-nine cell-based PCAs were used to profile 107 different compounds [123]. Rapid implementation of these assays and quantification of the results were enabled by using a common reporter system for all forty-nine assays. This was accomplished with a split yellow fluorescent protein reporter (YFP; λ_{ex} = 513, λ_{em} = 527 nm) in which YFP fragments were fused to critical interaction proteins known to function in different signal transduction pathways. Reconstitution of the YFP upon interaction of the two fusion proteins yielded a signal that was twenty times brighter than either split YFP fragment alone [128]. For example, a PCA employed in this study fused one-half of the split YFP to the C-terminal tail of the β-adrenergic receptor and the other half to β-arrestin to report on receptor activation (Figure 13.5D). An interesting result of this project was the discovery of compounds with unrelated structures with similar activity profiles. Hierarchal clustering of compound activity revealed a supercluster of drugs that affected common pathways. Although the compounds showed little structural similarity, all of the compounds in the supercluster seemed to possess a "hidden phenotype" of antiproliferative activity. Therefore, the use of PCA to profile compound activities across multiple signaling pathways can identify important phenotypes associated with compounds or off-target activity.

Fluorescent proteins can be configured to create FRET-based assays that measure the association of two proteins within cells. Assays based on FRET signals can show low changes in signal (approximately two-fold), but ratiometric processing of the donor/acceptor emission signals provides a means to correct for assay variations such as cell number variations, and can thus yield a highly precise measurement. For example, a FRET assay that used YFP and CFP to report on signaling events proximal to the insulin receptor (IR) in 3456-well microtiter plate format has been described [130] (Aurora Biosciences, San Diego, Calif.). The protein kinase Akt2 is sequestered to the plasma membrane through a PH domain of the kinase upon IR activation. In the FRET assay, CFP and YFP fusions of Akt2 were constructed and expressed in the same cell so that a FRET signal resulted when both Akt2 fusions were recruited to the plasma membrane upon IR activation, thus placing the YFP and CFP in close proximity. The final optimized assay conditions showed an S/B of 1.26- to 1.29-fold, although it was extremely precise (coefficient of variation=2%, Z' factor = 0.6), enabling assays for both agonists and potentiators of IR to be rapidly screened against one million compounds. However, the low S/B in such FRET assays often makes these assays difficult to optimize for HTS and requires optimization of the FRET pairs (in this case, more than thirty cell pools containing different constructions of the CFP/YFP PH domain fusions were evaluated to develop the assay).

In a BRET-based assay, the RLuc reporter is genetically fused to a candidate protein and a GFP reporter is genetically fused to another candidate protein of interest, such that when the two reporters are brought in close proximity through interaction of the two candidate proteins, an assay signal is generated (Table 13.1) [131]. RLuc (or RLuc8, see section **1.1.3**) serves as the donor to produce luminescent light that excites either GFP or YFP acceptors in close proximity [132]. The BRET system provides an advantage over FRET-based systems because an excitation light source is not required, thereby eliminating autofluorescence from compounds or cells [133]. The BRET system enables cell-based protein-protein interaction assays and has been successfully applied in an HTS for inhibitors of the CCR5/β-arrestin [134]. BRET has also been extended to other reporter pairs such as FLuc and DsRed [135].

Beetle luciferases have also been used to construct cell-based protein-protein interaction assays. For example, a split CBLuc has been designed to monitor nuclear receptor (NR) co-regulator interactions [136]]. The CBLuc was engineered to exhibit red-shifted luminescence (λ_{max} = 615 nm), which limits cell and compound autofluorescence, rendering this an ideal reporter for luminescence imaging in whole tissues or animals. A considerable amount of development was required to construct the optimal split CBLuc reporter, and the optimal halves were found to recover approximately 1% of the full CBLuc activity. This low recovery of activity can be typical for split FLuc reporters, but because of the high S/B inherent in luminescence, this activity generally provides adequate signal for either in vitro or in vivo applications. In this study, the split reporter fused one half of the CBLuc reporter to the androgen receptor (AR) ligand binding domain (LBD) that was constructed to contain a co-regulator peptide fused with the complementary half of CBLuc. The limit of detection for the cognate ligand 5α-dihydroxytestosterone was 100 nM, and the rank order of steroids tested was in agreement with other transcriptional reporter gene assays.

2.3. Post-Translational Assays for Protein Redistribution

A large number of nuclear receptors translocate from the cytoplasm to the nucleus in response to ligands, which leads to transcriptional events that either enhance or repress gene transcription. The measurement of cytoplasmic-to-nuclear localization of a reporter has been implemented in HTS systems, although typically, imaging technologies employing either wide-field confocal microscopes and image processing algorithms are used [51, 137]. These formats are often not suitable for HTS because of the cell fixation and washing steps required and long image acquisition and processing times. However, protein translocation can also be measured using EFC [138–140]. For this purpose, the EA fragment is sequestered in the nucleus through the use of a proprietary set of nuclear targeting sequence modifications [141, 142], and a nuclear receptor that undergoes ligand-dependent translocation from the cytosol to the nucleus is fused to the ED fragment. A cell-based 1536-well microplate assay for the glucocorticoid receptor (GR) has been described in which translocation of an ED-GR fusion to the nucleus was measured through an EFC approach [143]. This provided a simple, homogeneous assay in which ligands promoting GR translocation were detected with a robust chemiluminescent signal.

2.4. Molecular Sensors of Second Messengers

If the goal is to identify compounds that directly affect the functional activity of the receptor, a sensor of a second messenger is desirable as the assay signal that is generated from events more proximal to the receptor. Assays that enable the detection of second messengers, such as cAMP, provide an alternative assay format to transcriptional-based assays that measure the activity of cell-surface receptors through distal DNA response elements (e.g.; CRE; see previous section). These assays have been applied to one of the largest classes of cell-surface receptors that have been exploited by the drug discovery industry – the GPCR family that modulate either cAMP or Ca^{2+} levels [144, 145], among other activities.

A wide variety of assays are available to measure cAMP [3]. These include assays that use fluorescence polarization [146], homogeneous time-resolved fluorescence (HTRF) [147], and EFC [148]. These approaches are considered "end-point assays" because the cells undergo lysis during detection. Intracellular changes in Ca^{2+} levels can be measured in real time using fluorescent calcium indicators [149] (e.g.; Fura-3 or Fluo-4; also see *Assay Guidance Manual Version 5.0* on the National Institutes of Health (NIH) Chemical Genomics Web site [150]) and fast kinetic plate readers such as the Fluorimetric Imaging Plate Reader (FLIPR) [144, 151] or the FDSS6000 [91]. Coupling cAMP modulation to Ca^{2+} has been achieved with cell lines containing promiscuous or engineered G proteins [152, 153] that act as adaptors to couple GPCRs that normally regulate cAMP levels to a Ca^{2+} response. However, a molecular sensor of cAMP in living cells has been developed using a cyclic nucleotide-gated (CNG) ion channel [154]. The CNG ion channels are members of the ligand-gated ion channel family containing retinal and olfactory members. Whereas retinal CNG ion channels are gated by cGMP, olfactory CNG ion channels open in response to either cGMP or cAMP. These ion channels are highly selective for Ca^{2+} under physiologic conditions and provide a means to connect cAMP detection to Ca^{2+} influx detection using fast kinetic readers. In the CNG system, membrane potential dyes [155, 156] can be employed to allow detection using conventional plate readers. This cAMP assay has been applied to 1536-well microtiter plates using a titration-based screening approach known as qHTS [20], in which more than seventy thousand compounds were assayed at seven concentrations, and small molecule agonists of the TSH receptor were identified [157].

Protein kinase A (PKA) has also been engineered as a cAMP sensor by taking advantage of the cAMP-dependent interaction of its regulatory and catalytic subunits; upon exposure to increased concentrations of cAMP, the two domains dissociate. An early proof-of-principle study on the use of PKA as a cAMP sensor employed fluorescently labeled catalytic and regulatory domains to create a FRET assay in which a rise in cAMP levels resulted in a reduction in FRET signal because of disassociation of the two domains [158, 159].

PCA systems based on PKA have also been developed for the detection of cAMP. In one example, split fragments of RLuc were fused to the regulatory (RegF1) and catalytic subunits (CatF2) of PKA [129]. This system yielded an active RLuc in the RegF1:CatF2 complex that became inactive upon complex disassociation because of rises in cAMP levels. The inactivation of RLuc showed a half-life of approximately thirty seconds. Additionally, treatment of cells with forskolin resulted in reversible disassociation of 80% of the complex, consistent with the known dynamics of PKA regulation. The reversibility of the PKA reporter was only observed using RLuc, and the authors noted that they could not measure disruption of PKA regulatory interaction by cAMP using an YFP-based PCA reporter. This represents one criticism of split YFP/GFP reporters: They may not allow for dynamic measurements of protein complex formation/disassociation because of the stable β-barrel structure found in these fluorescent proteins [160].

Numerous other systems based on FRET between two linked fluorescent proteins exist, including sensors for Ca^{2+} levels [161], cAMP [162], and cGMP [163]. However, as mentioned in section **2.2**, with the CFP/YFP-based sensor for the insulin receptor, these assays have not been widely implemented in HTS because of the effort involved in optimizing FRET assays. Additionally, an issue that encumbers developing optimal FRET pairs using GFP variants is heterodimerization of the linked fluorescent proteins, which can make the assay less sensitive to identifying inhibitors [164].

2.5. Fluorescent Proteins as Molecular Sensors of Cell-Based Protease Activity

Enzyme activity can be measured within cell-based assays with appropriate molecular sensors. For example, activity of proteases such as caspase 3 has been measured in cell-based assays using a FRET-based detection. In this case, GFP mutants (e.g., CFP/YFP) are linked by a protease recognition sequence that, upon cleavage by the protease, results in a loss of FRET signal (Figure 13.5E). The stable β-barrel structure of GFP protects the fluorescent protein from degradation by the protease. This assay used an HTS to search for compounds that induce apoptosis, with strongly active compounds demonstrating a greater than 70% change in FRET signal. The assay showed low false negatives and positives rates (~0.2%) [165].

2.6. Splicing Assays

Research in genomics has led to the revelation that alternative pre-mRNA splicing events play a major role in regulating protein function in mammalian cells [166]. Prototypical splicing involves intron excision; however, splice isoforms can be created by exon retention, truncation, extension, or intron retention. This leads to an average of three spliced isoforms per gene, with nearly 75% of human genes thought to undergo alternative splicing [167]. Aberrant splicing can lead to many genetic diseases, and abnormal splicing has also been implicated in cancer. Constructing assays that report on splicing events has been difficult, because the changes in isoform expression are usually less than ten-fold [167]. One attractive approach employs a tandem reporter system for splicing in which the splicing signal is augmented by transcriptional synergy [168]. In this system, a transcriptional activator that is known to act synergistically when appropriate promoters are present is expressed in a splicing-dependent fashion (Figure 13.5F). Expression of the transcriptional activator then drives GFP expression in a manner in which a four- to five-fold change in splicing yields a thirty- to fifty-fold change in GFP expression.

Two-color reporter systems have also been employed to construct splicing assays, in which splicing specific events can be distinguished from effects on transcription or translation. An assay suitable for HTS has been described that uses a mini-gene reporter in which alternative splicing resulted in either GFP or RFP expression [169].

3. GENERAL CONSIDERATIONS AND ISSUES ASSOCIATED WITH CELL-BASED MOLECULAR SENSOR ASSAYS

3.1. Cytotoxicity

In HTS, in which hundreds of thousands of samples are typically screened, even artifacts that occur at low frequency can be cumulatively problematic. One of the main artifacts to plague reporter gene assays are false positives in antagonist assays due to cytotoxicity of a compound [8]. In this case, the reporter readout is decreased not because of inhibition of signaling but because of cell death. Ratiometric assays, such as the β-lactamase assay, can help flag these instances, as outlined in section **1.1.1**. Alternatively, including a cytotoxicity marker in the assay can also be used as a measure against falsely identifying "inhibitory" compounds [14]. In general, whenever possible, efforts should be made to design the assay so that a signal increase is obtained for compounds that are active at the target of interest, particularly if a single reporter will be used to construct the assay. However, inhibitors that stabilize the reporter are still a concern (described in section **3.3**).

3.2. Autofluorescence

It has been reported that aggregates of compounds, fluorescent particulate matter such as dust or lint, and fluorescent or colored compounds can affect the fluorescent ratio measurements in β-lactamase assays [8, 9]. For example, yellow-colored compounds have been found to increase fluorescence in the green channel [8]. In addition, a significant part of small molecule libraries are composed of autofluorescent compounds. The fluorescence profile of seventy thousand compounds found that more than 5% fluoresced brighter than a commonly used fluorophore present at a concentration of 10 nM (4-methyl umbelliferone [4-MU]) [21]. In addition, the majority of these compounds fluoresce in the blue-green spectral region – the region collected in the β-lactamase and GFP assays. "Activity" due to autofluorescence artifacts also tends to be reproducible and to titrate [21]. False identification of actives or inhibitors due to blue or green small molecule autofluorescence, respectively, may be reduced with the use of plate readers that provide higher content data, such as the laser scanning fluorescence microplate cytometer, which can be set to measure only cellular fluorescence. Ultimately, however, the construction of fluorescent counterscreen databases of the chemical library is desirable in order to provide corroborative evidence on compound autofluorescence.

Another phenomenon that has been found to contribute greatly to the false positives in HTS assays is the aggregation of compounds. In fact, in a β-lactamase assay, nearly 2% of compounds initially identified as actives in a seventy thousand compound library were identified as forming aggregates [23, 170]. Although these statistics were determined using cell-free assay systems, the phenomenon has been shown to also increase fluorescence in the green spectral region of *bla*-based reporter gene assays [8].

3.3. Reporter Inhibitors

Another mechanism of false positives that can occur in reporter gene assays is specific inhibition of enzymatic reporters by test compounds. For assays geared at identifying decreases in signal, reporter protein inhibition may be confused with antagonistic effects on the receptor or signaling pathway. However, as mentioned in section **1.1.3**, inhibitors of FLuc reporters have been shown to lead to false activation in reporter gene assays because of inhibitor-based stabilization of the luciferase reporter in cells [22, 78, 79]. The best counter-screen to guard against such a counterintuitive artifact is to simply screen for inhibitors of the reporter using purified preparations of the enzyme with substrate levels near the K_M (the concentration of substrate that gives the half maximal rate for an enzyme; also a measure of the affinity of the substrate for the enzyme) so that inhibitors can be readily identified [171]. Often, pure preparations of reporter enzymes such as FLuc can be obtained from commercial suppliers and used to construct simple enzyme assays to check for specific inhibition. Inhibitor-based reporter stabilization is a likely cause of the apparent activation when it is found that inhibition of the reporter enzyme tracks with the activation response in the cell-based assay [172]. As well, several dark reactions are mediated by FLuc (e.g. enzyme catalyzed reactions that do not result in light production [79]) that can give rise to complex modes of inhibition as illustrated by PTC124, a compound discovered in a series of FLuc cell-based assays that was reported to have nonsense-codon suppression activity [173]. Experiments to better understand FLuc inhibition by PTC124 demonstrated that the enzyme catalyzes the formation of a PTC124- adenylate species [174] leading to a rare form of enzyme inhibition due to formation of a multisubstrate adduct inhibitor (MAI; [175]). The enzyme-bound MAI exhibits picomolar affinity against FLuc, potently stabilizes the enzyme, but can be subsequently removed from the enzyme during detection using reagents such as Bright-Glo which ultimately leads to a rise in luminescence activity in cell-based assays [174]. Although the mechanism of FLuc inhibition can be complex, once information on the structure-activity relationship (SAR) of inhibitors is understood, this knowledge can be used to filter compounds showing either positive or negative effects on the cell-based reporter gene assay signal. The ability to define the SAR for reporter inhibitors offers an advantage over other sources of artifacts, such as fluorescent interferences where SAR can be considerably more difficult to define [21].

4. SUMMARY

New molecular sensor assays continue to be developed using a variety of strategies. For example, FLuc continues to find new uses as a molecular sensor and new luciferases are now being explored including secreted forms such the luciferase from *Guassia princeps* [176]. Taking advantage of the conformational change that occurs upon substrate binding, allosterically regulated and noncovalent variants of FLuc have been recently engineered to develop sensors for proteases or analytes such as cAMP or rapamycin [177, 178]. With the advent of efforts such as the Molecular Libraries Initiative [179] managed by the NIH and the proliferation of academic screening centers that are applying HTS technologies to new areas of biology [180, 181], there will be an increased need to understand and develop new molecular sensor assays that enable the broader application of HTS to basic research questions.

ACKNOWLEDGMENTS

We thank Allison Peck for editing of this chapter prior to submission.

REFERENCES

1. Fox, S., Farr-Jones, S., Sopchak, L., Boggs, A., Nicely, H. W., Khoury, R., and Biros, M. (2006). High-throughput screening: update on practices and success. *J Biomol Screen 11*, 864–869.
2. Wood, K. V. (1995). Marker proteins for gene expression. *Curr Opin Biotechnol 6*, 50–58.
3. Inglese, J., Johnson, R. L., Simeonov, A., Xia, M., Zheng, W., Austin, C. P., and Auld, D. S. (2007). High-throughput

screening assays for the identification of chemical probes. *Nat Chem Biol 3*, 466.

4. Ambler, R. P. (1980). The structure of beta-lactamases. *Philosophical transactions of the Royal Society of London 289*, 321–331.
5. Zlokarnik, G. (2000). Fusions to beta-lactamase as a reporter for gene expression in live mammalian cells. *Methods Enzymol 326*, 221–244.
6. Sutcliffe, J. G. (1978). Nucleotide sequence of the ampicillin resistance gene of *Escherichia coli* plasmid pBR322. *Proc Natl Acad Sci U S A 75*, 3737–3741.
7. Zlokarnik, G., Negulescu, P. A., Knapp, T. E., Mere, L., Burres, N., Feng, L., Whitney, M., Roemer, K., and Tsien, R. Y. (1998). Quantitation of transcription and clonal selection of single living cells with beta-lactamase as reporter. *Science 279*, 84–88.
8. Oosterom, J., van Doornmalen, E. J., Lobregt, S., Blomenrohr, M., and Zaman, G. J. (2005). High-throughput screening using beta-lactamase reporter-gene technology for identification of low-molecular-weight antagonists of the human gonadotropin releasing hormone receptor. *Assay Drug Dev Technol 3*, 143–154.
9. Kunapuli, P., Ransom, R., Murphy, K. L., Pettibone, D., Kerby, J., Grimwood, S., Zuck, P., Hodder, P., Lacson, R., Hoffman, I., Inglese, J., and Strulovici, B. (2003). Development of an intact cell reporter gene beta-lactamase assay for G protein-coupled receptors for high-throughput screening. *Anal Biochem 314*, 16–29.
10. Knapp, T., Hare, E., Feng, L., Zlokarnik, G., and Negulescu, P. (2003). Detection of beta-lactamase reporter gene expression by flow cytometry. *Cytometry A 51*, 68–78.
11. Bowen, W. P., and Wylie, P. G. (2006). Application of Laser-Scanning Fluorescence Microplate Cytometry in High Content Screening. *Assay Drug Dev Technol 4*, 209–221.
12. Chin, J., Adams, A. D., Bouffard, A., Green, A., Lacson, R. G., Smith, T., Fischer, P. A., Menke, J. G., Sparrow, C. P., and Mitnaul, L. J. (2003). Miniaturization of cell-based beta-lactamase-dependent FRET assays to ultra-high throughput formats to identify agonists of human liver X receptors. *Assay Drug Dev Technol 1*, 777–787.
13. Peekhaus, N. T., Ferrer, M., Chang, T., Kornienko, O., Schneeweis, J. E., Smith, T. S., Hoffman, I., Mitnaul, L. J., Chin, J., Fischer, P. A., Blizzard, T. A., Birzin, E. T., Chan, W., Inglese, J., Strulovici, B., Rohrer, S. P., and Schaeffer, J. M. (2003). A beta-lactamase-dependent Gal4-estrogen receptor beta transactivation assay for the ultra-high throughput screening of estrogen receptor beta agonists in a 3456-well format. *Assay Drug Dev Technol 1*, 789–800.
14. Hallis, T. M., Kopp, A. L., Gibson, J., Lebakken, C. S., Hancock, M., Van Den Heuvel-Kramer, K., and Turek-Etienne, T. (2007). An improved beta-lactamase reporter assay: multiplexing with a cytotoxicity readout for enhanced accuracy of hit identification. *J Biomol Screen 12*, 635–644.
15. Iversen, P. W., Eastwood, B. J., Sittampalam, G. S., and Cox, K. L. (2006). A comparison of assay performance measures in screening assays: signal window, Z′ factor, and assay variability ratio. *J Biomol Screen 11*, 247–252.
16. Zhang, J. H., Chung, T. D., and Oldenburg, K. R. (1999). A Simple Statistical Parameter for Use in Evaluation and Validation of High Throughput Screening Assays. *J Biomol Screen 4*, 67–73.
17. Zuck, P., Murray, E. M., Stec, E., Grobler, J. A., Simon, A. J., Strulovici, B., Inglese, J., Flores, O. A., and Ferrer, M. (2004). A cell-based beta-lactamase reporter gene assay for the identification of inhibitors of hepatitis C virus replication. *Anal Biochem 334*, 344–355.
18. Bresnick, J. N., Skynner, H. A., Chapman, K. L., Jack, A. D., Zamiara, E., Negulescu, P., Beaumont, K., Patel, S., and McAllister, G. (2003). Identification of signal transduction pathways used by orphan G protein-coupled receptors. *Assay and Drug Development Technologies 1*, 239–249.
19. National Center for Biotechnology Information PubChem: http://pubchem.ncbi.nlm.nih.gov/.
20. Inglese, J., Auld, D. S., Jadhav, A., Johnson, R. L., Simeonov, A., Yasgar, A., Zheng, W., and Austin, C. P. (2006). Quantitative high-throughput screening: A titration-based approach that efficiently identifies biological activities in large chemical libraries. *Proc Natl Acad Sci U S A 103*, 11473–11478.
21. Simeonov, A., Jadhav, A., Thomas, C. J., Wang, Y., Huang, R., Southall, N. T., Shinn, P., Smith, J., Austin, C. P., Auld, D. S., and Inglese, J. (2008). Fluorescence spectroscopic profiling of compound libraries. *J Med Chem 51*, 2363–2371.
22. Thorne, N., Auld, D. S., and Inglese, J. (2010). Apparent activity in high-throughput screening: origins of compound-dependent assay interference. *Curr Opin Chem Biol 14*, 315–324.
23. Babaoglu, K., Simeonov, A., Irwin, J. J., Nelson, M. E., Feng, B., Thomas, C. J., Cancian, L., Costi, M. P., Maltby, D. A., Jadhav, A., Inglese, J., Austin, C. P., and Shoichet, B. K. (2008). Comprehensive Mechanistic Analysis of Hits from High-Throughput and Docking Screens against beta-Lactamase. *J Med Chem 51*, 2502–2511.
24. Glickman, J. F., Wu, X., Mercuri, R., Illy, C., Bowen, B. R., He, Y., and Sills, M. (2002). A comparison of ALPHAScreen, TR-FRET, and TRF as assay methods for FXR nuclear receptors. *J Biomol Screen* 7, 3–10.
25. Shimomura, O., Johnson, F. H., and Saiga, Y. (1962). Extraction, purification and properties of aequorin, a bioluminescent protein from the luminous hydromedusan, Aequorea. *J Cell Comp Physiol 59*, 223–239.
26. Tsien, R. Y. (1998). The green fluorescent protein. *Annual Review of Biochemistry* 67, 509–544.
27. Cody, C. W., Prasher, D. C., Westler, W. M., Prendergast, F. G., and Ward, W. W. (1993). Chemical structure of the hexapeptide chromophore of the Aequorea green-fluorescent protein. *Biochemistry 32*, 1212–1218.
28. Cormack, B. P., Valdivia, R. H., and Falkow, S. (1996). FACS-optimized mutants of the green fluorescent protein (GFP). *Gene 173*, 33–38.
29. Johnson, F. H., Shimomura, O., Saiga, Y., Gershman, L. C., Reynolds, G. T., and Waters, J. R. (1962). Quantum efficiency of Cypridina luminescence with a note on that of Aequorea. *J Cell Comp Physiol 60*, 85–103.
30. Chalfie, M., Tu, Y., Euskirchen, G., Ward, W. W., and Prasher, D. C. (1994). Green fluorescent protein as a marker for gene expression. *Science 263*, 802–805.
31. Tsien, R. Y., and Miyawaki, A. (1998). Seeing the machinery of live cells. *Science 280*, 1954–1955.

32. Tour, O. (2005). EGFP as your targeted 'hitman'. *Nat Methods 2*, 491–492.
33. Cubitt, A. B., Heim, R., Adams, S. R., Boyd, A. E., Gross, L. A., and Tsien, R. Y. (1995). Understanding, improving and using green fluorescent proteins. *Trends Biochem Sci 20*, 448–455.
34. Hadjantonakis, A. K., and Nagy, A. (2001). The color of mice: in the light of GFP-variant reporters. *Histochemistry and cell biology 115*, 49–58.
35. Heim, R., and Tsien, R. Y. (1996). Engineering green fluorescent protein for improved brightness, longer wavelengths and fluorescence resonance energy transfer. *Curr Biol 6*, 178–182.
36. Baird, G. S., Zacharias, D. A., and Tsien, R. Y. (2000). Biochemistry, mutagenesis, and oligomerization of DsRed, a red fluorescent protein from coral. *Proc Natl Acad Sci U S A 97*, 11984–11989.
37. Campbell, R. E., Tour, O., Palmer, A. E., Steinbach, P. A., Baird, G. S., Zacharias, D. A., and Tsien, R. Y. (2002). A monomeric red fluorescent protein. *Proc Natl Acad Sci U S A 99*, 7877–7882.
38. Kremers, G. J., Goedhart, J., van Munster, E. B., and Gadella, T. W., Jr. (2006). Cyan and yellow super fluorescent proteins with improved brightness, protein folding, and FRET Forster radius. *Biochemistry 45*, 6570–6580.
39. Shaner, N. C., Steinbach, P. A., and Tsien, R. Y. (2005). A guide to choosing fluorescent proteins. *Nat Methods 2*, 905–909.
40. Shaner, N. C., Lin, M. Z., McKeown, M. R., Steinbach, P. A., Hazelwood, K. L., Davidson, M. W., and Tsien, R. Y. (2008). Improving the photostability of bright monomeric orange and red fluorescent proteins. *Nat Methods 5*, 545–551.
41. Muller-Taubenberger, A., and Anderson, K. I. (2007). Recent advances using green and red fluorescent protein variants. *Applied microbiology and biotechnology* 77, 1–12.
42. Shaner, N. C., Patterson, G. H., and Davidson, M. W. (2007). Advances in fluorescent protein technology. *J Cell Sci 120*, 4247–4260.
43. Niswender, K. D., Blackman, S. M., Rohde, L., Magnuson, M. A., and Piston, D. W. (1995). Quantitative imaging of green fluorescent protein in cultured cells: comparison of microscopic techniques, use in fusion proteins and detection limits. *Journal of microscopy 180*, 109–116.
44. Corish, P., and Tyler-Smith, C. (1999). Attenuation of green fluorescent protein half-life in mammalian cells. *Protein engineering 12*, 1035–1040.
45. Rogers, S., Wells, R., and Rechsteiner, M. (1986). Amino acid sequences common to rapidly degraded proteins: the PEST hypothesis. *Science 234*, 364–368.
46. Li, X., Zhao, X., Fang, Y., Jiang, X., Duong, T., Fan, C., Huang, C. C., and Kain, S. R. (1998). Generation of destabilized green fluorescent protein as a transcription reporter. *J Biol Chem 273*, 34970–34975.
47. Broom, W. J., Auwarter, K. E., Ni, J., Russel, D. E., Yeh, L. A., Maxwell, M. M., Glicksman, M., Kazantsev, A. G., and Brown, R. H., Jr. (2006). Two approaches to drug discovery in SOD1-mediated ALS. *J Biomol Screen 11*, 729–735.
48. Jones, J., Rodgers, J., Heil, M., May, J., White, L., Maddry, J. A., Fletcher, T. M., III, Shaw, G. M., Hartman, J. L. IV, and Kutsch, O. (2007). High throughput drug screening for human immunodeficiency virus type 1 reactivating compounds. *Assay Drug Dev Technol 5*, 181–189.
49. Ochsenbauer-Jambor, C., Jones, J., Heil, M., Zammit, K. P., and Kutsch, O. (2006). T-cell line for HIV drug screening using EGFP as a quantitative marker of HIV-1 replication. *Biotechniques 40*, 91–100.
50. Hellweg, C. E., Baumstark-Khan, C., and Horneck, G. (2003). Generation of stably transfected Mammalian cell lines as fluorescent screening assay for NF-kappaB activation-dependent gene expression. *J Biomol Screen 8*, 511–521.
51. Inglese, J., (Ed.) (2006). *Measuring Biological Responses with Automated Microscopy*, vol. 414, Elsevier Academic Press, San Diego.
52. Auld, D. S., Johnson, R. L., Zhang, Y. Q., Veith, H., Jadhav, A., Yasgar, A., Simeonov, A., Zheng, W., Martinez, E. D., Westwick, J. K., Austin, C. P., and Inglese, J. (2006). Fluorescent protein-based cellular assays analyzed by laser-scanning microplate cytometry in 1536-well plate format. *Methods Enzymol 414*, 566–589.
53. Fan, F., and Wood, K. V. (2007). Bioluminescent assays for high-throughput screening. *Assay Drug Dev Technol 5*, 127–136.
54. DeLuca, M., and McElroy, W. D. (1974). Kinetics of the firefly luciferase catalyzed reactions. *Biochemistry 13*, 921–925.
55. Wood, K. V. (1998). The chemistry of bioluminescent reporter assays. *Promega Notes 65*, 14–20.
56. de Wet, J. R., Wood, K. V., Helinski, D. R., and DeLuca, M. (1985). Cloning of firefly luciferase cDNA and the expression of active luciferase in Escherichia coli. *Proc Natl Acad Sci U S A 82*, 7870–7873.
57. de Wet, J. R., Wood, K. V., DeLuca, M., Helinski, D. R., and Subramani, S. (1987). Firefly luciferase gene: structure and expression in mammalian cells. *Mol Cell Biol* 7, 725–737.
58. Almond, B., Zdanovsky, A., Zdanovskaia, M., Ma, D., Stecha, P., Paguio, A., Garvin, D., and Wood, K. (2004). Introducing the Rapid Response Reporter Vectors. *Promega Notes 87*, 18–22.
59. Thompson, J. F., Hayes, L. S., and Lloyd, D. B. (1991). Modulation of firefly luciferase stability and impact on studies of gene regulation. *Gene 103*, 171–177.
60. Leclerc, G. M., Boockfor, F. R., Faught, W. J., and Frawley, L. S. (2000). Development of a destabilized firefly luciferase enzyme for measurement of gene expression. *Biotechniques 29*, 590–591, 594–596, 598 passim.
61. Hirota, T., Lewis, W. G., Liu, A. C., Lee, J. W., Schultz, P. G., and Kay, S. A. (2008). A chemical biology approach reveals period shortening of the mammalian circadian clock by specific inhibition of GSK-3beta. *Proc Natl Acad Sci U S A 105*, 20746–20751.
62. Loening, A. M., Wu, A.M., and Gambhir, S.S. (2007). Red-shifted *Renilla reniformis* luciferase variants for imaging in living subjects. *Nat Methods 4*, 641–643.
63. Craig, F. F., Simmonds, A. C., Watmore, D., McCapra, F., and White, M. R. (1991). Membrane-permeable luciferin esters for assay of firefly luciferase in live intact cells. *Biochem J 276* (Pt 3), 637–641.
64. Ignowski, J. M., and Schaffer, D. V. (2004). Kinetic analysis and modeling of firefly luciferase as a quantitative reporter

gene in live mammalian cells. *Biotechnol Bioeng 86*, 827–834.
65. Greer, L. F., 3rd, and Szalay, A. A. (2002). Imaging of light emission from the expression of luciferases in living cells and organisms: a review. *Luminescence 17*, 43–74.
66. Hart, R. C., Stempel, K. E., Boyer, P. D., and Cormier, M. J. (1978). Mechanism of the enzyme-catalyzed bioluminescent oxidation of coelenterate-type luciferin. *Biochem Biophys Res Commun 81*, 980–986.
67. Ward, W. W., and Cormier, M. J. (1978). Energy transfer via protein-protein interaction in *Renilla* bioluminescence. *Photochemistry and Photobiology 27*, 389–396.
68. Loening, A. M., Fenn, T. D., Wu, A. M., and Gambhir, S. S. (2006). Consensus guided mutagenesis of *Renilla* luciferase yields enhanced stability and light output. *Protein Eng Des Sel 19*, 391–400.
69. Hannah, R. R., Jennens-Clough, M. L., and Wood, K. V. (1998). Rapid luciferase reporter assay systems for high throughput studies. *Promega Notes 65*, 9–14.
70. Stables, J., Scott, S., Brown, S., Roelant, C., Burns, D., Lee, M. G., and Rees, S. (1999). Development of a dual glow-signal firefly and Renilla luciferase assay reagent for the analysis of G-protein coupled receptor signalling. *J Recept Signal Transduct Res 19*, 395–410.
71. Schagat, T., Paguio, A., and Kopish, K. (2007). Normalizing genetic reporter assays: approaches and considerations for increasing consistency and statistical significance. *Cell Notes*, 9–12.
72. Dyer, B. W., Ferrer, F. A., Klinedinst, D. K., and Rodriguez, R. (2000). A noncommercial dual luciferase enzyme assay system for reporter gene analysis. *Anal Biochem 282*, 158–161.
73. Hampf, M., and Gossen, M. (2006). A protocol for combined *Photinus* and *Renilla* luciferase quantification compatible with protein assays. *Anal Biochem 356*, 94–99.
74. Almond, B., Hawkins, M., Stecha, P., Garvin, D., Paguio, A., Butler, B., Beck, M., Wood, M., and Wood, K. (2003). Introducing Chroma-Luc technology. *Promega Notes 85*, 11–14.
75. Wood, K. V., Lam, Y. A., Seliger, H. H., and McElroy, W. D. (1989). Complementary DNA coding click beetle luciferases can elicit bioluminescence of different colors. *Science 244*, 700–702.
76. Davis, R. E., Zhang, Y. Q., Southall, N., Staudt, L. M., Austin, C. P., Inglese, J., and Auld, D. S. (2007). A cell-based assay for IkappaBalpha stabilization using A two-color dual luciferase-based sensor. *Assay Drug Dev Technol 5*, 85–104.
77. Auld, D. S., Southall, N. T., Jadhav, A., Johnson, R. L., Diller, D. J., Simeonov, A., Austin, C. P., and Inglese, J. (2008). Characterization of chemical libraries for luciferase inhibitory activity. *J Med Chem 51*, 2372–2386.
78. Auld, D. S., Thorne, N., Nguyen, D. T., and Inglese, J. (2008). A specific mechanism for nonspecific activation in reporter-gene assays. *ACS Chem Biol 3*, 463–470.
79. Thorne, N., Inglese, J., and Auld, D. S. (2010). Illuminating insights into firefly luciferase and other bioluminescent reporters used in chemical biology. *Chem Biol 17*, 646–657.
80. Conti, E., Franks, N. P., and Brick, P. (1996). Crystal structure of firefly luciferase throws light on a superfamily of adenylate-forming enzymes. *Structure 4*, 287–298.
81. Nakatsu, T., Ichiyama, S., Hiratake, J., Saldanha, A., Kobashi, N., Sakata, K., and Kato, H. (2006). Structural basis for the spectral difference in luciferase bioluminescence. *Nature 440*, 372–376.
82. Thompson, J. F., Geoghegan, K. F., Lloyd, D. B., Lanzetti, A. J., Magyar, R. A., Anderson, S. M., and Branchini, B. R. (1997). Mutation of a protease-sensitive region in firefly luciferase alters light emission properties. *J Biol Chem 272*, 18766–18771.
83. Deo, S. K., and Daunert, S. (2001). Luminescent proteins from Aequorea victoria: applications in drug discovery and in high throughput analysis. *Fresenius J Anal Chem 369*, 258–266.
84. Kendall, J. M., and Badminton, M. N. (1998). *Aequorea victoria* bioluminescence moves into an exciting new era. *Trends Biotechnol 16*, 216–224.
85. Brini, M., Marsault, R., Bastianutto, C., Alvarez, J., Pozzan, T., and Rizzuto, R. (1995). Transfected aequorin in the measurement of cytosolic Ca^{2+} concentration ($[Ca^{2+}]c$). A critical evaluation. *J Biol Chem 270*, 9896–9903.
86. Inouye, S., Noguchi, M., Sakaki, Y., Takagi, Y., Miyata, T., Iwanaga, S., Miyata, T., and Tsuji, F. I. (1985). Cloning and sequence analysis of cDNA for the luminescent protein aequorin. *Proc Natl Acad Sci U S A 82*, 3154–3158.
87. Brini, M., Pinton, P., Pozzan, T., and Rizzuto, R. (1999). Targeted recombinant aequorins: tools for monitoring $[Ca^{2+}]$ in the various compartments of a living cell. *Microscopy research and technique 46*, 380–389.
88. Prasher, D., McCann, R. O., and Cormier, M. J. (1985). Cloning and expression of the cDNA coding for aequorin, a bioluminescent calcium-binding protein. *Biochem Biophys Res Commun 126*, 1259–1268.
89. Sheu, Y. A., Kricka, L. J., and Pritchett, D. B. (1993). Measurement of intracellular calcium using bioluminescent aequorin expressed in human cells. *Anal Biochem 209*, 343–347.
90. Tanahashi, H., Ito, T., Inouye, S., Tsuji, F. I., and Sakaki, Y. (1990). Photoprotein aequorin: use as a reporter enzyme in studying gene expression in mammalian cells. *Gene 96*, 249–255.
91. Le Poul, E., Hisada, S., Mizuguchi, Y., Dupriez, V. J., Burgeon, E., and Detheux, M. (2002). Adaptation of aequorin functional assay to high throughput screening. *Journal of Biomolecular Screening* 7, 57–65.
92. Bovolenta, S., Foti, M., Lohmer, S., and Corazza, S. (2007). Development of a Ca^{2+}-activated photoprotein, Photina, and its application to high-throughput screening. *J Biomol Screen 12*, 694–704.
93. George, S. E., Schaeffer, M. T., Cully, D., Beer, M. S., and McAllister, G. (2000). A high-throughput glow-type aequorin assay for measuring receptor-mediated changes in intracellular calcium levels. *Anal Biochem 286*, 231–237.
94. Shimomura, O., Kishi, Y., and Inouye, S. (1993). The relative rate of aequorin regeneration from apoaequorin and coelenterazine analogues. *Biochem J 296 (Pt 3)*, 549–551.
95. Serebriiskii, I. G., and Golemis, E. A. (2000). Uses of lacZ to study gene function: evaluation of beta-galactosidase assays employed in the yeast two-hybrid system. *Anal Biochem 285*, 1–15.

96. New, D. C., Miller-Martini, D. M., and Wong, Y. H. (2003). Reporter gene assays and their applications to bioassays of natural products. *Phytother Res 17*, 439–448.
97. Alam, J., and Cook, J. L. (1990). Reporter genes: application to the study of mammalian gene transcription. *Anal Biochem 188*, 245–254.
98. Martin, C. S., Wight, P. A., Dobretsova, A., and Bronstein, I. (1996). Dual luminescence-based reporter gene assay for luciferase and beta-galactosidase. *Biotechniques 21*, 520–524.
99. Eglen, R. M., and Singh, R. (2003). Beta galactosidase enzyme fragment complementation as a novel technology for high throughput screening. *Comb Chem High Throughput Screen 6*, 381–387.
100. Berger, J., Hauber, J., Hauber, R., Geiger, R., and Cullen, B. R. (1988). Secreted placental alkaline phosphatase: a powerful new quantitative indicator of gene expression in eukaryotic cells. *Gene 66*, 1–10.
101. Farr, A., and Roman, A. (1991). A pitfall of using a second plasmid to determine transfection efficiency. *Nucleic Acids Research 20*, 920.
102. Ghazawi, I., Cutler, S. J., Low, P., Mellick, A. S., and Ralph, S. J. (2005). Inhibitory effects associated with use of modified *Photinus pyralis* and *Renilla reniformis* luciferase vectors in dual reporter assays and implications for analysis of ISGs. *J Interferon Cytokine Res 25*, 92–102.
103. Boss, V., Talpade, D. J., and Murphy, T. J. (1996). Induction of NFAT-mediated transcription by Gq-coupled receptors in lymphoid and non-lymphoid cells. *J Biol Chem 271*, 10429–10432.
104. Hill, S. J., Baker, J. G., and Rees, S. (2001). Reporter-gene systems for the study of G-protein-coupled receptors. *Curr Opin Pharmacol 1*, 526–532.
105. Macian, F., Garcia-Rodriguez, C., and Rao, A. (2000). Gene expression elicited by NFAT in the presence or absence of cooperative recruitment of Fos and Jun. *EMBO J 19*, 4783–4795.
106. Bai, C., Schmidt, A., and Freedman, L. P. (2003). Steroid hormone receptors and drug discovery: therapeutic opportunities and assay designs. *Assay Drug Dev Technol 1*, 843–852.
107. Chen, T., Xie, W., Agler, M., and Banks, M. (2003). Coactivators in assay design for nuclear hormone receptor drug discovery. *Assay Drug Dev Technol 1*, 835–842.
108. Grover, G. S., Turner, B. A., Parker, C. N., Meier, J., Lala, D. S., and Lee, P. H. (2003). Multiplexing nuclear receptors for agonist identification in a cell-based reporter gene high-throughput screen. *J Biomol Screen 8*, 239–246.
109. Hill, C. S., and Treisman, R. (1995). Differential activation of c-fos promoter elements by serum, lysophosphatidic acid, G proteins and polypeptide growth factors. *EMBO J 14*, 5037–5047.
110. Lalli, E., and Sassone-Corsi, P. (1994). Signal transduction and gene regulation: the nuclear response to cAMP. *J Biol Chem 269*, 17359–17362.
111. Montminy, M. R., Gonzalez, G. A., and Yamamoto, K. K. (1990). Regulation of cAMP-inducible genes by CREB. *Trends in neurosciences 13*, 184–188.
112. Price, M. A., Hill, C., and Treisman, R. (1996). Integration of growth factor signals at the c-fos serum response element. *Philosophical transactions of the Royal Society of London 351*, 551–559.
113. McCarthy, S. E., Licata, J. M., and Harty, R. N. (2006). A luciferase-based budding assay for Ebola virus. *J Virol Methods 137*, 115–119.
114. Karin, M., Yamamoto, Y., and Wang, Q. M. (2004). The IKK NF-kappa B system: a treasure trove for drug development. *Nat Rev Drug Discov 3*, 17–26.
115. Li, X., Fang, Y., Zhao, X., Jiang, X., Duong, T., and Kain, S. R. (1999). Characterization of NFkappaB activation by detection of green fluorescent protein-tagged IkappaB degradation in living cells. *J Biol Chem 274*, 21244–21250.
116. Gross, S., and Piwnica-Worms, D. (2005). Real-time imaging of ligand-induced IKK activation in intact cells and in living mice. *Nat Methods 2*, 607–614.
117. Stack, J. H., Whitney, M., Rodems, S. M., and Pollok, B. A. (2000). A ubiquitin-based tagging system for controlled modulation of protein stability. *Nature Biotechnology 18*, 1298–1302.
118. Zaman, G. J., van der Lee, M. M., Kok, J. J., Nelissen, R. L., and Loomans, E. E. (2006). Enzyme fragment complementation binding assay for p38alpha mitogen-activated protein kinase to study the binding kinetics of enzyme inhibitors. *Assay Drug Dev Technol 4*, 411–420.
119. Eglen, R. M. (2002). Enzyme fragment complementation: a flexible high throughput screening assay technology. *Assay Drug Dev Technol 1*, 97–104.
120. Lee, H. K., Brown, S. J., Rosen, H., and Tobias, P. S. (2007). Application of beta-lactamase enzyme complementation to the high-throughput screening of toll-like receptor signaling inhibitors. *Mol Pharmacol 72*, 868–875.
121. Robers, M. B., Horton, R. A., Bercher, M. R., Vogel, K. W., and Machleidt, T. (2008). High-throughput cellular assays for regulated posttranslational modifications. *Anal Biochem 372*, 189–197.
122. Michnick, S. W., Ear, P. H., Manderson, E. N., Remy, I., and Stefan, E. (2007). Universal strategies in research and drug discovery based on protein-fragment complementation assays. *Nat Rev Drug Discov 6*, 569–582.
123. MacDonald, M. L., Lamerdin, J., Owens, S., Keon, B. H., Bilter, G. K., Shang, Z., Huang, Z., Yu, H., Dias, J., Minami, T., Michnick, S. W., and Westwick, J. K. (2006). Identifying off-target effects and hidden phenotypes of drugs in human cells. *Nat Chem Biol 2*, 329–337.
124. Michnick, S. W. (2003). Protein fragment complementation strategies for biochemical network mapping. *Curr Opin Biotechnol 14*, 610–617.
125. Remy, I., and Michnick, S. W. (2004). Mapping biochemical networks with protein-fragment complementation assays. *Methods Mol Biol 261*, 411–426.
126. Giepmans, B. N., Adams, S. R., Ellisman, M. H., and Tsien, R. Y. (2006). The fluorescent toolbox for assessing protein location and function. *Science 312*, 217–224.
127. Barnea, G., Strapps, W., Herrada, G., Berman, Y., Ong, J., Kloss, B., Axel, R., and Lee, K. J. (2008). The genetic design of signaling cascades to record receptor activation. *Proc Natl Acad Sci U S A 105*, 64–69.
128. Yu, H., West, M., Keon, B. H., Bilter, G. K., Owens, S., Lamerdin, J., and Westwick, J. K. (2003). Measuring drug

action in the cellular context using protein-fragment complementation assays. *Assay Drug Dev Technol 1*, 811–822.

129. Stefan, E., Aquin, S., Berger, N., Landry, C. R., Nyfeler, B., Bouvier, M., and Michnick, S. W. (2007). Quantification of dynamic protein complexes using *Renilla* luciferase fragment complementation applied to protein kinase A activities in vivo. *Proc Natl Acad Sci U S A 104*, 16916–16921.
130. Marine, S., Zamiara, E., Smith, S. T., Stec, E. M., McGarvey, J., Kornienko, O., Jiang, G., Wong, K. K., Stack, J. H., Zhang, B. B., Ferrer, M., and Strulovici, B. (2006). A miniaturized cell-based fluorescence resonance energy transfer assay for insulin-receptor activation. *Anal Biochem 355*, 267–277.
131. Pfleger, K. D., and Eidne, K. A. (2006). Illuminating insights into protein-protein interactions using bioluminescence resonance energy transfer (BRET). *Nat Methods 3*, 165–174.
132. Loening, A. M., Fenn, T. D., and Gambhir, S. S. (2007). Crystal structures of the luciferase and green fluorescent protein from *Renilla reniformis*. *J Mol Biol 374*, 1017–1028.
133. Boute, N., Jockers, R., and Issad, T. (2002). The use of resonance energy transfer in high-throughput screening: BRET versus FRET. *Trends Pharmacol Sci 23*, 351–354.
134. Hamdan, F. F., Audet, M., Garneau, P., Pelletier, J., and Bouvier, M. (2005). High-throughput screening of G protein-coupled receptor antagonists using a bioluminescence resonance energy transfer 1-based beta-arrestin2 recruitment assay. *J Biomol Screen 10*, 463–475.
135. Bacart, J., Corbel, C., Jockers, R., Bach, S., and Couturier, C. (2008). The BRET technology and its application to screening assays. *Biotechnol J 3*, 311–324.
136. Kim, S. B., Otani, Y., Umezawa, Y., and Tao, H. (2007). Bioluminescent indicator for determining protein-protein interactions using intramolecular complementation of split click beetle luciferase. *Anal Chem 79*, 4820–4826.
137. Taylor, D. L., Haskins, J. R., and Giuliano, K. A., eds. (2006). *High Content Screening* vol. 356, (Totowa, N.J.: Humana Press).
138. Fung, P., Peng, K., Kobel, P., Dotimas, H., Kauffman, L., Olson, K., and Eglen, R. M. (2006). A homogeneous cell-based assay to measure nuclear translocation using beta-galactosidase enzyme fragment complementation. *Assay Drug Dev Technol 4*, 263–272.
139. Patel, A., Murray, J., McElwee-Whitmer, S., Bai, C., Kunapuli, P., and Johnson, E. N. (2009). A combination of ultrahigh throughput PathHunter and cytokine secretion assays to identify glucocorticoid receptor agonists. *Anal Biochem 385*, 286–292.
140. Wehrman, T. S., Casipit, C. L., Gewertz, N. M., and Blau, H. M. (2005). Enzymatic detection of protein translocation. *Nat Methods 2*, 521–527.
141. Kalderon, D., Roberts, B. L., Richardson, W. D., and Smith, A. E. (1984). A short amino acid sequence able to specify nuclear location. *Cell 39*, 499–509.
142. Saphire, A. C., Bark, S. J., and Gerace, L. (1998). All four homochiral enantiomers of a nuclear localization sequence derived from c-Myc serve as functional import signals. *J Biol Chem 273*, 29764–29769.
143. Zhu, P. J., Zheng, W., Auld, D. S., Jadhav, A., Macarthur, R., Olson, K. R., Peng, K., Dotimas, H., Austin, C. P., and Inglese, J. (2008). A miniaturized glucocorticoid receptor translocation assay using enzymatic fragment complementation evaluated with qHTS. *Comb Chem High Throughput Screen 11*, 545–559.
144. Chambers, C., Smith, F., Williams, C., Marcos, S., Liu, Z. H., Hayter, P., Ciaramella, G., Keighley, W., Gribbon, P., and Sewing, A. (2003). Measuring intracellular calcium fluxes in high throughput mode. *Comb Chem High Throughput Screen 6*, 355–362.
145. Williams, C. (2004). cAMP detection methods in HTS: selecting the best from the rest. *Nat Rev Drug Discov 3*, 125–135.
146. Prystay, L., Gagne, A., Kasila, P., Yeh, L. A., and Banks, P. (2001). Homogeneous cell-based fluorescence polarization assay for the direct detection of cAMP. *J Biomol Screen 6*, 75–82.
147. Gabriel, D., Vernier, M., Pfeifer, M. J., Dasen, B., Tenaillon, L., and Bouhelal, R. (2003). High throughput screening technologies for direct cyclic AMP measurement. *Assay Drug Dev Technol 1*, 291–303.
148. Weber, M., Ferrer, M., Zheng, W., Inglese, J., Strulovici, B., and Kunapuli, P. (2004). A 1536-well cAMP assay for Gs- and Gi-coupled receptors using enzyme fragmentation complementation. *Assay Drug Dev Technol 2*, 39–49.
149. Kao, J. P. Y., and Tsien, R. Y. (1988). Calcium binding kinetics of fura-2 and azo-1 from temperature-jump relaxation measurements. *Biophysical Journal 53*, 635–639.
150. Assay Guidance Manual Version 5.0 (2008). Eli Lilly and Company and NIH Chemical Genomics Center. assay.nih.gov.
151. Sullivan, E., Tucker, E. M., and Dale, I. L. (1999). Measurement of [Ca^{2+}] using the Fluorometric Imaging Plate Reader (FLIPR). *Methods Mol Biol 114*, 125–133.
152. Kostenis, E. (2001). Is Galpha16 the optimal tool for fishing ligands of orphan G-protein-coupled receptors? *Trends Pharmacol Sci 22*, 560–564.
153. Liu, A. M., Ho, M. K., Wong, C. S., Chan, J. H., Pau, A. H., and Wong, Y. H. (2003). Galpha(16/z) chimeras efficiently link a wide range of G protein-coupled receptors to calcium mobilization. *J Biomol Screen 8*, 39–49.
154. Reinscheid, R. K., Kim, J., Zeng, J., and Civelli, O. (2003). High-throughput real-time monitoring of Gs-coupled receptor activation in intact cells using cyclic nucleotide-gated channels. *Eur J Pharmacol 478*, 27–34.
155. Whiteaker, K. L., Gopalakrishnan, S. M., Groebe, D., Shieh, C. C., Warrior, U., Burns, D. J., Coghlan, M. J., Scott, V. E., and Gopalakrishnan, M. (2001). Validation of FLIPR membrane potential dye for high throughput screening of potassium channel modulators. *J Biomol Screen 6*, 305–312.
156. Zheng, W., Spencer, R. H., and Kiss, L. (2004). High throughput assay technologies for ion channel drug discovery. *Assay Drug Dev Technol 2*, 543–552.
157. Titus, S., Neumann, S., Zheng, W., Southall, N., Michael, S., Klumpp, C., Yasgar, A., Shinn, P., Thomas, C. J., Inglese, J., Gershengorn, M. C., and Austin, C. P. (2008). Quantitative high-throughput screening using a live-cell cAMP assay identifies small-molecule agonists of the TSH receptor. *J Biomol Screen 13*, 120–127.

158. Adams, S. R., Harootunian, A. T., Buechler, Y. J., Taylor, S. S., and Tsien, R. Y. (1991). Fluorescence ratio imaging of cyclic AMP in single cells. *Nature 349*, 694–697.
159. Zaccolo, M., De Giorgi, F., Cho, C. Y., Feng, L., Knapp, T., Negulescu, P. A., Taylor, S. S., Tsien, R. Y., and Pozzan, T. (2000). A genetically encoded, fluorescent indicator for cyclic AMP in living cells. *Nature Cell Biology 2*, 25–29.
160. Hu, C. D., Chinenov, Y., and Kerppola, T. K. (2002). Visualization of interactions among bZIP and Rel family proteins in living cells using bimolecular fluorescence complementation. *Mol Cell 9*, 789–798.
161. Miyawaki, A., Llopis, J., Heim, R., McCaffery, J. M., Adams, J. A., Ikura, M., and Tsien, R. Y. (1997). Fluorescent indicators for Ca2+ based on green fluorescent proteins and calmodulin. *Nature 388*, 882–887.
162. Nagai, Y., Miyazaki, M., Aoki, R., Zama, T., Inouye, S., Hirose, K., Iino, M., and Hagiwara, M. (2000). A fluorescent indicator for visualizing cAMP-induced phosphorylation in vivo. *Nat Biotechnol 18*, 313–316.
163. Russwurm, M., Mullershausen, F., Friebe, A., Jager, R., Russwurm, C., and Koesling, D. (2007). Design of fluorescence resonance energy transfer (FRET)-based cGMP indicators: a systematic approach. *Biochem J 407*, 69–77.
164. Collinson, A. D., Bligh, S. W., Graham, D. L., Mott, H. R., Chalk, P. A., Korniotis, N., and Lowe, P. N. (2004). Fluorescence properties of green fluorescent protein FRET pairs concatenated with the small G protein, Rac, and its interacting domain of the kinase, p21-activated kinase. *Assay Drug Dev Technol 2*, 659–673.
165. Jones, J., Heim, R., Hare, E., Stack, J., and Pollok, B. A. (2000). Development and application of a GFP-FRET intracellular caspase assay for drug screening. *J Biomol Screen 5*, 307–317.
166. Johnson, J. M., Castle, J., Garrett-Engele, P., Kan, Z., Loerch, P. M., Armour, C. D., Santos, R., Schadt, E. E., Stoughton, R., and Shoemaker, D. D. (2003). Genome-wide survey of human alternative pre-mRNA splicing with exon junction microarrays. *Science 302*, 2141–2144.
167. Black, D. L. (2003). Mechanisms of alternative pre-messenger RNA splicing. *Annu Rev Biochem 72*, 291–336.
168. Levinson, N., Hinman, R., Patil, A., Stephenson, C. R., Werner, S., Woo, G. H., Xiao, J., Wipf, P., and Lynch, K. W. (2006). Use of transcriptional synergy to augment sensitivity of a splicing reporter assay. *RNA 12*, 925–930.
169. Stoilov, P., Lin, C. H., Damoiseaux, R., Nikolic, J., and Black, D. L. (2008). A high-throughput screening strategy identifies cardiotonic steroids as alternative splicing modulators. *Proc Natl Acad Sci U S A. 105*, 11218–11223.
170. Feng, B. Y., Shelat, A., Doman, T. N., Guy, R. K., and Shoichet, B. K. (2005). High-throughput assays for promiscuous inhibitors. *Nat Chem Biol 1*, 146–148.
171. Cheng, Y., and Prusoff, W. H. (1973). Relationship between the inhibition constant (K1) and the concentration of inhibitor which causes 50 per cent inhibition (I50) of an enzymatic reaction. *Biochem Pharmacol 22*, 3099–3108.
172. Auld, D. S., Thorne, N., Maguire, W. F., and Inglese, J. (2009). Mechanism of PTC124 activity in cell-based luciferase assays of nonsense codon suppression. *Proc Natl Acad Sci U S A 106*, 3585–3590.
173. Welch, E. M., Barton, E. R., Zhuo, J., Tomizawa, Y., Friesen, W. J., Trifillis, P., Paushkin, S., Patel, M., Trotta, C. R., Hwang, S., Wilde, R. G., Karp, G., Takasugi, J., Chen, G., Jones, S., Ren, H., Moon, Y. C., Corson, D., Turpoff, A. A., Campbell, J. A., Conn, M. M., Khan, A., Almstead, N. G., Hedrick, J., Mollin, A., Risher, N., Weetall, M., Yeh, S., Branstrom, A. A., Colacino, J. M., Babiak, J., Ju, W. D., Hirawat, S., Northcutt, V. J., Miller, L. L., Spatrick, P., He, F., Kawana, M., Feng, H., Jacobson, A., Peltz, S. W., and Sweeney, H. L. (2007). PTC124 targets genetic disorders caused by nonsense mutations. *Nature 447*, 87–91.
174. Auld, D. S., Lovell, S., Thorne, N., Lea, W. A., Maloney, D. J., Shen, M., Rai, G., Battaile, K. P., Thomas, C. J., Simeonov, A., Hanzlik, R. P., and Inglese, J. (2010). Molecular basis for the high-affinity binding and stabilization of firefly luciferase by PTC124. *Proc Natl Acad Sci U S A 107*, 4878–4883.
175. Inglese, J., Blatchly, R. A., and Benkovic, S. J. (1989). A multisubstrate adduct inhibitor of a purine biosynthetic enzyme with a picomolar dissociation constant. *J Med Chem 32*, 937–940.
176. Tannous, B. A., Kim, D. E., Fernandez, J. L., Weissleder, R., and Breakefield, X. O. (2005). Codon-optimized *Gaussia* luciferase cDNA for mammalian gene expression in culture and in vivo. *Mol Ther 11*, 435–443.
177. Fan, F., Binkowski, B. F., Butler, B. L., Stecha, P. F., Lewis, M. K., and Wood, K. V. (2008). Novel genetically encoded biosensors using firefly luciferase. *ACS Chem Biol 3*, 346–351.
178. Hodgson, L. (2008). New approaches to in-cell detection of protein activity: genetically encoded chemiluminescence probes pave the way to robust HTS assays. *ACS Chem Biol 3*, 335–337.
179. Austin, C. P., Brady, L. S., Insel, T. R., and Collins, F. S. (2004). NIH Molecular Libraries Initiative. *Science 306*, 1138–1139.
180. Gordon, E. J. (2007). Small-molecule screening: it takes a village. *ACS Chem Biol 2*, 9–16.
181. Inglese, J., and Auld, D. S. (2008) High throughput screening (HTS) techniques: overview of applications in chemical biology, In *Wiley Encylopedia of Chemical Biology*, vol 1 (Hoboken, N.J.: John Wiley & Sons, Inc).
182. Almond, B., H. E., Stecha, P., Garvin, D., Paguio, A., Butler, B., Beck, M., Wood, M., and Wood K. (2003). Introducing Chroma-Luc technology. *Promega Notes 85*, 11–14.
183. Bronstein, I., Fortin, J., Stanley, P. E., Stewart, G. S., and Kricka, L. J. (1994). Chemiluminescent and bioluminescent reporter gene assays. *Anal Biochem 219*, 169–181.
184. Lefevre, C. K., Singer, V. L., Kang, H. C., and Haugland, R. P. (1995). Quantitative nonradioactive CAT assays using fluorescent BODIPY 1-deoxychloramphenicol substrates. *Biotechniques 19*, 488–493.

CHAPTER 14

Time-Resolved Fluorescence Resonance Energy Transfer Technologies in HTS

Yuhong Du
Jonathan J. Havel

1. FRET PRINCIPLES

Fluorescence resonance energy transfer (FRET) is the non-radiative transfer of energy between two fluorophores, termed the donor and the acceptor, with overlapping emission/excitation spectra (Figure 14.1A). When a donor fluorophore absorbs light of a certain wavelength, it is promoted to an excited electronic state. Upon vibrational relaxation, this energy is typically released as heat and light of a longer wavelength; however, if the donor is in sufficient proximity to an acceptor with specific spectral properties, energy can be transferred directly to the acceptor through long-range dipole-dipole interactions, resulting in photon release from the acceptor [1–3] (Figure 14.2A). The rate of this nonradiative energy transfer, that is, FRET, is strictly dependent upon a number of physical and spectral properties of the fluorophore pair and the surrounding environment. Förster described the relationship of these factors to the rate of FRET in the following equation:

$$K_T = [1/\tau_D] \cdot [R_0/r]^6 \qquad [14.1]$$

where K_T is the rate of energy transfer, τ_{D} is the emission half-life of the donor in the absence of the acceptor, r is the distance between the donor and acceptor, and R_0 is Förster's critical distance. R_0 is determined by the physical properties of the surrounding environment and the fluorophore pair. More specifically, R_0 is inversely proportional to the refractive index of the surrounding media and directly proportional to the relative orientation of the fluorophore dipoles, the integrated area of overlap between donor emission and acceptor excitation spectra (Figure 14.1A), and the quantum yield of the donor fluorophore. As such, R_0 is different for every fluorophore pair; however, most R_0 values fall between 2 and 6 nm. The relationship between R_0, r, and the efficiency of energy transfer (E_T) is described by the following equation:

$$E_T = 1/[1 + (r/R_0)^6] \qquad [14.2]$$

Therefore, when $r = R_0$, the efficiency of nonradiative energy transfer is at 50% of the maximum. Because the efficiency of FRET is inversely proportional to the sixth power of the distance between the donor and acceptor (r) (Equation 14.2), detection of FRET is very sensitive to small changes in r. Practically speaking, FRET events are only detectable over distances a factor of approximately 2 away from the R_0 value (i.e., ~1.5% to 98% efficiency), or approximately 1 to 12 nm. Serendipitously, this also happens to be the distance over which many biomolecular interactions occur, making FRET an appealing approach for the study of intra- or intermolecular interactions in biology [1–3]. In a typical biomolecular FRET assay, a protein "A" is genetically fused to a donor fluorescent protein such as cyan fluorescent protein (CFP), and "Protein B" is fused to an acceptor such as yellow fluorescent protein (YFP). If A and B interact, CFP and YFP will be brought into sufficient proximity for FRET to occur. Therefore, a positive FRET signal is indicative of a protein-protein interaction between A and B.

There are three primary methods for the detection of FRET events – donor photobleaching, acceptor photobleaching, and measurement of sensitized emission. In the first method, the rate of donor photobleaching is measured in the presence and absence of the acceptor. If a FRET event is occurring, the half-life of the donor's excited state will be decreased because of energy transfer. Because fluorophores are susceptible to photobleaching only when in their excited state, a decreased rate of donor photobleaching in the presence of the acceptor is indicative of FRET. The second approach, acceptor photobleaching, requires only one sample containing both the donor and acceptor. In this method, donor emission intensity is measured before and after acceptor photobleaching. Because the nonradiative energy transfer of FRET quenches the emission of the donor, photobleaching the acceptor will disrupt the energy transfer and cause an increase in donor emission in the case of a positive FRET event. In the final method, detection of sensitized emission, acceptor emission is detected upon donor excitation. Although it seems straightforward, this method is subject to two major confounding factors – spectral bleedthrough and media autofluorescence. Because of

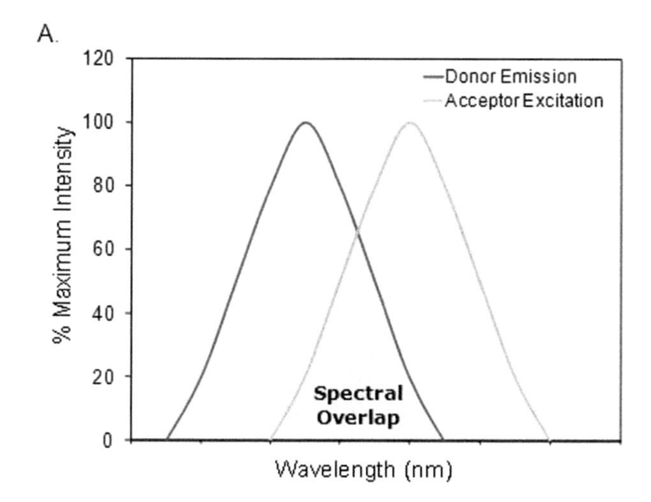

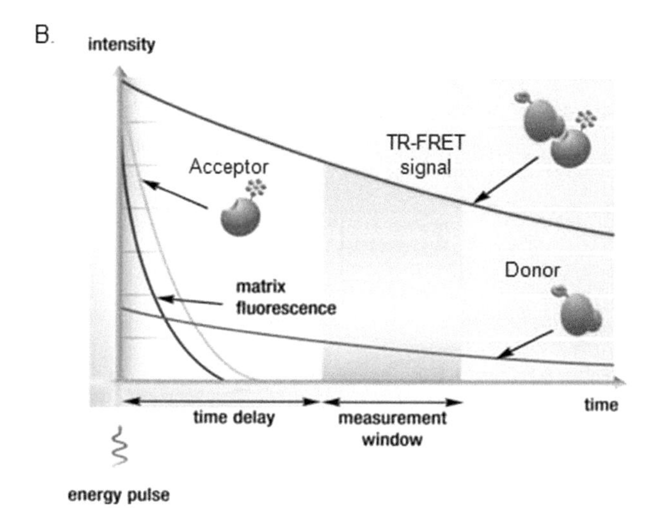

***FIGURE 14.1:** The basic principles of TR-FRET. **A.** FRET can only occur between two fluorophores with overlapping emission and excitation spectra. Specifically, the emission spectrum of the donor fluorophore (blue) must significantly overlap with the excitation spectrum of the acceptor (pink). The integrated area of spectral overlap (shaded area) is directly proportional to the distance over which FRET can be detected. **B.** The use of a donor with long-lived fluorescence emission, such as europium (Eu) or terbium, allows for a temporal delay (50 μs–150 μs) between donor excitation and measurement of the acceptor emission. The resultant FRET signal is therefore referred to as a time-resolved FRET (TR-FRET) signal. This temporal delay allows interfering short-lived fluorescence from three sources – 1) donor emission bleedthrough into acceptor detection ("donor"), 2) environmental or "matrix fluorescence," and 3) direct excitation of the acceptor by the donor excitation wavelength ("acceptor") – to decay, thereby increasing the signal-to-background ratio when the TR-FRET signal is recorded. Source: Adapted with minor modification from reference 7.*

the overlapping spectral properties of many FRET fluorophore pairs, the FRET channel (i.e., donor excitation, acceptor detection) is often contaminated by spectral bleedthrough from 1) a wide donor emission spectrum and 2) direct excitation of the acceptor by the donor's excitation wavelength. This spectral bleedthrough must be corrected by preparing donor-only and acceptor-only samples in addition to the FRET pair sample. Bleedthrough coefficients can be determined from the single fluorophore controls and used to correct bleedthrough in the FRET pair sample. The second confounding factor, media autofluorescence, is more difficult to avoid using traditional FRET fluorophore pairs [1 –3].

Although the first two FRET detection approaches are desirable because they lack spectral bleedthrough problems, they require advanced microscopy techniques and therefore are not amenable to high-throughput screening (HTS) applications. The "measurement of sensitized emission" method can be easily performed in a homogenous mix-and-read format by a high-throughput fluorometer; however, the use of this approach raises significant concerns about signal contamination from spectral bleedthrough and environmental autofluorescence. One effective solution to this problem is the use of rare earth metals, or lanthanides, as the FRET donor in a modification of traditional FRET known as time-resolved FRET (TR-FRET) [4–6]. In TR-FRET, the lanthanide ion Europium^{3+} (Eu) is typically used as the FRET donor because of its exceptionally long emission half-life (300–1500 μs). This long-lived donor emission allows for a temporal delay between donor excitation and detection of acceptor emission (Figure 14.1B) [7]. Because autofluorescence of biological media decays rapidly, this temporal delay virtually eliminates signal contamination due to media autofluorescence. Spectral bleedthrough from direct excitation of the acceptor is also avoided because the emission half-life of the acceptor is much shorter than that of the donor. As such, any acceptor emission due to direct excitation by the incident light will decay during the temporal delay and will be extinguished by the time acceptor emission is detected. Furthermore, the use of an acceptor with a red-shifted emission such as allophycocyanin or Alexa 680 obviates concerns about donor emission bleedthrough into acceptor emission detection [4, 5]. Another concern associated with the use of traditional FRET in HTS is the limited distances over which FRET occurs. The R_0 value, the distance at which 50% of the energy is transferred, is typically less than 50 Å for pairs of small organic dyes, thereby limiting the size of the complexes that can be assayed. The long-lived lanthanides have much larger R_0 values (~100 Å) [8], allowing the detection of larger complexes. Thus, TR-FRET assays are well suited for HTS applications because of their homogenous design, high signal-to-noise (S/N) ratios, and enhanced detection capabilities. The details of one such assay are described in section 4 of this chapter.

2. TR-FRET ASSAY PLATFORMS

TR-FRET assays are based on the proximity of two molecules, each labeled with one of two paired

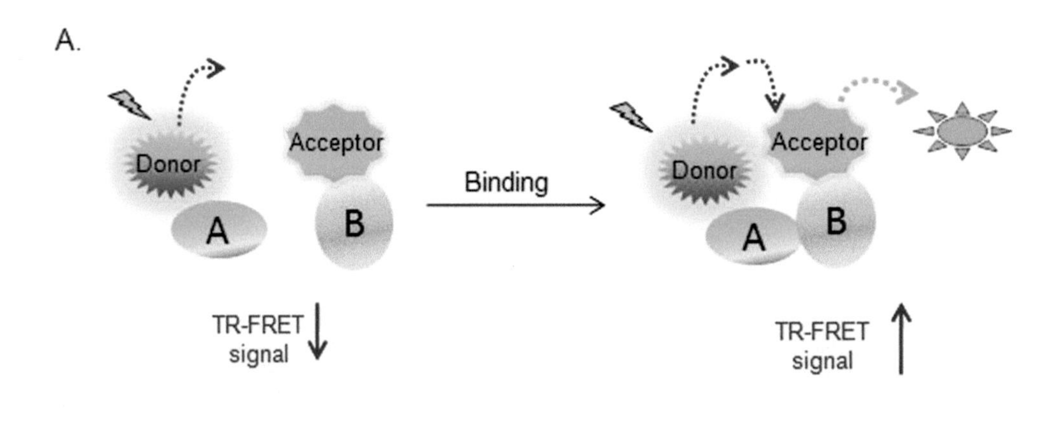

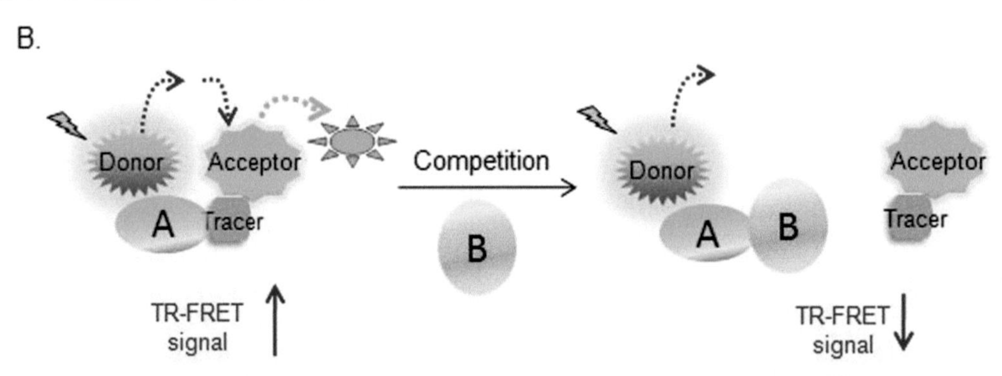

FIGURE 14.2: TR-FRET assay platform for monitoring the interaction of two biological molecules. Development of a TR-FRET assay requires the labeling of two molecules with two fluorophores serving as a FRET donor/acceptor pair. Two TR-FRET formats can be designed for monitoring bimolecular interactions. A. A classic binding assay in which the two binding partners are labeled. When there is no binding, emission signal from the excited donor cannot be transferred to the acceptor, and therefore there is no emission signal from the acceptor. Binding of two molecules brings the donor/acceptor into close proximity, enabling energy transfer from the excited donor to the acceptor. The emission signal of the acceptor can then be detected and defined as a FRET signal. B. A competition assay format. A TR-FRET signal is generated by the binding of a donor-conjugated molecule A to an acceptor-labeled tracer, which is known to bind to A. The binding of another molecule, B, to molecule A competes away the tracer, leading to a decrease in TR-FRET signal.

fluorophores – a donor (long-lived fluorescence) and an acceptor (short-lived fluorescence) with overlapping emission and excitation spectra, respectively. The feature of a time-delayed (typically 50–150 μs) measurement of acceptor emission makes the TR-FRET assay especially suitable for compound screening by minimizing the interference from short-lived background fluorescence, such as media or small molecule autofluorescence. TR-FRET based assays can be used to monitor both intra- and intermolecular status changes of molecules, for example, conformational changes, post-translational modifications, or binding/dissociation events. By varying the platforms used, TR-FRET assays can be applied to monitor both dual and single molecular events in biological systems.

Two major TR-FRET assay platforms are frequently used to detect the binding of one molecule to another: a direct binding assay or an indirect, competition-based assay. Direct binding assays require the labeling of the two molecules of interest with donor and acceptor fluorophores, respectively. Binding of the two molecules brings the donor and acceptor into close proximity, allowing generation of a TR-FRET signal (Figure 14.2A). However, in some cases, labeling the molecules can be problematic because of a lack of labeling tools or interference of the label with the natural function of the molecule being investigated. Therefore, indirect detection of binding determined by a competition assay may offer an alternative approach (Figure 14.2B). In such an assay format, only one interaction molecule (A) is labeled (e.g., with donor), and a TR-FRET signal is generated by the binding of the labeled molecule A to a known binding partner (referred to as the tracer) conjugated with an acceptor. The addition of another molecule (B), which also binds to molecule A, competes with the tracer for binding to A. The interaction of molecules A and B can therefore be determined by a decrease in TR-FRET signal (Figure 14.2B).

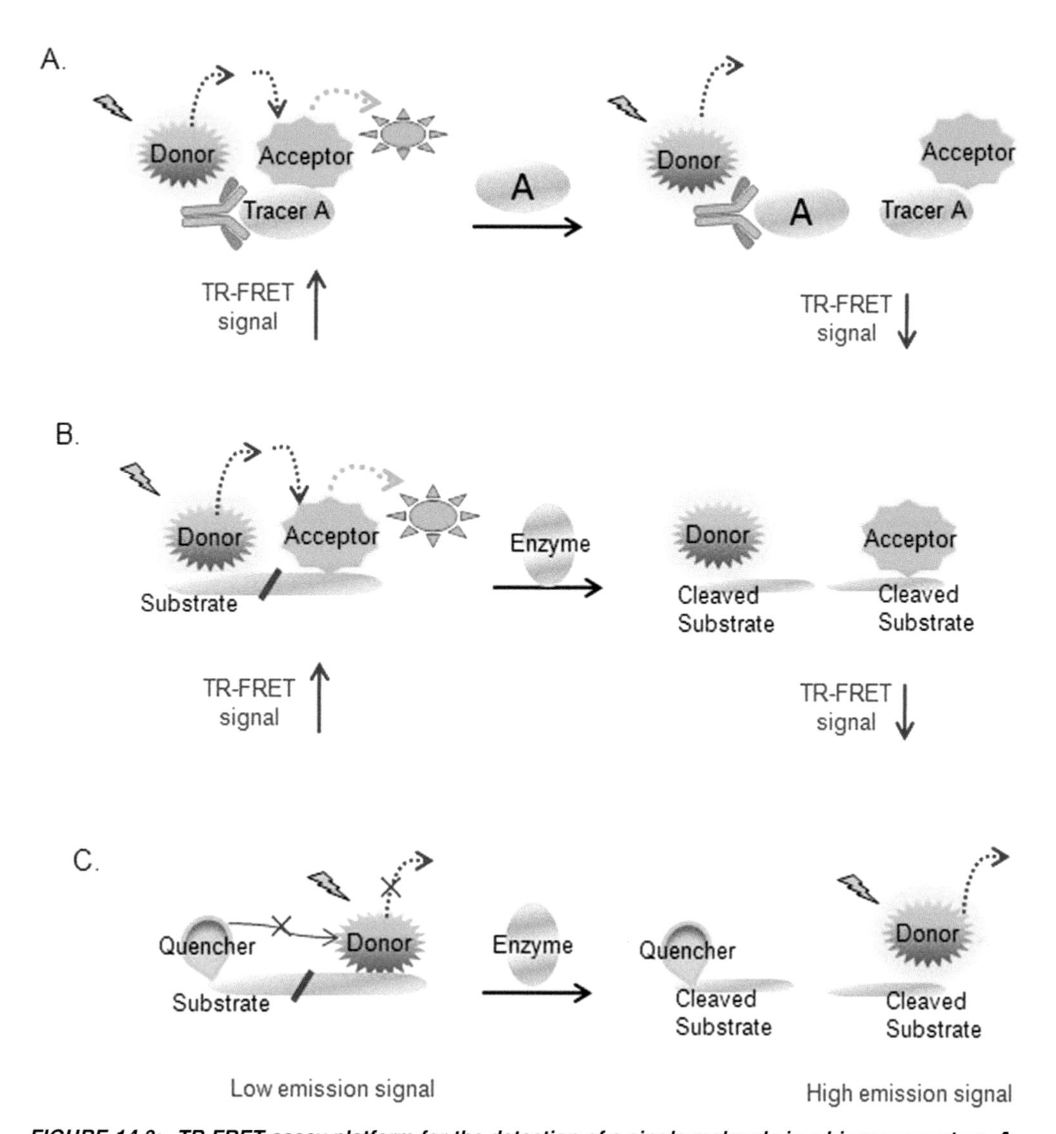

***FIGURE 14.3:** TR-FRET assay platform for the detection of a single molecule in a bioassay system. **A.** An antibody-based competition assay for measuring the concentration of a single molecule. A donor-labeled antibody directed against the molecule of interest (MOI) and an acceptor-labeled MOI (i.e. the tracer) interact to generate a TR-FRET signal. The addition of unlabeled MOI from an in vitro reaction or from cell lysate competes for antibody binding, resulting in a decrease in the TR-FRET signal. The amount of the unlabeled MOI is directly proportional to the measured decrease in TR-FRET signal. **B.** Competition assay for a single molecule with enzyme activity. The donor and acceptor fluorophores are labeled at two sites within the enzyme substrate molecule. Proteolytic enzyme activity cleaves the substrate molecule, resulting in separation of the donor from the acceptor and a decrease in TR-FRET signal. Therefore, proteolytic activity is directly proportional to the measured decrease in TR-FRET signal. **C.** An alternative format using only one fluorophore, a long-lived donor, for enzyme assays. A quencher of the donor is introduced to the labeled substrate and suppresses the donor emission signal. Cleavage of the substrate results in the recovery of donor emission. Therefore, time-delayed detection of donor emission is indicative of proteolytic enzyme activity.*

Using the same principle but a different platform, the TR-FRET assay can also be used to detect the level of a single molecule or activities of enzymes. Figure 14.3 presents three sample assay formats: 1) A competition assay can be adapted and used to monitor the level of a single molecule (Figure 14.3A). This approach requires the development of a specific antibody against the molecule of interest, which can be labeled with a FRET donor or acceptor. The interaction of labeled antibody with its paired TR-FRET fluorophore-conjugated tracer brings the donor and acceptor together and generates a TR-FRET signal. The addition of molecules from an experimental sample solution competes with the tracer for antibody binding. Therefore, the amount of molecule A can be determined from the decrease in TR-FRET signal; 2) For monitoring proteins with protease activity, the substrate can be labeled with both donor and acceptor at two sites on either side of the enzyme cleavage site, bringing the donor and acceptor into sufficient proximity for FRET to occur (Figure 3B). Addition of the enzyme cleaves the substrate, leading to physical separation of the donor and acceptor. The enzymatic activity is therefore determined by the resultant decrease in TR-FRET signal; 3). An alternative format of TR-FRET enzyme assays uses only the donor fluorophore, for example, Eu, and a quencher of Eu emission. The substrate being studied is conjugated to both the Eu and its quencher. Therefore, Eu emission is suppressed in the intact substrate (Figure 14.3C). The active enzyme cleaves the substrate, thereby releasing the donor from the quencher, resulting in increased fluorescence signal upon excitation. The use of long-lived Eu allows the measurement of emission signal after a time delay and increases assay sensitivity by reducing the background. With the development of specific antibodies in combination with labeling tools, the TR-FRET assay platform has been applied to a variety of assay systems, including kinase, protease, and second messenger (e.g., adenosine diphosphate [ADP], cyclic adenosine monophosphate [cAMP]) assays.

3. APPLICATIONS OF TR-FRET-BASED ASSAYS IN HTS

Because of its homogenous format, high sensitivity, flexibility and diversity of the fluorophore labeling tools, and relative ease of handling, TR-FRET based assays have been the preferred fluorescence assay technology for hit compound discovery through HTS. By varying the two components conjugated with a FRET donor and its acceptor, TR-FRET assays can be applied to monitor and characterize biomolecular interactions such as protein-protein, protein-compound, ligand-receptor, and protein-DNA/RNA interactions, as well as the enzymatic activity of kinases, proteases, and second-messenger levels. In this chapter, we briefly review the application of TR-FRET assays with selected examples in two major biological assay systems: in vitro biochemical assays and cell-based assays.

3.1. TR-FRET for In Vitro Biochemical Assays

3.1.1. Protein-Protein Interactions Protein-protein interactions play pivotal roles in many biological processes, such as cell growth, differentiation, intercellular communication, and the terminal event of programmed cell death. Therefore, identification of protein-protein interaction modulators can generate novel therapeutic agents as well as research tools to improve our understanding of biological events. Eukaryotic initiation factor (eIF) 4F plays a key role in recruiting 40S ribosomes and associated factors to mRNA templates during translation initiation. The function of this heterotrimeric complex is to deliver an RNA helicase to the 5′ cap proximal region of mRNAs in preparation for ribosome binding. This complex consists of three subunits: eIF4E, eIF4A, and eIF4G [9]. One mechanism by which translation initiation is regulated is via control of eIF4E availability for assembly into the eIF4F complex [9]. This control can be regulated by target of rapamycin (TOR) and is achieved by balancing the association of eIF4E with either eIF4G or with one of three negative regulatory proteins, known as the eIF4E-binding proteins (4EBPs) [9]. To study the interaction between subunits of this complex as well as to identify small molecules that could interfere with their association, Regina et al. have developed a TR-FRET assay to monitor the interactions between eIF4E and several of its interacting partners, eIF4GI, eIF4GII, and 4E-BP1, using epitope-tagged recombinant proteins [10]. The eIF4E protein with a 6xHis tag was indirectly labeled with a Eu chelate through a Eu-conjugated anti-6XHis antibody. GST-tagged eIF4G protein that spanned the eIF4E recognition domain was indirectly labeled with a Sure-Light Allophycocyanin (APC) conjugated anti-GST antibody. Eu and APC comprise a fluorescence energy transfer pair. Interaction of eIF4E and eIF4G brings the two conjugated fluorophores into sufficient proximity, leading to an energy transfer from Eu to APC and the generation of a FRET signal. This assay has been successfully optimized and used in a 1536-well plate format for uHTS. Such TR-FRET assays have also been used to monitor protein dimerization, such as Phox/Bem1p (PB1) domain heterodimerization [11], the dimerization of Bcl-2 family proteins [12], hepatitis C virus core dimerization [13], and α1-adrenoceptor and CXCR chemokine receptor dimerization [14].

3.1.2. Protein–Small Molecule Binding TR-FRET assays have also been used in a competition format for monitoring the binding of a protein, such as Hsp90, to a known small molecule inhibitor. The molecular chaperone heat-shock protein 90 (Hsp90) is required for maintaining the stability and function of a number of "client" proteins, some of which play important roles in promoting cancer cell growth and/or survival (e.g., Akt, Her2, HIF-1α) [15–17]. Interestingly, cancer cells appear to be more susceptible than normal cells to Hsp90 inhibition. Accordingly, Hsp90 has emerged as an important target in cancer and several other diseases, such as neurodegenerative diseases, nerve injuries, inflammation, and infection. Consequently, there has been increasing interest in developing novel inhibitors of this protein. Well-known Hsp90 inhibitors include natural products geldanamycin (GM) and its derivative, 17-allylamino-17-demethoxygeldanamycin (17AAG). For studying Hsp90 and its inhibitor interaction, a TR-FRET-based HTS assay was established that measures the binding of biotinylated geldanamycin (biotin-GM) to the 6xHis-tagged human Hsp90 N-terminal adenosine triphosphate (ATP)-binding domain (Hsp90N) [18]. The biotin-GM is indirectly labeled with APC (acceptor) through streptavidin (SA)-APC. The 6xHis-Hsp90N is labeled with a TR-FRET donor through an anti-6xHis antibody conjugated to Eu chelate. The binding of biotin-GM to His-Hsp90N brings APC and Eu into close proximity. After excitation with UV light (340 nm), the energy is transferred from the donor (Eu chelate) to the acceptor (APC) molecule and generates a TR-FRET signal. This TR-FRET assay format for monitoring protein–small molecule interactions has also been applied to the retinol-binding protein-4 (RBP4)/retinoid interaction [19].

3.1.3. Ligand-Receptor Interactions Compared to the commonly used radioligand binding assays for monitoring ligand-receptor interactions, TR-FRET-based assays provide a sensitive, homogeneous, nonradioactive, and faster assay platform that can be easily adapted to an HTS format. For example, the herpes virus entry mediator (HVEM) receptor and its ligand, HVEM-L, are involved in herpes simplex virus types 1 and 2 (HSV-1 and HSV-2) infection and in T-cell activation. Antagonists of this interaction are expected to have utility in the treatment of viral and inflammatory diseases. Using a Eu chelate-labeled HVEM receptor as a FRET donor and APC-conjugated ligand as the acceptor, a TR-FRET assay for monitoring HVEM receptor occupancy has been developed and validated for HTS [20]. Another example is a TR-FRET assay for monitoring G protein-coupled receptor ligand binding [21]. In this assay, purified human C5a ligand was labeled with terbium chelate as a fluorescence donor. Hemagglutinin (HA)-tagged human C5a receptor (C5aR) was obtained from the membrane fraction of transfected cells. Robust TR-FRET signal was obtained when the donor-labeled ligand and C5aR membrane fraction were mixed with the acceptor (AlexaFluor 488)-conjugated anti-HA antibody. The binding affinity of C5a to C5aR measured by this TR-FRET assay was consistent with the data determined by competition binding analysis using radiolabeled C5a. TR-FRET–based assays have also been used for detection of nuclear receptor activation [22], tumor necrosis factor receptor/ligand interaction [20], and estrogen receptor/ligand interactions [23, 24].

3.1.4. Enzyme Assays Enzymes promote the conversion of specific substrates into different molecules by altering the rates of chemical reactions. The competitive TR-FRET assay platform is often used to measure enzyme activities (Figure 14.3B), such as the enzymatic activity of heparanase [25]. Degradation of heparan sulfate proteoglycan by heparanase is an important process in tissue invasion by malignant tumor cells and inflammatory cells. Using heparan sulfate proteoglycan conjugated to biotin and Eu cryptate, a TR-FRET signal can be generated with the addition of XL665 (acceptor)-labeled streptavidin. Degradation of the substrate by heparanase in murine melanoma cell extract can then be detected as a decrease in TR-FRET signal due to increased distance of the donor from the acceptor fluorophore.

When designing a competitive TR-FRET assay format to monitor enzyme activity, one must keep in mind that it can be difficult to site-specifically incorporate two distinct labels into the substrate using chemical labeling. Direct labeling through biosynthesis, such as green fluorescent protein (GFP) fusion to the N-terminus of a substrate, offers an effective alternative approach. The enzyme reaction can be monitored in real time to obtain valuable kinetic information for studying the mechanism of enzyme activity and inhibition. Using this format, Invitrogen (www.invitrogen.com) has developed a LanthaScreen Terbium (Tb) deconjugation assay to monitor deubiquitinating enzymes (DUB) and ubiquitin-like (UBL) cleaving enzyme activity. Deubiquitinating (DUB) enzymes proteolytically cleave ubiquitin from proteins and are found to be overexpressed in cancer cells. Therefore, inhibition of DUBs is of interest as a potential therapeutic strategy for treating cancer. GFP fused to the N-terminus of ubiquitin (a DUB substrate) and a terbium chelate (Tb) conjugated to the C-terminal extension through an engineered cysteine residue comprise a TR-FRET pair, resulting in a higher TR-FRET signal in full-length versus cleaved ubiquitin. DUB-dependent cleavage causes separation of the donor (Tb) from the acceptor (GFP) and leads to a decrease in TR-FRET signal. In this assay format, the reaction can be monitored kinetically to detect the activation or inhibition of DUBs.

An additional TR-FRET assay format for monitoring enzyme activity is achieved by using an acceptor-conjugated antibody specific to the cleaved form of an enzyme's substrate. In this format, the substrate is also epitope tagged to allow constant donor association through a donor-conjugated epitope tag–specific antibody. The acceptor is only brought into sufficient proximity to the donor when the substrate is cleaved, allowing the acceptor-conjugated cleaved-substrate–specific antibody to bind. Therefore, proteolytic activity can be detected by an increase in the TR-FRET signal. This assay format has been applied to detect carboxypeptidase B activity for HTS [26].

A variant of TR-FRET termed time-resolved fluorescence quenching has been developed for enzyme assays using only donor labeling of the substrate, without involving the use of a fluorescent acceptor (Figure 14.3C). In this approach, as in other proteolysis assays, the high-energy emission Eu is conjugated to the substrate; however, in place of a FRET acceptor, the substrate is also conjugated at a different site to a Eu quenching dye. In the absence of enzyme, the short distance between Eu and its quencher allows rapid and efficient quenching of Eu. In this situation, there is no Eu emission signal upon excitation. The addition of enzyme cleaves the substrate, removes the quenching dye, and recovers Eu fluorescence. This assay system is suitable for the analysis of many different types of enzymes, such as proteases and helicases. As a model system, this homogeneous, time-resolved fluorescence quenching assay (LANCE, PerkinElmer Life Sciences/Wallac) has been developed to measure caspase-3 activity [27]. The assay utilizes a caspase-3 peptide labeled with both a Eu chelate and its quencher. Cleavage of the peptide by caspase-3 separates the quencher from Eu, and after a time delay, the caspase-3 activity can be detected by an increase in Eu fluorescence.

3.1.5. Kinase Assays Through phosphorylation-mediated signal transduction, kinases are involved in the regulation of many aspects of the cellular processes, including proliferation, differentiation, secretion, and apoptosis [2] [28, 29] Consequently, modulation of kinase function has potential therapeutic value in a wide range of diseases, including cancer and autoimmune disorders [30, 31]. The rapidly growing interest in kinases as potential targets for therapeutic intervention has prompted the development of many kinase assay technologies. Radioactive assays are a simple, classical method for measuring kinase activity. These assays utilize ATP that has been labeled with ^{32}P (or ^{33}P) on the γ-phosphate. Upon kinase catalysis, the γ-phosphate is transferred from ATP to a substrate, and by measuring the extent of isotope incorporation, the progress of the reaction can be monitored. Radioactive assays can be carried out in a variety of formats. However, those assays are generally labor intensive and are not routinely used in HTS because of the involvement of radioactive materials. With the development of antibodies that specifically recognize kinase reaction products, that is, phospho-substrates and ADP, the TR-FRET assay platform has been successfully applied to the kinase assay. Because of its ease of miniaturization for HTS and its robust, sensitive, and homogeneous nature, the TR-FRET kinase assay has rapidly become an important tool in kinase-based drug discovery as well as in target characterization and small molecule–aided mechanistic studies [32].

There are many labeling strategies that can be used to design a TR-FRET kinase assay; however, the general schemes are very similar. As illustrated in Figure 14.4, there are two steps in performing a TR-FRET kinase assay: 1) the kinase reaction followed by 2) the detection of the reaction products. The kinase reaction is carried out by mixing the kinase, substrate containing an affinity tag, and ATP. Two products are generated from the kinase reaction: phosphorylated substrate and ADP, which can be detected separately using the TR-FRET platform. The detection reagents for substrate phosphorylation usually include a donor-conjugated phospho-specific antibody and an acceptor-labeled anti-epitope tag antibody. The interaction of the phospho-specific antibody and the phosphorylated substrate brings the donor and acceptor together, allowing energy transfer from donor to acceptor to generate a TR-FRET signal (Figure 14.4A). Many kinase assays have been developed based upon the availability of phospho-specific antibodies. However, this platform requires the costly and time-consuming generation of antibodies targeting the phosphorylated form of specific substrates and presents challenges in some kinase assays because of specificity issues.

A generic TR-FRET kinase assay is used to detect ADP production rather than substrate phosphorylation. The detection reagents for ADP production include a Eu-labeled anti-ADP antibody developed by Bellbrook Labs (Madison, Wisc.) and an acceptor-conjugated ADP molecule as a tracer. The anti-ADP antibody interacts with tracer ADP and generates a TR-FRET signal. The ADP produced by a kinase reaction competes with acceptor-labeled ADP tracer for binding to the anti-ADP antibody, keeping the donor away from the acceptor and leading to a decrease in TR-FRET signal (Figure 14.4B). The strategy of detecting ADP formation makes this a universal and flexible assay for kinase profiling and HTS of kinases and ATPases.

Using a variety of FRET donor/acceptor pairs in different assay formats, TR-FRET–based kinase assays have been widely used in HTS for the discovery of kinase modulators. These target kinases are often implicated in the abnormal cellular signaling pathways in a wide range of diseases, including Cdc7/Dbf4 kinase [33], IGF-1R[34], Haspin/Gsg2 [35], Rho-kinase II (ROCK-II) [36], polo-like kinase 1[37], p38alpha protein kinase [38], cyclic GMP–dependent protein kinase (cGK) [39], Lck [40], insulin receptor tyrosine kinase [41], c-src [42], and protein kinase CK2 [43]. A noteworthy feature of the kinase assay with Eu cryptate as a donor fluorophore (Cisbio Bioassays) is that a very wide range of ATP concentrations can be used in the assay without affecting the assay performance. This allows the screening of kinase inhibitors under low or high ATP concentrations to identify ATP competitive inhibitors or non-ATP competitive inhibitors, respectively [32].

3.2. Profiling Cellular Signaling Pathways Using Cell-Based TR-FRET Assays

Dysregulated cell signaling pathways have been implicated in many human diseases. Identification of cellular signaling pathway modulators can not only greatly improve our understanding of the mechanisms of cellular regulation but also provide new tools for targeting dysregulated signaling pathways for therapeutic intervention. Cell signal transduction is the process of converting an extracellular signal to intracellular messages and is mediated by several types of transmembrane receptors. These receptors

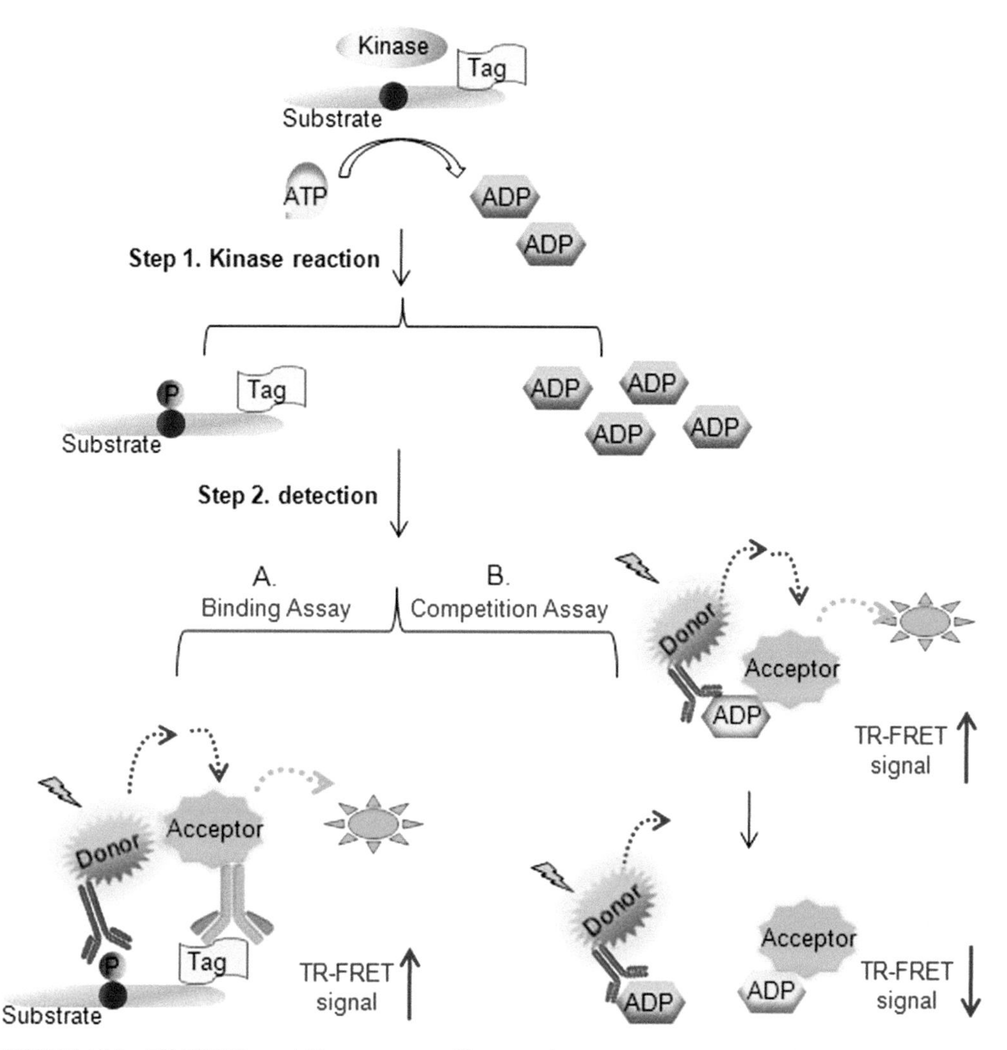

FIGURE 14.4: ***TR-FRET-based kinase assays. Kinase substrates are tagged with fusion proteins, peptides, or biotin. There are two steps involved in a TR-FRET kinase assay. The first step is the kinase reaction and is accomplished by mixing the kinase with ATP and a substrate, leading to substrate phosphorylation and ADP production. The second step is the detection of the two kinase products – the phosphorylated substrate and ADP. A. A donor (Eu)-labeled phospho-specific antibody binds a specific phosphorylated site within the substrate, while another antibody conjugated to an acceptor (XL665) binds a specific epitope tag fused to the substrate. The formation of this complex brings the donor and acceptor into sufficient proximity to allow the generation of a TR-FRET signal. B. A second, more generic detection method for kinase activity is the addition of a donor (Eu cryptate)-conjugated-anti-ADP antibody and an acceptor (d2)-labeled ADP tracer, which forms a FRET pair and generates a TR-FRET signal. Nonlabeled ADP produced from a kinase reaction competes with the tracer ADP for binding to the anti-ADP antibody, resulting in a decreased TR-FRET signal.***

all have extracellular domains that bind to specific ligands, transmembrane domains that span the plasma membrane, and cytoplasmic domains that participate in signal transduction. When bound to its ligand, a receptor is activated. The activated receptor can then initiate intracellular signaling either directly or indirectly, leading to the production of intracellular second messengers, thereby regulating cell function. Consequently, the activity of cell signaling pathways can be measured at different stages of a signal transduction process using TR-FRET–based technologies. Depending on the labeling strategy, one can design and use TR-FRET assays to profile cell signaling pathways in live cells by monitoring secreted cell products, receptor/ligand interactions, or membrane-protein interactions. TR-FRET can also be used to detect intracellular second messenger molecules in cell lysates (Figure 14.5).

3.2.1. TR-FRET Assays for Measuring Secreted Cell Products Alzheimer's disease (AD) is a debilitating, neurodegenerative disorder in the elderly affecting millions of individuals throughout the world. Evidence gathered

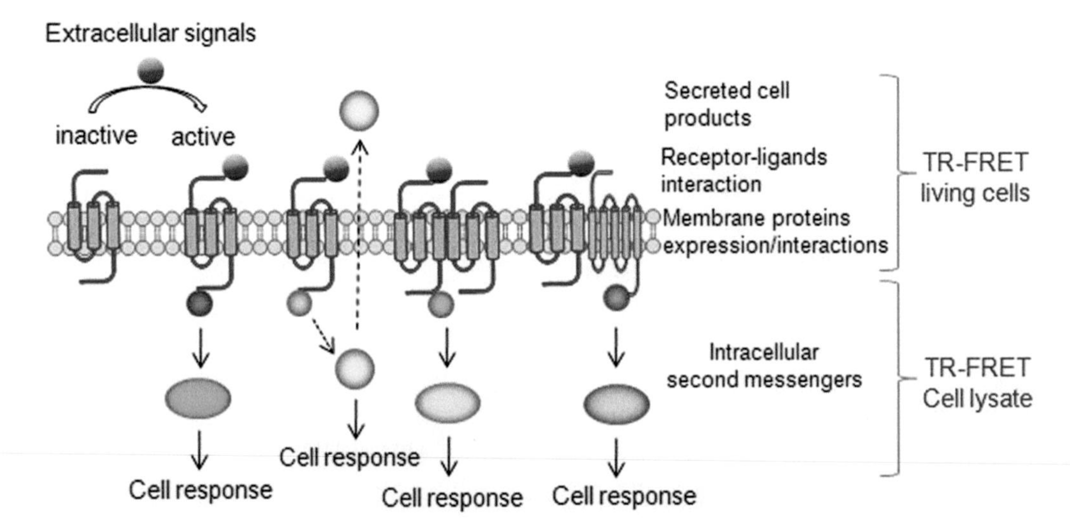

FIGURE 14.5: ***Applications of TR-FRET assays in profiling cellular signaling pathways. Binding of a ligand to its receptor leads to receptor activation. The activated receptor can then initiate intracellular signaling either directly or indirectly, through membrane protein interactions, leading to production of intracellular second messengers, thereby regulating cell function. In theory, depending on the labeling strategy, one can design and use TR-FRET assays to profile cell signaling pathways in live cells by monitoring secreted cell products, receptor-ligand interactions, and membrane-protein interactions. TR-FRET can also be used to detect intracellular second messenger molecules in crude cell lysates.***

suggests that β-amyloid (Aβ), the predominant proteinaceous component of senile plaques, plays an early and critical role in the etiology and pathogenesis of AD. Aβ is a secreted proteolytic product of a large transmembrane protein, the β-amyloid precursor protein (APP). Currently, the most pursued therapeutic strategies for the treatment of AD are aimed at decreasing the cerebral load of Aβ by altering Aβ production, deposition, or clearance [44]. Thus, compounds capable of reducing the accumulation of Aβ may be of value therapeutically as well as in research aimed at further dissecting the cellular pathways that regulate Aβ production and accumulation. A cell-based homogeneous TR-FRET assay has been developed for the detection of Aβ in cell supernatants to screen for inhibitors of Aβ production [45]. In this assay, total Aβ is detected by adding two commercially available antibody complexes. Antibodies conjugated with donor (Eu cryptate) or acceptor (allophycocyanin, XL665) bind the secreted Aβ at two different sites, leading to the generation of a TR-FRET signal. The assay has achieved a simple cell-based mix-and-read format with sensitivity of detection of approximately 1 nM.

For monitoring secreted cell products, traditional enzyme-linked immunosorbent assay (ELISA)-based assays have been widely used, but in general they are not suitable for HTS because multiple plate-washing steps are involved. These additional steps can introduce inaccuracies and lengthen assay times [46]. To overcome these limitations, a time-resolved fluoroimmunoassay (TR-FIA) has been developed with Eu as a fluorophore [47, 48]. The difference between this format and a traditional ELISA comes from using Eu fluorophore instead of horseradish peroxidase (HRP) for signal detection. For example, to detect the isotype of immunoglobulins (IgG, IdM, and IgA) secreted by lymphocytes, anti-mouse Ig was coated onto the bottom of a micotiter plate. After adding conditioned media from cells, the plate was washed, followed by the addition of biotinylated-secondary antibodies. Eu-labeled streptavidin was then added. The Eu fluorescence was measured after washing away the unbound Eu-streptavidin [48]. This format has broad applications in quantification of secreted cell products, such as immunoglobulins, cytokines, and growth factors. It can also be used to detect biomarkers in clinical samples for monitoring the progression of certain diseases.

3.2.2. Ligand-Receptor Interactions in Living Cells Cell-based assays are attractive for compound screening because they present the molecular targets in their cellular environment. Even though TR-FRET assays have been widely used for ligand-receptor interactions in vitro in a variety of formats, the development of ligand-receptor binding assays in living cells is limited, mostly because of the technical difficulty in labeling receptors in living cells without loss of their function.

Martikkala et al. have developed a HTS time-resolved fluorescence ligand binding assay in intact cells for the β2-adrenergic receptor (β2AR) [49]. This assay is based upon the detection of a Eu-labeled small molecule β2AR ligand, pindolol, which binds to intact, stably transfected human embryonic kidney (HEK) 293 cells overexpressing the human β2AR (hβ2AR). The system can be potentially adapted to other receptor-ligand binding assays through

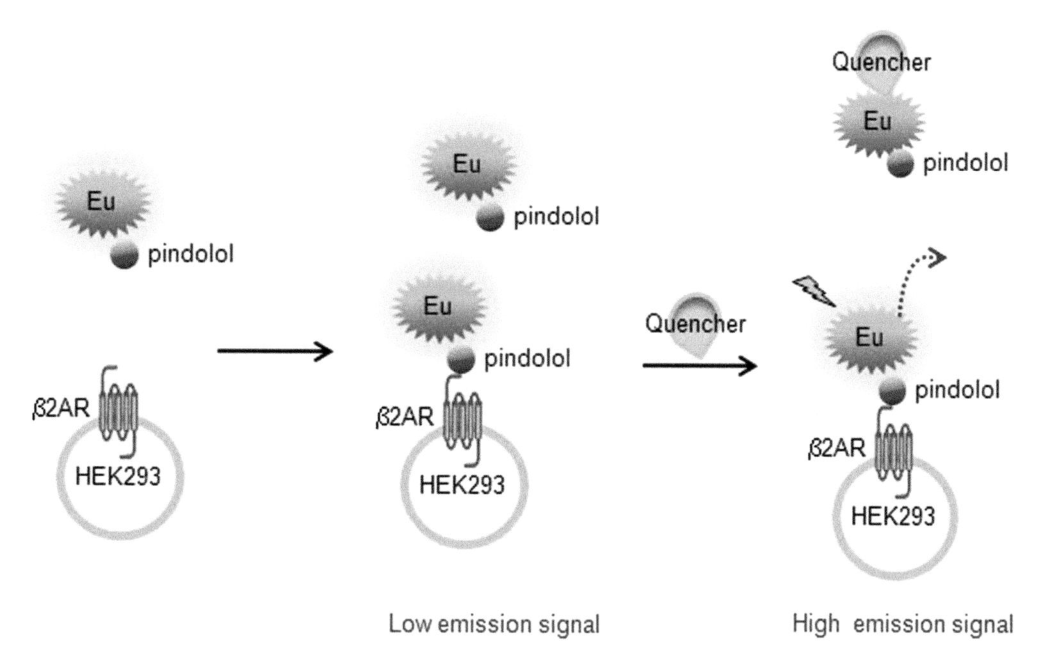

***FIGURE 14.6:** The principle of the homogeneous cell-based quenching resonance energy transfer technique (QRET). Intact HEK293 cells stably overexpressing the human β2AR are incubated with Eu-C12-pindolol. Before the measurement of Eu signal, soluble quencher is added. The quencher efficiently suppresses the fluorescence of the unbound Eu-ligand without affecting the fluorescence of receptor-bound ligand, thereby eliminating the need for a washing step to remove unbound ligand. Source: Adapted from reference 50.*

labeling other small molecule ligands with Eu. However, it requires a washing step to remove unbound Eu-ligand. To eliminate the washing step, researchers have developed a time-resolved fluorescence quenching resonance energy transfer (QRET) technology [50]. The technology is based on the use of a soluble quencher to quench the unbound fraction of Eu-pindolol (Figure 14.6) [50]. The quencher efficiently suppresses the fluorescence of the unbound Eu-labeled ligand, whereas the fluorescence of the receptor-bound fraction is not affected. This assay has been validated for HTS using β2AR antagonists and agonists. This technology has enabled the development of a single-label, separation-free ligand-receptor binding assay in intact cells. It simplifies current TR-FRET–based assay concepts by eliminating the need for dual labeling and potentially reduces overall cost and increases the efficiency of screening. The assay platform can be applied to other receptor-ligand interaction studies in intact cells and is theoretically applicable to immunoassays and non–cell-based receptor studies as long as the labeled ligand can be protected from quenching by receptor binding [50].

3.2.3. Membrane Protein Interactions Direct or indirect interactions between membrane proteins at the cell surface play a central role in numerous cell processes, including possible synergistic effects between different types of receptors. TR-FRET assays offer a new approach for studying membrane proteins and have been successfully applied to monitor the 1) expression and 2) interactions of membrane proteins in living cells [51 –53]. Examples are given below [51].

The expression of receptors in living cells can be measured using TR-FRET technology. Engineered cell surface proteins of interest can be directly or indirectly labeled with a fluorescence donor. After adding an acceptor-conjugated receptor-specific antibody, the TR-FRET signal generated by the bound antibody is directly proportional to the amount of receptors at the cell surface. For example, using Sf9 cells expressing low-affinity IL-2 receptors due to infection with a recombinant baculovirus, a cell-based TR-FRET assay has been developed to monitor interleukin (IL)-2 receptor expression [51]. The cells were biotinylated by using *N*-hydroxysuccinimide esters of biotin that bind covalently to free amine groups of proteins in the membrane. The biotinylated cells were pre-labeled with a streptavidin-Eu chelate, and excess donor molecules were removed. These prelabeled cells were then mixed with an hIL-2Ra-specific monoclonal antibody labeled with Cy5 in a homogenous TR-FRET assay. Binding of anti-hIL-2R-Cy5 to hIL-2R α expressed on the cell surface brings the donor and acceptor into sufficient proximity, allowing energy transfer and TR-FRET detection. This assay constitutes a general method because the entire cell surface is nonspecifically labeled with donor, thereby allowing a wide variety of binding studies to be performed on the cell membrane. In addition, because of the rapid fashion in which the cell TR-FRET assay is executed, it can be a valuable method for HTS applications.

Aided with well-designed antibodies raised against extracellular epitopes of two putative binding partners, homologous TR-FRET assays can be developed for monitoring membrane-protein interactions. Two cell surface receptors can be engineered and expressed with different tags. The interaction of two receptors in living cells can be monitored by the increased TR-FRET signal after adding donor- and acceptor-conjugated antibodies against the specific tags. For example, Maurel et al. [53] have developed a TR-FRET assay without washing steps to visualize membrane protein interactions at the surface of living cells. The γ-aminobutyrate B ($GABA_B$) receptor belongs to the G protein-coupled receptor (GPCR) family and has been reported to form obligatory heterodimers composed of the $GABA_{B1}$ and $GABA_{B2}$ subunits [53]. Using Eu cryptate-labeled anti-HA antibody and Alexa Fluor 647-labeled anti-myc antibody, they were able to detect oligomerization of HA-$GABA_{B1}$ and myc-$GABA_{B2}$ subunits in living cells in a 96-well plate format. This assay can be readily converted to a format for the detection of receptor expression by adding only one fluorophore-conjugated antibody followed by a washing step to remove the unbound antibody [53].

3.2.4. TR-FRET Assays for Detecting Cellular Second Messengers Using Cell Lysates In the past, receptor binding assays were the primary methods for monitoring receptor activities. These approaches are now complemented by methods to measure intracellular responses to receptor activation, such as the activation of cellular kinase cascades and accumulation of cAMP in cells.

Cell-based kinase assays offer some advantages over artificial in vitro biochemical assays, including measurement of activity in the native cellular environment and maintenance of native protein conformation. A generic approach to serine/threonine kinase cellular assays using TR-FRET has been described [54]. A cellular assay format, the LanthaScreen (Invitrogen), has been developed for monitoring the PI3K/AKT/mTOR pathway [55] and NFκB pathway [56]. The PI3K/AKT/mTOR pathway is central to cell growth control. Aberrant activation of this signaling cascade has been linked to several disease states. Therefore, major components of this pathway have become attractive targets for therapeutic intervention. The LanthaScreen cellular HTS assay format is based on the use of stable cell lines that express the kinase substrate of interest as a GFP fusion protein and a terbium (Tb)-labeled phospho-specific antibody to comprise a TR-FRET pair. Cells are stimulated to activate the endogenous signaling pathway to trigger phosphorylation of the substrate. Cell lysis and addition of a Tb-labeled phosphorylation site-specific antibody (in the same step) bring the Tb and GFP into close proximity, enabling energy transfer from Tb to GFP and generation of a TR-FRET signal. This assay format has been validated for key PI3K/AKT/mTOR pathway components, including Ser183 and Thr246 on the proline-rich AKT substrate of 40 kDa (PRAS40), Ser457 on programmed cell death protein 4 (PDCD4), and Thr308 and Ser473 on AKT. Because many intracellular proteins are post-translationally modified at more than one phosphorylation site or in more than one way (e.g., ubiquitination or acetylation), a single GFP-expressing cell line can be used to monitor multiple distinct changes if suitable modification state-specific antibodies are available. Regulation of intracellular concentrations of cAMP is one of the most ubiquitous mechanisms for regulating cellular functions. cAMP production via adenylyl cyclase (AC) activation has been extensively studied. cAMP is used by cells as a second messenger system for GPCR signaling. Beyond GPCR activation, increasing evidence suggests that cAMP synthesis is also regulated by many other mechanisms, such as ions (Ca^{2+}, Mg^{2+}), ATP concentrations, Raf kinase-mediated regulation of AC function [57], and GPCR-linked serine kinases, including PKC, PKA, and CaM kinase [58]. Cell-based TR-FRET cAMP assays have been developed for HTS to identify modulators of many GPCR-mediated signaling pathways [7]. For example, a cell-based TR-FRET assay was developed to measure intracellular cAMP levels in order to identify allosteric potentiators of EP_2 prostaglandin (PG) receptor [59]. EP_2 activation by its natural agonist (PGE_2) stimulates adenylate cyclase activity, resulting in elevated cAMP levels. A TR-FRET signal was generated between Eu cryptate (donor)-conjugated anti-cAMP antibody and cAMP-d2 (acceptor). Intracellular cAMP from the cell lysates competes with the cAMP-d2 for binding to Eu-anti-cAMP antibody, leading to decreased TR-FRET signals. Thus, decreased TR-FRET is proportional to increased cellular cAMP levels. This cell-based assay has been miniaturized to a 1536-well format for ultra-high-throughput screening. A total of 292,000 compounds have been screened, and 93 positive hits were identified that, at 20 uM, selectively increased the potency of PGE_2 on EP_2 receptors at least three-fold. The identified EP_2 allosteric potentiators show low cellular toxicity and are neuroprotective against *N*-methyl-D-aspartase (NMDA)-induced excitotoxicity on cultured hippocampal neurons [59].

4. A CASE STUDY: DEVELOPMENT OF TR-FRET ASSAYS FOR THE DISCOVERY OF ESTROGEN RECEPTOR-COACTIVATOR BINDING INHIBITORS

Antiestrogens, such as tamoxifen, displace the endogenous estrogen agonist estradiol and change the conformation of estrogen receptor (ER), thereby blocking many of its gene-regulatory functions [60]. Therapeutic blockade of estrogen action is typically achieved with conventional antagonists (CAs), compounds that displace estradiol from ER and induce formation of an ER conformation that cannot bind to coactivator proteins, such as the steroid receptor coactivators (SRCs). Antiestrogens, however, can have mixed agonist/antagonist activity, and their effectiveness in blocking ER activity in breast cancer can decrease with time, a phenomenon termed hormone resistance [61]. Thus, an alternative targeting strategy is desired. A set of TR-FRET assays has been developed for the discovery of estrogen receptor-coactivator binding inhibitors (CBIs) [24]. CBIs that compete directly with SRC for interaction with estradiol-bound ER would be interesting mechanistic probes of estrogen action and might also provide an alternative, more durable endocrine therapy for hormone-responsive breast cancer, in which cellular adaptations lead to resistance to CAs.

4.1. Assay Development

A TR-FRET assay for primary HTS of CBIs was developed based on the binding of Cy5-labeled SRC-1 peptide to the ligand binding domain (LBD) of ERα that is labeled (via a streptavidin-biotin interaction) with Eu in the presence of estradiol. The assay components include recombinant biotin-labeled ERα-LBD protein, SRC-Cy5 peptide, Lance Eu-W1024 labeled-streptavidin (SA-Eu) (PerkinElmer), and 17 β-estradiol in an assay buffer (20 mM Tris buffer, pH 7.5, 50 mM NaCl, and 0.01% NP40). A total volume of 30 μL is used for each well in a 384-well format. TR-FRET signal is measured with an Analyst HT reader (Molecular Devices). Eu excitation is set at 330/80 nm, and emission from the Eu donor and Cy5 acceptor is measured at 620/7.5 nm and 665/7.5 nm, respectively. All TR-FRET signals are expressed as F665 nm/F620 nm x 10^4.

To optimize the assay, each binding partner is evaluated through titration experiments. As shown in Figure 14.7A, the binding of increasing concentrations of ERα to SRC-1 peptide leads to energy transfer from Eu to Cy5 and results in increases in TR-FRET signals with a Kd of 0.3 nM. Similar results are obtained when increasing concentrations of SRC-Cy5 peptide are used (Kd of 1.1 nM; Figure 14.7B).

4.2. Assay Evaluation for HTS

To obtain the optimal conditions for HTS, a series of parameters are evaluated:

1) *Z′ factor and signal-to-background ratio (S/B)* for the evaluation of the robustness of the assay are calculated based on equations: $Z' = 1 - 3 \times (SDb + SDf)/(\mu b\text{-}\mu f)$ and $S/B = \mu b/\mu f$, where μb and μf are the FRET signals for bound (b) ERαSRC3-Cy5 and free (f) SRC3-Cy5 alone, respectively, and SDb and SDf are the standard deviations of bound and unbound FRET signal, respectively. The difference between mean signals for bound and free is represented by (μb-μf). A robust assay suitable for HTS should have a Z′ factor between 0.5 and 1 [62]. As shown in Figure 14.7C, in the ERα titration, the signal reaches its maximum at about 0.6 nM; the maximum S/B ratio is 140 at 0.625 nM ER, with a small decrease at higher concentrations, and Z′ values are approximately 0.9 in the entire concentration range.
2) *Dimethyl sulfoxide (DMSO) tolerance* of the assay is tested because the library compounds for screening are normally dissolved in DMSO (Figure 14.7D).
3) *Assay validation* is generally performed using positives controls if available. Such controls include known active compounds or peptides that can modulate the target activity. A small molecule CBI, pyrimidine, induces a dose-dependent decrease in TR-FRET signal as it dissociates SRC3-Cy5 from ERα (Figure 14.7E), validating the assay.
4) *Stability of the assay* was evaluated because the HTS campaign is generally carried out in different plates over a certain time period. As shown in Figure 14.7E, the inhibitory effect of pyrimidine on the ERα coactivator binding is stable for up to twenty hours, offering time-scheduling flexibility during HTS.

4.3. HTS for CBIs

After obtaining the assay conditions from assay optimization and evaluation, HTS is carried out in a mix-and-read format at a single concentration of test compound. Briefly, all of the assay components are mixed together, and 19 μL of the mixture is dispensed to a microtiter plate (384-well). Next, 0.5 μL of 1 mM compound in DMSO is added by a Sciclone liquid handler (Calpier LifeSciences), yielding a final compound concentration of 25 μM and a DMSO concentration of 2.6% (v/v). The plates are incubated for one hour at room temperature. The TR-FRET signals are recorded with the Analyst HT plate reader. Each plate contains three control groups: controls without one of the binding partners to determine the background signal, controls without the addition of testing compound but with vehicle control (DMSO) to determine the maximum signal, and positive controls using known inhibitors (small molecule or peptide). Using this homogenous TR-FRET assay, 86,106 compounds have been tested for the discovery of small molecule CBIs [24].

4.4. Data Analysis

HTS assay data are analyzed using the CambridgeSoft software. The percentage of inhibition is calculated with the following equation based on data from each plate:

$$\% \text{ of Inhibition} = 100 - [(FRETcompound - FRETpeptide\ only)/(FRETcontrol - FRET\ peptide\ only)] \times 100,$$

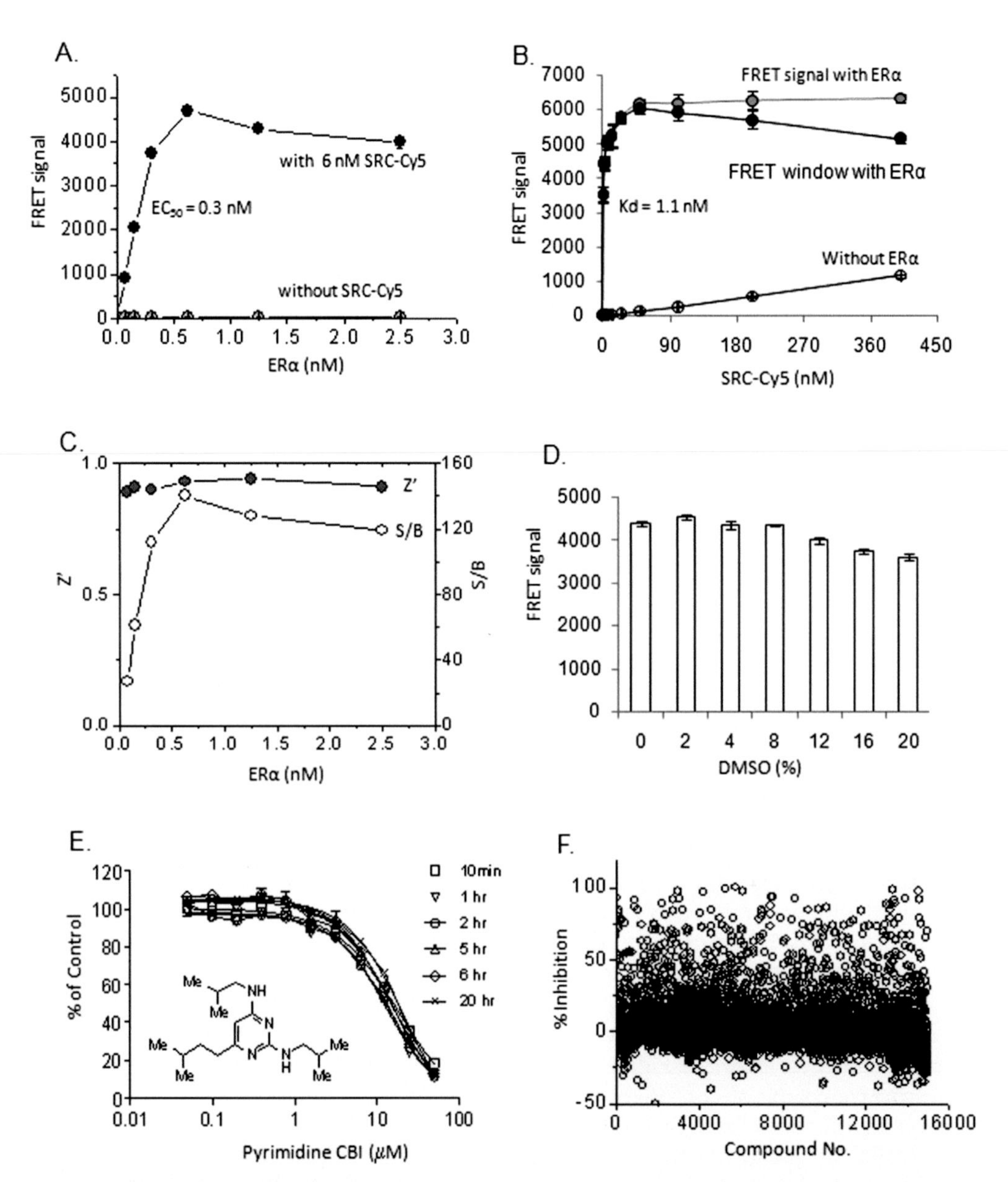

FIGURE 14.7: An example of TR-FRET HTS assay development for CBI identification. A. Titration of ERα-biotin proteins in the presence of SRC3-Cy5, SA-Eu, and estradiol. B. Titration of SRC-Cy5 in a solution of ERα-biotin, SA-Eu, and estradiol. C. Evaluation of assay performance for HTS (Z' and S/B). D. DMSO tolerance evaluation. E. Validation of the assay using known inhibitors and evaluation of assay stability. F. A sample HTS campaign of 15,000 compounds for ERα protein CBIs. **Source:** ***Adapted from reference 24.***

where FRETcompound is the FRET ratio from a well with a test compound, FRET peptide only is an average FRET signal from wells with SRC-Cy5 peptide only, and FRET-control is an average FRET ratio from wells containing Eu-conjugated ERα protein and SRC-Cy5 peptide, which defines maximal FRET signal. Compounds that inhibit the binding of ERα to SRC-Cy5 by more than 50% are considered potential hits, and a representative HTS result from a library of fifteen thousand compounds is shown in Figure 14.7F.

4.5. Conclusions

Using the optimized co-activator-ERα TR-FRET assay, a screen of 86,106 compounds has been carried out to identify putative ERα CBIs. A total of 1,442 compounds are identified as potential positives, which represents a hit rate of 1.67%. These positive hits require further secondary assays for validation. This study illustrates the use of TR-FRET assays in a simple "mix-and-measure" format for hit compound discovery through HTS [24].

5. RESOURCES FOR TR-FRET ASSAY DEVELOPMENT

5.1. Labeling Targets of Interest

There are a number of ways to label biological molecules. Direct labeling is achieved through biosynthesis, such as GFP fusion, or through chemical reactions to attach fluorescent probes to target molecules. Indirect labeling generally involves the use of secondary molecules that are labeled with fluorescent probes. The most frequently used secondary molecules are streptavidin or anti-epitope tag antibodies such as anti-GST, anti-FLAG, anti-6xHis, anti-HA, or anti-cMyc antibodies. These secondary molecules can react with molecules of interest conjugated to either biotin with streptavidin or a particular epitope tag. Indirect labeling requires that two binding molecules contain different tags in order to label one molecule with a FRET donor and the other with a FRET acceptor. Indirect labeling avoids direct chemical modification of important molecules to help maintain structure and activity. Another advantage of using indirect labeling is that the fluorescently labeled reagents are generic. For example, Eu-labeled anti-GST and APC-labeled streptavidin can be used to study the interactions of any paired biological molecules if one is fused with GST and another is labeled with biotin.

SNAP-tag technology from Covalys (www.covalys.com) offers an alternative tagging approach to covalently label proteins of interest. The SNAP-tag is a small, genetically engineered protein based on mammalian O6-alkylguanine-DNA-alkyltransferase (AGT). Derivatives of O6-benzylguanine (BG) bind specifically to the SNAP-tag. These SNAP ligands can carry a wide range of labels to serve as FRET assay donors or acceptors. In the SNAP-tag labeling reaction using a benzylguanine derivative carrying a FRET donor (long-lived cryptate, referred to as BG-TBP) or the BG-647 fluorescence acceptor for BG-TBP, the benzyl moiety is irreversibly transferred to a cysteine residue in the active site of the SNAP-tag, leading to covalent labeling and formation of stably labeled fusion proteins. SNAP-tag labeling reduces the size of the protein complex compared with labeling using antibodies, resulting in high-energy transfer efficiencies, thus streamlining assay development. This makes the SNAP-tag a valuable tool for TR-FRET assay development. The SNAP-tag–based TR-FRET assay has been used to study cell surface protein-protein interactions in living cells, such as GPCR oligomerization [52] and cell surface M3 muscarinic acetylcholine receptor interactions [63].

5.2. Commonly Used TR-FRET Pairs

Development of a TR-FRET assay requires two paired fluorophores, a donor (long-lived fluorophore) and an acceptor (short-lived fluorophore), which must fulfill certain compatibility criteria: 1) the emission spectra of the donor and acceptor must show nonoverlapping regions in order to individually measure each partner's emission, and 2) the excitation spectrum of the acceptor should overlap with the donor emission spectrum to allow efficient energy transfer from donor to acceptor when they are in close proximity. Currently, the two most commonly used lanthanide-based fluorescent donors for TR-FRET assays are Eu and terbium complex in the format of chelate or cryptate. Both are long-lived fluorophores with fluorescence lifetimes averaging between 200 μs to 1500 μs, whereas the fluorescence lifetime of most conventional fluorophores is 100 ns or less. Therefore, a delay of approximately 50 to 150 μs between excitation and fluorescence measurement allows the elimination of background noise from short-lived medium or test compound autofluorescence, bleedthrough of donor emission into acceptor emission, and direct acceptor excitation by donor excitation wavelengths (Figure 14.1B).

Resources for TR-FRET assay development are commercially available and are supplied either as ready-to-use kits for a specific assay (e.g., cAMP, tyrosine kinases, serine/threonine kinases) or as labeling toolboxes for customized assay development. Several major vendors offering TR-FRET assays and labeling reagents include Cisbio Bioassays (HTRF technology based on the use of Eu/terbium cryptate donor and XL665 or d2 acceptors. Its web site, www.htrf.com, provides detailed information on principles and chemistry of HTRF technology), PerkinElmer (Eu chelate donor and APC or Alexa Fluor 647 acceptors; www.perkinelmer.com), Invitrogen (terbium donor and fluorescein, Alexa Fluor 488, GFP, YFP, Rhodamine, or Alexa Fluor 546 as acceptors; www.invitrogen.com), and Molecules Devices (IMAP TR-FRET assay using terbium as donor; www.moleculardevices.com).

ACKNOWLEDGMENT

Y. D. is a recipient of Emory University's SPORE in Head and Neck Cancer Career Development award (P50 CA128613), Emory URC award and Kennedy seed grant at Emory Winship Cancer Instititue. J. J. H. is a recipient of a Ruth L. Kirschstein NRSA Pre-doctoral Fellowship (1F31NS067844–01A1) from the NIH NINDS and an NIH institutional training grant (T32GM008602).

REFERENCES

1. Centonze, V.E., Sun, M., Masuda, A., Gerritsen, H., and Herman, B. (2003). Fluorescence resonance energy transfer imaging microscopy. *Methods Enzymol 360*, 542–560.
2. Truong, K., and Ikura, M. (2001). The use of FRET imaging microscopy to detect protein-protein interactions and protein conformational changes in vivo. *Curr Opin Struct Biol 11*, 573–578.
3. Wu, P., and Brand, L. (1994). Resonance energy transfer: methods and applications. *Anal Biochem 218*, 1–13.

4. Bazin, H., Preaudat, M., Trinquet, E., and Mathis, G. (2001). Homogeneous time resolved fluorescence resonance energy transfer using rare earth cryptates as a tool for probing molecular interactions in biology. *Spectrochim Acta A Mol Biomol Spectrosc 57*, 2197–2211.
5. Hemmila, I., and Laitala, V. (2005). Progress in lanthanides as luminescent probes. *J Fluoresc 15*, 529–542.
6. Morrison, L.E. (1988). Time-resolved detection of energy transfer: theory and application to immunoassays. *Anal Biochem 174*, 101–120.
7. Degorce, F., Card, A., Soh, S., Trinquet, E., Knapik, G.P., and Xie, B. (2009). HTRF: A technology tailored for drug discovery – a review of theoretical aspects and recent applications. *Curr Chem Genomics 3*, 22–32.
8. Mathis, G. (1995). Probing molecular interactions with homogeneous techniques based on rare earth cryptates and fluorescence energy transfer. *Clin Chem 41*, 1391–1397.
9. Gingras, A.C., Raught, B., and Sonenberg, N. (2001). Regulation of translation initiation by FRAP/mTOR. *Genes Dev 15*, 807–826.
10. Cencic, R., Yan, Y., and Pelletier, J. (2007). Homogenous time resolved fluorescence assay to identify modulators of cap-dependent translation initiation. *Comb Chem High Throughput Screen 10*, 181–188.
11. Nakamura, K., Zawistowski, J.S., Hughes, M.A., Sexton, J.Z., Yeh, L.A., Johnson, G.L., and Scott, J.E. (2008). Homogeneous time-resolved fluorescence resonance energy transfer assay for measurement of Phox/Bem1p (PB1) domain heterodimerization. *J Biomol Screen 13*, 396–405.
12. Whitfield, J., Harada, K., Bardelle, C., and Staddon, J.M. (2003). High-throughput methods to detect dimerization of Bcl-2 family proteins. *Anal Biochem 322*, 170–178.
13. Kota, S., Scampavia, L., Spicer, T., Beeler, A.B., Takahashi, V., Snyder, J.K., Porco, J.A., Hodder, P., and Strosberg, A.D. A time-resolved fluorescence-resonance energy transfer assay for identifying inhibitors of hepatitis C virus core dimerization. *Assay Drug Dev Technol 8*, 96–105.
14. Milligan, G., Wilson, S., and Lopez-Gimenez, J.F. (2005). The specificity and molecular basis of alpha1-adrenoceptor and CXCR chemokine receptor dimerization. *J Mol Neurosci 26*, 161–168.
15. Neckers, L., and Ivy, S.P. (2003). Heat shock protein 90. *Curr Opin Oncol 15*, 419–424.
16. Workman, P. (2003). Overview: translating Hsp90 biology into Hsp90 drugs. *Curr Cancer Drug Targets 3*, 297–300.
17. Whitesell, L., Bagatell, R., and Falsey, R. (2003). The stress response: implications for the clinical development of hsp90 inhibitors. *Curr Cancer Drug Targets 3*, 349–358.
18. Zhou, V., Han, S., Brinker, A., Klock, H., Caldwell, J., and Gu, X.J. (2004). A time-resolved fluorescence resonance energy transfer-based HTS assay and a surface plasmon resonance-based binding assay for heat shock protein 90 inhibitors. *Anal Biochem 331*, 349–357.
19. Sharif, O., Hu, H., Klock, H., Hampton, E.N., Nigoghossian, E., Knuth, M.W., Matzen, J., Anderson, P., Trager, R., Uno, T., Glynne, R.J., Azarian, S.M., Caldwell, J.S., and Brinker, A. (2009). Time-resolved fluorescence resonance energy transfer and surface plasmon resonance-based assays for retinoid and transthyretin binding to retinol-binding protein 4. *Anal Biochem 392*, 162–168.
20. Moore, K.J., Turconi, S., Miles-Williams, A., Djaballah, H., Hurskainen, P., Harrop, J., Murray, K.J., and Pope, A.J. (1999). A homogenous 384-well high throughput screen for novel tumor necrosis factor receptor: ligand interactions using time resolved energy transfer. *J Biomol Screen 4*, 205–214.
21. Hu, L.A., Zhou, T., Hamman, B.D., and Liu, Q. (2008). A homogeneous G protein-coupled receptor ligand binding assay based on time-resolved fluorescence resonance energy transfer. *Assay Drug Dev Technol 6*, 543–550.
22. Zhou, G., Cummings, R., Hermes, J., and Moller, D.E. (2001). Use of homogeneous time-resolved fluorescence energy transfer in the measurement of nuclear receptor activation. *Methods 25*, 54–61.
23. Gowda, K., Marks, B.D., Zielinski, T.K., and Ozers, M.S. (2006). Development of a coactivator displacement assay for the orphan receptor estrogen-related receptor-gamma using time-resolved fluorescence resonance energy transfer. *Anal Biochem 357*, 105–115.
24. Gunther, J.R., Du, Y., Rhoden, E., Lewis, I., Revennaugh, B., Moore, T.W., Kim, S.H., Dingledine, R., Fu, H., and Katzenellenbogen, J.A. (2009). A set of time-resolved fluorescence resonance energy transfer assays for the discovery of inhibitors of estrogen receptor-coactivator binding. *J Biomol Screen 14*, 181–193.
25. Enomoto, K., Okamoto, H., Numata, Y., and Takemoto, H. (2006). A simple and rapid assay for heparanase activity using homogeneous time-resolved fluorescence. *J Pharm Biomed Anal 41*, 912–917.
26. Ferrer, M., Zuck, P., Kolodin, G., Mao, S.S., Peltier, R.R., Bailey, C., Gardell, S.J., Strulovici, B., and Inglese, J. (2003). Miniaturizable homogenous time-resolved fluorescence assay for carboxypeptidase B activity. *Anal Biochem 317*, 94–98.
27. Karvinen, J., Hurskainen, P., Gopalakrishnan, S., Burns, D., Warrior, U., and Hemmila, I. (2002). Homogeneous time-resolved fluorescence quenching assay (LANCE) for caspase-3. *J Biomol Screen 7*, 223–231.
28. Manning, G., Whyte, D.B., Martinez, R., Hunter, T., and Sudarsanam, S. (2002). The protein kinase complement of the human genome. *Science 298*, 1912–1934.
29. Hunter, T. (2000). Signaling – 2000 and beyond. *Cell 100*, 113–127.
30. Blume-Jensen, P., and Hunter, T. (2001). Oncogenic kinase signalling. *Nature 411*, 355–365.
31. Fantl, W.J., Johnson, D.E., and Williams, L.T. (1993). Signalling by receptor tyrosine kinases. *Annu Rev Biochem 62*, 453–481.
32. Jia, Y., Quinn, C.M., Gagnon, A.I., and Talanian, R. (2006). Homogeneous time-resolved fluorescence and its applications for kinase assays in drug discovery. *Anal Biochem 356*, 273–281.
33. Xu, K., Stern, A.S., Levin, W., Chua, A., and Vassilev, L.T. (2003). A generic time-resolved fluorescence assay for serine/threonine kinase activity: application to Cdc7/Dbf4. *J Biochem Mol Biol 36*, 421–425.
34. Wang, Y., Malkowski, M., Hailey, J., Turek-Etienne, T., Tripodi, T., and Pachter, J.A. (2004). Screening for small molecule inhibitors of insulin-like growth factor receptor (IGF-1R) kinase: comparison of homogeneous time-resolved fluorescence and 33P-ATP plate assay formats. *J Exp Ther Oncol 4*, 111–119.

35. Patnaik, D., Jun, X., Glicksman, M.A., Cuny, G.D., Stein, R.L., and Higgins, J.M. (2008). Identification of small molecule inhibitors of the mitotic kinase haspin by high-throughput screening using a homogeneous time-resolved fluorescence resonance energy transfer assay. *J Biomol Screen 13*, 1025–1034.
36. Schroter, T., Minond, D., Weiser, A., Dao, C., Habel, J., Spicer, T., Chase, P., Baillargeon, P., Scampavia, L., Schurer, S., Chung, C., Mader, C., Southern, M., Tsinoremas, N., LoGrasso, P., and Hodder, P. (2008). Comparison of miniaturized time-resolved fluorescence resonance energy transfer and enzyme-coupled luciferase high-throughput screening assays to discover inhibitors of Rho-kinase II (ROCK-II). *J Biomol Screen 13*, 17–28.
37. Sharlow, E.R., Leimgruber, S., Shun, T.Y., and Lazo, J.S. (2007). Development and implementation of a miniaturized high-throughput time-resolved fluorescence energy transfer assay to identify small molecule inhibitors of polo-like kinase 1. *Assay Drug Dev Technol 5*, 723–735.
38. Zhang, W.X., Wang, R., Wisniewski, D., Marcy, A.I., LoGrasso, P., Lisnock, J.M., Cummings, R.T., and Thompson, J.E. (2005). Time-resolved Forster resonance energy transfer assays for the binding of nucleotide and protein substrates to p38alpha protein kinase. *Anal Biochem 343*, 76–83.
39. Bader, B., Butt, E., Palmetshofer, A., Walter, U., Jarchau, T., and Drueckes, P. (2001). A cGMP-dependent protein kinase assay for high throughput screening based on time-resolved fluorescence resonance energy transfer. *J Biomol Screen 6*, 255–264.
40. Ohmi, N., Wingfield, J.M., Yazawa, H., and Inagaki, O. (2000). Development of a homogeneous time-resolved fluorescence assay for high throughput screening to identify Lck inhibitors: comparison with scintillation proximity assay and streptavidin-coated plate assay. *J Biomol Screen 5*, 463–470.
41. Biazzo-Ashnault, D.E., Park, Y.W., Cummings, R.T., Ding, V., Moller, D.E., Zhang, B.B., and Qureshi, S.A. (2001). Detection of insulin receptor tyrosine kinase activity using time-resolved fluorescence energy transfer technology. *Anal Biochem 291*, 155–158.
42. Braunwalder, A.F., Yarwood, D.R., Sills, M.A., and Lipson, K.E. (1996). Measurement of the protein tyrosine kinase activity of c-src using time-resolved fluorometry of europium chelates. *Anal Biochem 238*, 159–164.
43. Gratz, A., Gotz, C., and Jose, J. A FRET-based microplate assay for human protein kinase CK2, a target in neoplastic disease. *J Enzyme Inhib Med Chem 25*, 234–239.
44. Dominguez, D.I., and De Strooper, B. (2002). Novel therapeutic strategies provide the real test for the amyloid hypothesis of Alzheimer's disease. *Trends Pharmacol Sci 23*, 324–330.
45. Albrecht, H., Zbinden, P., Rizzi, A., Villetti, G., Riccardi, B., Puccini, P., Catinella, S., and Imbimbo, B.P. (2004). High throughput screening of beta-amyloid secretion inhibitors using homogenous time-resolved fluorescence. *Comb Chem High Throughput Screen* 7, 745–756.
46. Haugabook, S.J., Yager, D.M., Eckman, E.A., Golde, T.E., Younkin, S.G., and Eckman, C.B. (2001). High throughput screens for the identification of compounds that alter the accumulation of the Alzheimer's amyloid beta peptide (Abeta). *J Neurosci Methods 108*, 171–179.
47. Pennanen, N., Lapinjoki, S., Palander, A., Urtti, A., and Monkkonen, J. (1995). Macrophage-like RAW 264 cell line and time-resolved fluoroimmunoassay (TRFIA) as tools in screening drug effects on cytokine secretion. *Int J Immunopharmacol 17*, 475–480.
48. Ruedl, C., Wick, G., and Wolf, H. (1994). A novel and sensitive method for the detection of secreted cell products using time-resolved fluorescence. *J Immunol Methods 168*, 61–67.
49. Martikkala, E., Lehmusto, M., Lilja, M., Rozwandowicz-Jansen, A., Lunden, J., Tomohiro, T., Hanninen, P., Petaja-Repo, U., and Harma, H. (2009). Cell-based beta2-adrenergic receptor-ligand binding assay using synthesized europium-labeled ligands and time-resolved fluorescence. *Anal Biochem 392*, 103–109.
50. Harma, H., Rozwandowicz-Jansen, A., Martikkala, E., Frang, H., Hemmila, I., Sahlberg, N., Fey, V., Perala, M., and Hanninen, P. (2009). A new simple cell-based homogeneous time-resolved fluorescence QRET technique for receptor-ligand interaction screening. *J Biomol Screen 14*, 936–943.
51. Lundin, K., Blomberg, K., Nordstrom, T., and Lindqvist, C. (2001). Development of a time-resolved fluorescence resonance energy transfer assay (cell TR-FRET) for protein detection on intact cells. *Anal Biochem 299*, 92–97.
52. Maurel, D., Comps-Agrar, L., Brock, C., Rives, M.L., Bourrier, E., Ayoub, M.A., Bazin, H., Tinel, N., Durroux, T., Prezeau, L., Trinquet, E., and Pin, J.P. (2008). Cell-surface protein-protein interaction analysis with time-resolved FRET and snap-tag technologies: application to GPCR oligomerization. *Nat Methods 5*, 561–567.
53. Maurel, D., Kniazeff, J., Mathis, G., Trinquet, E., Pin, J.P., and Ansanay, H. (2004). Cell surface detection of membrane protein interaction with homogeneous time-resolved fluorescence resonance energy transfer technology. *Anal Biochem 329*, 253–262.
54. Adams, D.G., Wang, Y., Mak, P.A., Chyba, J., Shalizi, O., Matzen, J., Anderson, P., Smith, T.R., Garcia, M., Welch, G.L., Claret, E.J., Fink, M., Orth, A.P., Caldwell, J.S., and Brinker, A. (2008). Cellular Ser/Thr-kinase assays using generic peptide substrates. *Curr Chem Genomics 1*, 54–64.
55. Carlson, C.B., Robers, M.B., Vogel, K.W., and Machleidt, T. (2009). Development of LanthaScreen cellular assays for key components within the PI3K/AKT/mTOR pathway. *J Biomol Screen 14*, 121–132.
56. Robers, M.B., Horton, R.A., Bercher, M.R., Vogel, K.W., and Machleidt, T. (2008). High-throughput cellular assays for regulated posttranslational modifications. *Anal Biochem 372*, 189–197.
57. Ding, Q., Gros, R., Gray, I.D., Taussig, R., Ferguson, S.S., and Feldman, R.D. (2004). Raf kinase activation of adenylyl cyclases: isoform-selective regulation. *Mol Pharmacol 66*, 921–928.
58. Feldman, R.D., and Gros, R. (2007). New insights into the regulation of cAMP synthesis beyond GPCR/G protein activation: implications in cardiovascular regulation. *Life Sci 81*, 267–271.
59. Jiang, J., Ganesh, T., Du, Y., Thepchatri, P., Rojas, A., Lewis, I., Kurtkaya, S., Li, L., Qui, M., Serrano, G., Shaw, R., Sun, A., and Dingledine, R. Neuroprotection by selective allosteric potentiators of the EP2 prostaglandin receptor. *Proc Natl Acad Sci U S A 107*, 2307–2312.

60. Katzenellenbogen, B.S. (1996). Estrogen receptors: bioactivities and interactions with cell signaling pathways. *Biol Reprod 54*, 287–293.
61. Schiff, R., Massarweh, S.A., Shou, J., Bharwani, L., Arpino, G., Rimawi, M., and Osborne, C.K. (2005). Advanced concepts in estrogen receptor biology and breast cancer endocrine resistance: implicated role of growth factor signaling and estrogen receptor coregulators. *Cancer Chemother Pharmacol 56* Suppl 1, 10–20.
62. Zhang, J.H., Chung, T.D., and Oldenburg, K.R. (1999). A simple statistical parameter for use in evaluation and validation of high throughput screening assays. *J Biomol Screen 4*, 67–73.
63. Alvarez-Curto, E., Ward, R.J., Pediani, J.D., and Milligan, G. (2010) Ligand regulation of the quaternary organization of cell surface M3 muscarinic acetylcholine receptors analyzed by fluorescence resonance energy transfer (FRET) imaging and homogenous time-resolved FRET. *J Biol Chem. 285*, 23318–23330.

CHAPTER 15

Compound Profiling with High-Content Screening Methodology

Thomas Mayer
Stephan Schürer

Over the last decades, high-throughput screening (HTS) of small molecule libraries has developed into a very powerful tool for drug discovery. Although HTS capabilities initially were developed almost exclusively in the domain of pharmaceutical and biotech commercial ventures, they have more recently become available in academia. The Molecular Library Initiative of the National Institutes of Health (NIH) is one example that HTS has established itself as a routine technology in biomedical research [1].

The popularity of HTS is due to rapid developments in four different areas:

1) Robotic systems. Once exotic and expensive, robotic systems and automatic liquid handling have become more affordable and easier to operate. Turnkey systems are available for the most common applications.
2) Chemical libraries. Chemical libraries are becoming more readily available and affordable to the scientific community in academia and industry. Even large compound libraries, such as the one developed for the abovementioned Molecular Library Initiative, are accessible to the general scientific community.
3) Plate reader platforms. Reader platforms have undergone rapid development over the last two decades. As of today, high-speed microplate readers are available to very rapidly scan microplates in the 96-, 384-, and 1536-well formats or even higher densities. Automated confocal and conventional microscopes and other image-based reader platforms have evolved at an astonishing speed. Sophisticated image-based readers and corresponding image recognition and analysis software have led to the development of complex, high-content cell-based assays.
4) Detection technologies. A whole array of luminescence-based technologies have been developed and commercialized over the last few years. Among those are fluorescent proteins, high-yield fluorochromes, and luminescence substrates.

Advances in engineering and science have popularized HTS, and large libraries are now routinely screened against an increasing number of assays. Such screening campaigns typically identify numerous hit compounds. Confirmatory screens will eliminate false-positive results from the primary screen. Careful dose-response studies can further eliminate compounds with undesirable physicochemical properties.

Possibilities for assay and screen development are unlimited, but most screens fall into two categories: target-based or phenotypic assays. In a target-based assay, the biological target (often a protein) is known, and the assay measures how a compound interacts directly or indirectly with that target or how it affects its properties or activities. The most common types of target-based assays are based on enzyme activity or protein binding. In contrast to target-based assays, which are focused on the behavior of a known target, phenotypic assays study the behavior of a whole cell, microbe, or multicellular eukaryotic organism. Both types of assays have specific demands for result validation and for understanding of molecular target interaction mechanisms. Profiling has become a key technology for compound characterization. Because of differences between target-based and phenotypic assays, the demands on profiling are different for both types of assays. In this chapter, we focus on compound profiling for cell-based assays and give a short summary of profiling for target-based assays.

1. GENERAL APPLICATIONS OF COMPOUND PROFILING

1.1. Profiling for Target-Based Assays

Depending on the context, either individual compounds or whole libraries can be profiled. Today, technologies to profile individual compounds against a panel of targets are available; this is common practice in the case of kinases. Because of their highly conserved adenosine triphosphate

(ATP) binding sites, kinases are prone to promiscuity, and it is therefore desirable to profile development compounds across a panel of kinases. As one such technology, in vitro competition binding assays have been used to profile various known kinase inhibitors [2]. It is also technically feasible to profile compound libraries against panels of kinases; this may result in a paradigm shift away from the traditional linear, target-centric preclinical development approach [3]. In another example, to profile large libraries for important properties, the NIH Molecular Libraries Small Molecule Repository (MLSMR) has been tested for luciferase inhibitory activity [4]. Such information is relevant in the context of interpreting results of luciferase reporter gene assays.

1.2. Profiling Physical Properties of Compounds

One use of physical profiling is to flag those compounds that distort the readout without affecting the target. Fluorescence compounds often induce false readings in fluorescence-based assays. Those compounds can be flagged in libraries, which permits early detection of potential false positives. Assays have been developed to screen whole libraries to identify compounds with fluorescence assay interference at various wavelengths [5]. Fluorescence is only one example of how physical properties can affect an assay. There are multiple ways in which a compound can alter assay results – for instance, through changes in environmental parameters like pH or solubility. For example, some structural classes of small molecules form aggregates in solution, which may lead to promiscuous activity observed in biochemical enzyme assays [6]. High-throughput experimental profiling to identify aggregation-based inhibition in large compound libraries has been demonstrated for the MLSMR library [7].

1.3. Toxicology Profiling

Toxicologic screening permits information about toxic properties of compounds. Cells from different sources (species and organs) are exposed to test compounds, and cellular viability is then determined. The more tests that are performed early in compound development, the lower the risk that toxic effects will later emerge. In the process of lead optimization and chemical probe development, multiple parameters and readouts can be monitored to determine the presumptive toxicity of a compound [8]. Among these parameters are growth behavior, redox potential, ATP content, membrane integrity, and apoptotic markers like caspase activation.

1.4. Profiling for Compound Specificity

One of the goals of lead optimization (and similarly for the development of a chemical probe) often is specificity for a certain target. The potential target of a development project is usually a member of a much larger family of closely related proteins. Examples for extended protein families are proteases, kinases, G protein-coupled receptors (GPCRs), and phosphatases. As a popular example, more than five hundred kinases have so far been identified in the human genome, and all of these kinases use ATP as a substrate. The challenge of developing a compound specific to a certain kinase becomes evident (at least when targeting the ATP site). Kinase specificity profiling is therefore common. New development compounds are often tested against a panel of related proteins to ensure that the lead compound inhibits the desired target and a specific pathway with sufficient selectivity over closely related proteins, which may regulate a completely different pathway. For many of the druggable protein families, commercial profiling kits or profiling services are now available. Profiling is also used as a tool to narrow the specificity of a compound during its development.

2. CELL-BASED ASSAYS FOR COMPOUND PROFILING

Cell-based assays are used as primary as well as secondary screening assays. They are often an integral part of lead optimization and chemical probe development. Extensive cellular profiling is usually performed prior to expensive and time-consuming in vivo animal testing to evaluate toxic potential and bioavailability of a compound of interest. In target-based screening, specificity, toxicity, and physical properties of a hit compound are some obstacles to understanding the molecular interaction between target and compound. The same principles apply to cell-based screening, but cell-based assays are far more complex because the effect of a compound is measured in the context of a complex eukaryotic cell.

2.1. Assay Types

Cell-based assays can be simple proliferation assays to access cell growth or very elaborate assays using specifically engineered cell lines to study individual pathways in these cells. The readout can be as simple as counting nuclei as an indicator of cell numbers or as complicated as the redistribution of antigens to be quantified by sophisticated image analysis software. The same assay can be screened in different cells to determine cell type or tissue-specific effects. Using dividing or quiescent cells for the screen adds another layer of biological specificity and significance. Synchronized cell cultures allow investigations into cell cycle–specific effects.

Most high-content cell-based assays measure either signal translocation or a signal intensity modulation. Some examples of common types of high-content cell-based assays follow. Simple assays just monitor cell growth and count nuclei after staining the nuclear DNA with a fluorescent dye. Another class of assays measures the activation of pathways by studying transcription factor translocation to the nucleus. Transcription factor nuclear factor (NF)-κB translocation is a well-known example of a translocation event [9]. We discuss the NFκB assay in more detail later in paragraph 3 of this chapter. Another example of a translocation assay is the binding of β-arrestin to a G protein-coupled receptor after ligand binding [10]. Post-translational modification of proteins, such as Ser/Thr or Tyr phosphorylation, is often used as an indicator for the activity of individual proteins or pathways. Antibodies that specifically detect post-translational modifications of proteins can be used to design assays for monitoring the activity of those proteins by determining the amplitude and/or the location of a modification-specific signal [11]. From the few examples of cell-based assays just listed, it becomes clear that a challenge in interpreting assay results is to understand how a hit compound affects the cell. But again, cell-based assays can be used to deconvolute the activity of a compound by profiling it against a panel of cell-based assays monitoring a whole array of cellular functions.

2.2. Cautionary Notes

An important difference between a target and cell-based screen is the complexity of the underlying assay. A cell-based assay consists of the readout, the compound, and thousands of potential targets in the background of the complex network of intracellular pathways. Given the complexity of the mammalian cell and its pathways, it is hard to understand the effect of the compound at the molecular level. Visual inspection of images is a necessary step to exclude artifacts. Artifacts caused by physical properties of the compound or the solvent can mislead the image-analyzing software. Fluorescent compounds are known to interfere with fluorescence-based readouts. Dead or dying cells often give false readouts but can easily be spotted by visually inspecting the corresponding images [12].

3. THE NF-κB NUCLEAR IMPORT ASSAY: A CASE STUDY

3.1. NF-κB As an Anti-Inflammatory Target

The use of cell-based assays to profile compounds is a relatively new and still evolving field. In this chapter, a screen to measure the NF-κB nuclear import was used as a starting point for the discussion of profiling using an array of cell-based assays. This NF-κB assay is used to study and analyze early inflammatory processes at a cellular level. Chronic inflammatory diseases impose an enormous burden, both medical and financial, on the industrialized nations. Rheumatoid arthritis, atherosclerosis, chronic obstructive pulmonary disease, psoriasis, and inflammatory bowel disease are among the most common inflammatory diseases. Proinflammatory signals are detected and amplified by endothelial cells. Cytokines like tumor necrosis factor-alpha (TNFα), interleukin 1 (IL-1), and CD40 ligand (CD40L) are among the activators of the inflammatory response [13]. As a consequence, lymphocytes attach to the endothelial cells, become activated, and transmigrate the blood vessel wall at the site of the tissue damage [14].

A major step in the early inflammatory response is the translocation of the transcription factor NF-κB from the cytosol into the nucleus of the endothelial cell (Figure 15.1A), inducing the expression of an array of proinflammatory proteins [15].

Among these proteins are E-selectin, VCAM1, and P-selectin, which regulate the attachment of lymphocytes to the wall of the blood vessel. The NF-κB pathway is very promising for therapeutic intervention. Potential targets in this pathway are the cytokine receptors, the I kappa B kinase (IKK) complex, proteosome activity, and transcriptional activity of NF-κB [16].

Figure 15.2 summarizes the activation of the NF-κB pathway by TNFα. In the nonstimulated cell, the NF-κB dimer is associated with the inhibitor IκBα. The inhibitor binds to the p65 unit of the dimer and covers the nuclear import sequence of the transcription factor, keeping the transcription factor in the cytosol. Multiple events are known to activate NF-κB [17]. In the canonical pathway, NF-κB is activated by cytokines like TNFα (Figures 15.1 and 15.2). Binding of TNFα to its receptor results in the activation of the kinase complex, IKK. IKK then phosphorylates the inhibitor, IκBα. The phosphorylated form of IκBα undergoes proteosomal degradation. Once NF-κB is released from the complex, the nuclear localization sequence is exposed, and the transcription factor is imported into the nucleus of the cell (Figure 15.1). In our assays, we use TNFα or IL-1 to stimulate the nuclear accumulation of NF-κB.

3.2. Screening for NF-kB Inhibitors

The cells are incubated with the cytokine for twenty minutes and then fixed and permeabilized. Indirect immunostaining with mouse antibodies against the p65 subunit of the NF-κB transcription factor and secondary anti-mouse antibodies coupled to a fluorochrome are used to visualize intracellular distribution of the NF-κB transcription factor (Figure 15.1B). The nuclear-to-cytoplasmic ratio of the NF-κB concentration is used to quantify the translocation. After stimulation, the nuclear-to-cytoplasmic ratio of the NF-κB increases up to six times (Figure 15.1B). Inhibitors of IKKs, a major regulator of the NF-κB import pathway, prevent nuclear accumulation. Image acquisition and image analyses are done with a GE In Cell Analyzer 3000. The detailed experimental procedure is described in Mayer et al. (2009) [9].

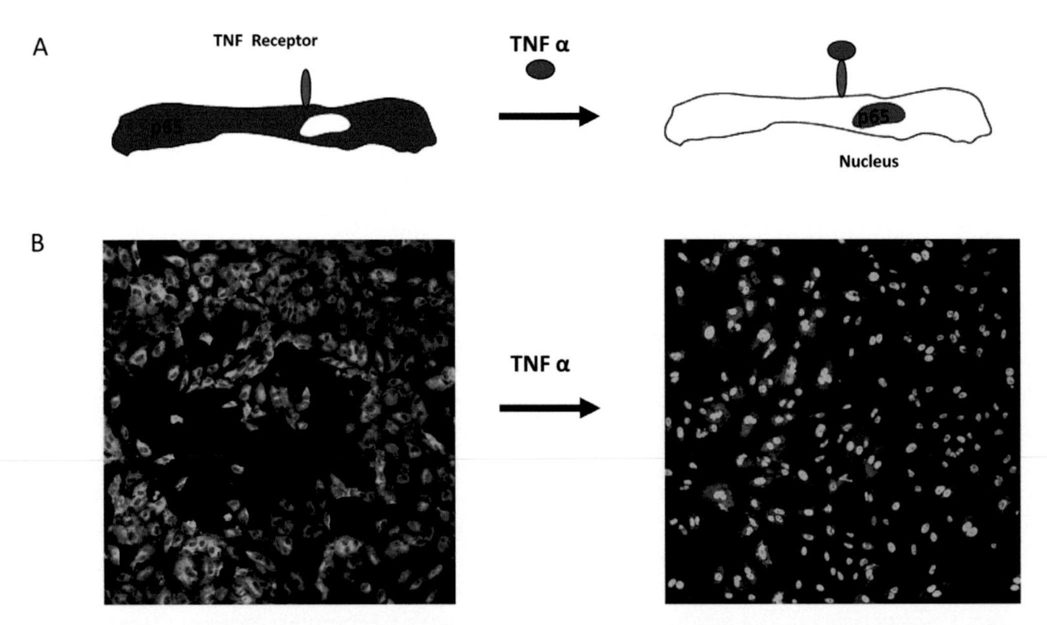

FIGURE 15.1: ***A. Distribution of NF-κB. In the nonstimulated, cell the NF-κB transcription factor is associated with IκBα and predominantly localized in the cytosol. Upon activation with cytokines like TNFα or IL-1, NF-κB is released from IκBα and translocated into the nucleus. B. Nuclear translocation of NF-κB in HUVECs. HUVECs were stimulated with TNFα for twenty minutes. To detect NF-κB distribution, cells were fixed, permeabilized, and stained with an antibody against the p65 part of NF-κB and a secondary antibody coupled to a fluorochrome. The left panel shows images of nonstimulated cells, and the right panel shows images of TNFα stimulated cells.***

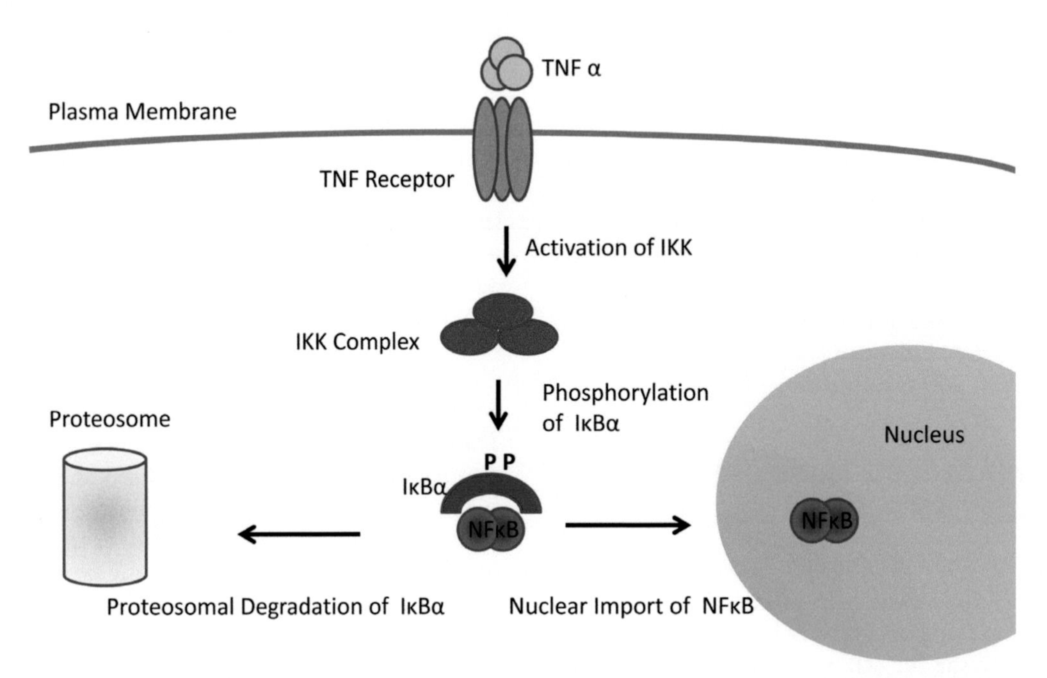

FIGURE 15.2: ***NF-κB pathway. NF-κB in the cytosol is bound to the inhibitor IκBα. Binding of TNFα to a TNF receptor activates the pathway. One event of this activation process is activation of the IKK complex. Activated IKK phosphorylates IκBα. Phosphorylated IκBα undergoes ubiquitination and is degraded by the proteosome. The free NF-κB dimer can now accumulate in the nucleus and activate transcription of its target genes.***

The cell type used for a cell-based assay and the assay conditions will determine the value and the biological relevance of the results obtained. The cells must represent a reliable model system for the biology that the assay is designed to analyze. For the NF-κB assay and the profiling assays described in this chapter, we used a primary cell type, human umbilical vein endothelial cells (HUVECs). HUVECs are often used as a tissue culture model for studying inflammation. HUVECs are readily available from commercial sources as well-characterized, pathogen-screened, and pooled preparations. HUVECs are nonimmortalized cells, but they can be expanded by growing them over several passages in vitro. This is an important consideration when planning cell-based assays. To keep the impact of batch-to-batch variations of the bovine serum at a minimum, obtain controlled growth conditions, and thereby increase assay robustness, we use growth medium formulations that contain low serum (2%) and a defined set of supplements.

Once the obvious artifacts are identified, the real task for the researcher is to understand how an active compound affects the assay. Initially, two simple assumptions can be made: 1) The compound affects a target unrelated to the targeted pathway and the effect is rather indirect (off-target effect); or 2) the compound directly hits in the pathway and shows at least some specificity for the desired pathway (on-target effect). Cheminformatics analysis of screening results and mining of databases like PubChem can help to elucidate effects of active compounds in cell-based assays. To generate a more comprehensive activity profile under standardized and comparable conditions, active compounds are rescreened against a panel of profiling or secondary assays. Results of such profiling panels can effectively be visualized as heat maps. One can assume that compounds affecting multiple assays and pathways are unlikely to have a specific effect on the targeted pathway.

3.3. A Case Study for Compound Profiling

Here we use the example of two compounds affecting the NF-κB assay to demonstrate how profiling in cell-based assays is an effective tool to validate a compound's activity and elucidate its mechanism of action (MOA).

The NF-κB assay was accepted into the MLP (Molecular Libraries Program, an NIH Roadmap Initiative) and screened against more than twelve thousand compounds in the MLSMR Library (PubChem Bioassay AID 438). In the primary screen, sixty compounds were described as active. In the confirmatory screen, eight compounds were verified as active (PubChem AID 462). Two of these compounds are used as an example of profiling. For demonstration purposes, we have selected two compounds that have very different structures and different known activities. This makes it likely that these two compounds inhibit the primary assay by different mechanisms. Compound **1** has the PubChem

1 (CID: 657314) **2** (CID: 659101)

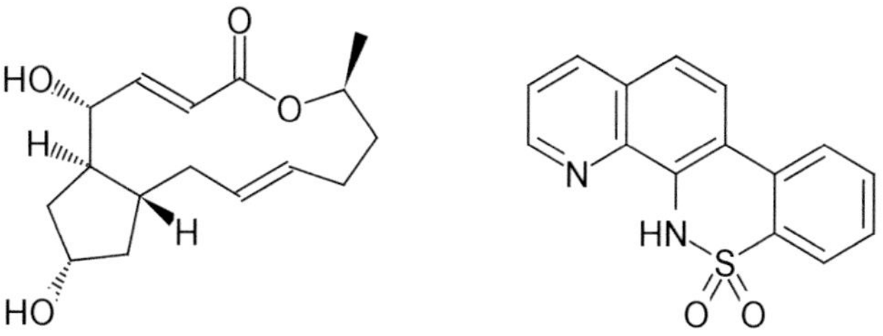

FIGURE 15.3: *Structure of compounds 1 (PubChem CID 657314) and 2 (PubChem CID 659101).*

Compound ID (CID) 657314, and Compound **2** has the CID 659101 (Figure 15.3). CID 657314 is also known as brefeldin A, a fungal metabolite that inhibits anterograde transport in the mammalian cell [18]. Compound CID 659101 is a C7-locked-*N*-(quinolin-8-yl) benzenesulfonamide. The reported IC_{50}s for CID 657314 and 659101 are 0.04 μM and 2.07 μM, respectively. At this point, it is known that both compounds are active in a cell-based NF-κB assay and are not obviously toxic; however, their specificity and MOA in the cell-based assay are unknown. There are several initial approaches to elucidate how these compounds affect the pathway:

1) Database mining of the hits and structurally similar compounds and analysis of their activity in various assays can give first clues. However, the diversity of experimental assay conditions, including cell types, readouts, assay time, assay conditions, and variations in data processing and reported end points, can make it impossible to analyze such data sets conclusively. At the time of the primary screen, the library of about twelve thousand compounds was selected based on structural diversity, and therefore no close analogs of **2** were among the screened compounds. Moreover, brefeldin A (**1**) is a natural product with no close structural analogs in the library.
2) The compounds may be tested against an array of possible targets. In this early stage of compound development, this is expensive and time-consuming because the targets have not been narrowed down.
3) The activity of compounds is tested against a panel of profiling assays to understand which principal cellular pathways and processes are affected. This is a relatively easy and fast way to generate an activity profile of the hit compounds.

For this purpose, a panel of twenty-five cell-based assays covering a wide variety of cellular processes has been developed. Table 15.1 summarizes the assays and lists the pathways and/or biological processes that are monitored by these assays. To simplify the interpretation of the results, the basic design of these assays is very similar to the design of the primary screening assay. Most importantly, all assays

TABLE 15.1: List of Profiling Assays
Table 15.1 summarizes the assays used to profile the hit compounds. The table lists the name of the assay, name of the activator (activator) when an activator was used and the time for which the assay was treated (time). The readout column specifies the event that was monitored.

Name of Assay	Activator	Time	Readout
NFκB	TNFα	30 min	NFκB Nuclear Translocation
NFκB	IL-1	30 min	NFκB Nuclear Translocation
E-Selection	TNFα	4 h	E-Selectin Expression
E-Selectin	IL-1	4 h	E-Selectin Expression
VCAM-1	TNFα	24 h	VCAM-1 Expression
VCAM-1	IL-1	24 h	VCAM-1 Expression
P38	Anisomycin	30 min	p38 Phosphorylation
P38	TNFα	30 min	p38 Phosphorylation
P38	IL-1	30 min	p38 Phosphorylation
ERK1/2	TPA	30 min	ERK1/2 Phosphorylation
STAT1	IFγ	30 min	Stat 1 Nuclear Translocation
STAT6	IL-4	30 min	Stat 6 Nuclear Translocation
JNK	Anisomycin	30 min	JNK Phosphorylation
JNK	TNFα	30 min	JNK Phosphorylation
JNK	IL-1	30 min	JNK Phosphorylation
HO1	–	–	HO1 Expression
CHK2	–	–	CHK2 Phosphorylation
P53	–	–	p53 Phosphorylation
P230 (A)	–	–	Golgin-230 Distribution
P230 (F)	–	–	Golgin-230 Distribution
Histone H3	–	–	Histone H3 Phosphorylation
Tubulin	–	–	Tubulin Distribution after 4 h
Tubulin	–	–	Tubulin Distribution after 24 h
Cell Number	–	–	Nuclear Count after 4 h
Cell Growth	–	–	Nuclear Count after 24 h

are run with HUVECs grown in a defined medium to study the effect of the hit compound in the same biological background. This is especially important, because different cell types may display different responses to the same compound. To allow simultaneous screening of multiple profiling assays, we have standardized and simplified the assay protocols. For example, cells are seeded at the same density for most of the assays. Cells are incubated with activators for thirty minutes, four hours, or twenty-four hours so that assays can be grouped and processed together.

A larger number of assays in the profiling set will generate a more complete picture of the compound's activity. However, for practical reasons, which include time and resources, we limit the number of assays to twenty-five. Each compound is tested in serial dilutions between 8 μM and 16 nM. Testing several compound concentrations against the same assay will allow computation of a dose-response curve for each compound if necessary. These dose-response curves will provide additional information on the compound's physical and biological properties.

The results of the profiling experiments are summarized in two heat maps (Figure 15.4). Red indicates an inhibitory effect of the tested compound and blue a stimulatory effect. The deeper the shade of the colors, the stronger the effect of the compound. White indicates that there is at best a weak effect. Comparison of the profiles of compounds **1** (CID 657314) and **2** (CID 659101) clearly indicates that **2** has a rather narrow activity spectrum. It inhibits mostly the inflammation-related readouts like NF-κB translocation, E-selectin, and VCAM1 expression. Inhibitory activity can be seen when the assays were activated with TNFα or IL-1. IL-1 and TNFα pathways converge on the level of the IKK complex. This indicates that the compound may act at or downstream of the IKK complex. Only small effects can be observed on p38, JNK, and Erk mitogen-activated protein kinase (MAP kinase) assays. Similar findings were observed for the STAT1 and STAT6 assays. These facts make a general inhibition of receptor signaling or nuclear import as an MOA unlikely. However, some assays monitoring cell growth and cell cycle were affected. After twenty-four hours of treatment with **2**, a lower cell count shows an inhibition of cell division. In addition to a lower cell count, p53 undergoes phosphorylation at serine 15, which points to a block in the cell cycle, most likely at the G1/S interphase. This is consistent with the fact that histone H3 phosphorylation (Ser10) is inhibited. Histone H3 undergoes phosphorylation in Ser10 as early as in the late G2 phase. The results from the golgin-230 and tubulin assays indicate that the overall integrity of the cell seems to be intact. The results from the profiling assays of **2** show that the compound mostly inhibits inflammation-related assays. Because the compound inhibits both arms of the pathway, the TNFα

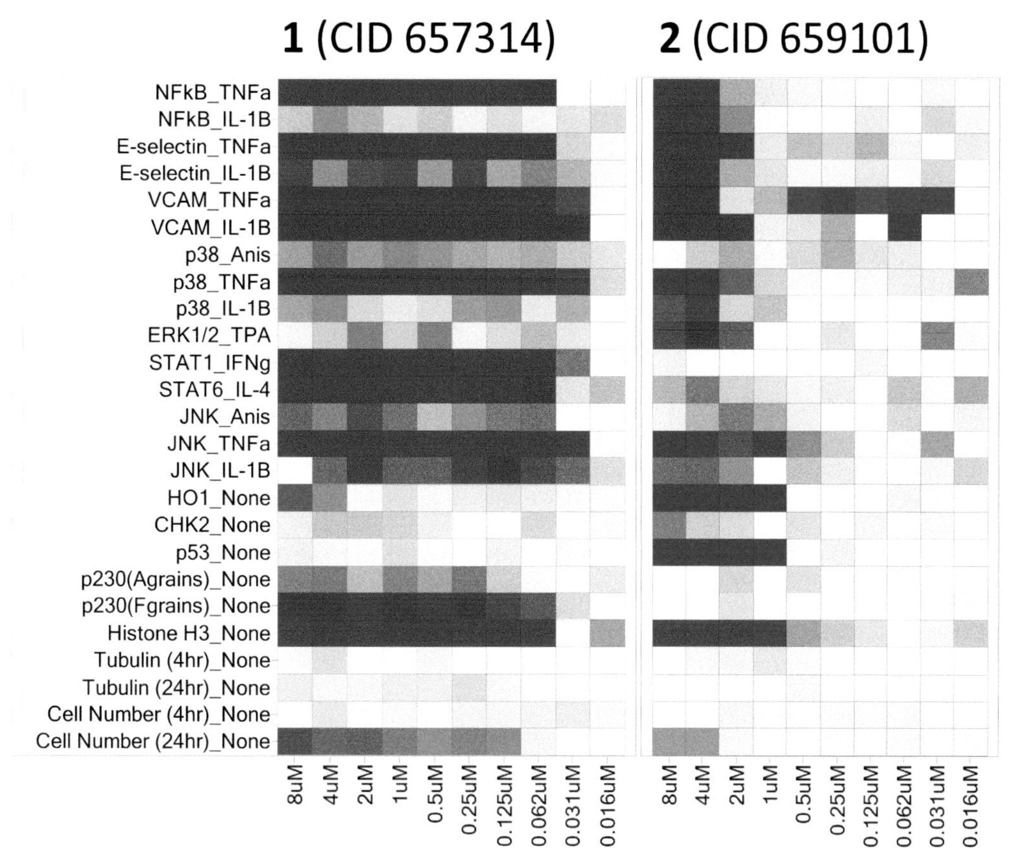

FIGURE 15.4: ***Activity maps of the profiled compounds. Activity maps of the hit compounds are displayed. Each compound was profiled against a set of 25 cell-based assays. Each assay was tested against 10 different concentrations. The lowest concentration was 0.016 μM and the highest was 8 μM. The arrow indicates increasing concentrations of the compound.***

and IL-1 arm, it must act on IKK complex, the conversion points of both arms, or somewhere downstream from the IKK complex in the pathway. It is likely that by blocking the NF-κB pathway, the cell undergoes cell cycle arrest.

A quite different picture emerges when compound 1 (CID 657314) is profiled. All assays that use cytokine/receptor interaction to activate the pathways (NF-κB, E-selectin, VCAM1, p38, STAT1, STAT6, and JNK) were inhibited by this compound. This fact makes it unlikely that the compound specifically interacts with proteins in the NF-κB pathway. The effect was much weaker when anisomycin, which does not use cell surface receptors, was used to activate the assays. The low cell numbers after twenty-four hours indicate that the cell division was inhibited. The inhibition of histone H3 phosphorylation is an indicator that the cell cycle is blocked. A closer look at the golgin-p230 and tubulin assays reveals that the tubulin cytoskeleton seems to be intact, but the integrity of the Golgi complex was affected. Figure 15.5 shows HUVECs treated with **1** (brefeldin A). To label the Golgi apparatus, the cells were fixed and stained with an antibody against golgin-230, a Golgi marker. In the treated cells, the Golgi apparatus is mostly disintegrated. It is a well-known fact that the Golgi apparatus is dissembled after brefeldin A treatment and that Golgi marker appears to be relocated to the endoplasmic reticulum (ER) membranes. In such cells, intracellular membrane transport is disrupted [19]. This defect in membrane transport may explain the broad inhibitory effect of compound 1. The treated cells are no longer able to transport the receptors for IL-1, TNFα, STAT1, and STAT 6 to the plasma membrane. With no receptor at the cell surface, the cytokines can no longer activate the pathways. Figure 15.6 summarizes the MOA of both compounds. In untreated HUVECs, NF-κB (Figure 15.6A) is located in the cytosol. Upon activation with TNFα, IKK becomes activated, IκBα is degraded, and NF-κB accumulates in the nucleus (Figure 15.6B). Treatment with CID 657314 (**1**) disassembled the Golgi apparatus and thus interrupted the transport of receptor to the cell surface (Figure 15.6C). As show by Xie et al., CID 659101 (**2**) inhibits IKK, and thus IκBα remains complexed with NF-κB, inhibiting nuclear accumulation (Figure 15.6D) [20].

The value of cell-based assays as a tool to profile compounds is well illustrated by the two examples described previously. In the early stage of the development of drug leads or chemical probes, it is necessary to get some

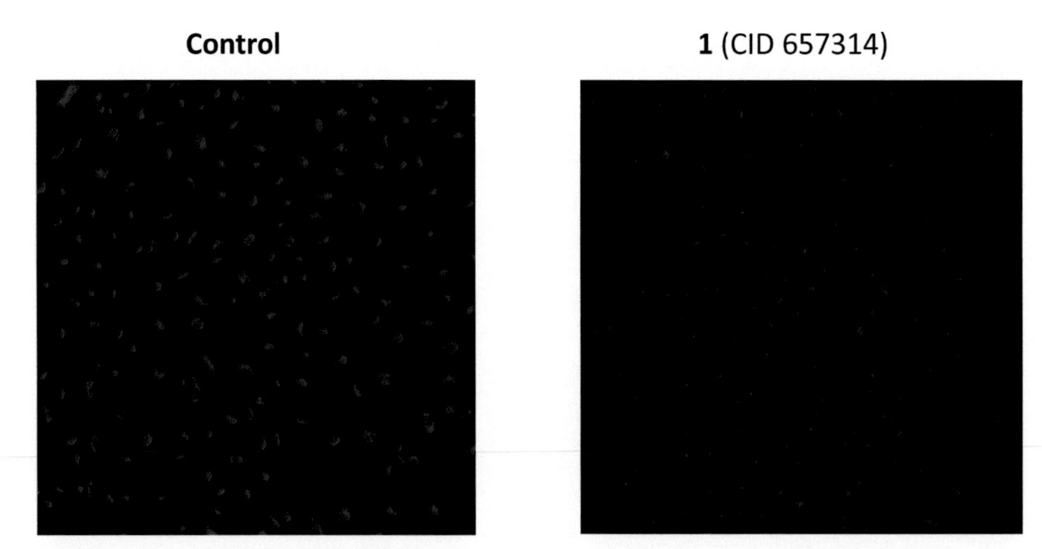

*FIGURE 15.5: CID 657314 disassembled the Golgi apparatus. Figure shows cells from the p230 assay. Cells in right panel were treated with CID 657314 (**1**). Cells shown in the left panel are the non treated control. Cells were then fixed and labeled with an antibody against golgin-230 and a secondary fluorescence labeled antibody. Pictures were taken with the In Cell Analyzer 3000.*

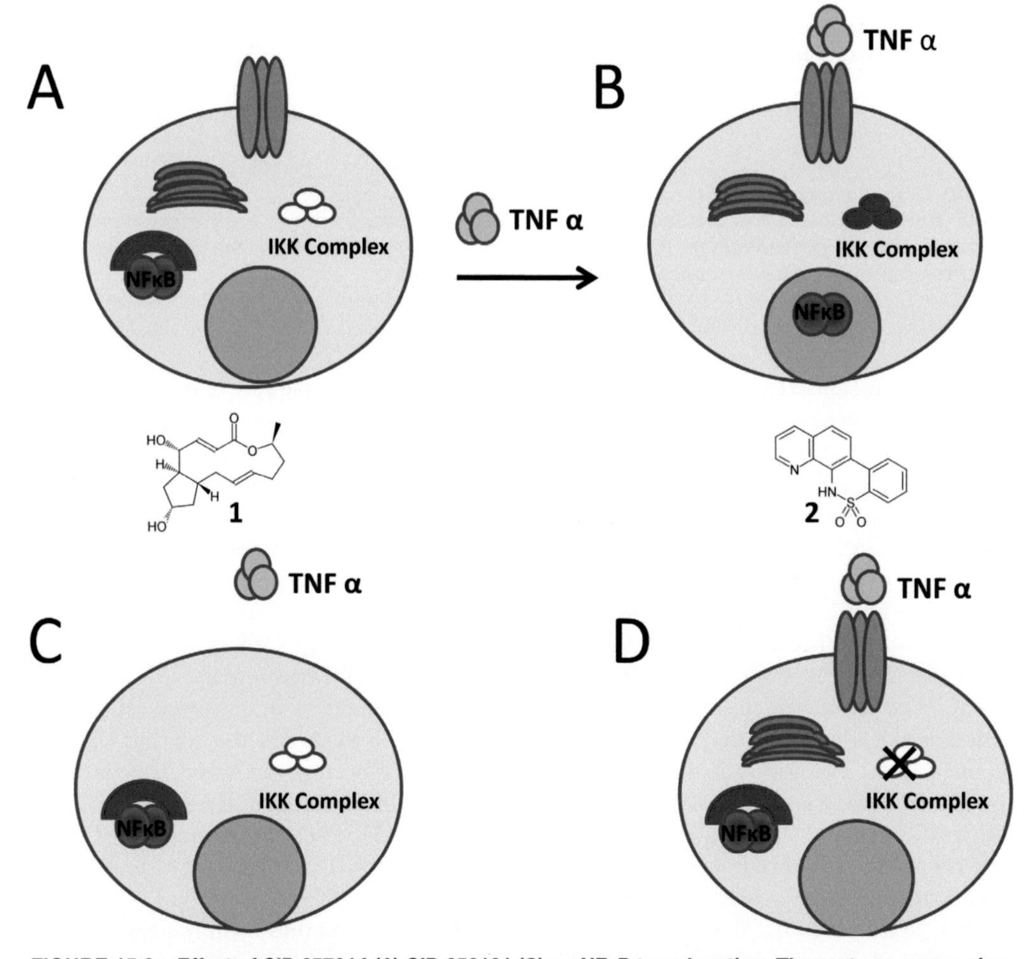

*FIGURE 15.6: Effect of CID 657314 (**1**) CID 659101 (**2**) on NFκB translocation. The cartoon summarizes the effects of both compounds on the NF-κB translocation assay. NF-κB is located in the cytosol (**A**). Upon activation with TNFα, IKK becomes activated and phosphorylates IκBα. Phosphorylated IκBα is degraded, and NF-κB is translocated into the nucleus (**B**). After treatment with CID 657314 (**1**), the Golgi complex is disassembled, intracellular transport is inhibited, and receptors cannot reach the cell surface (**C**). In cells incubated with CID 659101 (**2**) the IKK activity is blocked, and NF-κB remains associated with the inhibitor IκBα (**D**).*

information on the molecular activities of a confirmed hit compound to exclude those that have widespread activities and affect basic cellular processes rather than the desired pathways and targets. In the case of the two compounds described here, one (**2**, CID 659101) showed promising properties and was further developed into a molecular probe [20]. The example mentioned in paragraph 3.3 is a simple illustration of how cell-based assays can be employed to profile compounds. Future developments in two areas will allow a much more detailed compound profiling: 1) More assays covering additional pathways and biological processes will generate a broader knowledge on the biological effects of a test compound. Multiplexing of assays is one way to expand the assay set; and 2) new software programs using analysis algorithms that refer to sophisticated maps of pathways and cellular processes can generate models to predict molecular actions of compounds in question. Further development of powerful informatics will be one of the keys for understanding cross talks between pathways in the complex cellular and intracellular biochemical network.

ACKNOWLEDGEMENT

We thank Michael Wyler and Dr. Efithia Cayanis for help with the profiling assays. The development of the NFκB assays was started at Memorial Sloan-Kettering Center in New York City as part of the Functional Proteomic Project. Screening and profiling were done at the MLSCN Center at Columbia University. We thank Dr. James E. Rothman for his support and Dr. Martin Wiedmann for his suggestions and input during the project. The work was supported by the Molecular Libraries Initiative of the National Institutes of Health and a grant (R03 MH076344–01) to T. M. for assay development.

REFERENCES

1. Austin, C. P., Brady, L. S., Insel, T. R., and Collins, F. S. (2004). NIH Molecular Libraries Initiative. *Science 306*, 1138–1139.
2. Karaman, M. W., Herrgard, S., Treiber, D. K., Gallant, P., Atteridge, C. E., Campbell, B. T., Chan, K. W., Ciceri, P., Davis, M. I., Edeen, P. T., Faraoni, R., Floyd, M., Hunt, J. P., Lockhart, D. J., Milanov, Z. V., Morrison, M. J., Pallares, G., Patel, H. K., Pritchard, S., Wodicka, L. M., and Zarrinkar, P. P. (2008). A quantitative analysis of kinase inhibitor selectivity. *Nat Biotechnol 26*, 127–132.
3. Goldstein, D. M., Gray, N. S., and Zarrinkar, P. P. (2008). High-throughput kinase profiling as a platform for drug discovery. *Nat Rev Drug Discov* 7, 391–397.
4. Auld, D. S., Southall, N. T., Jadhav, A., Johnson, R. L., Diller, D. J., Simeonov, A., Austin, C. P., and Inglese, J. (2008). Characterization of chemical libraries for luciferase inhibitory activity. *J Med Chem 51*, 2372–2386.
5. Simeonov, A., Jadhav, A., Thomas, C. J., Wang, Y., Huang, R., Southall, N. T., Shinn, P., Smith, J., Austin, C. P., Auld, D. S., and Inglese, J. (2008). Fluorescence spectroscopic profiling of compound libraries. *J Med Chem 51*, 2363–2371.
6. Seidler, J., McGovern, S. L., Doman, T. N., and Shoichet, B. K. (2003). Identification and prediction of promiscuous aggregating inhibitors among known drugs. *J Med Chem 46*, 4477–4486.
7. Feng, B. Y., Simeonov, A., Jadhav, A., Babaoglu, K., Inglese, J., Shoichet, B. K., and Austin, C. P. (2007). A high-throughput screen for aggregation-based inhibition in a large compound library. *J Med Chem 50*, 2385–2390.
8. Kramer, J. A., Sagartz, J. E., and Morris, D. L. (2007). The application of discovery toxicology and pathology towards the design of safer pharmaceutical lead candidates. *Nat Rev Drug Discov 6*, 636–649.
9. Mayer, T., Jagla, B., Wyler, M. R., Kelly, P. D., Aulner, N., Beard, M., Barger, G., Tobben, U., Smith, D. H., Branden, L., and Rothman, J. E. (2006). Cell-based assays using primary endothelial cells to study multiple steps in inflammation. *Methods Enzymol 414*, 266–283.
10. Oakley, R. H., Hudson, C. C., Cruickshank, R. D., Meyers, D. M., Payne, R. E., Jr., Rhem, S. M., and Loomis, C. R. (2002). The cellular distribution of fluorescently labeled arrestins provides a robust, sensitive, and universal assay for screening G protein-coupled receptors. *Assay Drug Dev Technol 1*, 21–30.
11. Wyler, M. R., Smith, D. H., Cayanis, E., Tobben, U., Aulner, N., and Mayer, T. (2009). Cell-based assays to probe the ERK MAP kinase pathway in endothelial cells. *Methods Mol Biol 486*, 29–41.
12. Schmid, I., Hausner, M. A., Cole, S. W., Uittenbogaart, C. H., Giorgi, J. V., and Jamieson, B.D. (2001). Simultaneous flow cytometric measurement of viability and lymphocyte subset proliferation. *J Immunol Methods 247*, 175–186.
13. Senftleben, U., and Karin, M. (2002). The IKK/NF-kappaB pathway. *Crit Care Med 30*, S18-S26.
14. Karaman, M. W., Herrgard, S., Treiber, D. K., Gallant, P., Atteridge, C. E., Campbell, B. T., Chan, K. W., Ciceri, P., Davis, M. I., Edeen, P. T., Faraoni, R., Floyd, M., Hunt, J. P., Lockhart, D. J., Milanov, Z. V., Morrison, M. J., Pallares, G., Patel, H. K., Pritchard, S., Wodicka, L. M., and Zarrinkar, P. P. (2008). A quantitative analysis of kinase inhibitor selectivity. *Nat Biotechnol 26*, 127–132.
15. Ding, G. J., Fischer, P. A., Boltz, R. C., Schmidt, J. A., Colaianne, J. J., Gough, A., Rubin, R. A., and Miller, D. K. (1998). Characterization and quantitation of NF-kappaB nuclear translocation induced by interleukin-1 and tumor necrosis factor-alpha. Development and use of a high capacity fluorescence cytometric system. *J Biol Chem 273*, 28897–28905.
16. Gilmore, T. D., and Herscovitch, M. (2006). Inhibitors of NF-kappaB signaling: 785 and counting. *Oncogene 25*, 6887–6899.
17. Gilmore, T. D. (2006). Introduction to NF-kappaB: players, pathways, perspectives. *Oncogene 25*, 6680–6684.
18. Klausner, R. D., Donaldson, J. G., and Lippincott-Schwartz, J. (1992). Brefeldin A: insights into the control of membrane traffic and organelle structure. *J Cell Biol 116*, 1071–1080.

19. Orci, L., Palmer, D. J., Ravazzola, M., Perrelet, A., Amherdt, M., and Rothman, J. E. (1993). Budding from Golgi membranes requires the coatomer complex of non-clathrin coat proteins. *Nature 362*, 648–652.
20. Xie, Y., Deng, S., Thomas, C. J., Liu, Y., Zhang, Y. Q., Rinderspacher, A., Huang, W., Gong, G., Wyler, M., Cayanis, E., Aulner, N., Tobben, U., Chung, C., Pampou, S., Southall, N., Vidovic, D., Schurer, S., Branden, L., Davis, R. E., Staudt, L. M., Inglese, J., Austin, C. P., Landry, D. W., Smith, D. H., and Auld, D. S. (2008). Identification of N-(quinolin-8-yl)benzenesulfonamides as agents capable of down-regulating NFkappaB activity within two separate high-throughput screens of NFkappaB activation. *Bioorg Med Chem Lett 18*, 329–335.

CHAPTER 16

Use of Transgenic Zebrafish in a Phenotypic Screen for Angiogenesis Inhibitors

Jaeki Min

Yuhong Du

Brenda Bondesen

Brian Revennaugh

Peter Eimon

Ray Dingledine

Angiogenesis inhibitors have become essential tools in the treatment of diseases such as cancer, age-related macular degeneration, psoriasis, and diabetic retinopathy (reviewed in [1]), making angiogenesis a clinically relevant target process for drug discovery. For example, solid tumors require an adequate supply of blood vessels in order to survive, grow, and metastasize [2, 3]. Recently, a link between angiogenesis and Alzheimer's disease has also been postulated [4], possibly highlighting another clinical use for antiangiogenesis drugs.

Multiple in vitro and in vivo angiogenesis assays are commonly used for drug discovery. Each of these assays has distinct advantages and disadvantages [5 –7]. In vitro endothelial cell models of migration, proliferation, apoptosis, and tube formation are popular because of their simplicity and throughput. However, most of these models address only the effects of compounds on endothelial cells and not other tissues in the vascular bed, such as smooth muscle cells, fibroblasts, and endothelial progenitor cells. Because angiogenesis involves the proliferation and migration of endothelial cells in the context of a living organism, current in vitro models employed in screening campaigns may prove vulnerable to a variety of unanticipated limitations.

In vivo angiogenesis models include the corneal micropocket assay, the chick embryo chorioallantoic membrane (CAM) assay, the Matrigel plug assay, and the femoral artery ligation assay (reviewed in [7]). These models enable the study of endothelial cells in the context of their native environment but are expensive, labor intensive, technically challenging, inherently low-throughput, and require comparatively large quantities of test compound. In addition, performing genetic manipulations such as gene knockdown and overexpression in most in vivo models either is not feasible or requires the added complication of generating transgenic/knockout mice. Therefore, there is a growing demand for novel angiogenesis assays that combine the biological complexity of an in vivo environment with the potential for moderate- to high-throughput drug screening and rapid gene knockdown.

Zygogen, LLC has developed a rapid in vivo angiogenesis assay suitable for screening large compound libraries. This assay utilizes transgenic zebrafish (*Danio rerio*) embryos with fluorescent blood vessels to facilitate automated image analysis. The zebrafish is a small, freshwater fish of the minnow family native to the streams and rivers of India. Zebrafish have become a leading model organism for the study of vertebrate biology because of their fecundity, optical clarity, physiologic similarity to mammals, and adaptability to multiple experimental techniques [8]. Embryonic angiogenesis has been extensively characterized in zebrafish, and the molecular pathways underlying this process are well conserved with mammals (recently reviewed in [9]). Blood flow begins in the major vessels of the embryo approximately twenty-four hours postfertilization. These vessels are formed by vasculogenesis (the de novo production of new blood vessels from progenitor cells) and include the dorsal aorta and caudal veins. At about this time, additional vessels begin to form by sprouting from preexisting vasculature, a process termed angiogenesis. The first angiogenic vessels to form in the trunk of the embryo are the intersegmental vessels (ISVs), which sprout from the dorsal aorta. Several studies have shown that treatment of zebrafish embryos with clinical-stage antiangiogenic compounds inhibits ISV growth [10, 11].

Transgenic zebrafish lines that express a fluorescent protein throughout the vasculature [11] provide a convenient in vivo readout that facilitates automated visualization and quantification of angiogenic vessel development. Stable transgenic lines with fluorescent vessels have been developed by utilizing the vascular endothelial growth factor receptor 2 (*vegfr2*; also known as *kdr* or *flk-1*) promoter to drive expression of green reef coral fluorescent protein (GRCFP) or *Aequorea coerulescens* green fluorescent protein (AcGFP). *Vegfr2* is one of several receptors for the VEGF family of growth factors and is expressed specifically in blood vessels. The fluorescent blood vessels within these transgenic zebrafish allow robotic visualization of blood vessel growth to enable a medium- to high-throughput compound screen.

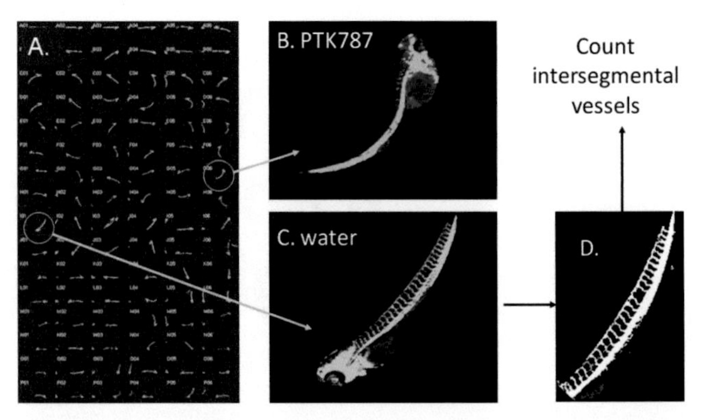

FIGURE 16.1:* *Assay for angiogenesis in zebrafish embryos expressing green fluorescent protein in the endothelial cells that line blood vessels. A. Partial view of a 384-well plate with embryos arrayed one per well. B. After overnight treatment with a VEGF receptor antagonist, there are no observable angiogenic vessels outside the spinal area. C. Embryo treated with water overnight shows robust angiogenesis. D. The spinal area is isolated with a box cursor, and the number of intersegmental vessels is counted as a measure of angiogenesis.

1. METHODS

Live zebrafish embryos can be arrayed in 384-well microplates, making them ideal for screening midsized compound libraries or as a secondary assay to follow up on hits from high-throughput in vitro screens. To demonstrate this, we used transgenic zebrafish embryos with fluorescent blood vessels to screen the commercially available LOPAC1280 compound library (Sigma-Aldrich) for antiangiogenic activity. The LOPAC library is a collection of 1,280 known pharmacologically active compounds. A schematic of the assay is shown in Figure 16.1. To determine whether the screening parameters and overall screening experience are comparable across multiple sites in this whole-animal assay, Zygogen and Emory each screened half of the LOPAC library (640 compounds each).

Transgenic zebrafish with fluorescent blood vessels – *tg(vegfr2:GRCFP)* – were mated with wild type zebrafish (Tübingen strain) to produce fertilized eggs for the screen. Embryos were treated with pronase (1 mg/mL) to remove the chorions, placed into Petri dishes containing Holtfreter's solution (60 mM sodium chloride, 2.4 mM sodium bicarbonate, 0.8 mM calcium chloride, 0.67 mM potassium chloride, 10 mM HEPES, pH 7.0), and incubated at 28°C. Embryos were allowed to develop until they had reached the prim-5 stage, about twenty-four hours postfertilization, at which time retinal and skin pigmentation becomes apparent and the heart begins beating. At this stage, ISVs are just beginning to sprout from the dorsal aorta. Embryos were arrayed into black 384-well polystyrene clear bottom plates (Greiner) with one embryo per well in 30 μL of Holtfreter's solution.

Compound stocks were diluted with Holtfreter's solution to produce a 2 × working stock (60 μM), and 30 μL was added to each microplate well using a Caliper Sciclone ALH 3000 (Zygogen) or a Biomek NX (Emory) robotic liquid handler. The final screening concentration was 30 μM in 60 μL of media (1% dimethyl sulfoxide [DMSO]). The antiangiogenic compound vatalanib (PTK787/ZK-222584) was used at 10 μM as a positive control, and 1% DMSO in Holtfreter's solution was used as the negative control. After the compounds were added, plates were centrifuged at 32 × g for sixty seconds and then incubated at 28°C for approximately twenty-four hours. At two days postfertilization, zebrafish embryos were anesthetized with 0.016% tricaine for thirty minutes to inhibit movement. A fluorescent image of each embryo in the 384-well plate was captured using an automated high-content imaging system (Discovery-1 at Zygogen or ImageXpress 5000 at Emory).

Image analysis software was used to increase assay throughput and quantify the effect of angiogenesis inhibitors. In each image, the trunk of the embryo was manually selected using MetaMorph software (Molecular Devices; Figure 16.1D). The angiogenic blood vessels (ISVs and their branching arteries) were then counted using an automated algorithm derived from the MetaMorph Neurite Outgrowth application drop-in. With the appropriate application settings, the vasculogenic vessels were identified as one or more cell bodies, and the angiogenic vessels that sprout from them were identified as outgrowths of these cell bodies. Because the trunk isolation step was interactive, each image was viewed by the user during the analysis, allowing exclusion of embryos that were unhealthy or in an improper orientation for analysis. A 384-well plate required approximately thirty minutes for image acquisition and approximately twenty minutes for data analysis.

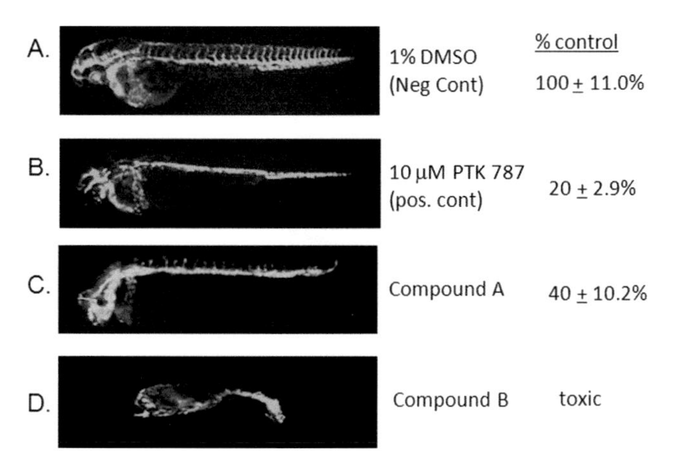

FIGURE 16.2: ***Visual confirmation of algorithm performance. Representative images of one-day-old zebrafish embryos treated for twenty-four hours with four compounds are shown in the left panels. The output of the angiogenesis measurement algorithm (mean ± SEM, n = 5 embryos) is shown for comparison on the right.***

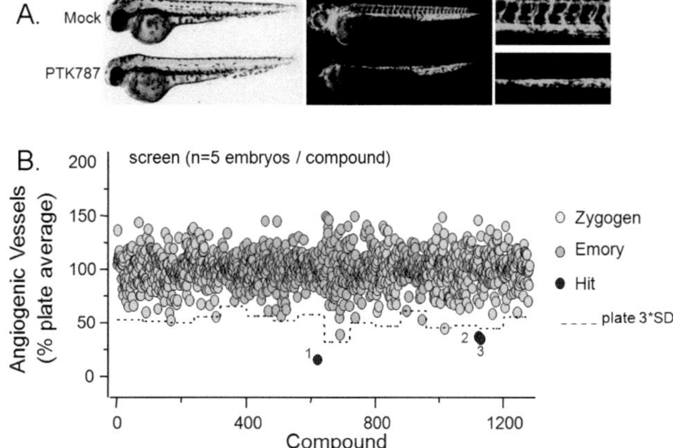

FIGURE 16.3: ***Screen of the LOPAC1280 compound library for antiangiogenic compounds. A. Representative images of zebrafish embryos treated for twenty-four hours with 1% DMSO (Mock) or 10 μM PTK787. B. Results of the screen. Each point represents the mean angiogenesis count of up to five embryos. Data were normalized as percentage of plate average. The three positive hits (red points) lie outside 3 standard deviations of the plate average.*** **Source:** ***Modified from Tran et al., 2007 [19].***

2. RESULTS

Many high-throughput screens expose each well to a different compound. In preliminary experiments, we found it necessary to dose five embryos per compound and average the results to achieve adequate screening robustness parameters. Screening data for each test compound were normalized as percent control and subjected to statistical analysis. Percent control was defined as 100 × [mean (compound) − mean (PTK787)]/[mean (DMSO) − mean (PTK787)], where mean (compound) is the mean vessel count of up to five embryos treated with the compound; mean (PTK787) is the mean vessel count of PTK787-treated embryos from the plate, and mean (DMSO) is the average blood vessel count of all embryos from the negative control wells. Figure 16.2 illustrates representative compound-treated embryos and the associated percent control values measured by the algorithm. Embryos exposed to a saturating concentration of the potent angiogenesis inhibitor PTK787 returned a value of 20% of control, even though few or no ISVs were apparent on visual inspection. The maximum assay window is thus approximately 5 (= 100%/20%). Forty compounds, representing 3.1% of the total library, were severely toxic or lethal at the screening dose (e.g., Figure 16.2, compound B) and were not analyzed further. The screen had robust quality control measures, with mean plate Z′ factor of 0.58, signal-to-background ratio of 4.8, and signal-to-noise (S/N) ratio of 10.2.

Compounds causing a reduction in the angiogenic vessel count in excess of 3 standard deviations from the plate mean were defined as positive hits. All hits were reconfirmed by visual inspection. Three hits were identified using these criteria, representing a 0.23% hit rate. (Figure 16.3B). Two are known angiogenesis inhibitors (SU4312, a VEGFR and platelet-derived growth factor receptor antagonist, and AG1478, an epidermal growth factor receptor antagonist). The third is novel: indirubin-3′-monoxime (IRO). IRO inhibits angiogenesis in zebrafish embryos with an IC_{50} of 0.31 μM (Figure 16.4). IRO is a cell-permeable derivative of indirubin, the active component of Dang Gui Longhui Wan, a mixture of plants used in traditional Chinese medicine for the treatment of chronic diseases, particularly chronic myelocytic leukemia.

To determine whether IRO could block vasculogenesis in addition to angiogenesis, we examined its effect on the dorsal aorta and posterior cardinal vein of zebrafish embryos treated with IRO ten hours postfertilization

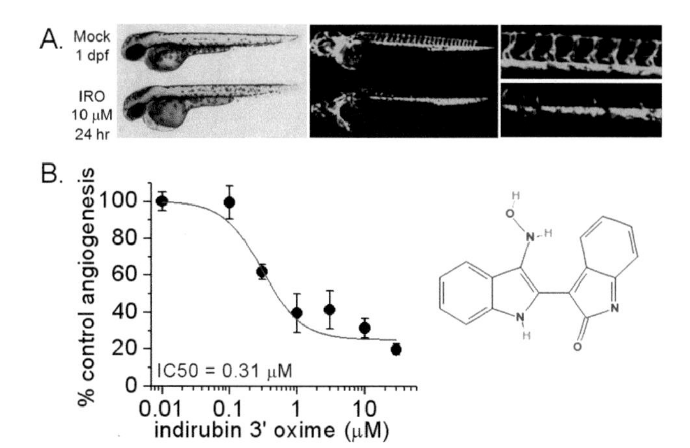

FIGURE 16.4: ***Indirubin 3′-monoxime (IRO) inhibits angiogenesis in zebrafish embryos. A. brightfield and fluorescent images of zebrafish embryos treated for twenty-four hours with 1% DMSO (Mock) or 10 μM IRO. B. The antiangiogenic effects of IRO are concentration dependent. The structure of IRO is presented to the right.*** **Source:** ***Modified from Tran et al., 2007 [19].***

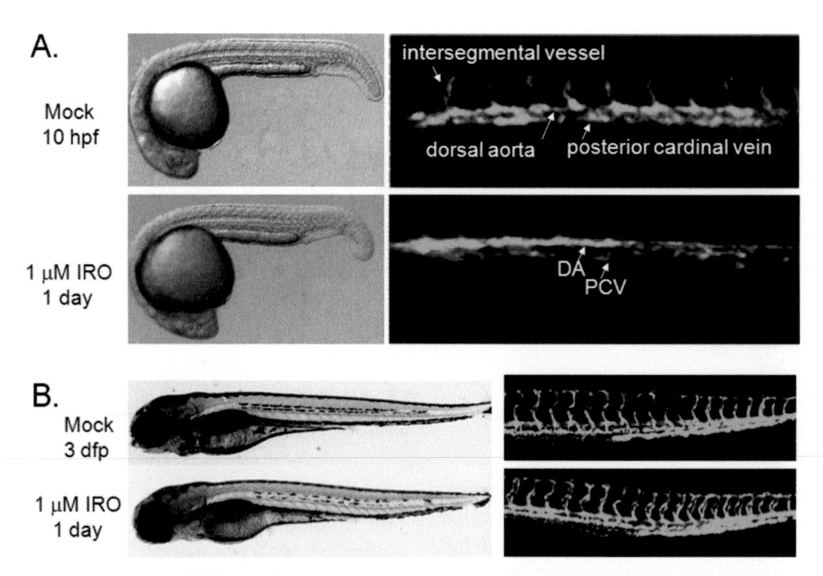

FIGURE 16.5: ***Lack of effect of IRO on vasculogenesis and established blood vessels. A. Overnight treatment of embryos ten hours postfertilization (hpf) with 1 μM IRO does not prevent growth of the dorsal aorta of posterior cardinal vein. B. Treatment of three-day-old embryos with IRO does not cause regression of existing vasculature.*** Source: ***Modified from Tran et al., 2007 [19].***

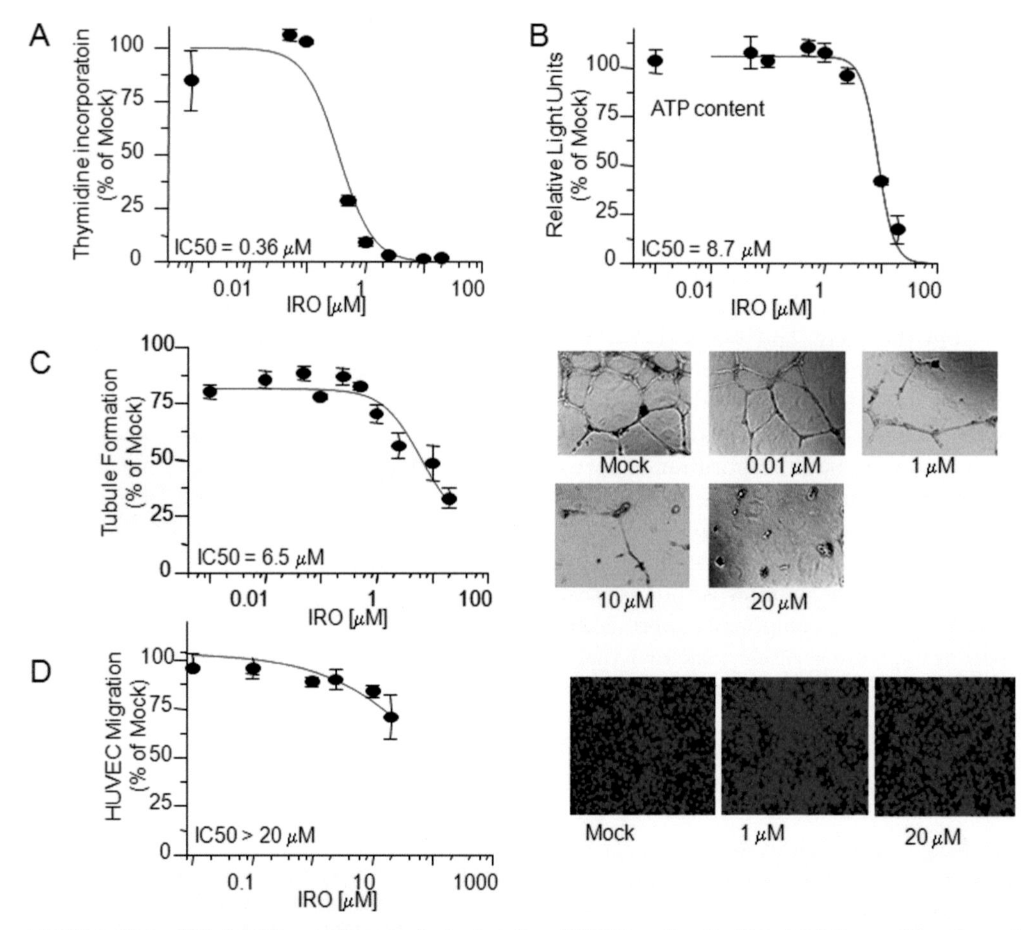

FIGURE 16.6: ***IRO inhibits angiogenesis in in vitro HUVEC cells. A. IRO inhibits proliferation of HUVECs in a concentration-dependent manner as assessed by the incorporation of ^{3}H-thymidine into newly dividing cells. B. The Cell-Titre Glow assay was used to evaluate the cellular toxicity of IRO. C. The effect of IRO on vascular tube formation is shown, with representative images on the right. D. Lack of effect of IRO on migration of HUVECs.*** Source: ***Modified from Tran et al., 2007 [19].***

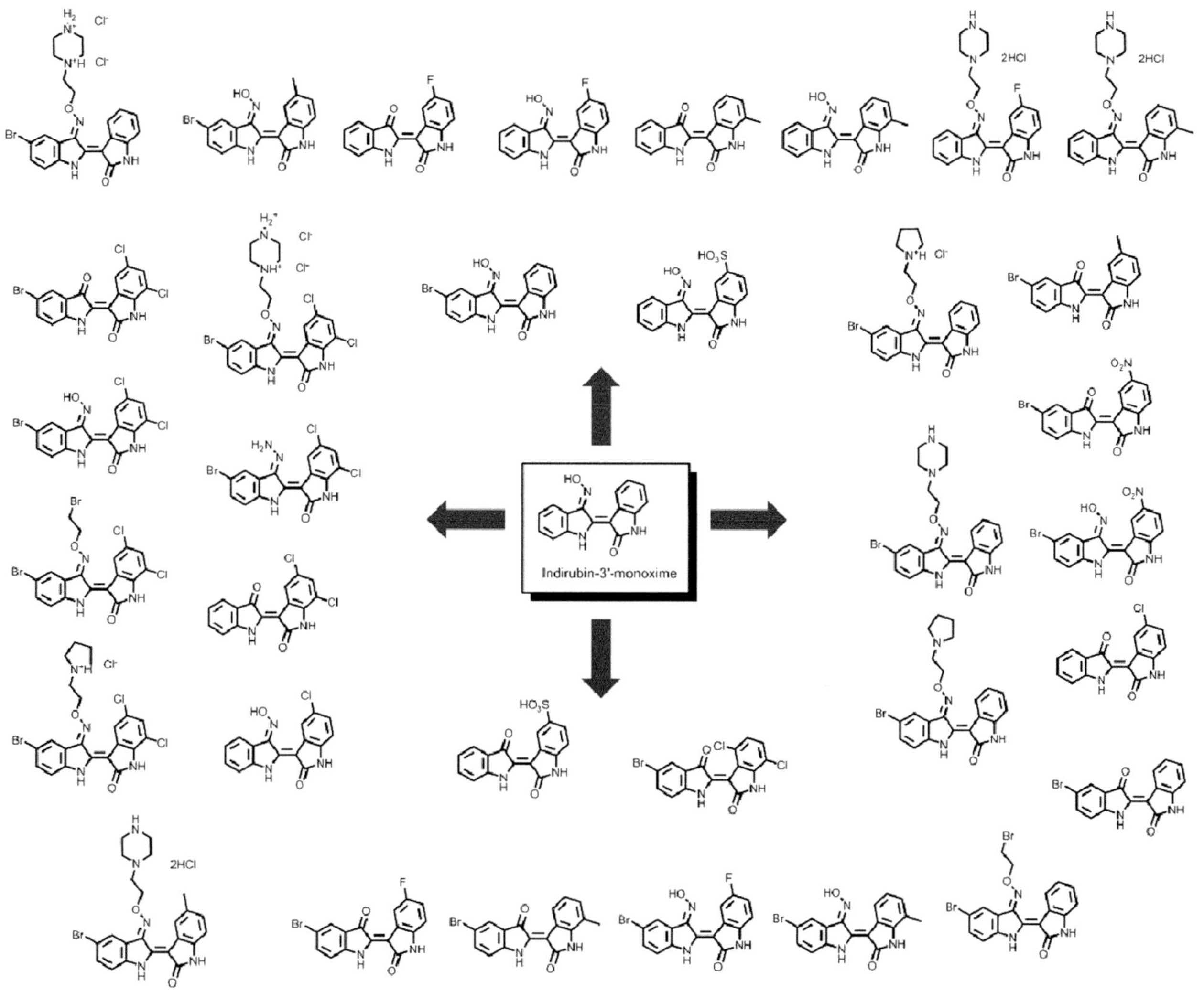

FIGURE 16.7: ***IRO itself and thirty-three analogs were synthesized for testing in the zebrafish angiogenesis assay.***

(Figure 16.5A). This developmental time point precedes the appearance of angioblasts, which differentiate to form the vasculogenic vessels. IRO did not block the formation of the aortic vessels under these conditions, although it continued to exert a strong antiangiogenic effect on the ISVs. Moreover, when IRO was applied to older embryos (three days postfertilization), it did not cause regression of preexisting ISVs (Figure 16.5B). These data demonstrate that IRO is a selective, antiangiogenic compound in zebrafish that neither inhibits vasculogenesis nor targets established angiogenic blood vessels.

To determine if IRO is antiangiogenic in human endothelial cells and to elucidate its mechanism of action, we studied its effects on human umbilical cord vein endothelial cell (HUVEC) assays. HUVEC proliferation was assessed using a ^{3}H-thymidine incorporation assay. Endothelial cells were seeded into a 96-well plate and treated with different concentrations of IRO, and ^{3}H-thymidine incorporation was measured. IRO inhibited endothelial cell proliferation with an IC_{50} of 0.36 μM (Figure 16.6A). Proliferation was inhibited at an approximately twenty-four-fold lower concentration than that needed to produce cytotoxicity (IC_{50} = 8.7μM), which was monitored by reduction in cellular adenosine triphosphate (ATP) levels (Figure 16.6B). Endothelial tube formation was assessed using HUVECs cultured on extracellular matrix. Full-growth media were used to stimulate tube formation. IRO was added to the growth medium, and tube formation was allowed to proceed for eighteen hours. Tube length was measured using Image Pro Plus software. IRO inhibited endothelial tube formation with an IC_{50} of 6.5 μM (Figure 16.6C). Finally, endothelial cell migration was assessed using a transwell migration assay (BD Biosciences). HUVECs were serum starved for four hours, then seeded into the top wells of migration chambers in endothelial basal medium (without growth factors). Complete medium–containing growth factor was placed in the bottom well as a chemoattractant. Migration was

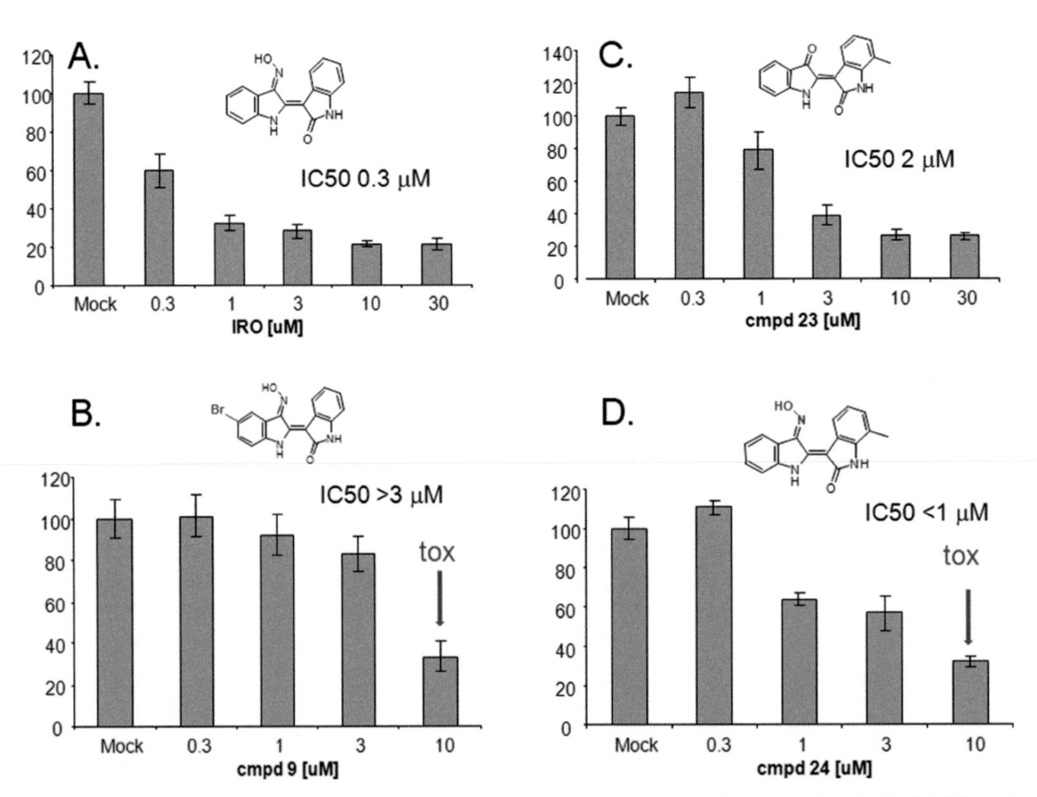

***FIGURE 16.8:** Some structural features of IRO that confer antiangiogenic properties. **A, B.** Addition of a bromo group to the 5-position of IRO reduced activity. **C, D.** Substitution of a ketone for the monoxime group preserved activity.*

allowed to proceed for twenty-two hours, at which point migrated cells were fixed, stained, and quantified. IRO did not significantly affect endothelial cell migration (Figure 16.6D). Collectively, these results indicate that the antiangiogenic effect of IRO is achieved mainly by inhibiting cell proliferation, with no significant effect on endothelial cell migration.

Indirubin and some of its derived analogs are reported to be inhibitors of cyclin-dependent kinases, glucogen synthase kinase 3 (GSK-3), and glycogen phosphorylase b and to be activators of the aryl hydrocarbon receptor (AhR), which was previously known as the dioxin receptor [12–17]. Thirty-four IRO analogs were synthesized, and an additional forty-three compounds were purchased for testing in the zebrafish angiogenesis assay. The purpose was to explore the structural features of IRO that are important for antiangiogenic activity. Interestingly, indirubin itself was inactive up to 30 μM, whereas the indirubin-3′-oxime inhibited angiogenesis with 0.3 μM IC_{50}. Based on the significance of the 3′-oxime group in the indirubin scaffold, we designed derivatives that were modified around both the phenyl ring and 3′-oxime. A new class of 3′-substituted 7-halogenoindirubins is known to induce cell death in a diversity of human tumor cell lines; some indirubins therefore may interact with new, unidentified target proteins, distinct from their well-characterized targets such as cyclin dependent kinases, GSK-3, and AhR [18]. We applied this strategy to construct a small, focused library of indirubins. We started by designing a method to synthesize 5- or 7-substituted indirubins with various substitutions on position 3′ using reported synthetic procedures [18]. The chemical structures of synthesized IRO derivatives and representative structure-activity relationships are shown in Figures 16.7 and 16.8. Similar to the observation that indirubin is inactive, the same substitution of ketone for oxime lost activity in the zebrafish angiogenesis assay (not shown). However, in other pairs of compounds (e.g., Figure 16.7C, D), the effect of the ketone substitution was less marked. Adding a bromo group on the 5′-position also reduced activity compared with IRO itself (Figure 16.8A, B). In the meantime, the substitution of methyl on the 7-position (PubChem SID 26676134) and fluoride on the 5-position (SID 26676136) of indirubin-3′-monoxime exhibited similar levels of activity compared with IRO itself. The introduction of other functional groups such as chloride, nitro, or sulphonic acid on the same 5- or 7-position reduced inhibition. In sum, we discovered several novel synthetic analogs of IRO that showed similar levels of inhibition against zebrafish angiogenesis.

3. SUMMARY AND OUTLOOK

A moderate-throughput assay (~5,000 embryos per week) was developed to study angiogenesis in zebrafish. Compared with other animal models, screening in zebrafish can be performed more rapidly on a larger number of animals using much less compound. The optimized assay displayed acceptable robustness parameters for screening ($Z' = 0.58$; S/N = 10). A novel antiangiogenic compound, indirubin 3′-monoxime ($IC_{50} = 0.31$ μM), was identified from a screen of the LOPAC1280 library. In addition to blocking angiogenesis in zebrafish, IRO also blocked proliferation of human vascular endothelial cells at a twenty-four-fold lower concentration ($IC_{50} = 0.36$ μM) than the concentration at which it caused cellular toxicity (IC_{50}=8.7μM). Importantly, the relatively low level of IRO-induced endothelial cytotoxicity shows that its antiproliferative activity is not due to broad cytotoxic effects. This is consistent with data from the zebrafish assay, which revealed no indications of overt toxicity in embryos treated with effective doses of IRO and no effect of IRO on preexisting vasculature.

Transgenic zebrafish combine the physiologic complexity of in vivo vertebrate models with the speed and convenience required for medium- to high-throughput assays and are thus well suited to screen small libraries or support medicinal chemistry campaigns. As such, the zebrafish angiogenesis assay may be used as a bridge between biochemical or cell-based screening assays and expensive, labor-intensive mammalian systems.

REFERENCES

1. Carmeliet, P., and Jain, R. K. (2000). Angiogenesis in cancer and other diseases. *Nature 407*, 249–257.
2. Hanahan, D., and Folkman, J. (1996). Patterns and emerging mechanisms of the angiogenic switch during tumorigenesis. *Cell 86*, 353–364.
3. Li, C. Y., Shan, S., Cao, Y., and Dewhirst, M. W. (2000). Role of incipient angiogenesis in cancer metastasis. *Cancer Metastasis Rev 19*, 7–11.
4. Vagnucci, A. H., Jr., and Li, W. W. (2003). Alzheimer's disease and angiogenesis. *Lancet 361*, 605–608.
5. Auerbach, R., Akhtar, N., Lewis, R. L., and Shinners, B. L. (2000). Angiogenesis assays: problems and pitfalls. *Cancer Metastasis Rev 19*, 167–172.
6. Jain, R. K., Schlenger, K., Hockel, M., and Yuan, F. (1997). Quantitative angiogenesis assays: progress and problems. *Nat Med 3*, 1203–1208.
7. Norrby, K. (2006). In vivo models of angiogenesis. *J Cell Mol Med 10*, 588–612.
8. Doan, T. N., Eilertson, C. D., and Rubinstein, A. L. (2004). High-throughput target validation in model organisms. *Drug Discovery Today: TARGETS 3*, 191–197.
9. Baldessari, D., and Mione, M. (2008). How to create the vascular tree? (Latest) help from the zebrafish. *Pharmacol Ther 118*, 206–230.
10. Chan, J., Bayliss, P. E., Wood, J. M., and Roberts, T. M. (2002). Dissection of angiogenic signaling in zebrafish using a chemical genetic approach. *Cancer Cell 1*, 257–267.
11. Cross, L. M., Cook, M. A., Lin, S., Chen, J. N., and Rubinstein, A. L. (2003). Rapid analysis of angiogenesis drugs in a live fluorescent zebrafish assay. *Arterioscler Thromb Vasc Biol 23*, 911–912.
12. Adachi, J., Mori, Y., Matsui, S., Takigami, H., Fujino, J., Kitagawa, H., Miller, C. A., 3rd, Kato, T., Saeki, K., and Matsuda, T. (2001). Indirubin and indigo are potent aryl hydrocarbon receptor ligands present in human urine. *J Biol Chem 276*, 31475–31478.
13. Guengerich, P. F., Martin, M. V., McCormick, W. A., Nguyen, L. P., Glover, E., and Bradfield, C. A. (2004). Aryl hydrocarbon receptor response to indigoids in vitro and in vivo. *Arch Biochem Biophys 423*, 309–316.
14. Hoessel, R., Leclerc, S., Endicott, J. A., Nobel, M. E., Lawrie, A., Tunnah, P., Leost, M., Damiens, E., Marie, D., Marko, D., Niederberger, E., Tang, W., Eisenbrand, G., and Meijer, L. (1999). Indirubin, the active constituent of a Chinese antileukaemia medicine, inhibits cyclin-dependent kinases. *Nat Cell Biol 1*, 60–67.
15. Knockaert, M., Blondel, M., Bach, S., Leost, M., Elbi, C., Hager, G. L., Nagy, S. R., Han, D., Denison, M., Ffrench, M., Ryan, X. P., Magiatis, P., Polychronopoulos, P., Greengard, P., Skaltsounis, L., and Meijer, L. (2004). Independent actions on cyclin-dependent kinases and aryl hydrocarbon receptor mediate the antiproliferative effects of indirubins. *Oncogene 23*, 4400–4412.
16. Kosmopoulou, M. N., Leonidas, D. D., Chrysina, E. D., Bischler, N., Eisenbrand, G., Sakarellos, C. E., Pauptit, R., and Oikonomakos, N. G. (2004). Binding of the potential antitumour agent indirubin-5-sulphonate at the inhibitor site of rabbit muscle glycogen phosphorylase b. Comparison with ligand binding to pCDK2-cyclin A complex. *Eur J Biochem 271*, 2280–2290.
17. Leclerc, S., Garnier, M., Hoessel, R., Marko, D., Bibb, J. A., Snyder, G. L., Greengard, P., Biernat, J., Wu, Y. Z., Mandelkow, E. M., Eisenbrand, G., and Meijer, L. (2001). Indirubins inhibit glycogen synthase kinase-3 beta and CDK5/p25, two protein kinases involved in abnormal tau phosphorylation in Alzheimer's disease. A property common to most cyclin-dependent kinase inhibitors? *J Biol Chem 276*, 251–260.
18. Ferandin, Y., Bettayeb, K., Kritsanida, M., Lozach, O., Polychronopoulos, P., Magiatis, P., Skaltsounis, A. L., and Meijer, L. (2006). 3′-Substituted 7-halogenoindirubins, a new class of cell death inducing agents. *J Med Chem 49*, 4638–4649.
19. Tran, T. C., Sneed, B., Haider, J., Blavo, D., White, A., Aiyejorun, T., Baranowski, T. C., Rubinstein, A. L., Doan, T. N., Dingledine, R., and Sandberg, EM. (2007). Automated, quantitative screening assay for antiangiogenic compounds using transgenic zebrafish. *Cancer Res. 67*, 11386–11392.

CHAPTER 17

Flow Cytometry Multiplexed Screening Methodologies

Virginia M. Salas
J. Jacob Strouse
Zurab Surviladze
Irena Ivnitski-Steele
Bruce S. Edwards
Larry A. Sklar

High-content analytical methodologies are increasingly becoming integrated into the screening phase of the discovery pipeline for small molecule modulators of biological targets. The iterative analysis among targets with related functions or among a series of interacting targets can be replaced with multiplexing approaches whereby multiple targets are analyzed simultaneously in the same well against the same compound or even against a mixture of compounds in the same well. It has become practical to use the multiparameter optical capabilities of the flow cytometer for the discovery of active small molecules from compound libraries by high-throughput screening (HTS). In this chapter, we present the fundamental principles of flow cytometry–based multiplex methods, including assay design and data analysis, with specific examples of cell-based and bead-based assays and a discussion of the added value of multiplex data sets.

1. OVERVIEW

Over the past two decades, HTS methodologies (generally defined as testing ten thousand to one hundred thousand compounds per day, depending on the technology) have become an essential part of discovery science for novel drugs and biological probes within the pharmaceutical, biotech, and academic research communities [1–3]. Technologies that are inherently capable of multiparametric measurements, including microscopy and flow cytometry, are well suited to produce high-content, high-complexity data by obtaining multiple data points from each well in a microtiter plate or other high-density format. In multiplex applications, information content is further enriched by obtaining data for each parameter on multiple targets. The advantages, compared with single-point iterative assays, include a substantial savings in time and reagents and the ability to produce rich data sets under conditions that can be optimized for statistical rigor and data quality for multiple targets simultaneously. For example, multiplex analysis among a family of related targets, which often exhibit similar biochemical and pharmacologic properties, produces selectivity screens in a single well with inherent consistency. Furthermore, the conditions in each well are normalized across targets for any number of parameters, and counterscreens for artifacts can be rigorously examined. Table 17.1 is a comparison of characteristics between single-target and multiplexed assays that summarizes the advantages for each approach.

1.1. High-Content Cell-Based Assays

High-content cellular assays were initially applied to secondary screening and have now become integrated throughout the discovery pipeline, including screening and beyond, and into systems cell biology research [4–6]. Primarily in the realm of high-content imaging, automated high-resolution microscopy platforms obtain topographical content via location of fluorescence signals for multiple cell functions. High throughput combined with multiparametric biological measurements in situ have enabled complex functional analysis of, for example, signaling pathways in a single cell, including inflammatory cell signaling [7] and G protein-coupled receptor (GPCR) activation and internalization [8]. High-content image analysis has been used in multiplexing formats for profiling of cell signaling pathways of as many as thirteen molecules simultaneously in myeloid cells [9]. In principle, high-content

TABLE 17.1: Single Target Analysis Versus Multiplex HTS

Parameter	Single Target	Multiplex
Format	High density, low volume, single target per well	High density, low volume, multiple targets per well
Assay development	Optimized for single target per screen; straightforward assay development and optimization	Optimized for each target initially; followed by optimization in multiplex; complex assay development; dynamic range of assay often compromised
Data acquisition	Single data point per target per well	Multiple data points per target per well
Data content	Multiple screens required for counterscreen information, SAR for related targets acquired from multiple screens	Single screen provides counterscreen information, rich SAR data for multiple targets from same screen
SAR analysis	Family of compounds against a single target	Compound activity across a target family
Statistical robustness	Assay variability must be normalized across screens for multiple target analysis	Controls as part of multiplex normalized data across several targets simultaneously
Cost (data point per compound)	One compound and assay per target: cost relatively high	One compound and assay for multiple targets: cost relatively low (as little as pennies per compound per target for bead-based multiplexes); must also consider cost of beads for bead arrays
Project management	Throughput of HTS is usually rate-limiting	Operations bottlenecks more likely at assay development/optimization and data analysis
Throughput (data points per compound)	Upwards of 100,000 samples analyzed per day	Same rate of screening throughput, multiple data points from each well increases data acquisition by factor of multiplex

cell imaging for obtaining both cell morphology (phenotypic) and subcellular (biochemical) events would provide a breadth of data unachievable by any other single technology.

1.2. High-Content Flow Cytometry

The other major multiparametric analytical platform is flow cytometry. The essential features of flow cytometry include a fluidics component for hydrodynamic focusing to create a continuous stream of particles that can be interrogated by the optical components (lasers or other light sources) and interpreted by the electronic components (detectors). Other electronics features control the data acquisition in a qualitative manner (e.g., time delays, gating determinations, and detector controls). The computational components enable the analytical functions for discerning multiparametric optical measurements, multiple particle populations, and time-dependent data acquisition, for example, in kinetic analysis applications. Intrinsic cellular parameters (light-scatter properties) and extrinsic cellular properties, measured by the use of fluorescent dyes, are the basis of flow cytometric detection. Because the flow cytometer measures a pulse of light associated with each cell or particle above a background signal, flow cytometry is typically able to discriminate in homogeneous assays and without wash steps the association of large or small labeled molecules with the particles [10].

For decades, flow cytometry has been the tool of choice for single-cell analysis within cell populations, and it continues to be a central platform in diagnostic and clinical settings [11, 12], where the technology is becoming increasingly accessible, even in resource-poor settings [13–15]. Flow cytometry has matured as a research technology for phenotypic and biochemical cell-based assays [16, 17], and assays using beads as a solid support matrix for molecular reactions have also become a mainstay of flow cytometry fluorescence-based detection for many applications [18–21].

Bead matrices can accommodate a highly diverse collection of materials, including biological reagents, such as purified antibody fragments [22], and bacteriophage display libraries [23]. Beads are amenable to a variety of surface chemistries for covalent coupling (via carboxyl, amine/hydrazide, and maleimide groups), noncovalent linkages, species-specific anti-immunoglobulin G (IgG), and high/low-density streptavidin-coated fluorophores. Other affinity tags for capturing proteins onto microspheres include gluthathione *S*-transferase (GST), Ni^{2+}-6x-histidine, and proteins A and G. Suspension arrays (fluorescence-encoded microsphere sets), largely pioneered by the Luminex Corporation [24], provide a large-capacity analytical substrate format that complements the analytical power of flow cytometry. Fluorescence-encoded microsphere kits are available for multi-analyte profiling of cytokines, micro (mi)RNAs, and infectious disease biomarkers and for conducting large-scale genotyping, as well as for the analysis of molecular assemblies [reviewed in 25].

1.3. The Challenge of Informatics in Multiplex Analysis

The development of informatics for integrating data collection and analysis presents an ongoing opportunity for fully realizing the potential of flow cytometry–based multiplex assays [26, 27]. Our group has made substantial contributions in the area of data acquisition and processing from high-density (384-well) formats [28, 29]. Processing of output from a single well may represent ten or more target data sets. Creating compatibility with downstream analysis, for example, in prioritization of actives, display of dose-response data, and cheminformatics analysis, are features of informatics for multiplex screening that will become critical for effectively transforming large data sets into useful, biologically rich information. A growing number of commercial packages are available for everything from management strategies for compound registration and tracking to data quality analysis (tracking and correction for systematic error) [30] to customizable programs for data output from a variety of assay platforms.

Multiplexed assays are well suited to address the demands of HTS in discovery science as it evolves from a focus on throughput and speed to one of high biological content for integration of the recent advancements in genomics and proteomics. This chapter focuses on flow cytometry–based, high-throughput, multiplexed screening methodologies with examples of assay design and implementation for both cell-based and bead-based assays.

1.4. Multiplex Data Sets and Structure-Activity Relationship Analysis

With regard to structure-activity relationship (SAR) analysis, multiplex data sets provide advantages over single point approaches for exploring the relationship between chemical and biological space simultaneously, thus enhancing the predictive value of comprehensive chemical biology modeling [31]. Multiplex screening among related targets in a common assay volume is likelier to provide reliable information about chemical cross-reactivity across targets and thus more efficiently lead to the discovery of multiple probes with selective reactivity for individual targets. For example, one could utilize biological and chemical data mining of existing databases such as PubChem to find structures active on a target of interest [32]. However, differences in experimental protocols can greatly influence the interpretation and analysis between screening campaigns. Furthermore, outside global screening databases such as those found in PubChem, the likelihood of similar small molecule chemotypes existing in the screening libraries across several research groups is low.

Understanding cross-target activity relationships against families of biological targets increases the breadth of information obtained at early-stage discovery, producing vital details about ligand/substrate specificity [33]. The practical limitations for translating SAR data from one target to a similar target are determined by the availability and quality of structural and biological data. For example, predictive strategies have shown high utility across multiple targets in a few well-characterized major protein families, including protein kinases [34, 35], and GPCRs [36, 37], including nuclear hormone receptors [38].

Integration of high-content screening data with SAR has predominantly been in the context of cellular imaging assays [39]. A novel analysis method has recently been described in which phenotypes with assigned biological meanings (e.g., nuclear morphology or chromosome condensation during cell cycle) are quantified and profiled. SARs are constructed by comparison of phenotypic profiles with chemical similarities and predicted protein binding of active compounds [40]. Using this technique, cell cycle analysis using fluorescent markers was used to profile a 6,500-member compound library and revealed active compounds that could be clustered into seven phenotypic classes. As predicted, phenotypic clustering correlates more highly with target prediction than with specific chemical structures. However, the phenotypic classes demonstrated good correlation between phenotypic activity and compound structural similarity. In this scenario, multiplexed data sets between targets are likely to demonstrate the correlation between phenotypic results with small molecules as well as offer more insight into subtle molecular scaffold changes that correlate with specificity.

Reverse chemical genomics has been used as an analytical approach for processing of large cell-based assay data sets from high-throughput physiologic screens [41]. In this application, protein function is characterized by identifying a small molecule ligand and then examining the cellular phenotype resulting from the treatment with said ligand. This is in contrast with the classical genetics approach, in which the phenotype is associated with a gene, and orthogonal ligand-protein pairs are evaluated to determine selectivity profiles [42, 43]. Muliplexed flow cytometry has the potential to explore both forward and reverse approaches through the analysis of whole cell phenotypic responses or through biochemical bead-based methods, respectively. To our knowledge, our group is among the first to explore SAR using data from a multiplexed HTS flow cytometry assay platform.

2. HIGH-THROUGHPUT FLOW CYTOMETRY MULTIPLEXED ASSAY PROTOCOLS

It has become practical to use the multiparameter optical capabilities of the flow cytometer for the discovery of

TABLE 17.2: When Is an Assay a Candidate for Multiplex?

Bead-based	Cell-based
Availability of a set of fluorescently tagged proteins in a well-defined pathway	A set of cell-lines expressing related ligands or proteins (preferably suspension lines)
Panels of wild type and mutants within protein families	Compatible growth conditions, similar doubling times; no effector functions between target cell types
Availability of a set of proteins involved in a molecular assembly; well-characterized with respect to binding characteristics to a fluorescent reporter molecule (less than two-fold range of K_d values, depending on the assay readout)	Availability of control cell lines (e.g., parental cell lines)
Reliable source of quality controlled, stable protein reagents	Stable uptake and retention of labeling dyes to discriminate between cell populations
Reliable source of bead arrays characterized for surface	Stable and robust expression of protein of interest

active small molecules from compound libraries by HTS. The efficiency of HTS has been boosted by increasing biological content through analysis of multiplexed targets. Multiplexing is achieved with mixtures of targets that are color coded as bead suspension arrays or mixtures of cells. Several receptors or transporters on distinct cell populations or several molecular assemblies on distinct microsphere populations are analyzed in a single assay well. This approach has been successfully applied to GPCR receptor families [44], efflux transporters [45], low molecular weight G proteins [PubChem (http://pubchem.ncbi.nlm.nih.gov/) AIDs 757–761, 764; 1333–1337, 1339–1341], Bcl-2 family interactions [PubChem AIDs 950–952; 1320, 1322, 1324, 1327–1333], and regulators of G-protein signaling (RGS) proteins [PubChem AIDs 1415, 1423, 1439–1441, 1836–1838, 1840–1841, 1869, 1871–1872, 1888]. Table 17.2 is a summary of factors to consider when evaluating an assay for multiplexing potential.

2.1. Multiplex Assay Optimization Considerations

There are inherent challenges in constructing multiplexed high-content screens for flow cytometry. The practical limitations of fluorescence imaging reagents, throughput capabilities, and data analysis are shifting as labeling technologies, integrated robotics HTS platforms, and commercial data analysis packages are being developed in response to production bottlenecks in each of these areas. The reader is referred to a series of technology reviews and reports of fluorescence-based tools for imaging applications [46–48], HTS platforms and fundamentals of assay design [49, 50], and data analysis tools for interfacing workflow, data mining, and decision making in lead development [51–56].

Assay design considerations for fluorescence-based multiplex experiments may include differences in physiologic features even among related targets, which may present issues of reagent compatibility, fluorescence signal overlap, and other potential artifacts. In principle, assay optimization begins with characterization of the component targets independently and then in multiplex. Typically, individual targets are displayed on sets of cells or particles that are encoded as suspension arrays. Principles for suspension arrays for cells have been described by Krutzik and Nolan [57] and for beads by Nolan and Sklar [58]. The color codes must be stable and chosen so that the target assays are resolved from the color coding scheme of the arrays. In the simplest case, if a binding assay involves a green fluorescent ligand, then the particles may be coded in multiple intensities of red fluorescence. Configuring an HTS assay in multiplex requires that the individual targets remain resolved in the array and the targets assays themselves retain their individual characteristics, such as binding affinity of the components. It is useful in this regard that target families be chosen in which a common fluorescent ligand binds with similar affinity to each of the individual targets associated with the particles, as in the case of the multiplexes for formylpeptide receptor (FPR) family (with the common ligand H_2N-Trp-[Fluoresceinyl-Lys]-Tyr-Met-Val-[DMet]-amide), low molecular weight guanosine triphosphate (GTP)-ase family (with a common fluorescent GTP ligand), Bcl-2 family (with a common Bim-FITC ligand), and RGS family (with a common fluorescent $G\alpha o$) [59–61].

Bead-based assays require special consideration with regard to factors affecting bead surface chemistry interactions with test compounds [62]. For example, glutathione *S*-transferase–glutathione (GST-GSH) interactions can be used to construct bead-based arrays for families of proteins (discussed in section 2.4.1 on bead-based multiplexing: HTS to identify specific small molecule inhibitors of Ras and Ras-related GTPases); however, the GST-GSH interaction can also be a target for small molecule interaction.

Autofluorescence properties of test compounds can be considerable and are another important assessment for fluorescence-based assay optimization and analysis. Autofluorescence profiling for a given compound collection is useful; however, findings may not be directly applicable to assays conducted in different buffer conditions, in the presence of other potential reagent interactions, or among assays with different dynamic range, depending upon the fluorescence intensity of the compound. We have found it useful to add control populations of beads to the multiplex. These populations can be configured to test interaction of the compounds with the bead surface itself or with control proteins that mimic the targets but are themselves not targets for the fluorescent probe.

A series of reports addresses other sources of variation in multiplex suspension arrays, including interinstrument variability and carryover of microspheres between sample wells [63, 64] and variation in fluorescence detection due to an observed range of microsphere size in any given lot of microspheres [65]. With the exception of assessment for sample-to-sample carryover, the impact of these effects on an otherwise robust HTS assay has not been evaluated in a systematic way. The sampling technology utilized in our laboratory (see section 2.2 on sample handling and data analysis) has an inherently low level of particle carryover because of sequential sampling from wells without a requirement for an intersample probe rinse step.

2.2. Sample Handling and Data Analysis

Until recently, flow cytometry has been limited in high-throughput applications to processing fewer than ten samples per minute. Sample handling technology is now available that uses a peristaltic pump in combination with an autosampler to boost assay performance into the range of one to two samples per second (HyperCyt, IntelliCyt, Albuquerque, N.M.). Sample delivery occurs in a continuous stream, and data acquisition is obtained in a single, time-resolved data file (Figure 17.1B, C, E, F). Proprietary software (IDLeQuery, developed by BSE) is used to merge the raw instrument data and work lists associated with the compound library. Raw data are parsed and analyzed by the software into annotated fluorescence summary data for each sample well. The results are then processed through a spreadsheet template file that segregates target specific data from each well, automatically calculates assay quality statistics (Z and Z′ analyses) [66–68], and normalizes data with respect to control wells and correction of systematic error trends across plates.

2.3. Cell-Based Multiplexing

Cell-based multiplexes have been conducted for both FPR and adenosine triphosphate (ATP)-binding cassette (ABC) transporter families. Assay details have been described in PubChem [PubChem (http://pubchem.ncbi.nlm.nih.gov/; AID 1325, 1326] and in publications [69, 70]. The FPR assay is based on the binding of a common green fluorescent ligand to individual receptor targets on two cell populations, one coded with a red fluorescent stain and the other unstained. Because this assay involved relatively high levels of fluorescence from several hundred thousand receptors per cell, the major technical challenges for the assay, once 384-well capabilities had been established, involved cell handling rather than cytometric sampling or data analysis.

2.3.1. High-Throughput Flow Cytometry to Detect Selective Inhibitors of ABCB1, ABCC1, and ABCG2 Transporters The general strategy for multiplexed cellular assays is based on discrimination of target cell populations by morphological (light scatter) or fluorescent staining. Cellular responses are assessed by staining characteristics imparted by a fluorescent probe or fluorescent ligand distinct from the fluorescent staining utilized to discriminate cell populations (Figure 17.1). In the protocol presented here, two cell-based assays were constructed to assess the specificity of small molecules that inhibit three transporters involved in cancer multidrug resistance [71].

2.3.2. Cells In one duplex, ABCB1 and ABCG2 transporters were evaluated using fluorescent JC1 as a common transporter substrate and cell lines Jurkat-DNR (a suspension cell line expressing ABCB1) and IgMXP3 (an adherent cell line expressing ABCG2). The second duplex examined suspension cell lines CCRF-Adr (ABCB1) and SupT1-Vin (ABCC1) using CaAM as a common fluorescent substrate. The cell lines used in this assay were selected on the basis of their overexpression of the transporters of interest. The ABCB1-expressing cell lines (CCRF-Adr and Jurkat-DNR) were color-coded to allow distinction from the alternate target in each assay. Cells were stained with 0.5 ng/mL of FarRed DDAO CellTrace SE (Invitrogen) for fifteen minutes at room temperature, washed twice by centrifugation, and then combined with unlabeled IgMXP3 cells (ABCG2) or unlabeled SupT1-Vin (ABCC1) in the assay buffer. The DDAO label binds covalently to amine groups in cells and is detected at red fluorescence emission wavelengths of 665 ± 10 nm upon excitation at 635 nm. Final in-well concentration of cells was 3×10^6 cells/mL at a 1:1 ratio of the two cell types. This concentration was sufficient to provide analysis of approximately one thousand cells of each type from each well (sampling rate forty wells per minute, sample volume $\sim$2 μL).

2.3.3. Fluorescence Marker Substrates and Controls JC1 and CaAM are fluorescence markers for active transport activity and are detected in this assay as retained fluorescence in cells. JC1 is a cationic dye that is retained in viable cells, producing increasing red fluorescence as it aggregates in mitochondria. Monomeric JC1 produces green fluorescence with mitochondrial depolarization and is detected in the FL 1 channel in this assay. CaAM is converted in live cells to green-fluorescent calcein after hydrolysis by

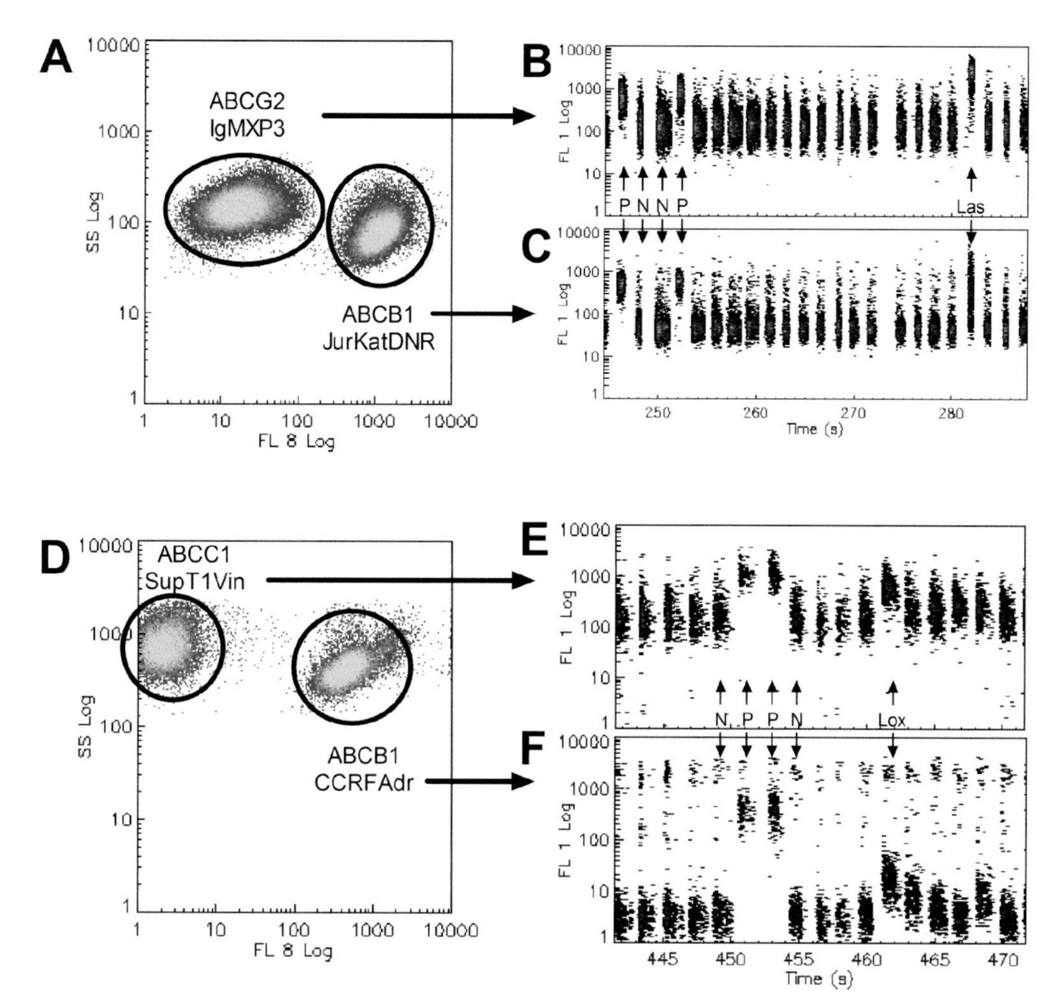

***FIGURE 17.1:** Cell-based multiplex for HTS: analysis of ABC transporter activity. **(A–C)**. JC1 duplex assay showing IgMXP3 cells (ABCG2, dim-red/FL8 fluorescent) and Jurkat DNR cells (ABCB1, bright-red FL8 fluorescent) distinguished in separate electronic gates (circles, panel A) to perform separate analysis of the green JC1 fluorescence (FL) intensity response (FL1) of cells sampled from a 384-well plate (panels B and C, respectively). In plots of time versus FL1 (right), each discrete cluster of dots represents cells sampled from a single well. Arrows indicate cells from DMSO vehicle negative control wells (N), wells containing the transporter block controls (P), and the well containing a compound displaying a significant assay response for both transporters (Las). **(D–F)** CaAM duplex assay showing SupT1-Vin cells (ABCC1, dim-red/FL8 fluorescent) and CCRF-ADR. Five thousand cells (ABCB1, bright-red/FL8 fluorescent) are distinguished in gates (panel D) to allow separate analysis of the green CaAM FL1 of cells sampled from a 384-well plate (panels E and F, respectively). Arrows indicate cells from negative (N) and positive (P) control wells as in (A–C), and cells treated with a compound that selectively produced a significant assay response in ABCC1-expressing cells (Lox). In the time-resolved dot plots (B,C, E, and F) a 30–40 second segment of the full 384-well sequence (11 min total) is shown.*

intracellular esterases. JC1 was found to be a substrate for both ABCB1 and ABCG2 transporters and a poor substrate for the ABCC1 transporter. CaAM is a good substrate for ABCB1 and ABCC1 but not ABCG2.

The assay response range was defined by control wells containing Nicardipine or Verapamil as pump inhibitors (transporter blockers) for JC1 and CaAM, respectively. The fluorescence readout is high with pump inhibitors (inhibitor controls) and low with active pumps (dimethyl sulfoxide [DMSO] control). Assay components for each duplex assay are summarized in Table 17.3.

2.3.4. Assay Format The screen was conducted in 384-well plates in a total assay volume of 100 μL per well. Each assay plate was configured with 64 control wells as follows: Outer columns 1, 2 and 23, 24 contained negative unblocked transporter activity controls (1% DMSO) and positive blocking transporter inhibition controls (nicardipine for JC1 or verapimil for CaAM, each at 50 μM final concentration). Test compound plates were prepared from DMSO stocks formatted in columns 3 to 22 of 384-well plates for a total of 320 compounds per assay plate. Test compounds in this assay were tested at approximately 5 μM.

TABLE 17.3: Reagent and Cell Components for Transporter Duplex Assays

Transporter	Cell Label	Substrate	Cell Type	Growth Conditions	Transporter Block Controls
			Duplex 1		
ABCB1	DDAO Far Red	JC1	Jurkat DNR	Suspension	Nicardipine
ABCG2	No label		IgMXP3	Adherent	
			Duplex 2		
ABCC1	No label	CaAM	SupT1-Vin	Suspension	Verapamil
ABCB1	DDAO Far Red		CCRF-Adr	Suspension	

For the JC1 duplex assay, Jurkat-DNR cells (ABCB1) were labeled with Far Red DDAO, as described in section 2.3.2. on cells, and mixed with unlabeled IgMXP3 cells (ABCG2), following release by trypsin, in JC1 assay buffer (provided by supplier). JC1 at a final concentration of 0.2% was added to the cell suspension, and mixture was dispensed into all wells of a 384-well plate at 100 μL per well. Test and control compounds were added to wells (1 μL per well). After a fifteen-minute incubation at room temperature, cells were evaluated by flow cytometry as described in section 2.2 on sampling handling and data analysis.

2.3.5. Assay Evaluation Efflux pump activity was evaluated for percent inhibition in response to each test compound on the basis of the median fluorescence intensity (MFI) of fluorescent substrate detected in the green fluorescence channel. The data reveal compounds that produce an assay response for both transporters and an assay response for a second compound that is selective for one transporter (Figure 17.1). Compound autofluorescence was a problem for this assay, and in retrospect there may have been value in defining additional cell populations without pumps in which the autofluorescence signal associated with cell accumulation of compounds could have been independently evaluated from their effects on the pumps.

2.4. Bead-Based Multiplexing

We have now performed four sets of bead-based multiplexes. Descriptions of three sets have been uploaded to PubChem (GTPase family, Bcl-2 family, RGS family), and a fourth (substrate families for botulinum toxin) is in preparation for publication. A detailed protocol for the Bcl-2 family has been described [72], which serves a step by step model for bead-based multiplexing.

2.4.1. HTS to Identify Specific Small Molecule Inhibitors of Ras and Ras-Related GTPases Assay formats for bead-based multiplexing rely on the binding of a fluorescent reporter to color-coded bead sets, each coupled with individual ligands or proteins. Bead sets are discriminated by varying color intensities (each with a unique optical address), and each set represents a unique target bound to the bead (Figure 17.2).

The protocol described here is a no-wash fluorescent GTP-binding assay in which multiple GTPases are simultaneously screened against a small molecule library. In the multiplex, fluorescent GTP binds to G protein–GST fusion proteins that are arrayed on a set of GSH beads (Figure 17.2). A set of six G proteins (Rac1 wt, Rab7 wt, Rac 1 act, Ras wt, Rab 2 wt, and Cdc wt) are assessed under conditions of divalent molecule depletion.

2.4.2. Assay Optimization: Preparation of Beads GSH bead set collections were acquired from Duke Scientific (Thermo Fisher Scientific). Individual sets of beads are distinguished by varying magnitude of emission at 655 ± 10 nm with excitation at 635 nm. The binding characteristics for each member of the bead array may not be consistent between bead sets or between bead lots. Assay optimization includes an evaluation of GSH binding site densities for bead sets by obtaining GST-GFP binding curves for each color level (Figure 17.3). The interaction between the bead-bound proteins and their fluorescent ligand binding partners need to be verified for consistency in terms of the affinity of the interaction as well as the maximal level of binding. Assays work best if the affinities are within a factor of two and the binding levels are such that the signal of the lowest target is about three times higher than blank beads. Proteins should be supplied in concentrations from 10 to 50 ng/mL, and the optimal concentrations used for coating beads are based on obtaining a fluorescence signal for each protein target that is comparable to that of more than thirty thousand molecules of equivalent soluble fluorophores (MESF) using an optimized concentration of fluorescence reporter.

Prior to coating with proteins, beads are treated in ionic buffers, usually intracellular mimics (30 mM HEPES pH 7.5, 100 mM KCl, 20 mM NaCl, 1 mM EDTA) containing detergent (e.g., 0.01% Tween-20 or NP-40) and protein (0.1% BSA) to eliminate nonspecific binding and to condition the binding surface of the beads. Bead-protein binding optimization is performed in single target reactions and then mixed to assess and optimize binding with

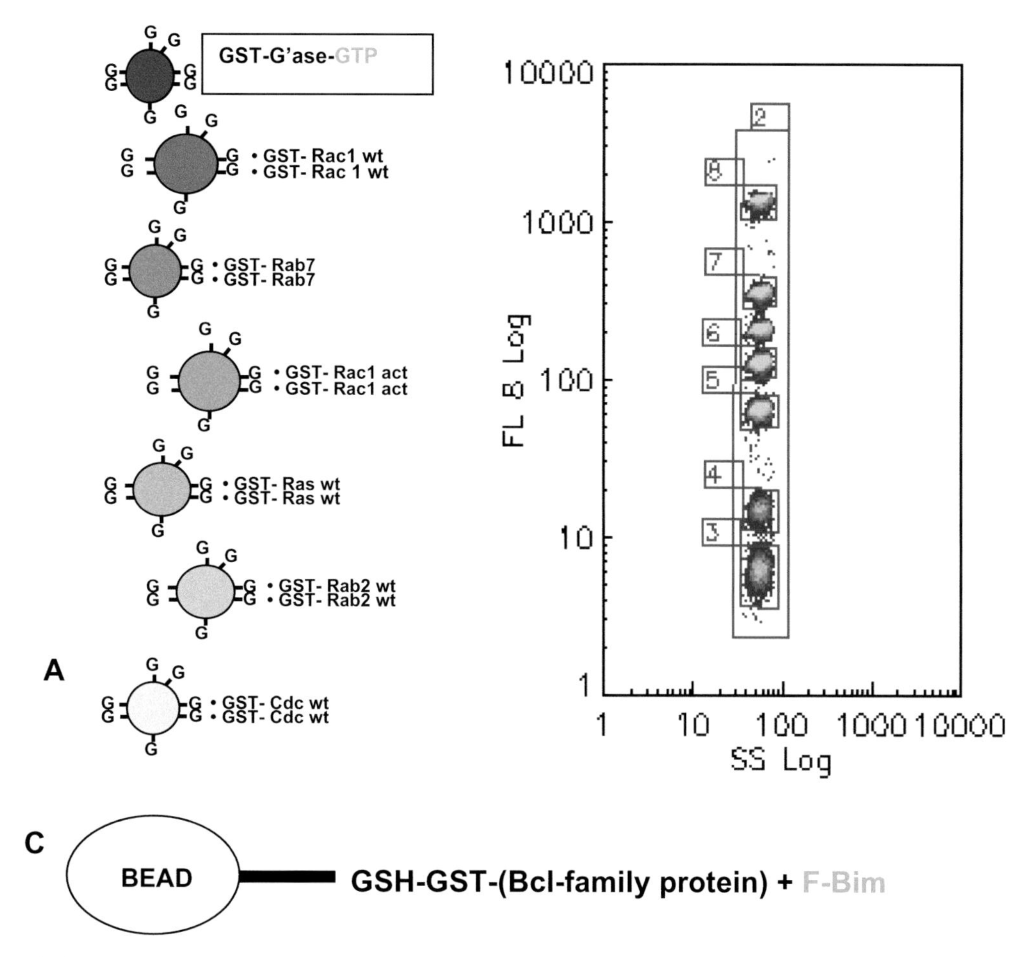

***FIGURE 17.2:** Bead-based multiplex for HTS: experimental design principles for exploring protein-protein interactions. The assay design is based on the binding of fluorescent GTP to G protein–GST fusion proteins on GSH beads (panel A boxed figure, upper left). A set of six G proteins (Rac 1 wt, Rab7 wt, Rac 1 activated, Ras wt, Rab 2 wt, Cdc wt) are arrayed under conditions of divalent molecule depletion. **(A)** Each component of the multiplex consists of a glutathione-labeled bead (glutathione indicted by G), a GST G protein family fusion target (six total), and a fluorescent peptide probe, F-GTP. Bead sets are coated with GST-conjugated proteins. The multiplex is constructed by using beads for each protein target that have been labeled with varying intensities of red color, such that each target is assayed on a unique bead set. **(B)** Each bead set is associated with a unique optical address. An example of a side scatter versus fluorescence intensity plot demonstrates uniformly sized bead populations discriminated by fluorescence intensity. Electronic gates (red boxes) allow for separate analysis of each bead population. **(C)** A similar HTS for small molecule regulators of Bcl-2 family protein interactions was constructed based on the binding of fluorescent Bim (F-Bim) to Bcl-2 family proteins (six total: Mcl-1, Bfl-1, Bcl-b, Bcl-W, Bcl-2, Bcl-XL) bound to GSH beads.*

the fluorescent reporter in multiplex. Ideally, for assays that are designed to detect both increased and decreased binding (e.g., activators and inhibitors), the differences in dissociation constant (K_d) values among the targets with the fluorescent reporter should be minimal, so as not to bias the assay results in favor of the target with the highest binding affinity. Characterization of fluorescent reporters is discussed further in section 2.4.4 on fluorescent reporters and controls.

2.4.3. Coating Bead Sets with Target Proteins Following optimization of protein concentration, each protein is bound to a bead set by incubating beads with protein at 4°C overnight on a rotating platform. Bead sets are then combined with the fluorescence scavenger bead set (see section 2.4.4 on fluorescent reporters and controls) in equal ratios such that two thousand beads per set are dispensed in a total volume of 5 mL per assay well.

2.4.4. Fluorescent Reporters and Controls The binding specificities among the targets used in this assay are based on the observation that individual GTPases, including wild type (wt) and activated (act) forms, exhibit measurably distinct affinities for Bodipy-FI-GTP versus GTP.

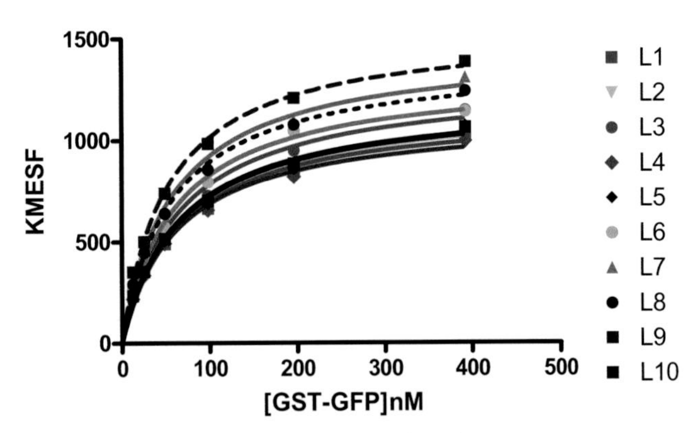

FIGURE 17.3: Characterization of GSH binding sites on bead arrays. A ten-color level (L1–10) bead array was evaluated for GSH binding site densities by obtaining binding curves of GFP fluorescence (KMESF) against increasing concentrations of labeled GST (GST-GFP). Binding site densities between bead sets is variable (range 1.1 x 10^6 sites (L4) to 1.5 x 10^6 sites (L9).

The Bodipy-GTP analogs were developed to measure nucleotide hydrolysis by $G\alpha_o$. High-affinity (11 nM K_d) binding to $G\alpha_o$ increases analog fluorescence by approximately six-fold [73].

Controls in this assay include a GST-only bead set (no associated GST protein) in each well as a scavenger for dissociated GST proteins and to visualize artifact from test compound fluorescence (see section 2.4.6 on assay evaluation for discussion of test compound fluorescence). The positive control includes the protein-coated bead mixture with the fluoresceinated GTP and no test compound (bright fluorescence). The negative control includes the bead mixture plus fluoresceinated GTP plus unlabeled GTP (dim fluorescence) as a control for the specificity of the fluorescent GTP binding.

2.4.5. Assay Format The assay is conducted in 384-well microtiter plates in a total well volume of 10.1 mL: 5 mL of protein-coated bead mixture, 0.1 mL of test compound at 1 mM in DMSO, and 5 mL of 200 nM Bodipy-FL-GTP in buffer containing BSA and DTT for a final concentration of GTP of 100 nM. Positive and negative controls are located in columns 1, 2 and 23, 24, respectively. Test compounds

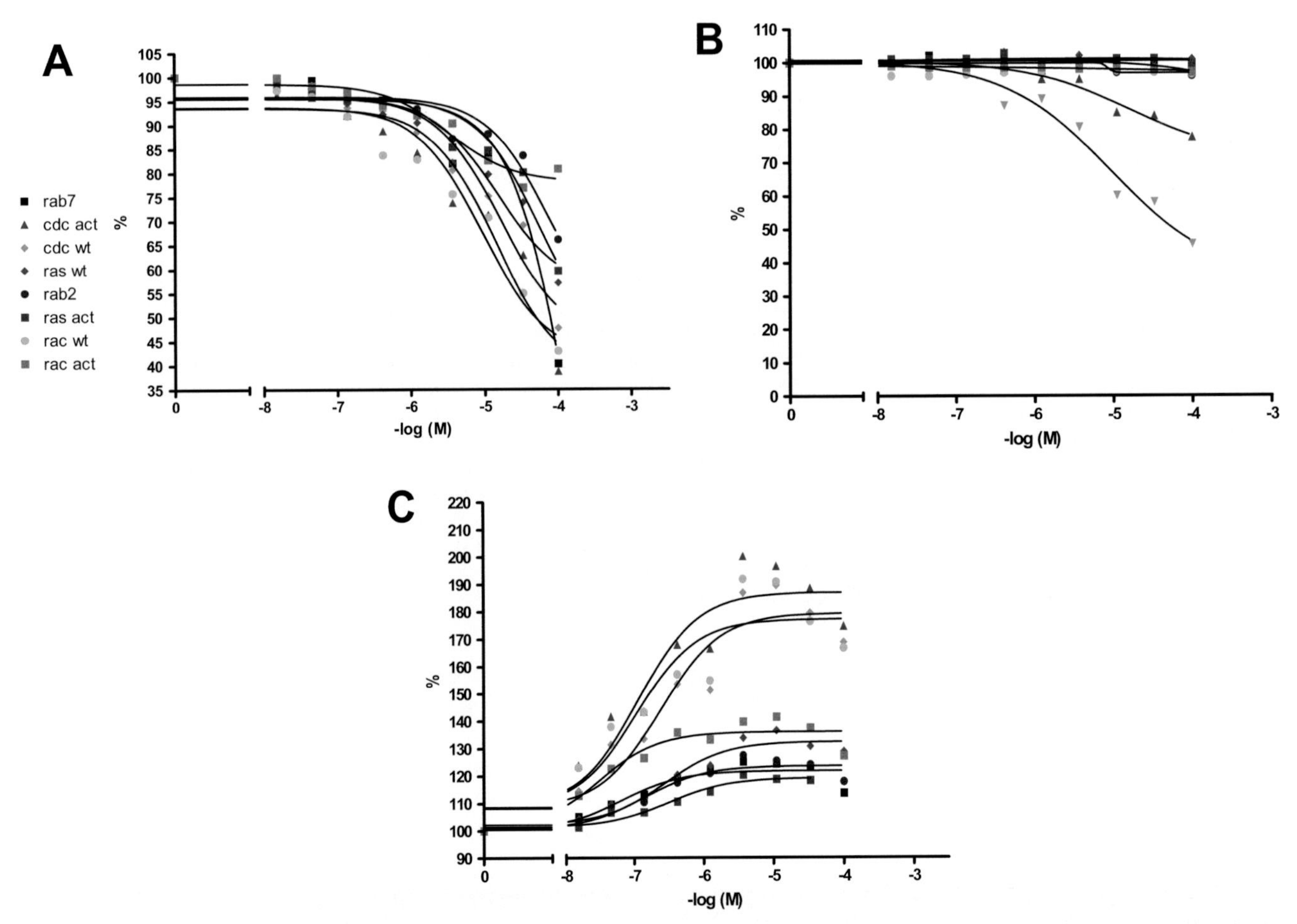

FIGURE 17.4: Multiplex dose response: confirmation of active wells identified in HTS multiplexed primary screening for small molecule activators/inhibitors of Ras and Ras-related GTPases. Eight protein targets were analyzed simultaneously against a 6-log concentration gradient of test compound. Results demonstrate the ability of multiplexing to identify: nonselective inhibition (panel A), selective inhibition (panel B), and activation (panel C) for one compound among related targets.

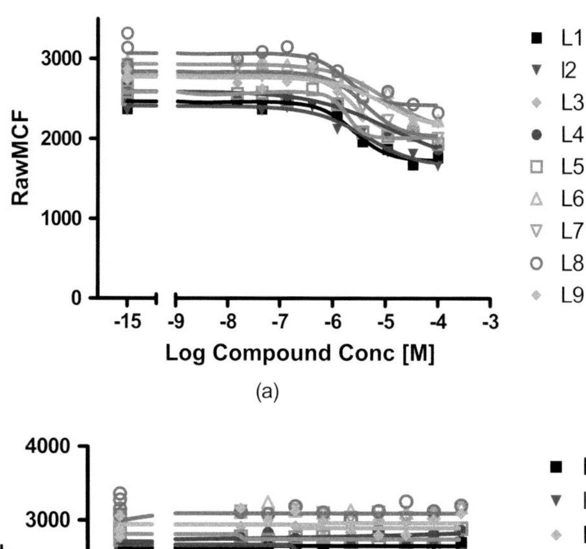

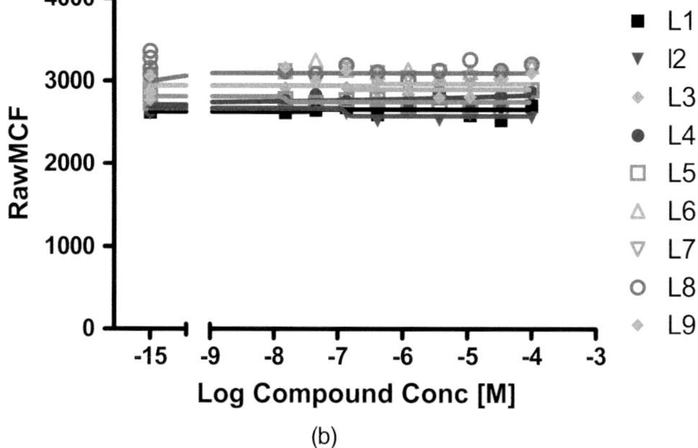

***FIGURE 17.5:* Evaluating bead-based assay artifact: compound interaction with GST-GSH. A bead-based multiplex assay to evaluate small molecule inhibitors of Ras and Ras-related GTPases is based on binding of fluorescent GTP to G-protein–GST fusion proteins on GSH beads. To evaluate the effect of compounds on the binding of GST fusion proteins to beads, bead sets were primed with GST-GFP and evaluated with compounds of interest (active wells from HTS) in a dose-response assay. Graphs represent raw mean channel fluorescence (MCF) value versus concentration of test compound for a 9-level (L1–9) bead array. (A) Concentration-dependent compound interaction with GST-GFP across all bead-protein sets demonstrating assay artifact. (B) Dose-response analysis demonstrates no interaction between compound of interest and GST-GFP binding.**

are dispensed into columns 3 to 22 (320 compounds tested per target per assay plate). Plates are placed on rotators and incubated for forty to forty-five minutes at 4°C. The rotators periodically move the plates between inverted and upright positions to maintain beads in suspension. Surface tension keeps the well contents from spilling.

2.4.6. Assay Evaluation Sample analysis is conducted using the HyperCyt sampling and analytical platform described in section 4.2 on sample handling and data analysis. Flow cytometric data of light scatter and fluorescence emission at 530 ± 20 nm (FL1) and 665 ± 10 nm (FL8) are collected. Electronic gating (Figure 17.2B) is based on FL8 emission and distinguishes the bead populations associated with each protein target, and the median fluorescence for each bead population is calculated on a minimum of twenty-five events (beads) for each compound.

Screening results from an inherently fluorescent test compound are considered to be inconclusive. Fluorescence of a compound is assessed by comparing the influence of compound fluorescence on all of the target proteins in one well with that on the positive control wells. Differences between sample fluorescence and positive control fluorescence are calculated for each protein target. The coefficient of variation (CV) of the entire protein target set is calculated. If the CV is less than 30%, indicating that the compound-attributed fluorescence is very similar between all of the different proteins, the compound is flagged as a potential green fluorescent compound.

Potential systematic trend over an entire plate (whole plate trends) require normalization to calculate percent activities of test compounds. Whole plate trends of the positive controls are evaluated by linear regression techniques. The data reveal small molecules with selectivity and specificity in both HTS primary screens and dose-response analysis (Figure 17.4). Additional controls show that small molecules do not impact the binding of GST-GSH (Figure 17.5).

3. SUMMARY AND FUTURE DIRECTIONS

The efficiency of flow cytometry–based HTS has been greatly improved by increasing biological content through analysis of multiplexed targets. Multiplexing is achieved with mixtures of targets that are color coded as suspension arrays of beads or cells. Several receptors or transporters on distinct cell populations or several molecular assemblies on distinct microsphere populations can be analyzed using multiplexing approaches. The data reveal small molecules with selectivity and specificity in both primary screens and dose-response analysis.

The contribution of multiplexing approaches to the areas of functional genomics or proteomics likely will depend on successful application in the areas of protein-protein, protein-ligand (small molecule), and protein–nucleic acid (DNA or RNA) interactions. Arrays can be constructed to reveal interactions between members of protein families as well as specificities, affinities, and kinetics of molecular interactions between related targets. Taken together with the advances in small molecule library design for broadly exploring chemical diversity or for targeted library collections and even combinatorial chemistry libraries, multiplexing approaches have the potential to improve both the efficiency of HTS and the quality of leads for therapeutics, small molecule biological probes, and even imaging agents for diagnostics.

ACKNOWLEDGMENTS

This work was supported by NIH grants 1X01 MH079850–01, 1X01 MH081231–01, 1X01 MH 078946–01, NS 057014–01, and U54 MH074425–01 and by the University of New Mexico Center for Molecular Discovery (formerly the New Mexico Molecular Libraries Screening Center) at the University of New Mexico Cancer Research and Treatment Center.

ABBREVIATIONS

ABC (B1, C1, G2)	ATP binding cassette (B1, C1, G2)
BODIPYFL-GTP	Guanosine 5′-triphosphate BODIPYFL 2′-O-(N-2-aminoethyl) urethane
CaAM	Calcein acetoxymethyl ester
DMSO	Dimethyl sulfoxide
DNR	Daunorubicin
GSH	Glutathione
GST	Glutathione-*S*-transferase
HTS	High-throughput screening
JC1	J-aggregate-forming lipophilic cation 5,5',6,6′-tetrachloro-1,1′,3,3′-tetraethylbenzimidazolcarbocyanine iodide)
SAR	Structure-activity relationship

REFERENCES

1. Pereira, D. A., and Williams, J. A. (2007). Review: Origin and evolution of high throughput screening. *Br J Pharmacol 152*, 53–61.
2. Macarron, R. (2006). Critical review of the role of HTS in drug discovery. *Drug Discov Today 11*, 277–279.
3. Bender, A., Bojanic, D., Davies, J. W., Crisman, T. J., Mikhailov, D., Scheiber, J., Jenkins, J. L., Deng, Z., Hill, W. A., Popov, M., Jacxoby, E., and Glick, M. (2008). Which aspects of HTS are empirically correlated with downstream success? *Curr Opin Drug Discov Devel 11*, 327–37.
4. Rausch, O. (2006). High content cellular screening. *Curr Opin Chem Biol 10*, 316–320.
5. Lang, P., Yeow, K., Nichols, A., and Scheer, A. . (2006). Cellular imaging in drug discovery. *Nat Rev Drug Discov 5*, 343–352.
6. Korn, K., and Krausz, E. (2007). Cell-based high-content screening of small-molecule libraries. *Curr Opin Chem Biol 11*, 503–510.
7. Bertelsen, M. (2006). Multiplex analysis of inflammatory signaling pathways using a high-content imaging system. *Methods Enzymol 414*, 348–363.
8. Ross, D. A., Lee, S., Reiser, V., Xue, J., Alves, K., Vaidya, S., Kreamer, A., Mull, R., Hudak, E., Hare, T., Detmers, P.A., Lingham, R., Ferrer, M., Strulovici, B., and Santini, F. (2008). Multiplexed assays by high-content imaging for assessment of GPCR activity. *J Biomol Screen 13*, 449–455.
9. Irish, J. M., Hovland, R., Krutzik, P. O., Perez, O. D., Bruserud, O., Gjertsen B. T., and Nolan G. P. (2004). Single cell profiling of potentiated phosphorylation networks in cancer cells. *Cell 118*, 217–228.
10. Sklar, L. A., Carter, M. B., and Edwards, B. S. (2007). Flow cytometry for drug discovery, receptor pharmacology and high-throughput screening. *Curr Opin Pharmacol* 7, 527–534.
11. Ryningen, A., and Bruserud, O. (2007). Epigenetic targeting in acute myeloid leukemia: use of flow cytometry in monitoring therapeutic effects. *Curr Pharm Biotechnol 6*, 401–411.
12. Kellar, K. L., and Iannone, M. A. (2002). Multiplexed microsphere-based flow cytometric assays. *Exp Hematol 30*, 1227–1237.
13. Larsen, C. H. (2008). The fragile environments of inexpensive CD4+ T-cell enumeration in the last developed countries: strategies for accessible support. *Cytometry B Clin Cytom 74S*, 107–116.
14. Pattanapanyasat, K., Phuang-Ngern, Y., Lerdwana, S., Wasinrapee, P., Sakulploy, N., Noulsri, E., Thepthai, C., and McNicholl, J. M. (2007). Evaluation of a single-platform microcapillary flow cytometer for enumeration of absolute CD4+ T-lymphocyte counts in HIV-1 infected Thai patients. *Cytometry B Clin Cytom* 72, 387–396.
15. Li, X., Ymeti, A., Lunter, B., Breukers, C., Tibbe, A. G., Terstappen, L. W., and Greve, J. (2007). CD4+ T lymphocytes enumeration by an easy-to-use single platform image cytometer for HIV monitoring in resource-constrained settings. *Cytometry B Clin Cytom* 72, 397–407.
16. Edwards, B. S., Oprea, T., Prossnitz, E. R., and Sklar, L. A. (2004). Flow cytometry for high-throughput, high-content screening. *Curr Opin Chem Biol 8*, 392–398.
17. Smith, R. A., and Giorgio, T. D. (2004). Cell-based screening: a high throughput flow cytometry platform for identification of cell-specific targeting molecules. *Comb Chem High Throughput Screen* 7, 141–151.
18. Salas, V. M., Edwards, B. S., and Sklar, L. A. (2008). Advances in multiple analyte profiling. *Adv Clin Chem 45*, 47–74.
19. Waller, A., Simons, P. C., Biggs, S. M., Edwards, B. S., Prossnitz, E. R., and Sklar, L. A. (2004). Techniques: GPCR assembly, pharmacology and screening by flow cytometry. *Trends Pharmacol Sci 25*, 663–669.
20. Nolan, J. P., and Mandy, F. (2006). Multiplexed and microparticle-based analyses: quantitative tools for the large-scale analysis of biological systems. *Cytometry A 69*, 318–325.
21. Mandy, F. F., Nakamura, T., Bergeron, M., and Sekiguchi, K. (2001). Overview and application of suspension array technology. *Clin Lab Med 4*, 713–729.
22. Ayriss, J., Woods, T., Bradbury, A., and Pavlik, P. (2007). High-throughput screening of single chain antibodies using multiplexed flow cytometry. *J Proteome Res 6*, 1072–1082.
23. Yang, L., and Nolan, J. P. (2007). High-throughput screening and characterization of clones selected from phage display libraries. *Cytometry A 71*, 625–631.
24. Dunbar, S. A. (2006). Applications of Luminex xMAP technology for rapid, high-throughput multiplexed nucleic acid detection. *Clin Chim Acta 363*, 71–82.

25. Nolan, J. P., and Yang L. (2007). The flow of cytometry into systems biology. *Brief Funct Genomic Proteomic 6*, 81–90.
26. Durr, O., Duval, F., Nichols, A., Lang, P., Brodte, A., Heyse, S., and Besson, D. (2007). Robust hit identification by quality assurance and multivariate data analysis of a high-content, cell-based assay. *J Biomol Screen 12*, 1042–1049.
27. Dunlay, R. T., Czekalaski, W. J., and Collins, M. A. (2007). Overview of informatics for high content screening. *Methods Mol Biol 356*, 269–280.
28. Edwards, B. S., Oprea, T., Prossnitz, E. R., and Sklar, L. A. (2004). Flow cytometry for high-throughput, high-content screening. *Curr Opin Chem Biol 8*, 392–398.
29. Sklar, L. A., Carter, M. B., and Edwards, B. S. (2007). Flow cytometry for drug discovery, receptor pharmacology and high-throughput screening. *Curr Opin Pharmacol* 7, 527–534.
30. Makarenkov, V., Zentilli, P., Kevorkov, D., Gagarin, A., Malo, N., and Nadon, R. (2007). An efficient method for the detection and elimination of systematic error in high-throughput screening. *Bioinformatics 23*, 1648–1657.
31. Haggerty, S. J. (2005). The principle of complementarity: chemical versus biological space. *Curr Opin Chem Biol 9*, 296–303.
32. Rosania, G. R., Crippen, G., Woolf, P., States, D., and Shedden, K. (2007). A cheminformatic toolkit for mining biomedical knowledge. *Pharm Res 24*, 1791–1802.
33. Harris, C. J., and Stevens, A. P. (2006). Chemogenomics: structuring the drug discovery process to gene families. *Drug Discov Today 11*, 880–888.
34. Noble, M. E., Endicott, J. A., and Johnson, L. N. (2004). Protein kinase inhibitors: insights into drug design from structure. *Science 303*, 1800–1805.
35. Cherry, M., and Williams, D. H. (2004). Recent kinase and kinase inhibitor x-ray structures: mechanisms of inhibition and selectivity insights. *Curr Med Chem 11*, 663–673.
36. Kristiansen, K. (2004). Molecular mechanisms of ligand binding, signaling and regulation within G-protein-coupled receptors: molecular modeling mutagenesis approaches to receptor structure and function. *Pharmacol Ther 103*, 21–80.
37. Jacob L. and Vert J. P. (2008). Protein-ligand interaction prediction: an improved chemogenomics approach. *Bioinformatics 24*, 2149–2156.
38. Klabunde, T., and Evers, A. (2005). GPCR antitarget modeling: pharmacophore models for biogenic amine binding GPCRs to avoid GPCR-mediated side effects. *Chembiochem 6*, 876–889.
39. Low, J., Stancato, L., Lee, J., and Sutherland, J. J. (2008). Prioritizing hits from phenotypic high-content screens. *Curr Opin Drug Discov Devel 11*, 338–345.
40. Young, D. W., Bender, A., Hoyt, J., McWhinnie, E., Chirn, G. W., Tao, C. Y., Tallarico, J. A., Labow, M., Jenkins, J. L., Mitchison, T. J., and Feng, Y. (2008). Integrating high-content screening and ligand target prediction to identify mechanism of action. *Nat Chem Biol 4*, 59–68.
41. Ross-Macdonald, P. (2005). Forward in reverse: how reverse genetics complements chemical genetics. *Pharmacogenomics 6*, 429–434.
42. Bol, D., and Ebner, R. (2006). Gene expression profiling in the discovery, optimization and development of novel drugs: one universal screening platform. *Pharmacogenomics* 7, 227–235.
43. Kubinyi, H. (2006). Chemogenomics in drug discovery. *Ernst Schering Res Found Workshop 58*, 1–19.
44. Waller, A., Simons, P. C., Biggs, S. M.,Edwards, B. S., Prossnitz, E. R., and Sklar, L. A. (2004). Techniques: GPCR assembly, pharmacology and screening by flow cytometry. *Trends Pharmacol Sci 25*, 663–669.
45. Ivnitski-Steele, I., Larson, R. S., Lovato, D. M., Khawaja, H. M., Winter, S. S., Oprea, T. I., Sklar, L. A., and Edwards, B. S. (2008). High-throughput flow cytometry to detect selective inhibitors of ABCB1, ABCC1, and ABCG2 transporters. *Assay Drug Dev Technol 2*, 263–276.
46. Shapiro, H. M. (2005). Fluorescent probes. In *Flow Cytometry for Biotechnology*, L. A. Sklar, ed. (New York: Oxford University Press, 15–39).
47. Patsenker, L., Tatarets, A., Kolosova, O., Obukhova, O., Povrosin, Y., Fedyunyayeva, I., Yermolenko, I., and Terpetschnig, E. (2008). Fluorescent probes and labels for biomedical applications. *Ann NY Acad Sci 1130*, 179–187.
48. Chen, W. (2008). Nanoparticle fluorescence based technology for biological applications. *Nanosci Nanotechnol 8*, 1019–1051.
49. Inglese, J., Johnson, R. L., Simeonov, A., Xia, M., Zheng, W., Austin, C. P., and Auld, D. S. (2007). High-throughput screening assays for the identification of chemical probes. *Nat Chem Biol 3*, 466–479.
50. Mayr, L. M., and Fuerst, P. (2008). The future of high-throughput screening. *J Biomol Screen 13*, 443–448.
51. Ling, X. B. (2008). High throughput screening informatics. *Comb Chem High Throughput Screen 11*, 249–257.
52. Harper, G., and Pickett, S. D. (2008). Methods for mining HTS data. *Drug Discov Today 11*, 694–699.
53. Makarenkov, V., Zentilli, P., Kevorkov, D., Gagarin, A., Malo, N., and Nadon, R. (2007). An efficient method for the detection and elimination of systematic error in high-throughput screening. *Bioinformatics 23*, 1648–1657.
54. Durr, O., Duval, F., Nichols, A., Lang, P., Brodte, A., Heyse, S., and Besson, D. (2007). Robust hit identification by quality assurance and multivariate data analysis of a high-content, cell-based assay. *J Biomol Screen 12*, 1042–1049.
55. Dunlay, R. T., Czekalaski, W. J., and Collins, M. A. (2007). Overview of informatics for high content screening. *Methods Mol Biol 356*, 269–280.
56. Garfinkle, L. S. (2007). Large-scale data management for high content screening. *Methods Mol Biol 356*, 281–291.
57. Krutzik, P. and Nolan, G. (2006). Fluorescent cell barcoding allows high throughput drug screening by flow cytometry. *Nat Meth 3*, 361–368.
58. Nolan, J. P., and Sklar, L. A. (2002). Suspension array technology; evolution of the flat array paradigm. *TIBS 20*, 9–12.
59. Young, D. W., Bender, A., Hoyt, J., McWhinnie, E., Chirn, G. W., Tao, C. Y., Tallarico, J. A., Labow, M., Jenkins, J. L., Mitchison, T. J., and Feng, Y. (2008). Integrating high-content screening and ligand target prediction to identify mechanism of action. *Nat Chem Biol 4*, 59–68.
60. Schwartz, S. L., Tessema, M., Buranda, T., Pylypenko, O., Rak, A., Simons. P. C., Surviladze, S., Sklar, L.A., and Wandinger-Ness, A. (2008). Flow cytometry for real-time

measurement of guanine nucleotide binding and exchange by Ras-like GTPases. *Anal Biochem 381*, 258–266.

61. Roman, D. L., Talbot, J. N., Roof, R. A., Sunahara, R. K., Traynor, J. R., and Neubig, R. R. (2007). Identification of small-molecule inhibitors of RGS4 using a high-throughput flow cytometry protein interaction assay. *Mol Pharmacol 71*, 169–175.
62. Nolan, J. P., and Sklar, L. A. (1998). The emergence of flow cytometry for the sensitive real-time analysis of molecular assembly. *Nat Biotechnol 16*, 833–838.
63. Hanley, B. (2007). Variance in multiplex suspension array assays: carryover of microspheres between sample wells. *J Negat Results Biomed 6*, 6.
64. Hanley, B. (2007). Variance in multiplex suspension array assays: intraplex method improves reliability. *Theor Biol Med Model 4*, 32–32.
65. Hanley, B., Xing, L., and Cheng, R. H. (2007). Variance in multiplex suspension array assays: microsphere size variation impact. *Theor Biol Med Model 4*, 31–31.
66. Zhang, J., Chung, T. D., Oldenburg, K. R. (1999). A simple statistical parameter for use in evaluation and validation of high throughput screening assays. *J Biomol Screen 4*, 67–73.
67. Glickman, J. F., and Auld, D. (2005). Another Z factor. *Assay Drug Devel Technol 3*, 339–350.
68. Malo, N., Hanley, J. A., Cerquozzi, S., Pelletier, J., and Nadon, R. (2006). Statistical practice in high-throughput screening data analysis. *Nat Biotechnol 24*, 167–175.
69. Ivnitski-Steele, I., Larson, R. S., Lovato, D. M., Khawaja, H. M., Winter, S. S., Oprea, T. I., Sklar, L. A., and Edwards, B. S. (2008). High-throughput flow cytometry to detect selective inhibitors of ABCB1, ABCC1, and ABCG2 transporters. *Assay Drug Dev Technol 2*, 263–276.
70. Young, S. M., Bologa, C. M., Fara, D., Bryant, B.K., Strouse, J. J., Arterburn, J. B., Ye, R. D., Oprea, T. I., Prossnitz, E. R., Sklar, L. A., and Edwards, B.S. (2008). Duplex high-throughput flow cytometry screen identifies two novel formylpeptide receptor family probes. *Cytometry A 75*, 253–263.
71. Ivnitski-Steele, I., Larson, R. S., Lovato, D. M., Khawaja, H. M., Winter, S. S., Oprea, T. I., Sklar, L. A., and Edwards, B.S. (2008). High-throughput flow cytometry to detect selective inhibitors of ABCB1, ABCC1, and ABCG2 transporters. *Assay Drug Dev Technol 2*, 263–276.
72. Curpan, R. F., Simons, P. C., Zhai, D., Young, S. M., Carter, M. B., Bologa, C.G., Oprea, T. I., Satterthwait, A. C., Reed, J. C., Edwards, B.S., and Skklar, L. A. (2011). High-throughput screen for the chemical inhibitors of antiapoptotic Bcl-2 family proteins by multiplex flow cytometry. *Assay Drug Dev Technol* . E-pub ahead of print May 11, 2011.
73. McEwen, D. P., Gee, K. R., Kang, H. C., and Neubig, R. (2001). Fluorescent BODIPY-GTP analogs: real-time measurement of nucleotide binding to G proteins. *Anal Biochem 291*, 109–117.

CHAPTER 18

Label-Free Biosensor Technologies in Small Molecule Modulator Discovery

Yuhong Du
Jie Xu
Haian Fu
Arron S. Xu

Label-free biosensor technologies allow the detection of molecular interactions and cell-based biological readouts without the need for conventional labels or engineered cell lines with overexpressed targets. A number of label-free detection technologies have been used for monitoring biomolecular interactions, most notably surface plasmon resonance spectroscopy (SPR) and isothermocalorimetry (ITC) [1–4]. With the rapid development of label-free based detection instrumentation in recent years, the power of various label-free technologies for studying biomolecular interactions and endogenous cellular events has been recognized by biomedical investigators, especially scientists involved in high-throughput screening (HTS). This, in turn, has led to a rapidly expanding list of various assay formats for HTS and allowed, for the first time, label-free screening for drug discovery in an HTS format [3, 5]. This chapter reviews basic principles of several major label-free biosensor technologies, with emphasis on microplate-based technologies and their practical applications in small molecule modulator discovery.

1. LABEL-BASED AND LABEL-FREE–BASED ASSAY TECHNOLOGIES

1.1. Traditional Label-Based Technologies for HTS

Various in vitro biochemical assays and cell-based methods are frequently employed in HTS for small molecule modulator discovery. These assays are often performed in a microplate format (96/384/1536 wells) in order to meet the demand of handling large sample numbers in primary HTS. In vitro biochemical assays rely on the use of an isolated or purified target that is often coupled with a tag (e.g., a fluorophor) for detection. These targets include purified adaptor proteins, enzymes, receptors, or cellular extract containing the target(s) of interest. Common detection methods in biochemical HTS assays include fluorescence polarization (FP), fluorescence resonance energy transfer (FRET), fluorescence intensity (FI), luminescence, and absorbance. FP measures the binding of a macromolecule to a relatively small peptide or ligand labeled with a fluorophore [6]. Binding of a macromolecule to the fluorophore-labeled small ligand/peptide slows down its rotation in solution and generates an FP signal. FRET allows the measurement of interacting molecules within a certain distance (<100 Å) (see Chapter 14 for details). The target and its binding partner are labeled separately, with a donor and a paired acceptor fluorophore containing matching excitation and emission spectra. Binding of the two molecules brings the donor and acceptor into close proximity, allowing energy transfer from the donor to the acceptor and FRET signal production. FRET signals can be derived in order to provide structural proximity information of molecular interactions [7]. FI-based assays measure changes in fluorescence intensity resulting from biochemical reactions. For example, cleavage of an appropriately fluorophore-labeled substrate by an enzyme can lead to a change in fluorescence intensity [8]. Biochemical assays are routinely used for monitoring receptor-ligand interactions, protein-protein interactions, and enzymatic activities (e.g., kinases and proteases).

Cell-based assays typically involve the use of established cell lines and occasionally primary cells. Examples of cell-based assays include reporter assays, cell viability and growth assays, ion translocation assays, and image-based assays for high-content screening (HCS). Cell-based assays generally involve the use of fluorescence-labeled or engineered cell lines. For example, fluorescent proteins such as green fluorescent proteins (GFPs) have been used to monitor calcium binding and translocation of proteins of interest. Luciferase proteins are widely used as reporters of gene expression and protein stability. Luciferase-coupled luminescence assays can also be used to measure cellular adenosine triphosphate (ATP) levels as well as

G protein-coupled receptor (GPCR) activation [9]. In contrast to biochemical assays, in which the mechanism is well-defined, cell-based assays generally provide complex and integrated cellular response information under a more physiologic condition.

These label-dependent assays offer mature technological platforms for a wide range of biological targets in HTS or HCS and have been extensively used for large-scale compound screenings. Because of their technical simplicity and low cost, it is unlikely that their use will diminish in the future. However, labels themselves may interfere with the functionality of the target molecules. Cellular responses measured by a label-based assay may reflect only a specific signaling pathway or coupled reaction, resulting in a lack of the detection of more comprehensive cellular responses. In addition, the presence of autofluorescence or fluorescence quencher compounds in a compound library may interfere with fluorescence-based readouts. Alternative assay technologies without labeling may offer viable alternative approaches to complement the existing label-dependent assays for compound screening.

1.2. Label-Free Biosensor Technologies

Label-free detection technologies represent a major new development in the bioassay technologies, which complements the existing label-based bioassays. Label-free technologies offer a number of distinct advantages over conventional label-dependent or artificial reporter assay platforms [3, 10]. Because they do not require fluorescence labeling of the target or engineering of cells, label-free technologies can possibly shorten the time needed for assay development. Targets of interest in biomolecule interactions or cellular signaling can be studied in their native states [11, 12]. Similarly, the problem of fluorescence interference from autofluorescent or fluorescence quenching compounds in the compound library is eliminated. Orphan target assay development may become easier as the lack of suitable labels for detection may be overcome by the use of label-free technologies. The noninvasive feature of label-free technologies in cell-based assays allows direct and real-time detection of complex and integrated cellular signaling events in native cells. This offers significant advantages over the use of engineered cell lines with overexpressed receptors by providing drug response data under more physiologically and pharmacologically relevant conditions [13–16].

Label-free detection requires the use of a biosensor, an analytical device that converts a biological response into a measurable optical, electrical, or mechanical signal. All label-free systems for monitoring bimolecular interactions have two fundamental components: a transducer coupled with a bioreceptor and a reader [17] (Figure 18.1). The bioreceptor is used to specifically capture biological recognition molecules, for example, proteins, small molecules, or living cells. The other binding partner (target analyte) can then be recognized by the bioreceptor, and corresponding biological responses are then converted into equivalent electrical signals by the transducer. A reader, another component of the label-free system, is then used to collect, amplify, and process signals from the transducer. Depending on the transducing mechanism used, biosensors can be of many types, such as resonant biosensors, optical detection biosensors, thermo-detection biosensors, ion-sensitive filed-effect transistor (ISFET) biosensors, and electrochemical biosensors [18].

In the past, label-free technologies have not been employed in primary assays for HTS because of their limited sensitivity and throughput. Instead, these techniques have been used primarily as the basis for secondary assays to confirm lead compounds and characterize the affinities, kinetics, and other properties of molecular binding interactions [3, 10, 19]. In recent years, because of technological advances and improved instrument design – especially the development of microplate (96/384/1536-well)-based label-free technologies with enhanced sensitivity and throughput – label-free detection has emerged as a promising complement to traditional HTS platforms not only in secondary follow-up assays but also in small molecule discovery in an HTS format.

2. THE USE OF LABEL-FREE BIOSENSOR TECHNOLOGIES FOR IN VITRO BIOCHEMICAL ASSAYS

2.1. Surface Plasmon Resonance (SPR) Technologies

2.1.1. Basic Principle of SRP. Among label-free biosensor technologies, SPR has been the most extensively explored and developed tool used to study bimolecular interactions between an analyte in solution and a bimolecular recognition element immobilized on the SPR sensor surface [20, 21]. The surface plasmon is a special type of the electromagnetic field propagating along a metal and dielectric interface, and its electromagnetic field decays evanescently into both the metal and dielectric. Metals such as gold and silver support the surface plasmons at visible frequencies. Because of its chemical stability, gold is the most frequently used metal in SPR studies.

The surface plasmon can be excited by attenuated total reflection in a prism coupler. A totally reflected light wave is obtained when a light wave travels through an optical prism and hits the metal film at an angle of incidence larger than the critical angle for the dielectric-prism system. As a result, an evanescent field is generated along the metal film. If the propagation constant of the evanescent field matches that of the surface plasmon of the metal film, coupling between the light wave and the surface plasmon occurs in such a way that part of the energy from the incident light is transferred into the energy of the surface plasmon, causing a reduction

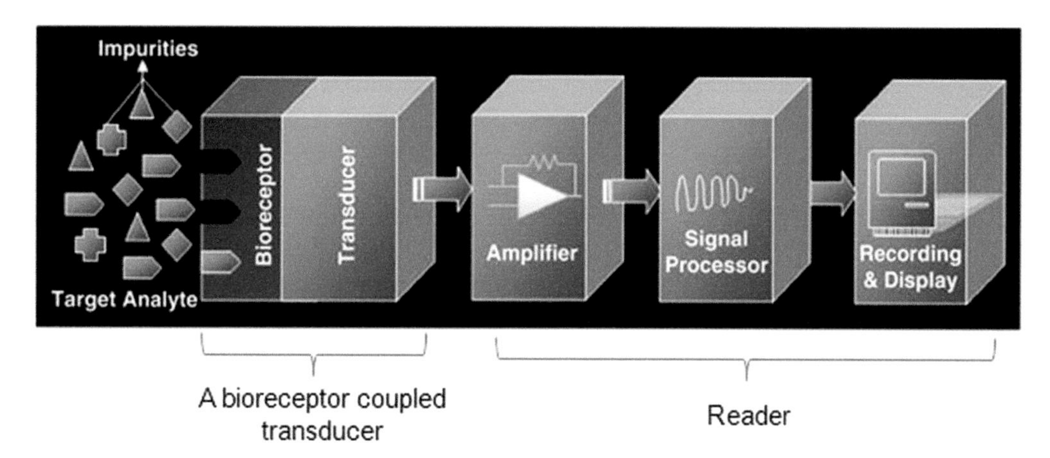

***FIGURE 18.1:** The two fundamental components of the label-free system: a bioreceptor coupled transducer and a reader. 1) The bioreceptor can capture specific biological molecules and is coupled with a transducer. The specific captured biomolecule recognizes the target analyte or its binding partner. The corresponding biological response (e.g., binding) is converted to electrical signals by the transducer. 2) The converted signal from transducer is then amplified, processed, stored, displayed, and analyzed by a "reader." (Modified from reference 17 with copyright permission from Elsevier.)*

in the intensity of reflected light. The coupling angle at which this intensity minimum occurs is a function of the refractive index of the solution close to the gold layer on the opposing face of the sensor surface (Figure 18.2A). Binding of the target of interest at the metal surface changes the local refractive index, which results in a shift of the SPR coupling angle of incidence.

SPR technologies have the advantage of label-free and real-time monitoring of biomolecular interactions in solution, so binding kinetics (association and dissociation), affinity, and concentration of targets can be quantified. Combining this technology with microfluidic design enables measurement of the interaction of an immobilized target with multiple analytes in a sequential mode. In addition, versatile surface chemistries on SPR surface and instrument configurations allow for flexibility in assay development, making it a promising tool both for characterizing biomolecular interactions in basic biology research and in screening and identifying small molecules in drug discovery efforts.

2.1.2. Applications of SPR A broad spectrum of molecular interaction assays using SPR have been reported, such as protein-protein interactions [22], antigen-antibody interactions [23], protein-peptide interactions [24, 25], protein–small molecule interactions [26, 27], receptor-ligand interactions [28, 29] protein-DNA/RNA interactions [30], protein-DNA interactions [31], and RNA–small molecule interactions [32] in addition to enzyme binding and functional assays such as kinase assays [33 –35].

SPR technologies are routinely used for the characterization of molecular binding equilibrium [36] or as complementary secondary assays to validate target/compound interactions identified from primary screening [26]. For example, SPR has been used to validate leads for dipeptidyl peptidase IV (DPP-IV), the main glucagon-like peptide 1 (GLP1)-degrading enzyme [37]. Inhibition of DPP-IV has been proposed for the treatment of type 2 diabetes. In this assay, DPP-IV is immobilized on a CM5 SPR sensor (Biacore) by standard amide coupling chemistry. The binding and dissociation of the inhibitor to the immobilized DPP-IV is monitored in real time by SPR. SPR has also been used to detect and characterize promiscuous inhibitors [38].

Recent instrumentation and technological advances resulting in increased sensitivity and throughput make SPR-based small molecule modulator discovery possible. For example, an array-based SPR apparatus has been developed for a kinase assay [35] with the achievement of multiple sample processing (one thousand samples per day) [34, 35]. In this system, the substrates, phospho-tyrosine (pTyr) residues, are selectively captured by anti-pTyr antibodies immobilized on a sensor chip. The level of tyrosine phosphorylation is detected and calculated from the captured level of the substrates and the binding level of anti-pTyr antibody.

2.1.3. SRP-Based Commercial Instruments There are a number of commercially available SPR-based instruments on the market for biochemical assays. Biacore (www.biacore.com) pioneered commercialization of label-free SPR technology with the launch of the first commercial SPR instrument in 1990. Biacore continues to improve its instrument design, surface chemistry, and microfluidic systems as well as its user interface. The new A-100 and T-100 instruments have achieved sensitivity sufficient for the detection of protein binding by chemical fragments of a molecular mass as small as 150 Da. Graffinity Pharmaceutical developed an SPR imager to detect SPR changes

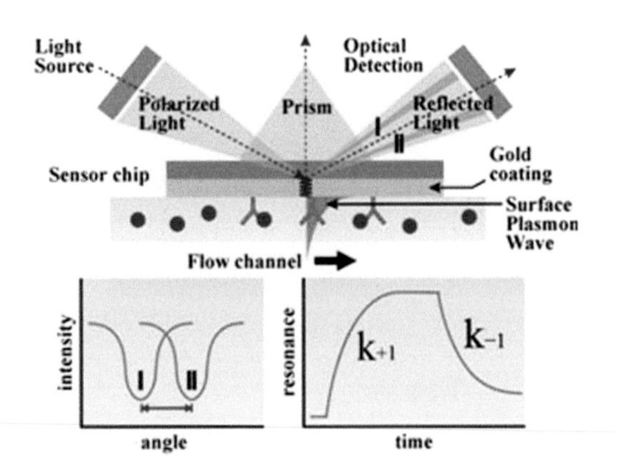

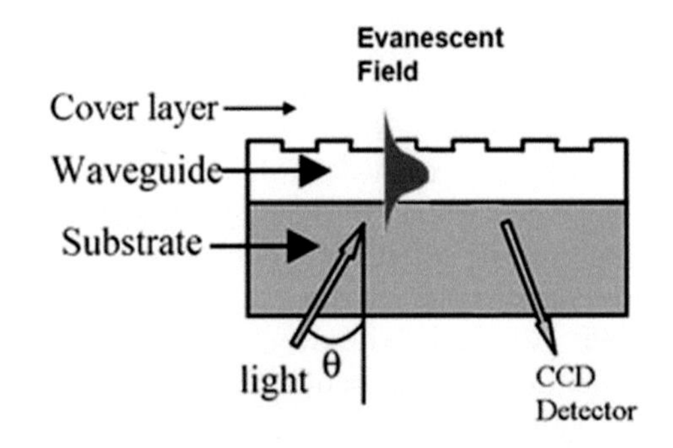

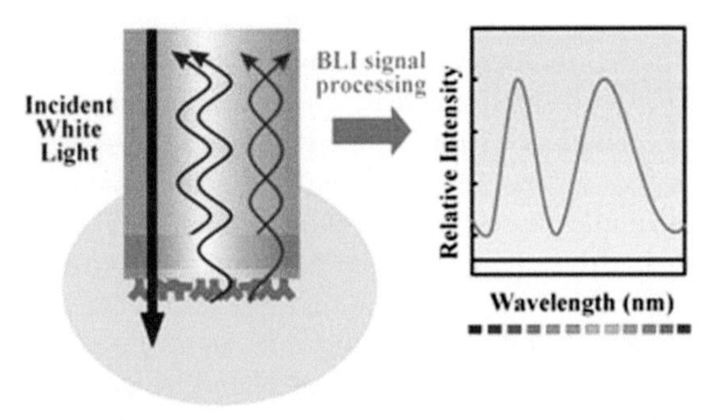

FIGURE 18.2: Biosensor systems for biochemical assays. (A) SPR biosensor system. In SPR, a surface plasmon is excited at the interface between the metal film and a dielectric medium (the cover layer). A change in the refractive index in this region creates a change in the propagation constant of the surface plasmon. This change in the propagation constant is determined by measuring a change in coupling angle or coupling wavelength. (B) In a grating coupler sensor, a thin waveguide layer with high refractive index is deposited on a glass substrate. Gratings are fabricated and used to couple light into or out of the waveguide. The evanescent field is very sensitive to local refractive index changes. Any surface binding will perturb the evanescent field, resulting in a change in the waveguide coupling condition. This change can be measured by monitoring the change in coupling angle or the coupling wavelength. (C) In Bio-Layer Interferometry (BLI), an interference pattern of white light back-reflected from two surfaces: a layer of immobilized protein on the biosensor

FIGURE 18.2 (continued) tip and an internal reference layer is obtained (Source: Adapted from www.fortebio.com). The binding between a ligand immobilized on the biosensor tip surface and an analyte in solution produces an increase in optical thickness at the biosensor tip, which results in a shift in the interference pattern that can be measured in real time.

in an array format (www.graffinity.com). Bio-Rad Laboratory (www.bio-rad.com) launched an SPR system (ProteOn XPR36) in 2007. The main advance of the ProteOn XPR36 system is its target array format, which utilizes a rotating microfluidic delivery system, thus effectively increasing the number of parallel assays performed on each array. Fujifilm launched the AP-3000 system in 2007 (www.fujifilm.com). The AP-3000 system uses a sensor stick (cartridge), which partly forms the internal fluidic system. Protein targets are delivered into the disposable sensor stick and immobilized onto the surface. According to Fujifilm, each sensor stick may be reused up to 160 times.

2.2. Resonant Waveguide Grating (RWG) Biosensor Technologies

*2.2.1. **Basic Principles of RWG Biosensor*** Another major label-free optical biosensor technology is based on resonant waveguide grating (RWG). Resonant waveguide sensors may be rendered in several formats or configurations [39]. In a typical RWG sensor, a thin waveguide layer with a high refractive index is deposited on a glass substrate. Light can be coupled into or out of the waveguide by a grating at the resonance condition, which is determined by Equation 18.1, where θ is the coupling angle, N is the effective refractive index of a waveguide layer, λ is the wavelength of the incident light beam, and Λ is the grating period. When light is excited inside the waveguide, an evanescent field generates and extends above the waveguide surface. Any refractive index change in this region will perturb the evanescent field, resulting in a change in the waveguide's effective refractive index. This change can be measured by monitoring changes in the coupling angle or coupling wavelength. When illuminated with broadband light, the RWG biosensor reflects only a specific wavelength. Binding of an analyte to the target attached to the biosensor surface leads to a change in the local index of refraction, which in turn leads to a change in wavelength of the reflected light to a higher wavelength. The measured wavelength shift, therefore, can be used as a readout for monitoring the bimolecular interactions (Figure 18.2B).

$$n\theta = N - \frac{\lambda}{\Lambda} \qquad (1)$$

*2.2.2. **Applications of RWG Biosensors*** RWG-based technologies like SPR require immobilization of the target onto the sensor surface and have been successfully applied to detect a variety of bimolecular interactions in solution as the following examples:

- direct binding of small molecules (4-carboxybezene sulfonamide, dansylamide, and acetazolamide) to immobilized targets (carbonic anhydrase II, 30 kDa) [40, 41]
- protein (warfarin, 308 Da) binding to human serum albumin (66 KDa) [41], and human IgG [5]
- biotin (244 Da) binding to avidin (60 KDa) [42]
- peptide-protein interactions with a signal trafficking phosphorylated SWTY peptide and 14-3-3 protein or anti-phosphopeptide antibodies [43]
- enzyme assays such as the functional lytic assays for HTS [44] and caspase-3 protease assay [42].

In addition, because an RWG-based surface has a distinct physical composition relative to the metal/glass sensor surface used in SPR, this offers the unique advantage of providing broader applicability by allowing both biochemical and cell-based assays to be carried out in a single platform [45]. The cell-based applications of RWG biosensors are discussed in a later session of this chapter.

2.2.3. Commercial RWG Sensor Instruments Two major RWG-based instruments are the Corning Epic system (Corning; www.Corning.com) and the BIND system (SRU Biosystems; www.srubiosystems.com). These instruments utilize various sensor microplates in industry-standard 96-well, 384-well, or 1536-well formats with different surface chemistries optimized for either biochemical assays or cell-based assays. The Epic microplate for biochemical assay incorporates a unique intrawell reference sensor design that compensates for small variations in refractive index due to temperature variations and bulk refractive index changes, such as the presence of dimethyl sulfoxide (DMSO). The Epic reader provides the capability for both low-throughput assay development and an automated high-throughput screening operation. The internal robotic plate handling design enables a throughput of approximately forty thousand wells per eight hours when the system is integrated with an external liquid handling device for sample delivery. The Epic system also provides sufficient sensitivity to detect both small and large molecular interactions. At a target immobilization density of 5 pg/mm^2, the Epic system is able to detect binding of a small molecule of approximately 150 Da to a 25 kDa protein target.

The SRU BIND system is a versatile, bench-top, label-free detection system based on the RWG principle. It has a variety of sensor surface chemistries for different biochemical targets and cell-based assays, respectively. The BIND system has been used in a number of biochemical binding assays as well as cell-based functional assays. With the efforts of developing 384/1536-well format and possibly cost reduction, the RWG biosensors have an advantage of higher throughput compared with other label-free technologies and the potential for rapidly increasing applications in an HTS format.

2.3. Optical Bilayer Interferometry Technology

Optical bilayer interferometry is another label-free detection technology available for molecular interaction analysis. Its operating principle is based on the optical wave interference phenomenon that occurs upon the combination of two optical beams back reflected from two surfaces: a internal reference layer and a layer of immobilized bioreceptors [39] (Figure 18.2C). The binding of an analyte to the biosensor surface leads to a phase change of the sample beam because of an increase in optical thickness. This phase change, in turn, generates a shift in the interference pattern. By monitoring this shift, binding information can be obtained. One commercially available bilayer interferometry instrument is the Octet system from ForteBio (ForteBio, www.fortebio.com). The biosensor of the Octet system is embedded at the tip of an optical fiber, creating a "dip and read" biosensor that can be configured for up to sixteen parallel sensors/channels in a 384-well microplate format. This system is primarily used for studying protein-protein interactions either in a simple buffer system or in a complex biological sample. It can be used to study protein–small molecule binding, antibody clone selection and epitope mapping, process optimization for protein expression, and affinity and kinetics measurements of protein-protein interaction.

2.4. Development of a Biochemical Label-Free Assay

Studying bimolecular interactions at a sensor surface requires one binding partner to be immobilized on the transducer surface. Therefore, the sensitivity of the method is essentially limited by the availability of the soluble binding partner and the inherent sensitivity of the mode of signal detection. The higher the affinity of the interaction, the lower the concentration of the analyte that can be detected, which makes this method very attractive for high-affinity interactions. As illustrated in Figure 18.3, development of a label-free biochemical assay generally involves two steps. The first step is the immobilization of a target-specific bioreceptor (e.g., a small molecule, peptide, enzyme, or antibody) on the biosensor surface. The second step is the binding reaction, which is initiated by adding an analyte (binding partner, e.g., small molecules, peptides, proteins, or DNA fragments) to a solution. The signal is measured in real time or a selected time point by different detection instruments throughout each step to obtain binding information.

2.4.1. Immobilization of Targets to the Biosensor Surface Usually, bioreceptor immobilization is performed through direct covalent binding or affinity capturing.

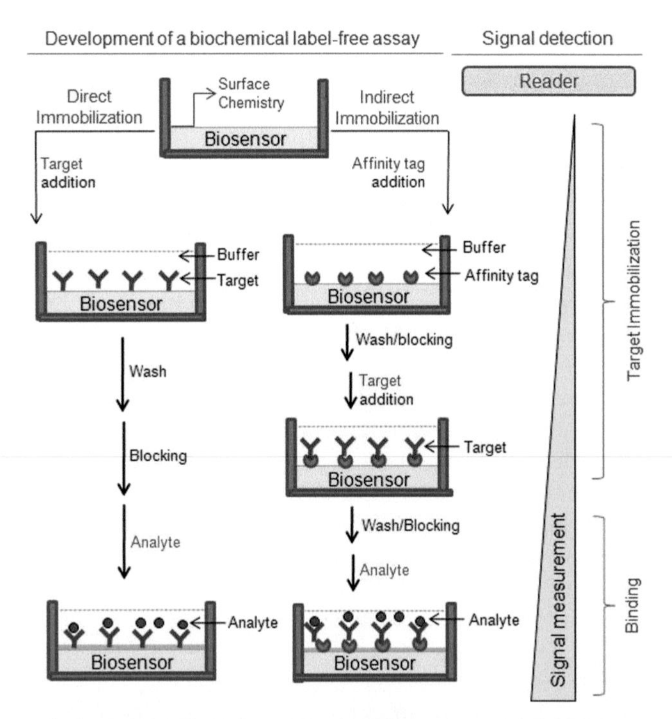

***FIGURE 18.3:* Procedures for developing a label-free biochemical assay. Setting up a label-free biochemical assays generally includes two steps: target immobilization to the biosensor surface followed by the addition of an analyte in solution (such as a small molecule, peptide, or a protein). The binding reaction is initiated upon adding the analyte, and the binding signal is then monitored in real time.**

Amine coupling to surface lysine residues or the N-terminal residue of a protein (or peptide) is the most commonly employed surface chemistry for directly coupling amine-containing targets to a biosensor surface. Heterobifuncational reagents containing an amine-reactive *N*-hydroxysuccinimide (NHS) ester are often used. However, the amine coupling process competes with the hydrolysis of the NHS ester in aqueous solution. The optimal pH of immobilization buffer should be evaluated by carrying out the immobilization step using buffers with varying pHs. A buffer with 1 to 2 pH units below the protein's isoelectric point allows the protein to be positively charged and is generally used for immobilization. However, consideration must be given to the stability of the protein at low pH during immobilization. The buffer used should also have low ionic strength and contain no primary amines (e.g., Tris). Acidic proteins (isoelectric point (pI) <3.5) are difficult to immobilize through amine coupling chemistry. Although direct covalent coupling of bioreceptors produces a viable surface for a wide range of proteins, it can introduce a heterogeneous population of target proteins with random orientations on the sensor surface and, in some cases, can inactivate the bioreceptor. The efficiency and effectiveness of covalent coupling is evaluated by measuring binding with its binding partner. An optimal immobilization condition is established when a high immobilization efficiency is correlated with high efficiency of binding of a positive control analyte.

For targets containing affinity tags, indirect affinity coupling via these tags is often employed. Avidin or streptavidin may be immobilized for affinity capture of biotinylated targets, whereas tag-specific antibodies (e.g., anti–glutathione *S*-transferase [GST], anti-6xHis, or anti-Flag) can be immobilized on the biosensor surface to capture targets containing the corresponding epitopes. The affinity of avidin to biotin is strong enough so that the binding is almost irreversible. The avidin/streptavidin immobilized surface is "sticky" and might increase nonspecific binding of the analyte to the surface. Optimization of buffer conditions for binding assays is generally required to reduce nonspecific binding of proteins or hydrophobic/lipophilic compounds to the sensor surface. Indirect immobilization through an affinity tag has the advantage of producing a more homogenous population of oriented targets on the surface and therefore creates a uniform surface. This provides an alternative strategy for coupling a target to the biosensor surface without the need to subject the protein or ligand to a low pH. In this way, target activity may be better preserved. This tag-based indirect coupling interaction is often highly specific and can be used for low-purity samples.

2.4.2. Binding Reaction Binding of an analyte to an immobilized target is initiated by delivery of the analyte in solution to the target on a sensor. The binding signal is obtained by subtracting the immobilization signal from the measured signal after analyte addition. Almost all label-free readers are capable of measuring the binding signal in real time after analyte addition. A time course is generally plotted using the recorded binding signals against time for kinetic analysis. To determine the equilibrium binding affinity of a molecular interaction, binding may be measured at different analyte concentrations in a concentration titration assay from which the dissociation constant, kd, can be determined. There are several experimental considerations for developing and optimizing a binding assay using label-free technologies: 1) An optimal binding buffer should preserve the activity of the target. Co-factors and buffer salts required to maintain the physiologic activity of a target are generally included; 2) Protein targets that exhibit significant nonspecific binding to sensor surface may require the inclusion of additional assay components to reduce nonspecific bindings. For example, surfactants (e.g., 0.05% 3-[(3-cholamidopropyl)dimethylammonio]-1-propanesulfonate (CHAPS) or P-20) could be added to the binding buffers; 3) Appropriate controls should be included to evaluate the specificity of a binding assay. Proper negative controls with target immobilized or captured but without analyte may be used. Generally, completely blank wells (wells without target present) with binding buffer alone are not optimal background controls for nonspecific interactions. Binding of an analyte to an unrelated protein with respect to the target of interest may also be considered as a negative control to evaluate binding specificity; 4) Analyte concentration–dependent binding allows for measurement of equilibrium binding or the dissociation constant and, thus, affinity ranking of binding partners; 5) Optical refractive index sensors are highly sensitive to refractive index change in samples. A small change of DMSO concentration in compound addition may result in buffer mismatch and introduce significant artifacts in optical response. An initial read prior to analyte addition must be performed in the same assay buffer as the one containing the analyte of interest; 5) Analyte aggregation may lead to an abnormally high response. This may, in some cases, be considered as a unique advantage of an optical label-free detection platform that reveals promiscuous or problematic compounds in the compound library; and 5). The detected binding response may be directly affected by the mass changes when a binding event occurs. Generally, smaller amounts of immobilized targets on the biosensor surface and larger amounts of analyte in the solution may lead to a larger signal output and, hence, is a more preferred assay design.

3. THE USE OF LABEL-FREE TECHNOLOGIES FOR CELL-BASED FUNCTIONAL ASSAYS

Label-free technologies for cell-based assays have received considerable attention because of advances in instrumentation, sensor, and experimental design that have significantly improved their capabilities [10, 12, 15, 16]. The current label-free technologies for cell-based assays make use of various aspects of integrated cellular and morphological changes in response to an environmental signal to obtain a multiparametric and biologically relevant readout [16, 46–49]. These integrated cellular changes may be associated with changes in electrogenic and impedance properties as well as intracellular mass redistribution-related optical properties [16, 50, 51]. The advanced label-free cell-based detection technologies enable multiparametric and real-time monitoring of cellular responses with sensitivity that is sufficient to detect a biological response without the need for the overexpression of a cellular target [50]. The current technologies allow for real-time monitoring of the biological events in a kinetic mode. There are a number of label-free technologies available for cell-based assays [15, 17]. Among them, two main technologies, impedance and RWG biosensors, have been successfully applied to study cell biology, particularly using cells with endogenous receptors [6, 10, 12, 18 –20].

3.1. Impedance-Based Biosensor Technologies

3.1.1. Basic Principles. Impedance biosensors are electrical biosensors and rely on the measurement of the electrical impedance (Z), which is defined as the ratio of voltage to current ($Z = V/I$) as described by Ohm's law. Recent instrumentation advances make this technology applicable in a microplate format for cell-based label-free HTS assay platform. Cells are seeded onto a 96-well microplate that contains electrodes at the bottom of wells. Cells and electrodes form an electrical circuit attached to an impedance analyzing system that supplies constant voltage, producing current that flows around and between cells (extracellular current, or Iec) and through cells (transcellular current, Itc) [16]. When cells are exposed to a stimulus such as a receptor ligand, signal transduction events occur that lead to cellular changes, such as changes in morphology, volume, adherence, or alternations in cell-to-cell contact and cell number [48]. These cellular changes individually or collectively affect the flow of extracellular and transcellular current and therefore affect the magnitude and characteristics of the measured impedance. The change of impedance signal can be monitored in real time after ligand addition (Figure 18.4A). These response profiles may be characteristics of a particular signaling pathway or a compound action [16].

(A) Impedance cell-based biosensor system

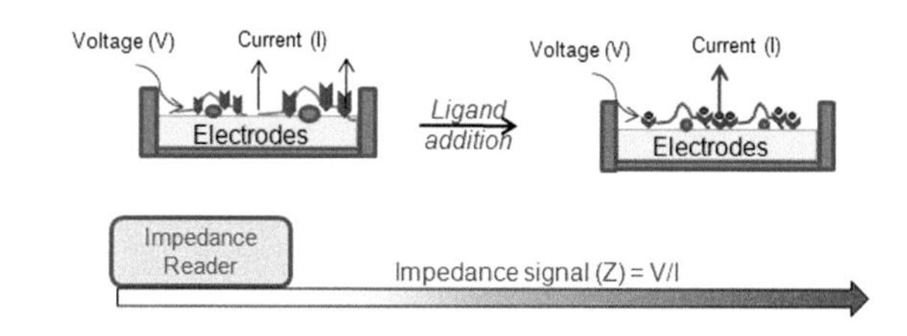

(B) Resonant waveguide grating (RWG) cell-based biosensor system

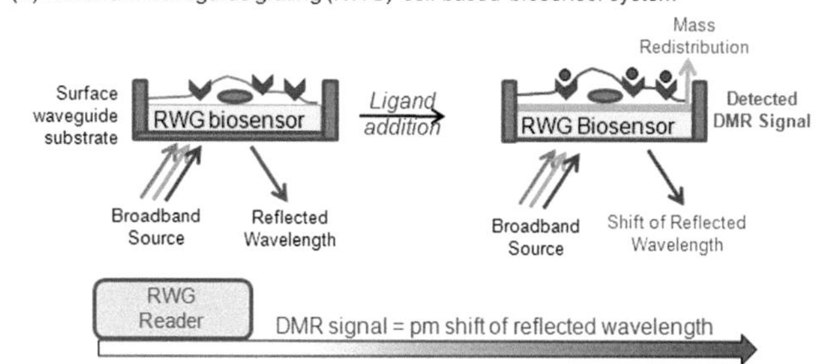

FIGURE 18.4: Biosensor systems for cell-based functional assays. (A) Impedance cell-based biosensor system. Cells are seeded onto electrodes coupled at the bottom of the well. When a constant voltage (V) is applied by an impedance reader, an electrical circuit between the cells and electrodes is generated and produces current (I), which is measured by the impedance reader as the impedance signal (Z = V/I). Change in cell number, morphology, volume, adherence, or alterations in cell-to-cell contact will result in impedance changes. (B) RWG cell-base biosensor system. An RWG biosensor measures changes in the local index of refraction resulting from a ligand-induced dynamic mass redistribution of the cell monolayer, as shown by a shift in the reflected resonant wavelength.

3.1.2. Applications of Impedance Biosensors The recent development of microplate-based impedance biosensors has allowed this technology to be successfully applied to cell-based assays for drug discovery. This technology has been utilized to monitor GPCR signaling [21, 22], cellular functions of receptor tyrosine kinases (RTKs) [46], cell proliferation, adhesion [52], viability, apoptosis, growth inhibition [53], cell migration, and invasion assays [15, 54]. Other applications include signaling pathway identification and deconvolution [50, 55] as well as hit identification and pharmacologic profiling [5, 55].

3.1.3. Commercial Impedance-Biosensor Instruments There are several commercially available impedance-based instruments for cell-based assays in a 96-well format for medium-throughput screenings, such as Cell-Key (MDS, www.cellkey.com), xCELLigence (ACEA Bioscience, www.aceabio.com), and ECIS (Applied Biophysics, www.biophysics.com).

3.2. RWG Biosensors for Cell-Based Assays

The RWG biosensor technology represents another major label-free detection technology for cell-based assays. Cell-based functional assays using optical RWG biosensors are based on monitoring changes of refractive index, which reflects cellular dynamic mass redistribution (DMR) as a result of cellular responses to extracellular signals such as small molecule ligands or drugs (Figure 18.4B). The shift of the reflected wavelength after ligand stimulation can be monitored in real time by a reader. The reflected wavelength difference before (baseline measurement) and after ligand stimulation (signal measurement) is defined as a DMR signal [4, 24]. Because the energy of the broadband evanescent electromagnetic field on the RWG sensor surface decays over distance from the surface, only the DMR change within limited depth from the biosensor surface is detected. Although the sensing depth of current RWG sensors may not penetrate through the entire volume of the cells attached to the sensor surface, the presence of integrated cellular signaling pathways enables the signal propagation within the entire cytoplasm and permits the detection of a cellular response within a partial volume of the cells. This is demonstrated by the agreement of well-characterized actions of known drugs detected with RWG technologies and other label or nonlabel technologies [55]. Current major commercial RWG-based instruments for cell-based assays in a microplate format include the Epic system and BIND system. Both systems are capable of performing cell-based assays as well as biochemical binding assays.

A number of applications of RWG biosensors for cell-based functional assays have been described. Cell-based RWG-biosensor assays include functional assays for the activation of receptors such as GPCRs [6, 18, 20] and RTKs

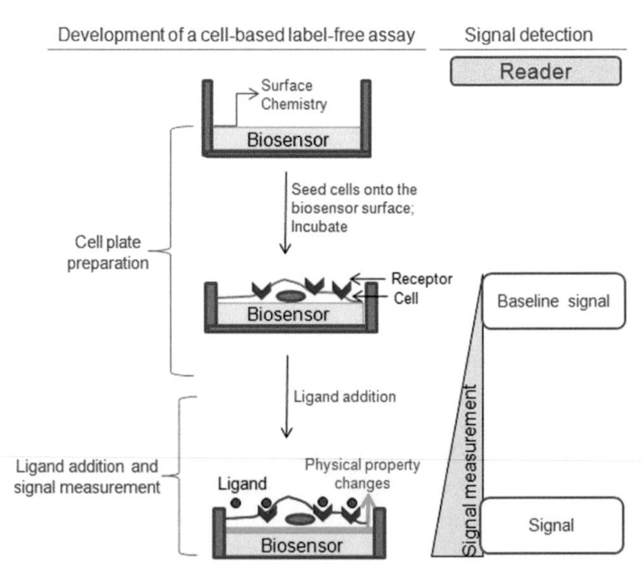

FIGURE 18.5: Procedures for developing a cell-based label-free assay in a microplate format. Cells are seeded onto biosensors coated at the bottom of microplates. After cells are attached to the biosensor surface, the baseline signal is recorded by the reader. A ligand is then added to the cells, and the signal is continuously monitored in real time.

[56] and the measurement of cell proliferation, viability, cytotoxicity, and growth inhibition assays [19, 23] [57]. It is also used for signaling pathway identification and deconvolution [4, 26, 27], probing cytoskeleton modulations [58], monitoring viral infection [59] and ion channel activities [60], and HTS to discover specific ligand-receptor induced pathway modulators [61].

3.2. Development of a Label-Free Cell-Based Biosensor Assay

The advent of the microplate-based label-free technologies has provided a major enabling path for cell-based assays in basic research and drug discovery. Cells can be cultured using conventional techniques on the biosensor-coated microplates. Label-free detections can then be performed directly on the same microplates. As illustrated in Figure 18.5, there are several steps in setting up a label-free cell-based assay. Depending on the mode of detection, for example, agonist or antagonist assays, multiple steps of ligand additions may be incorporated into a homogeneous assay design that allows the continuous monitoring of cellular response in a kinetic or an end-point mode.

As an example of developing a label-free cell-based assay, we have established the distinct growth factor–induced DMR profiles for monitoring oncogenic signaling pathways in various cancer cells [61]. Given the prominent importance of epidermal growth factor receptor (EGFR) signaling in cancer, targeting EGFR represents a new paradigm in cancer therapy. EGFR inhibitors have shown therapeutic promise in a wide range of tumor types,

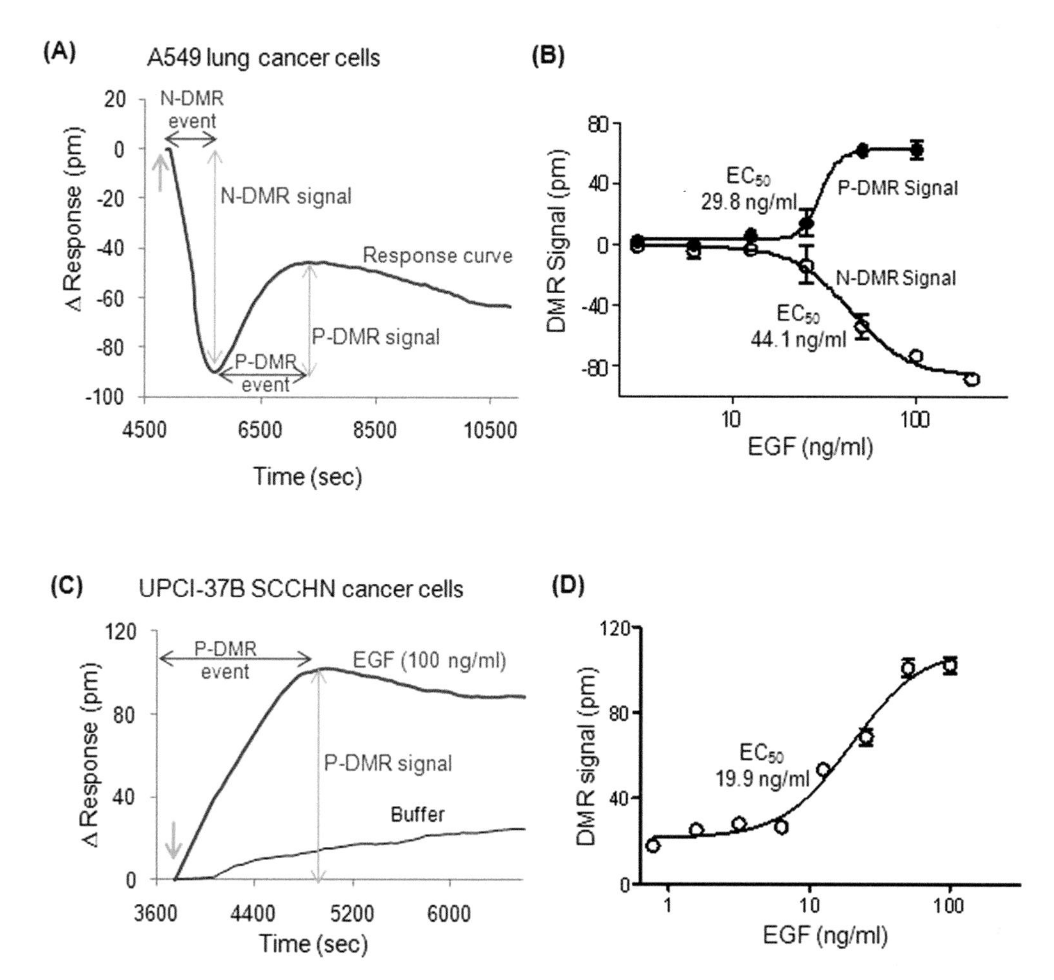

FIGURE 18.6: EGF induces DMR response in lung and SCCHN cancer cells. (A) EGF (100 ng/mL) induced a unique DMR response in A549 cells. (B) The dose-response of A549 cells to EGF. (C) UPCI-37B SCCHN cells display a distinct DMR response to EGF (100 ng/mL). Unlike the response of A549 cells, its response contains only one major event: P-DMR. (D) EGF dose-dependently induces DMR signal (P-DMR) in SCCHN cancer cells. The EC_{50} value was obtained using Prism 4.0 software. The solid green arrow indicates the time of EGF addition. (adapted from reference 61 with copyright permission.)

including squamous cell carcinoma of the head and neck (SCCHN) and non-small cell lung cancer (NSCLC). Using SCCHN and NSCLC cells as model systems, we have identified and characterized distinct optical signatures of EGF-induced signaling events [61], which are used to illustrate the utility of a RWG biosensor for cellular pathway profiling.

3.4.1. Experimental Procedure

1) *Cell plate preparation:* UPCI-37B SCCHN and lung adenocarcinoma A549 cancer cells are dispensed into a 384-well fibronectin-coated Epic biosensor plate at 10,000 cells per well and incubated overnight. After serum starvation for twenty-four hours, cells are washed twice with assay buffer (Hank's Balanced Salt Solution [HBSS] with 20 mM HEPES). The plates are delivered to the Corning Epic microplate reader;
2) *Baseline measurement:* After cell plates are thermo-equilibrated in the reader for one hour, basal DMR signals are recorded;
3) *Compound addition:* Test compounds or vehicle (DMSO) control are added to cells in each well and incubated for thirty minutes;
4) *Ligand stimulation:* EGF diluted in a defined assay buffer is added to each well containing cells;
5) *Signal measurement:* DMR signals are monitored and recorded continuously after ligand addition. Alternatively, the response may be monitored at a defined time point of the cellular response in an "end-point" mode. The end-point mode of measurement allows greater throughput for HTS; and
6) *Data analysis:* Depending on the technology platform, differential signals prior to and poststimulation are analyzed by the magnitude and temporal signal changes as a result of the integrated cellular response. The DMR

signal detected by the Epic reader is reported as picometer unit (pm) of wavelength shift (ΔResponse), which represents the shift of the reflected wavelength relative to an initial (time zero) measurement. Response curves are plotted based on ΔResponse after ligand addition versus time and used to define the DMR events [45]. An increase of DMR signal is defined as a positive DMR (P-DMR) event, and a decrease of DMR signal is referred as a negative-DMR (N-DMR) event. The characteristic cellular DMR in response to a specific stimulus (such as ligands of GPCRs, growth factor receptors, and other cell surface receptors) is defined by the shape of the response curve, which contains P-DMR events and/or N-DMR events (Figure 18.6A).

3.4.2. Results and Discussion

3.4.2.1. EGF induces unique optical DMR response signatures in lung and SCCHN cancer cells. A distinct DMR signal was obtained when serum-starved A549 lung cancer cells were stimulated with EGF (Figure 18.6A). The DMR response profile contained two events: a rapid initial decrease in the detected DMR signal, the N-DMR event [57], followed by an increase in the detected DMR signal, the P-DMR event. Stimulation with increasing concentrations of EGF led to dose-dependent increases in both the N-DMR and P-DMR signals (Figure 18.6B). However, in response to EGF, SCCHN cells displayed a completely different DMR profile compared with that of A549 cells (Figure 18.6C). Upon EGF addition, there was a rapid increase in the detected DMR response (P-DMR events), which was dose-dependently increased with increasing concentrations of EGF (Figure 18.6D). The unique DMR signatures in response to EGF stimulation obtained from lung and SCCHN cancer cells are consistent with the fact that pathways and mechanisms involved in the regulation of cell signaling are cell type specific. These data suggest that the DMR response of cells to EGF monitored by the Epic biosensor may serve as a promising approach for cell line characterization and for monitoring cell-specific oncogenic signaling pathways.

3.4.2.2. Identification of EGFR and PI3K as critical mediators of EGF-induced signaling in SCCHN cells. To test whether the DMR signal induced by EGF occurs specifically via the EGFR pathway and to determine the mechanisms by which the DMR signals were induced by EGF, the effect of a panel of small molecule signaling inhibitors on DMR signals were tested in SCCHN cells. Preincubation with AG 1478, an EGFR tyrosine kinase inhibitor, resulted in a delayed and decreased DMR signal (Figure 18.7A, B); whereas the VEGFR inhibitor SU 1498 did not attenuate the EGF-induced DMR signal (Figure 18.7C). The PI3K kinase inhibitors LY 294002 and wortmannin significantly blocked the EGF-induced DMR signal (Figure 18.8A, C), but the MEK1/MEK2 inhibitors U0126 or PD98059 did not (Figure 18.8B, C). These data

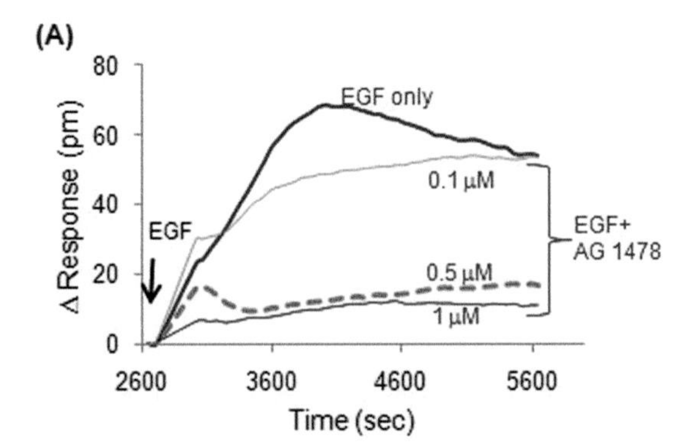

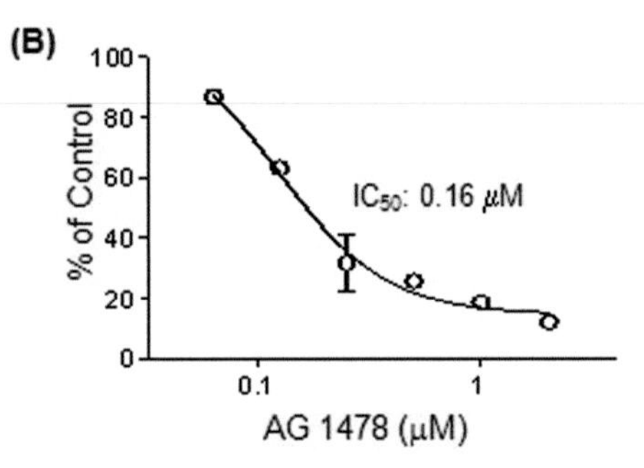

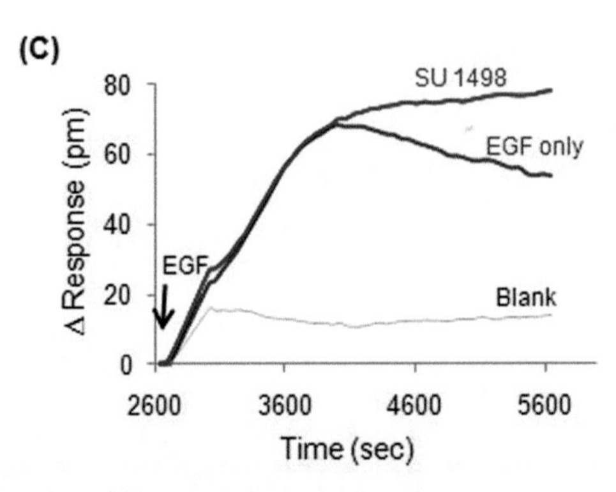

***FIGURE 18.7:** The EGF-induced DMR optical signature in SCCHN cells requires EGFR. **(A)** EGF-induced DMR response is does-dependently blocked by the tyrosine kinase inhibitor, AG 1478. **(B)** Dose-response effect of AG 1478 on EGF-induced DMR response. The IC_{50} was calculated from the dose-response curve using Prism 4.0 software. **(C)** The VEGFR inhibitor, SU 1498 (10 μM), has no effect on the DMR response induced by EGF (100 ng/mL). (adapted from reference 61 with copyright permission.)*

suggest that the EGF-induced DMR signals are dependent on EGFR tyrosine kinase activity and are specific to EGFR activation in SCCHN cells. The EGF-induced DMR signal in UPCI-37B SCCHN cells is mediated primarily by the PI3K pathway but not MAPK pathways.

3.4.2.3. Evaluation of the use of EGF-responsive DMR signal profiles for HTS to identify EGFR pathway inhibitors. EGFR is frequently activated in SCCHN, lung, and other cancer types and is a validated cancer target. The potential use of the Epic biosensor for the discovery of new EGFR signaling pathway targeting agents was evaluated by testing the effect of two known small

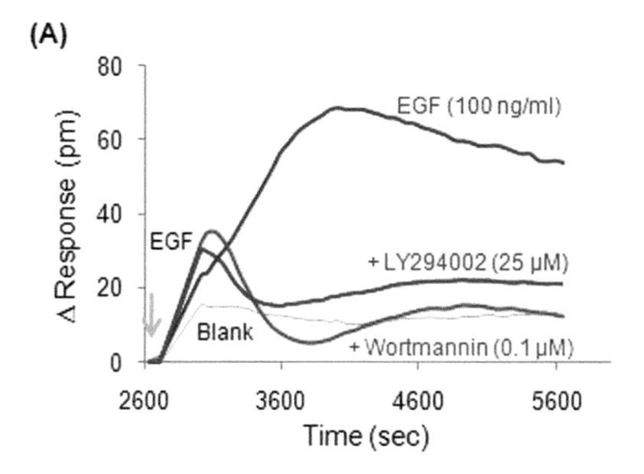

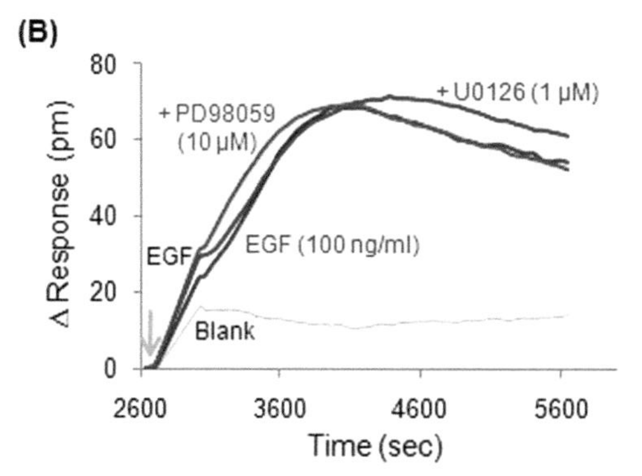

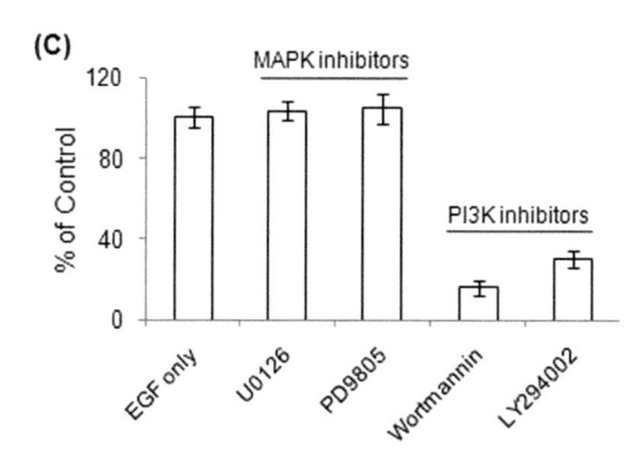

FIGURE 18.8:* *PI3K is a critical mediator of EGF-induced DMR in SCCHN cells. (A) The PI3K kinase inhibitors, LY 294002 (25 μM) and wortmannin (0.1 μM), significantly reduce the EGF-induced DMR response. (B) MEK1/MEK2 inhibitors U0126 (1 μM) and PD98059 (10 μM) do not block the EGF-induced DMR signal. (C) Normalized data. PI3K inhibitors, but not MAPK inhibitors, block the EGF-induced DMR signal in SCCHN cells. (adapted from reference 61 with copyright permission.)

molecule EGFR kinase inhibitors, gefitinib (Iressa) and erlotinib (Tarceva), on the EGF-induced DMR response. Treatment of cells with gefitinib or erlotinib showed significant attenuation of the DMR signal (Figure 18.9A, B). The inhibitory effects were dose dependent, with an estimated IC_{50} value of 1.42 μM and 0.57 μM for gefitinib and erlotinib, respectively (Figure 18.9C). The performance of the assay for HTS was further evaluated by estimating Z′ factors and signal-to-background ratios, which indicated a robust assay for HTS [61, 62]. These results demonstrate that the Epic biosensor can be used as an effective tool to monitor the efficacy of EGF-targeted therapy and as a novel approach to discover new therapeutic agents targeting abnormal pathways involved in human diseases (Figure 18.10).

4. SUMMARY

Label-free biosensor technologies offer a number of distinct advantages over label-dependent detection technologies. With direct detection principles, label-free assays preclude the need for labeling or modifying the molecules of interest, thus enabling the study of bimolecular interactions without potential interference from probe labeling. Compound fluorescence interference, a common problem in fluorescence label-based HTS assays, can be diminished with the label-free technology. Additionally, the ability to detect integrated cellular responses from endogenous receptors eliminates potential pitfalls of altering cellular biology with overexpressed cellular systems. The ability to detect cellular response with native cells by label-free technologies greatly broadens their applications in diverse cell types, including primary cells and stem cells. Furthermore, the pathway-unbiased detection by label-free technologies may enable the detection of an integrated cellular response from multiple signaling pathways that otherwise may be missed with the use of specific reporter-based assays for single pathways.

Among the current label-free technologies, SPR- and ITC-based technologies have been used extensively for molecular interaction studies. They have not been employed routinely in primary HTS campaigns, partly because of limitations of throughput and cost. The increased availability of the microplate-based label-free technologies addresses some of these limitations and thus offers the possibility of employing these technologies for small molecule modulator discoveries through HTS. However, the current label-free technologies are still in the early stage of development. Further improvements in reducing the cost of biosensor plates and in instrument optimization and software support are expected to accelerate its widespread technology adoption. There is also an important need for enhanced understanding of the biological and pharmacologic meaning and significance of the integrated cellular signals detected with label-free technologies. Such an understanding will be essential to help unlock the promising potential of current label-free biosensor technologies for both basic research and drug discovery.

ACKNOWLEDGEMENTS

We thank Mary Puckett for her critical reading and editing of the manuscript. Y. D. is a recipient of Emory University's SPORE in Head and Neck Cancer Career Development award (P50 CA128613), Emory URC award and Kennedy seed grant at Emory Winship Cancer Institute.

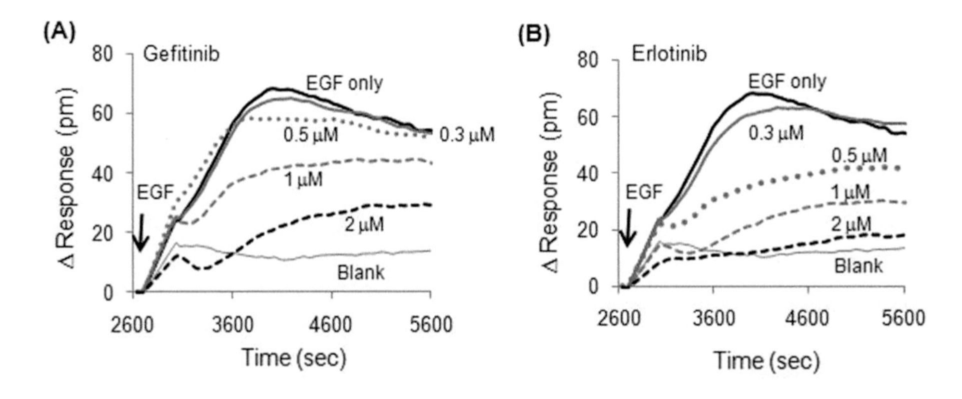

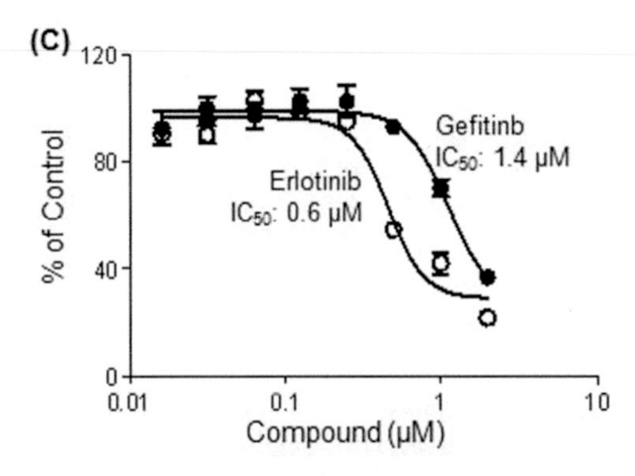

FIGURE 18.9: Using the established DMR signature to evaluate the efficacy of known EGFR targeting drugs, gefitinib and erlotinib. The effect of gefitinib (A) and erlotinib (B) on EGF-induced DMR response in UPCI-37B SCCHN cells is shown. (C) The dose-response effect of gefitinib and erlotinib on the P-DMR signal. The inhibitory effects of gefitinib and erlotinib are normalized to the EGF vehicle control wells. IC_{50} values are calculated using Prism 4.0 software. (adapted from reference 61 with copyright permission.)

H. F. is a Georgia Research Alliance Distinguished Investigator and Georgia Cancer Coalition Distinguished Cancer Scholars and is supported by the National Institutes of Health (P01CA116676).

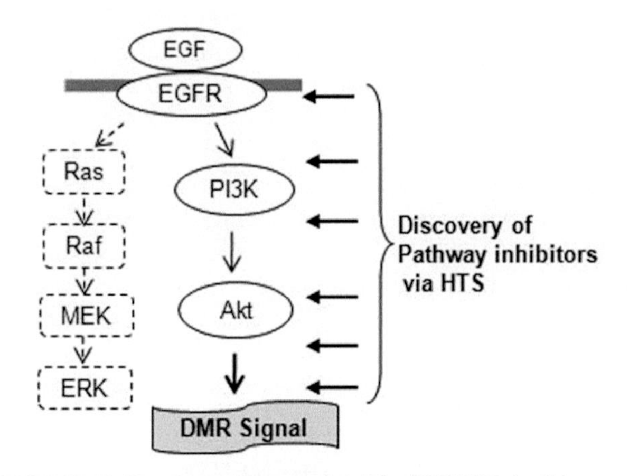

FIGURE 18.10: A working model. The PI3K/Akt (solid arrow), but not Ras/Raf/MEK/ERK (dotted arrow) pathway, mediates the EGF-induced DMR response in UPCI-37B SCCHN cells. The established cell-based optical signature assay enables a novel HTS approach for the discovery of inhibitors targeting the EGFR signaling pathway. (adapted from reference 61 with copyright permission.)

REFERENCES

1. Jason-Moller, L., Murphy, M., Bruno, J. (2006). Overview of Biacore systems and their applications. *Curr Protoc Protein Sci*, chap. 19, unit 19.*13*.
2. Gauglitz, G., and Proll, G. (2008). Strategies for label-free optical detection. *Adv Biochem Eng Biotechnol 109*, 395–432.
3. Rich, R. L., and Myszka, D. G. (2007). Higher-throughput, label-free, real-time molecular interaction analysis. *Anal Biochem 361*, 1–6.
4. Piliarik, M., Vaisocherova, H., and Homola, J. (2009). Surface plasmon resonance biosensing. *Methods Mol Biol 503*, 65–88.
5. Dodgson, K., Gedge, L., Murray, D. C., Coldwell, M. (2009). A 100K well screen for a muscarinic receptor using the Epic label-free system – a reflection on the benefits of the label-free approach to screening seven-transmembrane receptors. *J Recept Signal Transduct Res 29*, 163–172.
6. Smith, D. S., and Eremin, S. A. (2008). Fluorescence polarization immunoassays and related methods for simple, high-throughput screening of small molecules. *Anal Bioanal Chem 391*, 1499–1507.
7. Klostermeier, D., and Millar, D. P. (2001). Time-resolved fluorescence resonance energy transfer: a versatile tool for the analysis of nucleic acids. *Biopolymers 61*, 159–179.
8. Glickman, J. F., Wu, X., Mercuri, R., Illy, C., Bowen, B. R., He, Y., and Sills, M. (2002). A comparison of ALPHAScreen,

TR-FRET, and TRF as assay methods for FXR nuclear receptors. *J Biomol Screen* 7, 3–10.
9. Eglen, R. M. (2002). Enzyme fragment complementation: a flexible high throughput screening assay technology. *Assay Drug Dev Technol 1*, 97–104.
10. Cooper, M. A. (2006). Non-optical screening platforms: the next wave in label-free screening? *Drug Discov Today 11*, 1068–1074.
11. Cooper, M. A. (2006). Optical biosensors: where next and how soon? *Drug Discov Today 11*, 1061–1067.
12. Kenakin, T. P. (2009). Cellular assays as portals to seven-transmembrane receptor-based drug discovery. *Nat Rev Drug Discov 8*, 617–626.
13. Fang, Y. (2007). Non-invasive optical biosensor for probing cell signalling. *Sensors 4*, 2316–2329.
14. Fang, Y., and Ferrie, A. M. (2008). Label-free optical biosensor for ligand-directed functional selectivity acting on beta(2) adrenoceptor in living cells. *FEBS Lett 582*, 558–564.
15. Atienza, J. M., Yu, N., Kirstein, S. L., Xi, B., Wang, X., Xu, X., and Abassi, Y. A. (2006). Dynamic and label-free cell-based assays using the real-time cell electronic sensing system. *Assay Drug Dev Technol 4*, 597–607.
16. McGuinness, R. (2007). Impedance-based cellular assay technologies: recent advances, future promise. *Curr Opin Pharmacol* 7, 535–540.
17. Velusamy, V., Arshak, K., Korostynska, O., Oliwa, K., and Adley, C. An overview of foodborne pathogen detection: in the perspective of biosensors. *Biotechnol Adv 28*, 232–254.
18. Bosch, M. E., Sanchez, A. J., Rojas, F. S., and Ojeda, C. B. (2007). Optical chemical biosensors for high-throughput screening of drugs. *Comb Chem High Throughput Screen 10*, 413–432.
19. Shiau, A. K., Massari, M. E., and Ozbal, C. C. (2008). Back to basics: label-free technologies for small molecule screening. *Comb Chem High Throughput Screen 11*, 231–237.
20. Homola, J. (2003). Present and future of surface plasmon resonance biosensors. *Anal Bioanal Chem* 377, 528–539.
21. Hoa, X. D., Kirk, A. G., and Tabrizian, M. (2007). Towards integrated and sensitive surface plasmon resonance biosensors: a review of recent progress. *Biosens Bioelectron 23*, 151–160.
22. Crona, M., Furrer, E., Torrents, E., Edgell, D. R., and Sjoberg, B. M. (2010). Subunit and small-molecule interaction of ribonucleotide reductases via surface plasmon resonance biosensor analyses. *Protein Eng Des Sel 23*, 633–641.
23. Adamczyk, M., Moore, J. A., and Yu, Z. (2000). Application of surface plasmon resonance toward studies of low-molecular-weight antigen-antibody binding interactions. *Methods 20*, 319–328.
24. Shliom, O., Huang, M., Sachais, B., Kuo, A., Weisel, J. W., Nagaswami, C., Nassar, T., Bdeir, K., Hiss, E., Gawlak, S., Harris, S., Mazar, A., and Higazi, A. A. (2000). Novel interactions between urokinase and its receptor. *J Biol Chem 275*, 24304–24312.
25. Kawamoto, S. A., Thompson, A. D., Coleska, A., Nikolovska-Coleska, Z., Yi, H., and Wang, S. (2009). Analysis of the interaction of BCL9 with beta-catenin and development of fluorescence polarization and surface plasmon resonance binding assays for this interaction. *Biochemistry 48*, 9534–9541.
26. Huber, W. (2005). A new strategy for improved secondary screening and lead optimization using high-resolution SPR characterization of compound-target interactions. *J Mol Recognit 18*, 273–281.
27. Mattu, M., Di Giovine, P., Steinkuhler, C., De Francesco, R., Altamura, S., Cecchetti, O., Carfi, A., and Barbato, G. (2008). Development and optimization of a binding assay for histone deacetylase 4 using surface plasmon resonance. *Anal Biochem 377*, 267–269.
28. Cochran, S., Li, C. P., and Ferro, V. (2009). A surface plasmon resonance-based solution affinity assay for heparan sulfate-binding proteins. *Glycoconj J 26*, 577–587.
29. Mandine, E., Gofflo, D., Jean-Baptiste, V., Sarubbi, E., Touyer, G., Deprez, P., and Lesuisse, D. (2001). Src homology-2 domain binding assays by scintillation proximity and surface plasmon resonance. *J Mol Recognit 14*, 254–260.
30. Shumaker-Parry, J. S., Aebersold, R., and Campbell, C. T. (2004). Parallel, quantitative measurement of protein binding to a 120-element double-stranded DNA array in real time using surface plasmon resonance microscopy. *Anal Chem 76*, 2071–2082.
31. Buckle, M. (2001). Surface plasmon resonance applied to DNA-protein complexes. *Methods Mol Biol 148*, 535–546.
32. Davis, T. M., and Wilson, W. D. (2001). Surface plasmon resonance biosensor analysis of RNA-small molecule interactions. *Methods Enzymol 340*, 22–51.
33. Dickopf, S., Frank, M., Junker, H. D., Maier, S., Metz, G., Ottleben, H., Rau, H., Schellhaas, N., Schmidt, K., Sekul, R., Vanier, C., Vetter, D., Czech, J., Lorenz, M., Matter, H., Schudok, M., Schreuder, H., Will, D. W., and Nestler, H. P. (2004). Custom chemical microarray production and affinity fingerprinting for the S1 pocket of factor VIIa. *Anal Biochem 335*, 50–57.
34. Takeda, H., Fukumoto, A., Miura, A., Goshima, N., and Nomura, N. (2006). High-throughput kinase assay based on surface plasmon resonance suitable for native protein substrates. *Anal Biochem 357*, 262–271.
35. Takeda, H., Goshima, N., and Nomura, N. High-throughput kinase assay based on surface plasmon resonance. *Methods Mol Biol 627*, 131–145.
36. Boozer, C., Kim, G., Cong, S., Guan, H., and Londergan, T. (2006). Looking towards label-free biomolecular interaction analysis in a high-throughput format: a review of new surface plasmon resonance technologies. *Curr Opin Biotechnol 17*, 400–405.
37. Thoma, R., Loffler, B., Stihle, M., Huber, W., Ruf, A., and Hennig, M. (2003). Structural basis of proline-specific exopeptidase activity as observed in human dipeptidyl peptidase-IV. *Structure 11*, 947–959.
38. Giannetti, A. M., Koch, B. D., and Browner, M. F. (2008). Surface plasmon resonance based assay for the detection and characterization of promiscuous inhibitors. *J Med Chem 51*, 574–580.
39. Fan, X., White, I. M., Shopova, S. I., Zhu, H., Suter, J. D., and Sun, Y. (2008). Sensitive optical biosensors for unlabeled targets: a review. *Anal Chim Acta 620*, 8–26.
40. Cunningham, B. T., and Laing, L. (2006). Microplate-based, label-free detection of biomolecular interactions: applications in proteomics. *Expert Rev Proteomics 3*, 271–281.

41. Cunningham, B. T., Li, P., Schulz, S., Lin, B., Baird, C., Gerstenmaier, J., Genick, C., Wang, F., Fine, E., and Laing, L. (2004). Label-free assays on the BIND system. *J Biomol Screen 9*, 481–490.
42. Lin, B., Qiu, J., Gerstenmeier, J., Li, P., Pien, H., Pepper, J., and Cunningham, B. (2002). A label-free optical technique for detecting small molecule interactions. *Biosens Bioelectron 17*, 827–834.
43. Wu, M., Coblitz, B., Shikano, S., Long, S., Spieker, M., Frutos, A. G., Mukhopadhyay, S., and Li, M. (2006). Phosphospecific recognition by 14–3-3 proteins and antibodies monitored by a high throughput label-free optical biosensor. *FEBS Lett 580*, 5681–5689.
44. O'Malley, S. M., Xie, X., and Frutos, A. G. (2007). Label-free high-throughput functional lytic assays. *J Biomol Screen 12*, 117–125.
45. Fang, Y., Ferrie, A. M., Fontaine, N. H., Mauro, J., and Balakrishnan, J. (2006). Resonant waveguide grating biosensor for living cell sensing. *Biophys J 91*, 1925–1940.
46. Atienza, J. M., Yu, N., Wang, X., Xu, X., and Abassi, Y. (2006). Label-free and real-time cell-based kinase assay for screening selective and potent receptor tyrosine kinase inhibitors using microelectronic sensor array. *J Biomol Screen 11*, 634–643.
47. Fang, Y. (2006). Label-free cell-based assays with optical biosensors in drug discovery. *Assay Drug Dev Technol 4*, 583–595.
48. Giaever, I., and Keese, C. R. (1993). A morphological biosensor for mammalian cells. *Nature 366*, 591–592.
49. Xi, B., Yu, N., Wang, X., Xu, X., and Abassi, Y. A. (2008). The application of cell-based label-free technology in drug discovery. *Biotechnol J 3*, 484–495.
50. Fang, Y., Frutos, A. G., and Verklereen R. (2008). Label-free cell-based assays for GPCR screening. *Comb Chem High Throughput Screen 11*, 357–369.
51. Velasco-Garcia, M. N. (2009). Optical biosensors for probing at the cellular level: a review of recent progress and future prospects. *Semin Cell Dev Biol 20*, 27–33.
52. Ponti, J., Ceriotti, L., Munaro, B., Farina, M., Munari, A., Whelan, M., Colpo, P., Sabbioni, E., and Rossi, F. (2006). Comparison of impedance-based sensors for cell adhesion monitoring and in vitro methods for detecting cytotoxicity induced by chemicals. *Altern Lab Anim 34*, 515–525.
53. Xiao, C., Lachance, B., Sunahara, G., and Luong, J. H. (2002). Assessment of cytotoxicity using electric cell-substrate impedance sensing: concentration and time response function approach. *Anal Chem 74*, 5748–5753.
54. Wang, L., Zhu, J., Deng, C., Xing, W. L., and Cheng, J. (2008). An automatic and quantitative on-chip cell migration assay using self-assembled monolayers combined with real-time cellular impedance sensing. *Lab Chip 8*, 872–878.
55. Lee, P. H., Gao, A., van Staden, C., Ly, J., Salon, J., Xu, A., Fang, Y., and Verkleeren, R. (2008). Evaluation of dynamic mass redistribution technology for pharmacological studies of recombinant and endogenously expressed G protein-coupled receptors. *Assay Drug Dev Technol 6*, 83–94.
56. Fang, Y., Ferrie, A. M., Fontaine, N. H., and Yuen, P. K. (2005). Characteristics of dynamic mass redistribution of epidermal growth factor receptor signaling in living cells measured with label-free optical biosensors. *Anal Chem* 77, 5720–5725.
57. Chan, L. L., Gosangari, S. L., Watkin, K. L., and Cunningham, B. T. (2007). A label-free photonic crystal biosensor imaging method for detection of cancer cell cytotoxicity and proliferation. *Apoptosis 12*, 1061–1068.
58. Fang, Y., Ferrie, A. M., and Li, G. (2005). Probing cytoskeleton modulation by optical biosensors. *FEBS Lett 579*, 4175–4180.
59. Owens, R. M., Wang, C., You, J. A., Jiambutr, J., Xu, A. S., Marala, R. B., and Jin, M. M. (2009). Real-time quantitation of viral replication and inhibitor potency using a label-free optical biosensor. *J Recept Signal Transduct Res 29*, 195–201.
60. Fleming, M. R., and Kaczmarek, L. K. (2009). Use of optical biosensors to detect modulation of Slack potassium channels by G protein-coupled receptors. *J Recept Signal Transduct Res 29*, 173–181.
61. Du, Y., Li, Z., Li, L., Chen, Z. G., Sun, S. Y., Chen, P., Shin, D. M., Khuri, F. R., and Fu, H. (2009). Distinct growth factor-induced dynamic mass redistribution (DMR) profiles for monitoring oncogenic signaling pathways in various cancer cells. *J Recept Signal Transduct Res 29*, 182–194.
62. Zhang, J. H., Chung, T. D., Oldenburg, K. R. (1999). A simple statistical parameter for use in valuation and validation of high throughput screening assays. *J Biomol Screen 4*, 67–73.

CHAPTER 19

Basic Principles and Practices of Computer-Aided Drug Design

Chao-Yie Yang
Denzil Bernard
Shaomeng Wang

Technological advances in pharmaceutical research during the past few decades have transformed drug discovery and development from empirical trial-and-error methods to development of mechanism-based compounds, often referred to as targeted therapy. In the targeted therapy approach, the scientific endeavor is to find drugs that can act on specific targets, the majority of which are proteins. Functions, or more typically dysfunctions, of these targeted cellular proteins typically underlie diseases. The therapeutic concept assumes that binding of drugs to the target proteins can alter the proteins' function in the pathological states of cells. A favorable outcome of drug administration is to nullify or at least mitigate the disease. Development of a drug for treatment of a disease requires many resources as well as much time and collaborative effort of scientists from different disciplines, including chemistry, biology, and pharmacology. In the past decade, the computer has emerged as a powerful tool that can facilitate drug discovery and development. The initial step in the process calls for a correlation of the structures of known compounds with their activities in order to begin the search for new classes of molecules with the requisite activity. In recent years, more sophisticated computational molecular modeling methods have been developed and applied to the development of drugs, from initial discovery of hits and lead optimization to prediction of absorption, distribution, metabolism, and excretion (ADME) properties to toxicity (TOX) evaluation. In this chapter, we focus on the well-established computational methods applied to the identification and optimization of lead compounds. We first discuss the computational methodologies currently used in drug discovery. The strategies used for discovery and optimization of novel ligands by combining different computational tools are described next, and we then present case studies in which computational methods have been employed in drug design. We conclude by highlighting the current challenges and future perspectives of computer-aided drug design (CADD).

1. COMPUTATIONAL METHODOLOGIES

The ultimate goal of developing drugs that target a specific cellular protein is to find compounds or ligands that bind to the protein and modulate its function. The process typically progresses from the initial discovery of hits with weak to moderate potency against the target protein to the final optimized and highly potent compounds suitable for clinical trials. To achieve this goal, computational methods have been applied at various stages in the process that can be classified into four major areas:

1) Protein structure preparation: Identify and generate models of the target protein structure;
2) Ligand structure preparation: Identify and generate models of candidate ligands that may be available for experimental validation;
3) Docking: Identify and generate the structure of the complex formed by the protein and the ligand; and
4) Scoring: Estimate or predict the binding affinity of the ligand for the protein.

Next we discuss related computational methods in these four areas. Finally, protein flexibility, currently a very challenging issue in computational drug discovery, is discussed in some detail.

1.1. Protein Structure Preparation

Validation of the protein target is one of the most important steps in drug development. Comprehensive preliminary biochemical role and disease prevalence must be assessed before a protein target is identified [1]. Once such a protein is identified in a drug design project, structural information of the protein is sought. A common source for such information is the Protein Data Bank (PDB, www.rcsb.org) [2]. Currently, 75,594 structures have been deposited in the PDB and include 70,010 proteins, 2,294 nucleic acids, 3,266 protein/DNA complexes, and 24 other structures. The experimental methods employed to determine these structures are primarily x-ray crystallography, nuclear magnetic resonance (NMR) spectroscopy, and electron microscopy. Of these, x-ray crystallography provides the most accurate atomic resolution of the structures. NMR spectroscopy, however, can determine structures in solution [3]. Electron microscopy is widely used to determine a coarser model for large structural complexes that involve multiple protein units, such as viral capsids and ribosomes.

In a typical receptor-based structure design, the optimum starting point is a structure determined by x-ray crystallography and preferably that of the protein in complex with a ligand, because a protein may change its conformation upon binding to a ligand.

Although the PDB contains more than 50,000 protein structures, a number that increases each year, high levels of redundancy among the structures are found. If a less than 90% sequence identity is used as an indicator to select nonredundant proteins deposited in the PDB, only 5,290 structures are obtained from the database as of December 2008 [4]. This number is further limited by the fact that for various reasons, not all of the 5,290 proteins are suitable drug targets [5], namely disease related.

When no structural information of the protein is available in the PDB for a selected target, a homology modeling approach can be used to generate model structures for purposes of drug design. Two types of computational methods have been developed for protein structure modeling: template-based homology modeling [6, 7] and de novo protein structure prediction. The sequence of the target protein is aligned with homologous proteins with known structures, also called template proteins. Partial structures of the template proteins with identical or similar sequence to that of the target are patched together by satisfying spatial constraints. For sequence segments with unknown structures, model secondary structures are constructed, and the rotameric states of the side chain atoms are defined and used to augment the structure. Finally, the predicted structures are optimized using force field–based parameters, and their relative energetic stabilities are ranked. The de novo protein structure prediction [8] employs a much more sophisticated method to sample the protein backbone conformation [9]. The rotameric states of the side chain atoms are searched with a bias toward probabilistic observations from known protein structures. Intramolecular interactions in the predicted protein structure are assessed and include hydrogen bond and van der Waals interactions, solvation effects, and entropic penalties associated with the protein side chain conformations [10].

Of the two methods, comparative homology modeling, for example, with the program MODELLER [11–14], has been widely used with reasonable success [15]. The de novo protein structure method is the most computationally expensive approach, and its accuracy has improved recently because of the advances of computational technology [16]. Recently, a coarse model structure of the nuclear pore complex (NPC), a 50 MDa assembly of 456 proteins, was constructed using a collection of diverse, high-quality data as restraints during structure optimization [17]. This represents a state-of-the-art example of the insights that computer modeling can provide into the architecture of macromolecular complexes, which no single experimental method can provide.

1.2. Ligand Structure Preparation

Two file formats commonly used to describe a small molecule are the SMILES string [12, 13] and the SDF format from MDL [18]. Both of these formats encode the chemical composition and bonding information, and most commercial vendors provide their compound libraries in either or both formats. In most computer-aided drug design projects, three-dimensional (3D) structures of the small molecules are needed, so a computational conversion of the two-dimensional (2D) information to 3D structures is necessary. Many commercially available programs can perform such a task, and they include Sybyl [19], CORINA [20], CACTVS [21], MOE [22], and Omega from OpenEye [23]. The publicly accessible ZINC Web server (www.zinc.org) [24] has converted and made available more than one million commercially available compound libraries in 3D small molecule structures in the Mol2 format. The different conversion programs usually generate reasonably correct structures, but predefined rules are required to correctly obtain the tautomers of the small molecules [25]. Another important issue in 3D molecular structure generation is the assignment of protonation states in the small molecules. Common practice is to assume neutrality or to assign the protonation state of the small molecule according to a pH value of 7.0. In the small molecules libraries provided by the ZINC server, databases based on a different protonation state assumption are provided as well.

Several commercial programs also implement algorithms that compute the physicochemical properties of small molecule compounds and are designated as molecular descriptors. For example, MOE can calculate more than three hundred descriptors, including LogP, number of rotatable bonds, polar accessible surface areas, and number of hydrogen bond donors and acceptors, among others. Depending on one's purposes, molecular descriptors can be used in a filter function to exclude undesirable small molecules from a large database.

1.3. Docking

The fundamental concept of protein-ligand binding is depicted in Figure 19.1. Docking simulation is an *in silico* simulation evaluating and determining the structures of the protein and the ligand with necessary approximations.

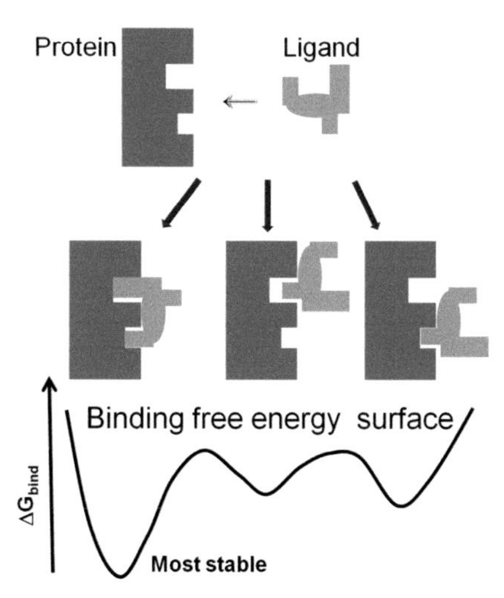

FIGURE 19.1: ***Protein-ligand binding process and the associated binding free energy surface.***

Computational examination of the structure of the complex formed by the receptor protein and the docked ligand can be broken down into two calculations, viz. docking and scoring. Docking serves to identify possible ligand conformations that can satisfactorily fit into the binding site of the target protein, and scoring is a means of evaluating the strength of the interaction between the target protein and the ligand based on the model protein-ligand structures. Computational methods required for docking include the sampling of ligand conformations and the translational/orientational positioning of the ligands in the protein binding site. Current implementations of the conformational sampling of ligands can be classified into two types of algorithms. The algorithm used in DOCK [26] and FlexX [27] breaks the ligand into chemical fragments connected by rotatable bonds, performs orientational sampling individually, and then combines the chemical fragments to generate the complete ligand. The second type of algorithm, used in Autodock [28], GOLD [29], and Glide [30, 31], performs angular sampling of the rotatable bonds of the ligands without invoking ligand fragmentation.

Exhaustive sampling of the angles of all rotatable bonds of the ligands and the translational and rotational freedom of ligands is, however, a computationally demanding task. A ligand with five rotatable bonds using a sampling step of 30 degrees for each rotatable bond, for instance, will require generation of 248,832 (12^5) conformations. If the translation/orientation sampling of the ligand at each of these conformations is included, several million conformations of each ligand will have to be generated for evaluation using such a systematic approach. To expedite the sampling process, a genetic algorithm approach has been incorporated in the Autodock [32, 33] and GOLD programs [34]. The genetic algorithm was inspired by evolutionary biology, and the guiding principles follow the mechanisms of mutation and crossover of genes, with the gene best adapted to the conditions surviving [35]. By analogy, each of the translational and orientational degrees of freedom and the angle of each rotatable bond is treated as a gene. Therefore, each ligand conformation is represented as a collection of these genes, or a chromosome. The mutation and crossover of the genes correspond to the change of variables in the ligand conformations randomly or by swapping. The adaptation of the chromosome to the environment (or the "fitness") is monitored by a scoring function that assesses the interaction – that is, the approximate binding free energy – between the protein and the ligand.

For docking of the ligand in the binding site, a box region around the site is first defined, and spaces occupied by the protein atoms are defined as excluded regions, that is, regions from which the ligand molecule is prohibited. Ligand placement in the protein binding site is then achieved in one of two ways. DOCK uses a maximum subgraph clique detection algorithm to identify matches between receptor cavities and ligand-heavy atoms [36], and chemical property matches between the receptor sites and ligands may also be considered. The ligand conformational search and fitting in this program and also in FlexX involve the placement of a primary fragment of the ligand – the anchor, or root – at the binding site and incremental addition of subsequent chemical fragments to the root via a "growth" method with local conformational sampling. If the growth on a certain path reaches a dead end, alternative paths are chosen for evaluation. Similarly, the algorithm used by FRED [37] approximates the ligand atoms by Gaussian shape functions, and the docking calculation proceeds by identifying shape complementarity between the ligands and the protein binding site [37]. A different approach used in Autodock and the GOLD program incorporates the translational and orientational variables together with the rotatable bond angles into the genes in a genetic algorithm. In this case, either a Monte Carlo search or Lamarckian genetic algorithm is used for variable sampling. The strength of the interactions between the protein and the ligand conformations obtained from the docking procedure are then assessed by some form of a scoring function, discussed in the following section.

1.4. Scoring

A very important component in receptor-based computational drug design is the function that evaluates the binding affinity between the receptor and the ligands, viz. the scoring function. Presumably, a ligand in solution can find different ways of binding to the target protein with varying strength of interaction between the two (Figure 19.1). The most stable protein-ligand complex structure yields the highest binding affinity and is often assumed to be the one observed in the crystal structure. In thermodynamic terms, the binding affinity is associated with the binding free energy, $\Delta G^0{}_{bind}$, between the protein and the ligand.

At equilibrium, the mathematical expression of $\Delta G^0{}_{bind}$ can be written as:

$$\Delta G^0_{bind} = G^0(\text{protein-ligand}) - G^0(\text{protein}) - G^0(\text{ligand}) \\ = -RT\ln(K_{eq}), \quad (19.1)$$

where $\Delta G^0{}_{bind}$ is the standard-state free energy change of the protein-ligand binding, G^0 is the standard-state free energy, R is the universal gas constant, T is the temperature, and K_{eq} is the equilibrium constant of the protein-ligand binding [38]. Various computational approaches have been developed to calculate the $\Delta G^0{}_{bind}$ for a given protein-ligand pair. A direct theoretical calculation based on Equation 19.1 requires calculation of the three standard-state free energies, the G^0 terms, of the protein-ligand complex, the protein, and the ligand. This then gives the absolute free energy of binding, $\Delta G^0{}_{bind}$, and the associated equilibrium constant, K_{eq} [39].

Such a direct calculation is not realizable in practice, however, because of the expensive computation required and the necessary approximation of parameters for the theoretical models. Instead, the relative binding free energy of a ligand to that of a reference ligand is usually calculated. The most accurate theoretical estimate for the relative protein-ligand binding free energy generally used is based on free energy perturbation theory [40]. In this calculation, a perturbation by mutating a chemical group between two partners is stipulated, and the resulting free energy difference of the protein with the initial- and final-state ligands is calculated. Less sophisticated but still computationally expensive approximations, such as the molecular-mechanics/Poisson-Boltzman Surface Area or Generalized Born Surface Area (MM-PBSA/MM-GBSA) [41] and linear interaction energy (LIE) [42] methods, have also been introduced. These expensive calculations are most suitable when more accurate estimates of the protein-ligand binding free energy are needed.

In many practical applications, less expensive calculations employing approximate models, namely scoring functions, are often applied to estimate the relative binding affinities between a protein and its ligands. In a typical computational virtual screening experiment, millions of compounds with multiple conformations for each compound are generated and evaluated against the target protein. This is computationally demanding, and the efficiency of the binding affinity evaluation becomes critical. When, by virtual screening, the number of candidate ligands is narrowed down to a more manageable size, perhaps about a few thousand, more elaborate scoring functions may be used to provide increasingly accurate assessment of the ligand potency.

Currently, three approaches have been adopted in scoring function development. They are the empirical [43, 44], force field [26, 30, 37, 45], and knowledge-based [29, 46–48] scoring functions. Empirical scoring functions involve a sum of several terms representing the physical interaction and properties between the protein and the ligand. These include van der Waals, electrostatic and hydrogen bond interactions, solvation/desolvation of the protein and the ligand due to complex formation, the ligand entropic penalty due to binding, and an offset term covering all unaccounted contributions. The coefficients of each term are regressed against a training set of protein-ligand structures with known binding affinities.

The force field–based scoring function is similar in form to the empirical scoring functions. However, some major differences are that each term is derived from the force field parameters used frequently in molecular dynamics simulations, and no regression is used to fit the data. The force field functions used in molecular docking give only the interaction energy between the protein and the ligand. They do not include solvation/desolvation effects or ligand entropic penalty upon binding, both of which should be considered in estimates of the binding free energy between a protein and a ligand. Additional approximations to these terms are needed in the force field–based scoring function. One method is to use solvation models to account for the solvation/desolvation effects of the protein and the ligand upon binding. Such solvation models include Poisson-Boltzman [49] and generalized Born [50] Surface Area models, which approximate the electrostatic contribution to solvation, whereas the nonpolar contribution is estimated by a surface accessible area model. Force field–based scoring functions have been implemented in DOCK, Autodock, and Glide.

Knowledge-based scoring functions employ a very different approach for assessment of the interaction between the protein and the ligand. The approach is built on the statistical observation of finding two atoms, one from the protein and one from the ligand, at a given distance found to be most probable from a set of protein-ligand crystal structures. For any given distance, a higher frequency of identifying two atoms than a predefined reference atom pair is correlated with a more favorable interaction between them. Through a Boltzman-like equation, the ratio of probabilistic observation of a protein-ligand atom pair and the reference atom pair is converted into a function that mimics a free energy (or potential energy) curve between these two atoms. Scores calculated based on the approximate free energy curves of all atom pairs between the protein and the ligand give an estimate of how likely it is that the designated protein-ligand structure will be found. Docking simulation using this statistical likelihood method as a scoring function has been found to identify complex structures very similar to the crystal structure [46]. Such scoring functions have also been shown to deliver good estimates of the binding affinity between proteins and the ligands in several studies [47, 48, 51]. Examples of this type of scoring function include GOLD [25], SMog96 [52], BLEEP [53, 54], PMF [48], Drugscore [46, 51], and M-Score [47].

Different scoring functions use different design philosophies to approximate the binding free energy efficiently. Empirical and force field–based scoring functions may provide a better evaluation of hydrogen bond interactions, being sensitive to the geometry of hydrogen bond donors and acceptors. Knowledge-based scoring functions developed based on statistical distribution between different atom/fragment pairs do not suffer from the regression biases employed in empirical scoring functions. The drawback of force field–based scoring functions, however, is the inaccuracy of the parameters used, which are mostly derived from high-level calculations of model systems in the gas phase and need to include empirical terms to be robust [55]. One strategy, consensus scoring [56], has been proposed to overcome the weaknesses of different types of scoring functions. In this scoring strategy, different types of scoring functions are used simultaneously to associate ligands with either scores or ranks. The consensus of scores from these scoring functions yields the final judgment of relative potencies of the ligands tested.

1.5. Protein Flexibility

Proteins undergo conformational changes to provide an optimized fit with bound ligands. Prediction of protein conformational change upon ligand binding is important yet cannot be accurately achieved even by state-of-the-art computational methods [57]. However, practical strategies have been developed and implemented in efforts to account for this fundamental phenomenon. Currently, there are at least two models to explain structural changes of proteins when they bind with partners [58]. One model proposes an induced-fit mechanism in which the protein undergoes conformational changes by adapting to the ligand when it docks into the protein's binding site. The other model proposes that the ligand-bound protein conformations exist in solution but constitute a far smaller conformational population than the more stable native conformational states [59]. During binding, the ligand may selectively bind to less common protein conformation that provides the highest binding free energy. Practical applications harnessing these two mechanisms have led to several computational methods that seek to account for protein flexibility in structure-based drug design.

In the induced-fit model, the ligand-bound protein conformations are defined only when ligands bind to the target site. In a simple scenario, it is possible to anticipate minor side chain changes of residues around the binding site in response to different ligands [60]. To incorporate this factor into drug design, Knegtel et al. [61] have combined multiple protein conformations, each with a different ligand, to generate average potential grids in DOCK to approximate the binding site flexibility for docking. In a different implementation, Ferrari et al. [62] changed the parameters of the repulsive potential in DOCK to permit greater latitude for ligands to fit in the binding site. Selective side chain flexibility has also been implemented in Autodock 4.0. In Glide, both side chain and backbone flexibilities of the proteins are implemented [63].

In the second model, the ligand-bound protein conformations may be observed in solution in the absence of the ligand. Frembgen-Kesner and Elcock [28] performed a 390-ns simulation on adenosine triphosphate (ATP)-bound p38 MAP kinase and identified the cryptic site next to the ATP binding site. They found that the binding sites of the five thousand selected protein conformations from the simulation were complementary to the inhibitor BIRB796. By docking, they were able to identify poses similar to those in the crystallographic structures. Although this is an appealing computational application, issues of whether sufficiently extensive protein conformational sampling is performed and how to identify representative protein conformations so as to predict ligand binding models are nontrivial. Further validation of this rationale and approach is needed.

Methods based on direct computational simulation of the protein, with or without ligands, to sample the protein conformations have been developed recently. They include the dynamic pharmacophore method [64] and the relaxed complex method [65]. In the dynamic pharmacophore method, the ligand free receptor was subjected to a lengthy simulation in which receptor conformations were sampled. Small molecule probes used as pharmacophores were then docked into the binding site to identify putative binding regions. The constructed model of the binding site based on these pharmacophores was then used in a small molecule library search. Inhibitors targeting MDM2 protein have been identified using this approach [66]. In the relaxed complex method, a receptor with a fragment in the binding site was subjected to lengthy simulations. The method sought to identify potential binding sites near the first fragment. Subsequent docking of small ligands to the nearby binding pocket was performed to identify potent fragments that could be tethered to the first fragment and improve the potency of the first fragment. This approach may be more suitable for ligand optimization where location of a potential nearby binding site unoccupied by a lead compound is known.

2. COMPUTATIONAL STRATEGIES IN DRUG DESIGN AND DISCOVERY

Identification of a suitable lead is one of the prominent bottlenecks in the drug discovery process [67]. Although high-throughput screening (HTS) has been extensively pursued as a means to overcome this, the role of CADD has become increasingly important as an efficient, rational, and economical approach. Recent findings concerning compound aggregation in HTS clearly emphasize a supplementary value provided by computational virtual screening

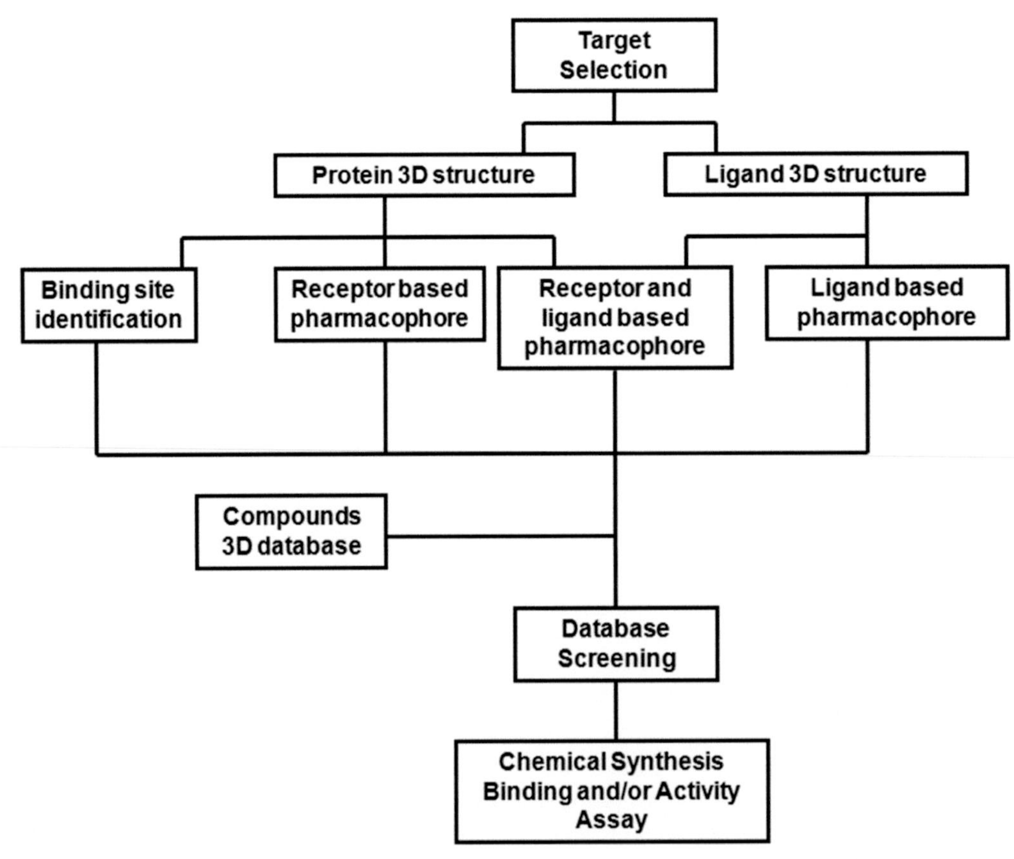

SCHEME 1: CADD flowchart.

(VS) [68]. In CADD, the lead can be identified via the simplified flowchart drawn in Scheme 1. The actual scheme can be tailored or modified depending on the available information. To carry out VS, either of two approaches can be adopted depending on the amount of information at hand: 1) If the 3D structure of the target protein binding site is known, structure-based VS by docking can be attempted; and 2) if compounds known to bind to the target are available, ligand-based VS is possible. When knowledge of both the target protein and the ligands binding to it is available, both approaches may be used in conjunction.

2.1. Receptor-Based Lead Discovery through VS

Current practice of structure-based VS integrates several computational methods to improve the identification of hits. The hierarchical scheme commonly adopted is depicted in Scheme 2. Large small molecule libraries from commercial providers or corporate sources typically store hundreds of thousands to millions of compounds, synthesized previously for many different purposes. The first step in screening preparation is to apply filters to the databases to remove undesirable compounds. Filters commonly used include the Lipinski "Rule of Five" [69] to select drug-like candidates, a filter to remove chemically active compounds, and a filter to select compounds with desirable physicochemical properties such as solubility. Applications of sets of filters can reduce the number of compounds in the library by as much as 30% to 40%. In a second step, target-specific filters can be applied. These types of filters can be built upon specific pharmacophore models derived from compounds known to be biologically active against the target. Alternatively, they can be filters for focused libraries such as those tailored to the kinase family of proteins. Similarity analysis of the compounds can also be used to increase the structural diversity of the tested compounds from the database. After the filtering steps, some number of compounds, typically on the order of ten thousand to one hundred thousand, will be subject to docking simulation against the target. This is a computationally demanding step, and program efficiency must be considered. After the docking simulation, the top few poses of each putative ligand can be used for rescoring using more accurate but computationally more expensive scoring functions. The scores so obtained can then be used for reranking, and the top 40% to 50% of the ligands can be subject to a second, more elaborate, and more time consuming docking simulation to obtain a more accurate prediction of the binding poses. Rescoring can be applied again to fine-tune the binding affinity prediction of these compounds, after which the binding poses of the top 5% to

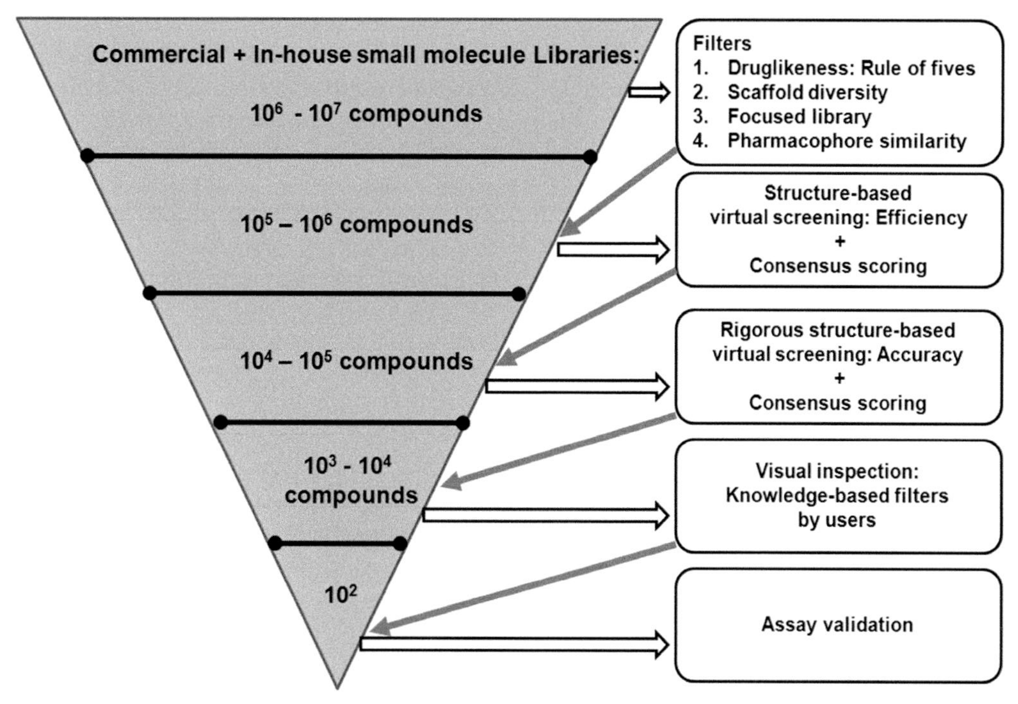

SCHEME 2: Hierarchical scheme used in virtual screening.

20% of the compounds may be subjected to visual inspection. At this stage, the user's knowledge of the target can be carefully applied during inspection to enhance the hit rate. Finally, a calculation of ADME/TOX properties [70] can be performed to improve the drug likeness of the candidate compounds that are to be chosen for subsequent validation by experimental assays. The recommended number of compounds for experimental validation by actual assay is typically between a few hundred to thousands, but the reduction in the number of compounds shown in Scheme 2 can be modified depending on the practitioner's budgetary capability.

2.2. Ligand-Based Lead Discovery

In the absence of the 3D structure of the target protein, drug design efforts can be assisted by ligand-based approaches. Three approaches are commonly used and include pharmacophore modeling, pseudoreceptor modeling, and quantitative structure-activity relationship (QSAR) modeling.

Pharmacophore models are based on a set of compounds with known activities and typically comprise a set of points in space defining the location of groups with specific properties and the distances between them, thus representing the groups contributing to the interaction with the target. The initial application of the pharmacophore model is generally for database mining, in which structures that satisfy the specified criteria are extracted for evaluation. Pseudoreceptor models, on the one hand, define a model of the binding site for screening but have not found much application because, at best, they are poor representations and unreliable. QSAR models, on the other hand, involve the determination of a mathematical relationship between the biological activities of a set of molecules and their physicochemical properties. Ligand-based methods have been useful for assisting drug discovery in the absence of structural information for the target, and they may also be employed in conjunction with structure-based methods.

The application of the QSAR to drug design is based on an understanding of the relationship between the chemical structure or properties of a molecule and its biological activity. This relationship was first described in the nineteenth century with regard to the toxicity of alcohols, which was known to increase with a decrease in water solubility. Quantitative methods building upon the initial relationship developed by Hammett correlating electronic properties of organic acids with their reactivity were developed [71]. The classical Hammett equation derived for the substituted benzenes *m*-XC6H4Y and *p*-XC6H4Y is given by Equation 19.2,

$$\log(k/k_0) = \sigma\rho \tag{19.2}$$

where k_0 is the rate constant for X = H, k is the rate constant for group X, ρ is a constant for a given reaction, and σ is a constant characteristic of the group X.

The transition from a chemical reactivity equation to one relating with biological activity was made by Hansch in formulating the first 2D QSAR equation [72]. The Hammett σ parameter accounting for the electronic effect of substituents was found to relate to biological activity inadequately, but Hansch obtained a meaningful relation utilizing the compound's lipophilicity, expressed as the *n*-octanol-water coefficient *P*, in Equation 19.3,

$$\log(1/C) = a \log P + b, \qquad (19.3)$$

where C is the molar concentration of a compound that produces a standard biological response measured as ED_{50}, LD_{50}, and so forth. Hansch inferred that the drug molecules should possess optimal lipophilicity; if the lipophilicity is too low, the molecule will not partition into the cell membrane, but if it is too high, it will remain there and not reach the target. The role of lipophilicity is well accepted today as an important factor in determining the bioavailability of a drug. It was quickly realized that combinations of the Hammett and Hansch equations and other forms of statistical relationships could be applied to obtain a prediction of the biological activity based on the structural properties of compounds, establishing the field of QSAR.

In recent years, many different descriptors and analytical methods have been employed in QSAR studies. Commonly used descriptors represent the hydrophobic, electronic, and steric characteristics of a molecule. Whereas classical QSAR refers to the use of regression analysis and is generally applied to a series of congeneric molecules, the more popular 3D QSAR methods have used principal component analysis (PCA) or partial least squares methods (PLS) and are commonly used for the analysis of chemically diverse molecules.

QSAR

The development of a QSAR model depends primarily on having a series of compounds with known biological activities. Ideally, a range of compounds covering a broad chemical space and having a significantly broad range of activities is required. The choice of descriptors is the next important criterion. Determination of the correlation between the individual descriptors and the biological activity is performed next. Although models are classically linear, parabolic relationships may often be found to be more suitable in the field of drug design because of the existence of an optimal value for a descriptor.

In drug design, it is preferable to have a model that can also predict the 3D molecular structure required for activity, even in the absence of knowledge of the specific target. 3D QSAR methods therefore have found more application where the spatial orientation of molecules is utilized and may provide insight into which interaction sites in the binding pocket are most important for activity. The applicability of this approach to a structurally diverse set of molecules also makes it preferable to 2D QSAR models. The first step in the generation of a 3D QSAR model, given the availability of a set of molecules with known activity, would be the alignment of these molecules. This relies on the fact that the biological activity is due to a common set of interactions with a particular protein target. Hence, the generation of a reliable pharmacophore model is a prerequisite to such analysis. The requirement for 3D alignment introduces the complexity of conformational flexibility of molecules. The common approach is to align all molecules to the most rigid molecule. If all molecules in the data set are flexible, then the lowest-energy conformers may often be utilized for alignment. Once alignment is achieved, a molecular field is computed on a grid composed of points in space around the molecule. These field descriptors typically capture spatial properties such as steric factors and electrostatic potentials. Fitting of the field points is then carried out by PCA or PLS techniques to obtain a predictive model. Models obtained from PLS analysis provide information on how activity correlates with a region in space and the characteristic of the group in that space, such as electron donating or electron withdrawing, similar to that from pharmacophore models. The first 3D QSAR method proposed by Cramer [73], which became popular, was termed comparative molecular field analysis (CoMFA). The method relies on the principle that differences in biological activity are related to changes in the shapes and strengths of electrostatic and steric fields surrounding the molecules. Thus, in CoMFA studies, the terms computed are generally steric in nature, based on the Lennard-Jones potential, and electrostatic, based on a Coulomb function. The molecular fields are evaluated by probes placed at regular points of a grid surrounding the molecules, where the grid spacing used and the probe types are particularly important. The resulting data are in the form of a matrix, with the rows representing individual molecules and the columns representing the energy value at each grid point. The biological activity is then correlated with the field values by Equation 19.4,

$$activity = C + \sum_{i=1}^{N} \sum_{j=1}^{P} c_{ij} \, S_{ij}, \qquad (19.4)$$

where c_{ij} is the coefficient for the column in the matrix S_{ij} that corresponds to placing probe group *j* at point *i*.

Instead of the grid-based method, a similarity index–based CoMFA method that utilizes steric, electrostatic, and hydrophobic potential has been developed, called CoMSIA [comparative molecular similarity indices analysis]. More recently, a four-dimensional (4D) QSAR method [74], in which an ensemble of conformations for each molecule serves as the fourth dimension, and five-dimensional (5D) QSAR [75], which considers hypotheses for changes in conformation of the receptor due to ligand binding, have been introduced. Validation of a QSAR model is generally performed by various statistical tests such as bootstrapping or leave-one-out.

A primary assumption in QSAR is that the molecules bind in a similar manner to the target site, which may or may not be true, and the method relies on alignment and

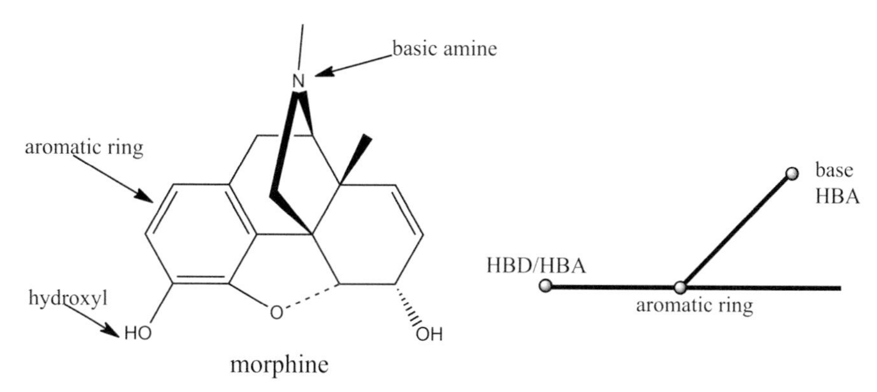

***FIGURE 19.2:** Opiate pharmacophore.*

superposition of molecules, which depend on the reference compound and conformers used. The reliability of the model also depends on the availability of an adequate number of compounds with biological data spanning at least 3 orders of magnitude. The method does not distinguish between the contribution of binding and transport properties in drug activity but takes into account the entire complex biological system without any information of the binding site. It thereby permits analysis of the SAR in the absence of a protein structure. It can be applied for both screening and lead optimization by predicting the activity of molecules yet to be synthesized and can be used along with structure-based methods.

PHARMACOPHORE MODEL

The term pharmacophore is thought to have been coined by Ehrlich in 1909 [76] to describe a drug molecule's molecular framework or substructure that is responsible for its biological activity. Since then, the term has evolved to the rigorous International Union of Pure and Applied Chemistry (IUPAC) definition: "A pharmacophore is the ensemble of steric and electronic features that are necessary to ensure the optimal supramolecular interactions with a specific biological target structure and to trigger (or to block) its biological response." Consequently, the pharmacophore does not represent a real molecule or a real association of functional groups but a purely abstract concept that accounts for the common molecular interaction capacities toward their target structure of a group of compounds. Simple functional groups such as guanidines or sulfonamides, or typical structural skeletons such as flavones or dihydropyridines, do not constitute a pharmacophore. The pharmacophore can be considered as the largest common denominator shared by a set of molecules with similar biological activities and recognized by the same site on the target protein. Simply put, the pharmacophore is a spatial arrangement of functional groups essential for biological activity. In a more generic sense, one may think of the lock-and-key analogy to enzyme substrate interactions or of a hand in a glove that accounts for steric complementarity as well as receptor flexibility and chirality.

The functional groups that impart the properties essential for interaction with a target protein are called the pharmacophoric descriptors. These include hydrogen bonding, hydrophobic, and electrostatic interaction sites defined by atoms, ring centers, and even virtual points. As an example, the important functional groups in the opiate analgesics are the basic amine, the phenol group, and the aromatic ring, shown in Figure 19.2. A simple pharmacophore is then defined using these groups.

In general, pharmacophoric features may be classified as: 1) polar – including hydrogen bond donors (HBD) and hydrogen bond acceptors (HBA); 2) ionic – including positively or negatively charged species; and 3) nonpolar – including aromatic and hydrophobic centers. These features do not rely on atom types; hence, different functionalities could represent the same feature. For instance, a basic amino group or a hydroxyl group could serve as a hydrogen bond donor. Similarly, the same functional group may have multiple properties, such as an amide group, which has both an HBA (carbonyl) and an HBD (NH group), or the benzene ring, which has both aromatic and hydrophobic character.

Simple 2D pharmacophores were initially described based on structure-activity relationships (SARs) before the advent of computer-aided methods, and one of the earliest such applications was in the classification of estrogenic molecules based on the 8.5 Å separation between the hydroxyl or keto groups determined from crystallographic structures of estrone and diethylstilbestrol (Figure 19.3) [77].

However, it was necessary to propose a 3D model for the difference in the interactions at the receptor of the *R*- and *S*-isomers of epinephrine, whose chiral center placed the hydroxyl group in a position interacting with the receptor in the *R*-isomer but not in the *S*-isomer (Figure 19.4) [78].

Insight into the nature of a protein binding site can thus be obtained by understanding the structure of ligands that

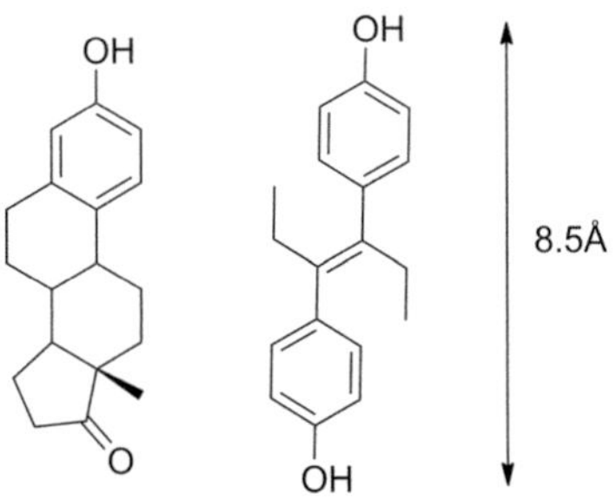

FIGURE 19.3: Comparison of estrone and diethylstilbestrol.

interact with it. Pharmacophores may be derived either from molecules analogous to a known ligand or from a diverse series of molecules known to have similar interactions with the target of interest or having the same biological effect because of interactions with the target protein. Alternatively, if the structure of the protein is known, a pharmacophore could be defined by analysis of the target site.

Pharmacophores have typically relied on simple geometrical descriptors describing the spatial relationship between important functional groups but could also use descriptors such as receptor-excluded volumes, which indicate regions occupied by the receptor and inaccessible to ligands.

An important consideration in the development of a pharmacophore model is the bioactive conformation of the known ligands. The most rigid compound usually serves as a template for comparison of other ligands. The general approach involves modeling of the molecules and creation of multiple conformations, followed by some alignment of the conformations that places important functional groups of each molecule in similar spatial orientations. Low-energy conformers of molecules are typically used in the process. A single definition may not include all known ligands, and the selection of a pharmacophore from several that may be generated by various alignments could be on the basis of known SAR or by statistical tests such as randomization. The lack of knowledge of the bioactive conformation and use of alignment techniques could result in incorrect models, and more recently, methods utilizing all ensemble conformers of the molecules in a reference set have been used to create conformationally sampled pharmacophores (CSP) that do not rely on alignment methods or require the assumption of a bioactive conformation [79, 80].

Once defined, a pharmacophore model can be used to design new molecules and to identify new lead compounds from a database of molecules through VS. Compounds obtained by VS could identify diverse classes of compounds with properties absent from the reference set of compounds. As the amount of available quantitative information increases, QSAR models can be developed using either traditional physicochemical or structural descriptors [81].

2.3. De Novo and Peptidomimetic Design

The ability to visualize both ligands and protein molecules, to describe the interactions between them, and to provide quantitative predictions of the activity of new molecules permits a rational approach to the discovery, chemical synthesis, and development of new therapeutic agents. The increasing number of resolved 3D structures of complexes of proteins and ligands provides structural insights that can be utilized in the rational design of new ligands from an entirely different perspective: the de novo design of ligands based on knowledge of the protein binding site [82, 83]. CADD permits the spontaneous creation of new molecules and an evaluation of their potential to bind to a target of interest. Such de novo design is often implemented based on fragment growth strategy, which involves the docking of initial core fragments into the target site followed by joining of additional fragments. Programs such as LUDI [84] and Glide [31] utilize this strategy in their de novo design algorithms. Visualization of molecules in the protein binding site also provides novel ideas pertaining

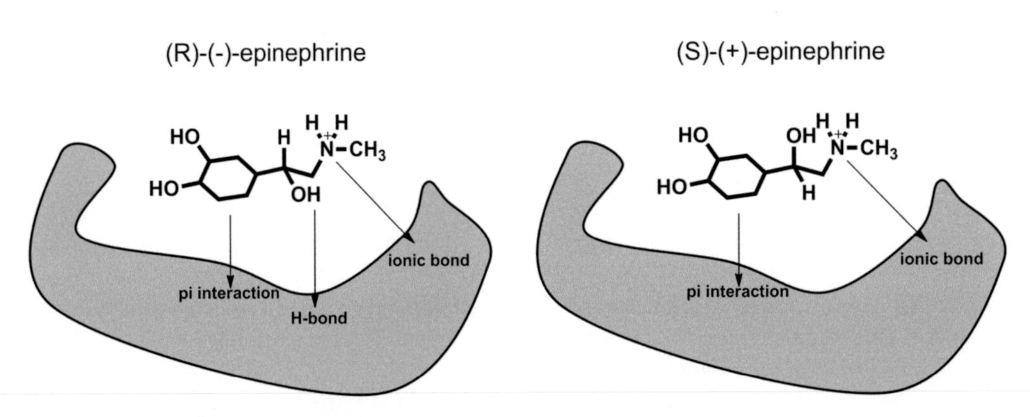

FIGURE 19.4: Difference in interactions of R(−)- and S(+)-epinephrine.

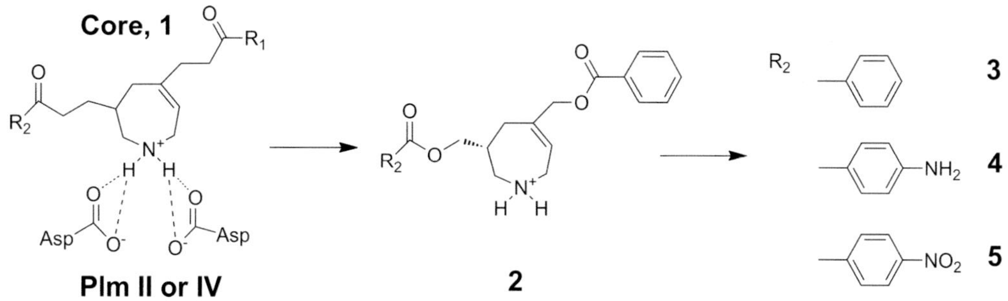

FIGURE 19.5: ***Plasmepsin II and IV inhibitors.***

to detailed chemical design of existing molecules in lead optimization.

De novo ligand design is a very challenging task. Its success depends highly on whether reasonable and accurate binding models of the designed compounds and the target protein can be obtained, on the synthetic feasibility of the designed molecules, and on the availability of robust biochemical assays with which to validate the hypothesis of the design. In this approach, iterative communication between computer modelers and chemists is necessary to generate potential synthetic modifications of the lead template. Docking of the compounds to the target protein is pursued to generate binding models. When no structural information is available for the template, a series of analogs based on the core scaffold can be used to produce and validate the hypothetical binding models.

A completely different approach to de novo design is peptidomimetic design [85]. The goal is to improve the potency between a known peptide template and the target protein so that the resulting compound has therapeutic value. In this strategy, the starting ligand template is a peptide segment that binds to the target protein binding site with known affinity. If the peptide segment is short and has a molecular weight of less than 500 to 600, it could be optimized and potentially become an ideal drug candidate. The optimization efforts are focused on introducing additional chemical components or restraints to the peptide template to improve potency, improving the cell permeability and the stability of the peptide in cells where the amide bond can be prone to catalytic cleavage.

3. APPLICATIONS OF COMPUTATIONAL METHODS IN DRUG DESIGN

Because drug discovery and development are multidisplinary in nature, it is perhaps not surprising that there is no commercially marketed drug that was purely designed using the computer [86]. Computational approaches, however, have played an important role in many successful drug design projects. In this section, we discuss several cases in which computational methods were applied to drug design.

3.1. Receptor-Based Virtual Screening

Thymidylate Synthase Thymidylate synthase (TS) is essential for DNA replication and has been an important target in oncology. It utilizes a co-factor to convert a substrate, deoxy-uridine 5′-monophosphate (dUMP), into deoxy-thymidine 5′-monophosphate (dTMP). Early design of inhibitors by Agouron scientists has led to potent inhibitors resembling a partial fragment of the co-factor. A virtual screening campaign conducted by DesJarlais et al. [87, 88] used a shape-matching algorithm in the early DOCK program and a library of compounds from the Fine Chemicals Directory to identify several TS inhibitors that were effective at high micromolar ranges. Follow-up study based on the initial hits identified phenolphthalein (K_i = 15 μM) bound to a site distinct from that utilized by either the co-factor or the substrate. The binding mode of the compound with TS was later verified by x-ray crystallography. Although the identified "hits" are clearly weak inhibitors of the target, this is one of the early examples in which *in silico* screening led to active compounds.

3.2. Virtual Screening and Optimization

Inhibitors of Plasmepsin II and IV The plasmepsins (Plm) are aspartic proteases responsible for hemoglobin degradation in the host cells in malaria. Most inhibitors developed to date target Plm II and are transition-state analogs of the substrates. This research was recently carried out by Luksch et al. [89]. As shown in Figure 19.5, docking simulation studies on Plm II using the crystal structure between Plm II and a different inhibitor, RS367, suggested that a 2,3,4,7-tetrahydro-1H-azepine core scaffold (compound **1**) should be an ideal starting point for structural modifications. A library of 2,083 esters was used as fragments that were combined with the core structure (compound **2**) and docked to the binding site using a combinatorial docking method with FlexX. Utilization of the library of esters in the docking simulation and evaluation served to identify putative candidates that could be synthesized and verified by bioassay. Analyses of the docked ligands led to the realization that an NO2- or NH2- substituted aryl fragment (compounds **5** and **4**) was preferred, and the three analogs shown in Figure 19.5 were synthesized for verification. The experimentally determined K_i values of these three compounds and their theoretical scores, predicted based on docking, were in agreement with expected increased potency when the unsubstituted phenyl group (compound **3**) was converted to the *p*-aminophenyl group (compound **4**). However, the prediction that the *p*-nitro substituted aryl compound (compound **5**) would be less potent than the unsubstituted compound as a result of steric hindrance in the docked pose, was found to be incorrect when compared with experimental K_i values. This unexpected result was attributed to a rotameric change of the residue in the protein identified from the molecular dynamics simulation.

3.3. 3D QSAR in the Design of Antifungal Agents

3D QSAR analysis was performed with a set of sixteen cyclopropanecarboxide antifungals to obtain insight into the nature of substituents suitable for design of new compounds. After alignment of the molecules using the common template skeleton, compound **6** in Figure 19.6, CoMFA steric and electrostatic fields were calculated using a spacing of 2 Å for the grid points, with a carbon probe atom having a charge of +1 and a van der Waals radius of 1.52 Å. Using a PLS method, a well correlated 3D-QSAR model was obtained with q^2 of 0.516 (cross-validated r^2) and r^2 of 0.800 (non–cross-validated r^2).

From the contour maps of the CoMFA interaction fields, it was found that an electron donating group at the 4 position on the R = Ph could improve the activity. Accordingly, a compound with R = p-CH_3Oph (compound **7**) was synthesized and found to have good antifungal activity. The model also indicated that larger groups may be tolerated at the 2 and the 3 positions of the phenyl ring but not at the 4 position. This is an example of the way in which QSAR can be used to gain an understanding of the structural properties that are important for the activity as well as for optimization of lead compounds.

A. **B.**

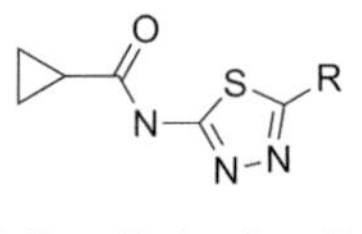

R=H, CH_3, C_2H_5, n-Pr, iso-Pr, n-Bu, Ph, *o*-CH_3Ph, *m*-CH_3Ph, *p*-ClPh, *o*-ClPh, *o*-FPh, *p*-NO_2Ph, Py, furan

6 7

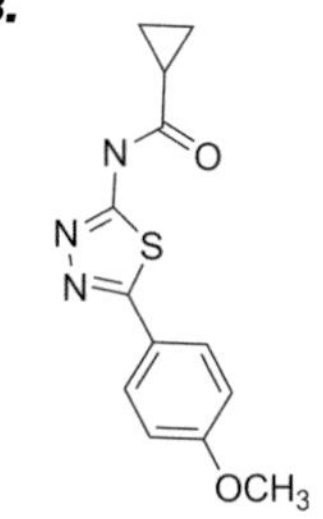

***FIGURE 19.6:* A. Compounds used in the generation of the 3D-QSAR model. B. Active compound synthesized based on the CoMFA contours.**

3.4. Pharmacophore-Based Screening

Small Molecule Inhibitors of the cMyc-Max Protein-Protein Interaction The literature is rich with examples of applications of pharmacophores in drug discovery. A recent example concerns the application of a pharmacophore model to obtain novel Myc-Max heterodimer disruptors [90]. A 3D pharmacophore was generated from a set of six known inhibitors of the c-Myc-Max heterodimer formation, the parent compound **8**, and its derivatives, such as **9** and **10** in Figure 19.7.

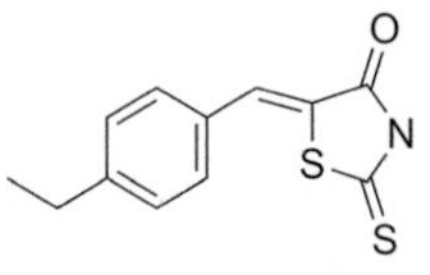

8 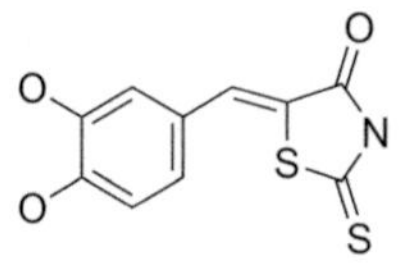 9 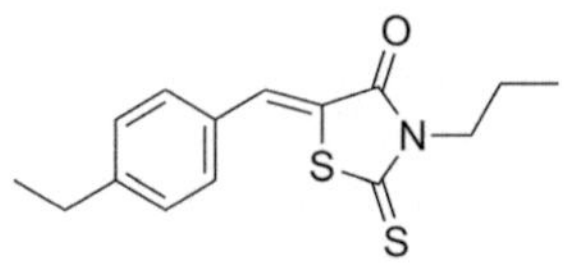 10

A. **B.**

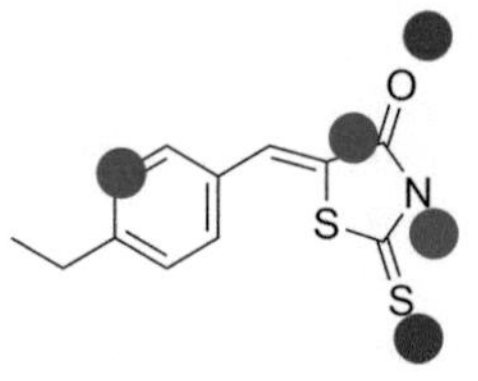

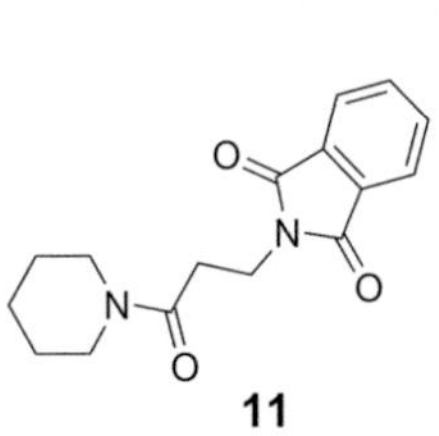

11 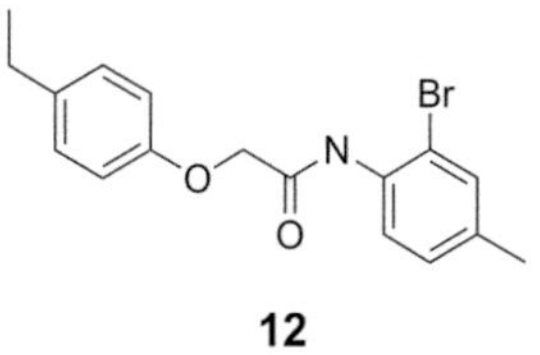 12 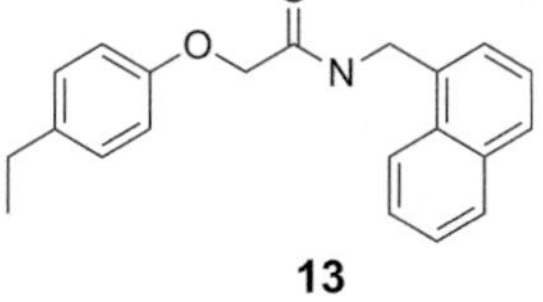 13

C.

***FIGURE 19.7:* A. Subset of compounds used in pharmacophore determination. B. Representation of the generated pharmacophore model is shown with one of the compounds used for model development (hydrophobic regions in green, hydrogen bond acceptors in red and hydrogen bond donors in blue). C. Subset of active compounds identified from database screening.**

A model of this pharmacophore was refined and validated by classifying a set of ten compounds with six active and four inactive compounds. All but one of the active compounds were clustered together along with one inactive compound. This hypothesis was used for screening the ZINC 7.0 database of about 5 million compounds, some 15,822 of which were regarded as hits. After applying a similarity screen and filtering the top hundred compounds through a screen for desirable ADME properties, the thirty best compounds were selected. Following this, another screen to select compounds that were unlikely to inhibit the metabolic enzyme cytochrome P450 3A4 (CYP3A4) was applied, and finally, nine compounds were tested for the ability to inhibit c-Myc and Max binding while not inhibiting CYP3A4. Just two compounds were found to be inactive, and the remaining seven (e.g., compounds **11**, **12**, and **13**), Figure 19.7, were found to possess affinities in the micromolar range.

HIV-1 Protease Inhibitors Human immunodeficiency virus (HIV)-1 protease, an important protein that is responsible for the maturation of the AIDS virus, is a homodimer. The binding site recognizes a four-amino acid motif and cleaves the peptide bond between the second and third amino acids. Early inhibitor design focused on developing peptidomimetic ligands that are bound to the target site and thus block the proteolytic function of the protease. In the 1990s, Dupont Merck initiated a rational design by performing a 3D ligand database search based on a pharmacophore hypothesis and identified compound **14** as a nonpeptidic HIV-1 protease inhibitor (Figure 19.8). One of the methoxyl groups in compound **14** was found to displace a key water molecule (flap water, Wat301) in the binding site. This led to several candidate scaffolds (compounds **15–17**) and to a series of cyclic urea scaffold compounds, including the clinical candidate DMP323 [91]. This was a classic example of computational methods being applied successfully to a series of compounds entering clinical trials.

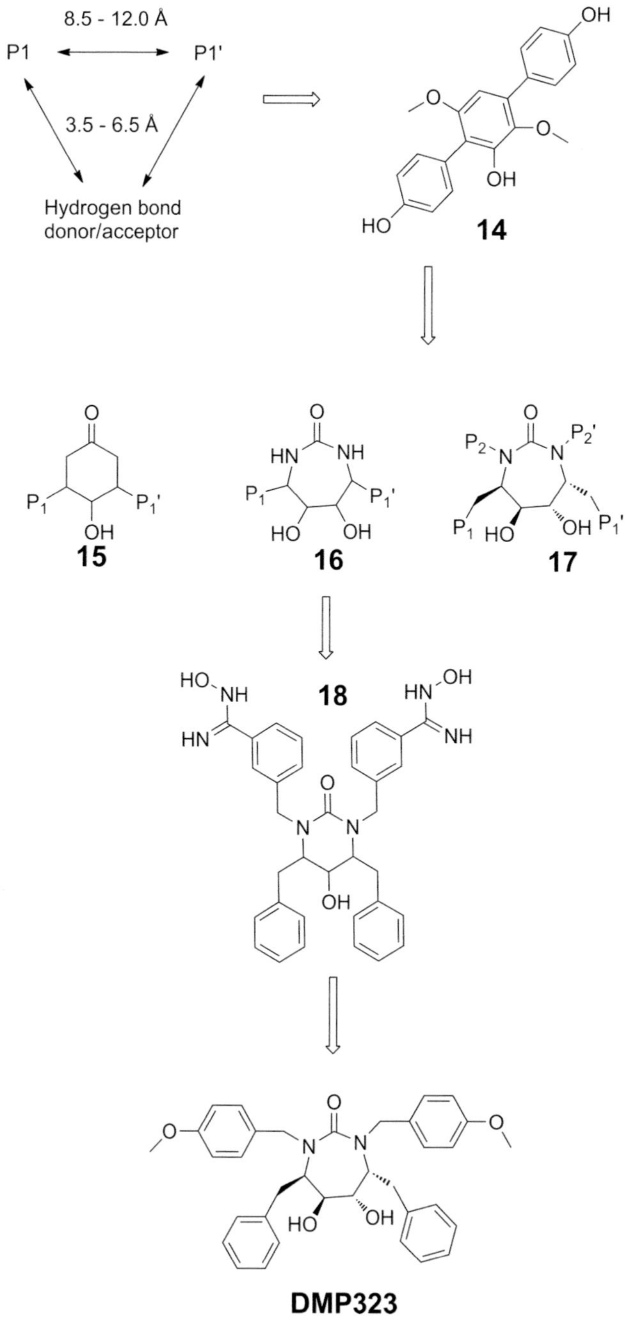

FIGURE 19.8: ***HIV-1 protease inhibitors.***

3.5. Combined Receptor- and Ligand-Based Virtual Screening

MDM2-p53 Interaction Inhibitors Although structure-based database screening forms the mainstay of VS, the availability of the biologically relevant conformer of the ligand, in the form of a structure of the protein-ligand complex, enables targeted screening employing both approaches in VS. Utilizing the available x-ray structures of the MDM2 protein complexed with several ligands, Lu et al. employed pharmacophore searching combined with structure-based screening for the discovery of novel small molecule inhibitors of the MDM2-p53 interaction [92].

In this study, a database of drug-like compounds from the National Cancer Institute (NCI) database [92–94] was first screened using a pharmacophore model based on the crystal structures of MDM2 in a complex with p53 peptide or with several small molecule inhibitors. Compounds that were identified from the pharmacophore search were further subjected to screening for drug-like properties such as hydrophobicity and the number of rotatable bonds, and non–drug-like compounds were discarded. The remaining compounds were docked into the MDM2 binding pocket using computational docking with the GOLD program [29, 95] to predict the binding pose and rank order of compounds. Compounds were then rescored, and the top

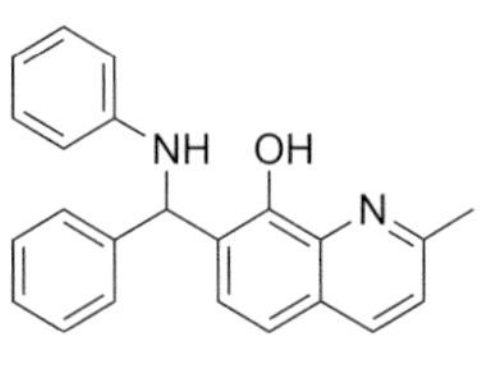

NH OH N

19 K_i = 120 nM

***FIGURE 19.9:** Compound obtained from combined pharmacophore and structure-based screening for inhibitors of MDM2 p53 interaction.*

two hundred compounds, ranked by Chemscore [43] and Xscore [44], were visually inspected to ensure that they mimic the key binding residues in p53. A total of sixty-seven compounds were obtained and tested in a competitive fluorescence polarization FP-based MDM2 binding assay, which yielded ten hits. The most potent MDM2 inhibitor identified from this study was compound **19** (Figure 19.9), with a K_i of 120 nM [92].

3.6. De novo Design

MDM2-p53 Interaction Inhibitors Ding and colleagues employed a de novo design strategy for the discovery of a class of nonpeptide MDM2 inhibitors with a spiro-oxindole core [96, 97]. Analysis of the crystal structure of the MDM2-p53 complex showed that four key hydrophobic residues, Phe19, Leu22, Trp23, and Leu26, in p53, mediate its interaction with MDM2 [98]. The spiro(oxindole-3,3′-pyrrolidine) structure was selected as the core for the design of new MDM2 inhibitors, with the oxindole moiety mimicking the key interaction between Trp23 in p53 and MDM2 in both hydrogen-bond formation and hydrophobic interactions with MDM2, and the spiropyrrolidine ring providing a rigid scaffold from which two hydrophobic groups could be projected to mimic the side chains of Phe19 and Leu26, as shown in Figure 19.10A. Candidate compounds were designed using a structure-based approach to mimic the p53 residues and were docked into the MDM2

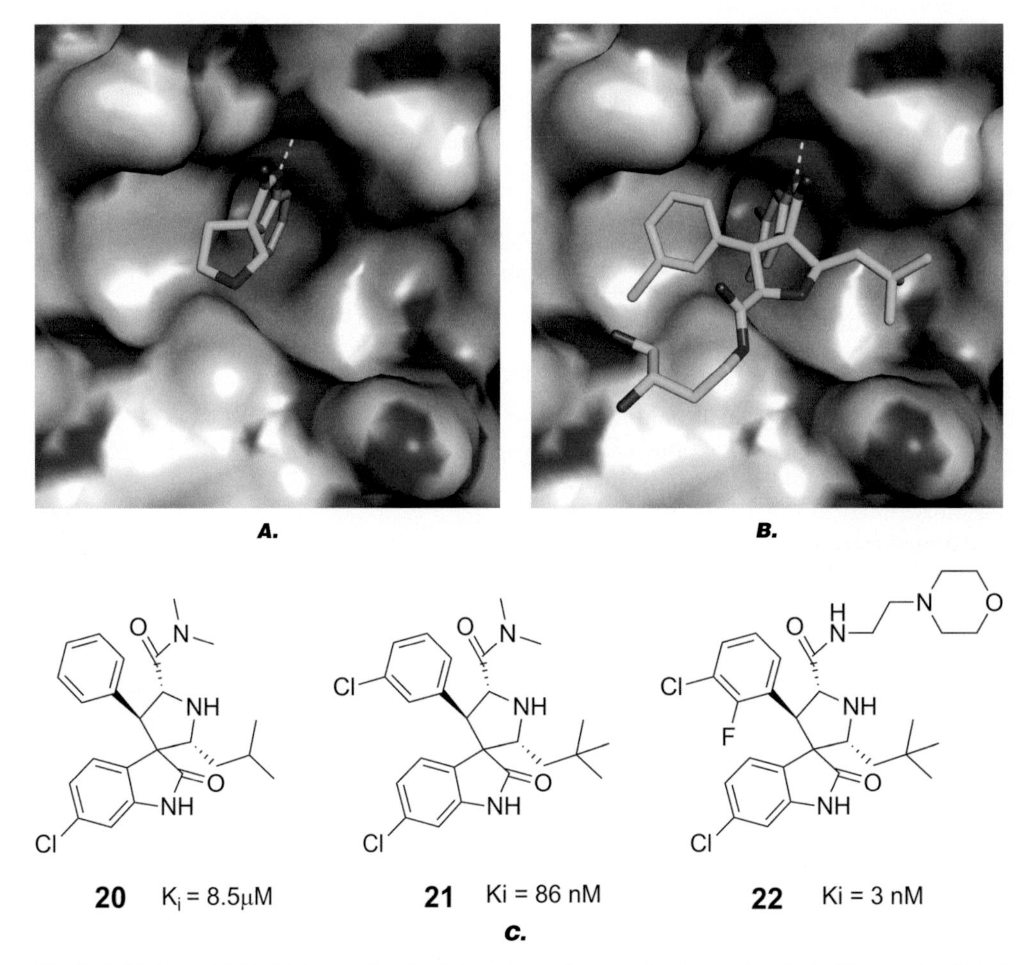

***FIGURE 19.10:** MDM2 p53 interaction inhibitors from de novo design and optimization. **A.** Predicted binding mode of spiro(oxindole-3,3′-pyrrolidine) to MDM2 by GOLD. Ligands are shown with carbons in cyan, nitrogen in blue, oxygen in red. The surface representation of MDM2 is shown with carbons in gray, nitrogen in blue, oxygen in red, and sulfur in yellow. Hydrogen bond is depicted by yellow dashed lines. This figure was generated by the program Pymol. **B.** MDM2-p53 inhibitors developed utilizing the spiro(oxindole-3,3′-pyrrolidine) scaffold.*

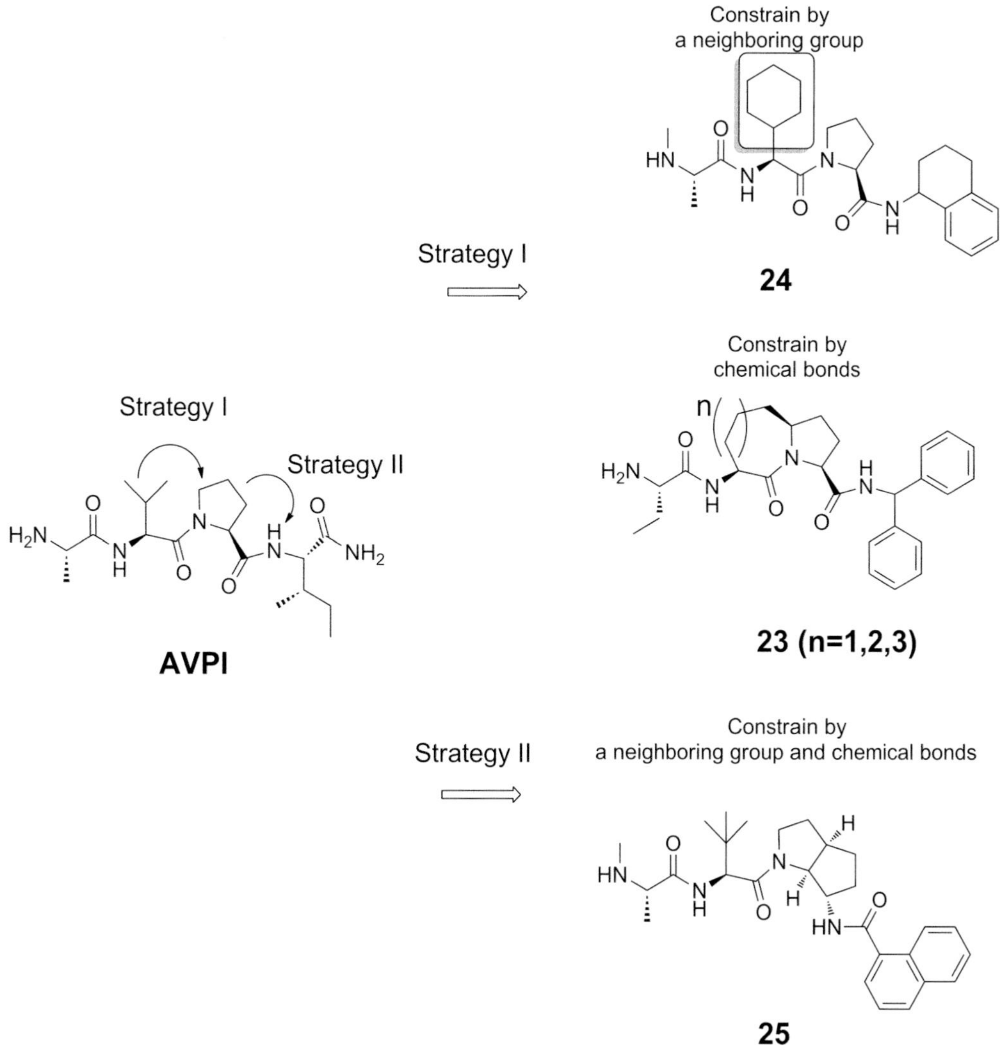

FIGURE 19.11: ***Design of potent, drug-like, small-molecule Smac mimetics as XIAP inhibitors.***

binding site. Compound **20** in Figure 19.10B was predicted to bind to MDM2 and, when tested, was found to have a K_i of 8.5 μM. This became the lead compound for subsequent optimization [96]. Modification of compound **20** yielded the spiro-oxindole compound **21**, which binds to MDM2 with K_i of 86 nM. Docking experiments predicted that compound **21** mimicked the Phe19, Trp23, and Leu26 residues in p53. Further modification to capture the interaction between Leu22 in p53 and MDM2 yielded compound **22**, MDM2 inhibitor-63, or MI-63 [97], which binds to MDM2 with a K_i of 3 nM, being one thousand times more potent than the natural p53 peptide.

3.7. Peptidomimetic Inhibitors Design

X-linked Inhibitor of Apoptosis Proteins (XIAP) One successful example of the peptidomimetic design approach is the development of potent, drug-like, small molecule mimetics that bind to the X-linked inhibitor of apoptosis protein (XIAP), a key apoptosis regulator [99]. XIAP blocks apoptosis by binding to caspase-3/-7 and caspase-9. In cells, the inhibitory functions of XIAP to these caspases are antagonized by the second mitochondria-derived activator of caspase (Smac) [100, 101]. X-ray crystallography [102] and NMR solution [103] structures showed that the four AVPI residues at the N-terminus of Smac bind to a surface groove in the BIR3 domain of XIAP. The AVPI Smac peptide was used as the starting point for the design of small molecule Smac mimetics to target the XIAP. The structure of the complex of AVPI and XIAP BIR3 suggested at least two possible routes for design and optimization, shown in Figure 19.11. One route was to form a bicyclic lactam system (compound **23**) by connecting the side chain of the valine in AVPI to the proline. This bicyclic lactam system could serve as an ideal conformational constrained template with a goal of reducing the peptidic nature of the resulting compounds. One key question was the size of the larger ring system in the bicyclic lactam system. Computational docking of the designed Smac-mimetic compounds showed that the ligands could ideally mimic the native AVPI conformation if a ring size of seven or eight were used [104] [105]. A large deviation was found if a ring size of six was adopted. In a subsequent study, it was also found that the docking pose of one of the [5, 8] lactam ring Smac mimetic compounds was predicted to be extremely close to the crystal structure, the two structures having a heavy atom root-mean-squared-deviation (RMSD) of only 0.7 Å [105]. The successful prediction of the docking poses in this case may be attributed to the fact that the residues around the binding site undergo similar conformational changes when XIAP BIR3 binds with the Smac mimetic compounds. Based on the crystal structure of the complex of XIAP BIR3 and the Smac peptide, a different strategy, applying a single non-natural amino acid substitution, has yielded compound **24** [106]. Another example in the design of Smac mimetics was reported by Genentech using the azabicyclooctane scaffold in compound **25** [107].

4. CHALLENGES AND FUTURE PERSPECTIVES

Computational methods have been widely used in the discovery of "hits" or initial lead compounds in many drug design projects. The success of computational methods in lead optimization has been limited. It is due to the fact that although current computational methods are capable of reproducing reasonably well the binding poses of ligands to their protein targets observed in the crystal structure [108, 109], the success of accurate "prediction" of ligand binding poses to proteins is still quite rare At least two major factors are attributed to the limited success of computational methods in lead optimization efforts. One, the conformational changes of the protein binding site, and the second, the poor performance of the current scoring functions in predicting ligand binding affinity to their target proteins [110].

Although drug discovery has focused on monomeric protein targets with relatively compact and small binding sites, an emerging area is to design small molecules to target the interface between proteins [111–113]. Protein-protein interaction interfaces are generally large and devoid of significant topography [112], but recent studies have shown that "hot spots" exist in the interface between proteins, which account for the bulk of the binding affinity [114, 115]. These relatively small hot spots in the protein-protein interaction interface make it possible to design potent drug-like small molecules, and some successful examples have been reported in recent years [111]. However, the challenging issue of protein flexibility will be more pronounced when designing inhibitors targeting protein-protein interactions. The research in protein flexibility and the improvement in the development of scoring functions will be a major focal point for improving drug design capability for computational scientists in the future.

ACKNOWLEDGMENT

We thank Dr. G. W. A. Milne and Ms. Karen Kreutzer for careful and critical reading of the manuscript and many useful suggestions.

APPENDIX: AVAILABLE WEB RESOURCES

1. www.ebi.ac.uk/Tools/clustalw2/: ClustalW2 is a general purpose program for performing multiple sequence alignment for DNA or proteins. This web site is hosted by the European Bioinformatics Institute (EBI) from the European Molecular Biology Laboratory (EMBL), where sequences of DNA or proteins can be uploaded for alignment.
2. www.rcsb.org Protein Data Bank: Collection of experimental determined protein, nucleic acid, protein-ligand, and nucleic acid-ligand structures.
3. www.salilab.org/modeller/: Modeller is a homology modeling program for comparative protein structure modeling by satisfaction of spatial restraints created by Dr. A. Sali's laboratory.
4. www.boinc.bakerlab.org/rosetta/: Rossetta@home, sets of programs for ab initio protein structure prediction created by Dr. D. Baker's laboratory.
5. www.zinc.org: Web site hosted by Dr. B. Shoichet's laboratory, where millions of commercial available small molecule libraries are provided. Some simple classification of the compounds is provided, and different protonation states of the compounds are prepared. The compounds are prepared in Mol2 format.
6. www.molecular-networks.com/software/corina: A commercial program to convert 2D chemical structures into 3D molecular structures.
7. pubchem.ncbi.nlm.nih.gov: a component of NIH's Molecular Libraries Roadmap Initiative, which hosts a collection of 48,019,951 compounds from 91 commercial and academic sources with 3,838,028 of them having associated bioassay information determined from 1,417 assays as of 2009.
8. www.drugbank.ca : A bioinformatics and cheminformatics resource that combines drug and drug target information. As of 2009, nearly 4,800 drug entries were curated. They include more than 1,350 FDA-approved small molecule drugs, 123 FDA-approved biotech (protein/peptide) drugs, 71 nutraceuticals, and more than 3,243 experimental drugs.
9. sw16.im.med.umich.edu/chmis-c/: A Comprehensive Herbal Medicine Information System for Cancer, a database for Chinese herbal medicine created by Dr. S. Wang's laboratory. This database integrates information on 203 cancer-related molecular targets, 527 anti-cancer herbal recipes that have been used for

the treatment of different types of cancer in clinics, 937 individual ingredients, and 9,366 small organic molecules isolated from herbal medicines.

10. Databases for protein-ligand structures and binding affinity information: www.pdbbind.org, www.pdbbind.org.cn, www.bindingmoad.org, www.csardock.org, and www.bindingdb.org.

REFERENCES

1. Smith, C. (2003). Drug target validation: Hitting the target. *Nature 422*, 341–347.
2. Berman, H. M., Westbrook, J., Feng, Z., Gilliland, G., Bhat, T. N., Weissig, H., Shindyalov, I. N., and Bourne, P. E. (2000). *The Protein Data Bank. Nucl Acids Res 28*, 235–242.
3. Eliezer, D. (2009). Biophysical characterization of intrinsically disordered proteins. *Curr Opin Struct Biol 19*, 23–30.
4. Target DB Statistics. Summary report. http://targetdb.pdb.org/statistics/TargetStatistics.html#redundancy.
5. Hopkins, A. L., and Groom, C. R. (2002). The druggable genome. *Nat Rev Drug Discov 1*, 727–730.
6. Zhang, Y., and Skolnick, J. (2004). Automated structure prediction of weakly homologous proteins on a genomic scale. *Proc Natl Acad Sci USA 101*, 7594–7599.
7. Zhang, Y. (2008). Progress and challenges in protein structure prediction. *Curr Opin Struct Biol 18*, 342–348.
8. Bonneau, R., Strauss, C. E. M., Rohl, C. A., Chivian, D., Bradley, P., Malmström, L., Robertson, T., and Baker, D. (2002). De novo prediction of three-dimensional structures for major protein families. *J Mol Biol 322*, 65–78.
9. Simons, K. T., Kooperberg, C., Huang, E., and Baker, D. (1997). Assembly of protein tertiary structures from fragments with similar local sequences using simulated annealing and bayesian scoring functions. *J Mol Biol 268*, 209–225.
10. Bonneau, R., and Baker, D. (2001). Ab Initio Protein Structure Prediction: Progress and Prospects. *Annu Rev Biophys Biomol Struct 30*, 173–189.
11. Fiser, A., Do, R. K. G., and Sali, A. (2000). Modeling of loops in protein structures. *Prot Sci 9*, 1753–1773.
12. Eswar, N., Marti-Renom, M. A., Webb, B., Madhusudhan, M. S., Eramian, D., Shen, M., Pieper, U., and Sali, A. (2007). Comparative protein structure modeling with MODELLER. In *Current Protocols in Protein Science*, Volume Supplement 50. (John Wiley & Sons, Inc.), pp. 2.9.1–2.9.31.
13. Sali, A., and Blundell, T. L. (1993). Comparative protein modelling by satisfaction of spatial restraints. *JMol Biol 234*, 779–815.
14. Marti-Renom, M. A., Stuart, A. C., Fiser, A., Sanchez, R., Melo, F., and Sali, A. (2000). Comparative protein structure modeling of genes and genomes. *Ann Rev Biophys Biomol Struct 29*, 291–325.
15. Kryshtafovych, A., and Fidelis, K. (2009). Protein structure prediction and model quality assessment. *Drug Disc Today 14*, 386–393.
16. Bradley, P., Misura, K. M. S., and Baker, D. (2005). Toward High-Resolution de Novo Structure Prediction for Small Proteins. *Science 309*, 1868–1871.
17. Alber, F., Dokudovskaya, S., Veenhoff, L. M., Zhang, W., Kipper, J., Devos, D., Suprapto, A., Karni-Schmidt, O., Williams, R., Chait, B. T., Rout, M. P., and Sali, A. (2007). Determining the architectures of macromolecular assemblies. *Nature 450*, 683–694.
18. Originally MDL Information Systems; now called Symyx Technologies.
19. Sybyl, a molecular modeling system, is supplied by Tripos, Inc., St. Louis, Miss.
20. Sadowski, J., and Gasteiger, J. (1993). From atoms and bonds to three-dimensional atomic coordinates: automatic model builders. *Chem Rev 93*, 2567–2581.
21. Ihlenfeldt, W. D., Takahashi, Y., Abe, H., and Sasaki, S. (1994). Computation and management of chemical properties in CACTVS: An extensible networked approach toward modularity and compatibility. *J Chem Inf Comp Sci 34*, 109–116.
22. Molecular operating environment from the Chemical Computing Group, Inc.
23. Omega; OpenEye Scientific Software: Santa Fe, N. M.
24. Irwin, J. J., and Shoichet, B. K. (2005). ZINC: A Free Database of Commercially Available Compounds for Virtual Screening. *J Chem Inf Model 45*, 177–182.
25. Oellien, F., Cramer, J., Beyer, C., Ihlenfeldt, W.-D., and Selzer, P. M. (2006). The impact of tautomer forms on pharmacophore-based virtual screening. *J Chem Inf Model 46*, 2342–2354.
26. Ewing, T. J. A., Makino, S., Skillman, A.G., and Kuntz, I. D. (2001). DOCK 4.0: Search strategies for automated molecular docking of flexible molecule databases. *J Comput Aided Mol Des 15*, 411–428.
27. Rarey, M., Kramer, B., Lengauer, T., and Klebe, G. (1996). A fast flexible docking method using an incremental construction algorithm. *J Mol Biol 261*, 470–489.
28. Frembgen-Kesner, T., and Elcock, A. H. (2006). Computational sampling of a cryptic drug binding site in a protein receptor: Explicit solvent molecular dynamics and inhibitor docking to p38 MAP Kinase. *J Mol Biol 359*, 202–214.
29. Jones, G., Willett, P., Glen, R. C., Leach, A. R., and Taylor, R. (1997). Development and validation of a genetic algorithm for flexible docking. *J Mol Biol 267*, 727–748.
30. Friesner, R. A., Banks, J. L., Murphy, R. B., Halgren, T. A., Klicic, J. J., Mainz, D. T., Repasky, M. P., Knoll, E. H., Shelley, M., Perry, J. K., Shaw, D. E., Francis, P., and Shenkin, P. S. (2004). Glide: A new approach for rapid, accurate docking and scoring. 1. Method and assessment of docking accuracy. *J Med Chem 47*, 1739–1749.
31. Halgren, T. A., Murphy, R. B., Friesner, R. A., Beard, H. S., Frye, L. L., Pollard, W. T., and Banks, J. L. (2004). Glide: A new approach for rapid, accurate docking and scoring. 2. Enrichment factors in database screening. *J Med Chem 47*, 1750–1759.
32. Belew, R. K., and Mitchell, M. (1996). *Adaptive Individuals in Evolving Populations: Models and Algorithms* (Reading, Mass.: Addison-Wesley).
33. Hart, W. E., Kammeyer, T. E., and Belew, R. K. (1994). In *Foundations of Genetic Algorithms III*, D. Whitley and M. Vose, eds. (San Francisco, Calif.: Morgan Kauffman).

34. Jones, G., Willett, P., and Glen, R. C. (1995). Molecular recognition of receptor sites using a genetic algorithm with a description of desolvation. *J Mol Biol 245*, 43–53.
35. Goldberg, D. E. (1989). *Genetic Algorithms in Search, Optimization and Machine Learning* (Reading, Mass.: Addison-Wesley).
36. Bemis, G. W., and Kuntz, I. D. (1992). A fast and efficient method for 2D and 3D molecular shape description. *J Comput Aided Mol Des 6*, 607–628.
37. Mark R. McGann, H. R. A., Anthony Nicholls, J. Andrew Grant, Frank K. Brown. (2003). Gaussian docking functions. *Biopolymers 68*, 76–90.
38. Hammes, G. (2000). *Thermodynamics and Kinetics for the Biological Sciences* (New York: John Wiley Interscience).
39. Woo, H.-J., and Roux, B. (2005). Calculation of absolute protein-ligand binding free energy from computer simulations. *Proc Natl Acad Sci USA 102*, 6825–6830.
40. Kollman, P. (1993). Free energy calculations: Applications to chemical and biochemical phenomena. *Chem Rev 93*, 2395–2417.
41. Kollman, P. A., Massova, I., Reyes, C., Kuhn, B., Huo, S., Chong, L., Lee, M., Lee, T., Duan, Y., Wang, W., Donini, O., Cieplak, P., Srinivasan, J., Case, D. A., and Cheatham, T. E. (2000). Calculating structures and free energies of complex molecules: combining molecular mechanics and continuum models. *Acc Chem Res 33*, 889–897.
42. Aqvist, J., Medina, C., and Samuelsson, J.-E. (1994). A new method for predicting binding affinity in computer-aided drug design. *Prot Eng.* 7, 385–391.
43. Eldridge, M. D., Murray, C. W., Auton, T. R., Paolini, G. V., and Mee, R. P. (1997). Empirical scoring functions: I. The development of a fast empirical scoring function to estimate the binding affinity of ligands in receptor complexes. *J Comput Aided Mol Des 11*, 425–445.
44. Wang, R., Lai, L., and Wang, S. (2002). Further development and validation of empirical scoring functions for structure-based binding affinity prediction. *J Comput Aided Mol Des 16*, 11–26.
45. Morris, G. M., Goodsell, D. S., Halliday, R. S., Huey, R., Hart, W. E., Belew, R. K., and Arthur J. Olson (1998). Automated docking using a Lamarckian genetic algorithm and an empirical binding free energy function. *J Comput Chem 19*, 1639–1662.
46. Gohlke, H., Hendlich, M., and Klebe, G. (2000). Knowledge-based scoring function to predict protein-ligand interactions. *J Mol Biol 295*, 337–356.
47. Yang, C.-Y., Wang, R., and Wang, S. (2006). M-Score: A Knowledge-Based Potential Scoring Function Accounting for Protein Atom Mobility. *J Med Chem 49*, 5903–5911.
48. Muegge, I., and Martin, Y. C. (1999). A General and Fast Scoring Function for Protein-Ligand Interactions: A Simplified Potential Approach. *J Med Chem 42*, 791–804.
49. Sharp, K., Nicholls, A., Fine, R., and Honig, B. (1991). Reconciling the magnitude of the microscopic and macroscopic hydrophobic effects. *Science 252*, 106–109.
50. Bashford, D., and Case, D. A. (2000). Generalized born models of macromolecular solvation effects. *Ann Rev Phys Chem 51*, 129–152.
51. Velec, H. F. G., Gohlke, H., and Klebe, G. (2005). DrugScoreCSD Knowledge-Based Scoring Function Derived from Small Molecule Crystal Data with Superior Recognition Rate of Near-Native Ligand Poses and Better Affinity Prediction. *J Med Chem 48*, 6296–6303.
52. DeWitte, R. S., and Shakhnovich, E. I. (1996). SMoG: de novo design method based on simple, fast, and accurate free energy estimates. 1. Methodology and supporting evidence. *J Am Chem Soc 118*, 11733–11744.
53. John, B. O. Mitchell, R. A. L., Alexander Alex, Janet M. Thornton. (1999). BLEEP: potential of mean force describing protein-ligand interactions: I. Generating potential. *J Comput Chem 20*, 1165–1176.
54. John, B. O. Mitchell, R. A. L., Alexander Alex, Mark J. Forster, Janet M. Thornton. (1999). BLEEP-potential of mean force describing protein-ligand interactions: II. Calculation of binding energies and comparison with experimental data. *J Comput Chem 20*, 1177–1185.
55. Friesner, R. A., Murphy, R. B., Repasky, M. P., Frye, L. L., Greenwood, J. R., Halgren, T. A., Sanschagrin, P. C., and Mainz, D. T. (2006). Extra precision glide: Docking and scoring incorporating a model of hydrophobic enclosure for protein-ligand complexes. *J Med Chem 49*, 6177–6196.
56. Charifson, P. S., Corkery, J. J., Murcko, M. A., and Walters, W. P. (1999). Consensus scoring: A method for obtaining improved hit rates from docking databases of three-dimensional structures into proteins. *J Med Chem 42*, 5100–5109.
57. Cozzini, P., Kellogg, G. E., Spyrakis, F., Abraham, D. J., Costantino, G., Emerson, A., Fanelli, F., Gohlke, H., Kuhn, L. A., Morris, G. M., Orozco, M., Pertinhez, T. A., Rizzi, M., and Sotriffer, C. A. (2008). Target flexibility: An emerging consideration in drug discovery and design. *J Med Chem 51*, 6237–6255.
58. Grünberg, R., Nilges, M., and Leckner, J. (2006). Flexibility and conformational entropy in protein-protein binding. *Structure 14*, 683–693.
59. Lee, G. M., and Craik, C. S. (2009). Trapping moving targets with small molecules. *Science 324*, 213–215.
60. Lange, O. F., Lakomek, N.-A., Fares, C., Schroder, G. F., Walter, K. F. A., Becker, S., Meiler, J., Grubmuller, H., Griesinger, C., and de Groot, B. L. (2008). Recognition dynamics up to microseconds revealed from an rdc-derived ubiquitin ensemble in solution. *Science 320*, 1471–1475.
61. Maria I. Zavodszky, L. A. K. (2005). Side-chain flexibility in protein-ligand binding: the minimal rotation hypothesis. *Prot Sci 14*, 1104–1114.
62. Knegtel, R. M. A., Kuntz, I. D., and Oshiro, C. M. (1997). Molecular docking to ensembles of protein structures. *J Mol Biol 266*, 424–440.
63. Ferrari, A. M., Wei, B. Q., Costantino, L., and Shoichet, B. K. (2004). Soft docking and multiple receptor conformations in virtual screening. *J Med Chem 47*, 5076–5084.
64. Sherman, W., Day, T., Jacobson, M. P., Friesner, R. A., and Farid, R. (2006). Novel Procedure for Modeling Ligand/receptor induced fit effects. *J Med Chem 49*, 534–553.
65. Carlson, H. A., Masukawa, K. M., Rubins, K., Bushman, F.D., Jorgensen, W.L., Lins, R.D., Briggs, J.M., and McCammon, J. A. (2000). Developing a dynamic pharmacophore model for HIV-1 integrase. *J Med Chem 43*, 2100–2114.

66. Lin, J.-H., Perryman, A. L., Schames, J. R., and McCammon, J. A. (2002). Computational drug design accommodating receptor flexibility: The relaxed complex scheme. *J Am Chem Soc 124*, 5632–5633.
67. Bowman, A. L., Nikolovska-Coleska, Z., Zhong, H., Wang, S., and Carlson, H. A. (2007). Small molecule inhibitors of the MDM2-p53 interaction discovered by ensemble-based receptor models. *J Am Chem Soc 129*, 12809–12814.
68. Klebe, G. (2006). Virtual ligand screening: strategies, perspectives and limitations. *Drug Dis Today 11*, 580–594.
69. Feng, B. Y., Simeonov, A., Jadhav, A., Babaoglu, K., Inglese, J., Shoichet, B. K., and Austin, C. P. (2007). A high-throughput screen for aggregation-based inhibition in a large compound library. *J Med Chem 50*, 2385–2390.
70. Lipinski, C. A., Lombardo, F., Dominy, B. W., and Feeney, P. J. (1997). Experimental and computational approaches to estimate solubility and permeability in drug discovery and development settings. *Adv Drug Del Rev 23*, 3–25.
71. Davis, A. M., and Riley, R. J. (2004). Predictive ADMET studies, the challenges and the opportunities. *Curr Opin Chem Biol 8*, 378–386.
72. Hammett, L. P. (1937). The effect of structure upon the reactions of organic compounds benzene derivatives. *J Am Chem Soc 59*, 96–103.
73. Hansch, C. (1969). A Quantitative Approach to Biochemical Structure-Activity Relationships. *Acc Chem Res 2*, 232–239.
74. Cramer, R. D., Patterson, D. E., and Bunce, J. D. (1988). Comparative molecular-field analysis (CoMFA). 1. Effect of shape on binding of steroids to carrier proteins. *J Am Chem Soc 110*, 5959–5967.
75. Hopfinger, A. J., Wang, S., Tokarski, J. S., Jin, B. Q., Albuquerque, M., Madhav, P. J., and Duraiswami, C. (1997). Construction of 3D-QSAR models using the 4D-QSAR analysis formalism. *J Am Chem Soc 119*, 10509–10524.
76. Vedani, A., and Dobler, M. (2002). 5D-QSAR: The key for simulating induced fit? *J Med Chem 45*, 2139–2149.
77. Ehrlich, P. (1909). On the present state of chemotherapy. *Ber Der Deut Chem Ges 42*, 17–47.
78. Schueler, F. W. (1946). Sex hormonal action and chemical constitution. *Science 103*, 221–223.
79. Beckett, A. H. (1959). Stereochemical factors in biological activity. *Fort der Arzneimittelforschung 1959*, 455–530.
80. Bernard, D., Coop, A., and MacKerell, A. D., Jr. (2003). 2D conformationally sampled pharmacophore: a ligand-based pharmacophore to differentiate delta opioid agonists from antagonists. *J Am Chem Soc 125*, 3101–3107.
81. Bernard, D., Coop, A., and Mackerell, A. D. (2005). Conformationally sampled pharmacophore for peptidic delta opioid ligands. *J Med Chem 48*, 7773–7780.
82. Bernard, D., Coop, A., and MacKerell, A. D., Jr. (2007). Quantitative conformationally sampled pharmacophore for delta opioid ligands: reevaluation of hydrophobic moieties essential for biological activity. *J Med Chem 50*, 1799–1809.
83. Shoichet, B. K. (2004). Virtual screening of chemical libraries. *Nature 432*, 862–865.
84. Taylor, R. D., Jewsbury, P. J., and Essex, J. W. (2002). A review of protein-small molecule docking methods. *J Comput Aided Mol Des 16*, 151–166.
85. Bohm, H. J. (1992). The computer program LUDI: a new method for the de novo design of enzyme inhibitors. *J Comput Aided Mol Des 6*, 61–78.
86. Wu, Y.-D., and Gellman, S. (2008). Peptidomimetics. *Acc Chem Res 41*, 1231–1232.
87. Jorgensen, W. L. (2004). The Many Roles of Computation in Drug Discovery. *Science 303*, 1813–1818.
88. DesJarlais, R. L., Seibel, G. L., Kuntz, I. D., Furth, P. S., Alvarez, J. C., Ortiz de Montellano, P. R., DeCamp, D. L., Babe, L. M., and Craik, C. S. (1990). Structure-based design of nonpeptide inhibitors specific for the human immunodeficiency virus 1 protease. *Proc Natl Acad Sci USA 87*, 6644–6648.
89. DesJarlais, R. L., Sheridan, R. P., Seibel, G. L., Dixon, J. S., Kuntz, I. D., and Venkataraghavan, R. (1988). Using shape complementarity as an initial screen in designing ligands for a receptor binding site of known three-dimensional structure. *J Med Chem 31*, 722–729.
90. Luksch, T., Chan, N. S., Brass, S., Sotriffer, C. A., Klebe, G., and Diederich, W. E. (2008). Computer-aided design and synthesis of nonpeptidic plasmepsin II and IV inhibitors. *ChemMedChem 3*, 1323–1336.
91. Mustata, G., Follis, A. V., Hammoudeh, D. I., Metallo, S. J., Wang, H., Prochownik, E. V., Lazo, J. S., and Bahar, I. (2009). Discovery of novel myc-max heterodimer disruptors with a three-dimensional pharmacophore model. *J Med Chem 52*, 1247–1250.
92. Kubinyi, H. (1998). Structure-based design of enzyme inhibitors and receptor ligands. *Curr Opin Drug Dis Dev 1*, 4–15.
93. Lu, Y., Nikolovska-Coleska, Z., Fang, X., Gao, W., Shangary, S., Qiu, S., Qin, D., and Wang, S. (2006). Discovery of a nanomolar inhibitor of the human murine double minute 2 (MDM2)-p53 interaction through an integrated, virtual database screening strategy. *J Med Chem 49*, 3759–3762.
94. Milne, G. W., Nicklaus, M. C., Driscoll, J. S., Wang, S., and Zaharevitz, D. (1994). National cancer institute drug information system 3D database. *J Chem Inf Comput Sci 34*, 1219–1224.
95. Voigt, J. H., Bienfait, B., Wang, S., and Nicklaus, M. C. (2001). Comparison of the NCI open database with seven large chemical structural databases. *J Chem Inf Comput Sci 41*, 702–712.
96. Verdonk, M. L., Cole, J. C., Hartshorn, M. J., Murray, C. W., and Taylor, R. D. (2003). Improved protein-ligand docking using GOLD. *Proteins 52*, 609–623.
97. Ding, K., Lu, Y., Nikolovska-Coleska, Z., Qiu, S., Ding, Y., Gao, W., Stuckey, J., Krajewski, K., Roller, P. P., Tomita, Y., Parrish, D. A., Deschamps, J. R., and Wang, S. (2005). Structure-based design of potent non-peptide MDM2 inhibitors. *J Am Chem Soc 127*, 10130–10131.
98. Ding, K., Lu, Y., Nikolovska-Coleska, Z., Wang, G., Qiu, S., Shangary, S., Gao, W., Qin, D., Stuckey, J., Krajewski, K., Roller, P. P., and Wang, S. (2006). Structure-based design of spiro-oxindoles as potent, specific small-molecule inhibitors of the MDM2-p53 interaction. *J Med Chem 49*, 3432–3435.
99. Kussie, P. H., Gorina, S., Marechal, V., Elenbaas, B., Moreau, J., Levine, A. J., and Pavletich, N. P. (1996). Structure of the MDM2 oncoprotein bound to the p53 tumor suppressor transactivation domain. *Science 274*, 948–953.

100. Holcik, M., Gibson, H., and Korneluk, R. (2001). XIAP: Apoptotic brake and promising therapeutic target. *Apoptosis 6*, 253–261.
101. Deveraux, Q. L., and Reed, J. C. (1999). IAP family proteins-suppressors of apoptosis. *Genes & Dev 13*, 239–252.
102. Salvesen, G. S., and Duckett, C. S. (2002). IAP proteins: blocking the road to death's door. *Nat Rev Mol Cell Biol 3*, 401–410.
103. Wu, G., Chai, J., Suber, T. L., Wu, J.-W., Du, C., Wang, X., and Shi, Y. (2000). Structural basis of IAP recognition by Smac/DIABLO. *Nature 408*, 1008–1012.
104. Liu, Z., Sun, C., Olejniczak, E. T., Meadows, R. P., Betz, S. F., Oost, T., Herrmann, J., Wu, J. C., and Fesik, S. W. (2000). Structural basis for binding of Smac/DIABLO to the XIAP BIR3 domain. *Nature 408*, 1004–1008.
105. Sun, H., Nikolovska-Coleska, Z., Yang, C.-Y., Xu, L., Tomita, Y., Krajewski, K., Roller, P. P., and Wang, S. (2004). Structure-based design, synthesis, and evaluation of conformationally constrained mimetics of the second mitochondria-derived activator of caspase that target the X-linked inhibitor of apoptosis protein/caspase-9 interaction site. *J Med Chem 47*, 4147–4150.
106. Sun, H., Stuckey, J. A., Nikolovska-Coleska, Z., Qin, D., Meagher, J. L., Qiu, S., Lu, J., Yang, C.-Y., Saito, N. G., and Wang, S. (2008). Structure-based design, synthesis, evaluation, and crystallographic studies of conformationally constrained smac mimetics as inhibitors of the X-linked inhibitor of apoptosis protein (XIAP). *J Med Chem 51*, 7169–7180.
107. Oost, T. K., Sun, C., Armstrong, R. C., Al-Assaad, A.-S., Betz, S. F., Deckwerth, T. L., Ding, H., Elmore, S. W., Meadows, R. P., Olejniczak, E. T., Oleksijew, A., Oltersdorf, T., Rosenberg, S. H., Shoemaker, A. R., Tomaselli, K. J., Zou, H., and Fesik, S. W. (2004). Discovery of potent antagonists of the antiapoptotic protein XIAP for the treatment of cancer. *J Med Chem 47*, 4417–4426.
108. Cohen, F., Alicke, B., Elliott, L. O., Flygare, J. A., Goncharov, T., Keteltas, S. F., Franklin, M. C., Frankovitz, S., Stephan, J.-P., Tsui, V., Vucic, D., Wong, H., and Fairbrother, W. J. (2009) Orally bioavailable antagonists of inhibitor of apoptosis proteins based on an azabicyclooctane scaffold. *J Med Chem 49*, 1723-1730.
109. Warren, G. L., Andrews, C. W., Capelli, A.-M., Clarke, B., LaLonde, J., Lambert, M. H., Lindvall, M., Nevins, N., Semus, S. F., Senger, S., Tedesco, G., Wall, I. D., Woolven, J. M., Peishoff, C. E., and Head, M. S. (2006). A critical assessment of docking programs and scoring functions. *J Med Chem 49*, 5912–5931.
110. Zentgraf, M., Steuber, H., Koch, C., La Motta, C., Sartini, S., Sotriffer, Christoph A., and Klebe, G. (2007). How reliable are current docking approaches for structure-based drug design? lessons from aldose reductase. *Angew Chem Int Ed 46*, 3575–3578.
111. Leach, A. R., Shoichet, B. K., and Peishoff, C. E. (2006). Prediction of protein-ligand interactions. docking and scoring: successes and gaps. *J Med Chem 49*, 5851–5855.
112. Wells, J. A., and McClendon, C. L. (2007). Reaching for high-hanging fruit in drug discovery at protein-protein interfaces. *Nature 450*, 1001–1009.
113. Yang, C.-Y., and Wang, S. (2006). Recent advances in design of small-molecule ligands to target protein-protein interactions. In *Ann Rep Comput Chem*, Volume 2, C. S. David, ed. (Elsevier), pp. 197–219.
114. Fry, David C. (2006). Protein-protein interactions as targets for small molecule drug discovery. *Pept Sci 84*, 535–552.
115. Clackson, T., and Wells, J. (1995). A hot spot of binding energy in a hormone-receptor interface. *Science 267*, 383–386.
116. Wells, J. A. (1996). Binding in the growth hormone receptor complex. *Proc Natl Acad Sci USA 93*, 1–6.

CHAPTER 20

Computational Approach for Drug Target Identification

Honglin Li

Mingyue Zheng

Xiaofeng Liu

Hualiang Jiang

Drug research and development (R&D) is a comprehensive, expensive, and time-consuming enterprise, full of risk throughout the process [1]. In general, the pipeline for drug discovery is composed of three major steps: drug target identification and validation, lead compound discovery and optimization, and preclinical research (Figure 20.1). In the last decades, a lot of new technologies have been developed and applied in drug R&D to shorten the research cycle and reduce the expenses. Among them, computational approaches have revolutionized the pipeline of the discovery and development [2]. In the past forty years, computational technologies for drug R&D have been evolving very quickly, especially in recent decades with the unprecedented development of biology, biomedicine, and computer capabilities. In the postgenomic era, because of the dramatic increase of small molecule and biomacromolecule information, computational tools have been applied in almost every stage of drug R&D, greatly changing the strategy and pipeline for drug discovery [2]. Computational approaches span almost all stages in the discovery and development pipeline, from target identification to lead discovery and from lead optimization to preclinical or clinical trials (Figure 20.1).

As shown in Figure 20.1, the target identification and validation are the first two key stages in the drug discovery pipeline. By 2000, only about 500 drug targets had been reported [3, 4]. The completion of human genome project and numerous pathogen genomes unveiled that there are thirty thousand to forty thousand genes and at least the same number of proteins, and many of these proteins are potential targets for drug discovery. However, it is still a challenging task to identify and validate those druggable targets from thousands of candidate macromolecules [5].

Numerous technologies for identifying targets have recently been developed. Experimental approaches such as genomic and proteomic techniques are the major tools for target identification [6, 7]. However, these methods have been proven inefficient in target discovery because of the laborious and time-consuming procedures. In addition, it is difficult to get relatively clear information related to drug targets from the enormous amount of information produced by genomics, expression profiling, and proteomics [5]. As a complement to the experimental methods, a series of computational (in silico) tools have also been developed for target identification in the past two decades [6, 8, 9]. In general, they can be categorized into sequence-based and structure-based approaches. Sequence-based approaches contribute to the processes of the target identification by providing functional information about target candidates and positioning information to biological networks. Such methods include sequence alignment for gene selection, prioritization of protein families, gene and protein annotation, and expression data analysis for microarray or gene chip. Here, we only introduce a special structure-based method for target identification: reverse docking.

1. A BRIEF INTRODUCTION OF MOLECULAR DOCKING

1.1. The Principle of Molecular Docking

Molecular recognition (binding) plays a central role in biological systems. A wide variety of physiologic processes are the reflection of ligand binding with macromolecules, especially with proteins. The most common examples are the binding between enzymes and substrates, between hormones and receptors, between protein receptors and their signal-inducing ligands, between antigens and their antibodies, and so on [10]. A detailed understanding of molecular recognition mechanisms is of particular interest in drug discovery, because most drugs interact with protein

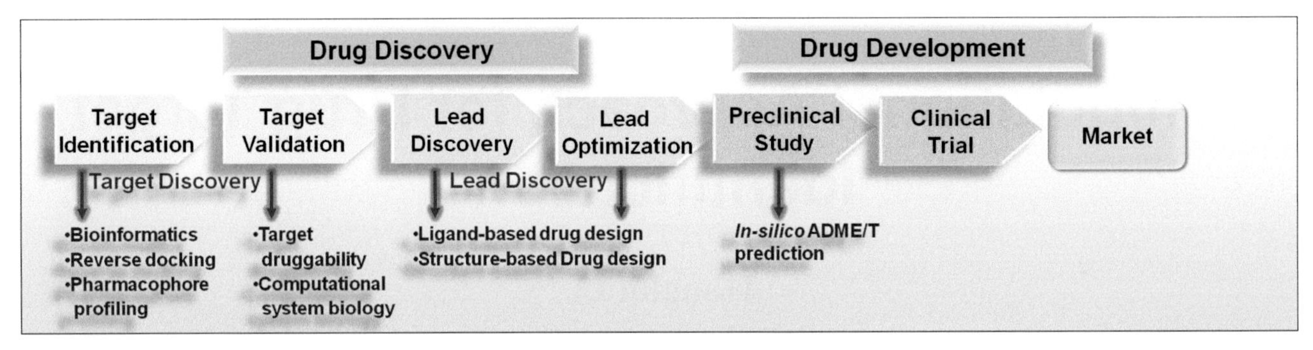

FIGURE 20.1: Computational approaches in line with the drug discovery pipeline.

targets such as enzymes or receptors [11]. The basic premise behind all approaches of rational drug design, either target structure-based drug design or small molecule structure-based drug design, is the theory of ligand-receptor recognition [12], the so-called lock-and-key principle [13]. Since Paul Ehrlich originated the concept (although he attributed it to Emil Fische in his Nobel Prize address), the lock-and-key principle has captivated drug discovery scientists through much of the twentieth century [14]. In 1982, Kuntz et al. [15] developed the first docking algorithm (DOCK), with which one can, for the first time, simulate ligand-receptor binding or interaction directly based on the lock-and-key principle.

The principle of molecular docking is straightforward. It is just a process to place a small molecule ligand into the binding site of a protein receptor or other biological macromolecules, optimize the relative orientation and conformation for a ligand interacting with the receptor, and calculate the binding affinity between the ligand and receptor using appropriate computational methods (Figure 20.2).

However, a real docking simulation is not as easy as represented in Figure 20.2. Obtaining the correct binding model of a ligand with a receptor is a core step of molecular docking. There are many factors to determine the binding pose of a ligand with its receptor, such as interaction components (e.g., hydrogen bonding, electrostatics, and hydrophobic interaction) and the flexibilities of both ligand and receptor. This implies that, during a docking simulation, the degrees of freedom to be searched include translational and rotational degrees of freedom of the ligand as a whole and the residues composed of the binding site of the receptor. To this end, a number of different optimization algorithms have been developed [16, 17]. Most molecular docking programs perform conformational optimization at the predefined binding sites, which can be obtained by comparing protein crystal structures with different ligands or by comparing proteins with similar functions. By conformation sampling of small molecules with three-dimensional (3D) structures and positioning them in the binding sites of target molecules, molecular docking methods optimize the orientation and conformation of a small molecule and search for binding poses with the most favorable ligand-protein interactions.

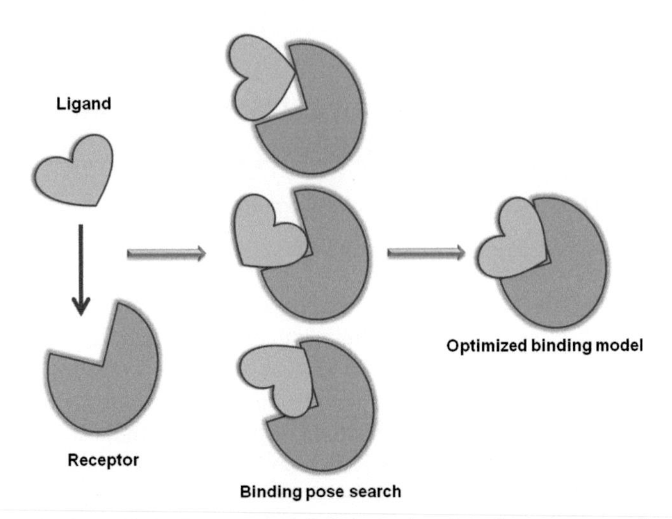

FIGURE 20.2: A cartoon representation for the process of molecular docking simulation.

Predicting the binding affinity between ligand and receptor is another important issue of docking, which is usually performed simultaneously with the docking simulation. Under the equilibrium condition, the noncovalent reversible reaction between receptor protein (P) and ligand (L) in solution can be expressed as:

$$[\mathrm{P}]_{aq} + [\mathrm{L}]_{aq} \rightleftharpoons [\mathrm{P'L'}]_{aq} \tag{20.1}$$

The receptor-ligand binding is determined by the variation of free energy in the binding process [17, 18],

$$\Delta G = -RT \ln K_A \, K_A = K_i^{-1} = \frac{[\mathrm{P'L'}]}{[\mathrm{P}][\mathrm{L}]}, \tag{20.2}$$

where T is the absolute temperature, R is the molar gas constant (8.3145 J $\cdot$ mol^{-1} $\cdot$ K^{-1}), K_{A} is the binding constant, and K_{i} is the dissociation constant of ligand-protein interaction [19]. Under the conditions of thermodynamic equilibrium, the binding free energy is usually estimated through separate determination of the enthalpy and entropy contributions to the complex stability:

$$\Delta G_{bind} = \Delta H - T\Delta S = G_{complex} - (G_{receptor} + G_{ligand}) \quad (20.3)$$

Free energy perturbation (FEP) and thermodynamic integration (TI) are the most accurate methods to estimate the binding free energy [20]. However, they are not suitable to drug design because of the time-consuming process. Unlike the accurate free energy calculation, docking programs need quick estimation of the binding affinity and usually only consider the complexed state explicitly. Some simplified scoring functions were proposed to approximately estimate the free energy. The scoring functions used in molecular docking can be classified into three types: force field–based, knowledge-based, and empirical scoring functions [10, 19]. Most scoring functions ignore or roughly estimate the entropy effect and only consider the ligand-receptor interaction energy in enthalpy change. The non-bond interaction $E_{\text{interaction}}$ of ligand and receptor consists of van de Waals (VDW) interaction E_{vdw}, electrostatic interaction E_{ele}, and hydrogen bonding (H-bond) interaction $E_{\text{H-bond}}$:

$$E_{\text{interaciton}} = E_{vdw} + E_{ele} + E_{H-bond} \quad (20.4)$$

In most cases, hydrogen bond (H-bond) interaction will be implicitly involved in electrostatic interaction energy, whereas the empirical functions, obtained by the method of fitting experimental data, have multiple polynomial expressions such as VDW, H-bond interaction, ionic interaction, hydrophobic or lipophilic interaction, and binding entropy effect.

1.2. The Application of Molecular Docking in Drug Discovery

In the last few decades, molecular docking has been developed as a useful tool for hit and lead discovery, called virtual screening [17]. In general, virtual screening is defined as researchers using computational methods (e.g., docking) in the in silico laboratory to evaluate virtual libraries (databases) against virtual receptors (targets) with the goal of speeding up drug discovery process. In essence, virtual screening is designed for searching large-scale hypothetical databases of chemical structures or virtual libraries by using computational analysis and for selecting a limited number of candidate molecules likely to be active against a chosen biological receptor [16, 21]. Recent promising advancements in virtual screening have demonstrated the efficiency of this approach in discovering leading (active) compounds [22–30]. Doman et al. [23] compared the performance of random high-throughput screening (HTS) with molecular docking-based virtual screening in a search for the inhibitors of protein tyrosine phosphatase 1B (PTP1B) – a target for the type 2 diabetes. A corporate library of approximately four hundred thousand compounds was screened using HTS. The bioassay showed that eighty-five compounds (0.021%) inhibited the enzyme at the level of IC_{50} less than 100 μM. Concurrently, virtual screening technology was used to screen approximately 235,000 commercially available compounds in the Available Chemicals Directory (ACD) database against the x-ray crystallographic structure of PTP1B. Of the 365 high-scoring molecules derived from virtual screening, 127 (34.8%) compounds showed the activity with IC_{50} values less than 100 μM. This study demonstrated that virtual screening enriched the hit rate by about 1,700-fold over random screening.

Liu et al. have established a 3D structural model for eukaryotic Shaker K^+ channels using a homology modeling method and performed virtual screening against ACD and Chinese Natural Product Database (CNPD) databases. By using docking-based virtual screening in conjunction with electrophysiologic assay, they have discovered ten new blockers of the eukaryotic Shaker K^+ channels, including four natural product blockers and six synthetic compounds [31]. Results show that virtual screening can recognize blockers of voltage-regulated potassium ion channel that are novel and diversified in structures and can overcome the disadvantage of HTS, which cannot be used in the discovery of blockers of potassium ion channel. Therefore, virtual screening, which is used complementarily with HTS, will play an increasingly important role in drug discovery.

2. REVERSE DOCKING

2.1. The Rise of the Reverse Docking Idea

After the completion of the human genome and numerous pathogen genomes, efforts are under way to understand the role of gene products in biological pathways and human disease and to exploit their functions for the sake of discovering new drug targets [32]. Genomic technologies, including proteomics and structural genomics, have been used to discover new drug targets that may lead to the best small molecule drugs for many major diseases [33]. However, how to identify druggable targets from thousands of candidate macromolecules and validate them is still challenging. Small and cell-permeable chemical ligands are used increasingly in genomics approaches to understand the global functions of proteins [34]. A frontier called *chemogenomics* (or *chemical genomics*) has emerged in the pipeline of "from genome to drug" [35]. Chemogenomics has been defined as the discovery and description of all possible drugs to all possible drug targets [35]. Using small, active chemical ligands as probes to reveal the functions of a proteome or a series of proteins in a biological pathway has been proven an effective way of finding clues for new targets [35–37].

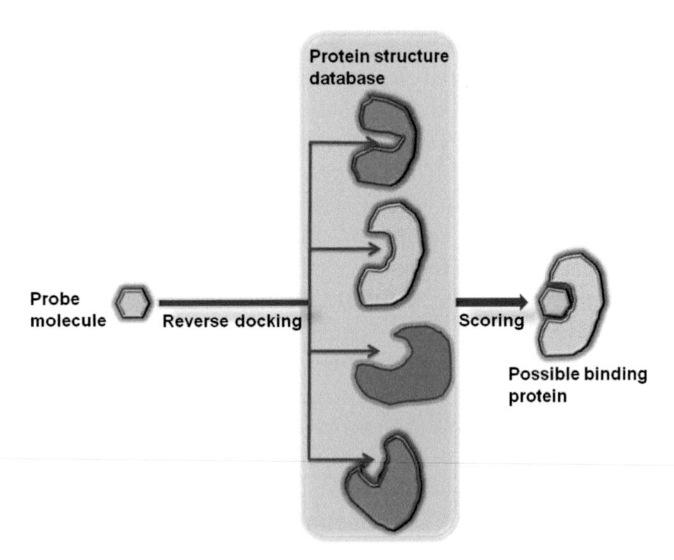

FIGURE 20.3: A cartoon representation for the process of reverse docking simulation.

To this end, proteomics may be an appropriate approach for identifying particular binding proteins of the small molecules by comparing the difference between protein expression profiles of pathogenic cells (or tissues) and compound treated cells (or tissues). However, this method has not been very successful in target discovery because it is time-consuming and has slower rates of reproducing experiments [5]. An alternative approach that has been promising in recent years is to use computational methods to find the probable binding protein(s) for an active compound, a natural product, or an old drug from the genomic or protein databases and then use traditional molecular and/or cell biology methods to validate the computational result [38–41]. Reverse to the virtual screening, the strategy of this computational approach docks known active compound(s) into the binding sites of all of the 3D structures of a protein database and selects potential binding protein(s) according to the scoring results for further biological tests. Accordingly, this approach is referred as reverse docking.

2.2. The Principle and Process of Reverse Docking

Like molecular docking, the lock-and-key principle still works in reverse docking simulation. Unlike the virtual screening based on conventional molecular docking, reverse docking uses a small molecule as a probe to search a protein database, fishing out potential binding proteins of this small molecule (Figure 20.3).

The general procedure of target identification by means of reverse docking is shown schematically in Figure 20.4. It includes four steps: probe (small molecule) structure preparation, potential drug target database (PDTD) modeling (protein structure preparation), reverse docking screening, and hit targets postprocessing.

The first step is to build up the structure of small molecules that act as probes to identify drug targets. Usually, we can construct a probe database containing useful small molecules, such as existing drugs, natural products, and active compounds reported in the publications. Therefore, such database preparation begins with the two-dimensional (2D) database construction, because information on available small molecules is stored in 2D form with atom types and connectivity (e.g., Chemical Abstract [CA]* and Beilstein databases†). The format of MDL databases‡ frequently serves as data repositories for the database creations. By checking the atom and bond types and assigning the protonation states and charges of the molecules in a 2D database, the 3D structures can be generated for the 3D database using the programs such as CORINA [42] or CONCORD‖.

Potential drug target database preparation is an important step in reverse docking. Adequate and diversified protein molecules with known 3D structures are demanded as the prerequisite of reverse docking methods. The cornerstone of reverse molecular docking is the establishment of related databases, based on which virtual screening against drug targets is performed. So far, 75,594 known crystal structures are deposited in the Protein Data Bank (PDB) [43], which provides a basic source of protein structures for reverse docking searches. Homology modeling protein structure is another reliable source. Typically, the primary structure of a protein target can be taken from PDB, in which the protein structures were determined by x-ray crystallography or nuclear magnetic resonance (NMR) spectroscopy. Alternatively, the initial structure of a protein target can also be constructed by homology modeling based on both sequence and structural information from known families of proteins. The next step is to calculate protonation states and assign atomic charges; then the entire structure of the protein should be optimized by a proper force field. The following step is to determine the space volume of a binding site. This is a critical step because a too-small binding site limits the available search space, whereas a larger receptor site uses more computing time [11]. Two methods can be used in this regard. Mostly, a binding site can be selected directly from the 3D structure of a ligand-receptor complex. For the receptor structures without bound ligands, the binding sites sometimes need to be manually assigned according to the biological function studies such as mutagenesis. Some programs (e.g., DOCK [15, 44, 45] and AutoDock [46–48]) need an extra step to construct the grid of the binding site, but whether this step is necessary depends on the docking algorithms employed.

* http://www.cas.org/ONLINE/CD/CACD/cover.html.
† http://www.beilstein.com.
‡ http://www.mdli.com.
‖ http://www.tripos.com.

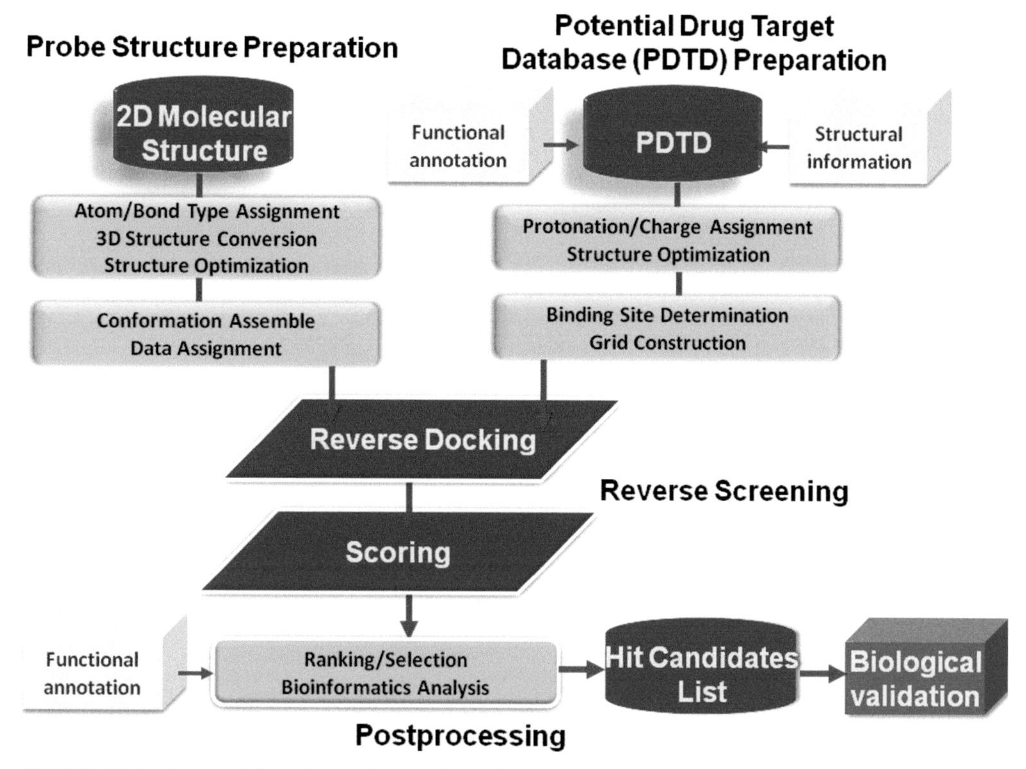

***FIGURE 20.4:** General flow chart of target identification by reverse docking.*

To fit the requirement of reverse docking, we constructed a potential drug target database (PDTD). PDTD is a dual function database that associates an informatics database with a structural database of known and potential drug targets [49]. PDTD is a comprehensive, Web-accessible database of drug targets* that focuses on those drug targets with known 3D-structures. It contains more than 1,200 entries covering more than 840 known and potential drug targets with structures from the PDB. In particular, PDTD is designed and integrated with a reverse docking program, which can be used to identify binding proteins for active compounds or existing drugs.

sc-PDB is another collection of 3D ligand-protein structure derived from the PDB, in which the ligand binding sites were extracted from the crystal structure. Each binding site is formed by all protein residues with at least one atom within 6.5 Å of any ligand atom. sc-PDB can also be used in reverse docking for target prediction of small molecules and is available at http://bioinfo-pharma.u-strasbg.fr/scPDB/. In addition, TTD, another drug target database, currently contains more than 1,500 known and explored therapeutic protein and nucleic acid targets and provides information about the targeted disease, pathway information, and the corresponding drugs/ligands directed at each of these targets [50]. DrugBank is a unique bioinformatics and cheminformatics resource that combines detailed drug (i.e., chemical, pharmacologic, and pharmaceutical) data with comprehensive drug target (i.e., sequence, structure, and pathway) information. Composed of more than 4,500 drug targets, DrugBank provides various information for each category of drugs, including chemical structures, protein and DNA sequences, related Web links, description of characteristics, and detailed pathology [51]. All of these databases offer abundant resources for the annotation of drug targets.

The core step of computational target identification is reverse docking and scoring. As indicated in Figure 20.3, reverse docking is a process to place a molecule (or a series of molecules) in the binding sites of receptors in the target database and optimize the relative orientation and conformation for the ligand interacting with these proteins and proteins or other kinds of macromolecules from the target database that possibly bind tightly to the small molecule. In principle, all docking programs can be modified into reverse docking programs. In 2001, Chen et al. developed a reverse molecular docking program, INVDOCK, based on the DOCK program [52]. INVDOCK uses small molecules to screen a series of target proteins, which is applied in the search for potential binding proteins of small molecules and furthermore explores the mechanisms of drug interactions or toxic side effects [39]. Paul et al. performed reverse virtual screening in a protein screening database (sc-PDB) containing 2,148 proteins by using the docking program GOLD [41] with a perl script. The complexes with ligands of sc-PDB were selected from the PDB database, which were saved in three databases: one containing

* http://www.dddc.ac.cn/pdtd.

protein atomic coordinates, one containing ligand 3D coordinates, and another that is the residue database of active sites obtained by using S(O,r) as the probe. Screening was performed against biotin, 4H-tamoxifen, 6-hydroxyl-1,6-dihydroprurine ribonucleoside, and methotrexate in the sc-PDB in an attempt to recover the original target proteins, and results demonstrate that their crystal structures are all recognized at the top 1% of the database. Normally, the computation of one ligand will be completed within sixty-four hours.

On the basis of DOCK4.0, we developed a reverse docking method, TarFisDock (Target Fishing Docking) [53]. TarFisDock makes use of ligand-protein reverse docking strategy to search out all possible binding proteins for small molecules from the PDTD (Figure 20.3). The small molecules could be active compounds screened out by cell- or animal-based bioassay, natural products, or existing drugs whose molecular targets are unknown. TarFisDock docks the small molecules into the protein targets in PDTD and outputs the top 2%, 5%, or 10% candidates ranked by the energy score, including their binding conformations and a table of the related target information. Therefore, TarFisDock may be a useful tool for target identification, and it can also be used in mechanism study of old drugs and natural products. Based on TarFisDock and PDTD, we also developed a reverse docking Web server.* TarFisDock has proven to be a tool of great potential value for identifying targets by more than seven hundred users from more than thirty countries and regions. One typical example of TarFisDock is that we found that peptide deformylase (PDF) is a target for anti–*Helicobacter pylori* drugs [54].

Ranking the binding targets of a target database is performed simultaneously with the reverse docking simulation. In general, all scoring functions described previously can be used in reverse docking.

The last step of reverse docking is the potential target hit list postprocessing. This step is generally integrated with bioinformatics analysis. Usually, the candidate proteins isolated by reverse docking will be submitted to further functional annotations to search their biological functions from the public databases or literatures and make possible connections to existing biological networks or pathways.

Using the aforementioned four steps, most of the undesired proteins or nucleic acids should be eliminated and the large target database was reduced to a manageable size for the experimental validation (Figure 20.4).

2.2. An Example of Reverse Docking in Target Identification

In the following, we use an example of our own research results to demonstrate how to discover drug target using the reverse docking approach. This example shows the finding

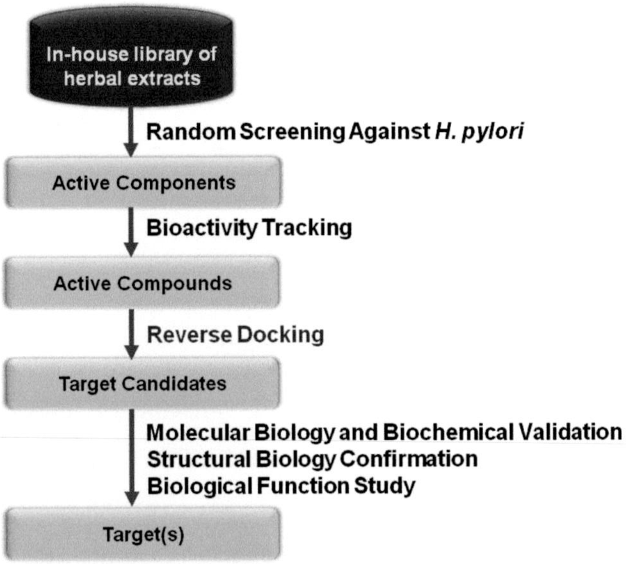

***FIGURE 20.5:** The workflow of target identification for anti-H. pylori drugs, from the active components of herbs to the binding proteins of natural products.*

of PDF as a potential target for anti–*H. pylori* drugs. The result was discovered using a computational method and verified with bioassay and x-ray crystallography [54].

The discovery procedure is shown in Figure 20.5. A random screening of a diverse small molecule and herbal extracts library used an agar dilution method to identify active components against *H. pylori*. Compound **1**, isolated from *Ceratostigma willmottianum*, a folk medicine used to remedy rheumatism, traumatic injury, and parotitis [55], showed inhibitory activity against *H. pylori* with a minimum inhibitory concentration (MIC) value of 180 μg/mL. Chemical modification on compound **1** afforded a number of analogs, and compound **2** is the most active one with an improved MIC value of 100 μg/mL against *H. pylori*.

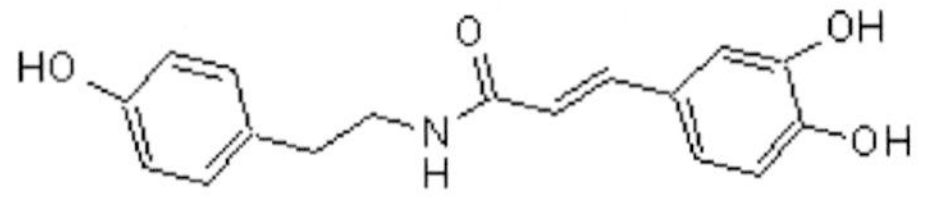

1

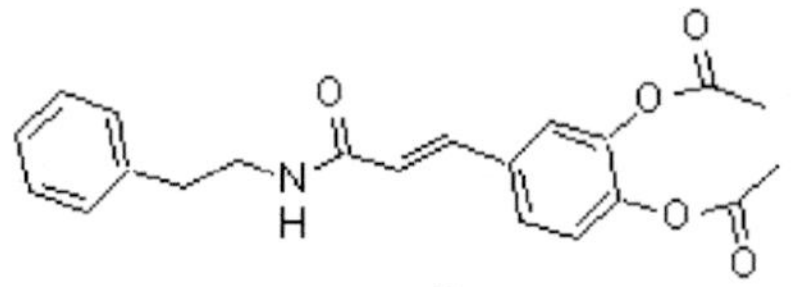

2

Potential binding proteins of compound **1** were screened from our in-house potential drug target database (PDTD)* using the reverse docking approach TarFisDock. Fifteen protein targets (Table 20.1) with interaction energies with compound **1** less than −35.0 kcal/mol were identified.

* http://www.dddc.ac.cn/tarfisdock/.

* http://www.dddc.ac.cn/pdtd/.

TABLE 20.1: Possible Binding Protein Candidates of Compound 1 Searched from The PDTD Using Reverse Docking

ID	Interaction Energy (kcal/mol)[a]	Protein Name	PDB Entry	Brief Annotation	Homology Protein of *H. pylori*
1	−41.33	Hormone binding domain of hormone/growth factor receptor	1R1K	A ligand-inducible nuclear transcription factor, ecdysone receptor (EcR), in noctuid and moth	No
2	−41.26	Diaminopimelate decarboxylase	1HKV	A pyridoxal-5′-phosphate (PLP)-dependent enzyme converting meso-diaminopimelic acid (DAP) to l-lysine in bacteria	Yes
3	−39.7	GppNHp nuclear transport	1QBK	A transport factor mediating the movement of macromolecules in nuclear-cytoplasmic transport pathways in the human being	No
4	−39.59	Prophospholipase	1HN4	An anion-assisted dimmers of prophospholipase A2 in pig	No
5	−38.89	Porcine factor Ixa	1PFX	A key enzyme in blood coagulation related to hemophilia B, a severe X chromosome–linked bleeding disorder afflicting pig	No
6	−38.79	Cholesterol oxidase	1IJH	A oxidoreductase catalyzing the oxidation and isomerization of cholesterol to cholest-4-en-3-one in bacteria	No
7	−38.74	T catalitic domain of matrix metalloproteinase-8	1I76	Human neutrophil collagenase (HNC, MMP-8)	No
8	−38.43	Lac repressor dimer	1EFA	Lac repressor bound to operator in bacteria	No
9	−37.89	Thiamin pyrophosphokinase	1IG0	Thiamin pyrophosphokinase, catalyzing the transfer of a pyrophosphate group from ATP to vitamin B_1 (thiamin) to form the coenzyme thiamin pyrophosphate (TPP) in yeast	No
10	−37.89	Acetolactate synthase	1OZF	A thiamine diphosphate (ThDP)-dependent enzyme, catalyzing the decarboxylation of pyruvate to give a co-factor–bound hydroxyethyl group in plants, fungi, and bacteria	No
11	−37.46	Peptide deformylase	1BS6	Peptide deformylase	Yes
12	−37.35	Methyltransferase	1JQE	Histamine N-methyltransferase (HNMT) plays the dominant role in histamine biotransformation in bronchial epithelial and endothelial cells of the human airways and gut and is the only enzyme responsible for termination of the neurotransmitter action of histamine in the mammalian brain	No
13	−35.88	Nitroreductase	1ICR	*Escherichia coli* nitroreductase is a flavoprotein that reduces a variety of quinone and nitroaromatic substrates. Its ability to convert relatively nontoxic prodrugs such as CB1954 (5-[aziridin-1-yl]-2,4-dinitrobenzamide) into highly cytotoxic derivatives has led to interest in its potential for cancer gene therapy	No
14	−35.86	Cyclooxygenase-2	1CX2	Cox-2 is an inducible form that plays a major role in prostaglandin biosynthesis in inflammatory cells (monocytes/macrophages) and in the central nervous system	No
15	−35.66	Cytochrome P450	1BVY	HEME and FMN binding domain of cytochrome P450 BM-3	No

[a] Interaction energies were calculated using the scoring function of DOCK4.0.

A homology search indicated that only two proteins of the fifteen candidates, diaminopimelate decarboxylase (DC) and PDF, have homologous proteins in the genome of *H. pylori*. DC is encoded by the *lysA* gene that converts *meso*-DAP to L-lysine, which is the final step in the bacterial lysine biosynthetic pathway. Lysine is essential for bacterial viability and development. Unlike bacteria, humans lack the lysine biosynthetic pathway and obtain lysine from dietary sources. Thus, DC could be a potential antibacterial drug target. In prokaryotes and eukaryotic organelles (e.g., mitochondria and chloroplast), protein synthesis is initiated with *N*-formyl-methionyl-tRNA$_i$, resulting in N-terminal formylation of all nascent polypeptides. During elongation of the polypeptide chain, PDF catalyzes the removal of a formyl group at the N-terminal. PDF is essential for bacterial cell growth [56, 57], and it has been proven that the inactivation of the *def* gene encoding PDF in *Streptococcus pneumoniae* cannot be achieved [58]. Although PDF has also been discovered in human beings [59, 60], it has been demonstrated that PDF has no effect on cytoplasmic protein synthesis in mammalian cells. Therefore, PDF is also an attractive target for discovering novel antibiotics [61], and we chose DC and PDF as probable binding proteins of compound **1** for target validation.

The inhibitory activities of compounds **1** and **2** on *Hp*PDF were measured by using the formate dehydrogenase (FDH)-coupled assay [62]. The IC_{50} values of compounds **1** and **2** inhibiting the catalytic activity of Co-*Hp*PDF were estimated as 10.8 and 1.25 μM, respectively, which is in agreement with the MIC values. This result indicates that *Hp*PDF is really a target of compound **1** and its analogs compound **2**. In parallel, the inhibitory activities of compounds **1** and **2** to *H. pylori* DC (*Hp*DC) were also measured using the double-enzyme coupled assay [63, 64]. However, compounds **1** and **2** have no inhibitory effect on *Hp*DC, indicating that *Hp*DC is not a target of compounds **1** and **2**.

To verify that *Hp*PDF is a binding protein of compounds **1** and **2** at the atomic level, the x-ray structures of both *Hp*PDF and *Hp*PDF-inhibitor complexes were determined by using the crystallographic method. Analysis of the diffraction data indicated that the crystal structure of *Hp*PDF belongs to space group $P2_12_12_1$ and contains one *Hp*PDF molecule per asymmetric unit. All of the x-ray crystal structures were refined to a resolution of 2.2 Å. The crystal structures indicated that both of these inhibitors bind with the binding pocket of *Hp*PDF (Figure 20.6).

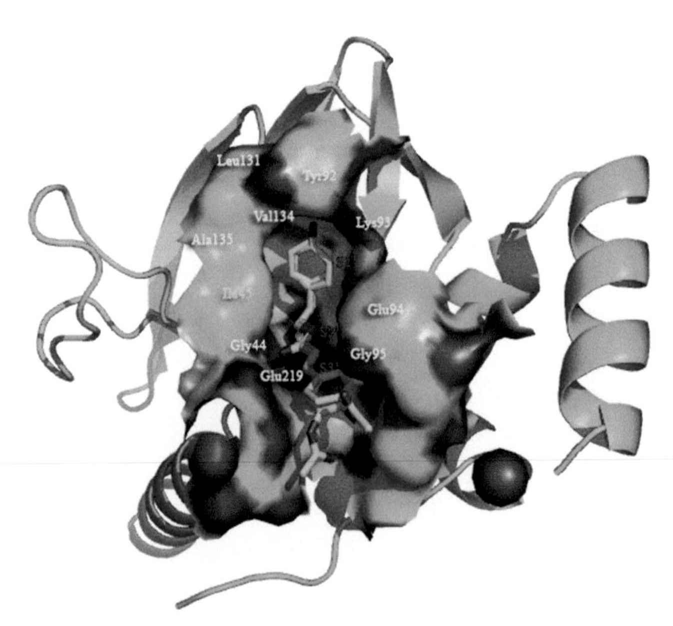

FIGURE 20.6: ***Binding models of compounds 1 and 2 with HpPDF derived from the X-ray crystal structure determination.***

3. PERSPECTIVE OF COMPUTATIONAL CHEMICAL BIOLOGY FOR TARGET IDENTIFICATION AND DRUG DISCOVERY

In drug development, it always takes too long and costs too much to bring new drugs to market. The identification of drug target and undesirable secondary effects are among the main challenges in developing new drugs [6]. To ease this problem, one possible solution is to identify new usages of the existing drugs, including traditional Chinese medicine (TCM). This approach is a proven shortcut between the basic research and the clinic study and will be a new paradigm for drug discovery [65]. Discovering new or alternative binding proteins or other macromolecules for the existing drugs is the first step to find their new usages. Therefore, computational chemical biology methods have numerous potential applications in this area.

On one hand, for those drugs or drug candidates with unknown mechanisms of action, such as some well-known antibacterial agents, computational approaches can help to determine their potential binding target proteins quickly and effectively, which will provide clues for the elucidation of the drug action mechanisms.

On the other hand, for the known drugs or drug leads with clear mechanisms, computational approaches can provide secondary binding protein information on compounds, which definitely will play a significant role in developing the new usage of old drugs [6, 66]. Furthermore, computational approaches can be applied to predict the interaction with potential protein involving the toxic side effects, which provides a fast and effective tool with low cost for safety evaluation of early-stage candidate molecules [40, 67]. After the toxic side effects have been validated, computational approaches such as similarity analysis can be performed to facilitate another round of target finding [68].

TCM is one of the most productive sources of drug leads. However, because of its ambiguous active constituents and implicated mechanisms, TCM is not acceptable from the viewpoint of the Western reductionist and mechanistic approach to diseases. Moreover, it is unarguable that many herbs from TCM, especially the active components of some natural products, have hold more and more interests because of their excellent pharmacological and safety profile [69–71]. Therefore, the development of TCM should shift its focus to how to determine the active components of herbs and elucidate the mechanism [72, 73]. Considering the vast amount of components in TCM and the difficulty of isolation, a simple but effective means is to predict potential targets of isolated compounds using the aforementioned computational approaches before biological and pharmacological experiments. Some success in real practice can be found elsewhere [54, 74].

In addition to the above-mentioned computational approaches, scientists have made great efforts to elucidate the role of gene products in biological pathways and human diseases and to explore their functions in an attempt to discover new drug targets [75]. Chemical genomics, which uses small and cell-permeable chemical ligands in genomics approaches to understand the global functions of genome and proteome, has been widely used experimentally [76]. Another emerging field is systems biology to model or simulate intracellular and intercellular events using data gathered from genomic, proteomic, or metabolomic experiments [77], which has thrown light on drug discovery and development. Along this direction, more corresponding computational biology methods will be appearing.

In this chapter, we discussed only reverse docking and its application in target identification. In addition to identifying target candidates for active compounds, the advantage of reverse docking is represented by its potential to identify the off-target effects of a drug [40]. However, reverse docking still has certain limitations. The major one is that there are not enough protein entries in the protein structure databases like PDB to cover all of the protein information of disease-related genomes. The second one is that this approach has not considered the flexibility of proteins during docking simulation. These two aspects will produce false negatives. A third limitation is that the scoring function for reverse docking is not accurate enough, which will produce false positives [53]. One strategy to overcome these shortages is to develop new docking programs that consider protein flexibility and use accurate scoring function. Another is to integrate sequence-based and structure-based approaches together.

REFERENCES

1. DiMasi, J. A., Hansen, R. W., and Grabowski, H. G. (2003). The price of innovation: new estimates of drug development costs. *J Health Econ 22*, 151–185.
2. Jorgensen, W. L. (2004). The many roles of computation in drug discovery. *Science 303*, 1813–1818.
3. Drews, J. (2000). Drug discovery: a historical perspective. *Science 287*, 1960–1964.
4. Leader, B., Baca, Q. J., and Golan, D. E. (2008). Protein therapeutics: a summary and pharmacological classification. *Nat Rev Drug Discov* 7, 21–39.
5. Huang, C. M., Elmets, C. A., Tang, D. C., Li, F., and Yusuf, N. (2004). Proteomics reveals that proteins expressed during the early stage of Bacillus anthracis infection are potential targets for the development of vaccines and drugs. *Genomics Proteomics Bioinformatics 2*, 143–151.
6. Marton, M. J., DeRisi, J. L., Bennett, H. A., Iyer, V. R., Meyer, M. R., Roberts, C. J., Stoughton, R., Burchard, J., Slade, D., Dai, H., Bassett, D. E., Jr., Hartwell, L. H., Brown, P. O., and Friend, S. H. (1998). Drug target validation and identification of secondary drug target effects using DNA microarrays. *Nat Med 4*, 1293–1301.
7. Lindsay, M. A. (2003). Target discovery. *Nat Rev Drug Discov 2*, 831–838.
8. Garcia-Lara, J., Masalha, M., and Foster, S. J. (2005). *Staphylococcus aureus*: the search for novel targets. *Drug Discov Today 10*, 643–651.
9. Orth, A. P., Batalov, S., Perrone, M., and Chanda, S. K. (2004). The promise of genomics to identify novel therapeutic targets. *Expert Opin Ther Targets 8*, 587–596.
10. Brooijmans, N., and Kuntz, I. D. (2003). Molecular recognition and docking algorithms. *Annu Rev Biophys Biomol Struct 32*, 335–373.
11. Ingo Muegge, M. R. (2001). Small molecule docking and scoring. In *Reviews in Computational Chemistry*, vol. 17, K. B. Lipkowitz and D. B. Boyd, eds. (John Wiley & Sons, Inc.), 1–60.
12. Carlson, H. A. (2002). Protein flexibility and drug design: how to hit a moving target. *Curr Opin Chem Biol 6*, 447–452.
13. Zeelen, F. J. (1995). The lock-and-key principle. *Trends Pharm Sci 16*, 322–322.
14. Abraham, D. J., John, B. T., and David, J. T. (2007). Structure-based drug design – a historical perspective and the future. In *Comprehensive Medicinal Chemistry II*, D. J. Triggle and J. B. Taylor, eds. (Oxford: Elsevier), 65–86.
15. Kuntz, I. D., Blaney, J. M., Oatley, S. J., Langridge, R., and Ferrin, T. E. (1982). A geometric approach to macromolecule-ligand interactions. *J Mol Biol 161*, 269–288.
16. Ekins, S., Mestres, J., and Testa, B. (2007). In silico pharmacology for drug discovery: methods for virtual ligand screening and profiling. *Br J Pharmacol 152*, 9–20.
17. Kitchen, D. B., Decornez, H., Furr, J. R., and Bajorath, J. (2004). Docking and scoring in virtual screening for drug discovery: methods and applications. *Nat Rev Drug Discov 3*, 935–949.
18. Teague, S. J. (2003). Implications of protein flexibility for drug discovery. *Nat Rev Drug Discov 2*, 527–541.
19. Böhm, H.-J., and Stahl, M. (2003). The use of scoring functions in drug discovery applications. In *Reviews in Computational Chemistry*, vol. 18, K. B. Lipkowitz and D. B. Boyd, eds. (John Wiley & Sons, Inc.), 41–87.
20. Kollman, P. A. (1993). Free energy calculations: applications to chemical and biochemical phenomena. *Chem Rev 93*, 2395–2417.
21. Ekins, S., Mestres, J., and Testa, B. (2007). In silico pharmacology for drug discovery: applications to targets and beyond. *Br J Pharmacol 152*, 21–37.

22. Waszkowycz, B., Perkins, T. D. J., Sykes, R. A., and Li, J. (2001). Large-scale virtual screening for discovering leads in the postgenomic era. *IBM Systems J 40*, 360–376.
23. Doman, T. N., McGovern, S. L., Witherbee, B. J., Kasten, T. P., Kurumbail, R., Stallings, W. C., Connolly, D. T., and Shoichet, B. K. (2002). Molecular docking and high-throughput screening for novel inhibitors of protein tyrosine phosphatase-1B. *J Med Chem 45*, 2213–2221.
24. Rochu, D., Georges, C., Repiton, J., Viguie, N., Saliou, B., Bon, C., and Masson, P. (2001). Thermal stability of acetylcholinesterase from Bungarus fasciatus venom as investigated by capillary electrophoresis. *Biochim Biophys Acta 1545*, 216–226.
25. Honma, T., Hayashi, K., Aoyama, T., Hashimoto, N., Machida, T., Fukasawa, K., Iwama, T., Ikeura, C., Ikuta, M., Suzuki-Takahashi, I., Iwasawa, Y., Hayama, T., Nishimura, S., and Morishima, H. (2001). Structure-based generation of a new class of potent Cdk4 inhibitors: new de novo design strategy and library design. *J Med Chem 44*, 4615–4627.
26. Enyedy, I. J., Ling, Y., Nacro, K., Tomita, Y., Wu, X., Cao, Y., Guo, R., Li, B., Zhu, X., Huang, Y., Long, Y. Q., Roller, P. P., Yang, D., and Wang, S. (2001). Discovery of small-molecule inhibitors of Bcl-2 through structure-based computer screening. *J Med Chem 44*, 4313–4324.
27. Iwata, Y., Arisawa, M., Hamada, R., Kita, Y., Mizutani, M. Y., Tomioka, N., Itai, A., and Miyamoto, S. (2001). Discovery of novel aldose reductase inhibitors using a protein structure-based approach: 3D-database search followed by design and synthesis. *J Med Chem 44*, 1718–1728.
28. Enyedy, I. J., Lee, S. L., Kuo, A. H., Dickson, R. B., Lin, C. Y., and Wang, S. (2001). Structure-based approach for the discovery of bis-benzamidines as novel inhibitors of matriptase. *J Med Chem 44*, 1349–1355.
29. Pang, Y. P., Xu, K., Kollmeyer, T. M., Perola, E., McGrath, W. J., Green, D. T., and Mangel, W. F. (2001). Discovery of a new inhibitor lead of adenovirus proteinase: steps toward selective, irreversible inhibitors of cysteine proteinases. *FEBS Lett 502*, 93–97.
30. Baxter, C. A., Murray, C. W., Waszkowycz, B., Li, J., Sykes, R. A., Bone, R. G., Perkins, T. D., and Wylie, W. (2000). New approach to molecular docking and its application to virtual screening of chemical databases. *J Chem Inf Comput Sci 40*, 254–262.
31. Liu, H., Gao, Z. B., Yao, Z., Zheng, S., Li, Y., Zhu, W., Tan, X., Luo, X., Shen, J., Chen, K., Hu, G. Y., and Jiang, H. (2007). Discovering potassium channel blockers from synthetic compound database by using structure-based virtual screening in conjunction with electrophysiological assay. *J Med Chem 50*, 83–93.
32. Maggio, E. T., and Ramnarayan, K. (2001). Recent developments in computational proteomics. *Drug Discov Today 6*, 996–1004.
33. Lenz, G. R., Nash, H. M., and Jindal, S. (2000). Chemical ligands, genomics and drug discovery. *Drug Discov Today 5*, 145–156.
34. Giese, K., Kaufmann, J., Pronk, G. J., and Klippel, A. (2002). Unravelling novel intracellular pathways in cell-based assays. *Drug Discov Today* 7, 179–186.
35. Caron, P. R., Mullican, M. D., Mashal, R. D., Wilson, K. P., Su, M. S., and Murcko, M. A. (2001). Chemogenomic approaches to drug discovery. *Curr Opin Chem Biol 5*, 464–470.
36. Zheng, X. F., and Chan, T. F. (2002). Chemical genomics in the global study of protein functions. *Drug Discov Today* 7, 197–205.
37. Oi, H., Matsuura, D., Miyake, M., Ueno, M., Takai, I., Yamamoto, T., Kubo, M., Moss, J., and Noda, M. (2002). Identification in traditional herbal medications and confirmation by synthesis of factors that inhibit cholera toxin-induced fluid accumulation. *Proc Natl Acad Sci U S A 99*, 3042–3046.
38. Rockey, W. M., and Elcock, A. H. (2005). Rapid computational identification of the targets of protein kinase inhibitors. *J Med Chem 48*, 4138–4152.
39. Chen, Y. Z., and Zhi, D. G. (2001). Ligand-protein inverse docking and its potential use in the computer search of protein targets of a small molecule. *Proteins 43*, 217–226.
40. Chen, Y. Z., and Ung, C. Y. (2001). Prediction of potential toxicity and side effect protein targets of a small molecule by a ligand-protein inverse docking approach. *J Mol Graph Model 20*, 199–218.
41. Paul, N., Kellenberger, E., Bret, G., Muller, P., and Rognan, D. (2004). Recovering the true targets of specific ligands by virtual screening of the protein data bank. *Proteins 54*, 671–680.
42. Gasteiger, J., Rudolph, C., and Sadowski, J. (1990). Automatic generation of 3D-atomic coordinates for organic molecules. *Tetrahedron Comput Methodol 3*, 537–547.
43. Berman, H. M., Westbrook, J., Feng, Z., Gilliland, G., Bhat, T. N., Weissig, H., Shindyalov, I. N., and Bourne, P. E. (2000). The Protein Data Bank. *Nucleic Acids Res 28*, 235–242.
44. Todd, J. A., and Ewing, I. D. K. (1997). Critical evaluation of search algorithms for automated molecular docking and database screening. *J Comput Chem 18*, 1175–1189.
45. Kuntz, I. D. (1992). Structure-based strategies for drug design and discovery. *Science 257*, 1078–1082.
46. Goodsell, D. S., and Olson, A. J. (1990). Automated docking of substrates to proteins by simulated annealing. *Proteins 8*, 195–202.
47. Morris, G. M., Goodsell, D. S., Huey, R., and Olson, A. J. (1996). Distributed automated docking of flexible ligands to proteins: parallel applications of AutoDock 2.4. *J Comput Aided Mol Des 10*, 293–304.
48. Morris, G. M., Goodsell, D. S., Halliday, R. S., Huey, R., Hart, W. E., Belew, R. K., and Olson, A. J. (1998). Automated docking using a Lamarckian genetic algorithm and an empirical binding free energy function. *J Comput Chem 19*, 1639–1662.
49. Gao, Z., Li, H., Zhang, H., Liu, X., Kang, L., Luo, X., Zhu, W., Chen, K., Wang, X., and Jiang, H. (2008). PDTD: a web-accessible protein database for drug target identification. *BMC Bioinformatics 9*, 104.
50. Chen, X., Ji, Z. L., and Chen, Y. Z. (2002). TTD: Therapeutic Target Database. *Nucleic Acids Res 30*, 412–415.
51. Wishart, D. S., Knox, C., Guo, A. C., Shrivastava, S., Hassanali, M., Stothard, P., Chang, Z., and Woolsey, J. (2006). DrugBank: a comprehensive resource for in silico drug discovery and exploration. *Nucleic Acids Res 34*, D668–D672.

52. Wang, J., Kollman, P. A., and Kuntz, I. D. (1999). Flexible ligand docking: a multistep strategy approach. *Proteins 36*, 1–19.
53. Li, H., Gao, Z., Kang, L., Zhang, H., Yang, K., Yu, K., Luo, X., Zhu, W., Chen, K., Shen, J., Wang, X., and Jiang, H. (2006). TarFisDock: a web server for identifying drug targets with docking approach. *Nucleic Acids Res 34*, W219–W224.
54. Cai, J., Han, C., Hu, T., Zhang, J., Wu, D., Wang, F., Liu, Y., Ding, J., Chen, K., Yue, J., Shen, X., and Jiang, H. (2006). Peptide deformylase is a potential target for anti-*Helicobacter pylori* drugs: reverse docking, enzymatic assay, and X-ray crystallography validation. *Protein Sci 15*, 2071–2081.
55. Yue, L., Zhu, D. N., Yan, Y. Q., and Yu, B. Y. (2006). Relationship between components changes and efficacy of Shengmaisan VII. Chemical dynamic change of schisandrin in Shengmaisan. *Zhongguo Zhong Yao Za Zhi 31*, 1010–1012.
56. Mazel, D., Pochet, S., and Marliere, P. (1994). Genetic characterization of polypeptide deformylase, a distinctive enzyme of eubacterial translation. *Embo J 13*, 914–923.
57. Meinnel, T., and Blanquet, S. (1994). Characterization of the Thermus thermophilus locus encoding peptide deformylase and methionyl-tRNA(fMet) formyltransferase. *J Bacteriol 176*, 7387–7390.
58. Margolis, P., Hackbarth, C., Lopez, S., Maniar, M., Wang, W., Yuan, Z., White, R., and Trias, J. (2001). Resistance of *Streptococcus pneumoniae* to deformylase inhibitors is due to mutations in *defB*. *Antimicrob Agents Chemother 45*, 2432–2435.
59. Lee, M. D., Antczak, C., Li, Y., Sirotnak, F. M., Bornmann, W. G., and Scheinberg, D. A. (2003). A new human peptide deformylase inhibitable by actinonin. *Biochem Biophys Res Commun 312*, 309–315.
60. Serero, A., Giglione, C., Sardini, A., Martinez-Sanz, J., and Meinnel, T. (2003). An unusual peptide deformylase features in the human mitochondrial N-terminal methionine excision pathway. *J Biol Chem 278*, 52953–52963.
61. Nguyen, K. T., Hu, X., Colton, C., Chakrabarti, R., Zhu, M. X., and Pei, D. (2003). Characterization of a human peptide deformylase: implications for antibacterial drug design. *Biochemistry 42*, 9952–9958.
62. Lazennec, C., and Meinnel, T. (1997). Formate dehydrogenase-coupled spectrophotometric assay of peptide deformylase. *Anal Biochem 244*, 180–182.
63. Ray, S. S., Bonanno, J. B., Rajashankar, K. R., Pinho, M. G., He, G., De Lencastre, H., Tomasz, A., and Burley, S. K. (2002). Cocrystal structures of diaminopimelate decarboxylase: mechanism, evolution, and inhibition of an antibiotic resistance accessory factor. *Structure 10*, 1499–1508.
64. Scriven, F., Wlasichuk, K. B., and Palcic, M. M. (1988). A continual spectrophotometric assay for amino acid decarboxylases. *Anal Biochem 170*, 367–371.
65. Chong, C. R., and Sullivan, D. J., Jr. (2007). New uses for old drugs. *Nature 448*, 645–646.
66. O'Connor, K. A., and Roth, B. L. (2005). Finding new tricks for old drugs: an efficient route for public-sector drug discovery. *Nat Rev Drug Discov 4*, 1005–1014.
67. Ni, B., Zhu, T., Jiang, Z. Z., Zhang, R., Zhang, T., and Zhang, L. Y. (2008). In vitro and in silico approaches for analyzing the toxicological effect of triptolide on Cx43 in Sertoli cells. *Toxicol Mech Meth 18*, 717–724.
68. Campillos, M., Kuhn, M., Gavin, A. C., Jensen, L. J., and Bork, P. (2008). Drug target identification using side-effect similarity. *Science 321*, 263–266.
69. Tsang, I. K. (2007). Establishing the efficacy of traditional Chinese medicine. *Nat Clin Pract Rheumatol 3*, 60–61.
70. Hoessel, R., Leclerc, S., Endicott, J. A., Nobel, M. E., Lawrie, A., Tunnah, P., Leost, M., Damiens, E., Marie, D., Marko, D., Niederberger, E., Tang, W., Eisenbrand, G., and Meijer, L. (1999). Indirubin, the active constituent of a Chinese antileukaemia medicine, inhibits cyclin-dependent kinases. *Nat Cell Biol 1*, 60–67.
71. Chen, S. T., Dou, J., Temple, R., Agarwal, R., Wu, K. M., and Walker, S. (2008). New therapies from old medicines. *Nat Biotechnol 26*, 1077–1083.
72. Harvey, A. L. (2007). Natural products as a screening resource. *Curr Opin Chem Biol 11*, 480–484.
73. Newman, D. J. (2008). Natural products as leads to potential drugs: an old process or the new hope for drug discovery? *J Med Chem 51*, 2589–2599.
74. Chen, X., Ung, C. Y., and Chen, Y. (2003). Can an in silico drug-target search method be used to probe potential mechanisms of medicinal plant ingredients? *Nat Prod Rep 20*, 432–444.
75. Kopec, K. K., Bozyczko-Coyne, D., and Williams, M. (2005). Target identification and validation in drug discovery: the role of proteomics. *Biochem Pharmacol 69*, 1133–1139.
76. Stockwell, B. R. (2004). Exploring biology with small organic molecules. *Nature 432*, 846–854.
77. Materi, W., and Wishart, D. S. (2007). Computational systems biology in drug discovery and development: methods and applications. *Drug Discov Today 12*, 295–303.

SECTION FIVE

CHEMICAL GENOMICS AND MEDICINE

CHAPTER 21

Pharmacogenomics to Link Genetic Background with Therapeutic Efficacy and Safety

Mark M. Bouzyk
Weining Tang
Brian Leyland-Jones

Pharmacogenomics (PGx) is a rapidly evolving science with the potential to transform the way in which medicines are developed and prescribed. It represents a new era in patient treatment and will enable delivery of medicines in a safer and more efficacious manner. The aim of PGx is to optimize drug therapy with respect to the individual's genetic profile by associating genetic markers (e.g., single-nucleotide polymorphisms [SNPs] or gene expression) with the drug's safety or efficacy, thereby heralding the advent of *personalized medicine* (the right drug for the right person), whereby a drug may be selected or optimized for a specific individual based on his or her genetic characteristics. Recent advances, approaches, technologies, and resources in this field are highlighted in this chapter and interlaced with established and new exemplars.

1. DEFINITION

Often, the term *pharmacogenomics* is used interchangeably with *pharmacogenetics*, and it has proved difficult to define each concisely. However, it is generally accepted that pharmacogenomics takes into account the whole genome in a wide application of genomic technologies used to predict drug efficacy and safety in order to characterize current drugs or discover new drugs while considering genetic variation, whereas PGx is seen as the study of inherited genetic variation that may be associated with different response or adverse events to drugs by assessing typically one or a few candidate genes.

2. BACKGROUND

One of the greatest challenges facing all branches of modern medicine is to move from a reactive model of care, in which interventions are directed toward those already ill, to a proactive model aimed at preventing and treating disease in its earliest stages. Achieving that goal will require significant advances in knowledge of how genes, and the interaction of genes with the environment, confer vulnerability to disease.

There are more than one hundred thousand deaths and more than 2 million serious events caused by adverse drug reactions (ADRs) in the United States each year alone [1]. ADRs are responsible for 5% to 7% of hospital admissions in the United States and Europe, lead to the withdrawal of 4% of new medicines, and cost society an amount equivalent to the costs of drug treatment [2]. When comparing the list of drugs most commonly associated with ADRs with the list of metabolizing enzymes with known variants, it was found that drugs commonly involved in ADRs were also those that were metabolized by enzymes with known polymorphisms [3]. Drug safety is thus a top priority and a major clinical driver for PGx.

Methods of drug discovery and novel therapeutics have evolved greatly during the last decade, paralleled by advances in diagnostic tools that are enabling physicians to improve clinical care. Only recently, however, has the idea of molecular diagnosis and profiling gained traction and acceptance as an essential element in the eventual conquest of many diseases.

The emerging field of translational genomics research is now seeking to harness the power of new discoveries resulting from the Human Genome Project and applying them to the development of improved diagnostics, prognostics, and therapies for cancer, neurologic disorders, diabetes, and other complex diseases.

One of the most promising and exciting areas in the field of predictive medicine is the area of PGx.

As a result of the sequencing and mapping of the human genome, PGx is becoming the first drug discovery technology to affect the structure and economics of the pharmaceutical industry. Over the last few years, there has been a rapid increase in PGx studies by pharmaceutical and biotechnology companies. In particular, high-throughput SNP mapping is allowing human disease-associated drug targets to be identified and is reducing attrition in early-phase clinical trials. The application of genome-wide mapping and screening methods to clinical medicine is considerably increasing. There are a growing number of examples of PGx in the literature showing toxicity and efficacy of drugs linked to an individual's genotype. Improvements in genomic technologies, tools, and panels are considerably accelerating this discipline. For example, there are now novel and wide-ranging assay sets readily available to screen for drug-related safety issues for drug absorption, distribution, metabolism, and excretion (see below).

Today, many proof-of-concept projects that involve direct access to clinical trial patients and utilize their capabilities to apply genome-wide technologies are progressing in the pharmaceutical and biotechnology industries. Large and potential resources of genetic and biological material collected from clinical trials or biological repositories, with appropriate subsets stored under the strictest quality control protocols, will significantly benefit the scientific community and accelerate the development of PGx. What was predicted a few years ago is being actively pursued today, especially in the industrial setting.

Therefore, new improving technologies, economics, and the harsh realities of ADRs are some of the main drivers toward the development of PGx and are the focus of much clinical interest.

3. HISTORY

In 1902, Archibald Garrod first hypothesized that genetic variations could cause adverse biological reactions when chemical substances were ingested [4]. He also suggested that enzymes were responsible for detoxifying foreign substances and that genetic variation could prevent elimination of specific foreign substances from the body because of enzyme insufficiency required to break down these materials. Specifically, Garrod observed and hypothesized this in the study on alcaptonuria and later in his work "Inborn Errors of Metabolism" [5, 6].

In 1932, a study observed that the inability to taste the chemical compound phenylthiocarbamide was linked to an autosomal recessive trait. Participants with two recessive alleles were unable to produce an enzyme that allowed them to taste the phenylthiocarbamide chemical [4]. Drug reactions based on inherited traits were recorded during World War II, when some soldiers developed anemia after receiving doses of the antimalarial drug primaquine. Subsequent studies confirmed that the anemia was caused by a genetic deficiency of the glucose-6-phosphate dehydrogenase enzyme [7]. Other early advances included serum cholinesterase level variation that was demonstrated to be an inherited factor after observations that several individuals experienced extended anesthesia following administration of the muscle relaxant succinylcholine [8]. The molecular basis of the variation has now been determined [9]. In addition, reactions to isoniazid were studied and revealed that enzyme deficiencies in N-acetyl transferase also led to an inability to metabolize this drug, including the development of peripheral neuropathy [10]. In 1957, after studying ADRs to primaquine, succinylcholine, and isoniazid, Moltulsky proposed that inherited traits may not only lead to adverse drug reactions, but may also affect whether the drugs are really efficacious [11].

In recent decades, further progress has been made in isolating genetic variations in major drug-metabolizing enzymes, including the cytochrome P450 (CYP450) system (and identification of genetic variability in drug oxidation mediated by the P450 system). Scientists originally began to study CYP450 when some patients experienced a severe decline in blood pressure while taking the antihypertensive drug debrisoquine [12]. The study revealed that these patients had two recessive alleles for the enzyme, resulting in an inability to metabolize the drug. Debrisoquine was subsequently shown to be metabolized by a single polymorphic CYP450 enzyme, CYP2D6, known as debrisoquine hydroxylase [13, 14]. Major gene-inactivating mutations in CYP2D6 have now been identified, and initial DNA-based tests to identify individuals with impaired CYP2D6 metabolism were originally developed in the 1990s [15]. Approximately 10% of the population metabolizes cytochrome P450 poorly, experiencing adverse effects and reduced drug uptake when they take drugs in the family of chemicals metabolized by this enzyme [16]. The study of CYP450 has led to the identification and characterization of a whole series of drug-metabolizing enzymes [14].

Currently, research in PGx is focused on preventing ADRs through the analysis of the relationship between drug-metabolizing enzymes and the chemical compounds that those enzymes metabolize. In addition, PGx is being used to determine which receptors or other candidate genes are best equipped to transport particular chemical compounds into the cell for the purpose of treating a disease. Moreover, advanced genomic technologies are now allowing the possibility of screening whole genomes of patients to predict outcome of drugs in order to facilitate individualized drug treatment based on an individual's complete genetic constitution.

4. IMPORTANT EXAMPLES OF PHARMACOGENOMICS TODAY

There are now a growing number of diagnostics to guide prescribing and an increasing number of companies developing diagnostic/therapeutic pairs in clinical trials as part of the drug discovery process. The number of such tests is too large to permit detail in this chapter, but an excellent updated reference of the current large and diverse pharmacogenomic landscape with a comprehensive list of and information on more than thirty companies offering, developing, or partnering to develop companion diagnostics is well worth reading [17]. A growing personalized medicine cabinet is thus on the horizon, if not already here, in which some of these drugs require testing by the FDA, others are recommended for testing by the FDA, and some have information available. A selection of some well-known examples of PGx with appropriate background as necessary are given below.

4.1. Cancer Pharmacogenomics

One of the earliest successes in the field of pharmacogenomics is in the area of cancer, and this is still continuing today. Just as two random people may look different, different people may respond very differently to a potentially beneficial medication. This is based on their genetic makeup and environmental experience, but it can complicate the delivery of life-saving treatments and the design of clinical trials and can lead to dangerous ADRs. The emerging field of personalized medicine tries to consider the individual basis of disease origin and response by identifying correlations between genomic mutations and potential protein or small molecule biomarkers that correlate with disease and therapeutic outcome.

This individualized treatment approach is especially important for the treatment of cancer. Therapeutic selection is a challenge because morphologically similar cancers can often have different molecular mutations and responses to therapies. In addition, patients of different genetic makeup may respond differently and sometimes adversely to potentially beneficial treatment. When choosing a therapeutic regime, it therefore is important to consider molecular information about both normal tissue and tumor, where available.

Some of the critical goals of cancer research over the past several decades have been the identification of molecular mechanisms by which cancers progress, identification of biomarkers for early cancer detection, and specific targeting of the pathways that are disrupted during cancer progression. This concept of *tailored* or *individualized medicine* has seen a few remarkable developments toward this end. One of the first examples highlighted that individuals genetically deficient for thiopurine methylotransferase (TPMT) are at high risk for developing severe neutropenia following standard thiopurine therapy [18, 19]. These individuals may be safely and efficaciously treated with thiopurine drugs, but at one-tenth the dose. At many cancer centers, patients are now routinely genotyped for TPMT prior to initiation of thiopurine treatment.

Another example was the discovery of a chromosomal translocation in chronic myelogenous leukemia that led to the development of a drug (imatinib, or Gleevec) that targets the fusion enzyme BCR-ABL, which drives the proliferation and survival of these cancer cells [20]. In breast cancer, women who are likely to benefit from antiestrogen therapy can be identified if they express the estrogen receptor that serves as a biomarker for prognosis. Moreover, trastuzumab (or Herceptin), is a drug that targets HER2 (a growth factor receptor) function in women with breast cancer who overexpress that receptor. HER2 overexpression therefore serves as a biomarker for prognosis and for treatment with Herceptin [21].

Cancer, however, is a collection of more than 270 different diseases. For most cancers, the molecular mutations have not been fully characterized, and there are few known or validated biomarkers for early detection, treatment planning, or targeted therapy. The diagnosis of cancers is still based largely on morphological examination of tumor biopsy specimens, as it has been for decades, but this approach has significant limitations for predicting a given tumor's potential for progression and response to treatment.

Cancer and other disease reclassification in pathophysiologic terms is central to the future progress of medicine and critically important to pharmacogenomics. Progress has been significant in the molecular characterization of certain cancers, such as hematological malignancies. In one study, a large-scale analysis of genomic alterations in a small set of breast and prostate cancers revealed that individual tumors accumulate an average of approximately ninety mutant genes but that only a subset of these actually contribute to the neoplasia [22].

Drug development and consequently therapy determination for most diseases still lacks well-defined molecular targets and is very much empirical. Indeed, most drugs have been shown to be efficacious in fewer than approximately 60% of patients in the disease populations. In cancer, this figure is very poor, with an average drug response rate of less than 25%, likely because of the significant amount of heterogeneity among patients with a given type of cancer [23, 24]. In addition, most cancer drugs are toxic agents that affect cell growth, so they often have significant side effects because of their activity against normal tissues in the body.

In oncology, there is considerable opportunity for improving the drug development process and developing tailored treatments as well as improving prevention, early detection, and diagnosis.

A cancer biomarker is a biological or physiologic indicator that could include a broad range of biochemical entities, such as DNA, RNA, proteins, sugars, lipids, and small metabolites, as well as whole cells, in either specific tissues of interest or in the circulation. Biomarkers should in theory improve patient outcomes by ensuring that all patients receive appropriate medication that is likeliest to be effective for their specific disease, which would improve their drug response rate and limit the likelihood of adverse events. In addition to enhancing the effectiveness of treatment, biomarkers provide the opportunity to improve the cost-effectiveness of drugs by ensuring a more targeted approach. Such an approach would bypass the use of costly therapies to which a cancer will not respond and additionally overcome the need to manage the toxicity issues of these drugs. Early detection of cancers with biomarkers at the most treatable time points would very likely improve the cost-effectiveness of treatment as well as patient outcome.

Biomarker discovery and detection, either individually or as larger sets, is now performed by a wide variety of methods ranging from imaging to molecular and biochemical analysis of blood or tissue samples.

The importance of biomarkers ranges across the spectrum from basic research to pharmaceutical discovery and development to clinical trials and ultimately translates to patient care. Screening for disease; stratification of disease risk, diagnosis, prognosis, and prediction; and planning and monitoring of therapy as well as post-treatment surveillance are all important clinical applications of biomarkers. Biomarkers are now viewed as a tremendously useful tool to evaluate drug candidates for evidence of efficacy and toxicity for the drug development translational pipeline.

4.2. Warfarin Pharmacogenomics

In addition to cancer, there are a number of recently emerging noteworthy examples that are worth commenting on. Warfarin, a commonly prescribed drug taken by more than one million people in the United States alone who are at risk for harmful blood clots, has complex factors that can pose challenges for optimal dosing. Until now, standard dosing decisions for this drug have been based largely on trial and error. Its use is complex, and accurate dosing is required because of a narrow therapeutic window, variability in dose response, interactions with drugs and diet, and risk for serious hemorrhaging [25]. Warfarin blocks the vitamin K pathway, which is required to make active clotting factors. CYP2C9, another cytochrome P450, metabolizes warfarin, of which some variants are linked with increased risk for bleeding [26]. If an individual is homozygous for the *3 variant of the gene, this drug is poorly metabolized. Warfarin action is also affected by the gene vitamin K oxidoreductase C1 (*VKORC1*). The optimal maintenance doses of warfarin can vary considerably depending on the number of copies of the low-dose/high-dose *VKORC1* variant. Variants in *VKORC1* are reported to be responsible for about 30% of the variation in the final warfarin dose, with *CYP2C9* responsible for about 10% [27].

The scientific and clinical evidence that supports lower warfarin doses for patients with certain genetic variations in *CYP2C9* and *VKORC1* has recently moved the US Food and Drug Administration (FDA) to change the labeling of warfarin, recommending genetic testing to guide warfarin dosing. The new warfarin label is not compulsory and does not require the genetic testing prior to or during warfarin treatment [28]. However, to educate the medical community, the FDA has taken a large step foward to assert that certain variations in the two genes may increase the need for more frequent patient monitoring and require lower warfarin doses. It was reported at the American Society of Human Genetics Meeting in November 2008 that another gene, *CYP4F2*, accounts for 1.5% of dose variation. If one also takes into account the approximately 15% contribution of nongenetic factors such as gender and age, more than 50% of warfarin dose variation can now be predicted. A randomized trial to unequivocally show that genotyping for warfarin dosing improves efficacy or safety is being supported by the National Institutes of Health. If safety and efficacy can attain a new level by moving away from individualized clinical judgment for warfarin dosing, this will be a real paradigm shift toward pharmacogenomics and personalized medicine.

4.3. Abacavir HIV Pharmacogenomics

Abacavir, a human immunodeficiency virus (HIV) drug, can cause serious side effects in approximately 4% of patients. Since 2002, studies have been conducted to identify HIV-1 patients at increased risk for abacavir hypersensitivity, a treatment-limiting and potentially life-threatening adverse event, and it has been demonstrated and confirmed in independent studies that the human leukocyte antigen HLA-B*5701 allele causes hypersensitivity to the drug. It has also been shown that approximately 94% of patients who do not carry the HLA-B*5701 allele are at low risk for hypersensitivity reaction to this drug [29]. Thus, a PGx test can be helpful to prevent this toxic effect of abacavir. Prospective HLA-B*5701 screening should identify patients at high risk before they receive treatment and definitively reduce the incidence of the hypersensitivity reaction [30]. Testing for that allele should thus prevent cases of this rare but potentially fatal adverse event and is offered by LabCorp, Burlington, N.C.

5. PHARMACOGENOMIC PLATFORMS AND TECHNOLOGIES

There is a wide range of differing technologies that are employed today for pharmacogenomics in both discovery and clinical testing, and key technologies are described in the following sections:

5.1. Microarray Technology for Custom and Whole Genome Variation Profiling

To date, the majority of studies have focused on variations in human drug metabolism enzymes, but with the advent of more powerful genomic tools, it is now possible to routinely scan the human genome for SNP variation ranging from hundreds of thousands of SNPs to one million SNPs. Both the 6.0 chip from Affymetrix (Santa Clara, Calif.) and the 1M duo BeadChip from Illumina (San Diego, Calif.) – which incidentally contain many SNPs for drug transporters and drug metabolizing enzymes – make such large-scale scanning possible. These technologies have now opened the floodgates for non–hypothesis driven whole genome scanning to search for predictors of toxicity and efficacy in patients treated with drugs. The Affymetrix GeneChip is composed of short, single-stranded oligonucleotides and is manufactured with a combination of solid-phase DNA synthesis and photolithography. The technology has applications in SNP genotyping, expression, copy number variation, and resequencing. Illumina has a bead-based technology that is used for SNP genotyping, expression, and methylation profiling. The bead arrays are composed of millions of tiny, etched wells into which thousands to hundreds of thousands of 3-micron beads are randomly self-assembled onto fiber-optic bundles or planar silica slides. This is followed by 50-mer gene-specific probes linked with "address" or "zip code" oligonucleotide sequences that are immobilized on the surface of the bead and subsequently facilitate a decoding process to map a specific bead type (comprising a particular sequence) to a specific location on the BeadChip. Detailed descriptions of microarray technologies have been extensively reported [31], not to mention comparisons of the advantages and disadvantages of such technologies [32–34].

5.2. The Roche AmpliChip CYP450 Test

For many drugs, including some antidepressants, antipsychotics, and attention deficit hyperactivity disorder (ADHD) drugs, metabolism is accomplished by the CYP450 enzyme system in the liver. Approximately 90% of all drug metabolism results from the activity of the CYP450 enzymes, and CYP2D6 and CYP2C19 are responsible for the metabolism of a large number of widely prescribed drugs [35]. In 2005, the FDA cleared the AmpliChip CYP450 Test, which measures variations in two genes of the CYP450 enzyme system: *CYP2D6* and *CYP2C19*. This was the first FDA-cleared test for analysis of these two genes and the first pharmacogenomic microarray test approved for clinical use. (These two genes code for enzymes that metabolize many antidepressants, antipsychotics, and ADHD drugs, as well as other medications.) The AmpliChip CYP450 test determines the metabolizing phenotype. If one metabolizes drugs ultra rapidly, at an intermediate rate, extensively, or ultra slowly, this may indicate an alteration in drug dosage; thus, patient treatment can be individualized to get the best therapeutic results possible.

The technology is based on polymerase chain reaction (PCR) from Roche and microarray from Affymetrix. It measures both duplications and deletions and assays twenty-seven mutations and polymorphisms from the *2D6* gene and three variants from the *2C19* gene. DNA can be used from buccal brushes or whole blood, and at least 25 ng is required per test. After PCR is carried out to amplify the genes of interest, the amplicons are fragmented and biotin labeled at their 3′ ends with terminal transferase. Next, the biotin-labeled PCR product is hybridized to the AmpliChip DNA microarrary. Finally, the chip is scanned on an Affymetrix GeneChip Scanner after washing and staining via a streptavidin and phycoerythrin conjugate. After extracting and analyzing the data feature, the genotype calls are processed. Results from this molecular test can guide physicians in customizing patient treatment for appropriate drug dosing for drugs metabolized via these genes.

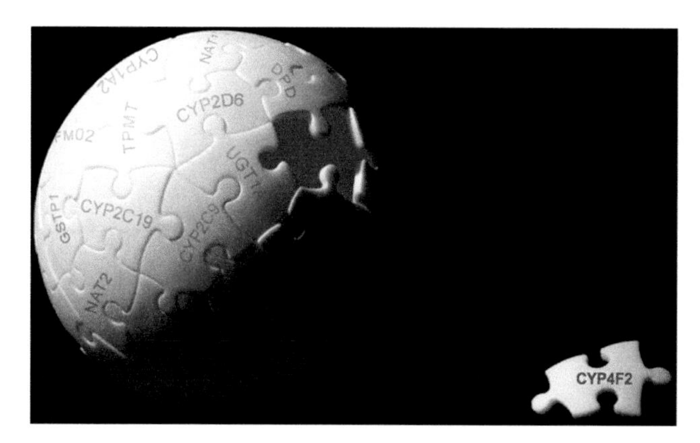

***FIGURE 21.1:* CYP4F2 was only recently discovered as associated with warfarin using the Affymetrix DMET product, and its contribution to Warfarin PGx was a highlight in the 58th American Society for Human Genetics meeting in Philadelphia in 2008. The DMET Plus product assays 1,936 drug metabolism biomarkers in 225 genes and contains markers in all FDA-validated genes to date.**

5.3. The Affymetrix Drug Metabolizing Enzymes and Transporters (DMET) Plus Chip

The DMET Plus microarray product profiles 1,936 drug metabolism biomarkers in 225 genes and automatically interprets data into a commonly used format that can be integrated into clinical trial work flows (see Figure 21.1 schematic). The information enables researchers to make more informed drug development decisions, which in turn significantly streamlines the drug development process and time to market.

This new product from Affymetrix contains the most comprehensive chip coverage of all adsorption, distribution, metabolism, and excretion (ADME) drug metabolism biomarkers, including common and rare variations, insertions, deletions, copy number, triallelic SNPs, and pseudogenes, many of which are not assayed by conventional microarray SNP methods.

Additionally, there are 169 core markers (known to have an effect on drug metabolism) that were selected in collaboration with pharmaceutical companies. Moreover, allelic frequencies for the core markers on DMET are very sensitive, typically below 0.09 on average. Additional details on DMET, including a complete list of all biomarkers assayed, can be found at http://www.affymetrix.com/products_services/arrays/specific/dmet.affx.

5.4. Applied Biosystems TaqMan Drug Metabolism Genotyping assays

Applied Biosystems has a large collection of TaqMan ADME genotyping assays. There are more than 2,600 unique validated assays to detect polymorphisms in 220 genes that code for various drug metabolism enzymes and drug transporters. A catalogue of these assays can be found at https://products.appliedbiosystems.com.

6. RESOURCES

There are numerous bioinformatics resources and other tools for the study of PGx as well as the design, execution, and subsequent validation of PGx experiments. However, the resources described below should act as an excellent starting point and window into this growing field.

6.1. The NIH Pharmacogenetics Research Network (PGRN)

The PGRN was established in 2000 to build a network of multidisciplinary research groups to conduct studies in PGx and pharmacogenomics as well as to populate a knowledge base (PharmGKB). PharmGKB is being used as a research tool to enable future PGx studies and is a world-class knowledge base in this field. The long-term goal is to translate this knowledge and identify safe and effective drug therapies designed for individual patients.

The PGRN consists of twelve independently funded, interactive research groups, including the PharmGKB. Each research group specializes in an identified area of PGx. PGRN research groups and their specific interests can be found at www.pharmgkb.org/network/pharmacogenetics_research_network.jsp.

The PGRN performs studies of variation in human genes relevant to pharmacokinetics (drug disposition) and pharmacodynamics (drug action) and the relationship of such variation to drug response phenotypes. Both raw and curated data from network (and non-network) researchers is deposited into PharmGKB.

6.2. PharmGKB (The Pharmacogenetics and Pharmacogenomics Knowledgebase)

PharmGKB (www.pharmgkb.org/index.jsp) is a publicly available, NIH-funded Internet research tool developed by Stanford University and is part of PGRN [36]. The PharmGKB database is a central repository for genetic, genomic, molecular, and cellular phenotype data and clinical information about people who have participated in pharmacogenomics research studies. The data include, but are not limited to, clinical and basic pharmacokinetic and pharmacogenomic research in the cardiovascular, pulmonary, cancer, pathways, metabolic, and transporter domains. Any pharmacogenomic investigator is welcome to submit data. There are currently thirty-nine genes that are considered very important for drug response, which are annotated and described as Very Important Pharmacogene (VIP) summaries. There are also fifty-six pathway diagrams. PharmGKB currently contains information on 545 drugs, 542 diseases, 608 genotyped genes, and 975 variants of interest. Recent enhancements associate genotypes with phenotypes, improve the depth of curated pharmacogenomics literature, and show SNPs on a gene's three-dimensional protein structure [37]. Figure 21.2 shows the Web site of PharmGKB, noted above, which is highly interactive, with useful features including a search function for gene, drug, or disease of interest as well as an interactive flowchart of PGx information. Moreover, to meet the needs of whole genome studies, PharmGKB has added new enhancements to browse the variant display by chromosome and cytogenetic locations outside the gene. It is also possible to compare across other data sources, such as dbSNP and HapMap data, as well as to store, display, and query large volumes of SNP microarray information [38].

7. FUTURE DEVELOPMENT AND CONCLUSIONS

Recent technological advances that enable examination of many potential biomarkers have fueled renewed interest in and optimism about developing PGx biomarkers, and it is widely believed that biomarkers can and will be used to improve disease screening and detection, to improve the drug development process, and to enhance the effectiveness and safety of clinical care by allowing physicians to customize treatment for individual patients and diseases.

The ongoing development of PGx not only will improve medical care for many patients, but it will also act as an example to institutions by striving to become a model for individualizing therapies through rational, molecularly guided approaches. The promise and development of PGx will enable it to serve a much broader patient base, attracting a much wider array of collaborating partners and investors, and may promote more rapid adoption of new diagnostic tools and create new revenue streams for the developers and manufacturers to ultimately benefit end users of these new products. The advent of new multidimensional approaches such as whole genome methylation profiling and other new technologies such as "next generation" sequencing will no doubt rapidly advance this field to a whole new level.

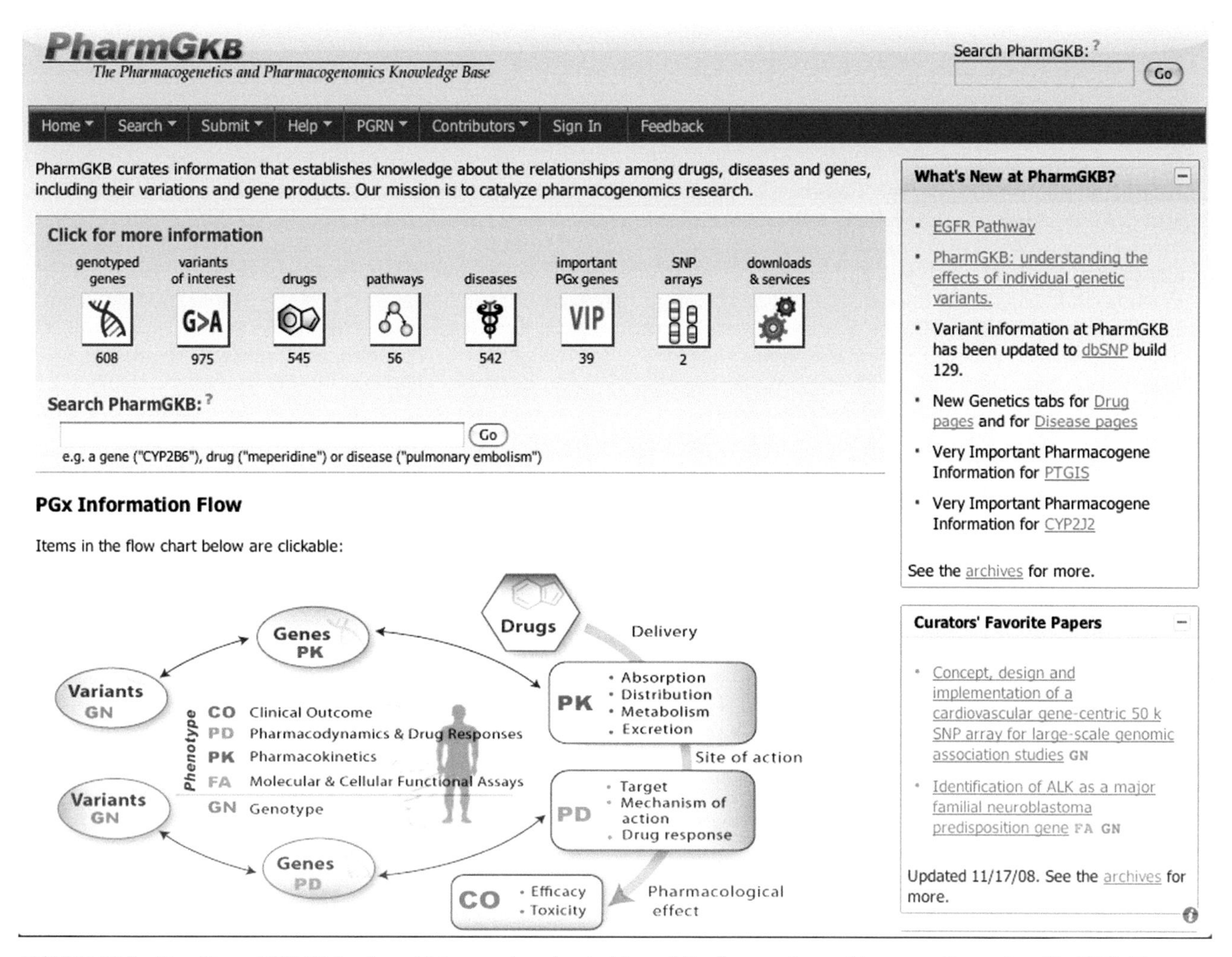

FIGURE 21.2: ***The PharmGKB Web site, which was downloaded from http://www. pharmgkb.org on November 23, 2008. The page acts as an excellent Web portal and starting point for educational and key up-to-date information on PGx.***

ACKNOWLEDGMENTS

The authors would like to sincerely thank Kurt Donner for providing the data on the Affymetrix DMET product as well as Figure 21.1 and Drs. Russ Altman and Teri Klein for allowing reproduction of Figure 21.2.

REFERENCES

1. Lazarou, J., Pomeranz, B. H., and Corey, P.N. (1998). Incidence of adverse drug reactions in hospitalized patients: a meta-analysis of prospective studies. *JAMA 279*, 1200–1205.
2. Ingelman-Sundberg, M., and Rodriguez-Antona, C. (2005). Pharmacogenetics of drug-metabolizing enzymes: implications for a safer and more effective drug therapy. *Philos Trans R Soc Lond B Biol Sci 360*, 1563–1570.
3. Phillips, K. A., Veenstra, D. L., Oren, E., Lee, J. K., and Sadee, W. (2001). Potential role of pharmacogenomics in reducing adverse drug reactions: a systematic review. *JAMA 286*, 2270–2279.
4. Mancinelli, L., Cronin, M., and Sadee, W. (2000). Pharmacogenomics: the promise of personalized medicine. *AAPS PharmSci 2*, E4.
5. Garrod, A. E. (1902). The incidence of alcaptonuria: a study in chemical individuality. *Lancet ii*, 1616–1620.
6. Garrod, A. E. (1909). *Inborn Errors of Metabolism* (New York: Oxford University Press).
7. Brown, S. M. (2003). *Essentials of Medical Genomics*. (New York: Wiley-Liss), 253.
8. Forbat, A., Lehmann, H., and Silk, E. (1953). Prolonged apnoea following injection of succinyldicholine. *Lancet 265*, 1067–1068.
9. La Du, B. N., Bartels, C. F., Nogueira, C. P., Hajra, A., Lightstone, H., Van der Spek, A., and Lockridge, O. (1990). Phenotypic and molecular biological analysis of human butyrylcholinesterase variants. *Clin Biochem 23*, 423–431.
10. Bonicke, R., and Lisboa, B. P. (1956). Uber die erbbedingtheit der intraindividuellen konstanz der isoniaszidausscheidung beim menschem (untersuchungen an eineiigen und zweieiigen zwiuingen). *Naturwissenschaften 44*.

11. Motulsky, A. G. (1957). Drug reactions enzymes, and biochemical genetics. *J Am Med Assoc 165*, 835–837.
12. Mahgoub, A., Idle, J. R., Dring, L. G., Lancaster, R., and Smith, R. L. (1977). Polymorphic hydroxylation of Debrisoquine in man. *Lancet 2*, 584–586.
13. Gonzalez, F. J., Skoda, R. C., Kimura, S., Umeno, M., Zanger, U. M., Nebert, D. W., Gelboin, H. V., Hardwick, J. P., and Meyer, U. A. (1988). Characterization of the common genetic defect in humans deficient in debrisoquine metabolism. *Nature 331*, 442–446.
14. Veenstra, D. L. (2004). The interface between epidemiology and pharmacogenomics. In *Human Genome Epidemiology: A Scientific Foundation for Using Genetic Information to Improve Health and Prevent Disease*, M. K. Khoury, J. Little, and W. Burke, eds. (New York: Oxford University Press), 234.
15. Gough, A. C., Miles, J. S., Spurr, N. K., Moss, J. E., Gaedigk, A., Eichelbaum, M., and Wolf, C. R. (1990). Identification of the primary gene defect at the cytochrome P450 CYP2D locus. *Nature 347*, 773–776.
16. Shargel, L., Wu-Pong, S., and Yu, A. (2005). *Applied Biopharmaceuticals & Pharmacokinetics*, 5th ed. (New York: McGraw-Hill), 355.
17. Allison, M. (2008). Is personalized medicine finally arriving? *Nat Biotechnol 26*, 509–517.
18. Weinshilboum, R. (2003). Inheritance and drug response. *N Engl J Med 348*, 529–537.
19. Evans, W. E., and McLeod, H. L. (2003). Pharmacogenomics–drug disposition, drug targets, and side effects. *N Engl J Med 348*, 538–549.
20. Baselga, J. (2006). Targeting tyrosine kinases in cancer: the second wave. *Science 312*, 1175–1178.
21. Duffy, M. J. (2005). Predictive markers in breast and other cancers: a review. *Clin Chem 51*, 494–503.
22. Sjoblom, T., Jones, S., Wood, L. D., Parsons, D. W., Lin, J., Barber, T. D., Mandelker, D., Leary, R. J., Ptak, J., Silliman, N., Szabo, S., Buckhaults, P., Farrell, C., Meeh, P., Markowitz, S. D., Willis, J., Dawson, D., Willson, J. K., Gazdar, A. F., Hartigan, J., Wu, L., Liu, C., Parmigiani, G., Park, B. H., Bachman, K. E., Papadopoulos, N., Vogelstein, B., Kinzler, K. W., and Velculescu, V. E. (2006). The consensus coding sequences of human breast and colorectal cancers. *Science 314*, 268–274.
23. Spear, B. B., Heath-Chiozzi, M., and Huff, J. (2001). Clinical application of pharmacogenetics. *Trends Mol Med* 7, 201–204.
24. Austin, M., and Babbiss, L. (2006). Commentary: when and how biomarkers could be used in 2016. *AAPS J 8*, E185–E189.
25. Ansell, J., Hirsh, J., Poller, L., Bussey, H., Jacobson, A., and Hylek, E. (2004). The pharmacology and management of the vitamin K antagonists: the Seventh ACCP Conference on Antithrombotic and Thrombolytic Therapy. *Chest 126*, 204S–233S.
26. Higashi, M. K., Veenstra, D. L., Kondo, L. M., Wittkowsky, A. K., Srinouanprachanh, S. L., Farin, F. M., and Rettie, A. E. (2002). Association between CYP2C9 genetic variants and anticoagulation-related outcomes during warfarin therapy. *JAMA 287*, 1690–1698.
27. Rieder, M. J., Reiner, A. P., Gage, B. F., Nickerson, D. A., Eby, C. S., McLeod, H. L., Blough, D. K., Thummel, K. E., Veenstra, D. L., and Rettie, A. E. (2005). Effect of VKORC1 haplotypes on transcriptional regulation and warfarin dose. *N Engl J Med 352*, 2285–2293.
28. Matthews, A. (2007). In milestone, FDA pushes genetic tests tied to drug. *Wall Street Journal*, August 16.
29. Mallal, S., Phillips, E., Carosi, G., Molina, J. M., Workman, C., Tomazic, J., Jagel-Guedes, E., Rugina, S., Kozyrev, O., Cid, J. F., Hay, P., Nolan, D., Hughes, S., Hughes, A., Ryan, S., Fitch, N., Thorborn, D., and Benbow, A. (2008). HLA-B*5701 screening for hypersensitivity to abacavir. *N Engl J Med 358*, 568–579.
30. Hughes, A. R., Spreen, W. R., Mosteller, M., Warren, L. L., Lai, E. H., Brothers, C. H., Cox, C., Nelsen, A. J., Hughes, S., Thorborn, D. E., Stancil, B., Hetherington, S. V., Burns, D. K., and Roses, A. D. (2008). Pharmacogenetics of hypersensitivity to abacavir: from PGx hypothesis to confirmation to clinical utility. *Pharmacogenomics J 8*, 365–374.
31. Gunderson, K. L., Steemers, F. J., Lee, G., Mendoza, L. G., and Chee, M. S. (2005). A genome-wide scalable SNP genotyping assay using microarray technology. *Nat Genet 37*, 549–554.
32. Hardiman, G. (2004). Microarray platforms – comparisons and contrasts. *Pharmacogenomics 5*, 487–502.
33. Wick, I., and Hardiman, G. (2005). Biochip platforms as functional genomics tools for drug discovery. *Curr Opin Drug Discov Devel 8*, 347–354.
34. Li, M., Li, C., and Guan, W. (2008). Evaluation of coverage variation of SNP chips for genome-wide association studies. *Eur J Hum Genet 16*, 635–643.
35. Michalets, E. L. (1998). Update: clinically significant cytochrome P-450 drug interactions. *Pharmacotherapy 18*, 84–112.
36. Klein, T. E., Chang, J. T., Cho, M. K., Easton, K. L., Fergerson, R., Hewett, M., Lin, Z., Liu, Y., Liu, S., Oliver, D. E., Rubin, D. L., Shafa, F., Stuart, J. M., and Altman, R. B. (2001). Integrating genotype and phenotype information: an overview of the PharmGKB project. Pharmacogenetics Research Network and Knowledge Base. *Pharmacogenomics J 1*, 167–170.
37. Sangkuhl, K., Berlin, D. S., Altman, R. B., and Klein, T. E. (2008). PharmGKB: understanding the effects of individual genetic variants. *Drug Metab Rev 40*, 539–551.
38. Hernandez-Boussard, T., Whirl-Carrillo, M., Hebert, J. M., Gong, L., Owen, R., Gong, M., Gor, W., Liu, F., Truong, C., Whaley, R., Woon, M., Zhou, T., Altman, R. B., and Klein, T. E. (2008). The pharmacogenetics and pharmacogenomics knowledge base: accentuating the knowledge. *Nucleic Acids Res 36*, D913–D918.

CHAPTER 22

Drugs, Genomic Response Signatures, and Customized Cancer Therapy

Rafael Rosell

Teresa Moran

Miguel Taron

Aberrant signaling generated by the activation of multiple pathways occurs in cancers and contributes to their growth, invasion, and survival. For example, lung cancer–specific epidermal growth factor receptor (*EGFR*) mutations result in constitutive activation of *EGFR* and downstream signaling components, such as *Akt*, with dramatic response to tyrosine kinase inhibitors (TKIs) (gefitinib or erlotinib). In patients with multiple metastases, a single pill of erlotinib daily can attain a 70% response rate, a twelve-month time to progression, and a twenty-two-month median survival. In fact, in some subgroups of patients, survival has not been reached. However, results can still be further improved. A growing body of evidence suggests a role for lateral signaling or cross talk between various receptor tyrosine kinases (TKs), with subsequent signaling through multiple receptors. This knowledge can facilitate the design of new therapeutic strategies. New data on the role of prognostic markers will help both to identify the high-risk group of patients who will relapse after surgery and to customize treatment for all patients. DNA damage response is a global signaling network that – among multiple functions – facilitates DNA repair processes and thus determines the sensitivity or resistance to different cytotoxic drugs. Central to DNA damage response is the breast cancer gene 1 (*BRCA1*). We have examined the predictive value of the *BRCA1-RAP80-Abraxas* complex in patients with metastatic lung cancer who were receiving customized treatment according to *BRCA1* messenger RNA (mRNA) levels. Our results indicate that a subgroup of patients with low levels of *BRCA1* and *RAP80* respond dramatically to cisplatin-based chemotherapy, with a time to progression of fourteen months and median survival not reached. These clinical findings require additional validation but can be even further improved with the proper combination with targeted therapy based on identification of the right molecular targets. This model of customization can be extrapolated to multiple different primary tumors.

1. SIMPLE FACTS AND CURRENT CANCER TREATMENT APPROACHES

It is estimated that about 565,650 Americans will die from cancer in the coming year – more than 1,500 deaths per day. Cancer is the leading cause of death among women aged forty to seventy-nine years and among men aged sixty to seventy-nine years. Cancers of the lung and bronchus, prostate, and colon and rectum in men and cancers of the lung and bronchus, breast, and colon and rectum in women continue to be the most common fatal cancers, accounting for half of the total cancer deaths. Lung cancer surpassed breast cancer as the leading cause of cancer death in women in 1987 and is expected to account for 26% of all female cancer deaths in 2008 [1]. Among men younger than forty years, leukemia is the most common fatal cancer, whereas cancer of the lung and bronchus predominates in men aged forty years and older. The second most common cause of cancer death is colorectal cancer among men aged forty to seventy-nine years and prostate cancer among men aged eighty years and older. Among women, leukemia is the leading cause of cancer death before age twenty; breast cancer ranks first at age twenty to fifty-nine; and lung cancer ranks first at age sixty and older [2].

1.1. Limited Cancer Biology Concepts Used by Medical Oncologists for Treatment Decisions

Understanding the molecular underpinnings of cancer is crucial for targeted intervention strategies. Identification of the right targets, however, is notoriously difficult and unpredictable. Malignant cell transformation requires the cooperation of a few oncogenic mutations that cause substantial reorganization of many cell features [3] and induce complex changes in gene expression patterns [4]. Genes critical to this multifaceted cellular phenotype have only been identified after signaling pathway analysis [5, 6] or on an ad hoc basis [7–9]. Cell transformation by cooperating oncogenic lesions depends on the synergistic modulation of downstream signaling circuitry, suggesting that malignant transformation involves synergy at multiple levels of regulation, including gene expression. For example, a large proportion of genes controlled synergistically by loss-of-function *p53* and *Ras* activation are critical to the malignant state of murine and human colon cells [10]. Targeted disruption of oncogenic mutations downstream may allow selective cancer deconstruction, yielding intervention strategies with high specificity for cancer cells.

However, no significant improvement in outcome has been reported in phase III trials of chemotherapy plus targeted agents in non-small cell lung cancer (NSCLC). Several studies combining platinum-based chemotherapy with *EGFR* TKIs, such as gefitinib and erlotinib, had negative findings [11–13]. (For further review, see Wheatley-Price and Shepherd [14].)

Angiogenesis inhibitors targeting the vascular endothelial growth factor (*VEGF*) signaling pathways have demonstrated therapeutic efficacy in mouse models of cancer and in an increasing number of human cancers. However, in both the preclinical and clinical settings, the benefits are at best transitory and are followed by a restoration of tumor growth and progression [15]. Recognition of the *VEGF* pathway as a key regulator of angiogenesis has led to the development of several *VEGF*-targeted agents, including neutralizing antibodies to *VEGF* or *VEGFRs*, soluble *VEGF* receptors or receptor hybrids, and TKIs with selectivity for *VEGFRs* [16]. Bevacizumab (Avastin, Genentech/Roche), the first *VEGF*-targeted agent, is an anti-*VEGF* monoclonal antibody that has shown clinical benefit in patients with metastatic colorectal cancer when combined with chemotherapy [17]. Bevacizumab and the *VEGFR* TKIs sorafenib (Nexavar, Bayer/Onyx) and sunitinib (Sutent, Pfizer) are currently approved by the U.S. Food and Drug Administration (FDA) for clinical use [16]. Bevacizumab is currently approved by the FDA for patients with metastatic colorectal cancer, NSCLC, and metastatic breast cancer in combination with chemotherapy [16]. The neutralization of *VEGF* by an antibody or soluble receptor construct (*VEGF* trap) can prevent its binding to and activation of *VEGFR1* and *VEGFR2* as well as *neuropilin 1* and *neuropilin 2*, which act as co-receptors for the *VEGFRs*. The TKI sorafenib has shown single-agent efficacy in patients with advanced renal cell carcinoma and hepatocellular carcinoma. Sunitinib has also been shown to be efficacious as a single agent in patients with renal cell carcinoma. It should be stated that because of their mode of action at the adenosine triphosphate (ATP) binding pocket, TKIs are selective rather than specific for a particular kinase(s). Therefore, TKIs designed to target *VEGF* receptors are actually considered multi-kinase inhibitors. For example, sorafenib and sunitinib also have significant activity against Raf, platelet-derived growth factor receptor (*PDGFRB*), fibroblast growth factor receptor (*FGFR*), *FLT3*, *KIT*, and *FMS* (also known as *CSF1R*) receptors [16].

In patients for whom *VEGF*-targeted therapy is found to be efficacious, the duration of activity is relatively short. Median survival was 12.3 months in stage IV NSCLC patients treated with paclitaxel plus carboplatin and bevacizumab compared with 10.3 months for those treated only with paclitaxel plus carboplatin (hazard ratio [HR] for death, 0.79; $P = .003$) [18]. However, HRs differed substantially according to the pattern of metastases, with the maximum benefit for bevacizumab seen in patients with liver metastases (HR, 0.68) whereas it was nil in patients with adrenal metastases (HR, 0.97) [18]. Intriguingly, in patients with metastatic colorectal cancer, median survival was 20.3 months for patients receiving irinotecan, bolus fluorouracil, and leucovorin (IFL) plus bevacizumab compared with 15.6 months for patients receiving IFL plus placebo (HR, 0.66; $P < .001$) [17]. Liver metastases are predominant in colorectal cancer, and the HR for these patients is similar to that observed in patients with NSCLC with liver metastases [18]. (In NSCLC, the incidence of liver metastases is as high as 35% [19].) In metastatic breast cancer, median survival was similar in patients receiving paclitaxel or paclitaxel plus bevacizumab (26.7 vs. 25.2 months; HR, 0.88; $P = .16$ [20]. The authors speculate that angiogenesis inhibitors could behave better in earlier phases of breast cancer in patients with micrometastatic disease in the adjuvant setting [20].

The *HIF* regulatory network plays an important role in angiogenesis, and metastatic renal cell carcinomas with high expression of carbonic anhidrase IX (*CA IX*) are most likely to benefit from interleukin-2 therapy (reviewed in Michaelson et al. [21]). Interestingly, low *CA IX* mRNA expression conferred poor prognosis in early resected NSCLC [22], raising the hypothesis that even in early stages of cancer, tumor growth could become independent from common angiogenesis mechanisms and thus be resistant to angiogenesis inhibitors. Hence, one of the most important shortcomings in the field of VEGF-targeted therapy is the failure to identify and validate predictive markers (a priori markers of drug activity that predict different degrees of benefit from a specific therapy).

Tumor growth encompasses many aspects of normal wound healing, and a wound response gene expression signature is reactivated in many types of human cancers, including breast and lung [23]. The wound response signature is composed of 512 genes that define the transcriptional response of fibroblasts to serum, the soluble fraction of clotted blood. In early breast cancer and lung adenocarcinoma, the wound response signature provides prognostic risk stratification of metastases development [23]. Several gene expression signatures, generally containing nonoverlapping genes, provide similar predictive information on clinical outcome, and a model combining several signatures did not perform better than each of the signatures separately. These signatures may be largely different from one another with regard to gene identity, but they occupy overlapping prognostic space [24] (further reviewed in Rosell et al. [25]). The special AT-rich binding protein (*SATB1*) was originally identified as a protein that recognized double-stranded DNA with a high degree of base-unpairing (referred to as base-unpairing regions [BURs]) [26]. *SATB1* was defined as a transcriptional regulator that functions as a "landing platform" for several chromatin remodeling enzymes and hence regulates large chromatin domains [27]. *SATB1* is a genome organizer that tethers multiple genomic loci and recruits chromatin-remodeling enzymes to regulate chromatin structure and gene expression. *SATB1* is expressed by aggressive breast cancer cells, and its expression level has high prognostic significance. RNA-interference-mediated knockdown of *SATB1* in highly aggressive cancer cells altered the expression of more than one thousand genes, inhibiting tumor growth and metastasis in vivo [28]. *ERBB1*, *ERBB2*, *ERBB3*, and *ERBB4* are among the upregulated genes, together with genes that stimulate invasion and mediate angiogenesis and bone metastasis [29], such as connective tissue growth factor (*CTGF*) [28]. *SATB1* nuclear staining was significantly correlated with survival in 985 patients with ductal breast carcinoma stratified by *SATB1* expression protein level [28]. Importantly, because *SATB1* tethers multiple genomic loci and regulates chromatin structure and gene expression [27], the analysis of *SATB1* mRNA or protein expression could provide potential prognostic and/or predictive information in clinical trials of angiogenesis inhibitors.

Translational clinical trials for validation of predictive markers are mandatory for understanding and optimizing treatment with the myriad of small molecule signal transduction inhibitors that are being tested in multiple tumors. Small molecules are being designed to interfere with specific steps along the deregulated signaling cascade from the cytoplasm membrane to the nucleus (reviewed in Leary and Johnston [30]). We focus here on *BRCA1* mRNA as a promising marker for customizing chemotherapy. Although the identification of patients who will respond to novel targeted therapies remains elusive, because current cancer management includes the combination of chemotherapy with targeted drugs, optimizing chemotherapy is a cardinal first step. Moreover, a causal relationship has been identified between *EGFR* TK domain mutations and a subgroup of NSCLC (as seen in section 5).

Mitogen-activated protein kinase (*MAPK*) cascades are key signaling pathways involved in the regulation of normal cell proliferation, survival, and differentiation. Aberrant regulation of *MAPK* cascades contributes to cancer. The extracellular signal-regulated kinase (*ERK*) *MAPK* pathway has been the subject of research leading to the development of pharmacologic inhibitors (reviewed in Roberts and Der [31]). *ERK* is a downstream component of an evolutionarily conserved signaling module that is activated by the *Raf* serine/threonine kinases. *Raf* activates the *MAPK/ERK* kinase (*MEK*) 1/2 dual-specific protein kinases, which then

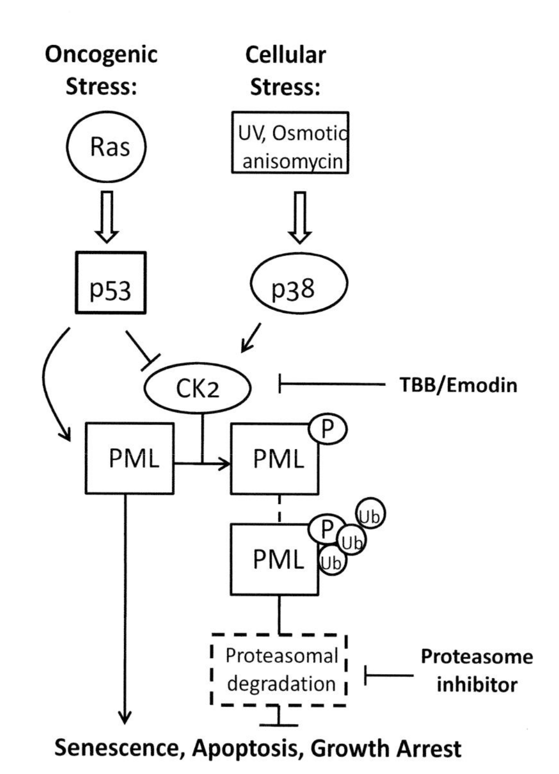

***FIGURE 22.1:** The molecular mechanisms controlling* **PML** *polyubiquitylation. During the cellular response to stress, casein kinase 2* **(CK2)** *controls* **PML** *protein levels through integration of upstream* **p53** *and p38* **MAPK** *signals. Therapy with* **CK2** *or proteasome inhibitors will abrogate aberrant* **PML** *protein degradation, leading to restoration of* **PML** *tumor-suppressor properties [48]. We hypothesize that because* **CK2** *can upregulate heterochromatin protein 1*(**HP1-β**)*, initiating DNA damage response [51] – and consequently* **BRCA1** *upregulation – loss of* **PML** *could be a surrogate of increased* **BRCA1** *mRNA observed in the clinical setting (see text).* (**Source:** *Figure 22.1 adapted with permission from Elsevier.)*

activate *ERK*1/2. The *Raf-MEK-ERK* pathway is a key downstream effector of the *Ras* small guanosine triphosphate (GTP)-ase, the most frequently mutated oncogene in human cancers (see Figure 1 in [31]). *B-Raf* mutant tumors are sensitive to *MEK* inhibition [32]. Tumor cell lines with *B-Raf* mutations were much more sensitive to *MEK* inhibition than cells with wild type *B-Raf* or mutant *K-ras*. In addition, *B-Raf*–positive tumor xenografts had completely abrogated tumor growth, whereas xenografts of *Ras* mutation-positive tumor cells were only partially inhibited [32]. Intriguingly, in melanoma with *Ras* mutations, cells switch their signaling from *B-Raf* to *C-Raf* [33]. Elevated *C-Raf* protein levels reflect a post-transcriptional regulatory mechanism of acquired resistance to *B-Raf* inhibition in melanoma harboring the recurrent V600E-activating *B-Raf* mutation [34]. The increased expression of *C-Raf* in resistant cells was not associated with *C-Raf* gene amplification or elevated mRNA levels of *C-Raf* [34]. Interestingly, the resistant cells exhibited exquisite sensitivity to the HSP90 inhibitor geldanamycin [34].

Tobacco smoking is the major cause of lung cancer, and nicotine in tobacco smoke leads to both addiction and further metabolism into potent carcinogens. Functional nicotinic acetylcholine receptors have been discovered in lung tumors, raising the question of whether exposure to nicotine could also be involved in lung cancer development. Three studies have found an association between single nucleotide polymorphisms (*SNPs*) on chromosome 15 and risk for lung cancer. Among the genes located in this region are those that encode subunits of nicotinic acetylcholine receptors, including *CHRNA3*, the α3 nicotinic receptor subunit gene, and *CHRNA5*, the α5 nicotinic receptor subunit gene [35–37]. Two of the most noteworthy of the SNPs associated with risk of lung cancer are rs1051730 (in exon 5 of *CHRNA3*) and rs16969968 (in exon 5 of *CHRNA5*). A third SNP, rs8034191, is also associated with predisposition to lung cancer. Nicotine may facilitate tumor transformation by stimulating angiogenesis and tumor growth mediated through its interaction with nicotinic acetylcholine receptors. However, an increased risk for developing lung cancer was also found in nonsmokers who harbored the SNPs of the *nicotinic receptor subunit* genes. There is a different *nicotinic acetylcholine receptor* gene expression pattern in lung cancers in smokers and nonsmokers. A sixty-five-gene-expression signature was associated with lung tumors of nonsmokers that overexpressed the mRNA of the α6 and β3 subunits [38]. The clinical usefulness of SNPs needs to be further elucidated, but the discovery of the association between these nicotinic acetylcholine receptor SNPs and lung cancer risk could pave the way for their use both in complementing the tools of lung cancer screening and in assessing new treatment strategies. Analysis of squamous cell lung cancers revealed increased mRNA levels of *α5* and *β3 nicotinic acetylcholine receptors* and increased levels of acetylcholine, associated with increased levels of choline acetyltransferase mRNA and decreased cholinesterase mRNA [39]. One approach to blocking the proliferative effects of nicotine and acetylcholine on lung cancers may be through M3 muscarinic acetylcholine receptor antagonists, which can limit the activation of *MAPK* that is caused by both nicotine and muscarinic signaling [39, 40]. The M3-selective muscarinic antagonist darifenacin blocked nicotine-stimulated H520 growth in vitro and also blocked H520 growth in nude mice in vivo [39].

2. POTENTIAL PREDICTIVE ROLE OF PROMYELOCYTIC LEUKEMIA PROTEIN

Loss of the promyelocytic leukemia (PML) tumor suppressor [41] has been observed in several cancers of various histologic origins. The *PML* gene is fused to the retinoic acid receptor alpha (*RARα*) gene in the vast majority of acute promyelocytic leukemias and has been implicated in the control of key tumor-suppressor pathways. *PML* expression was reduced or abolished in 63% of prostate adenocarcinomas, 28% of colon adenocarcinomas, 17% of breast carcinomas, 31% of NSCLC, 69% of brain tumors, 21% of lymphomas, and 49% of germ cell tumors, whereas normal control tissue samples revealed intense *PML* staining. In some tumor types (thyroid and adrenocortical carcinomas), no loss of *PML* protein expression was detected [42]. Intriguingly, *PML* mRNA was expressed in virtually all of the tumors, and *PML* mRNA expression was retained in samples negative for *PML* protein expression. The loss of *PML* protein may be due to proteasome-dependent mechanisms [42]. *PML* is typically found in multiprotein speckled subnuclear structures termed *PML* nuclear bodies. More than fifty proteins have been reported to co-localize with *PML* in the nuclear body, either transiently or constitutively, including *p53*, *pRB*, and *BRCA1* [43]. Therefore, *PML* could affect the function of these proteins by regulating their localization into the *PML* nuclear body (reviewed in Gurrieri et al. [42]). *PML* inactivation in mouse models renders cells resistant to the pro-apoptotic effects of ionizing radiation and chemotherapy [44].

PML has also been identified as a critical inhibitor of angiogenesis in vivo, in both ischemic and neoplastic conditions, through the control of protein translation. In hypoxic conditions, *PML* acts as a negative regulator of the synthesis rate of hypoxia-inducible factor 1 alpha (*HIF-1α*) by repressing mammalian target of rapamycin (*mTOR*) [45]. *PML* physically interacts with *mTOR* and negatively regulates its association with the small GTPase *Rheb* by favoring *mTOR* nuclear accumulation. Lack of *PML* inversely correlates with phosphorylation of *ribosomal protein S6* and tumor angiogenesis in mouse and human tumors [45]. Therefore, analysis of *PML* expression levels in human cancer could be used as a predictive marker of the efficacy of treatment with *mTOR* (further reviewed in Faivre et al. [46]) or angiogenesis inhibitors. Similarly, the transcription factor *ATF6* is a pivotal survival factor for quiescent – but not proliferative – squamous carcinoma cells. *ATF6α* induces survival through upregulation of *Rheb* and activation of *mTOR* signaling. Downregulation of *ATF6α* or *Rheb* reverted dormant tumor cell resistance to rapamycin [47].

In addition, casein kinase 2 (*CK2*) regulates *PML* protein levels by promoting ubiquitin-mediated degradation dependent on direct phosphorylation at Ser 517 [48]. *CK2* is frequently upregulated in human cancers and has been reported to confer poor prognosis in squamous cell cancer of the lung [49]. It is postulated that under conditions of oncogenic stress, such as those triggered by oncogenic *Ras*, *PML* is activated and exerts its tumor-suppressor function in conjunction with several partners, including *p53*. In this context, *p53* may inhibit *CK2* to achieve maximal *PML* activity and tumor-suppressive effects. Conversely, when *CK2* kinase activity is upregulated, *PML* is polyubiquitylated and degraded (Figure 22.1) [48]. The observation that *PML* loss cooperates with oncogenic *Ras* in a transgenic mouse model of NSCLC provides further support for this model; interestingly, *CK2* activity and *Ras* mutations were independent variables in the eight NSCLC cell lines examined [48]. (For further review of *ras* and *mTOR* signaling, see Shaw and Cantley [50].)

Finally, it has recently been found that *CK2* is involved in a signaling cascade that helps to initiate DNA damage response (DDR) in the presence of DNA breaks, altering chromatin by modifying a histone-code mediator protein, heterochromatin protein 1 *(HP1)-β*, that promotes *H2AX* (H2A histone family member X) phosphorylation [51].

PML plays an important prognostic role in chronic myeloid leukemia (CML). Patients with CML with low *PML* expression displayed higher complete molecular response and complete cytogenetic response than patients with high *PML* expression. Furthermore, low *PML* expression was strikingly predictive of better overall survival in CML [52].

3. DNA DAMAGE RESPONSE AND CUSTOMIZED CHEMOTHERAPY BASED ON BRCA1 mRNA LEVELS

Experimental findings suggest that both DNA double-strand breaks (DSBs) and DDR can be induced by ionizing radiation [53], hypoxia [54], DNA-damaging agents, or activated oncogenes [55]. DDR is a global signaling network that senses different types of damage and coordinates a response that includes activation of transcription, cell cycle control, apoptosis, senescence, and DNA repair processes. At the core of the DNA damage signaling apparatus are a pair of related protein kinases – *ATM* (ataxia telangiectasia mutated) and *ATR* (ATM and Rad3-related) – that are activated by DNA damage (reviewed in Harper and Elledge [56]). A large-scale proteomic analysis of proteins phosphorylated in response to DNA damage on consensus sites recognized by *ATM* and *ATR* identified more than nine hundred regulated phosphorylation sites encompassing more than seven hundred proteins [57]. This set of proteins is highly interconnected, with a large number of protein modules and networks not previously linked to DDR [57]. In our opinion, a module that is central to DDR is the breast cancer gene 1 (*BRCA1*), which includes *BRCA1*-associated ring domain protein (*BARD1*), *BRCA2*, partner and localizer of *BRCA2* (*PALPB2*), *ATM*, and *E2F1*. This module also contains *RAD51* and *XRCC3* (which are related to homologous recombination repair) and replication protein A (*RPA*) and excision repair cross-complementing 1 (*ERCC1*) (belonging to the nucleotide excision repair pathway) (see Figure 5H in Matsuoka et al. [57]). By dimerizing with *BARD1* through the RING domain, *BRCA1* forms an E3 ubiquitin ligase [58]. Another module involved in DDR is the *COP9 signasolome CSN*. This complex is involved in the *SCF* (SKp1-Cullin-F-box protein) pathway, where it participates in a neddylation-deneddylation cycle important in *SCF E3 ligase* function and alters its association with Cockayne syndrome type A (*CSA*) and damage-specific DNA binding protein 2 (*DDB2*), two DDB1-CUL4-associated factors (*DCAF*) proteins in *Cul4* ligases [57] (see Figure 5I in Matsuoka et al. [57]). Data suggest that the role of *DDB1-CUL4* may be to promote *DDB2* degradation once damage has been recognized, thereby facilitating the recruitment of the *XPC-Rad23* complex to initiate nucleotide excision repair [59]. *CUL4-DDB1-DDB2* complexes also promote *histone H2A* and *XPC* ubiquitination at damage sites, suggesting that *DDB2* serves multiple functions in this pathway. *XPC* ubiquitination promotes recruitment of other components of the XP repair system to chromatin [59]. It has also recently been reported that ultraviolet-DDB assembled on the *CUL4*A or *CUL4B*-RING platforms of ubiquitin ligases is a link between DNA repair, chromatin, and ubiquitination [60]. The association of both *CUL1A* and *CUL1B* with damaged DNA and the co-immunoprecipitation of each with *DDB1*, *DDB2*, and *RING box 1* (*RBX1*) argue for an early recruitment during the initiation of nucleotide excision repair [60]. (For further review, see Sugasawa [61], in which the human nucleotide excision repair pathway is fully described.)

Coordinated amplification of *CSN5* (also known as *JAB1* or *COPS5*) and *MYC* is the cause of wound response signature activation in human breast cancers. *CSN5* and *MYC* mRNA levels predict the wound response signature [62]. *CSN5* encodes the catalytic component of the *COP9* signasolome. The *COP9* signasolome, composed of eight proteins named *CSN1* to *CSN8*, maintains the activity of S-phase kinase protein 1 (*SKp1*), Cul 1 and F-box (*SCF*), and other cullin-RING family E3 ubiquitin ligases. The *COP9* signasolome deneddylates *Cul1*, and the catalytic activity resides in a metalloproteinase-like motif – termed the *JAMM* motif – in *CSN5*. Deneddylation by *CSN5* maintains *SCF* complexes in an active state, leading to the ubiquitination and proteosomal degradation of many proteins. It has been found that *CSN5* enhances the co-transcriptional ubiquitination of *MYC* that activates the transcriptional activity of *MYC* on a set of target genes, promoting cell proliferation, invasion, and angiogenesis (Figure 22.2) [62]. Macrophage migration inhibitory factor (*MIF*) is secreted by immune and parenchymal cells upon inflammatory and stress stimulation, but it is also expressed intracellularly, where it likely serves regulatory functions that are mediated by protein-protein interactions. *MIF* plays a pivotal role in the pathogenesis of acute and chronic inflammatory diseases including septic shock, rheumatoid arthritis, and atherosclerosis. Interestingly, *CSN5* acts as an intracellular *MIF*-binding protein that serves as a "molecular sink" to prevent the secretion of *MIF* [63]. *CSN5* is

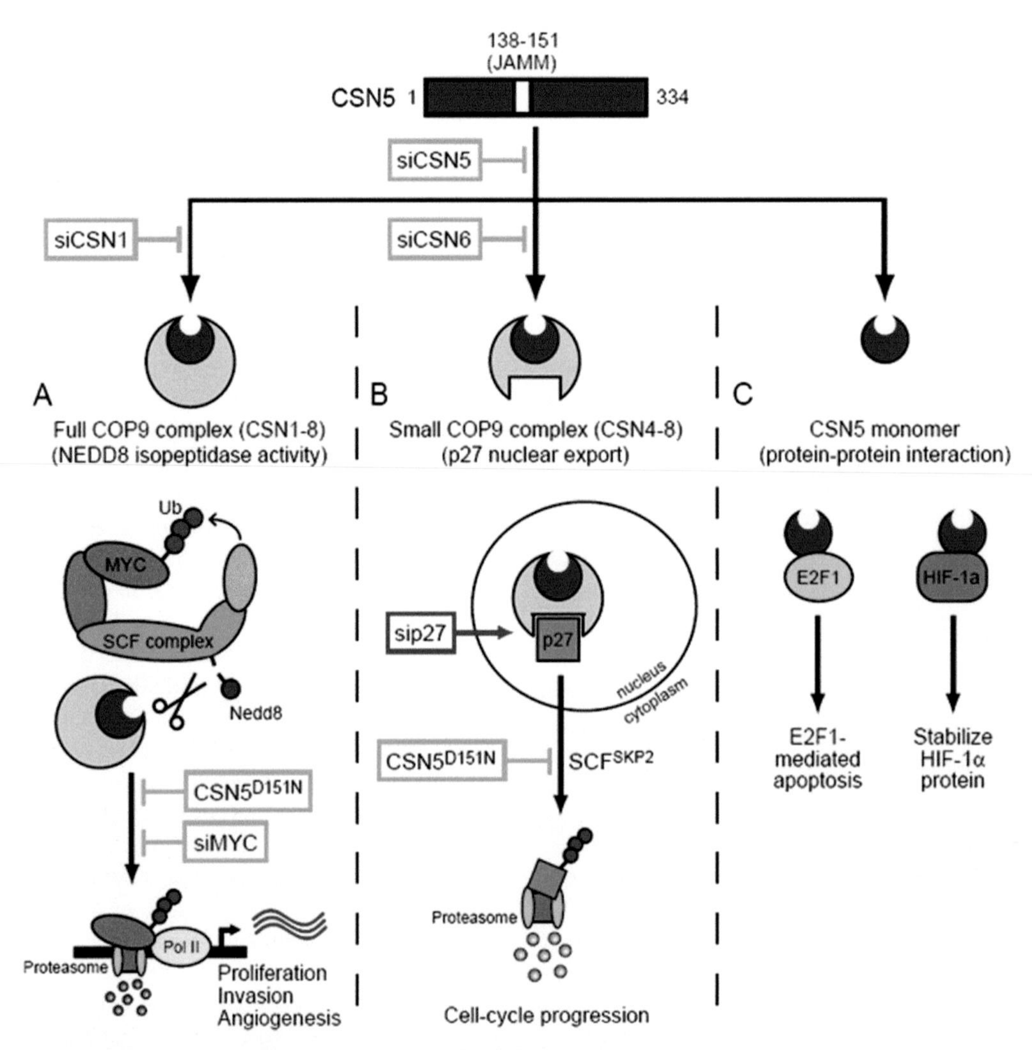

FIGURE 22.2:** Three modes of action of* **CSN5**. **CSN5** *encodes the catalytic component of the* **COP9** *signasolome. A main function of* **COP9** *is to maintain the activity of* **SCF** *(SKP1**, **CUL1**, *and* **F-box***) and other cullin-RING family E3 ubiquitin ligases.* **CSN5** *enhances the cotranscriptional ubiquitination of* **MYC** *that activates the transcriptional activity of* **MYC** *on a set of target genes promoting cell proliferation, invasion, and angiogenesis [64]. Other modes of action of* **CSN5** *are described in the text, including its implication in DNA damage response (DDR) (reviewed in Matsuoka et al. [57]). Polyubiquitination and subsequent proteasomal degradation is an ancient eukaryotic regulatory system by which the cell removes unwanted proteins. The sequential activity of three classes of enzymes marks proteins for proteasomal degradation. The ubiquitin activating enzyme (***E1***) facilitates the ATP-dependent activation of ubiquitin, which is then transferred to an ubiquitin conjugase (***E2***). Ubiquitin ligases (***E3***) then facilitate the transfer of ubiquitin onto specific lysine residues of the target protein. Specificity in this process is provided by the large family of* **E3** *enzymes, each of which recognizes particular substrates. A major group of the* **E3** *family is formed by the cullin-RING complexes. Of this group, the SCF1 complexes have been the most extensively studied. SCF1 complexes contain the RING-domain protein* **RBX1** **(RING-box 1)**, **CUL1** **(cullin 1)**, *and* **SKP1** *(S-phase-kinase associated protein 1). They also include one of a family of proteins with a distinctive motif, the F-box, that binds* **SKP1** *(reviewed in Sonnberg et al. [130]). An example of F-box protein is* **FBW7** *(F-box and WD repeat domain-containing 7), which is also involved in one of the modules involved in DDR (reviewed in Matsuoka et al. [57]. (***Source:** *Reprinted with permission of* **Cancer Research.***)*

overexpressed in many tumors and is indirectly linked to prosurvival effects in several carcinomas via *p27*. In a small *CSN* complex composed of five proteins (*CSN4*, *CSN5*, *CSN6*, *CSN7, and CSN8*), *CSN5* induces the cytoplasmic export and degradation of *p27* (Figure 22.2). Importantly, monomeric *CSN5* protein interacts with *HIF-1α*, leading to *HIF-1α* protein stabilization and increased angiogenic activity (Figure 22.2) [64]. *CSN5* regulates cellular signaling via the *MAPK* pathway – and indirectly via the *Akt* pathway – by inhibiting *MIF* secretion and its autocrine

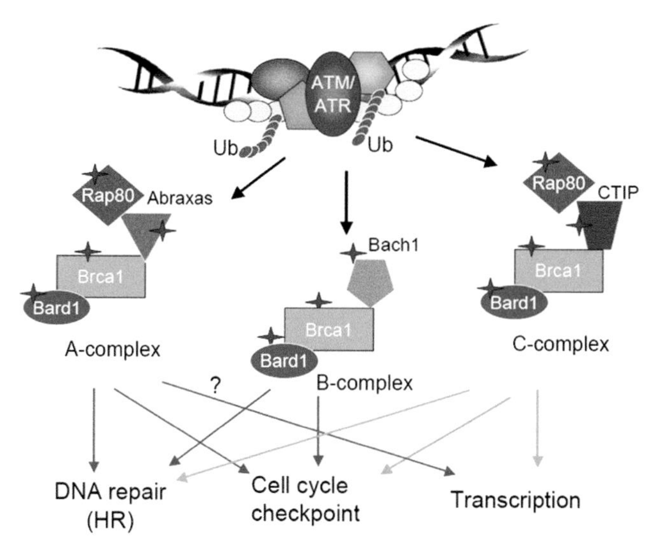

FIGURE 22.3: ***A model for breast cancer gene 1 (*****BRCA1*****) intervention in DNA damage response (DDR) has been proposed [58]. It is suggested that*** **BRCA1 BRCT** ***domains form mutually exclusive complexes with*** **Abraxas** ***1,*** **BRCA1*****-associated C-terminal helicase (*****BACH1*****), and CtBP-interacting protein (*****CtIP*****) through the pSXXF motif. These proteins may serve as adaptor proteins to recruit the*** **BRCA1-Bard1** ***E3 ubiquitin ligase to specific target proteins analogous to F-box protein's role in the SCF complexes. The*** **BRCA1-Abraxas-RAP80** ***complex is central to DDR, in which the function of*** **RAP80** ***is crucial for effective DDR and resistance to cisplatin-based chemotherapy, as epitomized in the first phase II Spanish Lung Cancer Group (SLCG) customized trial. This trial identified record time to progression and survival in a subgroup of patients with advanced NSCLC receiving cisplatin-based chemotherapy whose tumors showed very low mRNA expression levels of both*** **BRCA1** ***and*** **RAP80** ***(see text). (Source: Reprinted with permission from AAAS.)***

prosurvival activities. *MIF*-induced *Akt* activation leads to the phosphorylation and inactivation of two key proapoptotic proteins, *BAD* and *Foxo3a* [63]. *CSN5* is also involved in the regulation of *ERK MAPK* activity by *MIF* (see Figure 8 in Lue et al. [65]). It has also been found that the wound response signature (expressing *CSN5* plus *MYC*) highly correlates with a proteasome signature, both being strong predictors of response to bortezomib (a proteasome inhibitor); therefore, the wound response signature could be a critical determinant of bortezomib response in breast cancer [66]. Furthermore, bortezomib restored induction of the proapoptotic *BH3*-only *Bcl2*-interacting mediator of cell death (*BIM*), abrogated the *H-ras/MAPK* pathway-dependent paclitaxel resistance, and promoted *BIM*-dependent tumor regression [67].

Another module implicated in DDR is composed of *cyclin E* and two negative regulators of *cyclin E*: F-box and WD repeat domain-containing 7 (*FBW7*, also known as *FBXW7*, *CDC4*, *AGO*, and *SEL10*), which targets *cyclin E* for ubiquitination and destruction, and the cyclin-dependent kinase inhibitor *p27*, which binds to *cyclin E-Cdk2* complexes and inhibits their activity (see Figure 5J in Matsuoka et al. [57]). *Cyclin E* is autophosphorylated in cis by cyclin-dependent kinase 2 (*CDK2*) on S384 and by glycogen synthase kinase 3 (*GSK3*) on *T380*. This hyperphosphorylated degron generates a high-affinity binding site for *FBW7*, which makes contacts with both phosphorylation sites. *FBW7* recruits the remainder of the *SCF* ubiquitin ligase (including *RBX1*, *CUL1*, and *SKP1*), leading to *cyclin E* ubiquitylation and its subsequent degradation by the proteasome (see Figure 3 in Welcker and Clurman [68]). The ubiquitin-proteasome system is a major regulatory pathway of protein degradation and plays an important role in cellular division. *FBW7* is a member of the F-box family of proteins, which are substrate recognition components of the multisubunit ubiquitin ligase *SCF* (Figure 22.2); *FBW7* mediates the ubiquitin-dependent proteolysis of several oncoproteins, including – in addition to cyclin *E1* – *MYC*, *C-JUN*, and *NOTCH* [69] (also see Figure 5 in Welcker and Clurman [68]). All substrates of *FBW7* share a consensus phospho-degron motif called Cdc4 phospho-degron (*CPD*). In most cases, the central phospho-threonine is followed by a phosphor-serine in the +4 position that serves as a primer for *GSK3*. Interestingly, the SV40 large T oncoprotein inhibits the *FBW7* function through a decoy phospho-degron that stably binds to *FBW7* in a nondestructive manner (see Figure 5 in Welker and Clurman [68]). Interestingly, analysis of the SV40 T/t-antigen signature, which confers aggressive behavior in human breast, lung, and prostate cancers [70], revealed that *BRCA1* is overexpressed [70]. *MYC* is required for the induction of *BRCA1* [71].

Prognostic and Predictive Role of BRCA1 mRNA in Early Resected NSCLC We performed reverse-transcription quantitative PCR (RT-QPCR) in frozen lung cancer tissue specimens from 126 patients with early NSCLC who had undergone surgical resection and evaluated the association between survival and expression levels of several genes involved in DNA repair pathways. For validation, we used paraffin-embedded specimens from fifty-eight other NSCLC patients. A strong intergene correlation was observed between expression levels of *ERCC1*, *RRM1*, and *BRCA1*. Along with disease stage (stage I vs. II vs. III), *BRCA1* mRNA expression was significantly correlated with survival (HR, 1.98; $P = .02$). In the independent cohort of fifty-eight patients, *BRCA1* mRNA expression was also significantly correlated with survival (HR, 2.4; $P = .04$) [72]. Our findings indicate that although *BRCA1* is related to *ERCC1* and other genes, it stands out as the most significant prognostic marker of relapse. Patients whose tumors had high *BRCA1* expression had significantly worse survival and should be candidates for adjuvant chemotherapy. It is plausible that patients with the highest *BRCA1* levels would obtain more benefit from antimicrotubule, non–platinum-based chemotherapy (see Figure S5 in Rosell et al. [72]; also reviewed in Rosell et al. [25]).

TABLE 22.1: Potential Biomarkers for Customizing Treatment

Progression	Longer Survival	Shorter Survival	Shortest Survival
T/t antigen signature	Low *BRCA1* and *RAP80* mRNA	Intermediate *BRCA1* and *RAP80* mRNA	High *BRCA1* and *RAP80* mRNA
Wound response signature	Low *MYC* and *CSN5* mRNA	Intermediate *MYC* and *CSN5* mRNA	High *MYC* and *CSN5* mRNA
Proteasome signature	High *BIM* mRNA	Intermediate *BIM* mRNA	Low *BIM* mRNA
HSP90 mRNA	Low	High	High
EGFR mutations[a]	L858R // exon 19del	H1650 (exon 19del)	H1975 (T790M)
FBW7 ubiquitin ligase	*FBW7* intact	*FBW7* loss	*FBW7* loss
Drug response	cisplatin +++ antimicrotubules – antiangiogenic – [a]erlotinib +++	cisplatin + antimicrotubules ++ antiangiogenic + [a]erlotinib +	cisplatin – antimicrotubules +++ antiangiogenic +++ [a]erlotinib –

Moreover, bearing in mind that despite the lack of commonality of many genes identified between the different prognostic signatures, these gene expression signatures occupy overlapping prognostic space [24], we can extend the prognostic assessment of *BRCA1*. We compared the prognostic value of several gene signatures and individual genes in early resected NSCLC: a three-gene signature (*CSF1*, *EGFR*, and *CA IX*) [22]; a five-gene signature (*DUSP6*, *MMD*, *STAT1*, *ERBB3*, and *LCK*) [73]; the wound response signature (*CSN5* and *MYC*) [62]; *FBW7*; *RRM2*; and *BRCA1*. A close correlation was found between the expression levels of *CSN5* and *MYC* ($\rho = 0.40$; $P = .001$). In a risk score model used to classify patients into high or low risk for metastasis and death, patients with low *CSN5* and *MYC* had a median survival of sixty-two months, in contrast with thirty-two months for patients with high levels ($P = .01$). However, when all gene signatures, as well as *FBW7*, *RRM1*, and *BRCA1*, were included in the survival model, only *BRCA1* surfaced as a significant prognostic marker. Patients with low *BRCA1* levels had a median survival of sixty-two months, compared with thirty-three months for those with high *BRCA1* ($P = .01$). When *BRCA1* levels were analyzed as a continuous variable, the risk score was highly significant ($P = .009$) (R Rosell et al. unpublished).

We postulate that *BRCA1* mRNA expression may be a relevant prognostic and predictive marker in several tumors, including NSCLC. We propose a model (Table 22.1) in which low *BRCA1* mRNA expression – in correlation with normal expression of *FBW7* and low levels of *MYC* and *CSN5* – will confer better prognosis and predict sensitivity to platinum-based chemotherapy. At the other extreme, high *BRCA1* mRNA expression – in correlation with loss of *FBW7* and high levels of *MYC* and *CSN5* – can predict better response to antimicrotubule, non–platinum-based chemotherapy, as explained in further detail in section 4. In addition, high expression of *CSN5* may serve as a potential predictive marker for angiogenesis inhibitors (further reviewed in Rosell et al. [25]).

4. BRCA1 FOR CHEMOTHERAPY CUSTOMIZATION AND THE RELEVANCE OF NEW GENES IN BRCA1-MEDIATED DNA DAMAGE RESPONSE

The limited efficacy of current chemotherapy approaches is epitomized in metastatic (stage IV) NSCLC, in which median survival is only ten to eleven months with either noncustomized platinum-based chemotherapy [74, 75] or customized cisplatin-based chemotherapy based on *ERCC1* mRNA expression [76], and the two-year survival rate is only 14% to 21% [74–76].

Receptor-associated protein 80 (*RAP80*) or ubiquitin-interacting motif containing 1 (*UIMC1*) is a nuclear protein containing two functional UIMs at its amino terminus. Recently, it has been shown that *RAP80* plays a critical role in DNA damage response signaling [58, 77–79]. These studies report that *RAP80* translocates to ionizing radiation (IR)-induced foci (IRIF) after IR and that the UIMs are essential for relocalization. It was further found that *RAP80* forms a complex with *BRCA1* and that this association is dependent on the *BRCA1* COOH-terminal (*BRCT*) repeats of *BRCA1*. *BRCA1* plays a critical role in DNA repair and activation of cell cycle checkpoints. *RAP80* depletion disrupts the translocation of *BRCA1* to IRIF and causes defects in G2-M checkpoint activation after IR [58, 77, 78]. In addition, knockdown of *RAP80* expression by small interfering RNA reduces DSB-induced homology-directed recombination and increases the sensitivity of cells to IR-induced cytotoxicity [58, 79].

BRCT motifs are found in a number of proteins with functions in DNA repair response. Mutations in the *BRCT* motifs of *BRCA1* have been linked to elevated risk for breast and ovarian cancer. Structural analysis showed that many of the cancer-related mutations in the *BRCT* repeats disrupt the structure of the pSer (Thr)-X-X-Phe binding pocket, abolish the interaction of *BRCT* with its partners, and prevent *BRCA1* from translocating to DNA damage foci. The *BRCT* motifs of *BRCA1* are essential for its association with *RAP80*, and this interaction is required for the translocation of *BRCA1* to DNA damage foci [58, 77–79]. Along the same lines, the *BRCT* missense mutation R1699W of *BRCA1* abolishes the interaction of *BRCA1* with *RAP80* [80]. Interestingly, although *RAP80* has one potential pSer (Thr)-X-X-Phe motif, this motif is not required for interaction with *BRCA1*, suggesting that *RAP80* and *BRCA1* may interact indirectly by binding an intermediary protein. This concept is supported by recent studies showing that this interaction is mediated by coiled-coil domain-containing protein (*CCDC98*) or *Abraxas* (Figure 22.3) [58, 81, 82].

Abraxas binds *BRCA1* to the mutual exclusion of *BRCA1*-associated C-terminal helicase (*BACH*)/*BRCA1*-interacting protein (*Brip1*) and CtBP-interacting protein (*CTIP*) through the pSer-X-X-Phe motif. The *BRCA1-RAP80-Abraxas* complex (Figure 22.3) is clearly involved in DDR [58]. It is likely that different *BRCA1* complexes play redundant roles or promote multiple distinct steps in DDR. For example, the three complexes (Figure 22.3) are required for homologous recombination repair [58]. Furthermore, both the *BRCA1-RAP80-Abraxas* complex and the *BRCA1-RAP80-CTIP* complex are required for the G2-M checkpoint. These two complexes are also involved in transcription through their association with *RAP80* (Figure 22.3) [58]. It is also interesting to note that *RAP80* binds the estrogen receptor, suggesting that the *BRCA1-RAP80* complexes might mediate the *BRCA1* role in estrogen signaling in breast cancer [83]. Similar to our findings of poor prognosis in patients with NSCLC with elevated *BRCA1* expression [72], high *BACH1/Brip1* transcript levels found in more aggressive breast cancers with an estrogen receptor–negative, progesterone receptor–negative, or *HER-2*–positive status [84] seem to contradict their function as tumor suppressors.

In addition to *BRCA1*, *RAP 80*, and *Abraxas*, a deubiquitinating enzyme – *BRCC36* – has been shown to be present in the *RAP80-BRCA1* complex [77]. Interestingly, *BRCC36* is also aberrantly expressed in many breast cancers. Along the same lines, downregulation of *BRCC36* expression impairs the DNA repair pathway activated in response to IR by inhibiting *BRCA1* activation, thereby sensitizing breast cancer cells to IR-induced apoptosis [85]. *BRCC36* displays sequence homology to *CSN5* (the fifth subunit of the *COP9* signasolome), including a conserved *JAMM* motif (reviewed in Yan and Jetten [86]). How the *BRCA1/BARD1-RAP80-Abraxas-BRCC36* complex localizes to sites of DNA damage has been actively investigated. The two UIM motifs of *RAP80* are required for foci formation of *BRCA1* and *Abraxas* [58, 78]. It has been also shown that the UIM domains of *RAP80* bind to ubiquitin chains assembled through K63 linkages [77]. In addition, *RNF8* transduces the DNA-damage signal via histone ubiquitylation and checkpoint protein assembly [87–89]. *RNF8* ubiquitylates histones at DNA DSBs and promotes assembly of repair proteins. A model has been proposed for IRIF formation of the *BRCA1/BARD1-RAP80-Abraxas-BRCC36* complex. In response to DNA damage, *ATM* and *ATR* phosphorylate *H2AX* on Ser-139, which serves to recruit the mediator of DNA damage checkpoint protein 1 (*MDC1*) protein to chromatin, where it is also phosphorylated [90]. In an *H2AX*-and *MDC1*-dependent manner, *RNF8-Ubc13* complexes go to sites of DNA damage through their FHA domain and initiate the synthesis of K63 polyubiquitin chains on chromatin, which recruits the *BRCA1/BARD1-RAP80-Abraxas-BRCC36* complex through the UIM domains of *RAP80* (see Figure 5 in Wang and Elledge [87]). Other similar models have also been proposed (see Figure 1 in Kim and Chen [91], Figure 7 in Mailand et al. [89], and Figure 1 in Yan and Jetten [86]).

Protein complexes formed at DNA damage sites have to be removed after the damage is repaired. Although little is understood about this process, it likely involves dephosphorylation of *gamma-H2AX* and other proteins, K63-liked deubiquitination, and ubiquitination-proteasome-dependent degradation. It is therefore intriguing that *BRCA1/BARD1*, which functions as an E3 ligase, is associated with the deubiquitinase *BRCC36* [77, 92]. It can be speculated that the deubiquitinase activity of *BRCC36* might play a role in terminating the DNA repair signaling at a later stage of DDR [86].

Based on the roles that *RNF8*, *RAP80*, and *CCDC98/Abraxas* have in DDR, it could be expected that mutations in genes that would compromise *BRCA1* activity are involved in breast and ovarian cancer. From 5% to 10% of patients with different types of cancer were found to contain specific antibodies against *RAP80*, indicating that *RAP80* might function as a new cancer-associated gene. However, mutational analysis of *RAP80* and *CCCDC98/Abraxas* genes in 168 multiple-case breast/ovarian cancer families who were negative for mutations in *BRCA1* or *BRCA2* showed only few variants – most of which are considered common [93]. Another study analyzing *RAP80* mutations in 152 women with familial breast cancer who were negative for *BRCA1* and *BRCA2* mutations detected several missense but no truncating mutations [94]. Two cases of lung adenocarcinomas with *EGFR* mutations in the kinase domain also had germline mutations in *BRCA2*; the daughter, mother, and maternal aunt of one of these patients all had breast or ovarian cancer. DNA from the patient and daughter harbored the same *BRCA2* mutation. However, among 110 Jewish patients with lung cancer, only three harbored germline BRCA mutations (2.7%). This rate is similar to the 2.5% expected in the general Ashkenazi population. None of the lung tumors harbored *EGFR* mutations. No association was found between mutant lung cancers and BRCA mutations carriers [95].

4.1. BRCA1 as a Differential Modulator of Chemosensitivity

A growing body of evidence indicates that *BRCA1* confers sensitivity to apoptosis induced by antimicrotubule drugs (paclitaxel and vincristine) but induces resistance to DNA-damaging agents (cisplatin and etoposide) and radiotherapy [96–99]. These preclinical findings are supported by a variety of experimental models in breast and ovarian cancer cells: Inducible expression of *BRCA1* enhanced paclitaxel sensitivity [100]; a short interfering RNA-mediated inactivation of endogenous *BRCA1* led to paclitaxel and docetaxel resistance [101–103]; and reconstitution of *BRCA1*-deficient cells with wild type *BRCA1* enhanced sensitivity to paclitaxel and vinorelbine [101]. (See Figure 3 in Quinn et al. [101] and Figure 2 in Quinn et al. [103].) This differential modulating effect of *BRCA1* mRNA expression was also observed in tumor cells isolated from malignant effusions of NSCLC and gastric cancer patients, in which *BRCA1* mRNA levels correlated negatively with cisplatin sensitivity and positively with docetaxel sensitivity [104]. We have found a good correlation between *BRCA1* – but not *ERCC1* – mRNA expression and cisplatin sensitivity in six NSCLC cell lines with wild type *K-ras*. Intriguingly, the A549 and H23 NSCLC cell lines, harboring *K-ras* codon 12 mutations, were relatively sensitive to cisplatin in spite of having relatively high levels of *BRCA1* and *ERCC1* (M. A. Molina, personal communication). Three retrospective studies – in patients with NSCLC [105] and ovarian cancer [103, 106] – found that low or intermediate *BRCA1* mRNA levels were correlated with a significantly longer survival following platinum-based chemotherapy [103, 105], whereas survival in patients with higher *BRCA1* expression increased following taxane-based chemotherapy [103].

4.2. Customized Therapy in Advanced NSCLC Based on BRCA1 mRNA Expression

We recruited patients to this phase II, prospective, multicenter trial based on screening of *EGFR* mutations followed by *BRCA1* mRNA expression analysis in paraffin-embedded tumor tissue (Spanish Lung Cancer Group (SLCG, unpublished data). Patients with *EGFR* mutations – either the exon 19 deletion or the L858R mutation – received 150 mg of daily oral erlotinib continuously until progression or intolerable adverse effects. Each cycle was twenty-eight days. Patients with wild type *EGFR* received customized chemotherapy based on *BRCA1* mRNA levels. Patients in the lowest tercile of *BRCA1* expression received cisplatin 75 mg/m^2 on day 1 plus gemcitabine 1250 mg/m^2 on days one and eight. Patients in the intermediate tercile received cisplatin 75 mg/m^2 on day 1 plus docetaxel 75 mg/m^2 on day 1. Patients in the highest tercile received docetaxel 75 mg/m^2 on day 1. All chemotherapy was repeated every three weeks for a maximum of six cycles unless there was earlier evidence of disease progression or intolerable adverse effects. Between March 2005 and July 2007, a total of 123 patients from 25 centers were enrolled in the study. Median follow-up was eight months (range, one to twenty-eight months). Twelve patients had *EGFR* mutations and were assigned to receive erlotinib (*EGFR* group). Of the 111 patients with wild type *EGFR*, 38 were in the lowest tercile of *BRCA1* expression and were assigned to receive cisplatin plus gemcitabine (low *BRCA1* group), 40 were in the intermediate tercile and were assigned to receive cisplatin plus docetaxel (intermediate *BRCA1* group), and 33 were in the highest tercile and were assigned to receive docetaxel alone (high *BRCA1* group). Median survival was not reached for the *EGFR* group, compared with ten months (95% confidence interval [CI], 8.5 to 15.5) for patients in all three *BRCA1* groups. Two-year survival for patients in the *BRCA1* groups was 26.7%. For patients in the low *BRCA1* group, median survival was eleven months (95% CI, 1.1–20.9), and two-year survival was 41.2%. For those in the intermediate *BRCA1* group, median survival was nine months (95% CI, 5.4–12.6), and two-year survival was 16.2%. For patients in the high *BRCA1* group, median survival was eleven months (95% CI, 8.2–13.9), and two-year survival was 0%. Median survival was influenced by *RAP80* levels. In patients with low *BRCA1* levels, median survival was not reached in patients with low *RAP80* levels, whereas it was eight months for patients with intermediate *RAP80* and seven months for those with high *RAP80*. In patients with low *BRCA1* levels, time to progression was fourteen months in patients with low *RAP80* levels, whereas it was four months for patients with intermediate *RAP80* levels and six months for those with high *RAP80* levels (Figure 22.4A). In patients with intermediate *BRCA1* levels, time to progression was four months in patients with low *RAP80* levels, whereas it was nine months for patients with intermediate *RAP80* levels and six months for those with high *RAP80* levels. In patients with high *BRCA1* levels, time to progression was two months in patients with low *RAP80* levels, ten months in patients with intermediate *RAP80* levels, and four months for those with high *RAP80* levels (Figure 22.4A). An exploratory multivariate analysis in the eighty-six patients, with the use of a Cox proportional-hazards model, identified Eastern Cooperative Oncology Group (ECOG) performance status and *RAP80* as significant variables for survival (HRs: performance status 1, 2.72; $P = .005$; *RAP80*, 1.3; $P = .05$). No clinical characteristics, type of metastases, second-line chemotherapy (thirty-six patients), or *Abraxas* levels influenced survival. The Cox model for time to progression also showed that only performance status and *RAP80* were significant variables (SLCG, unpublished data).

5. EGFR MUTATIONS IN LUNG CANCER

Mutations in the TK domain of the *EGFR* have been identified as a cause of NSCLC, particularly in adenocarcinoma and bronchoalveolar carcinoma (BAC) [107–111]. The most common oncogenic mutations are small in-frame deletions in exon 19 and a point mutation that substitutes leucine 858 with arginine (L858R) in exon 21. These mutations likely cause constitutive activation of the kinase by destabilizing the autoinhibited conformation, which is normally maintained in the absence of ligand stimulation [112]. The activating mutations confer dramatic sensitivity to the small molecule TKIs gefitinib and erlotinib [107–109], with a reported median survival of 17.5 months in a prospective study of first-line gefitinib in advanced NSCLC [113].

The SLCG has carried out a prospective study of first- and second-line erlotinib in advanced NSCLC harboring *EGFR* mutations. Response was observed in 71% of 193 patients with NSCLC, including 24 patients with a complete response. The probability of response was double in patients aged sixty-one to seventy-one years and in patients with the exon 19 deletion. No other variables influenced response, including gender, smoking status, histology, brain or bone metastases, and prior chemotherapy. Median time to progression for all 193 patients was twelve months, and it was longer in women (fifteen months) than in men (eight months; $P = .001$). The multivariate analysis for time to progression showed that the exon 19 deletion and the absence of brain and bone metastases were significantly associated with a lower HR of progression. Overall median survival for all 193 patients was twenty-two months; it was twenty-eight months for 139 women and seventeen months for 54 men ($P = .04$). In the multivariate analysis for survival, PS 0, female gender, the presence of the exon 19 deletion, and the absence of brain metastases predicted a favorable outcome (SLCG, unpublished data). The fact that the exon 19 deletion surfaces as an independent factor for response, time to progression, and survival supports the finding in kinetic studies of erlotinib that the deletion – more so than the L858R mutation – markedly decreases the affinity of the kinase for ATP, with which erlotinib competes for binding [114]. Previous studies have also shown that the exon 19 deletion is associated with longer survival in patients treated with gefitinib or erlotinib [115]. (Further information on potential mechanisms of resistance to EGFR TKIs in patients harboring EGFR mutations is reviewed in Rosell et al. [25].)

6. NEW AVENUES IN CLINICAL TRIALS AND CONCLUSIONS

The vast majority of clinical trials combining chemotherapy plus targeted agents are based on empirical criteria and in general yield short-lived responses and no substantial benefit in prolonging survival in a meaningful way for the patient. Moreover, some chemotherapy drugs may well be antagonistic when combined with others and even with targeted drugs. Therefore, the first endeavor should focus on the correct choice of chemotherapy drugs and combinations. If further validated, the assessment of *BRCA1* together with *RAP80* mRNA expression levels could become a bona fide predictive marker for customizing chemotherapy, based on the appealing differential modulating effect of *BRCA1* on platinums and antimicrotubule drugs (docetaxel, paclitaxel, or vinorelbine) (Table 22.1). In the SLCG prospective customized study based on the baseline analysis of *BRCA1* mRNA expression in advanced NSCLC (nonsquamous histology), a record time to progression of fourteen months was observed in the subgroup of patients expressing low levels of both *BRCA1* and *RAP80* mRNA (lowest tercile). The length of this time to progression is three times that attained in trials with docetaxel plus cisplatin [74], with pemetrexed (a potent inhibitor of thymidylate synthase) plus cisplatin [75], or with customized docetaxel plus cisplatin based on *ERCC1* mRNA levels [76]. Interestingly, time to progression was only 5.2 months in NSCLC (nonsquamous histology) treated with pemetrexed plus cisplatin [75]. A time to progression of fourteen months in patients receiving gemcitabine plus cisplatin based on low expression of *BRCA1* in conjunction with low expression of *RAP80* is a milestone in NSCLC chemotherapy (Table 22.1, Figure 22.4A). In the only prospective study reported of gefitinib, time to progression was 9.2 months in a subgroup of thirty-one patients with NSCLC (nonsquamous histology) harboring *EGFR* TK mutations [113]. In the customized SLCG phase II study, time to progression for patients with *EGFR* mutations treated with erlotinib was thirteen months, and in the SLCG prospective trial of erlotinib in NSCLC patients with *EGFR* mutations, time to progression was twelve months.

A crucial clinical discovery emerging from this *BRCA1* mRNA phase II customized study, which concurs with preclinical findings, is that time to progression and survival decline abruptly in patients in the lowest tercile of *BRCA1* mRNA expression treated with cisplatin plus gemcitabine when *RAP80* mRNA is elevated (intermediate or in the highest-expression tercile (Figure 22.4A). This concurs with the observation that the UIM-containing protein *RAP80* plays a key role in the recruitment of *BRCA1* protein complexes to DNA DSBs [58, 77–79]. Conversely, because *RAP80* by itself is able to translocate to IRIF in HCC1937 cells that express a truncated *BRCA1* unable to migrate to IRIF, *RAP80* alone could replace the *BRCA1* function in cells lacking *BRCA1* [79]. This can explain the fact that patients with high levels of both *BRCA1* and *RAP80* mRNA show significant benefit when treated with single-agent docetaxel, which showed no significant benefit in spite of high levels of *BRCA1* when *RAP80* levels were low. Similarly, the best time to progression and

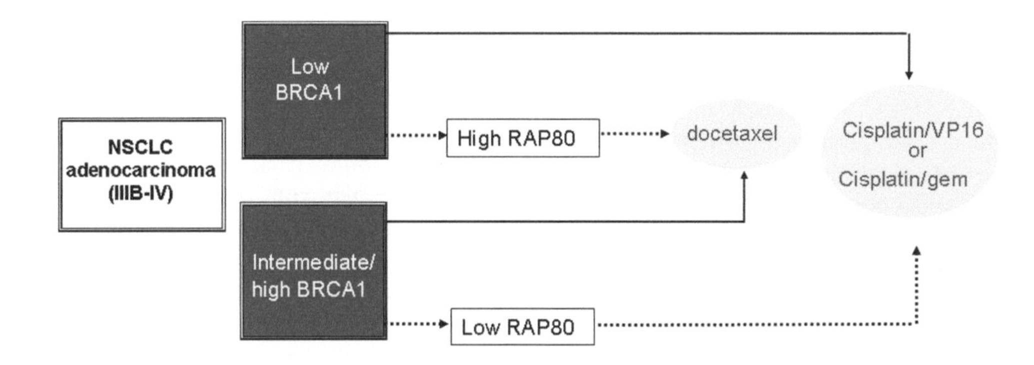

		RAP80 LEVELS					
		≤0.79		0.79-1.41		<1.41	
BRCA1 LEVELS	N	mo (95% CI)	N	mo (95% CI)	N	mo (95% CI)	P
Low	11	14 (5-22.9)	9	4 (2.8-5.1)	5	6 (-)	0.08
Intermediate	11	4 (3.1-4.9)	7	9 (2.5-15.5)	9	6 (3.1-8.9)	0.42
High	5	2 (0-4.1)	9	10 (7.3-12.6)	12	4 (1.7-6.6)	0.006

(a)

(b)

***FIGURE 22.4:** The proposed customized approach based on* BRCA1 *mRNA expression in advanced NSCLC based on the SLCG phase II customized chemotherapy trial. Patients whose tumors have very low* BRCA1 *mRNA could receive the maximum benefit from cisplatin alone or in combination with etoposide (VP16), gemcitabine, or pemetrexed. However, the* BRCA1 *customization model is strongly influenced by the levels of* RAP80 *mRNA. If both* BRCA1 *and* RAP80 *mRNA levels are in the lowest tercile, time to progression with cisplatin plus gemcitabine is fourteen months (time to progression is only five months with any type of chemotherapy combination). Dramatic decline in time to progression was observed in patients with low* BRCA1 *levels when* RAP80 *was elevated (Figure 22.4A). Similarly, in patients with intermediate levels of* BRCA1 *who received customized docetaxel plus cisplatin, the maximum benefit was observed in those who also had intermediate levels of* RAP80 *(time to progression of nine months). Figure 22.4B illustrates a customized phase II study in patients with advanced gastric cancer and non-small-cell lung cancer (NSCLC) to be undertaken by the SLCG together with Nanjing University. Patients will receive customized chemotherapy according to their mRNA levels of both* BRCA1 *and* RAP80*. (For further information, see the text.)*

survival were attained in patients in the intermediate terciles of both *BRCA1* and *RAP80* treated with docetaxel plus cisplatin (Figure 22.4A). A study of a common customizing model for NSCLC and gastric cancer will be carried out in collaboration with Nanjing Drum Tower Hospital (China) (Figure 22.4B). Patients with the lowest levels of *BRCA1* and *RAP80* will receive a platinum compound (either cisplatin or oxaliplatin [116], either alone or with non–*BRCA1*-interfering drugs (VP16, 5-fluorouracil, gemcitabine, or pemetrexed). Patients with the highest expression of both *BRCA1* and *RAP80* will receive single-agent docetaxel. Patients with low *BRCA1* but high *RAP80* will receive docetaxel or similar antimicrotubule drugs. Finally, for patients with high *BRCA1* but low *RAP80*, it can be inferred that the lack of *RAP80* hampered the recruitment of *BRCA1* for effective DDR, and the effect of platinum-based treatment in these patients should be tested in the setting of a clinical trial (Figure 22.4B). In patients with high *BRCA1* and high *RAP80*, *RRM1* levels will be examined, because low *RRM1* could indicate high sensitivity to the combination of docetaxel plus gemcitabine [117]. We postulate that appropriate chemotherapy customization could also optimize clinical trials combining chemotherapy with targeted drugs, like angiogenesis inhibitors (Table 22.1). *CSN5* and *MYC* mRNA expression could become a predictive marker of angiogenesis, as well as *SATB1* and *PML* protein expression.

In summary, a time to progression of fourteen months and median survival not reached in a subset of patients with advanced NSCLC displaying the lowest *BRCA1* and *RAP80* mRNA represents a new landmark in cancer chemotherapy. Could this outcome be further improved? Inhibitors of poly(ADP-ribose) polymerase (*PARP*) induce single-strand breaks that can result in DSBs as a consequence of stalled replication forks. Such lesions would normally be repaired by homologous recombination, but this is abrogated in *BRCA1*- or *BRCA2*-deficient cancer cells (reviewed in Ashworth [118]). The inhibition of PARP leads to the persistence of these DNA lesions [119]. In consequence, phase I-II studies in NSCLC will be undertaken in patients with low *BRCA1* mRNA expression being treated with pemetrexed plus cisplatin and a PARP inhibitor. Intriguingly, resistance to PARP inhibitors has been shown in patients with acquired *BRCA2* mutations. PARP inhibitor–resistant clones are resistant to cisplatin but not to docetaxel in tumors with secondary *BRCA2* mutations that restored the wild type *BRCA2* reading frame [120, 121]. The wound response signature that merges with the proteasome signature could also be predictive of resistance to PARP inhibitors (Table 22.1) (reviewed in Rosell et al. [25]).

The next question is, despite this impressive time to progression of fourteen months, what else can be done to delay acquired resistance in NSCLC or other tumors with low *BRCA1* and *RAP80* mRNA expression treated with platinum-based chemotherapy? Intriguingly, upregulation of *BRCA1* is observed in irradiated cells involving an *ERK* feedback regulatory loop that avoids *BRCA1* degradation by the proteasome (see Figure 6 in Yan et al. [80]). It has been shown that histone deacetylase (*HDAC*) inhibitors exhibit synergy with radiotherapy and cisplatin by inhibiting *HDAC* enzymes that are involved in DDR. Combination of PARP inhibitors with HDAC inhibitors produced a synergistic effect on apoptosis. Not surprisingly, the HDAC inhibitor decreased transcript mRNA levels of *BRCA1*, *BRCA2*, and *RAD51* [122]. In addition, in ovarian cancer cells expressing relatively high levels of *BRCA1* and *ERCC1* mRNA, a significant decrease in mRNA and protein levels of both *BRCA1* and *ERCC1* was observed with treatment with an HDACi [106] (see Figure 5H in Matsuoka et al. [57]).

High levels of thioredoxin-1 (*TRX*) are also associated with resistance to cisplatin [123]. We have found a close correlation between mRNA levels of *TRX* and *BRCA1* in NSCLC [72]. TRX binding protein 2 (*TBP-2*), which blocks *TRX* activity, is expressed at low levels in many tumors. Vorinostat (Zolinza, suberoylanilide hydroxamic acid) is a small molecule that inhibits *HDAC* activity and induces *TBP-2* mRNA [124, 125]. We postulate that the optimal use of vorinostat in combination with chemotherapy could be in tumors with low *BRCA1* and RAP 80 expression treated with platinum-based, non-antimicrotubule chemotherapy because of its potential in avoiding upregulation of *BRCA1*.

Finally, *BIM* abrogation is a potential mechanism of resistance to erlotinib and gefitinib in tumors containing EGFR TK mutations. EGFR cell lines with *EGFR* mutations interact with *HSP90*, and it has been observed that vorinostat induces *HSP90* acetylation, leading to *BIM* upregulation. Other mechanisms of *BIM* downregulation could be through the proteasome signature, and *CSN5* and *MYC* mRNA could be a potential predictive marker that warrants testing (Table 22.1). An inverse relationship between the expression of *BIM* and *BRCA1* in tumors with EGFR mutations has been described. Transcriptome analysis revealed the upregulation of *BIM* and downregulation of *BRCA1*, *MCL1*, and *HSP90*, among others, following effective treatment with a *panHER/EGFR* inhibitor in gefitinib-resistant H1975 cells expressing the T790M mutation [126] (reviewed in Rosell et al. [25]).

There is still room for further improvement, and some puzzling observations require further research. For example, why are NSCLC cell lines with *K-ras* mutations sensitive to cisplatin in spite of relatively high levels of *BRCA1* mRNA (M. A. Molina, personal communication)? One plausible explanation – not yet tested – is that these *K-ras* mutant cell lines do not express *RAP80*, hampering the *BRCA1* effect on DDR. We can also hypothesize that oncogenic ras may activate *p53*, leading to *CK2* inhibition (Figure 22.1) [48]. *CK2* inhibition suppresses *HP1*-β mobilization, attenuating the initiation of DDR via *H2AX* phosphorylation [51] and preventing the formation of the downstream *BRCA1* complexes. In short, *K-ras* status could be an additional marker to be examined, especially in lung adenocarcinomas.

Further investigation is also needed to understand the molecular basis of chromatin remodeling in which inaccessible, compact, and repressed chromatin is converted into an open, accessible form for active gene transcription or vice versa. Both the repressed and transcriptionally active states involve the recruitment of protein complexes that alter chromatin structure through covalent modifications of histone tails. These include histone H3 lysine methylation (H3Kme). As an added layer of regulation, lysine residues can be mono-(me1), di-(me2), or tri(Me3) methylated by ***histone methyl transferases*** (*HMTs*) with specificity for a particular lysine residue. The epigenetic modification H3K9me3, which is usually linked to gene repression, is catalyzed by the conserved mammalian ortholog of the Drosophila suppressor of position-effect variegation, *Su(var)3–9*-related *HMT*s, *SUV39H1* in humans [127]. H3K9m3 plays key roles in gene silencing by recruiting repressors and co-factors, including *HDACs* and *HP1-α*. Intriguingly, the absence of heterochromatin marks, such as *H3K9m3*, leads to the presence of alternative lengthening of telomeres–associated *PML* bodies [128]. Loss of *SUV39H1* decreases *H3K9m3* methylation and decreases the expression of *HP1-β* (also referred to as *CBX1*) (see Figure 4 in Garcia-Cao et al. [129]). The clinical relevance of these observations is the fact that loss of *SUV39H1* leads to the loss of *HP1-β* and can prevent DDR, presumably linked to low *BRCA1* levels because *H2AX* will not be activated [51]. In this situation, low *CK2* and high *PML* can confer better prognosis as well as low *BRCA1* expression. Low *BRCA1* mRNA enhances sensitivity to cisplatin and etoposide (VP16) (Table 22.1). Another possible scenario is that overexpression of *CK2* leads to the activation of *HP1-β* and high *BRCA1* mRNA. High *CK2* abrogates *PML* and confers poor prognosis. High *BRCA1* and other interacting partners, such as *BRCC36* and *BACH1/Brip1*, confer poor prognosis, and high *BRCA1* mRNA confers resistance to platinum compounds and hypersensitivity to docetaxel (Table 22.1).

REFERENCES

1. Jemal, A., Siegel, R., Ward, E., Hao, Y., Xu, J., Murray, T., and Thun, M. J. (2008). Cancer statistics, 2008. *CA Cancer J Clin 58*, 71–96.
2. Jemal, A., Siegel, R., Ward, E., Murray, T., Xu, J., and Thun, M. J. (2007). Cancer statistics, 2007. *CA Cancer J Clin 57*, 43–66.
3. Hanahan, D., and Weinberg, R. A. (2000). The hallmarks of cancer. *Cell 100*, 57–70.
4. Huang, E., Ishida, S., Pittman, J., Dressman, H., Bild, A., Kloos, M., D'Amico, M., Pestell, R. G., West, M., and Nevins, J. R. (2003). Gene expression phenotypic models that predict the activity of oncogenic pathways. *Nat Genet 34*, 226–230.
5. Vogelstein, B., Lane, D., and Levine, A. J. (2000). Surfing the p53 network. *Nature 408*, 307–310.
6. Downward, J. (2003). Targeting RAS signalling pathways in cancer therapy. *Nat Rev Cancer 3*, 11–22.
7. Okada, F., Rak, J. W., Croix, B. S., Lieubeau, B., Kaya, M., Roncari, L., Shirasawa, S., Sasazuki, T., and Kerbel, R. S. (1998). Impact of oncogenes in tumor angiogenesis: mutant K-ras up-regulation of vascular endothelial growth factor/vascular permeability factor is necessary, but not sufficient for tumorigenicity of human colorectal carcinoma cells. *Proc Natl Acad Sci U S A 95*, 3609–3614.
8. Yang, J., Mani, S. A., Donaher, J. L., Ramaswamy, S., Itzykson, R. A., Come, C., Savagner, P., Gitelman, I., Richardson, A., and Weinberg, R.A. (2004). Twist, a master regulator of morphogenesis, plays an essential role in tumor metastasis. *Cell 117*, 927–939.
9. Minn, A. J., Gupta, G. P., Siegel, P. M., Bos, P. D., Shu, W., Giri, D. D., Viale, A., Olshen, A. B., Gerald, W. L., and Massague, J. (2005). Genes that mediate breast cancer metastasis to lung. *Nature 436*, 518–524.
10. McMurray, H. R., Sampson, E. R., Compitello, G., Kinsey, C., Newman, L., Smith, B., Chen, S. R., Klebanov, L., Salzman, P., Yakovlev, A., and Land, H. (2008). Synergistic response to oncogenic mutations defines gene class critical to cancer phenotype. *Nature 453*, 1112–1116.
11. Giaccone, G., Herbst, R. S., Manegold, C., Scagliotti, G., Rosell, R., Miller, V., Natale, R. B., Schiller, J. H., Von Pawel, J., Pluzanska, A., Gatzemeier, U., Grous, J., Ochs, J. S., Averbuch, S. D., Wolf, M. K., Rennie, P., Fandi, A., and Johnson, D. H. (2004). Gefitinib in combination with gemcitabine and cisplatin in advanced non-small-cell lung cancer: a phase III trial – INTACT 1. *J Clin Oncol 22*, 777–784.
12. Herbst, R. S., Giaccone, G., Schiller, J. H., Natale, R. B., Miller, V., Manegold, C., Scagliotti, G., Rosell, R., Oliff, I., Reeves, J. A., Wolf, M. K., Krebs, A. D., Averbuch, S. D., Ochs, J. S., Grous, J., Fandi, A., and Johnson, D. H. (2004). Gefitinib in combination with paclitaxel and carboplatin in advanced non-small-cell lung cancer: a phase III trial – INTACT 2. *J Clin Oncol 22*, 785–794.
13. Gatzemeier, U., Pluzanska, A., Szczesna, A., Kaukel, E., Roubec, J., De Rosa, F., Milanowski, J., Karnicka-Mlodkowski, H., Pesek, M., Serwatowski, P., Ramlau, R., Janaskova, T., Vansteenkiste, J., Strausz, J., Manikhas, G. M., and Von Pawel, J. (2007). Phase III study of erlotinib in combination with cisplatin and gemcitabine in advanced non-small-cell lung cancer: the Tarceva Lung Cancer Investigation Trial. *J Clin Oncol 25*, 1545–1552.
14. Wheatley-Price, P., and Shepherd, F. A. (2008). Epidermal growth factor receptor inhibitors in the treatment of lung cancer: reality and hopes. *Curr Opin Oncol 20*, 162–175.
15. Bergers, G., and Hanahan, D. (2008). Modes of resistance to anti-angiogenic therapy. *Nat Rev Cancer 8*, 592–603.
16. Ellis, L. M., and Hicklin, D. J. (2008). VEGF-targeted therapy: mechanisms of anti-tumour activity. *Nat Rev Cancer 8*, 579–591.
17. Hurwitz, H., Fehrenbacher, L., Novotny, W., Cartwright, T., Hainsworth, J., Heim, W., Berlin, J., Baron, A., Griffing, S., Holmgren, E., Ferrara, N., Fyfe, G., Rogers, B., Ross, R.,

and Kabbinavar, F. (2004). Bevacizumab plus irinotecan, fluorouracil, and leucovorin for metastatic colorectal cancer. *N Engl J Med 350*, 2335–2342.
18. Sandler, A., Gray, R., Perry, M. C., Brahmer, J., Schiller, J. H., Dowlati, A., Lilenbaum, R., and Johnson, D. H. (2006). Paclitaxel-carboplatin alone or with bevacizumab for non-small-cell lung cancer. *N Engl J Med 355*, 2542–2550.
19. Hoang, T., Xu, R., Schiller, J. H., Bonomi, P., and Johnson, D. H. (2005). Clinical model to predict survival in chemonaive patients with advanced non-small-cell lung cancer treated with third-generation chemotherapy regimens based on eastern cooperative oncology group data. *J Clin Oncol 23*, 175–183.
20. Miller, K., Wang, M., Gralow, J., Dickler, M., Cobleigh, M., Perez, E. A., Shenkier, T., Cella, D., and Davidson, N. E. (2007). Paclitaxel plus bevacizumab versus paclitaxel alone for metastatic breast cancer. *N Engl J Med 357*, 2666–2676.
21. Michaelson, M. D., Iliopoulos, O., McDermott, D. F., McGovern, F. J., Harisinghani, M. G., and Oliva, E. (2008). Case records of the Massachusetts General Hospital. Case 17–2008. A 63-year-old man with metastatic renal-cell carcinoma. *N Engl J Med 358*, 2389–2396.
22. Skrzypski, M., Jassem, E., Taron, M., Sanchez, J. J., Mendez, P., Rzyman, W., Gulida, G., Raz, D., Jablons, D., Provencio, M., Massuti, B., Chaib, I., Perez-Roca, L., Jassem, J., and Rosell, R. (2008). Three-gene expression signature predicts survival in early-stage squamous cell carcinoma of the lung. *Clin Cancer Res 14*, 4794–4799.
23. Chang, H. Y., Sneddon, J. B., Alizadeh, A. A., Sood, R., West, R. B., Montgomery, K., Chi, J. T., van de Rijn, M., Botstein, D., and Brown, P. O. (2004). Gene expression signature of fibroblast serum response predicts human cancer progression: similarities between tumors and wounds. *PLoS Biol 2*, E7.
24. Massague, J. (2007). Sorting out breast-cancer gene signatures. *N Engl J Med 356*, 294–297.
25. Rosell, R., Taron, M., and Jablons, D. (2011). Lung cancer metastasis. In *Cancer Metastasis: Biologic Basis and Therapeutics*, D. R. Welch, D. C. Lyden, and B. Psaila, eds. (New York: Cambridge University Press).
26. Cai, S., Han, H. J., and Kohwi-Shigematsu, T. (2003). Tissue-specific nuclear architecture and gene expression regulated by SATB1. *Nat Genet 34*, 42–51.
27. Yasui, D., Miyano, M., Cai, S., Varga-Weisz, P., and Kohwi-Shigematsu, T. (2002). SATB1 targets chromatin remodelling to regulate genes over long distances. *Nature 419*, 641–645.
28. Han, H. J., Russo, J., Kohwi, Y., and Kohwi-Shigematsu, T. (2008). SATB1 reprogrammes gene expression to promote breast tumour growth and metastasis. *Nature 452*, 187–193.
29. Kang, Y., Siegel, P. M., Shu, W., Drobnjak, M., Kakonen, S. M., Cordon-Cardo, C., Guise, T. A., and Massague, J. (2003). A multigenic program mediating breast cancer metastasis to bone. *Cancer Cell 3*, 537–549.
30. Leary, A., and Johnston, S. R. (2007). Small molecule signal transduction inhibitors for the treatment of solid tumors. *Cancer Invest 25*, 347–365.
31. Roberts, P. J., and Der, C. J. (2007). Targeting the Raf-MEK-ERK mitogen-activated protein kinase cascade for the treatment of cancer. *Oncogene 26*, 3291–3310.
32. Solit, D. B., Garraway, L. A., Pratilas, C. A., Sawai, A., Getz, G., Basso, A., Ye, Q., Lobo, J. M., She, Y., Osman, I., Golub, T. R., Sebolt-Leopold, J., Sellers, W. R., and Rosen, N. (2006). BRAF mutation predicts sensitivity to MEK inhibition. *Nature 439*, 358–362.
33. Dumaz, N., Hayward, R., Martin, J., Ogilvie, L., Hedley, D., Curtin, J. A., Bastian, B. C., Springer, C., and Marais, R. (2006). In melanoma, RAS mutations are accompanied by switching signaling from BRAF to CRAF and disrupted cyclic AMP signaling. *Cancer Res 66*, 9483–9491.
34. Montagut, C., Sharma, S. V., Shioda, T., McDermott, U., Ulman, M., Ulkus, L. E., Dias-Santagata, D., Stubbs, H., Lee, D. Y., Singh, A., Drew, L., Haber, D. A., and Settleman, J. (2008). Elevated CRAF as a potential mechanism of acquired resistance to BRAF inhibition in melanoma. *Cancer Res 68*, 4853–4861.
35. Amos, C. I., Wu, X., Broderick, P., Gorlov, I. P., Gu, J., Eisen, T., Dong, Q., Zhang, Q., Gu, X., Vijayakrishnan, J., Sullivan, K., Matakidou, A., Wang, Y., Mills, G., Doheny, K., Tsai, Y. Y., Chen, W. V., Shete, S., Spitz, M. R., and Houlston, R. S. (2008). Genome-wide association scan of tag SNPs identifies a susceptibility locus for lung cancer at 15q25.1. *Nat Genet 40*, 616–622.
36. Hung, R. J., McKay, J. D., Gaborieau, V., Boffetta, P., Hashibe, M., Zaridze, D., Mukeria, A., Szeszenia-Dabrowska, N., Lissowska, J., Rudnai, P., Fabianova, E., Mates, D., Bencko, V., Foretova, L., Janout, V., Chen, C., Goodman, G., Field, J. K., Liloglou, T., Xinarianos, G., Cassidy, A., McLaughlin, J., Liu, G., Narod, S., Krokan, H. E., Skorpen, F., Elvestad, M. B., Hveem, K., Vatten, L., Linseisen, J., Clavel-Chapelon, F., Vineis, P., Bueno-de-Mesquita, H. B., Lund, E., Martinez, C., Bingham, S., Rasmuson, T., Hainaut, P., Riboli, E., Ahrens, W., Benhamou, S., Lagiou, P., Trichopoulos, D., Holcatova, I., Merletti, F., Kjaerheim, K., Agudo, A., Macfarlane, G., Talamini, R., Simonato, L., Lowry, R., Conway, D. I., Znaor, A., Healy, C., Zelenika, D., Boland, A., Delepine, M., Foglio, M., Lechner, D., Matsuda, F., Blanche, H., Gut, I., Heath, S., Lathrop, M., and Brennan, P. (2008). A susceptibility locus for lung cancer maps to nicotinic acetylcholine receptor subunit genes on 15q25. *Nature 452*, 633–637.
37. Thorgeirsson, T. E., Geller, F., Sulem, P., Rafnar, T., Wiste, A., Magnusson, K. P., Manolescu, A., Thorleifsson, G., Stefansson, H., Ingason, A., Stacey, S. N., Bergthorsson, J. T., Thorlacius, S., Gudmundsson, J., Jonsson, T., Jakobsdottir, M., Saemundsdottir, J., Olafsdottir, O., Gudmundsson, L. J., Bjornsdottir, G., Kristjansson, K., Skuladottir, H., Isaksson, H. J., Gudbjartsson, T., Jones, G. T., Mueller, T., Gottsater, A., Flex, A., Aben, K. K., de Vegt, F., Mulders, P. F., Isla, D., Vidal, M. J., Asin, L., Saez, B., Murillo, L., Blondal, T., Kolbeinsson, H., Stefansson, J. G., Hansdottir, I., Runarsdottir, V., Pola, R., Lindblad, B., van Rij, A. M., Dieplinger, B., Haltmayer, M., Mayordomo, J. I., Kiemeney, L. A., Matthiasson, S. E., Oskarsson, H., Tyrfingsson, T., Gudbjartsson, D. F., Gulcher, J. R., Jonsson, S., Thorsteinsdottir, U., Kong, A., and Stefansson, K. (2008). A variant associated with nicotine dependence, lung cancer and peripheral arterial disease. *Nature 452*, 638–642.
38. Lam, D. C., Girard, L., Ramirez, R., Chau, W. S., Suen, W. S., Sheridan, S., Tin, V. P., Chung, L. P., Wong, M. P.,

Shay, J. W., Gazdar, A. F., Lam, W. K., and Minna, J. D. (2007). Expression of nicotinic acetylcholine receptor subunit genes in non-small-cell lung cancer reveals differences between smokers and nonsmokers. *Cancer Res 67*, 4638–4647.
39. Song, P., Sekhon, H. S., Fu, X. W., Maier, M., Jia, Y., Duan, J., Proskosil, B. J., Gravett, C., Lindstrom, J., Mark, G. P., Saha, S., and Spindel, E. R. (2008). Activated cholinergic signaling provides a target in squamous cell lung carcinoma. *Cancer Res 68*, 4693–4700.
40. Song, P., Sekhon, H. S., Lu, A., Arredondo, J., Sauer, D., Gravett, C., Mark, G. P., Grando, S. A., and Spindel, E. R. (2007). M3 muscarinic receptor antagonists inhibit small cell lung carcinoma growth and mitogen-activated protein kinase phosphorylation induced by acetylcholine secretion. *Cancer Res 67*, 3936–3944.
41. Salomoni, P., and Pandolfi, P. P. (2002). The role of PML in tumor suppression. *Cell 108*, 165–170.
42. Gurrieri, C., Capodieci, P., Bernardi, R., Scaglioni, P. P., Nafa, K., Rush, L. J., Verbel, D. A., Cordon-Cardo, C., and Pandolfi, P. P. (2004). Loss of the tumor suppressor PML in human cancers of multiple histologic origins. *J Natl Cancer Inst 96*, 269–279.
43. Daniels, M. J., Marson, A., and Venkitaraman, A. R. (2004). PML bodies control the nuclear dynamics and function of the CHFR mitotic checkpoint protein. *Nat Struct Mol Biol 11*, 1114–1121.
44. Wang, Z. G., Ruggero, D., Ronchetti, S., Zhong, S., Gaboli, M., Rivi, R., and Pandolfi, P. P. (1998). PML is essential for multiple apoptotic pathways. *Nat Genet 20*, 266–272.
45. Bernardi, R., Guernah, I., Jin, D., Grisendi, S., Alimonti, A., Teruya-Feldstein, J., Cordon-Cardo, C., Simon, M. C., Rafii, S., and Pandolfi, P. P. (2006). PML inhibits HIF-1alpha translation and neoangiogenesis through repression of mTOR. *Nature 442*, 779–785.
46. Faivre, S., Kroemer, G., and Raymond, E. (2006). Current development of mTOR inhibitors as anticancer agents. *Nat Rev Drug Discov 5*, 671–688.
47. Schewe, D. M., and Aguirre-Ghiso, J. A. (2008). ATF6alpha-Rheb-mTOR signaling promotes survival of dormant tumor cells in vivo. *Proc Natl Acad Sci U S A 105*, 10519–10524.
48. Scaglioni, P. P., Yung, T. M., Cai, L. F., Erdjument-Bromage, H., Kaufman, A. J., Singh, B., Teruya-Feldstein, J., Tempst, P., and Pandolfi, P. P. (2006). A *CK2*-dependent mechanism for degradation of the PML tumor suppressor. *Cell 126*, 269–283.
49. O-charoenrat, P., Rusch, V., Talbot, S. G., Sarkaria, I., Viale, A., Socci, N., Ngai, I., Rao, P., and Singh, B. (2004). Casein kinase II alpha subunit and C1-inhibitor are independent predictors of outcome in patients with squamous cell carcinoma of the lung. *Clin Cancer Res 10*, 5792–5803.
50. Shaw, R. J., and Cantley, L. C. (2006). Ras, PI(3)K and mTOR signalling controls tumour cell growth. *Nature 441*, 424–430.
51. Ayoub, N., Jeyasekharan, A. D., Bernal, J. A., and Venkitaraman, A. R. (2008). HP1-beta mobilization promotes chromatin changes that initiate the DNA damage response. *Nature 453*, 682–686.
52. Ito, K., Bernardi, R., Morotti, A., Matsuoka, S., Saglio, G., Ikeda, Y., Rosenblatt, J., Avigan, D. E., Teruya-Feldstein, J., and Pandolfi, P. P. (2008). PML targeting eradicates quiescent leukaemia-initiating cells. *Nature 453*, 1072–1078.
53. Su, T. T. (2006). Cellular responses to DNA damage: one signal, multiple choices. *Annu Rev Genet 40*, 187–208.
54. Bristow, R. G., and Hill, R. P. (2008). Hypoxia and metabolism. Hypoxia, DNA repair and genetic instability. *Nat Rev Cancer 8*, 180–192.
55. Halazonetis, T. D., Gorgoulis, V. G., and Bartek, J. (2008). An oncogene-induced DNA damage model for cancer development. *Science 319*, 1352–1355.
56. Harper, J. W., and Elledge, S. J. (2007). The DNA damage response: ten years after. *Mol Cell 28*, 739–745.
57. Matsuoka, S., Ballif, B. A., Smogorzewska, A., McDonald, E. R., III, Hurov, K. E., Luo, J., Bakalarski, C. E., Zhao, Z., Solimini, N., Lerenthal, Y., Shiloh, Y., Gygi, S. P., and Elledge, S. J. (2007). ATM and ATR substrate analysis reveals extensive protein networks responsive to DNA damage. *Science 316*, 1160–1166.
58. Wang, B., Matsuoka, S., Ballif, B. A., Zhang, D., Smogorzewska, A., Gygi, S. P., and Elledge, S. J. (2007). Abraxas and RAP80 form a BRCA1 protein complex required for the DNA damage response. *Science 316*, 1194–1198.
59. O'Connell, B. C., and Harper, J. W. (2007). Ubiquitin proteasome system (UPS): what can chromatin do for you? *Curr Opin Cell Biol 19*, 206–214.
60. Guerrero-Santoro, J., Kapetanaki, M. G., Hsieh, C. L., Gorbachinsky, I., Levine, A. S., and Rapic-Otrin, V. (2008). The cullin 4B-based UV-damaged DNA-binding protein ligase binds to UV-damaged chromatin and ubiquitinates histone *H2A*. *Cancer Res 68*, 5014–5022.
61. Sugasawa, K. (2008). Xeroderma pigmentosum genes: functions inside and outside DNA repair. *Carcinogenesis 29*, 455–465.
62. Adler, A. S., Lin, M., Horlings, H., Nuyten, D. S., van de Vijver, M. J., and Chang, H. Y. (2006). Genetic regulators of large-scale transcriptional signatures in cancer. *Nat Genet 38*, 421–430.
63. Lue, H., Thiele, M., Franz, J., Dahl, E., Speckgens, S., Leng, L., Fingerle-Rowson, G., Bucala, R., Luscher, B., and Bernhagen, J. (2007). Macrophage migration inhibitory factor (MIF) promotes cell survival by activation of the Akt pathway and role for CSN5/JAB1 in the control of autocrine MIF activity. *Oncogene 26*, 5046–5059.
64. Adler, A. S., Littlepage, L. E., Lin, M., Kawahara, T. L., Wong, D. J., Werb, Z., and Chang, H. Y. (2008). CSN5 isopeptidase activity links COP9 signalosome activation to breast cancer progression. *Cancer Res 68*, 506–515.
65. Lue, H., Kapurniotu, A., Fingerle-Rowson, G., Roger, T., Leng, L., Thiele, M., Calandra, T., Bucala, R., and Bernhagen, J. (2006). Rapid and transient activation of the ERK MAPK signalling pathway by macrophage migration inhibitory factor (MIF) and dependence on JAB1/CSN5 and Src kinase activity. *Cell Signal 18*, 688–703.
66. Wong, D. J., Nuyten, D. S., Regev, A., Lin, M., Adler, A. S., Segal, E., van de Vijver, M. J., and Chang, H. Y. (2008). Revealing targeted therapy for human cancer by gene module maps. *Cancer Res 68*, 369–378.
67. Tan, T. T., Degenhardt, K., Nelson, D. A., Beaudoin, B., Nieves-Neira, W., Bouillet, P., Villunger, A., Adams, J. M., and White, E. (2005). Key roles of BIM-driven apoptosis in epithelial tumors and rational chemotherapy. *Cancer Cell 7*, 227–238.

68. Welcker, M., and Clurman, B. E. (2008). FBW7 ubiquitin ligase: a tumour suppressor at the crossroads of cell division, growth and differentiation. *Nat Rev Cancer 8*, 83–93.
69. Akhoondi, S., Sun, D., von der Lehr, N., Apostolidou, S., Klotz, K., Maljukova, A., Cepeda, D., Fiegl, H., Dafou, D., Marth, C., Mueller-Holzner, E., Corcoran, M., Dagnell, M., Nejad, S. Z., Nayer, B. N., Zali, M. R., Hansson, J., Egyhazi, S., Petersson, F., Sangfelt, P., Nordgren, H., Grander, D., Reed, S. I., Widschwendter, M., Sangfelt, O., and Spruck, C. (2007). FBXW7/hCDC4 is a general tumor suppressor in human cancer. *Cancer Res 67*, 9006–9012.
70. Deeb, K. K., Michalowska, A. M., Yoon, C. Y., Krummey, S. M., Hoenerhoff, M. J., Kavanaugh, C., Li, M. C., Demayo, F. J., Linnoila, I., Deng, C. X., Lee, E. Y., Medina, D., Shih, J. H., and Green, J. E. (2007). Identification of an integrated SV40 T/t-antigen cancer signature in aggressive human breast, prostate, and lung carcinomas with poor prognosis. *Cancer Res 67*, 8065–8080.
71. Menssen, A., and Hermeking, H. (2002). Characterization of the c-MYC-regulated transcriptome by SAGE: identification and analysis of c-MYC target genes. *Proc Natl Acad Sci U S A 99*, 6274–6279.
72. Rosell, R., Skrzypski, M., Jassem, E., Taron, M., Bartolucci, R., Sanchez, J. J., Mendez, P., Chaib, I., Perez-Roca, L., Szymanowska, A., Rzyman, W., Puma, F., Kobierska-Gulida, G., Farabi, R., and Jassem, J. (2007). BRCA1: a novel prognostic factor in resected non-small-cell lung cancer. *PLoS ONE 2*, e1129.
73. Chen, H. Y., Yu, S. L., Chen, C. H., Chang, G. C., Chen, C. Y., Yuan, A., Cheng, C. L., Wang, C. H., Terng, H. J., Kao, S. F., Chan, W. K., Li, H. N., Liu, C. C., Singh, S., Chen, W. J., Chen, J. J., and Yang, P. C. (2007). A five-gene signature and clinical outcome in non-small-cell lung cancer. *N Engl J Med 356*, 11–20.
74. Fossella, F., Pereira, J. R., von Pawel, J., Pluzanska, A., Gorbounova, V., Kaukel, E., Mattson, K. V., Ramlau, R., Szczesna, A., Fidias, P., Millward, M., and Belani, C. P. (2003). Randomized, multinational, phase III study of docetaxel plus platinum combinations versus vinorelbine plus cisplatin for advanced non-small-cell lung cancer: the TAX 326 study group. *J Clin Oncol 21*, 3016–3024.
75. Scagliotti, G. V., Parikh, P., von Pawel, J., Biesma, B., Vansteenkiste, J., Manegold, C., Serwatowski, P., Gatzemeier, U., Digumarti, R., Zukin, M., Lee, J. S., Mellemgaard, A., Park, K., Patil, S., Rolski, J., Goksel, T., de Marinis, F., Simms, L., Sugarman, K. P., and Gandara, D. (2008). Phase III study comparing cisplatin plus gemcitabine with cisplatin plus pemetrexed in chemotherapy-naive patients with advanced-stage non-small-cell lung cancer. *J Clin Oncol 26*, 3543–3551.
76. Cobo, M., Isla, D., Massuti, B., Montes, A., Sanchez, J. M., Provencio, M., Vinolas, N., Paz-Ares, L., Lopez-Vivanco, G., Munoz, M. A., Felip, E., Alberola, V., Camps, C., Domine, M., Sanchez, J. J., Sanchez-Ronco, M., Danenberg, K., Taron, M., Gandara, D., and Rosell, R. (2007). Customizing cisplatin based on quantitative excision repair cross-complementing 1 mRNA expression: a phase III trial in non-small-cell lung cancer. *J Clin Oncol 25*, 2747–2754.
77. Sobhian, B., Shao, G., Lilli, D. R., Culhane, A. C., Moreau, L. A., Xia, B., Livingston, D. M., and Greenberg, R. A. (2007). RAP80 targets BRCA1 to specific ubiquitin structures at DNA damage sites. *Science 316*, 1198–1202.
78. Kim, H., Chen, J., and Yu, X. (2007). Ubiquitin-binding protein RAP80 mediates BRCA1-dependent DNA damage response. *Science 316*, 1202–1205.
79. Yan, J., Kim, Y. S., Yang, X. P., Li, L. P., Liao, G., Xia, F., and Jetten, A. M. (2007). The ubiquitin-interacting motif containing protein RAP80 interacts with BRCA1 and functions in DNA damage repair response. *Cancer Res 67*, 6647–6656.
80. Yan, J., Yang, X. P., Kim, Y. S., and Jetten, A. M. (2008). RAP80 responds to DNA damage induced by both ionizing radiation and UV irradiation and is phosphorylated at Ser 205. *Cancer Res 68*, 4269–4276.
81. Kim, H., Huang, J., and Chen, J. (2007). CCDC98 is a BRCA1-BRCT domain-binding protein involved in the DNA damage response. *Nat Struct Mol Biol 14*, 710–715.
82. Liu, Z., Wu, J., and Yu, X. (2007). CCDC98 targets BRCA1 to DNA damage sites. *Nat Struct Mol Biol 14*, 716–720.
83. Yan, J., Kim, Y. S., Yang, X. P., Albers, M., Koegl, M., and Jetten, A. M. (2007). Ubiquitin-interaction motifs of RAP80 are critical in its regulation of estrogen receptor alpha. *Nucleic Acids Res 35*, 1673–1686.
84. Eelen, G., Vanden Bempt, I., Verlinden, L., Drijkoningen, M., Smeets, A., Neven, P., Christiaens, M. R., Marchal, K., Bouillon, R., and Verstuyf, A. (2008). Expression of the BRCA1-interacting protein Brip1/BACH1/FANCJ is driven by E2F and correlates with human breast cancer malignancy. *Oncogene 27*, 4233–4241.
85. Chen, X., Arciero, C. A., Wang, C., Broccoli, D., and Godwin, A. K. (2006). BRCC36 is essential for ionizing radiation-induced BRCA1 phosphorylation and nuclear foci formation. *Cancer Res 66*, 5039–5046.
86. Yan, J., and Jetten, A. M. (2008). RAP80 and RNF8, key players in the recruitment of repair proteins to DNA damage sites. *Cancer Lett 271*, 170–190.
87. Wang, B., and Elledge, S. J. (2007). Ubc13/Rnf8 ubiquitin ligases control foci formation of the Rap80/Abraxas/Brca1/Brcc36 complex in response to DNA damage. *Proc Natl Acad Sci U S A 104*, 20759–20763.
88. Huen, M. S., Grant, R., Manke, I., Minn, K., Yu, X., Yaffe, M. B., and Chen, J. (2007). RNF8 transduces the DNA-damage signal via histone ubiquitylation and checkpoint protein assembly. *Cell 131*, 901–914.
89. Mailand, N., Bekker-Jensen, S., Faustrup, H., Melander, F., Bartek, J., Lukas, C., and Lukas, J. (2007). RNF8 ubiquitylates histones at DNA double-strand breaks and promotes assembly of repair proteins. *Cell 131*, 887–900.
90. Stewart, G. S., Wang, B., Bignell, C. R., Taylor, A. M., and Elledge, S. J. (2003). MDC1 is a mediator of the mammalian DNA damage checkpoint. *Nature 421*, 961–966.
91. Kim, H., and Chen, J. (2008). New players in the BRCA1-mediated DNA damage responsive pathway. *Mol Cells 25*, 457–461.
92. Wu, W., Koike, A., Takeshita, T., and Ohta, T. (2008). The ubiquitin E3 ligase activity of BRCA1 and its biological functions. *Cell Div 3*, 1.
93. Osorio, A., Barroso, A., Garcia, M. J., Martinez-Delgado, B., Urioste, M., and Benitez, J. (2008). Evaluation of the BRCA1 interacting genes RAP80 and CCDC98 in familial breast cancer susceptibility. *Breast Cancer Res Treat 113*, 371–376.

94. Akbari, M. R., Ghadirian, P., Robidoux, A., Foumani, M., Sun, Y., Royer, R., Zandvakili, I., Lynch, H., and Narod, S. A. (2008). Germline RAP80 mutations and susceptibility to breast cancer. *Breast Cancer Res Treat 113*, 377–381.
95. Marks, J. L., Golas, B., Kirchoff, T., Miller, V. A., Riely, G. J., Offit, K., and Pao, W. (2008). EGFR mutant lung adenocarcinomas in patients with germline BRCA mutations. *J Thorac Oncol 3*, 805.
96. Lafarge, S., Sylvain, V., Ferrara, M., and Bignon, Y. J. (2001). Inhibition of BRCA1 leads to increased chemoresistance to microtubule-interfering agents, an effect that involves the JNK pathway. *Oncogene 20*, 6597–6606.
97. Husain, A., He, G., Venkatraman, E. S., and Spriggs, D. R. (1998). BRCA1 up-regulation is associated with repair-mediated resistance to cis-diamminedichloroplatinum(II). *Cancer Res 58*, 1120–1123.
98. Bhattacharyya, A., Ear, U. S., Koller, B. H., Weichselbaum, R. R., and Bishop, D. K. (2000). The breast cancer susceptibility gene BRCA1 is required for subnuclear assembly of *Rad51* and survival following treatment with the DNA cross-linking agent cisplatin. *J Biol Chem 275*, 23899–23903.
99. Abbott, D. W., Thompson, M. E., Robinson-Benion, C., Tomlinson, G., Jensen, R. A., and Holt, J. T. (1999). BRCA1 expression restores radiation resistance in BRCA1-defective cancer cells through enhancement of transcription-coupled DNA repair. *J Biol Chem 274*, 18808–18812.
100. Mullan, P. B., Quinn, J. E., Gilmore, P. M., McWilliams, S., Andrews, H., Gervin, C., McCabe, N., McKenna, S., White, P., Song, Y. H., Maheswaran, S., Liu, E., Haber, D. A., Johnston, P. G., and Harkin, D. P. (2001). BRCA1 and GADD45 mediated G2/M cell cycle arrest in response to antimicrotubule agents. *Oncogene 20*, 6123–6131.
101. Quinn, J. E., Kennedy, R. D., Mullan, P. B., Gilmore, P. M., Carty, M., Johnston, P. G., and Harkin, D. P. (2003). BRCA1 functions as a differential modulator of chemotherapy-induced apoptosis. *Cancer Res 63*, 6221–6228.
102. Chabalier, C., Lamare, C., Racca, C., Privat, M., Valette, A., and Larminat, F. (2006). BRCA1 downregulation leads to premature inactivation of spindle checkpoint and confers paclitaxel resistance. *Cell Cycle 5*, 1001–1007.
103. Quinn, J. E., James, C. R., Stewart, G. E., Mulligan, J. M., White, P., Chang, G. K., Mullan, P. B., Johnston, P. G., Wilson, R. H., and Harkin, D. P. (2007). BRCA1 mRNA expression levels predict for overall survival in ovarian cancer after chemotherapy. *Clin Cancer Res 13*, 7413–7420.
104. Wang, L., Wei, J., Qian, X., Yin, H., Zhao, Y., Yu, L., Wang, T., and Liu, B. (2008). *ERCC1* and BRCA1 mRNA expression levels in metastatic malignant effusions is associated with chemosensitivity to cisplatin and/or docetaxel. *BMC Cancer 8*, 97.
105. Taron, M., Rosell, R., Felip, E., Mendez, P., Souglakos, J., Ronco, M. S., Queralt, C., Majo, J., Sanchez, J. M., Sanchez, J. J., and Maestre, J. (2004). BRCA1 mRNA expression levels as an indicator of chemoresistance in lung cancer. *Hum Mol Genet 13*, 2443–2449.
106. Weberpals, J., Garbuio, K., O'Brien, A., Clark-Knowles, K., Doucette, S., Antoniouk, O., Goss, G., and Dimitroulakos, J. (2009). The DNA repair proteins BRCA1 and *ERCC1* as predictive markers in sporadic ovarian cancer. *Int J Cancer 124*, 806–815.
107. Lynch, T. J., Bell, D. W., Sordella, R., Gurubhagavatula, S., Okimoto, R. A., Brannigan, B. W., Harris, P. L., Haserlat, S. M., Supko, J. G., Haluska, F. G., Louis, D. N., Christiani, D. C., Settleman, J., and Haber, D. A. (2004). Activating mutations in the epidermal growth factor receptor underlying responsiveness of non-small-cell lung cancer to gefitinib. *N Engl J Med 350*, 2129–2139.
108. Paez, J. G., Janne, P. A., Lee, J. C., Tracy, S., Greulich, H., Gabriel, S., Herman, P., Kaye, F. J., Lindeman, N., Boggon, T. J., Naoki, K., Sasaki, H., Fujii, Y., Eck, M. J., Sellers, W. R., Johnson, B. E., and Meyerson, M. (2004). EGFR mutations in lung cancer: correlation with clinical response to gefitinib therapy. *Science 304*, 1497–1500.
109. Pao, W., Miller, V., Zakowski, M., Doherty, J., Politi, K., Sarkaria, I., Singh, B., Heelan, R., Rusch, V., Fulton, L., Mardis, E., Kupfer, D., Wilson, R., Kris, M., and Varmus, H. (2004). EGF receptor gene mutations are common in lung cancers from "never smokers" and are associated with sensitivity of tumors to gefitinib and erlotinib. *Proc Natl Acad Sci U S A 101*, 13306–13311.
110. Gazdar, A. F., Shigematsu, H., Herz, J., and Minna, J. D. (2004). Mutations and addiction to EGFR: the Achilles "heal" of lung cancers? *Trends Mol Med 10*, 481–486.
111. Johnson, B. E., and Janne, P. A. (2005). Epidermal growth factor receptor mutations in patients with non-small cell lung cancer. *Cancer Res 65*, 7525–7529.
112. Yun, C. H., Boggon, T. J., Li, Y., Woo, M. S., Greulich, H., Meyerson, M., and Eck, M. J. (2007). Structures of lung cancer-derived EGFR mutants and inhibitor complexes: mechanism of activation and insights into differential inhibitor sensitivity. *Cancer Cell 11*, 217–227.
113. Sequist, L. V., Martins, R. G., Spigel, D., Grunberg, S. M., Spira, A., Janne, P. A., Joshi, V. A., McCollum, D., Evans, T. L., Muzikansky, A., Kuhlmann, G. L., Han, M., Goldberg, J. S., Settleman, J., Iafrate, A. J., Engelman, J. A., Haber, D. A., Johnson, B. E., and Lynch, T. J. (2008). First-line gefitinib in patients with advanced non-small-cell lung cancer harboring somatic EGFR mutations. *J Clin Oncol 26*, 2442–2449.
114. Carey, K. D., Garton, A. J., Romero, M. S., Kahler, J., Thomson, S., Ross, S., Park, F., Haley, J. D., Gibson, N., and Sliwkowski, M. X. (2006). Kinetic analysis of epidermal growth factor receptor somatic mutant proteins shows increased sensitivity to the epidermal growth factor receptor tyrosine kinase inhibitor, erlotinib. *Cancer Res 66*, 8163–8171.
115. Jackman, D. M., Yeap, B. Y., Sequist, L. V., Lindeman, N., Holmes, A. J., Joshi, V. A., Bell, D. W., Huberman, M. S., Halmos, B., Rabin, M. S., Haber, D. A., Lynch, T. J., Meyerson, M., Johnson, B. E., and Janne, P. A. (2006). Exon 19 deletion mutations of epidermal growth factor receptor are associated with prolonged survival in non-small cell lung cancer patients treated with gefitinib or erlotinib. *Clin Cancer Res 12*, 3908–3914.
116. Wei, J., Zou, Z., Qian, X., Ding, Y., Xie, L., Sanchez, J. J., Zhao, Y., Feng, J., Ling, Y., Liu, Y., Yu, L., Rosell, R., and Liu, B. (2008). *ERCC1* mRNA levels and survival of advanced gastric cancer patients treated with a modified FOLFOX regimen. *Br J Cancer 98*, 1398–1402.

117. Souglakos, J., Boukovinas, I., Taron, M., Mendez, P., Mavroudis, D., Tripaki, M., Hatzidaki, D., Koutsopoulos, A., Stathopoulos, E., Georgoulias, V., and Rosell, R. (2008). Ribonucleotide reductase subunits M1 and M2 mRNA expression levels and clinical outcome of lung adenocarcinoma patients treated with docetaxel/gemcitabine. *Br J Cancer 98*, 1710–1715.
118. Ashworth, A. (2008). A synthetic lethal therapeutic approach: poly(ADP) ribose polymerase inhibitors for the treatment of cancers deficient in DNA double-strand break repair. *J Clin Oncol 26*, 3785–3790.
119. Farmer, H., McCabe, N., Lord, C. J., Tutt, A. N., Johnson, D. A., Richardson, T. B., Santarosa, M., Dillon, K. J., Hickson, I., Knights, C., Martin, N. M., Jackson, S. P., Smith, G. C., and Ashworth, A. (2005). Targeting the DNA repair defect in BRCA mutant cells as a therapeutic strategy. *Nature 434*, 917–921.
120. Edwards, S. L., Brough, R., Lord, C. J., Natrajan, R., Vatcheva, R., Levine, D. A., Boyd, J., Reis-Filho, J. S., and Ashworth, A. (2008). Resistance to therapy caused by intragenic deletion in *BRCA2*. *Nature 451*, 1111–1115.
121. Sakai, W., Swisher, E. M., Karlan, B. Y., Agarwal, M. K., Higgins, J., Friedman, C., Villegas, E., Jacquemont, C., Farrugia, D. J., Couch, F. J., Urban, N., and Taniguchi, T. (2008). Secondary mutations as a mechanism of cisplatin resistance in *BRCA2*-mutated cancers. *Nature 451*, 1116–1120.
122. Adimoolam, S., Sirisawad, M., Chen, J., Thiemann, P., Ford, J. M., and Buggy, J. J. (2007). HDAC inhibitor PCI-24781 decreases *RAD51* expression and inhibits homologous recombination. *Proc Natl Acad Sci U S A 104*, 19482–19487.
123. Arnold, N. B., Ketterer, K., Kleeff, J., Friess, H., Buchler, M. W., and Korc, M. (2004). Thioredoxin is downstream of Smad7 in a pathway that promotes growth and suppresses cisplatin-induced apoptosis in pancreatic cancer. *Cancer Res 64*, 3599–3606.
124. Ungerstedt, J. S., Sowa, Y., Xu, W. S., Shao, Y., Dokmanovic, M., Perez, G., Ngo, L., Holmgren, A., Jiang, X., and Marks, P. A. (2005). Role of thioredoxin in the response of normal and transformed cells to histone deacetylase inhibitors. *Proc Natl Acad Sci U S A 102*, 673–678.
125. Xu, W., Ngo, L., Perez, G., Dokmanovic, M., and Marks, P. A. (2006). Intrinsic apoptotic and thioredoxin pathways in human prostate cancer cell response to histone deacetylase inhibitor. *Proc Natl Acad Sci U S A 103*, 15540–15545.
126. de La Motte Rouge, T., Galluzzi, L., Olaussen, K. A., Zermati, Y., Tasdemir, E., Robert, T., Ripoche, H., Lazar, V., Dessen, P., Harper, F., Pierron, G., Pinna, G., Araujo, N., Harel-Belan, A., Armand, J. P., Wong, T. W., Soria, J. C., and Kroemer, G. (2007). A novel epidermal growth factor receptor inhibitor promotes apoptosis in non-small cell lung cancer cells resistant to erlotinib. *Cancer Res 67*, 6253–6262.
127. Villeneuve, L. M., Reddy, M. A., Lanting, L. L., Wang, M., Meng, L., and Natarajan, R. (2008). Epigenetic histone H3 lysine 9 methylation in metabolic memory and inflammatory phenotype of vascular smooth muscle cells in diabetes. *Proc Natl Acad Sci U S A 105*, 9047–9052.
128. Gonzalo, S., Jaco, I., Fraga, M. F., Chen, T., Li, E., Esteller, M., and Blasco, M. A. (2006). DNA methyltransferases control telomere length and telomere recombination in mammalian cells. *Nat Cell Biol 8*, 416–424.
129. Garcia-Cao, M., O'Sullivan, R., Peters, A. H., Jenuwein, T., and Blasco, M. A. (2004). Epigenetic regulation of telomere length in mammalian cells by the Suv39h1 and Suv39h2 histone methyltransferases. *Nat Genet 36*, 94–99.
130. Sonnberg, S., Seet, B. T., Pawson, T., Fleming, S. B., and Mercer, A. A. (2008). Poxvirus ankyrin repeat proteins are a unique class of F-box proteins that associate with cellular SCF1 ubiquitin ligase complexes. *Proc Natl Acad Sci U S A 105*, 10955–10960.

CHAPTER 23

Current Drug Targets and the Druggable Genome

Margaret A. Johns*
Andreas Russ*
Haian Fu

1. WHAT IS THE DRUGGABLE GENOME?

The importance of drug discovery is reflected in its long history, starting thousands of years ago with natural remedies and leading to the estimated $825 billion annual global pharmaceutical market that exists today[1]. Thus, the definition of the druggable genome, or defining those molecular entities within the human genome that can be manipulated to improve health, impacts the direction of drug discovery research around the globe. New advances in genomics studies and modern medicine have transformed both the process of drug discovery and the definition of the druggable genome. Historically, the drug discovery process used a "forward pharmacology" approach whereby an active compound was identified based on its efficacy, and then its mechanism of action was determined. This approach was limited by the availability of appropriate animal models but had the advantage of identification of drugs with efficacy in vivo. As understanding of the molecular basis of cellular function has increased, the drug discovery process has moved toward a "reverse pharmacology" approach, whereby drug targets are defined and then compounds affecting their function are identified.

The shift in focus of drug discovery into the reverse pharmacology approach has led to the development of a whole field of "biologicals," biologically created products as opposed to chemically synthesized products, for targeting proteins. For example, numerous therapeutic monoclonal antibodies have been successfully developed, including Enbrel, a monoclonal antibody that functions as a tumor necrosis factor antagonist and has proven successful for the treatment of rheumatoid arthritis (for review, see Silva et al. [1]) and Herceptin, a monoclonal antibody that interferes with the HER2/neu receptor and is used as an adjuvant treatment in patients with HER2-overexpressing breast cancer (for review of cancer immunotherapy and Herceptin, see Dillman [2]). Although biologicals have received increasing attention because of recent successes, they are limited by the restricted properties of target proteins, poor oral availability, and high cost of manufacture. As a result, the identification of therapeutic small molecules has remained the main focus of drug discovery.

One of the hurdles in the reverse pharmacology drug discovery process, for development of either biological or small molecule therapeutics, is the identification of a modifiable, biologically relevant target. With the ultimate goal being the identification of efficacious drugs and with the continued focus of drug discovery on small molecules, the natural evolution of the target identification process led to a narrowing down of the list of available genes to those that have the potential to be regulated by a small molecule. Thus, the druggable genome was first defined in the landmark paper by Hopkins and Groom as "the number of molecular targets that represent an opportunity for therapeutic intervention" [3]. The authors estimated the number of drug targets in the human genome to be 600 to 1,500, based on an intersection of the set of disease-modifying genes (~3,000) with the set of genes that contain a druggable domain (~3,000) (Figure 23.1). The authors defined a druggable domain as a ligand-binding site on a protein that is capable of binding a drug with high affinity. Follow-up papers seem to arrive at a consensus of approximately 300 known drug targets in the human genome and a predicted number of approximately 3000 druggable genes in the human genome [4–6]. Given the current estimate of 20,500 total genes in the human genome [7], these estimated numbers were surprisingly small.

In the decade since the original description of the druggable genome, technological advances have driven the development of an expanded scientific knowledge base and new methods for targeting protein expression and activity. Now, technologies for high-throughput sequencing and screening, as well as novel tools for targeting protein networks and gene expression, including protein-protein interaction inhibitors (PPIIs), small interfering RNAs (siRNA) and microRNAs (miRNA), and

* M. A. Johns and A. Russ contributed equally to this work.
[1] http://www.imshealth.com.

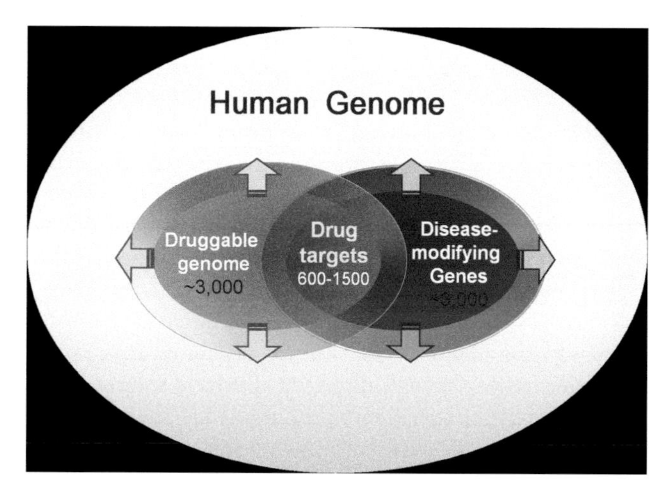

***FIGURE 23.1:* The expanding pool of drug targets. Hopkins and Groom originally defined the effective number of exploitable drug targets as the intersection of the number of genes linked to disease and the "druggable" subset of the human genome. In the last decade, the numbers of genes linked to disease has expanded with the expanding "omics" knowledge base. The "druggable" subset of the human genome has also expanded. The current druggable genome is evolving rapidly toward a new definition (as indicated by arrows). (Adapted by permission from Macmillan Publishers Ltd: Nature Reviews Drug Discovery, Hopkins A. & Groom C., Vol. 1, p. 727, 2002.)**

DNA/RNA/protein aptamers, have changed the landscape of the druggable genome and broadened the definition of a druggable target (Figure 23.1). Thus, the druggable genome is no longer limited to the individual genes mutated in diseases but has expanded to include indirect targets linked to disease genes in functional networks. Additionally, the druggable genome is no longer limited to the previously defined ligand-binding protein domains but has expanded to include protein-protein interaction interfaces and RNA interference (RNAi) and aptamer targets. In this chapter, we first review the characteristics of known druggable targets in the human genome. Next, we discuss the evolving state of the druggable genome in the "omics" age. Finally, we provide resources for exploring the druggable genome. The druggable genome is rapidly evolving and will lead to personalized, safe, and efficacious therapies in the future.

2. MEDICAL DRUGS AND THEIR TARGETS

2.1. The Known Druggable Genome

To define the known druggable genome, druggability has typically been investigated by analyzing the suitability of individual proteins as drug targets. A set of empirical rules is applied to predict domains that are privileged for drug interaction. The presence of binding pockets or protein folds are indicated that are likely to interact with candidate compounds with high affinity and high specificity [3]. Here, we present an overview of the inventory of currently approved drugs and their targets, and we review the characteristics of existing druggable binding domains.

2.1.1. Limitations on Defining the Druggable Genome

In creating a list of druggable targets, the first obstacle is to define the catalog of relevant drug compounds. This can be as stringent as the World Health Organization (WHO) essential medicines list or can be extended to all approved drugs, experimental drugs at various stages of development, or compounds of known biological activity that are not medically used. It also has to be considered whether substitution therapies, hormones, and anti-infective drugs (the targets of which would be encoded in the genome of pathogens rather than the host's) should be included in the catalog. Comprehensive and nonredundant data on drug-like compounds are very demanding to compile and curate; a number of public and private databases now provide this information (see Resources, section 6). Most published efforts have compiled information from several sources to suit the individual question asked.

Even when using the smallest common denominator of drugs with widely accepted clinical efficacy and well-investigated mechanisms of action, we encounter compounds for which the primary molecular target is only poorly defined. In addition, following the law of mass action, drugs will not interact with just a single species of protein target, and in many cases the target is not a uniquely distinct species but rather a clan or swarm of homologous proteins. Even if in vitro data indicate that a compound is highly selective to interact with only one member of a protein family, genetic redundancy makes it very difficult to demonstrate a clear one-to-one relationship of drug and efficacy target in vivo, for example, by gene knockout studies in mice. It therefore cannot be easily mapped to a single gene encoding the protein that is typically described as the *efficacy target*, the molecular partner through which the compound mediates its desired activity. The efficacy target is distinct from other proteins interacting with the compound, for example, by mediating its metabolism, activation of a pro-drug, or non–mechanism-related unwanted effects. In many cases, it is challenging to define an individual protein encoded by a single gene as an unambiguous efficacy target. This phenomenon, known as polypharmacology (or "dirty drugs"), confounds the characterization of the druggable genome but may be clinically advantageous. Morrow et al. recently reviewed the analysis and impact of polypharmacology in detail [8]. The dirty drug concept is well accepted in drug development and provides a conceptual limitation to the one-to-one mapping of drug and the gene encoding the target. Thus, although apparently straightforward as a concept, the practical implementation of a "reverse pharmacology" approach highlights the limitations of our definition of drug targets and the incomplete compatibility of pharmacologic and genetic entities.

Another difficulty in estimating the number of potential drug targets is the critical dependence on reliable annotation of the protein coding genes in the human genome. Because of alternative splicing, variable promoter usage, and the many other methods that have evolved to create protein diversity, a single gene often gives rise to multiple protein sequences. In a recent attempt to accurately annotate the genome, Clamp et al. describe curation of high-quality annotation of the human genome and report that the human genome contains approximately 20,500 bona fide protein coding loci, a number that is substantially lower than previous estimates [7]. A number of curated genome databases now exist (see Resources, section 6), but all are limited by the lack of experimental validation of predicted protein coding regions.

2.1.2. Inventories of Targets for Known Drugs Despite the limitations on defining the set of druggable targets, a number of researchers have made attempts to do so. Landry et al., Imming et al., Russ et al., and Overington et al. describe attempts to map the molecular targets of the current repertoire of medical drugs [4–6, 9]. The general consensus is that approximately three hundred targets from little more than one hundred protein families define the activity of the vast majority of known drugs. Swinney highlights important kinetic features of pathways that are suitable for modulation by drugs, whereas Drews discusses historical lessons in drug discovery [10–11]. In the most recent of these reviews, Overington et al. proposed a comprehensive, curated list of drug targets [5]. They identified a list of 1,204 unique small molecule FDA-approved drugs, analyzed the literature for each drug, and then associated a molecular target to 1,065 of the drugs. The drug targets were found to represent 266 distinct molecular entities in the human genome (Note: as this publication was going to press, a new review of drug targets was published by Rask-Andersen, et al. [12].)

An updated inventory of drug targets is maintained in the curated Therapeutic Target Database (TTD). As of the most recent publication in 2009, there were 348 successful, 292 clinical trial, and 1,254 research targets represented in the database [13]. Information about these targets was collected from a comprehensive literature search, including patents. The TTD list of targets is one of the most comprehensive and accessible inventories of all therapeutics and targets, including viral targets and antibody and antisense therapeutics. However, as a hand-curated data set, the annotation of targets and therapeutics in the database is not complete.

Despite limitations in the annotation of the TTD data, a summary of the biochemical classes associated with the current targets represented in the database provides a useful overview of the functions of current, and potentially future, targets (Figure 23.2). The most represented protein family among the current drug targets is that of receptors, specifically G protein-coupled receptors (GPCRs). In addition, enzymes, such as oxidoreductases, hydrolases (including phosphatases), and transferases (including kinases), and ion channels and membrane transporters are highly represented. Although it is clear that the majority of new targets in clinical trials fall into the same classes as current drug targets, there does appear to be an increase in the representation of kinases and growth factors for therapeutics in clinical development.

The enzymes, membrane transporters and ion channels, and receptors most commonly targeted by current drugs all have functions that can be regulated by a small molecule, either through direct or allosteric interactions. Many different enzymes serve as targets for therapeutics, including oxidoreductase, hydrolase, transferase, lyase, isomerase, protease, and other enzymatic activities. Drugs that interact with an enzyme active site typically lead to inhibition of enzyme activity, binding to the active site as either competitive or noncompetitive inhibitors. Using the TTD biochemical class annotation, approximately 50% of current targets are enzymes (Figure 23.2).

Ion channels and membrane transporters account for approximately 20% of current drug targets using the TTD data (Figure 23.2). Ion channels mediate the conductance of Na^+, Ca^{2+}, K^+, and Cl^- across the plasma membrane. They can be either voltage-gated, where the permeability is regulated by changes in the membrane potential, or ligand-gated. Membrane transporters typically contain a 12 transmembrane spanning region, and they mediate Na^+- or H^+-dependent transport of small molecules such as neurotransmitters, ions, and cationic amino acids. For both ion channels and membrane transporters, drugs typically either increase or decrease permeability.

Receptors form the largest family of drug targets, with GPCRs representing approximately 25% of current drug targets (Figure 23.2). Receptors interact with a diverse set of ligands (including growth factors, lipids, and ions) and mediate a large variety of cellular functions. GPCRs contain seven transmembrane domains, an amino-terminal extracellular domain with potential *N*-linked glycosylation sites that may be involved in determining ligand affinity, and a C-terminal cytoplasmic domain important for G protein coupling. Current drugs targeting receptors either mimic binding of the endogenous ligands (agonists) or interfere with the action of the natural ligand (antagonists).

Proteins within the druggable gene families contain "privileged druggable domains." These are conserved domains, or more specifically folds, which favor interaction with a compound. Often, these domains are also known to bind an endogenous ligand. Of a predicted ten thousand folds in the human genome, the proteins within the druggable families identified by Overington et al. were found to contain only 130 "privileged druggable domains" [5].

The identification of conserved signatures for privileged protein domains allows the characterization and then prediction of drug-binding pockets. Overall, the defining characteristics of drug-binding pockets include pocket volume

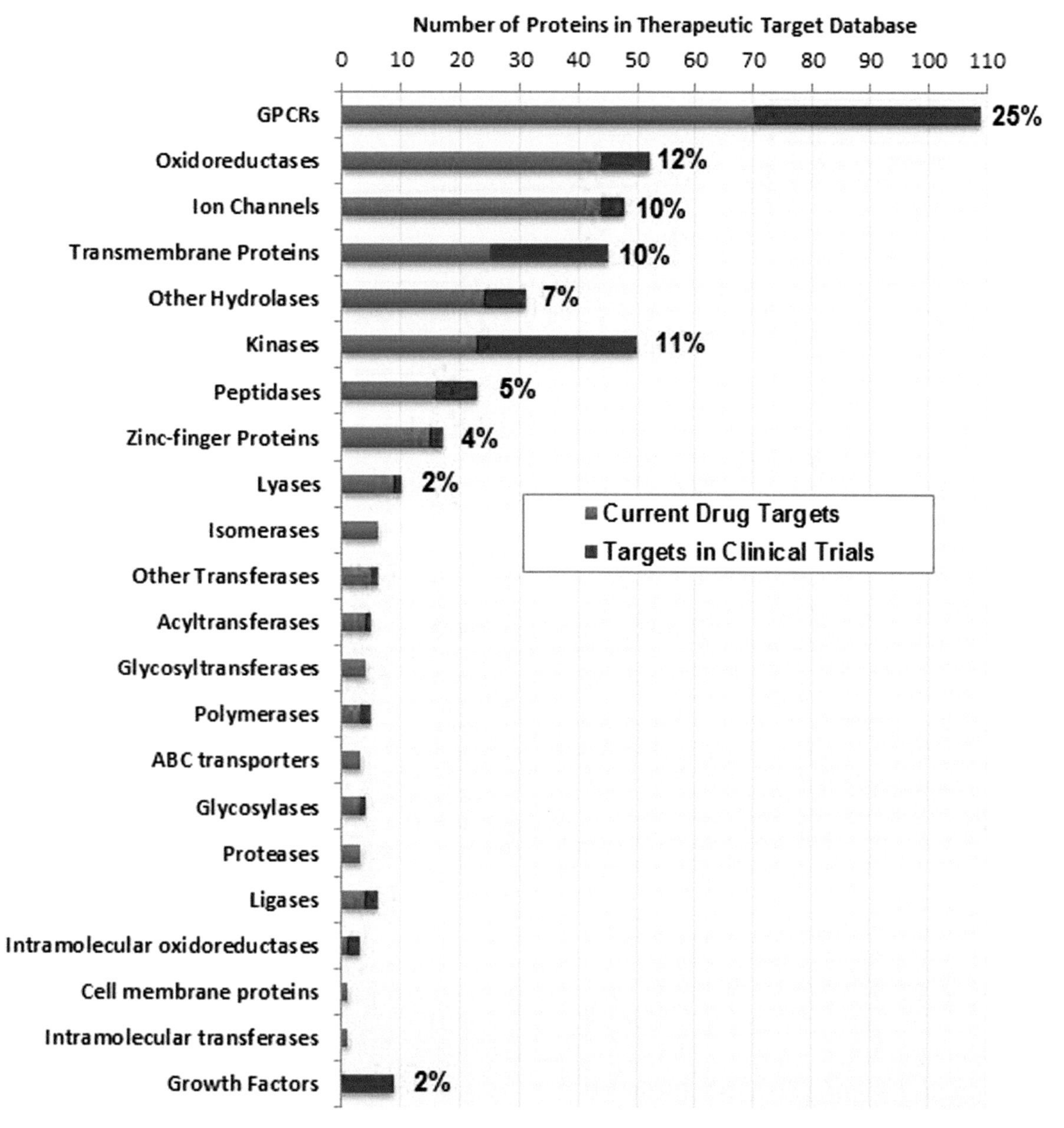

FIGURE 23.2: ***The gene family distribution of current drug targets and those involved in clinical trials. These data are derived from the annotated data in the hand-curated Therapeutic Targets Database (TTD). The data were downloaded on July 3, 2011. The blue bars represent the range of Biochemical Classes for the 308 total proteins in TTD annotated as a "Successful Target" and containing annotation for "Biochemical Class." The red bars represent the range of Biochemical Classes for the 133 total proteins in TTD annotated as in "Clinical Trial" and containing "Biochemical Class" annotation. G protein-coupled receptors (GPCRs) represent the largest target family for current and developing therapeutics. The number of new therapeutics targeting kinases and growth factors appears to be increasing.***

and shape and complementarity of features such as the hydrophobicity and ability to form hydrogen and ionic bonds, charge transfer complexes, and covalent bonds. The typical affinity of a pocket, as defined by Overington et al., for its ligand is 20 nM, with a range of 200 mM to 10 pM [5]. For a detailed review of drug-binding cavities, see Nayal and Honig [14].

The specific molecular characteristics of the binding pockets of druggable domains have been highly studied, and many programs have been developed using these characteristics for prediction of novel drug targets and for structure-based drug design. A range of strategies for the annotation, classification, and clustering of protein domain structures forms the basis for the protein family–based approach to drug target discovery. The InterPro database integrates all major sequence-based annotation algorithms into a single resource. The Pfam database, which describes protein domain signatures as hidden Markov models, provides a precise and flexible tool to analyze protein sequences with standard domain models but also empowers the researcher to develop models tailor-made for specific questions using the HMMER software package. The SCOP and CATH databases use structural rather than sequence homology as the basis for protein family assignment. Table 23.1 lists a range of druggable domain models using the PFAM annotation.

2.1.3. Prediction of the Complete Druggable Genome
Using the characteristics of known druggable domains, the consensus number of potential small molecule drug targets in the human genome hovers at approximately three thousand [3, 6]. These predictions are based on sequence or structure properties and require a starting data set that is appropriately annotated for druggable properties and an algorithm that is used to perform the classification of novel sequences. The accuracy of prediction is critically dependent on the quality of the training set. For a detailed review of different prediction methods, see [15, 16]. An important point for the existing estimates of the number of drug targets in the human genome is that they are limited to targets that are related by sequence or structure to known drug targets.

Work on the known druggable genome has provided research leads for identifying and targeting novel proteins for therapeutic ends. A surprisingly small list of approximately three hundred current drug targets has been identified, and approximately three thousand potential targets in the human genome have been predicted. The majority of these drug targets fall into relatively few protein families. Studies of the druggable genome, however, are changing with advances in science and technology, leading to a new definition of the druggable genome in the "omics" age.

3. DEFINITION OF DRUGGABLE TARGETS IN THE "OMICS" AGE

3.1. Advances in the Genome, Epigenome, Transcriptome, and Proteome

The drive into the "omics" age, marked by a rapid expansion in scientific data associated with the genome, epigenome, transcriptome, and proteome, has in large part occurred by advances in technology allowing high-throughput research. The new technologies range from next-generation sequencing techniques, which led to the production of the complete sequence of the human genome in 2003 [17], to high-throughput biology that now allows efficient performance of biological experiments, to small molecule compound screens in high density plates, to advanced mass spectrometry that accelerates the detection of a particular molecular entity from complex biological samples. Although the known druggable genome is

TABLE 23.1: Druggable Domain Models Using Pfam

PFAM Accession Number	Domain Description
PF00001	7 transmembrane receptor (rhodopsin family)
PF00002	7 transmembrane receptor (secretin family)
PF00003	7 transmembrane receptor (metabotropic glutamate family)
PF00005	ABC transporter
PF00026	Eukaryotic aspartyl protease
PF00044	Glyceraldehyde 3-phosphate dehydrogenase, NAD binding domain
PF00060	Ligand-gated ion channel
PF00067	Cytochrome P450
PF00068	Phospholipase A2
PF00069	Protein kinase domain
PF00070	Pyridine nucleotide-disulphide oxidoreductase
PF00077	Retroviral aspartyl protease
PF00078	Reverse transcriptase (RNA-dependent DNA polymerase)
PF00079	Serpin (serine protease inhibitor)
PF00080	Copper/zinc superoxide dismutase (SODC)
PF00083	Sugar (and other) transporter
PF00085	Thioredoxin
PF00089	Trypsin
PF00091	Tubulin/FtsZ family, GTPase domain
PF00092	von Willebrand factor type A domain
PF00102	Protein-tyrosine phosphatase
PF00104	Ligand-binding domain of nuclear hormone receptor
PF00106	short chain dehydrogenase
PF00107	Zinc-binding dehydrogenase
PF00112	Papain family cysteine protease
PF00128	α-amylase, catalytic domain
PF00133	tRNA synthetases class I (I, L, M, and V)
PF00135	Carboxylesterase
PF00147	Fibrinogen beta and gamma chains, C-terminal globular domain
PF00151	Lipase
PF00153	Mitochondrial carrier protein
PF00156	Phosphoribosyl transferase domain
PF00160	Cyclophilin type peptidyl-prolyl *cis-trans* isomerase
PF00171	Aldehyde dehydrogenase family
PF00174	Oxidoreductase molybdopterin binding domain
PF00175	Oxidoreductase NAD-binding domain
PF00185	Aspartate/ornithine carbamoyltransferase, Asp/Orn binding domain
PF00186	Dihydrofolate reductase
PF00194	Eukaryotic-type carbonic anhydrase
PF00200	Disintegrin
PF00202	Aminotransferase class-III
PF00204	DNA gyrase B
PF00209	Sodium:neurotransmitter symporter family
PF00211	Adenylate and guanylate cyclase catalytic domain
PF00233	3′5′-cyclic nucleotide phosphodiesterase
PF00246	Zinc carboxypeptidase
PF00248	Aldo/keto reductase family
PF00254	FKBP-type peptidyl-prolyl *cis-trans* isomerase
PF00255	Glutathione peroxidase
PF00264	Common central domain of tyrosinase
PF00265	Thymidine kinase
PF00278	Pyridoxal-dependent decarboxylase, C-terminal sheet domain

PFAM Accession Number	Domain Description
PF00282	Pyridoxal-dependent decarboxylase conserved domain
PF00294	pfkB family carbohydrate kinase
PF00303	Thymidylate synthase
PF00305	Lipoxygenase
PF00316	Fructose 1,6-bisphosphatase
PF00317	Ribonucleotide reductase, all-α domain
PF00324	Amino acid permease
PF00326	Prolyl oligopeptidase family
PF00343	Carbohydrate phosphorylase
PF00351	Biopterin-dependent aromatic amino acid hydroxylase
PF00357	Integrin α cytoplasmic region
PF00362	Integrin, β chain
PF00368	Hydroxymethylglutaryl-coenzyme A reductase
PF00387	Phosphatidylinositol-specific phospholipase C, Y domain
PF00388	Phosphatidylinositol-specific phospholipase C, X domain
PF00413	Matrixin
PF00432	Prenyltransferase and squalene oxidase repeat
PF00452	Apoptosis regulator proteins, Bcl-2 family
PF00454	Phosphatidylinositol 3- and 4-kinase
PF00474	Sodium:solute symporter family
PF00478	IMP dehydrogenase/GMP reductase domain
PF00494	Squalene/phytoene synthase
PF00508	E2 (early) protein, N-terminal
PF00520	Ion transport protein
PF00521	DNA gyrase/topoisomerase IV, subunit A
PF00530	Scavenger receptor cysteine-rich domain
PF00554	Rel homology domain (RHD)
PF00557	Metallopeptidase family M24
PF00581	Rhodanese-like domain
PF00591	Glycosyl transferase family, a/b domain
PF00641	Zn-finger in Ran binding protein and others
PF00644	Poly(ADP-ribose) polymerase catalytic domain
PF00648	Calpain family cysteine protease
PF00654	Voltage-gated chloride channel
PF00656	Caspase domain
PF00664	ABC transporter transmembrane region
PF00670	*S*-adenosyl-L-homocysteine hydrolase, NAD binding domain
PF00675	Insulinase (peptidase family M16)
PF00690	Cation transporter/ATPase, N-terminus
PF00703	Glycosyl hydrolases family 2, immunoglobulin-like beta-sandwich domain
PF00716	Assemblin (peptidase family S21)
PF00755	Choline/carnitine o-acyltransferase
PF00782	Dual specificity phosphatase, catalytic domain
PF00809	Pterin binding enzyme
PF00837	Iodothyronine deiodinase
PF00854	POT family
PF00858	Amiloride-sensitive sodium channel
PF00864	ATP P2X receptor
PF00868	Transglutaminase family
PF00870	P53 DNA-binding domain
PF00884	Sulfatase
PF00896	Phosphorylase family 2
PF00909	Ammonium transporter family
PF00927	Transglutaminase family, C-terminal Ig-like domain
PF00955	HCO3-transporter family
PF00962	Adenosine/AMP deaminase
PF00999	Sodium/hydrogen exchanger family
PF01000	RNA polymerase Rpb3/RpoA insert domain
PF01007	Inward rectifier potassium channel
PF01028	Eukaryotic DNA topoisomerase I, catalytic core
PF01039	Carboxyl transferase domain
PF01055	Glycosyl hydrolases family 31
PF01067	Calpain large subunit, domain III
PF01073	3-beta hydroxysteroid dehydrogenase/isomerase family
PF01074	Glycosyl hydrolases family 38 N-terminal domain
PF01082	Copper type II ascorbate-dependent monooxygenase, N-terminal domain
PF01094	Receptor family ligand binding region
PF01120	α-L-fucosidase
PF01124	MAPEG family
PF01154	Hydroxymethylglutaryl-coenzyme A synthase
PF01168	Alanine racemase, N-terminal domain
PF01180	Dihydroorotate dehydrogenase
PF01223	DNA/RNA nonspecific endonuclease
PF01233	Myristoyl-CoA:protein N-myristoyltransferase, N-terminal domain
PF01239	Protein prenyltransferase α-subunit repeat
PF01273	LBP/BPI/CETP family, N-terminal domain
PF01347	Lipoprotein amino terminal region
PF01360	Monooxygenase
PF01397	Terpene synthase, N-terminal domain
PF01400	Astacin (peptidase family M12A)
PF01401	Angiotensin-converting enzyme
PF01421	Reprolysin (M12B) family zinc metalloprotease
PF01431	Peptidase family M13
PF01432	Peptidase family M3
PF01433	Peptidase family M1
PF01447	Thermolysin metallopeptidase, catalytic domain
PF01457	Leishmanolysin
PF01531	Glycosyl transferase family 11
PF01536	Adenosylmethionine decarboxylase
PF01546	Peptidase family M20/M25/M40
PF01593	Flavin-containing amine oxidoreductase
PF01596	O-methyltransferase
PF01742	Clostridial neurotoxin zinc protease
PF01752	Collagenase
PF01758	Sodium bile acid symporter family
PF01839	FG-GAP repeat
PF01896	DNA primase small subunit
PF02031	Streptomyces extracellular neutral proteinase (M7) family
PF02055	O-glycosyl hydrolase family 30
PF02102	Deuterolysin metalloprotease (M35) family
PF02128	Fungalysin metallopeptidase (M36)
PF02518	Histidine kinase-, DNA gyrase B-, and HSP90-like ATPase
PF02729	Aspartate/ornithine carbamoyltransferase, carbamoyl-P binding domain
PF02743	Cache domain
PF02784	Pyridoxal-dependent decarboxylase, pyridoxal binding domain

(continued)

TABLE 23.1 ***(continued)***

PFAM Accession Number	Domain Description
PF02799	Myristoyl-CoA:protein N-myristoyltransferase, C-terminal domain
PF02800	Glyceraldehyde 3-phosphate dehydrogenase, C-terminal domain
PF02836	Glycosyl hydrolases family 2, TIM barrel domain
PF02837	Glycosyl hydrolases family 2, sugar binding domain
PF02867	Ribonucleotide reductase, barrel domain
PF02868	Thermolysin metallopeptidase, α-helical domain
PF02885	Glycosyl transferase family, helical bundle domain
PF02886	LBP/BPI/CETP family, C-terminal domain
PF02931	Neurotransmitter-gated ion-channel ligand binding domain
PF03062	MBOAT family
PF03073	TspO/MBR family
PF03098	Animal haem peroxidase
PF03104	DNA polymerase family B, exonuclease domain
PF03322	Gamma-butyrobetaine hydroxylase
PF03351	DOMON domain
PF03372	Endonuclease/exonuclease/phosphatase family
PF03712	Copper type II ascorbate-dependent monooxygenase, C-terminal domain
PF03926	Putative metallopeptidase (SprT family)
PF03933	Matrix metalloprotease, N-terminal domain
PF04228	Putative neutral zinc metallopeptidase
PF04298	Putative neutral zinc metallopeptidase
PF05090	Vitamin K–dependent gamma-carboxylase
PF05221	*S*-adenosyl-L-homocysteine hydrolase
PF05299	M61 glycyl aminopeptidase
PF05342	M26 IgA1-specific metallo-endopeptidase N-terminal region
PF05547	Immune inhibitor A peptidase M6
PF05572	Pregnancy-associated plasma protein-A
PF05574	Putative zincin metallopeptidase
PF05649	Peptidase family M13
PF06167	Protein of unknown function (DUF980)
PF06262	Domain of unknown function (DUF1025)
PF07108	PipA protein
PF07565	Band 3 cytoplasmic domain
PF07690	Major facilitator superfamily

surprisingly small, the increasing size of the scientific knowledge base and the availability of novel gene/protein targeting techniques make it clear that the druggable genome of the "omics" age will be much larger.

3.1.1. The Genome The increased scientific data available describing not only the human genome, but also the genomes of many additional species, has had a tremendous impact on the ability to identify and predict druggable targets. Knowledge of the complete sequence of the genome now allows identification and grouping of all human genes containing a protein domain. From this, potential novel drug targets can be identified based on sequence homology. In addition, sequences from other species are used to identify conserved amino acids important for structure and function. The availability of this data has strengthened the prediction abilities for structure-based drug design (SBDD) and off-target binding. Kirchmair et al. present a review of in silico screening, using genome data, and off-target binding prediction techniques [18].

The sequencing of multiple human genomes has also led to the identification of large numbers of single nucleotide polymorphisms (SNPs). SNPs are the most common type of genetic variation, occurring on average once in every three hundred nucleotides. Advances in the identification of SNPs have led to the field of pharmacogenomics, the study of how genetic variation affects pharmacokinetics and pharmacodynamics. SNPs not only can affect the binding of a drug to its receptor but also can impact how a drug is metabolized and the resulting side effects. In terms of defining the druggable genome, information about SNPs can be used to strengthen SBDD and increase the accuracy of predictions. The increasing data on SNPs are a critical part of the future of "personalized" medicine, in which genotypes will be used to provide individualized therapy.

3.1.2. The Epigenome Expression of the genome is regulated at many levels. Over the last decade, the identification of epigenetic modifications as an important regulatory component of genome expression has led to novel research avenues for drug discovery (for a recent review, see Yang et al. [19]). New technologies have been developed to study the epigenome. These technologies include the development of modification-specific antibodies for the various histones, which are then combined with chromatin immunoprecipitation (ChIP) coupled with microarrays (ChIP-chip) or used with direct sequencing technology (ChIP-seq). These new technologies allow global examination of the patterns of epigenetic modifications and have identified significant changes in the pattern of epigenetic modifications in disease.

Epigenetic regulation can be separated into three interrelated layers: DNA methylation, nucleosome positioning, and histone modifications [20]. DNA methylation is the result of the covalent addition of a methyl group to the cytosine contained in the CpG dinucleotide by the DNA methyltransferase (DNMT) enzyme. Most CpG dinucleotides are methylated in the genome, except for those contained in CpG islands. CpG islands are regions, often in the promoters of genes, containing a higher-than-expected CpG dinucleotide content. Methylation of CpG islands is associated with the silencing of gene expression by changing chromatin accessibility. Major changes in the pattern of methylation occur in cancer, including gene-specific promoter hypermethylation leading to the silencing of tumor suppressor genes; genome-wide and promoter hypomethylation that can result in the activation of oncoproteins; and loss of imprinting.

Nucleosome positioning is also an important feature of epigenetic regulation. Human DNA is tightly wrapped around histone cores into packages known as nucleosomes. The histone cores generally contain two copies of each of the four histone proteins H2A, H2B, H3, and H4. The histone composition of nucleosomes is regulated and varies dynamically; this variation leads to changes in chromatin structure and gene expression. The histone variations that occur include post-translational modifications, inclusion of specialized variant type of histone proteins, and removal of partial or complete histones from the nucleosome. The dynamic interplay of sequence-specific DNA-binding proteins, histone variants, histone-modifying enzymes, chromatin-associated proteins, histone chaperones, and adenosine triphosphate (ATP)-dependent nucleosome remodelers regulate the packing of regions of DNA and its expression [21].

Alterations in the epigenetic network have been identified in cancer, are associated with aging, and appear to be linked to mental illness and addiction. In recent years, drug discovery efforts focused on targeting proteins involved in regulating the epigenome, including DNMTs and histone deacetylases (HDACs), have led to promising clinical trials for cancer-related diseases. In clinical studies, 5-azacytadine and decitabine, both inhibitors of DNMTs, have shown promise for the treatment of hematological malignancies [22] and [23]. Numerous HDAC inhibitors, including the FDA-approved agent vorinostat, also have been developed (for review, see Minucci et al. [24]). Vorinostat has been approved for the treatment of cutaneous T-cell lymphoma in patients who have progressive, persistent, or recurrent disease.

3.1.3. The Transcriptome Beyond the genome, the complete transcriptome, which is the body of data encompassing the expression of genes at the RNA level, can now be studied in tissues or even single cells. Numerous high-throughput technologies have been developed, including hybridization with oligonucleotide microarrays, counting of sequence tags, DNA sequencing of ditags, and ChIP-Seq. These technologies provide a large data set of when and where individual genes are expressed. Expression can be tracked in tissues through development, disease processes, or even before and after drug treatment, leading to increased information about both the direct and indirect effects of drugs. The expression pattern of large sets of genes can be used to create gene signatures for normal and disease phenotypes, or before and after treatment with a drug. These gene signatures allow grouping of patients and individualized therapy based on gene expression patterns (for review in cancer, see Baehner et al. [25]).

With sequencing of the human genome, the importance of the non-coding regions of the genome has become clear. Contained in the non-coding region are ultra-conserved elements and non-coding RNAs (ncRNAs), important for the regulation of transcript and chromosome structure; RNA processing, stability, and modification; and translation, protein stability and transport. The ncRNAs include miRNAs and siRNAs. miRNAs are small, non-coding RNAs that bind to target mRNAs in their 3′ untranslated region and regulate translation of specific mRNAs into protein. So far, approximately one thousand miRNAs have been described, and they are thought to regulate the expression of up to 30% of all human genes. Studies have shown that miRNAs are dysregulated in tumors at every stage of tumor progression, and thus miRNAs may serve as novel targets [26].

Another type of ncRNA is the siRNAs. The siRNAs result from the processing of long, double-stranded RNA (dsRNA) into 21–24 nucleotide double-stranded siRNAs. These siRNA duplexes are incorporated into effector complexes known as the RNA-induced silencing complex (RISC), which regulates mRNA degradation. Together with the proteins in the RISC complex, the siRNAs form part of a catalytic process leading to the selective reduction of specific mRNA molecules and their corresponding proteins. The identification of microRNAs and siRNAs has led to the development of novel gene targeting strategies, as discussed in section 3.2.2.

3.1.3. The Proteome High-throughput technologies for studying the proteome, focused on the identification of proteins and mapping their interactions in a cellular context, have also impacted the landscape of druggable targets. High-throughput proteomics techniques include two-dimensional gel electrophoresis and high-throughput mass spectrometry. As with RNA expression data, protein expression signatures can be used to identify patterns associated with disease or drug treatment. Also, yeast two-hybrid screens, affinity-based assays, and mass spectrometry are being used for large-scale studies of protein-protein interactions. Furthermore, in vivo screens following protein expression in live cells or the whole organism (i.e., zebrafish embryos) can now be performed in high-throughput formats.

Study of the proteome is complex: a single gene often produces multiple different forms of a protein because of splicing and post-translational modifications; unlike the relatively static genome, the proteome varies dramatically in the lifetime of an organism, and proteins are interconnected into large protein networks or systems known as interactomes [27]. In the druggable genome of the "omics" age, scientists now look beyond individual genes and instead identify network nodes in the interactome, where multiple pathways converge on a single gene. This single gene becomes a target; affecting the function of this gene can cause substantial changes in cellular and organismal function. The distinction between features on the protein level versus the network or pathway level has been designated as the properties of the micro-target versus the properties of the macro-target [6]. Advances in knowledge of the network of genes will indicate whether a target protein will be amplified to yield the desired effect or will be minimized because of a compensating response of the network. Another issue is whether the target node is a homogenous entity; that is, can the target node be modulated by a single drug-like compound acting on a homogenous target, or is the target node a swarm of proteins with related functions but different structures? For the reverse pharmacology process of drug discovery, the identification of druggable target genes and network nodes is limited by our understanding of the interactome and the functional consequence of the compound intervention on these nodes.

As our knowledge of genomics, transcriptomics, and proteomics expands, the role of bioinformatics has become critical. The development of computational methods for coordinating vast amounts of scientific data and coalescing the data into defining the druggable genome is still in its infancy. Many new tools for data analysis have been developed, recently reviewed by Schneider and Orchard [28]. In addition, new algorithms, visualization tools, and applications form the horizon leading to the next steps of drug discovery.

3.2. New Tools for Targeting the Genome

Alongside the traditional approaches that have led to the discovery of many successful drugs, new tools and techniques for targeting genes have been developed. Protein domains that were previously thought to be undruggable because of their size or shape can now be targeted (for review, see [29]). New RNA interference strategies, based on advancing knowledge of ncRNAs, allow targeting of previously undruggable genes. In addition, there has been the discovery that small bits of DNA, RNA, or protein, known as aptamers, can be used to regulate drug targets. With these new abilities to regulate genes has arisen a new druggable genome.

3.2.1. Protein-Protein Interaction Inhibitors Previously, the definition of the druggable genome was limited to proteins with a ligand-binding pocket, as described above. Proteins such as the GPCRs and ion channels have well-described druggable pockets. It was thought that other domains, such as protein-protein interaction (PPI) domains, which are generally larger and flatter than ligand-binding domains, were undruggable. However, this view has been challenged [29]. Many PPIs have "hot spots" that are essential for high-affinity ligand binding. Small molecules that target PPIs, known as PPIIs, are now going through clinical trials. Some notable PPIIs that made it to clinical trial include nutlin-3, which inhibits the MDM-2/p53 interaction in tumors, and tramiprosate, an amyloid-beta protofibril formation inhibitor for Alzheimer's disease. Databases of small molecule PPIIs and their proteins are being built (see Resources). Some characteristics that distinguish PPI interfaces from traditional ligand-binding domains are being described, including larger size, more hydrophobic nature, and structures with more rigid aromatic scaffolds [30]. The advantage of targeting PPIs is that instead of just upregulating or downregulating protein activity, pathway-specific modulation of the activity of a target may be possible.

3.2.2. RNA Interference The identification of siRNAs and miRNAs and the use of small hairpin RNAs (shRNAs) has led to the development of RNAi as a mechanism for therapeutically targeting gene expression. As described in section 3.1.3, siRNAs and miRNAs are native to the human genome. By contrast, shRNAs are exogenous RNAs introduced into cells with a vector. The shRNAs form a hairpin structure that is cleaved into siRNA with a sequence homologous to the target mRNA. RNAi can only be used to decrease or silence gene expression, whereas small molecules can either activate or inhibit protein expression or activity. Unlike small molecules, the synthesis of RNAis is known and is straightforward, and because RNAis are chemically synthesized, they follow the same FDA approval process as small molecules. In addition, RNAis are natural parts of the cell with known breakdown products, so they are expected to have reduced toxicity compared with small molecules. Finally, RNAis are very potent compared with other antisense strategies (antisense DNA oligonucleotides and ribozymes), although the off-target effects are unknown. Because potentially any cell type, tissue, or gene can be targeted with RNAi, RNAis hold great potential for expanding the number of druggable targets.

3.2.3. Aptamers An additional genome-based targeting technique has evolved out of the discovery of aptamers. Aptamers are small oligonucleotide or protein molecules, usually selected from a large pool of random sequences that bind to a target protein with high affinity and specificity. Aptamers can be used to target protein function, or they can also be modified for use as delivery or imaging agents. Bunka et al. recently reviewed the development of RNA aptamers as therapeutics, and Borghouts et al. reviewed peptide aptamers [31, 32]. RNA aptamers, compared with peptide aptamers, are easier to administer therapeutically. However, an advantage of peptide aptamers is that they can be developed to selectively block specific protein domains. As with RNAi, the use of aptamers as a tool to therapeutically target genes and proteins promises to greatly expand the druggable genome of the future.

4. THE FUTURE OF GENOME DRUGGABILITY

Druggability is an operational definition and not a law of nature. It is a moving target and a function of our technical abilities at a given point in time. In the long term, as the coverage of genome annotation and structural and functional genomics is increasing, the analysis of druggability will shift from "Is it druggable?" to "How can we make it druggable?" Using new data from the genome, epigenome, transcriptome, and proteome as well as new tools for targeting coding and non-coding DNA, RNA, proteins, and pathways, the druggable genome of the future is expanding to a new definition.

5. RESOURCES

5.1. Publicly Available Databases of Drugs and Drug Targets

- DrugBank: a knowledgebase for drugs, drug actions and drug targets. http://www.drugbank.ca/
- ChEMBL: a database of bioactive drug-like small molecules, including chemical, target, and biochemical information. https://www.ebi.ac.uk/chembldb/index.php
- TTD: Therapeutic Target Database. http://bidd.nus.edu.sg/group/cjttd/TTD_ns.asp
- PDTD: a web-accessible protein database for drug target identification. http://www.dddc.ac.cn/pdtd/index.php
- SuperTarget and Matador: resources for exploring drug-target relationships. http://insilico.charite.de/supertarget http://matador.embl.de/
- IUPHAR-DB: the IUPHAR database of G protein-coupled receptors and ion channels. http://www.iuphar-db.org/index.jsp
- Fpocket: the Druggable Cavity Directory, an open-source for information related to protein pockets. http://fpocket.sourceforge.net/dcd
- SiteMap: protein binding site identification and analysis. https://www.schrodinger.com/products/14/20/
- sc-PDB: an annotated archive of druggable binding sites extracted from Protein Databank. http://bioinfo-pharma.u-strasbg.fr/scPDB

6.2. Exploring the Genome, Epigenome, Transcriptome, and Proteome

6.2.1. Databases and Resources for the Human Genome

- Genbank: an annotated collection of all publicly available DNA sequences. http://www.ncbi.nlm.nih.gov/genbank
- Ensembl: a genome database browser for vertebrates and other eukaryotic species. http://www.ensembl.org/
- The UCSC Genome Browser: a browser for a large collection of genomes. http://www.genome.ucsc.edu/
- HapMap: the International HapMap Consortium resource for genetic variations in human beings. http://www.hapmap.org
- 1000 Genomes: a deep catalog of human genetic variation. http://www.1000genomes.org

6.2.2. Epigenome Databases and Resources

- HEP: the Human Epigenome Project effort to identify, catalogue, and interpret genome-wide DNA methylation patterns of all human genes. http://www.epigenome.org
- Epigenesys: European research initiative on epigenetics with protocols and resources for epigenetic research. http://www.epigenesys.eu
- Histone Database: a curated collection of sequences and structures of histones and non-histone proteins containing histone folds. http://genome.nhgri.nih.gov/histones/

6.2.3. Transcriptome Databases and Resources

- GO Ontology: bioinformatics initiative with the aim of standardizing the representation of gene and gene product attributes across species and databases. http://www.geneontology.org
- ArrayExpress: a database of functional genomics experiments including gene expression. http://www.ebi.ac.uk/arrayexpress
- Gene Expression Omnibus: a public functional genomics data repository. http://www.ncbi.nlm.nih.gov/geo/
- CIBEX: a public database for microarray data. http://cibex.nig.ac.jp

6.2.4 Proteome Databases and Resources

- Pfam: protein families: http://pfam.sanger.ac.uk/
- Interpro: protein families, domains, and regions. http://www.ebi.ac.uk/interpro/
- PRIDE: a public repository for mass spectrometry–based proteomics. http://www.ebi.ac.uk/pride/
- InAct: freely available database system and analysis tools for molecular interactions. http://www.ebi.ac.uk/intact/main.xhtml
- Cytoscape: an open-source platform for complex network analysis and visualization. http://www.cytoscape.org
- ScanSite: a tool for identifying protein motifs. http://scansite.mit.edu/

6.2.5. Genetic Disease Databases and Resources

- OMIM: Online Mendelian Inheritance in Man. http://www.ncbi.nlm.nih.gov/omim
- TCGA: The Cancer Genome Atlas is a National Cancer Institute– and National Human Genome Research

Institute–funded project to create an atlas of genome alterations in a range of tumor types. http://cancergenome.nih.gov

- Genetic Association Database: an archive of human genetic association studies of complex diseases and disorders. http://geneticassociationdb.nih.gov

6.3. Resources and Databases for New Targeting Technologies

6.3.1. Databases for Protein-Protein Interactions

- ANCHOR: a eb server and database for analysis of protein-protein interaction hotspots for drug discovery. http://structure.pitt.edu/anchor/
- MIPS: mammalian protein-protein interaction database. http://mips.helmholtz-muenchen.de/proj/ppi/
- STRING: a database of known and predicted protein-protein interactions. http://string-db.org/
- BioGrid: a database of genetic and physical interactions. http://thebiogrid.org/
- DIP: database of interacting proteins. http://dip.doe-mbi.ucla.edu/dip/Main.cgi
- MINT: the molecular interaction database. http://mint.bio.uniroma2.it/mint/Welcome.do

6.3.2. RNA Interference Databases

- HuSiDa: a depository for sequences of published siRNA molecules. http://itb.biologie.hu-berlin.de/~nebulus/sirna/index.htm
- siRNA Database: a collection of siRNAs obtained from published data maintained by MIT. http://web.mit.edu/sirna/
- MicroRNA Target Collection: microRNA and their potential targets predicted with miRanda software. http://cbio.mskcc.org/mirnaviewer/
- MicroRNAdb: a comprehensive database for micro-RNAs. http://www.rnaiweb.com/RNAi/microRNA/miRNA_Databases/index.html
- miRBase: a searchable database of published miRNA sequences and annotation. http://www.rnaiweb.com/RNAi/microRNA/miRNA_Databases/index.html

6.3.3. Aptamer Resources

- Selex DB: experimental data on selected affinity enriched sequences from different combinatorial libraries. http://wwwmgs.bionet.nsc.ru/mgs/systems/selex/homepage.html
- Aptamer Database: a curated database of published aptamers. http://aptamer.icmb.utexas.edu/search.php

REFERENCES

1. Silva, L. C., Ortigosa, L. C., and Benard, G. (2010). Anti-TNF-alpha agents in the treatment of immune-mediated inflammatory diseases: mechanisms of action and pitfalls. *Immunotherapy 2*, 817–833.
2. Dillman, R. O. (2011). Cancer immunotherapy. *Cancer Biother Radiopharm 26*, 1–64.
3. Hopkins, A. L., and Groom, C. R. (2002). The druggable genome. *Nat Rev Drug Discov 1*, 727–730.
4. Russ, A. P., and Lampel, S. (2005). The druggable genome: an update. *Drug Discov Today 10*, 1607–1610.
5. Overington, J. P., Al-Lazikani, B., and Hopkins, A. L. (2006). How many drug targets are there? *Nat Rev Drug Discov 5*, 993–996.
6. Imming, P., Sinning, C., and Meyer, A. (2006). Drugs, their targets and the nature and number of drug targets. *Nat Rev Drug Discov 5*, 821–834.
7. Clamp, M., Fry, B., Kamal, M., Xie, X., Cuff, J., Lin, M., Kellis, M., Linblad-Toh, K., and Lander, E. S. (2007). Distinguishing protein-coding and noncoding genes in the human genome. *Proceedings of the National Academy of Sciences of the United States of America 104*, 19428–19433.
8. Morrow, J. K., Tian, L., and Zhang, S. (2010). Molecular networks in drug discovery. *Crit Rev Biomed Eng 38*, 143–156.
9. Landry, Y., and Gies, J. P. (2008). Drugs and their molecular targets: an updated overview. *Fundam Clin Pharmacol 22*, 1–18.
10. Swinney, D. C. (2004). Biochemical mechanisms of drug action: what does it take for success? *Nat Rev Drug Discov 3*, 801–808.
11. Drews, J. (2000). Drug discovery: a historical perspective. *Science 287*, 1960–1964.
12. Rask-Andersen, M., Almen, M. S., Schioth, H. B. (2011). Trends in the exploitation of novel drug targets. *Nat Rev Drug Discov 10*, 579–590.
13. Zhu, F., Han, B., Kumar, P., Liu, X., Ma, X., Wei, X., Huang, L., et al. (2010). Update of TTD: Therapeutic Target Database. *Nucleic Acids Research 38*, D787–D791.
14. Nayal, M., and Honig, B. (2006). On the nature of cavities on protein surfaces: application to the identification of drug-binding sites. *Proteins 63*, 892–906.
15. Egner, U., and Hillig, R. C. (2008). A structural biology view of target drugability. *Expert Opinion on Drug Discovery 3*, 391–401.
16. Laurie, A. T., and Jackson, R. M. (2006). Methods for the prediction of protein-ligand binding sites for structure-based drug design and virtual ligand screening. *Curr Protein Pept Sci 7*, 395–406.
17. International Human Genome Sequencing Consortium (2004). Finishing the euchromatic sequence of the human genome. *Nature 431*, 931–945.
18. Kirchmair, J., Distinto, S., Schuster, D., Spitzer, G., Langer, T., Wolber, G. (2008). Enhancing drug discovery through in silico screening: strategies to increase true positives retrieval rates. *Curr Med Chem 15*, 2040–2053.
19. Yang, X., Lay, F., Han, H., and Jones, P. A. (2010). Targeting DNA methylation for epigenetic therapy. *Trends Pharmacol Sci 31*, 536–546.

20. Jones, P. A., and Baylin, S. B. (2007). The epigenomics of cancer. *Cell 128*, 683–692.
21. Deal, R. B., and Henikoff, S. (2010). Capturing the dynamic epigenome. *Genome Biol 11*, 218.
22. Wijermans, P., Lubbert, M., Verhoef, G., Bosly, A., Ravoet, C., Andre, M., and Ferrant, A. (2000). Low-dose 5-aza-2′-deoxycytidine, a DNA hypomethylating agent, for the treatment of high-risk myelodysplastic syndrome: a multicenter phase II study in elderly patients. *J Clin Oncol 18*, 956–962.
23. Issa, J. P., Garcia-Manero, G., Giles, F. J., Mannari, R., Thomas, D., Faderl, S., Bayar, E., et al. (2004). Phase 1 study of low-dose prolonged exposure schedules of the hypomethylating agent 5-aza-2′-deoxycytidine (decitabine) in hematopoietic malignancies. *Blood 103*, 1635–1640.
24. Minucci, S., and Pelicci, P. G. (2006). Histone deacetylase inhibitors and the promise of epigenetic (and more) treatments for cancer. *Nat Rev Cancer 6*, 38–51.
25. Baehner, F. L., Lee, M., Demeure, M. J., Bussey, K. J., Kiefer, J. A., and Barrett, M. T. (2011). Genomic signatures of cancer: basis for individualized risk assessment, selective staging and therapy. *J Surg Oncol 103*, 563–573.
26. Lee, S. K., and Calin, G. A. (2011). Non-coding RNAs and cancer: new paradigms in oncology. *Discov Med 11*, 245–254.
27. Cusick, M. E., Klitgord, N., Vidal, M., and Hill, D. E. (2005). Interactome: gateway into systems biology. *Hum Mol Genet 14* Spec No. 2, R171–R181.
28. Schneider, M. V., and Orchard, S. (2011). Omics technologies, data and bioinformatics principles. *Methods Mol Biol 719*, 3–30.
29. Wells, J. A., and McClendon, C. L. (2007). Reaching for high-hanging fruit in drug discovery at protein-protein interfaces. *Nature 450*, 1001–1009.
30. Buchwald, P. (2010). Small-molecule protein-protein interaction inhibitors: therapeutic potential in light of molecular size, chemical space, and ligand binding efficiency considerations. *IUBMB Life 62*, 724–731.
31. Bunka, D. H., Platonova, O., and Stockley, P. G. (2010). Development of aptamer therapeutics. *Curr Opin Pharmacol 10*, 557–562.
32. Borghouts, C., Kunz, C., and Groner, B. (2008). Peptide aptamer libraries. *Comb Chem High Throughput Screen 11*, 135–145.

Index

Printed in the United States
by Baker & Taylor Publisher Services